# Understanding and improving the functic nutritional properties

BURLEIGH DODDS SERIES IN AGRICULTURAL SCIENCE
NUMBER 114

# Understanding and improving the functional and nutritional properties of milk

Edited by Professor Thom Huppertz, Wageningen University & Research, The Netherlands and Professor Todor Vasiljevic, Victoria University, Australia

burleigh dodds
SCIENCE PUBLISHING

Published by Burleigh Dodds Science Publishing Limited
82 High Street, Sawston, Cambridge CB22 3HJ, UK
www.bdspublishing.com

Burleigh Dodds Science Publishing, 1518 Walnut Street, Suite 900, Philadelphia, PA 19102-3406, USA

First published 2021 by Burleigh Dodds Science Publishing Limited


Library of Congress Control Number: 2021940817

British Library Cataloguing in Publication Data
A catalogue record for this book is available from the British Library

ISBN 978-1-78676-819-3 (Print)
ISBN 978-1-78676-822-3 (PDF)
ISBN 978-1-78676-821-6 (ePub)
ISSN 2059-6936 (print)
ISSN 2059-6944 (online)

DOI: 10.19103/AS.2022.0099

Typeset by Deanta Global Publishing Services, Dublin, Ireland

# Contents

# Series list

| Title | Series number |
|---|---|
| **Achieving sustainable cultivation of maize - Vol 1**<br>From improved varieties to local applications<br>Edited by: Dr Dave Watson, CGIAR Maize Research Program Manager, CIMMYT, Mexico | 001 |
| **Achieving sustainable cultivation of maize - Vol 2**<br>Cultivation techniques, pest and disease control<br>Edited by: Dr Dave Watson, CGIAR Maize Research Program Manager, CIMMYT, Mexico | 002 |
| **Achieving sustainable cultivation of rice - Vol 1**<br>Breeding for higher yield and quality<br>Edited by: Prof. Takuji Sasaki, Tokyo University of Agriculture, Japan | 003 |
| **Achieving sustainable cultivation of rice - Vol 2**<br>Cultivation, pest and disease management<br>Edited by: Prof. Takuji Sasaki, Tokyo University of Agriculture, Japan | 004 |
| **Achieving sustainable cultivation of wheat - Vol 1**<br>Breeding, quality traits, pests and diseases<br>Edited by: Prof. Peter Langridge, The University of Adelaide, Australia | 005 |
| **Achieving sustainable cultivation of wheat - Vol 2**<br>Cultivation techniques<br>Edited by: Prof. Peter Langridge, The University of Adelaide, Australia | 006 |
| **Achieving sustainable cultivation of tomatoes**<br>Edited by: Dr Autar Mattoo, USDA-ARS, USA and Prof. Avtar Handa, Purdue University, USA | 007 |
| **Achieving sustainable production of milk - Vol 1**<br>Milk composition, genetics and breeding<br>Edited by: Dr Nico van Belzen, International Dairy Federation (IDF), Belgium | 008 |
| **Achieving sustainable production of milk - Vol 2**<br>Safety, quality and sustainability<br>Edited by: Dr Nico van Belzen, International Dairy Federation (IDF), Belgium | 009 |
| **Achieving sustainable production of milk - Vol 3**<br>Dairy herd management and welfare<br>Edited by: Prof. John Webster, University of Bristol, UK | 010 |
| **Ensuring safety and quality in the production of beef - Vol 1**<br>Safety<br>Edited by: Prof. Gary Acuff, Texas A&M University, USA and Prof. James Dickson, Iowa State University, USA | 011 |
| **Ensuring safety and quality in the production of beef - Vol 2**<br>Quality<br>Edited by: Prof. Michael Dikeman, Kansas State University, USA | 012 |
| **Achieving sustainable production of poultry meat - Vol 1**<br>Safety, quality and sustainability<br>Edited by: Prof. Steven C. Ricke, University of Arkansas, USA | 013 |
| **Achieving sustainable production of poultry meat - Vol 2**<br>Breeding and nutrition<br>Edited by: Prof. Todd Applegate, University of Georgia, USA | 014 |

# Introduction

Milk and dairy products have long been one of the major food groups in diets in many parts of the world, particularly since the domestication of cattle approximately 10,000 years ago. Current annual global milk production exceeds 850 billion kg, equating more than 100 kg of milk per person per day on average. Undoubtedly one of the most important reasons why milk and dairy products have become such an important food group in diets worldwide is the richness of nutrients. For example, in the Netherlands, milk and dairy products provide close to 60% of total dietary calcium, close to 40% of dietary vitamin B2 and B12, >30% of dietary retinol and phosphorus and >20% of dietary iodide, protein and zinc, while providing only 15% of total energy intake. Milk and dairy products also provide proportionally more potassium, selenium and magnesium than energy to the Dutch diet. High contributions of milk and dairy products to total nutrient intake are also found in many other counties, particularly in Europe, North America, Oceania and parts of the middle East and of Asia. In addition to the nutrient richness and other nutritional properties, the relative ease with which milk can be converted, with relative ease, into a wide variety of safe and shelf-stable products with favored sensory properties has further fueled development and growth of the dairy sector. In addition, milk and by-products of dairy production also can be used as the basis for ingredient manufacture for application in a wide range of dairy and non-dairy food products, but also non-food products, including pharmaceuticals and cosmetics.

Despite the aforementioned great role of milk and dairy products in many diets worldwide, the dairy sector also faces challenges, as does the whole food system. The growing world population leads to increasing demands for food worldwide, and many areas in the world still suffer from malnutrition. However, there is also an urgent need to address issues relating to the environmental sustainability of the food system to combat climate change. Key areas of concern in terms of the environmental impact of the food system include greenhouse gas emissions, land use and water use. A sustainable food system should provide healthy diets for the global population and do this within planetary boundaries. For milk and dairy products, as well as for any other food product, it is thus crucial that environmental impact is limited as much as possible and is justified by contributions to human nutrition and health. For milk and dairy products, this requires an in-depth understanding of compositional and nutritional properties, which can subsequently be applied to design strategies for further optimization.

Understanding and improving the nutritional properties of milk is thus of critical important basis for the role of milk and dairy products in sustainable food systems. In this book, we cover essential elements of this, which together can form an important basis for not only understanding the nutritional properties of milk and dairy products, but also for improving them. In Part 1, the importance of milk and dairy products in the human diet is covered in Chapter 1, and distinctly outlines the many important contributions of milk and dairy products to human nutrition and health. In addition, Chapter 1 also highlights that milk and dairy products should not be considered as simply a collection of nutrients, but that the way in which these nutrients interact in the product, i.e. the product matrix, is very important for the nutritional and health benefits of milk and dairy products. This importance of matrix effects is also clearly highlighted in Chapter 2, which focusses on understanding the digestion of milk proteins and fat. From this Chapter, it is clear that compositional variation is important for digestion, but the structural changes induced by processing, e.g. heating and homogenization, also strongly affect digestion of milk proteins and milk fat. In addition, interspecies variation can also strongly affect milk properties, including those to nutrition and health. While most milk consumed worldwide is bovine milk, particularly buffalo, caprine and ovine milk make important contributions to diets in different parts of the world. To illustrate the importance of this interspecies variation, Chapter 3 covers interspecies variation of milk properties and composition, both in terms of macronutrients and micronutrients.

Many technological and nutritional functionalities of milk and dairy products are related to the milk proteins and milk lipids, which are the focus of Part 2 of this book. The caseins are the most abundant class of milk proteins, and are the focus of Chapter 4. This chapter focusses on advances of understanding of caseins and casein micelles, covering recent insights into genetic variation and posttranslational modification, but also casein-salt interactions and their effect on casein micelle structure, stability and functionality. Next to caseins, the whey proteins also provide key functionalities in milk and dairy products, but also dairy ingredients. Physicochemical and technological properties of whey proteins are covered in Chapter 5, which highlights a wide variety of opportunities for functional whey protein ingredients. The nutritional properties of whey proteins and whey protein ingredients are covered in Chapter 6 and illustrate valuable applications in many areas, including infant nutrition, nutrition for patients recovering from operations or other medical treatments and to prevent sarcopenia. In addition, bioactive properties of whey proteins are also covered in Chapter 6. There are cases where bioactivity of whey proteins can be enhanced by enzymatic hydrolysis. The application is widely applied in the dairy industry and the various functional milk protein hydrolysates are covered in Chapter 7. Next to milk proteins, milk fat also has important functional properties. Technological functionality is often related to fatty acid composition and stereospecificity of the fatty acids and their effect on crystallization behavior, which is covered in Chapter 8. Chapter 9 describes advances in the understanding of nutritional and bioactive properties of milk fat. A very important feature here is the organization of milk fat in the form of milk fat globules and the milk fat globule membrane on the surface of these milk fat globules.

Both nutritionally and functionally, proteins and lipids play an important role in milk and dairy products, but other constituents are equally important and are covered in Part 3 of this book. Chapter 10 covers the third macronutrient in milk, i.e. lactose. With comparatively limited current research attention to lactose and main focus often on lactose intolerance, important nutritional properties of lactose are at risk of being overlooked. For instances its low glycemic index and low cariogenic potential, which can be key contributors to human health. In addition, many other carbohydrates from milk or derived from lactose can have important nutritional benefits. These include for instance oligosaccharides but also glycans attached to proteins. The properties and benefits associated with these carbohydrates are covered in Chapter 11. Chapter 12 focusses on the enzymes in milk. Enzymatic processes in milk can affect nutritional quality, but also physical stability of milk and are thus always an important topic of consideration when looking at improving the properties of milk. Another key influencing factor for stability of milk and dairy products are the milk salts, which are the focus of Chapter 13. The milk salts often govern protein stability and are thus very important. Likewise, milk and dairy products are also a key source of e.g. calcium and phosphorus in many diets and this requires the salts to be bioavailable, which is also covered in Chapter 13.

The important role of milk and dairy products in many current diets, coupled with the necessity for more sustainable diets, drives much research towards optimizing milk composition and quality. Many key aspects of this are covered in Part 4 of this book. State-of-the-art instrumental techniques are a prerequisite for being able to monitor quality aspects of milk quickly, efficiently and accurately. The latest advances in these techniques, including e.g. infrared and nuclear magnetic resonance techniques are covered in Chapter 14. Such techniques are often at the basis of research into improving milk composition and quality. Such improvements may be achieved via either selection of animals based on genetics or via feed. Chapter 15 describes the opportunities for improving milk properties via genetic selection, either to improve gross composition or to improve functional or nutritional properties through selection of specific protein variants. Next to genetic selection, animal feed also is crucial in milk composition and offers opportunities to improve composition and nutritional properties of milk, which are covered in Chapter 16. Of course, improvements in milk composition and properties through genetics and feed become of true value when they can be applied in consumer products. This requires the milk to be converted into safe, stable and sensorially desired dairy products. Sensory aspects of milk, both in terms of understanding and analysis are therefore covered in Chapter 17, whereas Chapter 18 covers microbial quality and spoilage of milk. Both these aspects are crucial to create products which are accepted and favored by consumers. These products can be in various formats, with three main classes of dairy products being liquid milk products, fermented dairy products and cheese. For each product class, the constituents of the milk, their interactions and their changes during the shelf-life of the products have notable and quite different effects on product texture and sensory perception thereof. For this reason, the final three chapters of this book cover the relationship between the different milk constituents and their interactions in relation to the textual properties of liquid milk products (Chapter 19), fermented dairy products (Chapter 20) and cheese (Chapter 21).

Overall, we believe that the wealth and knowledge and insights provided by the authors throughout all the chapters in this book clearly highlights the depth of understanding that is available on the nutritional properties of milk, and how all milk constituents, and their interactions, play a crucial role in this. In addition, to understanding, the chapters also provide clear paths to further improving our understanding, and pathways to apply the understanding to improve milk and dairy products further, e.g. via improved milk production strategies or via improved processing. This understanding and these improvements will undoubtedly further strengthen the importance of milk and dairy products in human diets worldwide.

# Part 1

## General

# Chapter 1

# The role of the dairy matrix in the contribution of milk and dairy products to the human diet

*Jan Geurts, FrieslandCampina, The Netherlands*



## 1 Introduction

Milk and dairy products are an important part of the daily diet in many regions of the world due to their recognized contribution to the recommended intake of a variety of valuable nutrients, the amount of energy (calories) provided compared to other basic food categories and the natural power to compensate for foods in the diet with lesser nutritional value, such as grains.[1-4] An example of the nutrient contribution of dairy products to a typical diet is presented in Fig. 1. In this figure, the mean contribution of dairy products and substitutes to the intake of nutrients in the Dutch diet is given. Dairy products represent one of the five core food groups in most dietary guidelines worldwide. Data extracted from the Food and Agriculture Organization (FAO) database on food-based dietary guidelines from 80 countries[5,6] shows that 59 countries provide specific recommendations for dairy intake: 56 countries give daily recommendations and 3 countries advice weekly intake of dairy (Dominican Republic, Guatemala and Honduras). The other 21 countries mention dairy in various ways in their statements. The United Kingdom recommends 'some' milk and dairy every day, Canada no longer has recommendations to consume a certain number of servings of milk products and there are no serving sizes mentioned in the Canadian National guidance documents either. The Russian Federation recommends 325 kg of milk and dairy/person/year. Colombia's

http://dx.doi.org/10.19103/AS.2022.0099.02

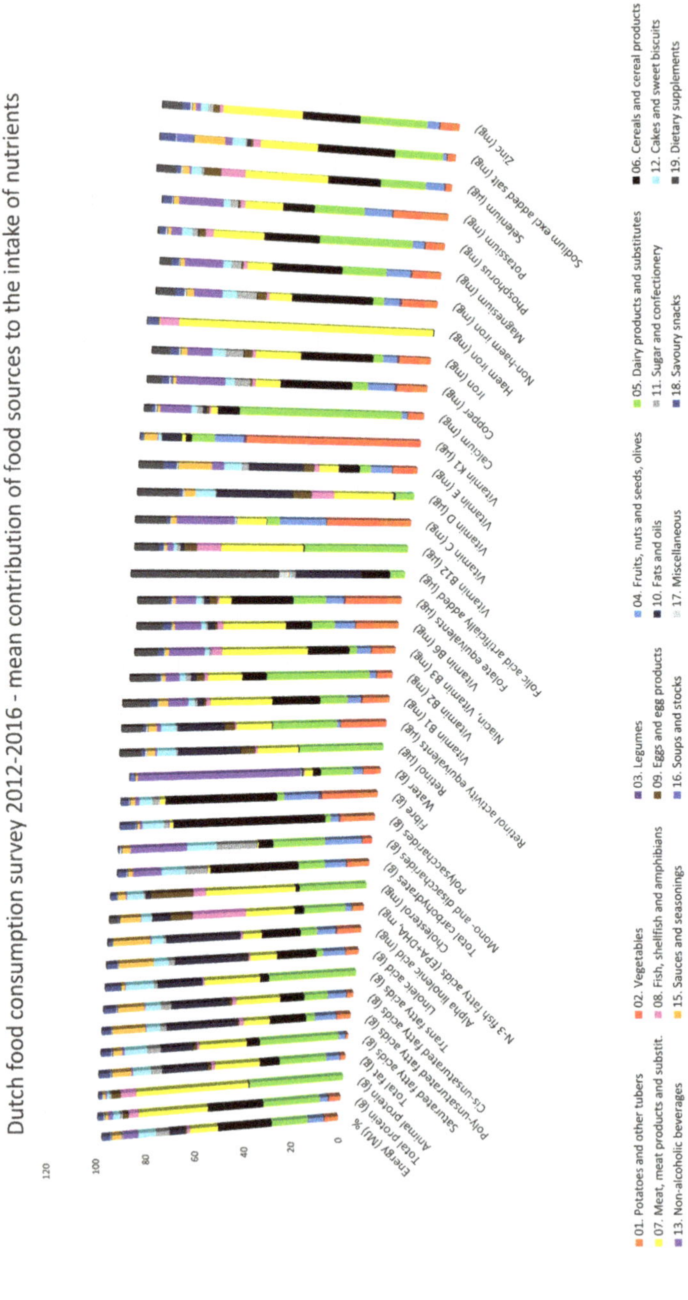


**Figure 1** Mean contribution of food sources to the intake of nutrients in the Dutch diet. Source: food consumption survey 2012–2016 (VCP 2012–2016), National Institute for Public Health and the Environment (RIVM), Bilthoven, the Netherlands.

recommendations are based on gender and are 1000 ml/day for men and 700 ml/day for women. An overview of the 80 countries that mention dairy products in their dietary recommendation is presented in Table 1.

On average, the worldwide recommendation is 2.3 servings per day (based on 44 countries, serving size varies from one country to another) which amount to, on average, 455 ml milk/day (based on 40 countries) and contributes, on average, 5% of energy, 10% of protein and 9% of fat intake (Table 2). In all food regulations milk and dairy products are present for their recognized contribution of substantial amounts of high-quality protein, minerals (calcium, zinc, magnesium and potassium), B-vitamins and other micronutrients such as iodine and selenium, as well as lower but still significant amounts of the fat-soluble vitamins A and D to the diet, with relatively modest energy (Table 3).

Dairy can play an important role in low- and middle-income countries (LMICs) where nutrient deficiencies are frequent. In these countries, animal source products, like milk, can help correct these deficiencies. Unfortunately, in LMICs, less than one in five children aged 6–23 months get animal-source foods[7] and most children consume only one animal-source food serving or less per day.[8] Children consuming food with inadequate amounts of animal-source protein are reported to have greater incidences of stunting.[9-13] Interestingly, among the various animal-source foods, evidence indicates a stronger favourable role for the dairy protein in promoting linear growth in children from LMIC and developed countries.[14-20] These effects may be mediated through insulin-like growth factor-1 (IGF-1), which is known to stimulate the proliferation of chondrocytes resulting in bone and muscle growth,[20] and higher levels of bioavailable essential amino acids (EAAs), such as tryptophan and leucine, involved in anabolic signaling via mTORC1.[21,22] Many biological processes are regulated by mTORC1 including skeletal muscle cell growth and homeostasis, myelination of neurons in the nervous system, chondral plate and bone growth and immune function. Milk proteins contain all (nine) EAAs in quantities that are above the age-specific amino acid requirement reference patterns and are easily digested, absorbed and available for various bodily functions. Milk proteins are therefore classified as high-quality proteins that can compensate for the often-low protein quality of the diets.[23,24]

The importance of dairy consumption in LMICs was also shown in the South East Asia Nutrition Survey (SEANUTS).[25] SEANUTS was a nationally representative multicentre survey in Indonesia, Malaysia, Thailand and Vietnam, conducted between 2010 and 2011, to assess the nutritional status and lifestyle factors of over 16 500 children aged 0.5–12 years old. The objective of SEANUTS was to determine whether children who consumed dairy products as part of their daily diet had a better nutritional status as measured by anthropometric indices and blood status for iron, vitamins A and D compared to children who did not. It was shown that children consuming one or more daily servings of dairy had higher

**Table 1** Overview of dietary guidelines worldwide that specifically mention dairy

| Country | Year | Model | Source |
|---|---|---|---|
| Albania | 2008 | Albanian Food Pyramid | http://www.fao.org/3/a-as658e.pdf |
| Antigua and Barbuda | 2013 | Pineapple | http://www.fao.org/3/a-as848e.pdf |
| Argentina | 2016 | Nutritional Circle | https://bancos.salud.gob.ar/sites/default/files/2020-08/guias-alimentarias-para-la-poblacion-argentina_manual-de-aplicacion_0.pdf |
| Australia | 2013 | Australian Guide to Healthy Eating | https://www.nhmrc.gov.au/about-us/publications/australian-dietary-guidelines |
| Austria | 2010 | Ernaehrungspyramide | http://www.fao.org/3/a-as659o.pdf |
| Bahamas | 2002 | Goat skin drum | http://www.fao.org/3/a-as849e.pdf |
| Bangladesh | 2013 | Pyramid | http://www.fao.org/3/a-as880e.pdf |
| Barbados | 2017 | Barbados map | http://www.fao.org/3/I9680EN/i9680en.pdf |
| Belgium (Flanders) | 2017 | Food Triangle | https://www.gezondleven.be/themas/voeding/voedingsdriehoek |
| Belgium (Federal Superior Health Council) | 2011 (reviewed in 2017) | Food Guide Pyramid | https://www.foodinaction.com/telechargez-pyramide-alimentaire/ |
| Belize | 2012 | Food Basket | http://www.fao.org/3/a-as852e.pdf |
| Benin | 2014 (reviewed in 2020) | African Food House | http://wiredhealthresources.net/presentations/82/story_content/external_files/Benin.pdf |
| Bolivia | 2014 | Arco de la alimentación | http://www.fao.org/nutrition/education/food-dietary-uidelines/regions/countries/bolivia-plurinational-state-of/en/ |
| Bosnia and Herzegovina | 2004 | Food Pyramid | http://www.fao.org/3/a-as669o.pdf |
| Brazil | 2014 | 10 steps to healthy eating | http://bvsms.saude.gov.br/bvs/publicacoes/dietary_guidelines_brazilian_population.pdf |

| | | | |
|---|---|---|---|
| Bulgaria | 2006 | Pyramid | Htttp://www.fao.org/nutrition/education/food-dietary-uidelines/regions/countries/bulgaria/en/ |
| Canada | 2019 | Canada's Food Guide | https://food-guide.canada.ca/en/ |
| Chile | 2013–2014–2019 | Eating guidelines for the Chilean population | https://inta.cl/material-informativo/page/2/ |
| China | 2016 | Food Guide Pagoda | http://dg.en.cnsoc.org |
| Colombia | 2015 | Plato saludable de la familia Colombia | http://www.icbf.gov.co/portal/page/portal/PortalICBF/bienestar/nutricion/educacion/DOCUMENTO%20TECNICO%20GABA_UV_WEB%20AGOSTO%202016.pdf |
| Croatia | 2002–2012 | Food Pyramid | http://www.fao.org/nutrition/education/food-dietary-guidelines/regions/countries/croatia/en/ |
| Cuba | 2009 | Food Guide | http://www.fao.org/3/a-as864s.pdf |
| Cyprus | 2007 | Food Pyramid | http://www.fao.org/3/a-as673o.pdf |
| Denmark | 2013 | Danish official dietary guidelines | http://www.fao.org/3/a-as675o.pdf |
| Dominican Republic | 2009 | Pilón de la alimentación y nutrición | http://www.fao.org/3/a-as866s.pdf |
| Ecuador | 2018 | Documento Técnico de las Guías Alimentarias Basadas en Alimentos (GABA) del Ecuador | http://www.fao.org/nutrition/education/food-dietary-guidelines/regions/countries/ecuador/en/ |
| El Salvador | 2012 | Guías alimentarias para las familias salvadoreñas | http://www.fao.org/3/a-as867s.pdf |
| Estonia | 2015 | Food Pyramid | https://intra.tai.ee/images/prints/documents/149019033869_eesti%20toitumis-%20ja%20liikumissoovitused.pdf |
| Finland | 2014 | Food triangle & Food plate | https://www.ruokavirasto.fi/en/themes/healthy-diet/nutrition-and-food-recommendations/ |
| Macedonia | 2014 | Dietary guidelines for the population in TRNM | http://iph.mk/wp-content/uploads/2014/11/vodic_ishrana_2013-2.pdf |

*(Continued)*

**Table 1** *(Continued)*

| Country | Year | Model | Source |
|---|---|---|---|
| France | 2016 | The French National Nutrition and Health Program's dietary guidelines | https://www.mangerbouger.fr/Les-recommandations/Aller-vers/Les-produits-laitiers-lait-yaourts-fromage-et-fromage-blanc |
| Georgia | 2005 | Healthy eating - the main key to health | http://www.fao.org/3/a-as682e.pdf |
| Germany | 2017 | 10 guidelines of the German Nutrition Society (DGE) for a wholesome diet | https://www.dge.de/ernaehrungspraxis/vollwertige-ernaehrung/10-regeln-der-dge/en/https://www.dge-ernaehrungskreis.de/lebensmittelgruppen/milch-und-milchprodukte/ |
| Greece | 1999-2014 | Food Pyramid | http://www.diatrofikoiodigoi.gr/?page=diatrofikes-sistaseishttp://www.fao.org/nutrition/education/food-dietary-guidelines/regions/countries/greece/en/ |
| Grenada | 2005 | Nutmeg | http://www.fao.org/3/a-as854e.pdf |
| Guatemala | 2012 | Family pot | http://www.fao.org/3/a-as870s.pdf |
| Honduras | 2013 | A pot | http://www.fao.org/3/a-as874s.pdf |
| Hungary | 2004 | House of healthy nutrition | http://www.fao.org/3/a-as684o.pdf |
| Iceland | 2014 | Food Circle & Fod Plate | http://www.landlaeknir.is/servlet/file/store93/item25796/radleggingar-um-mataraedi-2015.pdf |
| India | 2011 | Indian Food Pyramid | https://www.nin.res.in/downloads/DietaryGuidelinesforNIN website.pdf |
| Iran | 2015 | Food Pyramid | https://nut.behdasht.gov.ir |
| Ireland | 2016 | Irish Food Pyramid | https://www.healthpromotion.ie/hp-files/docs/HPM00796.pdf |
| Israel | 2017 | Food pyramid | https://www.health.gov.il/English/Topics/FoodAndNutrition/Pages/default.aspx |
| Italy | 2018 | Dietary Guidelines for Healthy Eating | https://www.crea.gov.it/en/web/alimenti-e-nutrizione/-/linee-guida-per-una-sana-alimentazione-2018 |
| Japan | 2010 | Food Guide Spinning Top | https://www.maff.go.jp/j/balance_guide/b_use/pdf/eng_reiari.pdf |

| | | | |
|---|---|---|---|
| Kenya | 2017 | National Guidelines for Healthy Diets and Physical Activity | http://nak.or.ke/wp-content/uploads/2017/12/NATIONAL-GUIDELINES-FOR-HEALTHY-DIETS-AND-PHYSICAL-ACTIVITY-2017-NEW-EDIT.pdf |
| Latvia | 2008 | Food Pyramid | http://www.fao.org/3/a-as687o.pdf |
| Lebanon | 2013 | The Lebanese cedar food guide | https://www.aub.edu.lb/fafs/nfsc/Documents/FBDG%20English%20Version.pdf |
| Lithuania | 2010 | Dietary Guideline | https://www.euro.who.int/__data/assets/pdf_file/0008/119672/E67884.PDF?ua=1 |
| Malaysia | 2010 | Food Pyramid | https://www.moh.gov.my/index.php/pages/view/370 |
| Malta | 2016 | Healthy Plate | https://deputyprimeminister.gov.mt//en/health-promotion/documents/library/publications/healthy%20plate%20en.pdf |
| Mexico | 2015 | El plato del bien comer | https://www.insp.mx/eppo/blog/3878-guias-alimentarias.html |
| Mongolia | 2010 | The ger (Mongolian tent) | http://www.fao.org/nutrition/education/food-dietary-guidelines/regions/countries/mongolia/en/ |
| Nepal | 2012 | Food-based dietary guidelines for Nepalese | http://www.fao.org/nutrition/education/food-dietary-guidelines/regions/countries/nepal/en/ |
| The Netherlands | 2015 | Sheel of five | https://www.voedingscentrum.nl/Assets/Uploads/voedingscentrum/Documents/Professionals/Schijf%20van%20Vijf/Richtlijnen%20Schijf%20van%20Vijf.pdf |
| New Zealand | 1990s | New Zealand food and nutrition guideline | https://www.health.govt.nz/our-work/eating-and-activity-guidelines/current-food-and-nutrition-guidelines |
| Nigeria | 2001 | Food Pyramid | http://www.fao.org/3/a-as841e.pdf |
| Norway | 2012 | Norwegian guidelines on diet, nutrition and physical activity | http://norden.diva-portal.org/smash/get/diva2:704251/FULLTEXT01.pdf |
| Oman | 2009 | Omani Healthy Plate | http://www.fao.org/3/a-as845e.pdf |
| Panama | 2013 | Plate of healthy eating | http://www.fao.org/3/a-as876s.pdf |

(Continued)

**Table 1** (*Continued*)

| Country | Year | Model | Source |
|---|---|---|---|
| Paraguay | 2015 | Nutritional pot | http://www.fao.org/3/a-ax401s.pdf |
| Philippines | 2012 | Healthy Food Plate + Daily Nutritional Guide Pyramid | https://www.fnri.dost.gov.ph/index.php/tools-and-standard/nutritional-guide-pyramid |
| Poland | 2009 | Pyramid Healthy Nutrition + Pyramid of Healthy Nutrition for children | http://www.izz.waw.pl/pl/?option=com_content&view=article&id=7&Itemid=5&lang=pl |
| Portugal | 2003 | Food Wheel Guide (A roda dos alimentos) | http://www.fao.org/3/a-ax434o.pdf |
| Qatar | 2015 | Shell | http://www.fao.org/3/a-az908e.pdf |
| Romania | 2006 | Food Pyramid | http://www.fao.org/3/a-as693o.pdf |
| Russian Federation | 2005 | Healthy Food Pyramid | https://www.euro.who.int/__data/assets/pdf_file/0020/150158/E73401.pdf |
| Seychelles | 2006 | Seychelles Food Guide | http://www.fao.org/3/a-as997e.pdf |
| Slovenia | 2007 | Food Guide Pyramid + Healthy plate | http://www.fao.org/3/a-az912o.pdf |
| South Africa | 2003 (revised in 2013) | Food Guide | http://www.fao.org/3/a-as842e.pdf |
| Spain | 2017 | Piramid +Mediterranean Pyramid | http://dietamediterranea.com/nutricion-saludable-ejercicio-fisico/ https://www.nutricioncomunitaria.org |
| Sri Lanka | 2011 | Food Pyramid | http://www.fao.org/3/a-as886e.pdf |
| Sweden | 2015 | Swedish dietary guidelines | https://www.livsmedelsverket.se/en/food-habits-health-and-environment/dietary-guidelines |
| Switzerland | 2011 | Swiss Food Pyramid | https://www.sge-ssn.ch/fr/toi-et-moi/boire-et-manger/aux-differents-ages/enfance/ |
| Thailand | 2010 | Nutrition flag | http://www.fao.org/3/a-as979e.pdfhttp://www.fao.org/3/a-as887e.pdf |

| | | | |
|---|---|---|---|
| Turkey | 2006 | Turkish Food guide (four leaves) | http://www.fao.org/3/a-as697e.pdf |
| United Kingdom | 2016 | Eatwell guide | https://www.gov.uk/government/publications/the-eatwell-guide |
| Uruguay | 2016 | Disk 5 food groups | https://www.gub.uy/ministerio-salud-publica/comunicacion/campanas/campana-de-alimentacion-saludable |
| USA | 2016 | Myplate | https://health.gov/sites/default/files/2019-09/2015-2020_Dietary_Guidelines.pdfhttps://www.choosemyplate.gov/dairy |
| Venezuela | 1991 | Food spinning top | http://www.fao.org/3/a-as878s.pdf |
| Vietnam | 2013 | Food Pyramid | http://www.fao.org/3/as981o/as981o.pdf |

**Table 2** Daily dairy recommendations per age group

| | All dairy (excl. cheese) | | | | Cheese | | | |
|---|---|---|---|---|---|---|---|---|
| Country | Children | Adults | Pregnant | Elderly | Children | Adults | Pregnant | Elderly |
| Albania | 2 bowls | 1 bowl | NS | NS | NS | 50 g | | NS |
| Antigua and Barbuda | NS | NS | NS | NS | NS | NS | NS | NS |
| Argentina | NS | 3 portions[1] | NS | NS | NS | 30 g | NS | NS |
| Australia | 1.5–3.5 servings[3] | 2.5 servings[4] | Adults (19–50 y): 2.5 servings/Girls (<18 y): 3.5 servings[5] | NS | NS | Low-salt cheese varieties are preferred | NS | NS |
| Austria | NS | 3 portions[6] | NS | NS | NS | NS | NS | NS |
| Bahamas | NS | NS | NS | NS | NS | NS | NS | NS |
| Bangladesh | 1–2 servings[7] | NS | NS | Fat-free of skimmed milk | NS | NS | NS | NS |
| Barbados | | 2–3 servings | NS | | NS | NS | NS | NS |
| Belgium (Flanders) | 350–600 g | 450 g | 600 g | 600 g | 10–40 g | 20 g | 20–40 g | 20–40 g |
| Belgium (Federal Superior Health Council) | NS | 2–3 portions[8] | NS | NS | NS | NS | NS | NS |
| Belize | NS | 5–7 servings[9] | NS | NS | NS | NS | NS | NS |
| Benin | 1 portion | 1-2 portions | 1-2 portions | NS | NS | NS | NS | NS |
| Bolivia | NS | 2-4 servings[10] | NS | NS | NS | NS | NS | NS |
| Bosnia and Herzegovina | NS | 2 portions | NS | NS | NS | NS | NS | NS |
| Brazil | NS | NS | NS | NS | NS | NS | NS | NS |

| | | | | | | | | |
|---|---|---|---|---|---|---|---|---|
| Bulgaria | NS | 1 glass[11] | NS | NS | NS | 50 g | NS | NS |
| Canada | NS | NS | NS | NS | NS | NS | NS | NS |
| Chile | NS | 3 servings[12] | NS | NS | NS | NS | NS | NS |
| China | NS | 300 g | NS | NS | NS | 25–30 g | NS | NS |
| Colombia | 2–2.5 portions | 3.5 portions (1) or 5 portions (1)[13] | NS | 3.5 portions | NS | NS | NS | NS |
| Croatia | NS | NS | NS | NS | NS | NS | NS | NS |
| Cuba | 1–2 portions[14] | 1–3 portions | 3 portions | 1.5 portions | NS | NS | NS | NS |
| Cyprus | NS | NS | NS | NS | NS | NS | NS | NS |
| Denmark | NS | 250–500 g of milk | NS | NS | NS | NS | NS | NS |
| Dominican Republic | NS | ~0.5 portions[15] | NS | NS | NS | NS | NS | NS |
| Ecuador | NS | NS | NS | NS | NS | NS | NS | NS |
| El Salvador | NS | NS | NS | NS | NS | NS[16] | NS | NS |
| Estonia | NS | NS[17] | NS | NS | NS | NS | NS | NS |
| Finland | NS | 500–600 g | NS | NS | NS | 20–40 g | NS | NS |
| Macedonia | NS | NS | NS | NS | NS | NS | NS | NS |
| France | 3–4 portions[18] | 2 portions | 3–4 portions | NS | NS | NS | NS | NS |
| Georgia | NS | 2–3 portions | NS | NS | NS | NS | NS | NS |
| Germany | NS | 200–250 ml | NS | NS | NS | 50–60 g | NS | NS |
| Greece | NS | 2 servings | NS | NS | NS | NS | NS | NS |
| Grenada | NS | 4 portions[19] | NS | NS | NS | NS | NS | NS |
| Guatemala | NS | ~0.5 portions[20] | NS | NS | NS | NS | NS | NS |
| Honduras | NS | ~0.5 portions[21] | NS | NS | NS | NS | NS | NS |
| Hungary | NS | 500 ml | NS | NS | NS | NS[22] | NS | NS |

(Continued)

**Table 2** (*Continued*)

| | All dairy (excl. cheese) | | | | Cheese | | | |
|---|---|---|---|---|---|---|---|---|
| Country | Children | Adults | Pregnant | Elderly | Children | Adults | Pregnant | Elderly |
| Iceland | NS | 500 ml[23] | NS | NS | NS | NS | NS | NS |
| India | NS | 300 g | NS | NS | NS | NS | NS | NS |
| Iran | NS | NS[24] | NS | NS | NS | NS | NS | NS |
| Ireland | 3 portions[25] | 3 portions | NS | 3 portions | NS | NS | NS[26] | NS |
| Israel | NS | 2–3 servings | NS | NS | NS | NS | NS | NS |
| Italy | NS | 3 portions[27] | NS | NS | NS | NS | NS[28] | NS[29] |
| Japan | NS | 2–3 servings[30] | NS | NS | NS | NS | NS | 40 g |
| Kenya | NS | NS | NS | NS | NS | NS | NS | NS |
| Latvia | NS | 3 portions | NS | NS | NS | NS | NS | NS |
| Lebanon | NS | 3 servings | 3–4 servings | NS | NS | NS | NS | NS |
| Lithuania | NS | 2–3 servings[31] | NS | NS | NS | NS | NS | NS |
| Malaysia | NS | 1–3 servings | NS | NS | NS | NS | NS | NS |
| Malta | NS | 2 servings[32] | NS | NS | NS | NS | NS | NS |
| Mexico | 2 portions | 2 portions | 2 portions | 2 portions | NS | NS | NS | NS |
| Mongolia | NS | 160–180 ml | NS | NS | NS | NS | NS | NS |
| Nepal | NS | NS | NS | NS | NS | NS | NS | NS |
| The Netherlands | 300–450 g | 300–350 g | NS | 600 g | 0–20 g | 40 g | NS | 40 g |
| New Zealand | 2-3 servings | ≥ 2 servings[33] | 3 servings | ≥ 3 servings | NS | NS | NS | NS |
| Nigeria | NS | NS | NS | NS | NS | NS | NS | NS |
| Norway | NS | 3 servings[34] | NS | NS | NS | NS | NS | NS |
| Oman | 0.3-0.5 servings | 0.5 servings[35] | 0.5 servings | 0.5 servings | NS | NS | NS | NS |

| | | | | | | | | |
|---|---|---|---|---|---|---|---|---|
| Panama | NS | NS | NS | NS | NS | NS | NS | NS |
| Paraguay | NS | 2 portions[36] | NS | NS | NS | 1 slice | NS | NS |
| Philippines | 1 serving[37] | 1 serving | 1 serving | 2–4 servings | NS | NS | NS | NS |
| Poland | NS | 2 large glasses of milk | NS | NS | NS | NS | NS | NS |
| Portugal | NS | 2-3 portions[38] | NS | NS | NS | NS | NS | NS |
| Qatar | NS | 2 calcium equivalents[39] | NS | NS | NS | NS | NS | NS |
| Romania | NS | 2–3 portions | NS | NS | NS | NS | NS | NS |
| Russian Federation | NS | 325 kg/yr/person | NS | NS | NS | NS | NS | NS |
| Seychelles | NS | 3 portions | NS | NS | NS | NS | NS | NS |
| Slovenia | NS | 2-4 portions[40] | NS | NS | NS | NS | NS | NS |
| South Africa | NS | 400–500 ml low-fat milk | NS | NS | 40 g | NS | NS | NS |
| Spain | 2–3 portions[41] | 2-4 portions | NS | NS | NS | NS | NS | NS |
| Sri Lanka | 2–4 servings | 1-2 servings | 2–4 servings | NS | NS | NS | NS | NS |
| Sweden | NS | 200-500 ml[42] | NS | NS | NS | NS | NS | NS |
| Switzerland | NS | 3 servings[43] | NS | NS | NS | NS | NS | NS |
| Thailand | 1-2 glasses[44] | 1-2 glasses | 1–2 glasses | 1–2 glasses | NS | NS | NS | NS |
| Turkey | 3–4 servings | 2 servings[45] | 3–4 servings | NS | NS | NS | NS | NS |
| United kindom | NS | NS[46] | NS | NS | NS | NS | NS | NS |
| Uruguay | NS | 500 ml | NS | NS | NS | NS | NS | NS |
| USA | 2–3 cups milk/ yogurt | 3 cups[47] | NS | NS | NS | NS | NS | NS |

(Continued)

**Table 2** (*Continued*)

| | All dairy (excl. cheese) | | | | Cheese | | | |
|---|---|---|---|---|---|---|---|---|
| Country | Children | Adults | Pregnant | Elderly | Children | Adults | Pregnant | Elderly |
| Venezuela | 1-2 portions | 1-2 portions | 1-2 portions | NS | NS | NS | NS | NS |
| Vietnam | NS | NS | NS | NS | NS | NS | NS | NS |

[1] In total, 500 ml milk or yogurt per day, [2] Soft cheese, [3] A standard serving is 1 cup (250 ml) of fresh, UHT long life, reconstituted powdered milk (500-600 kJ), [4] For women up to 50 years and men up to 70 years, [5] Advice is to consume reduced-fat milk, yogurt and cheese during pregnancy and breastfeeding and to avoid unpasteurized dairy products, and soft, semi-soft and surface-ripened cheeses, [6] Portion is defined as 200 ml milk/180-250 g yogurt/200 g cottage cheese/50-60 g of cheese. Recommended combination of two portions of liquid/soft dairy and 1 portion of cheese per day, [7] One cup of milk (150 ml) or 1 cup of dairy (100 ml), [8] No specific recommendations are given. In the guidelines, it is stated that drinking milk or eating dairy helps achieve the recommended intake of calcium, vitamin B2 and vitamin B12, [9] This corresponds to 1600 or 22 kcal energy intake, [10] One serving corresponds to 200 ml, [11] One glass corresponds to 200 ml, [12] One portion is defined as 1 cup of 200 ml low-fat milk or 175 g of yogurt or 50 g of soft cheese, [13] One portion is equivalent to 200 ml of milk, [14] One portion is equivalent to 1 cup of milk or yogurt (240 g) or 30 g of cheese, [15] Three portions per WEEK are recommended [16] Varieties such as quark, curd and cottage cheese are recommended as they are rich in protein and low fat, [17] General advice to consume more milk and milk products, [18] One serving is equivalent with 150 ml milk, 125 g of yogurt or 30 g of cheese, [19] For a diet of 1600 lcal/day, 4 portions per day, for a diet of 2200 kcal/day, 2 portions per day and for a diet of 2800 kcal/day, 4 portions per day are recommended, [20] The recommendation is three portions per WEEK. One portion corresponds to one glass of milk/yogurt or 30 g of cheese, [21] The recommendation is three portions per WEEK. One portion corresponds to one glass milk, a small glass of yogurt or 30 g of cottage cheese, [22] It is recommended to eat reduced-fat cheese regularly, [23] In the Icelandic recommendations, one dose of dairy is equivalent to 25 g of cheese, [24] The recommendation states dairy consumption every day, but no quantities are given [25] One portion is equivalent to one glass of milk (200 ml) or 125 g of yogurt or 25 g of cheese, [26] The recommendation states "enjoy cheese in small amounts", [27] One portion is defined as 125 g of milk (125 g), [28] It is recommended to enrich the diet with foods high in calcium such as milk and yogurt, [29] Cheese is recommended to contrast the lack of calcium, [30] One serving corresponds to 90 ml of milk or 100 g of yogurt or 20 g of cheese, [31] One serving corresponds to one glass of milk or sour milk, 0.5 glass of curd or 40g of cheese, [32] One serving is equivalent to 250 ml milk or 150 ml of yogurt or 30-40 g of cheese, [33] One serving is defined as one glass of milk (250 ml), one small bottle of yogurt (125-150 g), 40 g of cheese or 250 ml of fortified soy milk. It is recommended to choose low- or reduced-fat options, [34] One serving is equivalent to 200 ml of milk, 125 g of yogurt, 100 g of cottage cheese or 20 g of light cheese. To help cover the iodine intake, at least two of these should be milk or yogurt, [35] The recommended daily amount of the milk group for a 2000 kcal diet is about 0.5 servings (1 cup of milk of 30 ml or 45 g natural cheese/60 g processed cheese, [36] One portion corresponds to 200 ml of whole milk (200 ml) or 30 g of cheese, [37] One serving corresponds to one glass of whole milk or 1/2 cup of evaporated milk, [38] One portion is equivalent with 250 ml of milk or 200 g of yogurt or 40 g of cheese or 50 g of fresh cheese, [39] One equivalent corresponds to 1 cup of milk or yogurt, 14 tablespoons of labneh or 50 g of cheese, [40] One portion corresponds to one cup of partially skimmed milk or one cup of yogurt or 1/2 a slice of less fatty cheese, [41] One portion equals 200-250 ml of milk or 200-250 g of yogurt or 40-60 g of hard cheese or 80-125 g of soft cheese, [42] Milk, curdled milk or yogurt, [43] One serving equals 200 ml of milk or 150-200 g of yogurt, quark, cottage cheese or 30 g of semi-hard cheese or 60 g of soft cheese, [44] Milk or yogurt, [45] .One serving corresponds to one middle-sized glass (200 ml) of milk or yogurt with portions of matchbox-sized cheese, [46] Some milk and other dairy products are recommended every day. One serving equals 200 ml of milk (206 g), 125-150 g of yogurt or 30 g of hard cheese, [47] One cup of milk, yogurt, or soy beverage, 1 ½ ounce of natural cheese, or 2 ounces of processed cheese can be considered as 1 cup from the dairy group. The 3 cup or equivalent recommendation for adults equates to 360 ml.
NS: not specified

total dietary intakes than children consuming <1 daily dairy serving (Table 4). Specific analyses in Vietnam showed that higher dairy intake was associated with significantly higher intakes of all assessed nutrients[26] while analyses performed in Indonesia showed that adequate dairy intake can substantially add to the achievement of recommended dietary allowance RDA in Indonesian children.[27]

## 2 Dairy as part of a sustainable diet

To meet the needs of the ever-growing world population, natural resources are depleted and pressure on our world is growing. To feed the growing world population, it is important to ensure that the food produced also reaches people at the bottom of the pyramid. In 2019, the FAO stated:

> sustainable healthy diets are dietary patterns that promote all dimensions of individuals' health and wellbeing; have low environmental pressure and impact; are accessible, affordable, safe and equitable; and are culturally acceptable. The aims of sustainable healthy diets are to achieve optimal growth and development of all individuals and support functioning and physical, mental and social wellbeing at all life stages for present and future generations; contribute to preventing all forms of malnutrition (i.e. undernutrition, micronutrient deficiency, overweight and obesity); reduce the risk of diet-related-non-communicable diseases; and support the preservation of biodiversity and planetary health. Sustainable diets must combine all the dimensions of sustainability to avoid unintended consequences.

A global reference diet that considers the health and environmental sustainability aspects of eating patterns was recently published in the EAT-Lancet report,[28] which calls for restricting dairy consumption to a moderate level. However, dairy has an important role to play in food security and poverty reduction, especially in the developing world. Food security is defined as 'all people have always access to sufficient food for an active, healthy lifetime'.[29] By 2050, we face the challenge of feeding ten billion people while, at the same time, minimizing the impact of food production on the environment.[30] Food security and sustainable food production have been identified as the biggest challenges for the future.[31]

According to the FAO report World Livestock 2011,[32] dairy products can play an important role in reaching food security. Their natural nutrient density is seen as a key asset in this respect.[33] The concept of nutrient density was proposed by Drewnowski et al. to separate foods that are energy-dense from those that are nutrient-rich.[34] The nutrient-rich food (NRF) score X.Y is based on recommended daily allowances for adult populations, with X representing the number of nutrients to encourage and Y the number of nutrients to limit. Recently, Drewnowski et al. designed a hybrid nutrient density score, which also includes food groups. He showed that this score would also include the healthfulness of foods not readily captured by a score based on nutrients only.[35]

**Table 3** Recommendations regarding dairy including scientific underpinning

| Country | General recommendations | Does the dietary guideline mention scientific evidence? |
|---|---|---|
| Albania | NS | Milk and dairy are good sources of vitamins A and D, calcium and protein. Vitamin A helps eyesight, skin and hair. Vitamin D helps in absorbing calcium and nerve functioning. Fats serve as building blocks of membranes, play a key regulatory role in biological functions and provide fat-soluble vitamin absorption. Overuse of salt is related to high blood pressure and death from cardiovascular disease. |
| Anigua and Barburda | NS | No |
| Argentina | NS | Reducing salt intake helps avoid high blood pressure, cardiovascular disease and renal Insufficiency. Fruits and vegetable consumption protects against obesity, diabetes, colorectal cancer and cardiovascular disease. Excessive intake of fats, sugars and salt, such as butter, can lead to obesity, high blood pressure, diabetes and other cardiovascular diseases. |
| Australia | NS | No |
| Austria | NS | No |
| Bahamas | NS | Eating too many greasy foods is unhealthy and can cause heart problems and obesity. Choose foods with less sugar and salt, as they increase your risk of developing heart disease, high blood pressure and type 2 diabetes. |
| Bangladesh | Consume adequate amounts of milk and dairy. | Milk is a good source of calcium and phosphorus, essential for building strong bones and teeth. Also contains protein, lactose and vitamins (B12), which promote the growth and proper functioning of body tissues. Milk consumption also helps to prevent osteoporosis. |
| Barbados | Vegetable beverages can be substituted with whole milk. | Diets high in fat are linked to obesity, heart disease and cancer, especially saturated fatty acids and trans fats. |
| Belgium (Flanders) | | Thanks to milk you probably run less risk of colon cancer. Yogurt and milk or cheese with reduced fat content may help prevent diabetes. Milk products would have a beneficial effect on blood pressure. Milk products could protect you against cardiovascular disease, although there is only limited evidence for this. Also carefully proven: milk products may have an adverse effect on the development of prostate cancer. It has not been proven that milk in itself protects against osteoporosis. However, is it proven that calcium helps to get healthy bones. |

| | | |
|---|---|---|
| Belgium (Federal Superior Health Council) | NS | Milk and dairy products are important sources of protein, calcium, vitamin B2 and vitamin B12 and can contribute to a healthy and complete diet. Thanks to milk you probably run less risk of colon cancer. Yogurt and milk or cheese with reduced fat content may help prevent diabetes. Milk products can protect you against cardiovascular disease and prostate cancer. |
| Belize | NS | Reduce foods high in fat, such as cheese, butter, full cream milk are dangerous to your health, because they can lead to overweight and obesity, diabetes, cancer, high blood pressure, and risk of heart and blood vessel disease. |
| Benin | NS | Too much oil/fat, especially animal fat, may promote obesity and cardiovascular disease. Excessive consumption of salt increases the risk of high blood pressure. Carbonated and sugar-sweetened drinks promote obesity and diabetes. |
| Bolivia | Increase the consumption of milk and dairy. They are rich in proteins, vitamin D and calcium, which help growth in childhood and reduction of risk of osteoporosis. Cow's milk contains proteins (casein, lactoalbumin and lactoglobulin), lactose, minerals (phosphate) and fats, as well as vitamins A, B and D. | No |
| Bosnia and Herzegovina | NS | Milk and dairy are good sources of vitamins A, B, D, calcium and phosphorus. |
| Brazil | Full-fat milk is rich in calcium and vitamins but also in saturated fatty acids. The consumption of natural yogurt and other fermented milk foods is increasing in Brazil. | Saturated fatty acids can increase your risk of cardiovascular disease, obesity and other chronic diseases. Milk and dairy are rich in proteins, vitamin A and calcium. |

(Continued)

**Table 3** *(Continued)*

| Country | General recommendations | Does the dietary guideline mention scientific evidence? |
|---|---|---|
| Bulgaria | Milk and dairy are the food group rich in highly available calcium. They also contain proteins of high biological value, vitamins B12 and B6 and other vitamins of B group, vitamin A. The foods from this group are rich in fats with the exception of those with specially reduced fat content; brinded cheese contains also high amounts of salt. | Calcium is vitally important for building up and maintaining bone structure and teeth. Adequate calcium intake reduces the risk of osteoporosis and hypertension. Yogurt and other fermented milk products have special health effects. Lactic acid bacteria improve digestion, stimulate the motility of the intestines and reduce constipation. They suppress the reproduction of harmful bacteria in the intestines, thus decreasing the cases of intestinal disorders, their duration and severity. Lactic acid bacteria improve immunity, decrease the number of cases with lactose intolerance, reduce the atherosclerosis risk by decreasing blood cholesterol. They have the potential to decrease the risk of colon and rectum cancer. Bioactive peptides reducing the blood pressure and stimulating immunity are formed from milk proteins during the fermentation processes provided by lactic acid bacteria. |
| Canada | NS | Choose foods that have little to no added sodium, sugars or saturated fat. Saturated fatty acids are found in animal foods, such as cream, butter, cheeses. Reducing their intake can help lower cardiovascular risk factors such as LDL cholesterol. Reducing intake of foods containing mostly saturated fatty acids by replacing them with food containing mostly unsaturated fat decreases total and LDL cholesterol and lowers cardiovascular disease risk. Foods and beverages with added sugars are associated with an increased risk of weight gain, overweight and obesity and type 2 diabetes. |
| Chile | NS | Consuming too much salt can increase your blood pressure, and lead to cardiovascular disease. Milk and dairy contain calcium, which can help prevent osteoporosis later in life, as well as the loss of bone mass. |
| China | Consume plenty of vegetables, milk and soybeans. | No |
| Colombia | Take care of your heart: eat nuts, peanuts, and avocado; reduce the consumption of vegetable oils and margarine; and avoid animal fats like butter and lard. | Milk and dairy consumption can improve bone health in children and adults. Also, there is an inverse relationship between milk and dairy intake and high blood pressure and cardiovascular disease. The consumption of this group of foods can also reduce risk of diabetes. |

| | | |
|---|---|---|
| Croatia | Snacks should be chosen carefully: give preference to fermented milk products, fresh and dried fruits and nuts. | No |
| Cuba | Milk and dairy consumption is required in children, teenagers, pregnant and breastfeeding women and elderly people. Choose vegetable oils above the ones coming from animals. | Animal fats contain saturated fatty acids, which increase the risk of cardiovascular disease |
| Cyprus | NS | No |
| Denmark | NS | Dairy products are rich in protein, vitamins and minerals, such as calcium. They also contain saturated fatty acids, which can increase the risk of lifestyle diseases in excessive amounts. |
| Dominican Republic | NS | Calcium is essential for bone formation, muscle contraction and nervous system function. Zinc is essential for health. |
| Ecuador | NS | No |
| El Salvador | Include in your daily diet at least one of the following foods: milk, cheese, curd, cottage cheese or eggs. | Reduce your intake of oils and fats, as well as sugary foods, as they can increase your risk of overweight, obesity and diabetes. Milk and dairy contain lactose, proteins, vitamins, minerals and calcium, which can help form bone and prevent osteoporsis. Fats are important to transport fat-soluble vitamins and hormone formation but should be consumed in small amounts, as in excess can lead to atherosclerosis |
| Estonia | Unless there is a direct need, milk should not be replaced by plant-origin analog products. | Milk contains essential amino acids for the human body. It is a very good source of calcium and magnesium. Milk fats are an important building material and source of energy. |
| Finland | NS | No |
| Macedonia | NS | Eat milk and dairy products with low fat and salt content. |

(Continued)

**Table 3** (*Continued*)

| Country | General recommendations | Does the dietary guideline mention scientific evidence? |
|---|---|---|
| France | Consume foods that are rich in calcium (mainly dairy products, in addition to vegetables and mineral water rich in calcium, for those who consume mineral water). | Milk is rich in calcium, essential for bone and teeth formation. Calcium also helps muscle contraction, coagulation and the nervous system functioning. |
| Georgia | NS | Milk and dairy contain Ca and protein, which are important in bone and teeth formation. Also, help blood circulation. |
| Germany | Consume milk and dairy products such as yogurt and cheese daily. | Milk and dairy products are good sources of high-quality protein, vitamin B2 and calcium. |
| Greece | NS | No |
| Grenada | NS | Reducing fat intake, especially animal fat, can decrease the risk of heart and blood vessel problems, improve weight control and lead to a better quality of life. Therefore, change from whole milk to reduced fat, from reduced-fat to low fat, and from low fat to non-fat. |
| Guatemala | Could be cow's or goat's milk. Consume especially during pregnancy, childhood, and breastfeeding. | Milk and dairy are rich in calcium, essential for teeth and bone formation. Excessive consumption of full-fat milk can lead to cardiovascular disease. |
| Honduras | NS | No |
| Hungary | Drink milk (0.5 L) every day; choose fermented dairy products (curdled milk, kefir and yogurt) as often as possible. | No |
| Iceland | NS | No |

| | | |
|---|---|---|
| India | Milk which provides good quality proteins and calcium must be essential. | Avoid overeating to avoid overweight and obesity. |
| Iran | NS | No |
| Ireland | Fats, spreads and oils provide essential fats but use in very small amounts. Use as little as possible. | No |
| Israel | NS | No |
| Italy | NS | No |
| Japan | NS | No |
| Kenya | NS | No |
| Latvia | NS | No |
| Lebanon | Choose white cheeses over yellow, as they contain less total and saturated fat. Consume natural cheeses more often than processed cheeses, as they contain less salt. Consume calcium and vitamin D-enriched milk. Consume natural yogurt instead of salted yogurt-based drinks or sweetened yogurt. Soak high-salt cheeses to decrease salt content. | Milk and dairy products are the major source of calcium and vitamin D, two vital nutrients for promoting dental and bone health, and for attaining peak bone mass during childhood and adolescence, which is important for reducing the risk of fractures and osteoporosis later in life. They are also important sources of protein (vital for cell synthesis), riboflavin (necessary for normal cell function, growth, and energy production), vitamin A (important for vision and cellular growth), vitamin B12 (vital for maintaining healthy nerve cells and red blood cells) and zinc (which is important for the immune system and wound healing). |
| Lithuania | NS | Milk and dairy products provide the body with protein and calcium. Calcium is needed for children, adolescents and women in specific. It is required for the formation of bones and teeth, and the development of the nervous system, as well as internal and external gland secretion, blood coagulation and skeletal and cardiac muscle function. |

(*Continued*)

**Table 3** (*Continued*)

| Country | General recommendations | Does the dietary guideline mention scientific evidence? |
|---|---|---|
| Malaysia | Consume adequate amounts of milk and milk products. | No |
| Malta | NS | Cut down on the amount and frequency of sugary foods, a diet high in sugar will result in poor oral health and may contribute to heart and lung problems, diabetes and stroke. Full-fat milk and dairy can be high in saturated fatty acids. |
| Mexico | Include the three food groups: fruits and vegetables, cereals, legumes and animal-source foods in your breakfast, lunch and dinner. | No |
| Mongolia | Consume milk and milk products every day. | No |
| Nepal | Consume milk or milk products daily. | No |
| The Netherlands | Take a few portions of dairy products daily, including milk or yogurt. | No |
| New Zealand | Include milk and milk products in your diet. | No |
| Nigeria | The diet should contain as wide a variety of foods as possible, e.g. cereals, legumes, roots/tubers, fruits, vegetables, fish, lean meat, local cheese (wara). | In order to reduce the risk of cardiovascular disease, diabetes mellitus and obesity, reduce intake of full cream evaporated milk, butter, cheese, full cream powdered milk and condensed full cream milk. |
| Norway | NS | No |

| | | |
|---|---|---|
| Oman | Limit cheese intake. | Adequate amounts of dietary fat are essential for health. Polyunsaturated fatty acids lower the risk of coronary heart diseases. Saturated fatty acids and trans fats increase the risk of chronic diseases. Vitamin D is required to maintain normal blood levels of calcium and phosphate, which are needed for bone health, muscle contraction, nerve conduction and general cellular function. High sodium intake is associated with high blood pressure, cardiovascular disease and stomach cancer. Milk is rich in calcium, which may reduce the risk of osteoporosis. |
| Panama | NS | Milk and dairy contain vitamins A, B and D, minerals such as calcium and phosphate, as well as proteins, which help bone and tissue formation. A diet high in fat and sugars can lead to obesity, diabetes and cardiovascular disease. |
| Paraguay | NS | Use small quantities of iodized salt when preparing your meals to prevent high blood pressure. Increase daily intake of dairy products to maintain strong bones and teeth. |
| Philippines | Consume milk, milk products and other calcium-rich foods, such as small fish and shellfish, every day for healthy bones and teeth. Limit intake of salty, fried, fatty and sugar-rich foods to prevent cardiovascular disease. Be active to prevent obesity. | No |
| Poland | Drink at least two large glasses of milk a day. Milk could be substituted by yogurt, kefir and some cheese. | No |
| Portugal | NS | Saturated fatty acids and trans fats found in milk, butter or cheese increase blood cholesterol. Mono- and polyunsaturated fats reduce it. Milk also contains calcium, essential for bone and teeth formation. |

(*Continued*)

**Table 3** (*Continued*)

| Country | General recommendations | Does the dietary guideline mention scientific evidence? |
|---|---|---|
| Qatar | Choose vitamin D fortified milk. | Milk and dairy are high in calcium. Consuming low-fat dairy food is related to a reduced risk of heart diseases, stroke, high blood pressure, colorectal cancer, type 2 diabetes mellitus, metabolic syndrome and strengthening bones. Limit processed cheese and soak high salt white cheese in water before eating to remove excess salt. An excessive salt intake can increase blood pressure and the risk of cardiovascular disease, diabetes mellitus, chronic kidney disease and stomach cancer. |
| Romania | NS | No |
| Russian Federation | NS | No |
| Seychelles | NS | No |
| Slovenia | NS | No |
| South Africa | Have milk, maas or yogurt every day. | Milk has a low sodium to potassium ratio, as well as bioactive peptides, which may protect against the development of non-communicable diseases. Milk and dairy play an important role in the regulation of body weight and bone mineral content. Dairy can be protective against metabolic syndrome, obesity, high blood pressure and cardiovascular disease and cancer. |
| Spain | Eat milk and dairy products every day. | Milk and dairy are rich in calcium and vitamin D, which can improve bone formation in children. They can also prevent high blood pressure and metabolic syndrome. |
| Sri Lanka | Consume milk or milk products daily. Non-fat milk is preferable to adults, obese children and people with cardiovascular disease. | Milk and dairy help in growth and maintenance of your bones and teeth. |
| Sweden | It's fine to enjoy a little butter now and again. | Saturated fatty acids, like the ones from butter, have been associated with an increased risk of cardiovascular disease. Dairy products contain lots of calcium, important for bones and teeth, and they contain important vitamins and minerals, as well as protein. Low-fat dairy help reduce the risk of high blood pressure, stroke and type 2 diabetes. |

| | | |
|---|---|---|
| Switzerland | Consume three portions of dairy products and also one portion of meat/fish/eggs/tofu per day. Alternate between these protein-rich foods. | No |
| Thailand | NS | Milk is rich in calcium and phosphorus, which are essential for building strong bones and teeth. Also, it contains protein, lactose and vitamins (B12), which promotes growth and proper functioning of body tissues. Saturated fatty acids and cholesterol from milk can lead to high levels of blood cholesterol, and increase risk of developing cardiovascular disease. |
| Turkey | NS | Milk and dairy are rich in protein, calcium, phosphorus, vitamin B2, B12, which help bone and teeth development. They are also rich in fat, in particular SFA and cholesterol, as well as vitamin A. For individuals with high cholesterol levels we recommend consuming low-fat milk and dairy. Salt content in cheese is high, so prefer low in salt cheese varieties. Eating yogurt and drinking salty ayran for treatment of diarrhoea may save life. |
| United Kingdom | Have some dairy or dairy alternatives (such as soy drinks). | Milk and dairy food are good sources of protein and vitamins, and they are also important source of calcium, which helps to keep our bones strong. Some dairy food can be high in saturated fatty acids, but there are plenty of lower fat options to choose. |
| Uruguay | Dairy products are important at all ages, consume at least half a litre of milk every day. | No |
| USA | A healthy eating pattern limits saturated fats and trans fats, added sugars and sodium. | The dairy group contributes many nutrients, including calcium, phosphorus, vitamins A, D, B2, B12, K, protein, zinc, choline, magnesium and selenium. Butter contains saturated fatty acids, which are related to an increase in cardiovascular disease risk. |

(Continued)

**Table 3** *(Continued)*

| Country | General recommendations | Does the dietary guideline mention scientific evidence? |
|---|---|---|
| Venezuela | Milk provides essential nutrients for children, teenagers, pregnant and breastfeeding women and elderly people. It is not essential for adults, as can be substituted by other animal-originated foods. | Animal foods are rich in protein, calcium, vitamin B1, B2, zinc, and can reduce risk of cardiovascular disease. During pregnancy, lactation and childhood milk and dairy are good sources of calcium to meet the needs of this mineral. |
| Vietnam | Children over 6 months and adults are recommended to consume milk and dairy products appropriate to their age. | No |

NCD: non-communicable disease
NS: not specified

In 2014, the 'Rome Declaration on Nutrition', drawn up under the responsibility of the FAO and WHO, and embraced by 170 nations, raised an urgent appeal to everybody to battle hunger, achieve food security and encourage a healthy and sustainable diet.[36] The sustainability of food production and consumption is a hot topic for consumers, industries, governments and scientists alike. In Western societies, the debate frequently condenses to 'consume more plant-based foods and less animal-source foods and eat less in general', especially from the non-basic food categories. In this context, less animal-source foods are generally interpreted as less dairy products, less eggs and less meat, referring to the higher ecological impact associated with their production.

To combat global warming, it is important to reduce human-induced greenhouse gas emissions. Greenhouse gas emissions of products are expressed as carbon footprint ($CO_2$ eq.). The food system is held responsible for around 25% of greenhouse gas emissions worldwide, most of which is related to food production. Worldwide, the dairy sector is responsible for 3.2% of the total greenhouse gas emissions. It was recently shown that the global average carbon footprint of raw milk has decreased from 2.8 kg $CO_2$ eq./kg fat and protein corrected milk (FPCM) in 2005 to 2.5 kg $CO_2$ eq./kg FPCM in 2015.[37] Animal-based products have a higher carbon footprint than plant-based products. On a kg basis, the footprints of semi-skimmed milk (1.2 kg $CO_2$ eq.) is higher than of soy drink (0.3–0.5 kg $CO_2$ eq.),[38] almond drink (~0.15 kg $CO_2$ eq.)[39] and oat drink (0.3–0.4 kg $CO_2$ eq.).[40] However, per 100 g of protein the footprints of dairy and plant-based drinks are largely comparable, except

**Table 4** Total dietary intake as per cent of local RDA for three different dairy user groups in all SEANUTS countries combined

| | <1 Dairy consumption | | 1 Dairy consumption | | ≥2 Dairy consumptions | |
|---|---|---|---|---|---|---|
| Dairy per day | Mean | SE | Mean | SE | Mean | SE |
| Energy | 74[a] | 1 | 90[b] | 1 | 97[c] | 1 |
| Protein | 119[a] | 1 | 158[b] | 1 | 179[c] | 1 |
| Calcium | 55[a] | 1 | 96[b] | 1 | 83[c] | 1 |
| Iron | 76[a] | 1 | 117[b] | 2 | 101[b] | 2 |
| Zinc | 75[a] | 1 | 112[b] | 1 | 148[c] | 2 |
| Vitamin B1 | 71[a] | 1 | 114[b] | 1 | 141[c] | 2 |
| Vitamin B2 | 69[a] | 1 | 147[b] | 2 | 184[c] | 2 |
| Vitamin B3 | 63[a] | 1 | 92[b] | 1 | 103[c] | 1 |
| Vitamin C | 73[a] | 2 | 132[b] | 3 | 99[c] | 2 |
| Vitamin A | 63[a] | 1 | 96[b] | 1 | 88[c] | 2 |
| Vitamin D | 41[a] | 1 | 86[b] | 2 | 101[c] | 2 |

[a,b,c] Different letters in superscripts indicate the significant difference with other cells in the same row.

for soy drink (semi-skimmed milk 3.3 kg $CO_2$ eq., soy drink 1.3 kg $CO_2$ eq., almond drink 3.8 kg $CO_2$ eq. and oat drink 3.8 kg $CO_2$ eq., resp.). To make a 'fair comparison' one should account for both protein and calories, as well as vitamins and minerals.[41] With the introduction of plant-based alternatives, it is important to recall that the definition of milk refers to 'normal mammary secretion of milking animals obtained from one or more milkings without either addition to it or extraction from it, intended for consumption as liquid milk or for further processing'.[42]

In 2010, Smedman calculated the Nutrient Density to Climate Impact index, which accounted for protein quantity, as well as vitamins and minerals. Using data of unfortified products, milk scored substantially better on this index than other beverages, such as water, orange juice, oat drink and soy drink.[43] It should be noted that discussing sustainability only makes sense considering whole food systems instead of isolated products. A particular product cannot be sustainable on its own as it is always part of a diet. A sustainable diet provides safe and healthy foods, in an affordable and culturally acceptable way, to consumers. A single product with a low carbon footprint is not equal to a sustainable solution.

The higher environmental impact of animal-based products is not restricted to greenhouse gas emissions. For animal food production plant materials need to be converted. As a result, animal products score, on average, worse on exhaustion of natural and fossil resources (e.g. land usage), acidification (especially ammonia emissions), fertilization (especially nitrogen and phosphate) and greenhouse gas emissions (methane, nitric oxide, $CO_2$). On the other hand, it is interesting to realize that cows convert grass and other feed that's indigestible to humans into milk with its many nutrients and high-quality proteins. The efficacy of the protein conversion from the feed has been calculated for Dutch cows. The milk protein production is 4.38 times greater than the sum of protein added through the feed[44]!

In the scientific research that seeks to link 'healthy eating' with 'sustainable eating', we see various approaches, on both health and sustainability.[45-49] The indicator 'health' is regularly translated via the technique of linear programming as 'meeting the recommendations for nutrients and/or following guidelines for good nutrition'. From there, the desired changes toward a more sustainable and healthier diet are modelled. This is a useful exercise that clarifies what choices consumers could, respectively, should make to eat in a more sustainable manner. An example is Optimeal®, a quadratic calculation model developed by Blonk Consultants (Netherlands) in collaboration with the Netherlands Nutrition Centre.[41] Interestingly, this model shows that following a healthy and sustainable diet is not always as easy as it seems. The total environmental footprint of our diet is the result of production methods and consumer choices. Optimeal® uses nutritional and environmental data from over 200 products regularly consumed in the Netherlands and uses life-cycle assessments (LCAs) to calculate the environmental impact of each product. Modelling showed that lowering dairy consumption, unlike meat, has only negligible environmental effects. This can be explained by the nutrient density of dairy – when lowering or omitting dairy from the diet, nutrients (e.g. calcium, protein) must be provided by other products to retain a diet that still provides all required nutrients. To get the same amount of nutrients as provided by dairy, an individual must consume much more fruit and vegetables than the recommended daily portions. This not only affects portion sizes but also the total cost of the diet. Optimeal® is only a nutrient calculation model which does not consider the health effects of foods on e.g., non-communicable diseases.

There are however attempts to link sustainability indicators to health outcomes.[50,51] This adds additional complexity, as research must then determine whether more sustainable food patterns actually produce health benefits, other than through the usual analyses using scores for meeting healthy eating indexes, which *a priori* indicate what is healthy or less healthy. The update to the 2006 Dutch dietary guidelines[52] is a nice illustration of this approach. It considered the totality of evidence about the relationship between different whole foods and a range of health outcomes. A scientific panel evaluated the latest evidence in relation to ten different chronic disease outcomes for nutrients, foods and food patterns[52] and categorized the strength of the evidence. Food-based dietary guidelines were developed based on this systematic, evidence-based approach.

## 3 The dairy matrix

In most food products, components are neither homogenously dispersed nor present in free form. Instead, they are part of complex microstructures.[53] The term 'food matrix' is often used to capture this level of complexity and is frequently coined to explain why food components behave differently in isolated form than as part of food structures. Each food has its own matrix that is impossible to reduce to the sum of its elements.[2,54] A clear-cut example of the dairy matrix is dietary saturated fatty acids (SFAs). Lowering the consumption of saturated fat has been a continuous central theme of dietary goals and recommendations in many countries since the late 1970s. Since the 1980s, it has been recommended that SFA intake be limited to <10% of total calories as a means of reducing risk for cardiovascular disease (CVD).[55-57] Since then, many studies have produced evidence that contrasts this association.[58-63] These studies show that the physiological effects of SFAs are dependent on many factors, one important factor being the product via which they are consumed. From these studies, it becomes clear that the physiological effect cannot be reduced to the effects of SFAs in isolation. In cheese, for example, SFA content would be expected to increase CVD risk, but the opposite is actually found.[64] Similar effects were observed for other product groups as well, e.g., beef[65] and chocolate.[66]

The organization of lipids in milk is yet another illustration of the dairy matrix. In raw milk, the milk lipids are part of milk fat globules of, on average, 4 μm diameter that are surrounded by a trilayer structure known as the milk fat globule membrane or MFGM.[67] The MFGM contains many bioactive components like MGFM proteins and polar lipids. Processing during product production changes both the size of lipid droplets as well as the architecture of the triglycerides (TAG)/water interface. Homogenized lipid droplets are typically smaller (<1 μm in diameter) than milk fat globules and are (partially) covered by milk proteins instead of the MFGM. These changes affect the digestion and metabolism of the fat within the resultant dairy products.[67-69] The amount of TAG-water interface available for lipases directly influences digestion efficiency. Triglycerides from small native fat globules (~1.75 μm) are digested more efficiently than large native fat globules, both during the gastric and the intestinal phase. For processed lipid droplets the opposite effect is seen; the changes in interface composition (adsorption of caseins and whey proteins) upon decreasing globule size by homogenization lead to lower levels of lipolysis. Detailed structural characterizations performed during *in vitro* digestion experiments revealed differences in microstructure between native fat globules and processed lipid droplets due to coagulation of milk proteins at acidic pH. Interactions between the caseins adsorbed at the interface of homogenized lipid droplets and the protein network formed upon coagulation of milk in the stomach decrease accessibility for the gastric lipase to TAG and cause a decrease in the level of lipolysis.[67] Furthermore, the physicai stability of the emulsion formed in the stomach affects the gastric emptying rate.[70] It was also shown that human pancreatic lipase has a lower catalytic efficiency for processed lipid droplets than for native fat globules.[71] In conclusion, alterations in droplet size, interfacial composition and architecture in processed lipid droplets lead to different behaviour in the gastrointestinal tract compared to native milk fat globules. This impacts digestive lipolysis and thereby lipid bioavailability.

From a nutritional perspective, the dairy matrix can be described as *the structural organization of physically and chemically interacting food components that influence nutrient bioaccessibility,*[1] *nutrient bioavailability,*[2] *gut endocrine function and gut microbiota*.

Differences among food matrices are of considerable importance for the nutritional performance and health potential of products that have similar

1 Release of nutrients from the food matrix.
2 Absorption of nutrients in the gastrointestinal tract.

chemical composition.[72] This is defined as the 'food matrix effect'.[62,73] In the case of milk and dairy products, different dairy products have different matrices that result in different matrix effects. This was clearly shown in a study with elderly individuals, where it was found that consumption of different dairy products and dairy ingredients lead to clear differences in postprandial blood essential amino acid (EAA) concentrations. Upon the ingestion of dairy products containing 25 g of protein, a higher increase in EAA concentrations in blood was observed after consumption of yogurt, compared with milk and cheese[74] (Fig. 2).

A recent study showed another nice example of a matrix effect. In this study it was shown that consumption of Cheddar cheese leads to a slower but more sustained appearance of branched-chain amino acids in circulation over the postprandial period, peaking at ~120 min, yet with a lower glycemic profile, compared to an isonitrogenous amount of milk. Both milk and cheese stimulated a muscle anabolic program associated with mTORC1 signaling that was more evident with milk, but more persistent with cheese.[75] The authors speculate that '*drinking a small glass of milk with cheese, may result in a greater net protein balance over several hours in comparison to a bolus of milk alone, because of the ceiling for acute anabolic activation and subsequent oxidation of excess amino acids (coined the 'muscle-full effect') thereby limiting the anabolic benefit of the beverage on its own*'.

The recognition that foods are often more than the sum of their individual nutrients and possess a unique matrix and matrix effects has led to a new paradigm that is based on the understanding that nutrients act together in tissues, that all tissues need multiple nutrients and that inadequate intake impairs all body systems (though obviously some more prominently than others). This new paradigm has opened the way to scientific insights that would not have been expected based on a reductionistic single-nutrient approach. Figure 3 summarizes the findings related to dairy consumption and cardiovascular-related outcomes. As can be seen in this figure, mostly neutral or favourable associations have been found in observational and interventional trials. Drouin-Chartier et al.[76] studied the postprandial lipid profiles of volunteers fed 33 g of dairy fat from either a firm Cheddar cheese, cream cheese or butter, and observed different lipolysis profiles over an 8-h period. The triacylglycerol concentrations were greater at 2 h following the cream cheese compared to the butter or cheese, while at 6 h, the Cheddar cheese had a 42% increase in triglyceride response from baseline compared to 14% increase following the cream cheese.[76] These results clearly show that nutritional strategies should not remain focused on reductionist strategies, such as total fat or total saturated fat, thereby overlooking the importance of food matrix (food type and quality), processing methods and dietary patterns.[77] Labeling single nutrients as 'good' or 'bad' and, through association, any foods containing those nutrients also as 'good' or 'bad', regardless of their overall nutrient profile, is an oversimplification

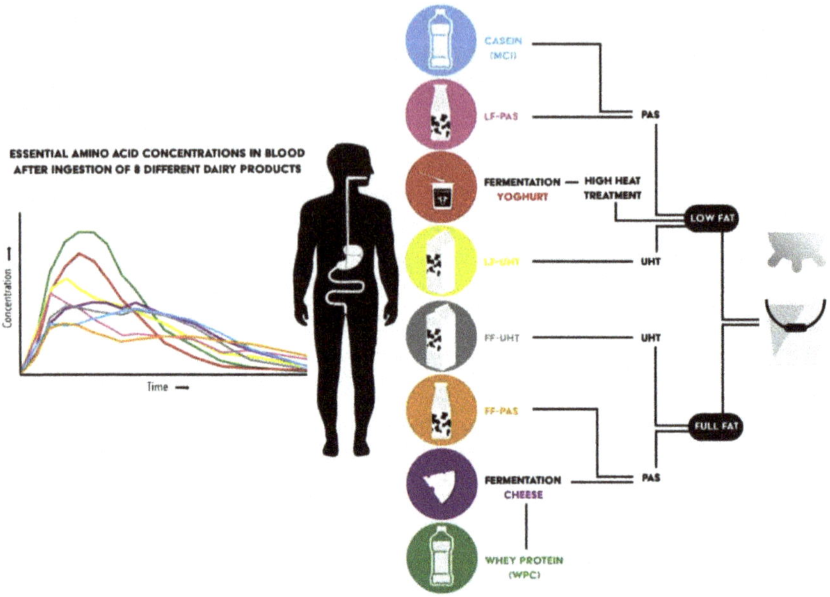


**Figure 2** Postprandial essential amino acid response after the consumption of eight different dairy products/ingredients. MCI: micellar casein isolate; LF-PAS: low-fat pasteurized milk; LF-UHT: low-fat UHT-treated milk; FF-UHT: full-fat UHT-treated milk; FF-PAS: full-fat pasteurized milk; WPC: whey protein concentrate. Source: Horstman et al.[74]

and does often no justice to the food items in question. Nutrition research should move beyond reductionism and toward the study of whole foods and dietary patterns to better inform dietary guidelines, by considering the effects of whole foods and whole diets on human wellbeing, rather than individual constituents independently.

## 4 Dairy matrix effects: case studies

Among animal-source foods, milk is one of the richest in diversity and complexity. Milk contains high-quality proteins, hundreds of different fatty acids and oligopeptides, lactose and almost all vitamins and minerals in significant proportions.[78] Many dairy products like yogurt, cheese, ice-cream and butter are made from milk. Each one has its own, unique dairy matrix. There are liquid structures like milk but also the semi-solid or gel-like structure of yogurt. Bovine milk is an oil-in-water emulsion consisting of lipid molecules, protein (mainly whey protein and casein) and lactose suspended in water.[79] A wide range of factors, including the specific breed of cow, the feeding regime, the age of the cow at milking and the time of year[80] can all influence the nutrient profile of unprocessed milk. Clearly exploring the health effects of dairy foods only makes sense if dairy is no longer treated as a homogeneous food group but instead as a diverse group of different products. In this section, some examples are given of how the dairy matrix influences human physiology.

### 4.1 Dairy and musculoskeletal health

Currently, around 200 million people worldwide show signs of osteoporosis. This number is expected to increase by circa 30% in the coming years. Nearly one in two women and one in three men, over 50 years of age, will suffer from at least one osteoporotic fracture during the second half of their lives.[81,82] According to the US National Osteoporosis Foundation, based on a thorough review of evidence-based literature, the best evidence is available for the positive effects of calcium intake and physical activity, especially during the late childhood and peripubertal years - a critical period for bone accretion. Further, they concluded that good evidence is also available for the role of vitamin D and dairy consumption.[83]

Another major clinical problem in the elderly population is sarcopenia, the age-related decline in muscle mass and function.[84] Sarcopenia may lead to an increased risk of falls and more bone fractures which will negatively impact mobility and overall quality of life. In addition, the risk for metabolic disorders like type 2 diabetes tends to increase due to the lower levels of physical activity and reduced muscle mass in the elderly.[85] The following sections will describe the role of dairy products and their specific matrix in relation to musculoskeletal health.

#### 4.1.1 Dairy and bone health

Bone is a rigid tissue that is made up of a network of proteins filled with hydroxyapatite, a complex calcium phosphate salt. Other fillers include sodium, magnesium, potassium and zinc. Together, these minerals, in combination with the protein scaffold, provide our bones their strength and rigidity.[86] In humans, bone acts as a calcium reserve, for which size is determined by the combination of mechanical loading and dietary calcium availability. From birth to adolescence, bones grow in length and width. Bones are continuously renewed in a process known as bone remodelling. Bone remodelling replaces old bone (i.e. the process of bone resorption) with new (i.e. the process of bone formation) without changing the shape of the bone. This remodelling allows for the preservation of skeletal mechanical integrity and the control of calcium homeostasis, i.e. the process of releasing calcium into the circulation when needed. Obligatory calcium losses must be compensated by sufficient calcium intakes and efficient intestinal absorption. Otherwise, bone is used as a source of calcium to maintain homeostasis in extracellular calcium concentration.

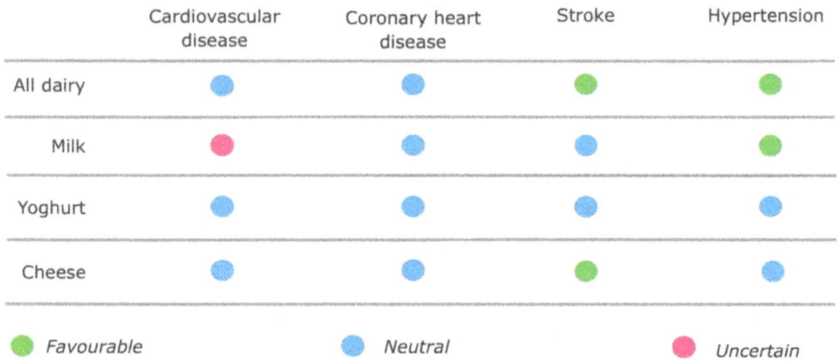


**Figure 3** Dairy product consumption and risk of cardiovascular-related clinical outcomes. Adapted from Drouin-Chartier et al.[184]

Several nutrients are required for proper bone health, irrespective of age. Next to the above-mentioned minerals and protein, vitamin D and vitamin K are important.[87] Dairy products are widely acknowledged for their significant contribution to bone health. Most dairy products contain all the nutrients needed for bone formation and maintenance, and it takes only a few servings to deliver up to about 70% of the recommended daily dietary calcium intake. Some vegetables and dried fruits are also considered good sources of calcium, yet the serving size needed to equal the dairy contribution is generally much higher. Also, the fractional absorption of calcium is often much lower than from dairy products as was shown for spinach.[88] So, whereas it is theoretically possible to achieve adequate calcium intake without dairy consumption, in practice, this is far from convenient.

Especially in the elderly, bone health is particularly sensitive to dietary inadequacies. Reduced vitamin D synthesis in the skin, reduced intestinal calcium absorption and failure to adapt to lower levels of calcium in their diet are thought to explain this. For women after menopause, changes in sex hormone levels and nutrition are associated with an increase in bone remodelling, and thus with a higher risk for bone fragility. Although the reduction in bone mass during ageing is a natural process, it should be restricted to a minimum to prevent bones from becoming too porous and brittle, e.g., to prevent osteoporosis. Osteoporosis is a serious health problem that affects one in three women over 50 and one in five men over 50. It has been shown that sufficient dietary calcium, protein and vitamin D are necessary for bone health of the elderly.[89] Dairy products are also associated with increased bone mass and decreased fractures rates in the elderly.[90]

As stated before, dairy products deliver a wide variety of bone-related nutrients to the diet like calcium, zinc, iodine, vitamins B2, B12, D, A and indispensable amino acids. The dairy matrix allows for far greater amounts of calcium and phosphate in milk to be present than would be possible if simple solutions of calcium phosphate salts would be used. In a simple solution, this would lead to an undesirable precipitate of calcium phosphate. In milk, the presence of both citrate, which forms soluble complexes with calcium, and caseins, which encapsulate calcium phosphate into the casein micelles, allows bovine milk to readily provide approximately ten times more calcium in stable and bioavailable form than would be possible with a simple solution of calcium phosphate.[91] To meet growth requirements, foods need to provide calcium and phosphorus in a certain ratio. The recommended intake ratio for children is 2:1 (Ca:P) and is 1:1 for adults.[92] The ratio (*w/w*) of Ca:P in bovine milk is 1.3:1. Interestingly, it has been found that in rats, calcium intakes in the form of milk increase bone strength more than calcium in the form of calcium carbonate supplements.[93] In another rat study, it was shown that calcium-supplemented almond drink was not capable of recovering bone mineralization and could be detrimental to bone mineralization and strength outcomes.[94] Two studies tried to replicate this effect in humans. In a 2-year randomized, placebo-controlled trial (RCT), 195 healthy girls (aged 10–12 years old), who had a habitual dietary calcium intake <900 mg/d were allocated to one of four groups: calcium (1000 mg) + vitamin D3 (200 IU/5 μg) supplement, calcium supplement (1000 mg), cheese (1000 mg calcium) or placebo. Cheese consumption resulted in the highest percentage change in cortical thickness of the tibia compared to the other groups. When compliance was considered in a per-protocol analysis, cheese consumption resulted in a higher whole-body bone mineral density in those with >50% compliance. Thus, cheese had a superior effect on bone mass accrual than calcium, or calcium and vitamin D, supplementation. With the use of a statistical model to control for growth velocity differences, these differences disappeared.[95] In another randomized controlled trial, the effect of dairy products or a calcium supplement, both providing 1200 mg calcium/d, on markers of bone metabolism and bone mineral density in 101 postmenopausal women was studied.[96] After 12 months, the dairy group had significantly greater improvements in the pelvis, total spine and total-body bone mineral density than the calcium supplement group or the control group.

There are only a small number of prospective studies and randomized controlled trials available that have compared plant-based alternatives to the dairy food group, but the available scientific evidence is quite revealing.[97,98] The nutritional content of plant-based alternatives may not compare at all to milk, especially in those who have very low protein content (Table 5). If cows' milk is exchanged for non-fortified and non-supplemented plant drinks consumers may risk various nutrient deficiencies including calcium, B12 and indispensable amino acids. The latter are very important for bone health as they stimulate the synthesis of growth factors such as insulin-like growth factor 1 (IGF-1) and promote the formation of bone.[99] When comparing milk to soy drink in a randomized controlled trial, milk proved to be superior in preventing loss of bone mineral density (BMD) in 141 postmenopausal Chinese women, at the same levels of calcium intake.[97] Differences in calcium bioavailability could be relevant for this observation. The bioavailability of minerals in plant sources can be quite variable and is dependent on many factors such as the type of mineral used for fortification (e.g. tricalcium phosphate, calcium carbonate etc.), the presence or absence of anti-nutritional factors (oxalates, phytates etc.), the processing of the product, etc.

At the moment, no RCT has yet found conclusive evidence for the protective effects of dairy foods on bone fracture risk. Two case-control studies have hinted at a protective effect,[100,101] however, the evidence remains inconsistent. Shortness of study length, lack of consistency in which dairy foods are included in analyses, number, timing and quality of dietary assessments and validations of fractures are some of the issues that help explain the scarcity of solid evidence in this area.

### *4.1.2 Dairy and muscle health*

Muscle tissue is constantly being synthesized and broken down, at a rate of approximately 1–2% per day.[102] The main anabolic stimuli for muscle synthesis are food intake and physical activity. Especially proteins are important in this respect. The postprandial rise in muscle protein synthesis is influenced by the level of protein digestion, amino acid absorption, plasma amino acid availability, hormone response, postprandial perfusion of the muscle, amino acid-induced anabolic signaling and myofibrillar protein synthesis. Ingestion of a meal-like amount of protein (~20 g) allows for ~55% of the protein-derived amino acids to become available in circulation. In total, ±20% of the protein-derived amino acids will be incorporated in *de novo* synthesized skeletal muscle during a 5-h postprandial period.[103] Postprandial protein handling and muscle protein synthetic response in relation to protein ingestion are dependent on the amount, type and amino acid composition of the protein ingested as well as the food matrix. The quality of a protein is defined by its amino acid composition (especially the essential amino acids) combined with its digestibility characteristics. Various methods have been developed to assess the quality of the dietary protein fraction. From this information, it is concluded that milk proteins are of very high quality.[104]

Whey protein stimulates postprandial muscle protein accretion even more than casein or casein hydrolysate, which may be explained by its faster digestion and absorption and higher leucine content.[105] Both types of proteins have a high digestibility, but the amino acids from casein are delivered more slowly than those from whey. In general, it is observed that approximately 4 h after consumption of whey protein, the level of circulating EAAs has returned to levels before eating or drinking. Upon consumption of casein, a relatively small

**Table 5** The nutritional profile of cow's milk and plant-based alternatives, with and without fortification

| | Milk | Soy drink | Soy drink, fortified | Almond drink | Rice drink | Rice drink, fortified | Oat drink | Oat drink, fortified |
|---|---|---|---|---|---|---|---|---|
| Energy (kcal) | 64 | 55 | 45 | 47 | 50 | 54 | 50 | 45 |
| Protein (g) | 3.4 | 3.1 | 3.3 | 0.7 | 0.1 | 0.1 | 0.6 | 0.5 |
| Total lipid (g) | 3.5 | 2.3 | 1.9 | 2.2 | 0.9 | 0.9 | 1 | 1.6 |
| Carbohydrate (g) | 4.9 | 5.3 | 3.6 | 5.9 | 10.3 | 11.4 | 9.7 | 7.2 |
| Vitamin A (RE) | 35.3 | 0 | 0 | 0 | 0 | 0 | 0.34 | 0 |
| Vitamin B2 (mg) | 0.18 | 0.01 | 0.02 | 0.02 | 0 | 0 | 0.01 | 0.01 |
| Vitamin B12 (µg) | 0.39 | 0 | NA | NA | NA | NA | NA | NA |
| Calcium (mg) | 119 | 9.86 | 74.5 | 8.8 | 1.85 | 84.3 | 6.56 | 126 |
| Zinc (mg) | 0.36 | 0.25 | 0.28 | 0.11 | 0.02 | 0.05 | 0.41 | 0.08 |
| Iron (mg) | 0.02 | 0.45 | 0.50 | 0.12 | 0.01 | 0.23 | 0.03 | 0.44 |
| Iodine (µg) | 16.5 | 1.3 | 9.45 | 0.89 | 1.04 | 2.5 | 0.418 | 5.9 |
| Phosphorus (mg) | 91 | 44.1 | 41.5 | 14.3 | 7.39 | 28 | 13.2 | 16.9 |

Nutritional content per 1 mL of beverage
RE: retinol equivalents
NA: not assessed
Adapted from Geiker et al.[185]

increase in the circulating levels is observed, but these levels are maintained over a longer period (more than 6 h).

When comparing plant-based protein sources to animal-based protein sources, the latter perform better in stimulating muscle protein accretion. This has been attributed to the relatively low EAA content, relatively low leucine content and the shortage of specific amino acids such as lysine and methionine in plant-based proteins.[106] These observations were also made in balance studies in rapidly growing animals and young children. These studies clearly showed that there are clear and predictable differences between protein sources in terms of digestibility and net protein utilization for tissue building, with values ranging from near-perfect utilization for animal proteins, too much lower values for some plant-based diets (perfect=100%).[107]

Next to high-quality proteins, dairy products deliver many other nutrients needed for healthy muscles like magnesium, potassium and calcium.

Another interesting example of the versatility of the dairy matrix can be seen in sports nutrition. With exercise and sports activities, acute nutrient needs will change. Fluid intake to ensure the replacement of water and minerals (electrolytes) lost in sweat is important. Energy needs also increase because of the elevated energy expenditure with physical activity. Interestingly, in recent years, several studies have hinted at milk, flavoured and chocolate milk, as being good supporting fluids for sports nutrition.[108-113] Various reasons are given: the water and minerals in milk are involved in rehydration; the lactose in milk helps to refill the used carbohydrate stores in the body (mainly present in the liver and muscle) quickly; the effective repair of the damaged proteins in muscles due to the superior quality and bioavailability of milk proteins. Recently, it was shown that ingested lactose can be readily oxidized during exercise and results in higher fat oxidation and reduced reliance on endogenous carbohydrate oxidation.[114,115]

In the situation that still additional growth of the skeletal muscle compartment is required, such as strength or endurance exercise, the availability of EAAs and the presence of anabolic signals must be synchronized to obtain a maximum result. The available literature and experiences by many athletes indicate that whey proteins outperform all other proteins in muscle anabolism. This benefit is primarily caused by its high level of EAAs and extremely good digestibility. In addition to these superior characteristics, the good-tasting whey proteins can provide the body with a high level of branched-chain amino acids (isoleucine, leucine and valine, collectively known as BCAAs). The requirement for this group of EAAs in muscle is relatively high, especially after physical exercise.

With adequate energy and carbohydrate intake, low- to moderate-intensity endurance activity has little impact on dietary protein requirements, and 1 g of protein per kg body weight per day is sufficient. The only situation where dietary protein requirements exceed those for relatively sedentary individuals is in top sport athletes where the maximal requirement is ~1.6 g/kg/d. Although most endurance athletes get enough protein to support any increased requirements, those with low energy or carbohydrate intakes may require nutritional advice to optimize dietary protein intake.[116]

Whey proteins are also a good protein source for the elderly.[117] They deliver higher peak values for their amino acids in the blood, which is beneficial for protein synthesis in the elderly as their muscle protein machinery seems less stimulated by the levels of circulating EAA from casein. Casein and whey proteins differently influence insulin, which is important for uptake of EAA in the muscle. Ageing is accompanied by a progressive loss of skeletal muscle mass and strength, a condition known as sarcopenia. The age-related loss of skeletal muscle mass is attributed to a disruption in the regulation of skeletal muscle protein turnover, resulting in an imbalance between muscle protein synthesis and degradation. The causes are multifactorial but include decreased physical activity, malnutrition, increased inflammatory cytokine production, oxidative stress and reductions in both growth and sex hormones. The dietary protein requirement for elderly individuals is dependent upon their health status; however, 0.8 g of protein per kg bodyweight per day is inadequate to maintain lean body mass in the average healthy elderly individual. It appears that an intake of up to 1.5 g of protein/kg BW/day, or about 15–20% of total caloric intake, is a reasonable target for elderly individuals wishing to optimize protein intake in terms of health and function.[118] Needless to say that high-quality protein is preferred.

Physical activity and/or exercise stimulate post-exercise muscle protein accretion in both the young and the elderly. However, with the elderly timing of administration seems more important. This means that consumption of milk protein should be distributed over the various meal moments during the day, breakfast, lunch and dinner.[119]

## 4.2 Dairy and blood pressure

Hypertension is a risk factor for the development of other cardiovascular diseases and remains one of the leading causes of death worldwide. In 2010, the US Dietary Guidelines Advisory Committee (DGAC) published an evidence-based review on the effects of milk and milk products (dairy) on blood pressure (BP). The DGAC's conclusion was that a moderate body of evidence suggests an inverse relationship between the intake of milk and milk products and BP.[120] Since then, more evidence has been found for a potential beneficial effect of dairy. In a Chinese study carried out in Singapore,[121] it was found that dairy food intake was inversely associated with the risk of hypertension in a dose-dependent manner. Hazard ratios (HR) across quartiles were 1.00 (lowest quartile, reference), 0.97 (95% CI: 0.92, 1.02), 0.98 (95% CI: 0.92, 1.03) and 0.93 (95% CI: 0.88, 0.98) (P-trend = 0.01). Milk accounted for ~80% of all dairy products consumed in this population. Daily milk drinkers had a lower risk of hypertension (HR: 0.94; 95% CI: 0.89, 0.99) than did non-drinkers. Also, a recent Dutch intervention study showed that a high intake of dairy products is associated with a reduction of BP in overweight middle-aged men and women.[122] Interestingly, it was found that both systolic and diastolic BP were lowered.

The Dietary Approaches to Stop Hypertension (DASH) study, in which the effect of a whole diet approach on blood pressure was investigated, provided some important insights into the role of low-fat dairy products in blood pressure regulation.[123] Participants who consumed the DASH diet (a diet including daily servings of fruit and vegetables and three daily servings of low-fat dairy products) produced the greatest reductions in blood pressure compared with all other groups. Systolic and diastolic blood pressures were reduced by 5.5 mmHg ($P < 0.001$) and 3.0 mmHg ($P < 0.001$), respectively, compared with the control diet. The effects of the DASH diet were apparent within 2 weeks of starting the diet and were maintained for the next 6 weeks. Blood pressure was lowered in both men and women, hypertensive as well as normotensive subjects. The greatest benefits were seen in those hypertensives at the start of the trial; in these subjects, systolic and diastolic blood pressures were reduced by 11.4 mmHg ($P < 0.001$) and 5.5 mmHg ($P < 0.001$), respectively. Substituting low-fat dairy products for full-fat variants in the DASH diet did not change the blood pressure-lowering effect.[124] Full-fat dairy also reduced the plasma triacylglycerol and very low-density lipoprotein concentrations without significantly increasing LDL cholesterol.

The study by Crippa et al.[125] is yet another example of a dairy matrix effect in relation to blood pressure. Intake of the Grana Padano cheese resulted in a significant decrease in systolic and diastolic blood pressure (decrease of 7–8 mmHg and 5–7 mmHg, respectively) compared to placebo. This is surprising as cheese might have been expected to increase blood pressure due to its high saturated fat and sodium content. But this study shows the opposite effect. The effect of certain nutrients may vary dramatically depending on the product matrix in which they are present and whether fermented or not.

The dairy matrix contains many nutrients that are related to blood pressure regulation like phosphorus, calcium, proteins and bioactive peptides.

In the combined ARIC and MESA study, after an average of 6.2 years of follow-up, phosphorus from dairy products, but not from other sources, was associated with lower baseline BP and reduced risk of incident hypertension. This may indicate that the benefits of phosphorous intake are dependent on interactions with other dairy components.[126]

The role of calcium in BP regulation was addressed by McGrane et al.[126] They refer to a systematic review from the US Agency for Healthcare Research and Quality (AHRQ) on (vitamin D and) calcium. The AHRQ report on calcium and BP covered six systematic reviews and 69 RCTs on calcium intake from foods and/or supplements. Meta-analyses within the reviews covered subjects with hypertension, without hypertension and the general public with and without hypertension. The review included five longitudinal cohort studies on calcium intake and hypertension incidence. Of these, only the Women's Health Study, with a 10-year follow-up, found a highly significant trend across quintiles of calcium consumption, with significantly higher rates of hypertension among women in the lowest quintile of calcium intake. Based on the five longitudinal studies, the AHRQ report concluded that evidence of an association between calcium intake and hypertension incidence was inconsistent.[126]

In 2008, a meta-analysis of 15 RCTs with a total of 826 subjects that examined fermented milk or milk-derived isoleucine-proline-proline (IPP) and valine-proline-proline (VPP) showed a positive association between lactotripeptides and lower BP. Most of the 15 trials were conducted with hypertensive individuals; however, two studies were conducted among prehypertensive individuals.[127] Another review of the evidence from 18 RCTs that tested lactotripeptides IPP and VPP (in either milk drinks or tablets) showed lactotripeptides were effective in reducing elevated BP but not normotensive BP. Effective doses for lactotripeptides ranged from 3.07 mg/d to 52.5 mg/d.[128]

Few studies have investigated the effects of intact whey protein and casein on BP and vascular function. Fekete et al.[129] found one study[130] that reported comparative effects of whey and casein. While both milk proteins were shown to reduce BP, only whey protein improved measures of arterial stiffness. A more recent study by Fekete et al.,[131] also compared the effects of both milk proteins on BP and vascular function. Significant reductions in 24-h BP were observed (23.9 mmHg for systolic BP and 22.5 mmHg for diastolic BP; $P = 0.050$ for both) after whey protein consumption, compared with control intake (=maltodextrin).

In 2017, a meta-analysis was published of seven RCTs (conducted in a range of adult populations) on the effects of milk proteins on BP. Overall, milk protein interventions (from whole foods or supplements) significantly lowered systolic BP by −3.33 mmHg (95% confidence interval −5.62, −1.03) and diastolic BP by −1.08 mmHg (95% confidence interval −3.38, −0.22). It is suggested that higher dietary protein intake (from milk as well as other foods) may contribute to increased tryptophan levels, which have been shown to reduce BP in animal studies.[132]

The inverse association that is observed for total dairy consumption, in mainly observational studies, appears to be largely attributable to low-fat dairy products. A recent meta-analysis of intervention studies shows no relation between total dairy intake and BP.[133]

For the specific effect of yogurt on blood pressure, several potential explanations have been suggested by Wang et al.[134] Yogurt contains approximately 50% more potassium (K), calcium (Ca) and magnesium (Mg) and 30% more protein per serving than milk. They suggest that nutrients in yogurt may also be more bioavailable than in other dairy products and probiotics abundant in yogurt but not in unfermented dairy products have been found to inhibit angiotensin-converting enzyme, and thus reduce BP.

Dairy foods are rich sources of various bioactive peptides, medium-chain fatty acids and micronutrients such as Ca, K and Mg. These nutrients may contribute to potentially underlying mechanisms for the BP-lowering effect of dairy products, such as regulating vascular resistance and promoting vasodilation by enhancing the synthesis of nitric oxide, reducing renal Na retention, blocking $Ca^{2+}$ channels and reducing intracellular $Ca^{2+}$.[134]

Hidayat et al.[132] suggest that higher dietary protein intake (particularly from tryptophan-rich foods such as milk) may also contribute to increased amino acid tryptophan levels, which have been shown to reduce BP in animal studies. The BP-lowering effect of lactotripeptides has consistently been demonstrated and is confirmed in meta-analyses. Whether ACE inhibition is indeed the major mechanism of action for these peptides, or whether other mechanisms may play a role as well, has not yet been elucidated.

High dairy consumption has also been linked with lower uric acid concentrations in the circulation.[135] Elevated serum uric acid predicts the development of obesity and hypertension because it reduces the bioavailability of endothelial nitric oxide (NO).[136] The reduction of uric acid that is associated with dairy intake may yet be another mechanism that helps to explain the blood pressure-lowering effect of the dairy matrix.

## 4.3 The yogurt matrix and type 2 diabetes

Almost half a billion people worldwide live with diabetes of which 90% or more suffer from type 2 diabetes. Diabetes leads to various microvascular and macrovascular complications that have a serious clinical impact on both patients and carers and profoundly negatively impact the healthcare systems. Type 2 diabetes is characterized by insulin deficiency caused by pancreatic β-cell dysfunction and insulin resistance in target organs.[137]

Several systematic reviews and meta-analyses have been published on the relation between dairy products and type 2 diabetes mellitus (T2DM).[138-143] Many reported a significant inverse association between total dairy consumption and T2DM risk.[139-141,144]

A nonlinear inverse association between yogurt consumption and T2DM risk at 80 g/d has been reported (RR: 0.86; 95% CI: 0.83, 0.90 - compared with 0 g/d),[140] although with significant heterogeneity existing between studies ($I^2$ = 73%), which was consistent with previous meta-analyses reporting yogurt consumption to be inversely associated with lower T2D risk.[139,141] The study by Soedamah-Muthu and de Goede[144] updated the meta-analysis of Gijsbers et al.[140] by adding the two new studies[145,146] and found a similar result. Again, a nonlinear significant inverse association was between yogurt consumption and T2D risk (RR: 0.86; 95% CI: 0.83, 0.90; $I^2 = 69\%$, at 80 g/d compared with 0 g/d). Both whole-fat and low-fat yogurt are associated with a lower risk of developing T2DM.[147,148]

The yogurt matrix seems to have a greater impact on metabolic health than non-fermented dairy, but the underlying mechanisms remain obscure.[149] Yogurt is nutrient-dense, may help to reduce body weight gain and to prevent T2DM by the concerted action of protein, calcium, bioactive lipids and several other micronutrients. It has also been suggested that fermentation with bacterial strains produces bioactive peptides that may contribute to the beneficial effect of yogurt on metabolic health.[149] Furthermore, fermented dairy may cause beneficial compositional shifts of the gut microbiota[150] upon its consumption.

Zeroing in on the gut microbiota, an interesting study was performed by Veiga et al.[151] who showed that consumption of a fermented milk product (containing *Bifidobacterium animalis subsp. lactis CNCM I-2494*, *Streptococcus thermophilus CNCM I-1630*, *Lactobacillus delbrueckii subsp. bulgaricus CNCM I-1632* and *CNCM I-1519* and *Lactococcus lactis CNCM I-1631*) stimulated colonic production of acetate, butyrate and total SCFAs. Earlier, it had already been shown that increased production of SCFAs arising from dietary resistant starch was associated with improved insulin sensitivity,[152] and that the concentration of serum acetate, but not propionate or butyrate, was negatively associated with fasting serum insulin and 2-h insulin concentrations after an oral glucose tolerance test (OGTT). Serum acetate was also inversely associated with visceral adipose tissue.[153]

## 4.4 Dairy and oral health

Establishing or maintaining oral health is important to protect teeth and periodontal tissues from damage, demineralization and inflammation.[154] Pathological conditions in the mouth may have a significant systemic impact.[155] Caries and changes in the oral microbiota may lead to inflamed periodontal tissues with a limited barrier function and increased passive transfer of oral bacteria into the bloodstream.[156] Systemic infections with oral bacteria are associated with several diseases including cardiovascular disease, diabetes, and Alzheimer's disease (AD).[157,158] Composition and frequency of meals, the make-up of the oral microbiota (dental plaque), the practice of mechanical oral hygiene (teeth brushing to control the size of dental plaque) and adequate saliva flow are important factors. The constant flow of saliva is not only important for oral cavity clearance but also for buffer capacity and pH control. The latter influences calcium phosphate homeostasis and affects bacterial metabolism.[159] Observational research has convincingly shown that consumption of milk and dairy products is inversely associated with dental caries,[160] especially for yogurt[161-163] and cheese[156,164,165] strong associations have been reported. Cheese increases enamel hardness more than saliva after partial demineralization through exposure to an acidic drink,[166] a finding that was later replicated for cream cheese.[167] Hard cheese was even shown to be able to reharden the dental enamel in patients who had received head and neck irradiation for malignancies and who were suffering from a dry mouth. The investigators suggested that hard cheese consumption in those patients 'may be regarded as effective to keep initial caries under control'.[168] The Danish European Youth Heart Study found that dairy intake among children and adolescents predicts the future risk of dental caries. Dairy intake at 9 years old was associated with fewer dental caries experiences from 9 to 12 years, as well as from 9 to 15 years while calcium, whey and casein intake were associated with lower caries incidence over 3 and 6 years of follow-ups.[169]

The cariogenic load of chocolate milk and other flavoured milk products (such as sugar and fruit juice-sweetened milk products) was shown to be negligible compared to similarly sweetened soft drinks.[170] One of the explanations for this may be the presence of lactose in dairy products. Dental caries occurs due to tooth demineralization by organic acids produced by the combination of orally fermentable carbohydrates (e.g. glucose, sucrose, digestible oligosaccharides and starches) and dental plaque bacteria.[171] Lactose is a disaccharide composed of galactose and glucose that is found exclusively in milk. In cow's milk, the lactose concentration is about 4.6%. Lactose can be fermented by the oral biofilm bacteria but to a significantly lesser extent than other sugars. Lactose is the least cariogenic sugar of all fermentable sugars.[172,173] Fermentable carbohydrates are seen as dietary risk factors for caries, in which the risk level is determined by the type of bacteria that ferment the carbohydrates, at what rate that occurs and to what level the pH drops.[174] Under normal conditions, sucrose fermentation lowers the pH to below 5.0 whereas lactose fermentation lowers it to around 6.0, a pH lower than 5.5. damages the tooth enamel and a pH lower than 6.2 damages root tissues.[172] Both chronic and acute studies show that lactose elicits a less detrimental effect on dental health than sucrose, glucose, fructose and maltose because of its lower acidogenicity.[173,175-178] Therefore, lactose is not thought to have significant adverse dental effects and is considered a low cariogenic sugar.[179,180]

Besides lactose, the dairy matrices of individual dairy products each contribute in a specific way to oral health and there are some common dominators. Cheese acts cariostatic by effectively increasing the amount of calcium and phosphate required for dental enamel remineralization and by buffering plaque pH.[168,181] Petti et al.[162] found a protective effect of milk and yogurt against dental decay, including lower numbers of *Streptococcus mutans*, which may be attributed to different milk components, such as calcium, phosphorus, casein and whey proteins as well as lipids. Caseinophosphopeptides (CPP) and glycomacropeptide (GMP) have been shown to inhibit the growth of several cariogenic bacteria including *Streptococcus mutans*. Furthermore, CPP can form nanoclusters with amorphous calcium phosphate (AMP) so that a calcium and phosphate ion reservoir is created at the tooth surface. This creates a state of super saturation with respect to tooth enamel. This buffers plaque pH and provides ions for tooth enamel remineralization. Glycosidic structures attached to GMP have been shown to be important for its anti-cariogenicity. Like CPP, GMP has shown inhibitory activity to enamel demineralization and promotes tooth enamel remineralization.[182] Milk contains several antibacterial proteins and peptides. The antibacterial protein bovine lactoferrin can compete with *Streptococcus mutans* for binding to the peptide domain of salivary agglutinin protein, thereby inhibiting the adherence of *Streptococcus mutans* to dental surfaces.[183]

The dairy matrix is of crucial importance for our understanding of the role of (unsweetened) milk and dairy products in oral health.

## 5 Conclusion

Dairy foods are an important part of the human diet, providing many key nutrients, including calcium, vitamins B2 and B12, phosphorus and high-quality protein. Dairy is not a homogeneous group as dairy products differ in physico-chemical attributes, taste and nutrient composition as well as functionality. Each dairy product has its own unique matrix with its own unique matrix effect(s) that, by influencing digestion and absorption parameters, results in specific health effects. These health effects cannot simply be replicated with isolated dairy components as has been shown in many studies. Clearly, a dairy product is more than the sum of its components, in this respect. These findings highlight the importance of food-based approaches over single-nutrient ones and stress that findings from single-nutrient studies in one (model) system cannot be simply extrapolated to whole foods based solely on concentrations of a particular nutrient. This should lead to more holistic, food-based dietary guidelines that are generally easier to follow by consumers/the general population.

Nutritional research should assess whole foods, not only single nutrients, in future observational and intervention studies of health outcomes. Insights and data from novel technologies and approaches like systems biology, prevention-oriented lifestyle interventions and digital health should become part of nutritional research. These novel technologies Create at a fast pace an enormous amount of data that can be analyzed in an integrated way and will help to further our understanding of the mechanisms behind health outcomes, either on an individual basis or for specific target groups. Such efforts will allow investigation of the health effects of dairy foods based on how they are actually consumed and may lead to more realistic and nuanced dietary guidelines in the future.

## 6 Where to look for further information

A good overview of the role of milk and dairy products in the human diet can be found in Milk and Dairy Products in Human Nutrition (2013), by E. Muehlhoff, A. Bennett and D. McMahon, Food and Agriculture Organisation of the United Nations (FAO), Rome. E-ISBN: 978-92-5-107864-8 (PDF). More information on the functional effects of dairy related to reducing the risk of key chronic diseases can be found in Milk and Dairy Foods: their functionality in human health and disease (2020), by I. Givens, Academic Press. ISBN: 978-0-12-815603-2. DOI: https://doi.org/10.1016/C2017-0-03582-6. General information about dairy products can be found on the following websites: American Dairy Science Association (http://www.adsa.org), Global Dairy Platform (https://globaldairyplatform.com) and International Dairy Federation (https://fil-idf.org).

Important conferences include IDF World Dairy Summit (https://fil-idf.org), which is held every year as well as dairy conferences, for example, from the American Dairy Science Association. Further, the International Milk Genomics Consortium (www.milkgenomics.org) and their annual symposium cover lactation biology, comparative genetics and the role of dairy products in health and disease.

## 7 References


1. Pereira, P. C. Milk nutritional composition and its role in human health. *Nutrition* 30(6), 619–627 (2014).
2. Zhang, X., Chen, X., Xu, Y., Yang, J., Du, L., Li, K. and Zhou, Y. Milk consumption and multiple health outcomes: umbrella review of systematic reviews and meta-analyses in humans. *Nutr. Metab. (Lond.)* 18(1), 7 (2021).
3. Thorning, T. K., Raben, A., Tholstrup, T., Soedamah-Muthu, S. S., Givens, I. and Astrup, A. Milk and dairy products: good or bad for human health? An assessment of the totality of scientific evidence. *Food Nutr. Res.* 60, 32527 (2016).
4. Godos, J., Tieri, M., Ghelfi, F., Titta, L., Marventano, S., Lafranconi, A., Gambera, A., Alonzo, E., Sciacca, S., Buscemi, S., Ray, S., Del Rio, D., Galvano, F. and Grosso, G. Dairy foods and health: an umbrella review of observational studies. *Int. J. Food Sci. Nutr.* 71(2), 138–151 (2020).
5. http://www.fao.org/nutrition/nutrition-education/food-dietary-guidelines/en/.
6. https://www.fil-idf.org.

7. Aguayo, V. M. Complementary feeding practices for infants and young children in South Asia. A review of evidence for action post-2015. *Matern. Child Nutr.* 13 (Suppl. 2), 12439 (2017).
8. White, J. M., Bégin, F., Kumapley, R., Murray, C. and Krasevec, J. Complementary feeding practices: current global and regional estimates. *Matern. Child Nutr.* 13 (Suppl. 2), e12505 (2017).
9. Arsenault, J. E. and Brown, K. H. Dietary protein intake in young children in selected low-income countries is generally adequate in relation to estimated requirements for healthy children, except when complementary food intake is low. *J. Nutr.* 147(5), 932–939 (2017).
10. Grasgruber, P., Sebera, M., Hrazdíra, E., Cacek, J. and Kalina, T. Major correlates of male height: a study of 105 countries. *Econ. Hum. Biol.* 21, 172–195 (2016).
11. Headey, D., Hirvonen, K. and Hoddinott, J. Animal sourced foods and child stunting. *Am. J. Agric. Econ.* 100(5), 1302–1319 (2018).
12. Moradi, A. Nutritional status and economic development in sub-Saharan Africa, 1950–1980. *Econ. Hum. Biol.* 8(1), 16–29 (2010).
13. Krasevec, J., An, X., Kumapley, R., Bégin, F. and Frongillo, E. A. Diet quality and risk of stunting among infants and young children in low- and middle-income countries. *Matern. Child Nutr.* 13 (Suppl. 2), e12430 (2017).
14. Dror, D. K. and Allen, L. H. The importance of milk and other animal-source foods for children in low-income countries. *Food Nutr. Bull.* 32(3), 227–243 (2011).
15. Allen, L. H. Global dietary patterns and diets in childhood: implications for health outcomes. *Ann. Nutr. Metab.* 61 (Suppl. 1), 29–37 (2012).
16. Cohen Kadosh, K., Muhardi, L., Parikh, P., Basso, M., Jan Mohamed, H. J., Prawitasari, T., Samuel, F., Ma, G. and Geurts, J. M. Nutritional support of neurodevelopment and cognitive function in infants and young children-an update and novel insights. *Nutrients* 13(1), 199 (2021).
17. Allen, L. Comparing the value of protein sources for maternal and child nutrition. *Food Nutr. Bull.* 34(2), 263–266 (2013).
18. Grenov, B., Larnkjær, A., Lee, R., Serena, A., Mølgaard, C., Michaelsen, K. F. and Manary, M. J. Circulating Insulin-like growth factor-1 is positively associated with growth and cognition in 6- to 9-year-old schoolchildren from Ghana. *J. Nutr.* 150(6), 1405–1412 (2020).
19. Michaelsen, K. F. Cow's milk in the prevention and treatment of stunting and wasting. *Food Nutr. Bull.* 34(2), 249–251 (2013).
20. Thomas, T., Singh, M., Swaminathan, S. and Kurpad, A. V. Age-related differences in height gain with dairy protein and micronutrient supplements in Indian primary school children. *Asia Pac. J. Clin. Nutr.* 29(2), 355–362 (2020).
21. Millward, D. J. An adaptive metabolic demand model for protein and amino acid requirements. *Br. J. Nutr.* 90(2), 249–260 (2003).
22. Millward, D. J. Nutrition, infection, and stunting: the roles of deficiencies of individual nutrients and foods, and of inflammation, as determinants of reduced linear growth of children. *Nutr. Res. Rev.* 30(1), 50–72 (2017).
23. FAO. Dietary protein quality evaluation in human nutrition. Report of an FAO expert consultation. *FAO Food Nutr. Pap.* 92, 1–66 (2013).
24. WHO. Protein and amino acid requirements in human nutrition. *World Health Organ. Tech. Rep. S.* (935), 1–265, back cover (2007).

25. Nguyen Bao, K. L., Sandjaja, S., Poh, B., Rojroongwasinkul, N., Huu, C., Sumedi, E., Aini, J., Senaprom, S., Deurenberg, P., Bragt, M. and Khouw, I. The consumption of dairy and its association with nutritional status in the South East Asian Nutrition Surveys (SEANUTS). *Nutrients* 10(6), 759 (2018).
26. Bao Khanh, le N., Burgers, M. R., Huu Chinh, N., Tuoc, B. V., Dinh Dung, N., Deurenberg, P., Schaafsma, A. and SEANUTS Study Group. Nutrient intake in Vietnamese preschool and school-aged children is not adequate: the role of dairy. *Food Nutr. Bull.* 37(1), 100–111 (2016).
27. Widodo, Y., Sandjaja, S., Sumedi, E., Khouw, I. and Deurenberg, P. The effect of socio-demographic variables and dairy use on the intake of essential macro- and micronutrients in 0.5–12-year-old Indonesian children. *Asia Pac. J. Clin. Nutr.* 25(2), 356–367 (2016).
28. Willett, W., Rockström, J., Loken, B., Springmann, M., Lang, T., Vermeulen, S., Garnett, T., Tilman, D., DeClerck, F., Wood, A., Jonell, M., Clark, M., Gordon, L. J., Fanzo, J., Hawkes, C., Zurayk, R., Rivera, J. A., De Vries, W., Majele Sibanda, L., Afshin, A., Chaudhary, A., Herrero, M., Agustina, R., Branca, F., Lartey, A., Fan, S., Crona, B., Fox, E., Bignet, V., Troell, M., Lindahl, T., Singh, S., Cornell, S. E., Srinath Reddy, K., Narain, S., Nishtar, S. and Murray, C. J. L. Food in the Anthropocene: the EAT-Lancet Commission on healthy diets from sustainable food systems. *Lancet* 393(10170), 447–492 (2019).
29. Mechlem, K. Food security and the right to food in the discourse of the United Nations. *Eur. Law J.* 10(5), 631–648 (2004).
30. Smith, P. and Gregory, P. J. Climate change and sustainable food production. *Proc. Nutr. Soc.* 72(1), 21–28 (2013).
31. Dewulf, K., Wever, R. and Brezet, H. Greening the design brief. *Des. Innov. Value Towards Sustain. Soc.*, 457–462 (2012).
32. FAO. *World livestock (2011).Livestock in Food Security*. Food and Agriculture Organization, Rome, 1–13.
33. FAO. *Milk and Dairy Products in Human Nutrition*. Willet-Blackwell. A John Wiley & Sons, Ltd., Publication, Rome, 1–376 (2013).
34. Drewnowski, A. Defining nutrient density: development and validation of the nutrient rich foods index. *J. Am. Coll. Nutr.* 28(4), 421S–426S (2009).
35. Drewnowski, A., Dwyer, J., King, J. C. and Weaver, C. M. A proposed nutrient density score that includes food groups and nutrients to better align with dietary guidance. *Nutr. Rev.* 77(6), 404–416 (2019).
36. FAO and WHO. Rome declaration on nutrition. *ICN* 2, 1–6 (2014).
37. GDP and FAO. *Climate Change and the Global Dairy Cattle Sector: The Role of the Dairy Sector in a Low-Carbon Future*, Rome. 1–36 (2018).
38. Broekema, R. and Kramer, G. LCA of Dutch semi-skimmed milk and semi-mature cheese. *Blonk Consult. Neth.*, 1–69 (2014).
39. https://www.bbc.com/news/science-environment-46654042.
40. https://www.oatly.com/nl/products/haverdrank-semi.
41. Weidema, B. P. and Stylianou, K. S. Nutrition in the life cycle assessment of foods–function or impact? *Int. J. Life Cycle Assess.* 25(7), 1210–1216 (2020).
42. WHO & FAO. Codex general standard for the use of dairy terms (CODEX STAN 206-1999). Joint FAO. *WHO Food Standard Programme*. Food and Agriculture Organization, Rome (1999).
43. Smedman, A., Lindmark-Månsson, H., Drewnowski, A. and Edman, A. K. Nutrient density of beverages in relation to climate impact. *Food Nutr. Res.* 54, 5170 (2010).

44. Dijkstra, J., France, J., Ellis, J. L., Strathe, A. B., Kebreab, E. and Bannink, A. Production efficiency of ruminants: feed, nitrogen and methane. In: *Sustainable Animal Agriculture* Kebreab, E. (Ed), 10–25 (2013).
45. DELTA model. https://sustainablenutritioninitiative.com/sustainable-nutrition-initiative/.
46. Kramer, G. F., Tyszler, M., Veer, P. V. and Blonk, H. Decreasing the overall environmental impact of the Dutch diet: how to find healthy and sustainable diets with limited changes. *Public Health Nutr.* 20(9), 1699–1709 (2017).
47. Heller, M. C., Keoleian, G. A. and Willett, W. C. Toward a life cycle-based, diet-level framework for food environmental impact and nutritional quality assessment: a critical review. *Environ. Sci. Technol.* 47(22), 12632–12647 (2013).
48. Drewnowski, A. Healthy diets for a healthy planet. *Am. J. Clin. Nutr.* 99(6), 1284-1285 (2014).
49. van Dooren, C., Marinussen, M., Blonk, H., Aiking, H. and Vellinga, P. Exploring dietary guidelines based on ecological and nutritional values: a comparison of six dietary patterns. *Food Policy* 44, 36–46 (2014).
50. NOR-eden. https://www.med.uio.no/imb/english/research/projects/nor-eden/.
51. Ernstoff, A., Fulgoni III, V., Heller, M. C., Keoleian, G. A., Fante, P and Jolliet, O. J. Integrating nutritional benefits and impacts in a life cycle assessment framework: a US dairy consumption case study. *Proceedings of the 9th International Conference on Life Cycle Assessment in the Agri-Food Sector (LCA Food 2014)*, San Francisco, California, USA, 8–10 October 2014 174, 339–347 (2014).
52. Kromhout, D., Spaaij, C. J., de Goede, J. and Weggemans, R. M. The 2015 Dutch food-based dietary guidelines. Eur. J. Clin. Nutr. 70, 869–878 (2016).
53. Aguilera, J. M. The food matrix: implications in processing, nutrition and health. *Crit. Rev. Food Sci. Nutr.* 59(22), 3612–3629 (2019).
54. Astrup, A., Geiker, N. R. W. and Magkos, F. Effects of full-fat and fermented dairy products on cardiometabolic disease: food is more than the sum of its parts. *Adv. Nutr.* 10(5), 924S–930S (2019).
55. Mozaffarian, D., Rosenberg, I. and Uauy, R. History of modern nutrition science-implications for current research, dietary guidelines, and food policy. *BMJ* 361, k2392 (2018).
56. Mozaffarian, D., Micha, R. and Wallace, S. Effects on coronary heart disease of increasing polyunsaturated fat in place of saturated fat: a systematic review and meta-analysis of randomized controlled trials. *PLoS Med.* 7(3), e1000252 (2010).
57. Hooper, L., Martin, N., Abdelhamid, A. and Davey Smith, G. Reduction in saturated fat intake for cardiovascular disease. *Cochrane Database Syst. Rev.* (6), CD011737 (2015).
58. de Souza, R. J., Mente, A., Maroleanu, A., Cozma, A. I., Ha, V., Kishibe, T., Uleryk, E., Budylowski, P., Schünemann, H., Beyene, J. and Anand, S. S. Intake of saturated and trans unsaturated fatty acids and risk of all-cause mortality, cardiovascular disease, and type 2 diabetes: systematic review and meta-analysis of observational studies. *BMJ* 351, h3978 (2015).
59. Harcombe, Z., Baker, J. S. and Davies, B. Evidence from prospective cohort studies did not support the introduction of dietary fat guidelines in 1977 and 1983: a systematic review. *Br. J. Sports Med.* 51(24), 1737–1742 (2017).
60. Ramsden, C. E., Zamora, D., Majchrzak-Hong, S., Faurot, K. R., Broste, S. K., Frantz, R. P., Davis, J. M., Ringel, A., Suchindran, C. M. and Hibbeln, J. R. Re-evaluation of the traditional diet-heart hypothesis: analysis of recovered data from Minnesota Coronary Experiment (1968–73). *BMJ* 353, i1246 (2016).

61. Siri-Tarino, P. W., Sun, Q., Hu, F. B. and Krauss, R. M. Meta-analysis of prospective cohort studies evaluating the association of saturated fat with cardiovascular disease. *Am. J. Clin. Nutr.* 91(3), 535–546 (2010).
62. Givens, D. I. Saturated fats, dairy foods and health: a curious paradox? *Nutr. Bull.* 42(3), 274–282 (2017).
63. Malhotra, A., Redberg, R. F. and Meier, P. Saturated fat does not clog the arteries: coronary heart disease is a chronic inflammatory condition, the risk of which can be effectively reduced from healthy lifestyle interventions. *Br. J. Sports Med.* 51(15), 1111–1112 (2017).
64. Astrup, A., Magkos, F., Bier, D. M., Brenna, J. T., de Oliveira Otto, M. C., Hill, J. O., King, J. C., Mente, A., Ordovas, J. M., Volek, J. S., Yusuf, S. and Krauss, R. M. Saturated fats and health: a reassessment and proposal for food-based recommendations: jacc state-of-the-art review. *J. Am. Coll. Cardiol.* 76(7), 844–857 (2020).
65. O'Connor, L. E., Kim, J. E. and Campbell, W. W. Total red meat intake of ≥0.5 servings/d does not negatively influence cardiovascular disease risk factors: a systemically searched meta-analysis of randomized controlled trials. *Am. J. Clin. Nutr.* 105(1), 57–69 (2017).
66. Gianfredi, V., Salvatori, T., Nucci, D., Villarini, M. and Moretti, M. Can chocolate consumption reduce cardio-cerebrovascular risk? A systematic review and meta-analysis. *Nutrition* 46, 103–114 (2018).
67. Lopez, C., Cauty, C. and Guyomarc'h, F. Organization of lipids in milks, infant milk formulas and various dairy products: role of technological processes and potential impacts. *Dairy Sci. Technol.* 95(6), 863–893 (2015).
68. Thorning, T. K., Bertram, H. C., Bonjour, J. P., de Groot, L., Dupont, D., Feeney, E., Ipsen, R., Lecerf, J. M., Mackie, A., McKinley, M. C., Michalski, M. C., Rémond, D., Risérus, U., Soedamah-Muthu, S. S., Tholstrup, T., Weaver, C., Astrup, A. and Givens, I. Whole dairy matrix or single nutrients in assessment of health effects: current evidence and knowledge gaps. *Am. J. Clin. Nutr.* 105(5), 1033–1045 (2017).
69. Rosqvist, F., Smedman, A., Lindmark-Månsson, H., Paulsson, M., Petrus, P., Straniero, S., Rudling, M., Dahlman, I. and Risérus, U. Potential role of milk fat globule membrane in modulating plasma lipoproteins, gene expression, and cholesterol metabolism in humans: a randomized study. *Am. J. Clin. Nutr.* 102(1), 20–30 (2015).
70. Golding, M., Wooster, T. J., Day, L., Xu, M., Lundin, L., Keogh, J. and Clifton, P. Impact of gastric structuring on the lipolysis of emulsified lipids. *Soft Matter* 7(7), 3513–3523 (2011).
71. Berton, A., Rouvellac, S., Robert, B., Rousseau, F., Lopez, C. and Crenon, I. Effect of the size and interface composition of milk fat globules on their in vitro digestion by the human pancreatic lipase: native versus homogenized milk fat globules. *Food Hydrocoll.* 29(1), 123–134 (2012).
72. Capuano, E., Oliviero, T. and van Boekel, M. A. J. S. Modeling food matrix effects on chemical reactivity: challenges and perspectives. *Crit. Rev. Food Sci. Nutr.* 58(16), 2814–2828 (2018).
73. Lecerf, J.-M. and Legrand, P. Les effets des nutriments dépendent-ils des aliments qui les portent? L'effet matrice. *Cah. nutr. diététique* 50(3), 158–164 (2015).
74. Horstman, A. M. H., Ganzevles, R. A., Kudla, U., Kardinaal, A. F. M., van den Borne, J. J. G. C. and Huppertz, T. Postprandial blood amino acid concentrations in older adults after consumption of dairy products: the role of the dairy matrix. *Int. Dairy J.* 113, 10.1016/j.idairyj.2020.104890 (2021).

75. de Hart, N. M. M. P., Mahmassani, Z. S., Reidy, P. T., Kelley, J. J., McKenzie, A. I., Petrocelli, J. J., Bridge, M. J., Baird, L. M., Bastian, E. D., Ward, L. S., Howard, M. T. and Drummond, M. J. Acute Effects of Cheddar cheese consumption on circulating amino acids and human skeletal muscle. *Nutrients* 13(2), 614 (2021).
76. Drouin-Chartier, J. P., Tremblay, A. J., Maltais-Giguère, J., Charest, A., Guinot, L., Rioux, L. E., Labrie, S., Britten, M., Lamarche, B., Turgeon, S. L. and Couture, P. Differential impact of the cheese matrix on the postprandial lipid response: a randomized, crossover, controlled trial. *Am. J. Clin. Nutr.* 106(6), 1358–1365 (2017).
77. Peters, S. The food matrix: food is more than the sum of its nutrients. *Voeding Magazine* 2, 1–4 (2017).
78. Jensen, R. G. *Handbook of Milk Composition*, Academic Press, London1-944 (1995).
79. Månsson, H. L. Fatty acids in bovine milk fat. *Food Nutr. Res.* 52, 1821 (2008).
80. Auldist, M. J., Walsh, B. J. and Thomson, N. A. Seasonal and lactational influences on bovine milk composition in New Zealand. *J. Dairy Res.* 65(3), 401–411 (1998).
81. Odén, A., McCloskey, E. V., Kanis, J. A., Harvey, N. C. and Johansson, H. Burden of high fracture probability worldwide: secular increases 2010–2040. *Osteoporos. Int.* 26(9), 2243–2248 (2015).
82. Curtis, E. M., Moon, R. J., Harvey, N. C. and Cooper, C. The impact of fragility fracture and approaches to osteoporosis risk assessment worldwide. *Bone* 104, 29–38 (2017).
83. Weaver, C. M., Gordon, C. M., Janz, K. F., Kalkwarf, H. J., Lappe, J. M., Lewis, R., O'Karma, M., Wallace, T. C. and Zemel, B. S. The National Osteoporosis Foundation's position statement on peak bone mass development and lifestyle factors: a systematic review and implementation recommendations. *Osteoporos. Int.* 27(4), 1281–1386 (2016).
84. Cruz-Jentoft, A. J., Bahat, G., Bauer, J., Boirie, Y., Bruyère, O., Cederholm, T., Cooper, C., Landi, F., Rolland, Y., Aihie Sayer, A., Schneider, S. M., Sieber, C. C., Topinkova, E., Vandewoude, M., Visser, M and Zamboni, M. Writing Group for the European Working Group on Sarcopenia in Older People 2 (EWGSOP2), and the Extended Group for EWGSOP2. Sarcopenia: revised European consensus on definition and diagnosis. *Age Ageing* 48, 16–31 (2019).
85. Hunter, G. R., Singh, H., Carter, S. J., Bryan, D. R. and Fisher, G. Sarcopenia and its implications for metabolic health. *J. Obes.* 2019, 8031705 (2019).
86. Heaney, R. P. Dairy and bone health. *J. Am. Coll. Nutr.* 28 (Suppl. 1), 82S–90S (2009).
87. Rizzoli, R. Nutrition: its role in bone health. *Best Pract. Res. Clin. Endocrinol. Metab.* 22(5), 813–829 (2008).
88. Heaney, R. P., Weaver, C. M. and Recker, R. R. Calcium absorbability from spinach. *Am. J. Clin. Nutr.* 47(4), 707–709 (1988).
89. Eastell, R. and Lambert, H. Plenary Lecture: strategies for skeletal health in the elderly. *Proc. Nutr. Soc.* 61(2), 173–180 (2002).
90. Heaney, R. P. Calcium, dairy products and osteoporosis. *J. Am. Coll. Nutr.* 19(2) (Suppl.), 83S–99S (2000).
91. Lenton, S., Nylander, T., Teixeira, S. C. and Holt, C. A review of the biology of calcium phosphate sequestration with special reference to milk. *Dairy Sci. Technol.* 95, 3–14 (2015).
92. Koletzko, B., Goulet, O., Hunt, J., Krohn, K. and Shamir, R. Guidelines on Paediatric parenteral Nutrition of the European Society of Paediatric Gastroenterology, Hepatology and Nutrition (ESPGHAN) and the European Society for Clinical

Nutrition and Metabolism (ESPEN), Supported by the European Society of Paediatric Research (ESPR): iron, minerals and trace elements. *J. Pedriatr Gastroenterol. Nutr.* 41, S1-87 (2005).

93. Weaver, C. M., Janle, E., Martin, B., Browne, S., Guiden, H., Lachcik, P. and Lee, W. H. Dairy versus calcium carbonate in promoting peak bone mass and bone maintenance during subsequent calcium deficiency. *J. Bone Miner. Res.* 24(8), 1411-1419 (2009).
94. Cakebread, J. A., Wallace, O. A. M., Kruger, M. C., Vickers, M. H. and Hodgkinson, A. J. Supplementation with bovine milk or soy beverages recovers bone mineralization in young growing rats fed an insufficient diet, in contrast to an almond beverage. *Curr. Dev. Nutr.* 3(11), nzz115 (2019).
95. Cheng, S., Lyytikäinen, A., Kröger, H., Lamberg-Allardt, C., Alén, M., Koistinen, A., Wang, Q. J., Suuriniemi, M., Suominen, H., Mahonen, A., Nicholson, P. H., Ivaska, K. K., Korpela, R., Ohlsson, C., Väänänen, K. H. and Tylavsky, F. Effects of calcium, dairy product, and vitamin D supplementation on bone mass accrual and body composition in 10-12-y-old girls: a 2-y randomized trial. *Am. J. Clin. Nutr.* 82(5), 1115-1126; quiz 1147 (2005).
96. Manios, Y., Moschonis, G., Trovas, G. and Lyritis, G. P. Changes in biochemical indexes of bone metabolism and bone mineral density after a 12-mo dietary intervention program: the postmenopausal health study. *Am. J. Clin. Nutr.* 86(3), 781-789 (2007).
97. Gui, J. C., Brašić, J. R., Liu, X. D., Gong, G. Y., Zhang, G. M., Liu, C. J. and Gao, G. Q. Bone mineral density in postmenopausal Chinese women treated with calcium fortification in soymilk and cow's milk. *Osteoporos. Int.* 23(5), 1563-1570 (2012).
98. Morency, M. E., Birken, C. S., Lebovic, G., Chen, Y., L'Abbé, M., Lee, G. J., Maguire, J. L. and TARGet Kids! Collaboration. Association between noncow milk beverage consumption and childhood height. *Am. J. Clin. Nutr.* 106(2), 597-602 (2017).
99. Scholz-Ahrens, K. E., Ahrens, F. and Barth, C. A. Nutritional and health attributes of milk and milk imitations. *Eur. J. Nutr.* 59(1), 19-34 (2020).
100. Konstantynowicz, J., Nguyen, T. V., Kaczmarski, M., Jamiolkowski, J., Piotrowska-Jastrzebska, J. and Seeman, E. Fractures during growth: potential role of a milk-free diet. *Osteoporos. Int.* 18(12), 1601-1607 (2007).
101. Dagnelie, P. C., Vergote, F. J., van Staveren, W. A., van den Berg, H., Dingjan, P. G. and Hautvast, J. G. High prevalence of rickets in infants on macrobiotic diets. *Am. J. Clin. Nutr.* 51(2), 202-208 (1990).
102. van Loon, L. J. Is there a need for protein ingestion during exercise? *Sports Med* 44 (Suppl. 1), S105-111 (2014).
103. Groen, B. B., Horstman, A. M., Hamer, H. M., de Haan, M., van Kranenburg, J., Bierau, J., Poeze, M., Wodzig, W. K., Rasmussen, B. B. and van Loon, L. J. Post-prandial protein handling: you are what you just ate. *PLoS ONE* 10(11), e0141582 (2015).
104. Millward, D. J., Layman, D. K., Tomé, D. and Schaafsma, G. Protein quality assessment: impact of expanding understanding of protein and amino acid needs for optimal health. *Am. J. Clin. Nutr.* 87(5), 1576S-1581S (2008).
105. Pennings, B., Boirie, Y., Senden, J. M., Gijsen, A. P., Kuipers, H. and van Loon, L. J. Whey protein stimulates postprandial muscle protein accretion more effectively than do casein and casein hydrolysate in older men. *Am. J. Clin. Nutr.* 93(5), 997-1005 (2011).
106. van Vliet, S., Burd, N. A. and van Loon, L. J. The skeletal muscle anabolic response to plant- versus animal-based protein consumption. *J. Nutr.* 145(9), 1981-1991 (2015).

107. FAO. Expert consultation on protein and energy requirements. Protein and amino acid requirements in human nutrition. *World Health Organ. Tech. Rep. S.* 935, 1-265 (2007).
108. Karp, J. R., Johnston, J. D., Tecklenburg, S., Mickleborough, T. D., Fly, A. D. and Stager, J. M. Chocolate Milk as a post-exercise recovery aid. *Int. J. Sport Nutr. Exerc. Metab.* 16(1), 78-91 (2006).
109. Gilson, S. F., Saunders, M. J., Moran, C. W., Moore, R. W., Womack, C. J. and Todd, M. K. Effects of chocolate milk consumption on markers of muscle recovery following soccer training: a randomized cross-over study. *J. Int. Soc. Sports Nutr.* 7, 19 (2010).
110. Watson, P., Love, T. D., Maughan, R. J. and Shirreffs, S. M. A comparison of the effects of milk and a carbohydrate-electrolyte drink on the restoration of fluid balance and exercise capacity in a hot, humid environment. *Eur. J. Appl. Physiol.* 104(4), 633-642 (2008).
111. Lee, J. K., Maughan, R. J., Shirreffs, S. M. and Watson, P. Effects of milk ingestion on prolonged exercise capacity in young, healthy men. *Nutrition* 24(4), 340-347 (2008).
112. Shirreffs, S. M., Watson, P. and Maughan, R. J. Milk as an effective post-exercise rehydration drink. *Br. J. Nutr.* 98(1), 173-180 (2007).
113. Thomas, K., Morris, P. and Stevenson, E. Improved endurance capacity following chocolate milk consumption compared with 2 commercially available sport drinks. *Appl. Physiol. Nutr. Metab.* 34(1), 78-82 (2009).
114. Odell, O. J., Podlogar, T. and Wallis, G. A. Comparable exogenous carbohydrate oxidation from lactose or sucrose during exercise. *Med. Sci. Sports Exerc.* 52(12), 2663-2672 (2020).
115. Odell, O. and Wallis, G. A. The application of lactose in sports nutrition. *Int. Dairy J.* 116, 104970 (2020).
116. Tarnopolsky, M. Protein requirements for endurance athletes. *Nutrition* 20(7-8), 662-668 (2004).
117. Dangin, M., Guillet, C., Garcia-Rodenas, C., Gachon, P., Bouteloup-Demange, C., Reiffers-Magnani, K., Fauquant, J., Ballèvre, O. and Beaufrère, B. The rate of protein digestion affects protein gain differently during aging in humans. *J. Physiol.* 549(2), 635-644 (2003).
118. Wolfe, R. R., Miller, S. L. and Miller, K. B. Optimal protein intake in the elderly. *Clin. Nutr.* 27(5), 675-684 (2008).
119. Layman, D. K. Dietary Guidelines should reflect new understandings about adult protein needs. *Nutr. Metab. (Lond.)* 6, 12 (2009).
120. Rice, B. H., Quann, E. E. and Miller, G. D. Meeting and exceeding dairy recommendations: effects of dairy consumption on nutrient intakes and risk of chronic disease. *Nutr. Rev.* 71(4), 209-223 (2013).
121. Talaei, M., Pan, A., Yuan, J. M. and Koh, W. P. Dairy food intake is inversely associated with risk of hypertension: the Singapore Chinese health study. *J. Nutr.* 147(2), 235-241 (2017).
122. Rietsema, S., Eelderink, C., Joustra, M. L., van Vliet, I. M. Y., van Londen, M., Corpeleijn, E., Singh-Povel, C. M., Geurts, J. M. W., Kootstra-Ros, J. E., Westerhuis, R., Navis, G. and Bakker, S. J. L. Effect of high compared with low dairy intake on blood pressure in overweight middle-aged adults: results of a randomized crossover intervention study. *Am. J. Clin. Nutr.* 110(2), 340-348 (2019).

123. Appel, L. J., Moore, T. J., Obarzanek, E., Vollmer, W. M., Svetkey, L. P., Sacks, F. M., Bray, G. A., Vogt, T. M., Cutler, J. A., Windhauser, M. M., Lin, P. H. and Karanja, N. A clinical trial of the effects of dietary patterns on blood pressure. DASH collaborative research group. *N. Engl. J. Med.* 336(16), 1117-1124 (1997).
124. Chiu, S., Bergeron, N., Williams, P. T., Bray, G. A., Sutherland, B. and Krauss, R. M. Comparison of the DASH (dietary approaches to stop hypertension) diet and a higher-fat DASH diet on blood pressure and lipids and lipoproteins: a randomized controlled trial. *Am. J. Clin. Nutr.* 103(2), 341-347 (2016).
125. Crippa, G., Zabzuni, D., Bravi, E., Piva, G., De Noni, I., Bighi, E. and Rossi, F. Randomized, double blind placebo-controlled pilot study of the antihypertensive effects of Grana Padano D.O.P. cheese consumption in mild - moderate hypertensive subjects. *Eur. Rev. Med. Pharmacol. Sci.* 22(21), 7573-7581 (2018).
126. McGrane, M. M., Essery, E., Obbagy, J., Lyon, J., Macneil, P., Spahn, J. and Van Horn, L. Dairy consumption, blood pressure, and risk of hypertension: an evidence-based review of recent literature. *Curr. Cardiovasc. Risk Rep.* 5(4), 287-298 (2011).
127. Pripp, A. H. Effect of peptides derived from food proteins on blood pressure: a meta-analysis of randomized controlled trials. *Food Nutr. Res.* 52 (2008).
128. Boelsma, E. and Kloek, J. Lactotripeptides and antihypertensive effects: a critical review. *Br. J. Nutr.* 101(6), 776-786 (2009).
129. Fekete, Á. A., Givens, D. I. and Lovegrove, J. A. The impact of milk proteins and peptides on blood pressure and vascular function: a review of evidence from human intervention studies. *Nutr. Res. Rev.* 26(2), 177-190 (2013).
130. Pal, S. and Ellis, V. The chronic effects of whey proteins on blood pressure, vascular function, and inflammatory markers in overweight individuals. *Obesity (Silver Spring)* 18(7), 1354-1359 (2010).
131. Fekete, Á. A., Giromini, C., Chatzidiakou, Y., Givens, D. I. and Lovegrove, J. A. Whey protein lowers blood pressure and improves endothelial function and lipid biomarkers in adults with prehypertension and mild hypertension: results from the chronic Whey2Go randomized controlled trial. *Am. J. Clin. Nutr.* 104(6), 1534-1544 (2016).
132. Hidayat, K., Du, H. Z., Yang, J., Chen, G. C., Zhang, Z., Li, Z. N. and Qin, L. Q. Effects of milk proteins on blood pressure: a meta-analysis of randomized control trials. *Hypertens. Res.* 40(3), 264-270 (2017).
133. Ding, M., Huang, T., Bergholdt, H. K., Nordestgaard, B. G., Ellervik, C., Qi, L. and CHARGE Consortium. Dairy consumption, systolic blood pressure, and risk of hypertension: Mendelian randomization study. *BMJ* 356, j1000 (2017).
134. Wang, H., Fox, C. S., Troy, L. M., Mckeown, N. M. and Jacques, P. F. Longitudinal association of dairy consumption with the changes in blood pressure and the risk of incident hypertension: the framingham heart study. *Br. J. Nutr.* 114(11), 1887-1899 (2015).
135. Eelderink, C., Rietsema, S., van Vliet, I. M. Y., Loef, L. C., Boer, T., Koehorst, M., Nolte, I. M., Westerhuis, R., Singh-Povel, C. M., Geurts, J. M. W., Corpeleijn, E. and Bakker, S. J. L. The effect of high compared with low dairy consumption on glucose metabolism, insulin sensitivity, and metabolic flexibility in overweight adults: a randomized crossover trial. *Am. J. Clin. Nutr.* 109(6), 1555-1568 (2019).
136. Nakagawa, T., Hu, H., Zharikov, S., Tuttle, K. R., Short, R. A., Glushakova, O., Ouyang, X., Feig, D. I., Block, E. R., Herrera-Acosta, J., Patel, J. M. and Johnson, R. J. A causal

role for uric acid in fructose-induced metabolic syndrome. *Am. J. Physiol. Ren. Physiol.* 290(3), F625–F631 (2006).
137. Chatterjee, S., Khunti, K. and Davies, M. J. Type 2 diabetes. *Lancet* 389(10085), 2239–2251 (2017).
138. Guo, J., Givens, D. I., Astrup, A., Bakker, S. J. L., Goossens, G. H., Kratz, M., Marette, A., Pijl, H. and Soedamah-Muthu, S. S. The impact of dairy products in the development of type 2 diabetes: where does the evidence stand in 2019? *Adv. Nutr.* 10(6), 1066–1075 (2019).
139. Aune, D., Norat, T., Romundstad, P. and Vatten, L. J. Dairy products and the risk of type 2 diabetes: a systematic review and dose-response meta-analysis of cohort studies. *Am. J. Clin. Nutr.* 98(4), 1066–1083 (2013).
140. Gijsbers, L., Ding, E. L., Malik, V. S., de Goede, J., Geleijnse, J. M. and Soedamah-Muthu, S. S. Consumption of dairy foods and diabetes incidence: a dose-response meta-analysis of observational studies. *Am. J. Clin. Nutr.* 103(4), 1111–1124 (2016).
141. Gao, D., Ning, N., Wang, C., Wang, Y., Li, Q., Meng, Z., Liu, Y. and Li, Q. Dairy products consumption and risk of type 2 diabetes: systematic review and dose-response meta-analysis. *PLoS ONE* 8(9), e73965 (2013).
142. Elwood, P. C., Pickering, J. E., Givens, D. I. and Gallacher, J. E. The consumption of milk and dairy foods and the incidence of vascular disease and diabetes: an overview of the evidence. *Lipids* 45(10), 925–939 (2010).
143. Tong, X., Dong, J. Y., Wu, Z. W., Li, W. and Qin, L. Q. Dairy consumption and risk of type 2 diabetes mellitus: a meta-analysis of cohort studies. *Eur. J. Clin. Nutr.* 65(9), 1027–1031 (2011).
144. Soedamah-Muthu, S. S. and de Goede, J. Dairy consumption and cardiometabolic diseases: systematic review and updated meta-analyses of prospective cohort studies. *Curr. Nutr. Rep.* 7(4), 171–182 (2018).
145. Hruby, A., Ma, J., Rogers, G., Meigs, J. B. and Jacques, P. F. Associations of dairy intake with incident prediabetes or diabetes in middle-aged adults vary by both dairy type and glycemic status. *J. Nutr.* 147(9), 1764–1775 (2017).
146. Brouwer-Brolsma, E. M., van Woudenbergh, G. J., Oude Elferink, S. J., Singh-Povel, C. M., Hofman, A., Dehghan, A., Franco, O. H. and Feskens, E. J. Intake of different types of dairy and its prospective association with risk of type 2 diabetes: the Rotterdam Study. *Nutr. Metab. Cardiovasc. Dis.* 26(11), 987–995 (2016).
147. Guasch-Ferré, M., Becerra-Tomás, N., Ruiz-Canela, M., Corella, D., Schröder, H., Estruch, R., Ros, E., Arós, F., Gómez-Gracia, E., Fiol, M., Serra-Majem, L., Lapetra, J., Basora, J., Martín-Calvo, N., Portoles, O., Fitó, M., Hu, F. B., Forga, L. and Salas-Salvadó, J. Total and subtypes of dietary fat intake and risk of type 2 diabetes mellitus in the Prevención con Dieta Mediterránea (PREDIMED) study. *Am. J. Clin. Nutr.* 105(3), 723–735 (2017).
148. Díaz-López, A., Bulló, M., Martínez-González, M. A., Corella, D., Estruch, R., Fitó, M., Gómez-Gracia, E., Fiol, M., García de la Corte, F. J., Ros, E., Babio, N., Serra-Majem, L., Pintó, X., Muñoz, M. Á, Francés, F., Buil-Cosiales, P. and Salas-Salvadó, J. Dairy product consumption and risk of type 2 diabetes in an elderly Spanish Mediterranean population at high cardiovascular risk. *Eur. J. Nutr.* 55(1), 349–360 (2016).
149. Fernandez, M. A., Panahi, S., Daniel, N., Tremblay, A. and Marette, A. Yogurt and cardiometabolic diseases: a critical review of potential mechanisms. *Adv. Nutr.* 8(6), 812–829 (2017).

150. Hirahatake, K. M., Slavin, J. L., Maki, K. C. and Adams, S. H. Associations between dairy foods, diabetes, and metabolic health: potential mechanisms and future directions. *Metabolism* 63(5), 618-627 (2014).
151. Veiga, P., Pons, N., Agrawal, A., Oozeer, R., Guyonnet, D., Brazeilles, R., Faurie, J. M., van Hylckama Vlieg, J. E., Houghton, L. A., Whorwell, P. J., Ehrlich, S. D. and Kennedy, S. P. Changes of the human gut microbiome induced by a fermented milk product. *Sci. Rep.* 4, 6328 (2014).
152. Robertson, M. D., Bickerton, A. S., Dennis, A. L., Vidal, H. and Frayn, K. N. Insulin-sensitizing effects of dietary resistant starch and effects on skeletal muscle and adipose tissue metabolism. *Am. J. Clin. Nutr.* 82(3), 559-567 (2005).
153. Layden, B. T., Yalamanchi, S. K., Wolever, T. M., Dunaif, A. and Lowe, W. L. Negative association of acetate with visceral adipose tissue and insulin levels. *Diabetes Metab. Syndr. Obes.* 5, 49-55 (2012).
154. Touger-Decker, R., Mobley, C. and Academy of Nutrition and Dietetics, O. N. A. D. position of the academy of nutrition and dietetics: oral health and nutrition. *J. Acad. Nutr. Diet.* 113(5), 693-701 (2013).
155. Kane, S. F. The effects of oral health on systemic health. *Gen. Dent.* 65(6), 30-34 (2017).
156. Kilian, M. The oral microbiome - friend or foe? *Eur. J. Oral Sci.* 126 (Suppl. 1), 5-12 (2018).
157. Olsen, I. and Singhrao, S. K. Can oral infection be a risk factor for Alzheimer's disease? *J. Oral Microbiol.* 7, 29143 (2015).
158. Liccardo, D., Cannavo, A., Spagnuolo, G., Ferrara, N., Cittadini, A., Rengo, C. and Rengo, G. Periodontal disease: a risk factor for diabetes and cardiovascular disease. *Int. J. Mol. Sci.* 20(6), 1414 (2019).
159. Dodds, M. W., Johnson, D. A. and Yeh, C. K. Health benefits of saliva: a review. *J. Dent.* 33(3), 223-233 (2005).
160. Dror, D. K. and Allen, L. H. Dairy product intake in children and adolescents in developed countries: trends, nutritional contribution, and a review of association with health outcomes. *Nutr. Rev.* 72(2), 68-81 (2014).
161. Tanaka, K., Miyake, Y. and Sasaki, S. Intake of dairy products and the prevalence of dental caries in young children. *J. Dent.* 38(7), 579-583 (2010).
162. Petti, S., Cairella, G. and Tarsitani, G. Rampant early childhood dental decay: an example from Italy. *J. Public Health Dent.* 60(3), 159-166 (2000).
163. Głabska, D., Sińska, B. and Remiszewski, A. Analysis of the dependence between milk and dairy products consumption, and dental caries observed in group of children and teenagers. *Roczniki Panstwowego Zakladu Hig.* 58(1), 69-75 (2007).
164. Llena, C. and Forner, L. Dietary habits in a child population in relation to caries experience. *Caries Res.* 42(5), 387-393 (2008).
165. Ohlund, I., Holgerson, P. L., Backman, B., Lind, T., Hernell, O. and Johansson, I. Diet intake and caries prevalence in four-year-old children living in a low-prevalence country. *Caries Res.* 41(1), 26-33 (2007).
166. Gedalia, I., Davidov, I., Lewinstein, I. and Shapira, L. Effect of hard cheese exposure, with and without fluoride prerinse, on the rehardening of softened human enamel. *Caries Res.* 26(4), 290-292 (1992).
167. Lewinstein, I., Ofek, L. and Gedalia, I. Enamel rehardening by soft cheeses. *Am. J. Dent.* 6(1), 46-48 (1993).

168. Kashket, S. and DePaola, D. P. Cheese consumption and the development and progression of dental caries. *Nutr. Rev.* 60(4), 97–103 (2002).
169. Lempert, S. M., Christensen, L. B., Froberg, K., Raymond, K. and Heitmann, B. L. Association between dairy intake and caries among children and adolescents. results from the Danish EYHS follow-up study. *Caries Res.* 49(3), 251–258 (2015).
170. Levine, R. S. Milk, flavoured milk products and caries. *Br. Dent. J.* 191(1), 20 (2001).
171. Moynihan, P., Tanner, L. M., Holmes, R. D., Hillier-Brown, F., Mashayekhi, A., Kelly, S. A. M. and Craig, D. Systematic review of evidence pertaining to factors that modify risk of early childhood caries. *JDR Clin. Trans. Res.* 4(3), 202–216 (2019).
172. Johansson, I. Milk and dairy products: possible effects on dental health. *Scand. J. Nutr.* 46(3), 119–122 (2002).
173. Rugg-Gunn, A. J., Roberts, G. J. and Wright, W. G. Effect of human milk on plaque pH in situ and enamel dissolution in vitro compared with bovine milk, lactose, and sucrose. *Caries Res.* 19(4), 327–334 (1985).
174. Lamothe, L. M., Lê, K. A., Samra, R. A., Roger, O., Green, H. and Macé, K. The scientific basis for healthful carbohydrate profile. *Crit. Rev. Food Sci. Nutr.* 59(7), 1058–1070 (2019).
175. Koulourides, T., Bodden, R., Keller, S., Manson-Hing, L., Lastra, J. and Housch, T. Cariogenicity of nine sugars tested with an intraoral device in man. *Caries Res.* 10(6), 427–441 (1976).
176. Birkhed, D., Imfeld, T. and Edwardsson, S. pH changes in human dental plaque from lactose and milk before and after adaptation. *Caries Res.* 27(1), 43–50 (1993).
177. Woodward, M. and Rugg-Gunn, A. J.. Chapter 8: milk, yoghurts and dental caries. *Monogr. Oral Sci.* 28, 77–90 (2020) Chapter 8.
178. Brudevold, F., Tehrani, A., Attarzadeh, F., van Houte, J. and Russo, J. Enamel demineralization potential of dietary carbohydrates. *J. Dent. Res.* 62(12), 1218–1220 (1983).
179. Aimutis, W. R. Lactose cariogenicity with an emphasis on childhood dental caries. *Int. Dairy J.* 22(2), 152–158 (2012).
180. Sheiham, A. Dietary effects on dental diseases. *Public Health Nutr.* 4(2B), 569–591 (2001).
181. Merritt, J., Qi, F. and Shi, W. Milk helps build strong teeth and promotes oral health. *J. Calif. Dent. Assoc.* 34(5), 361–366 (2006).
182. Aimutis, W. R. Bioactive properties of milk proteins with particular focus on anticariogenesis. *J. Nutr.* 134(4), 989S–95S (2004).
183. Oho, T., Bikker, F. J., Nieuw Amerongen, A. V. and Groenink, J. A peptide domain of bovine milk lactoferrin inhibits the interaction between streptococcal surface protein antigen and a salivary agglutinin peptide domain. *Infect. Immun.* 72(10), 6181–6184 (2004).
184. Drouin-Chartier, J. P., Brassard, D., Tessier-Grenier, M., Côté, J. A., Labonté, M. È, Desroches, S., Couture, P. and Lamarche, B. Systematic review of the association between dairy product consumption and risk of cardiovascular-related clinical outcomes. *Adv. Nutr.* 7(6), 1026–1040 (2016).
185. Geiker, N. R. W., Mølgaard, C., Iuliano, S., Rizzoli, R., Manios, Y., van Loon, L. J. C., Lecerf, J. M., Moschonis, G., Reginster, J. Y., Givens, I. and Astrup, A. Impact of whole dairy matrix on musculoskeletal health and aging-current knowledge and research gaps. *Osteoporos. Int.* 31(4), 601–615 (2020).

# Chapter 2

## Digestion of milk protein and milk fat

*Aiqian Ye, Riddet Institute, Massey University, New Zealand*



## 1 Introduction

Milk is considered to be the perfect food, especially for the youth and the elderly. It is nutrient rich and well balanced (Patton, 2017). In addition to high-quality proteins, lactose and fat, milk contains many minor constituents with biological properties, such as minerals, vitamins and the components of the milk fat globule membrane (MFGM). Many food products are developed from milk, including commercial liquid milk, cheese, yoghurt, milk powders, infant formulae, ice cream, cream and butter (O'Mahony and Fox, 2014). These dairy products produced from milk make an important contribution to the total dietary supply of nutrients, and in particular, a significant contribution to the daily diet of Asian populations. During the manufacture of dairy products, milk components undergo different processing treatments that can alter their macro- and microstructures and cause modifications at the molecular level, all of which may influence their digestion behaviour and their digestibility. The relationships between the structures that is formed by the milk proteins/fat and their digestion behaviour, and the physiological response, have recently been of interest and have been studied extensively using both in vivo and in vitro models; in particular, the phenomena observed during in vivo work have been explained by applying an advanced in vitro dynamic or semi-dynamic digestion model combined with modern analytical technologies and knowledge of protein interactions (Fardet et al., 2019; Huppertz and Chia, 2020; Mulet-Cabero et al., 2020; van Lieshout et al., 2020; Ye et al., 2016b; Ye et al., 2020).

http://dx.doi.org/10.19103/AS.2022.0099.23

Because dairy products have different matrices in their initial form and after restructuring during digestion, we must take into account the structure of the products and interactions relating to structure formation when investigating the digestion of milk components.

## 2 Digestibility of milk proteins

Milk proteins have many unique properties in function and digestion. Milk products containing proteins are sources of high-quality dietary protein for humans. Using both in vivo and in vitro evaluation methods, it has been found that all milk proteins and the milk proteins in dairy products are highly digestible compared with proteins from other sources (Gilani and Sepehr, 2003; Lacroix et al., 2006; Rutherfurd and Moughan, 2003, 2005; Rutherfurd and Moughan, 1998). Using the protein digestibility-corrected amino acid score in rats, Gilani and Sepehr (2003) reported that the digestibility scores of casein and whey protein concentrate were close to 100%. The true ileal reactive lysine digestibility using the rat model was high (> 91%) in all milk products; the highest in Ultra Heat Temperature (UHT) treated milk (100%) and the lowest in infant formulae (91–93%) (Rutherfurd and Moughan, 2005). The digestibility of reactive lysine was similar to the digestibility values obtained for most other amino acids. It is worth noting that the digestibility of milk protein is significantly higher than those of soy protein and other protein sources (Gilani and Sepehr, 2003; Rutherfurd and Moughan, 1998, 2003, 2005). In in vitro experiments, purified caseins could be hydrolysed by pepsin in 1 minute (Dupont et al., 2010). Unlike the caseins, the whey proteins, particularly native β-lactoglobulin (β-lg), are resistant to some proteases, especially pepsin, because most of the hydrophobic amino acids, which are potential cleavage sites for pepsin, are buried inside the hydrophobic core at low pH and are not readily accessible (Miranda and Pelissier, 1983; Reddy et al., 1988). Other whey proteins - α-lactalbumin (α-la), bovine serum albumin (BSA) and lactoferrin - are more susceptible to hydrolysis by pepsin (Furlund et al., 2013; Nik et al., 2010).

Although the milk proteins in most dairy products are highly digestible, differences among dairy products have been observed; for example, Rutherfurd and Moughan (2005) found that the digestibility of the proteins in infant formulae was lower than that of the proteins in UHT milk. This may be related to the differences in the digestion kinetics of the milk proteins. The kinetic differences in digestion and the delivery of amino acids among the milk proteins and even for the same protein in different milk products highlight the importance of the matrix in digestion and absorption. Lacroix et al. (2006) showed that the transfer of dietary nitrogen to urea occurred earlier after the ingestion of soluble milk proteins than after the ingestion of micellar casein and total milk protein, and that concentrations remained high for 8 h, together with higher but transient hyperaminoacidaemia and a higher incorporation of dietary nitrogen into plasma amino acids. In contrast, deamination, postprandial hyperaminoacidaemia and the incorporation of dietary nitrogen into plasma amino acids were lower in micellar casein and total milk protein. The slower digestion and absorption of casein, compared with whey protein, is thought to be related to its behaviour during gastric digestion (Boirie et al., 1997; Mahe et al., 1996). The delay in the delivery of amino acids to the gut is attributed to the phosphate content of the caseins; they precipitate and are retained as a clot under the acidic gastric pH conditions (Boirie et al., 1997; Mahe et al., 1996). In contrast, whey protein remains soluble and passes rapidly through the stomach, leading to a faster delivery of amino acids to the circulation (Boirie et al., 1997). This has elicited the concept of 'slow' digested caseins and 'fast' digested whey proteins (Boirie et al., 1997). However, this concept can be altered by differences in the previous processing of the ingested protein, for example heat treatment or no heat treatment, and the intensity of the heat treatment. These pre-treatments can alter the physical properties of the dietary proteins, like viscosity, osmolality, particle size and degree of hydrolysis by enzymes, because of the changes in their structure. These factors affect the gastric emptying rate (Azpiroz and Malagelada, 1990; Hunt and Stubbs, 1975; Vist and Maughan, 1995).

The influence of the processing treatment on the digestibility of the milk proteins has been demonstrated extensively. Wada and Lönnerdal (2014) showed that the protein bands in sodium dodecyl sulphate polyacrylamide gel electrophoresis (SDS-PAGE) gels of in-can-sterilised milk underwent a more rapid decrease in intensity than those of UHT milk during in vitro digestion. Proteomic analysis revealed that the level of lactulosyllysine, which reflects a decrease in the digestibility of protein in α-la, β-lg and caseins, was higher in in-can-sterilised milk than in UHT milk. Thus, industrial heating may improve the digestibility of milk proteins by denaturation, but the improvement is likely to be offset by heat-induced protein modifications, which may decrease the digestibility of the protein, for example, through the formation of cross-links such as lysinoalanine and lanthionine (Wada and Lönnerdal, 2014). Miranda and Pelissier (1987) reported that heat treatment of milk resulted in accelerated gastric emptying and an increase in the rate of hydrolysis of the caseins, in an in vivo rat experiment. Barbé et al. (2013) also reported a significant influence of heat treatment on the resistance and sensitivity of both casein and β-lg to hydrolysis in the human stomach, in an in vivo experiment. Tunick et al. (2016) investigated the effect of heat and homogenisation on the in vitro digestion of milk protein and fat and showed a faster release of the free fatty acids (FFAs) from homogenised milk into the intestinal fluid. Miranda and Pelissier (1987) reported that heat treatment results in accelerated gastric emptying during the digestion of milk and appears to increase the rate of hydrolysis of the caseins in the rat stomach. These phenomena have been attributed to the different coagulation behaviours of the milk proteins under gastric conditions, because of the interactions among the proteins and between the proteins and other components of milk during processing treatments.

## 3 Digestion behaviour of milk proteins

### *3.1 Coagulation/aggregation of milk proteins under gastric conditions*

Typically, the pH of human gastric fluid during the fasting state varies from about pH 1 to pH 3 because of the presence of hydrochloric acid. Such a strongly acidic environment prevents the growth of microorganisms (N'Goma et al., 2012). However, the pH may alter with the nature and the amount of ingested food and there may be significant differences between individuals (McClements et al., 2008). Generally, after consuming a meal, the pH value of the gastric content increases to about 5.5–7. After around 60 min of gastric digestion (i.e. the gastric half-emptying time), the pH reduces to about 4–5, and then drops to the initial acidic pH with further gastric emptying, when all digesta leave the stomach (N'Goma et al., 2012). In the fasting state, the typical ionic strength of the gastric fluid is around 100 mM; chloride is the dominant ion, with a concentration of about 100 ± 30 mM (Lindahl et al., 1997). The gastric secretion also contains different enzymes (i.e. pepsin and gastric lipase). Protein may be partially hydrolysed by pepsin in the stomach, and about 10–30% of the lipids may be digested to FFAs and diacylglycerol by the gastric lipase. The optimum pH for the activity/stability of pepsin is about 2.0. Furthermore, pepsin is able to induce milk coagulation at pH > 6, but will lose its milk-clotting ability at pH > 6.8 (Oberg et al., 1992).

#### *3.1.1 Casein and casein micelles*

The caseins can be precipitated by hydrophobic-interaction-induced aggregation when the pH decreases to their isoelectric point (pH ~ 5) during gastric digestion (Horne, 2020). Under gastric conditions, $\alpha_{s1}$-casein, $\alpha_{s2}$-casein and β-casein can be precipitated by calcium and other ions because of their ionic sensitivity. The precipitation of the caseins can be enhanced during gastric digestion at 37°C, because the hydrophobic interactions and the calcium sensitivity are temperature dependent. The caseins in caseinate are not coagulated by the milk-clotting enzyme (i.e. pepsin) in the stomach, despite κ-casein being hydrolysed by pepsin, as in casein micelles (Dalgleish and Corredig, 2012; Wang et al., 2018).

The casein micelles are very stable to processing treatments such as high temperature, commercial homogenisation, high ionic strength and calcium ions (O'Mahony and Fox, 2013). During gastric digestion, the casein micelles can be coagulated by both acidic conditions and proteases (e.g. pepsin) in the stomach. However, it has been shown that the casein micelles coagulate to form the structured curd before the pH in the stomach drops to the level that can lead to aggregation of the micelles during gastric digestion (Ye et al., 2016b). This suggests that the milk clotting induced by pepsin is the major mechanism for coagulation of the casein micelles during the gastric digestion of milk.

Milk clotting by proteolytic enzymes involves two stages: enzymatic hydrolysis, which alters the κ-casein and causes a loss in its ability to stabilise the casein micelle, and aggregation of the hydrolysed casein micelles. Whether aggregation of the casein micelles results in a firm clot or a soft fractured curd depends on the environmental conditions and the structure and composition of the formed curd. For the first stage, the rate and the extent of the enzymatic hydrolysis of κ-casein depend on the enzyme concentration and the pH. Chymosin uniquely cleaves the Phe105–Met106 peptide bond of κ-casein at an optimal pH (at pH ~ 5.6) (Carlson et al., 1987). Para-κ-casein remains on the surface of the casein micelles. The macropeptide leaves the casein micelle and moves into the serum phase. Pepsin is considered to be less specific (Fox and Walley, 1971), but there is little information on the κ-casein-cleaving properties of pepsin, especially under gastric conditions with a dynamic secretion of the enzyme and the gastric fluid. For the second stage, the aggregation of the milk micelle is dependent on the level of cleavage of the κ-casein because of the loss of electrostatic repulsion and steric repulsion of the κ-casein. The proteolysis of κ-casein also leads to a significant decrease in the zeta-potential of the casein micelles, which may enhance the sensitivity to the environmental stress in the stomach, and promote colloidal coagulation of the casein micelles during gastric digestion. Under gastric conditions in the presence of calcium ions and ionic strength, aggregation of the casein micelles, as induced by pepsin proteolysis, may be promoted. The presence of calcium ions facilitates the formation of the coagulum by connecting micelles as a bridge and inducing isoelectric conditions (Horne, 2020). Using dynamic low-amplitude oscillatory rheology, Roy et al. (2020b) reported that skimmed milk starts to gel at five units pepsin/100 mL at pH 5.

### *3.1.2 Whey protein*

Unlike the caseins, which are coagulable following specific, limited proteolysis, the whey proteins are not coagulated by proteolysis. It is well known that native β-lg is resistant to some proteases, particularly to pepsin; the highly hydrophobic β-barrel buries most hydrophobic sequences, making it very difficult for enzymes to gain access (Miranda and Pelissier, 1983; Reddy et al., 1988). In addition, whey protein will remain soluble at gastric pH (from around pH 5 to pH 2). Denaturation by processing treatments, such as heating, high pressure treatment, adsorption at the interface, addition of alcohols and esterification, can increase the susceptibility of β-lg to hydrolysis by pepsin (Chobert et al., 1995; Dalgalarrondo et al., 1995; Guo et al., 1995; Nik et al., 2010; Zeece et al., 2008). These treatments induce conformational changes in β-lg, resulting in increased exposure of peptic cleavage sites and thus increased susceptibility to pepsin action. In addition, as the changes in conformation of the proteins result in increased sensitivity to pH, aggregation could occur at low pH, close to the isoelectric point of the whey proteins, before hydrolysis by pepsin when the pH is lower than pH 4 under gastric conditions (Wang et al., 2018). In addition, several different kinds of intermediates and aggregates are formed during heat treatment, and it has been shown that different aggregates and modified monomers may behave differently under gastrointestinal conditions, particularly in the presence of pepsin (Peram et al., 2013). The β-lg fibrils formed by heating β-lg solutions at 80°C and pH 2.0 for 20 h were found to be digested completely by pepsin within 2 minutes (Bateman et al., 2010). The peptides in the fibrils (2000–8000 Da) could be digested to smaller peptides (mostly < 2000 Da) by pepsin. The peptides in the fibrils were believed to be more susceptible to access and attack by pepsin because of their hydrophobic nature.

Other whey proteins - α-la, BSA and lactoferrin - are more susceptible to hydrolysis by pepsin (Furlund et al., 2013; Nik et al., 2010). Although the hydrolysis of β-lg, α-la and BSA by pepsin has been studied extensively (Chobert et al., 1995; Dalgalarrondo et al., 1995; Guo et al., 1995; Nik et al., 2010; Sah et al., 2016; Zeece et al., 2008), the interactions of these proteins in the form of whey protein isolate (WPI) in dynamic gastric digestion have been studied only recently (Wang et al., 2018). In this study, hydrolysis of the proteins in WPI and heated WPI was investigated in a dynamic gastric digestion model, in which the pH was decreased from pH ~ 6 to pH ~ 2 at 240 min, the pepsin content was increased gradually by a consistent secretion to the digestion and the protein content was emptied to the intestine at a consistent rate. The β-lg in the unheated WPI remained intact without being hydrolysed during the whole digestion process and was released gradually to the intestine. The β-lg in the heated WPI formed aggregates at about pH 5 and was hydrolysed when the pH decreased to < 4, which led to a different composition of the gastric digesta for release to the intestine. The other major whey proteins such as α-la and BSA remained intact during the early digestion period when the pH was > 5, but were hydrolysed by pepsin at pH < 4 (Wang et al., 2018).

## 3.2 Gastric digestion of milk proteins in dairy products

### 3.2.1 Milk

The metabolic effects - such as the effects on body weight, cardiometabolic disease risk and bone health - of whole dairy products may be different from those of single dairy constituents (Thorning et al., 2017). In raw milk, the caseins are present in casein micelles and the whey proteins exist as individual proteins. After being processed into various milk products for the market, such as pasteurised milk and UHT milk, these proteins interact with each other to form aggregates, casein micelle–whey protein complexes and fat globules with surface casein and serum protein. These interactions between the milk proteins could significantly change their coagulation behaviour during gastric digestion and hence have an impact on the digestion of the milk constituents. In addition, the environmental conditions of the milk serum, such as calcium ions, pH, ionic strength and buffering capability of the casein and whey protein, should be considered in the coagulation of the milk proteins during gastric digestion.

As demonstrated with a dynamic digestion model – a human gastric simulator (HGS) – unheated milk formed a firm, dense, cheese-like clot with a porous network structure at about 10 min when the pH was > 6.2 (Fig. 1). The clot structures became tighter and less permeable at longer gastric digestion times (Ye et al., 2016b); this was confirmed in an in vivo rat study (Ye et al., 2019). The formation of the clot is considered to be the result of the hydrolysis

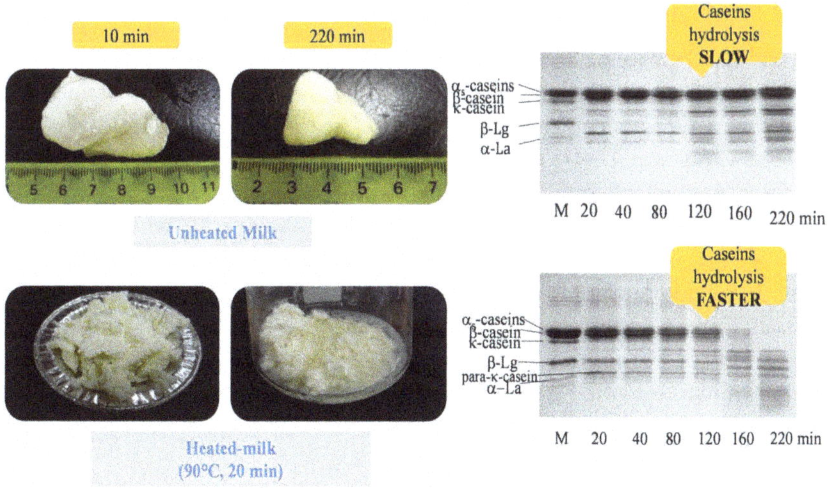


**Figure 1** Photographs of clots of unheated and heated milks formed during in vitro dynamic gastric digestion in the HGS and the SDS-PAGE patterns, under reducing conditions of proteins in the clots as a function of the digestion time. From Ye, Cui, Dalgleish, and Singh (2016a), reproduced with the permission of Elsevier Inc.

of κ-casein by pepsin, leading to destabilisation of the casein micelles, rather than the effect of low pH on the stability of the casein micelles, because the coagulation occurs at pH > 6.2. SDS-PAGE (Fig. 1) showed that the intensity of the κ-casein band decreased markedly even after 20 min of digestion, and was virtually absent after 40 min. In contrast, the intensities of the $\alpha_S$-casein ($\alpha_{S1}$- + $\alpha_{S2}$-casein) and β-casein bands did not appear to change with the digestion time, which confirmed the hypothesis that the pepsin in the simulated gastric fluid was involved in converting the milk into a clot. However, β-lg and α-la were not involved in the formation of the clot in the unheated milk.

As in unheated skim milk, this coagulation of the proteins and the formation of a clot occur at a similar digestion time or at a similar pepsin concentration under gastric conditions for all milk types with a variety of processing treatments, including heated skimmed milk, whole milk, pasteurised milk, UHT milk and homogenised milk (Ye et al., 2016a,b, 2017, 2019). Similar coagulation behaviour during gastric digestion has also been observed in the milks from other species such as goat milk and sheep milk (Roy et al., 2020b), although the composition of the casein micelles in these milks is different from that in bovine milk (Roy et al., 2020a). This suggests that altering the composition and the processing treatment is unlikely to alter the coagulation of the casein micelles by pepsin and the formation of a clot.

However, the structure of the formed clot is influenced by changes in the milk composition and the pre-processing treatment. For heated skimmed milk (90°C for 20 min), in which the whey protein had been fully denatured, a looser, fragmented-network-structured clot with numerous larger voids was formed under gastric conditions (Fig. 1) (Ye et al., 2016b). In contrast to the unheated milk, the whey proteins were involved in the formation of the clot during the gastric digestion of the heated milk, as both β-lg and α-la were observed in the clot and were hydrolysed, along with the caseins, into peptides during digestion. This was partly because of the association of the whey proteins with the casein micelles during heat treatment (Anema and Li, 2003b; Donato et al., 2007), and partly because the non-micelle-bound whey protein–κ-casein complexes formed during heating associated with the micelles during proteolysis (Kethireddipalli et al., 2011) and at pH values below ~ 5.3 (Donato et al., 2007; Vasbinder et al., 2003). However, the emptied digesta from the heated milk did not contain intact casein and whey proteins (Fig. 1), suggesting that the denatured whey proteins in the heated milk were hydrolysed rapidly in the digesta. In contrast, both β-lg and α-la remained intact during the whole digestion period in unheated milk (Fig. 1).

Commercially available milk is commonly homogenised to increase the stability of the milk fat by reducing the size of the milk fat globules. Homogenisation leads to the formation of coagula with fragmented and crumbled structures, compared with the coagulum formed from raw whole milk (Fig. 2), but a larger fraction of the protein and more fat globules are incorporated into the coagula, as induced by the action of the milk-clotting enzyme pepsin (Ye et al., 2017). Heat treatment of the homogenised milk enhances the looseness of the gastric coagula (Fig. 2). Homogenisation causes a marked increase in the number of fat globules by reducing the fat globule size, which leads to the caseins and whey proteins adsorbing on to the surface of the fat globules and reducing the amount of MFGM at the fat globule surface (Ye et al., 2008). The large number of fat globules that embed in the clots formed during digestion can affect the network of the protein matrix in the clot of homogenised milk, which results in a looser and fragmented structure (Ye et al., 2017). A greater reduction in the amount of protein at an earlier gastric digestion time in the clot of homogenised milk has been observed, suggesting greater hydrolysis of the proteins in the homogenised milk than in the untreated milk, in agreement with the more rapid breakdown of the clot (Ye et al., 2017).

### 3.2.2 Disintegration of clot and hydrolysis of protein during gastric digestion

Gastric fluid is secreted continuously and the pH gradually decreases to about pH 2 for adults or pH 3 for infants, after which the proteins coagulate and form the structured clot/curd. The clot structures become tighter and less permeable with increasing digestion time (Ye et al., 2016b). The different rates of pepsin hydrolysis for the caseins at the surface of the clot and for those in the inner part of the clot indicate that the clot structure prevents access of the pepsin to the interior of the clot, consequently influencing the hydrolysis of the proteins by pepsin (Ye et al., 2016b). This leads to the very

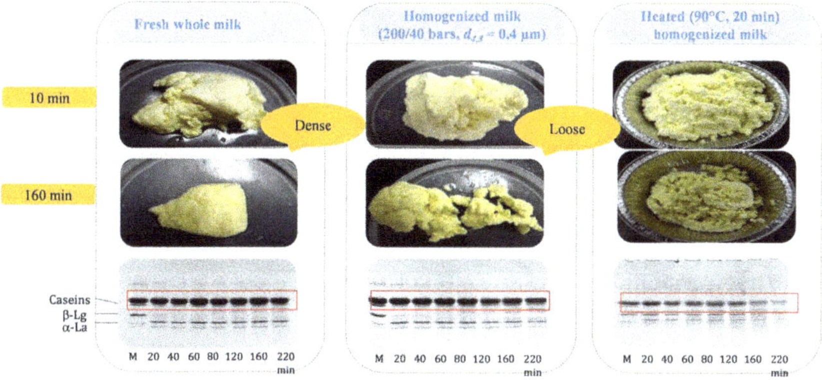


**Figure 2** Photographs of clots of raw whole milk, homogenised milk and heated homogenised milk formed during in vitro dynamic gastric digestion in the HGS and the SDS-PAGE patterns, under reducing conditions of proteins in the clots as a function of the digestion time. From Ye et al. (2017), reproduced with the permission of Elsevier Inc.

slow hydrolysis of the $\alpha_s$-casein ($\alpha_{s1}$- + $\alpha_{s2}$-casein), β-casein and para-κ-casein with digestion time in the clot of unheated milk (Fig. 1). However, there is greater hydrolysis by pepsin of the caseins and whey proteins during further gastric digestion of heated milk, because the open, fragmented structure of the clot/curd formed from heated milk increases the diffusion of gastric fluid within the clot and enlarges the effective contact area with pepsin. In addition, both β-lg and α-la are also hydrolysed gradually, along with the caseins, into peptides during digestion, and are emptied from the stomach into the intestine (Fig. 1). A greater hydrolysis of the proteins and a more rapid breakdown of the clot during gastric digestion are also observed in homogenised milk, because of its looser and more open-structured clot compared with untreated milk.

UHT-treated (140°C for 4 s) homogenised milk exhibited a more rapid rate of protein hydrolysis than unheated milk and pasteurised (72°C for 15 s or 85°C for 15 s) homogenised milk during gastric digestion, because of the looser structure of the clot and the larger voids within the curds formed, as observed in both an in vivo rat study and an in vitro dynamic gastric digestion model (Ye et al., 2019). This indicated that the heat intensity of the UHT treatment of milk was sufficient to cause a significant change in the digestion of the milk proteins. An in vivo pig study, (Lacroix et al., 2008) also reported that UHT treatment enhanced the rate of digestion of the milk proteins, causing a greater transfer of dietary nitrogen into serum amino acids and protein, but that pasteurisation did not alter the outcome. Using a semi-dynamic digestion model, Mulet-Cabero et al. (2019) also showed that the higher the temperature of the heat treatment, the softer the coagulum obtained and the greater the protein hydrolysis at the end of digestion for whole milk, with homogenisation, and heat treatments of pasteurisation at 72°C for 15 s and UHT treatment at 140°C for 3 s.

A possible mechanism for the difference in the structures of clots formed during the gastric digestion of unheated milk and heated milk has been suggested to be related to the involvement of the whey proteins with the casein micelles during heat treatment (Fig. 3) (Ye et al., 2019). Heat treatment leads to the association of denatured whey proteins with the casein micelles and with the surface of the fat globules (Anema and Li, 2003a,b; Ye et al., 2004), which could reduce casein–casein interactions and casein–fat globule interactions during coagulation. This may also alter the orientation of the casein strands and provide a less complete fusion rearrangement, which may result in a change in the structure of the curds to a more open and looser network. The casein-micelle-bound whey proteins may sterically hinder aggregation of the casein micelles, even after the removal of nearly all of the κ-casein and a decrease in the concentration of ionic calcium in heated milk (Kethireddipalli and Hill, 2015).

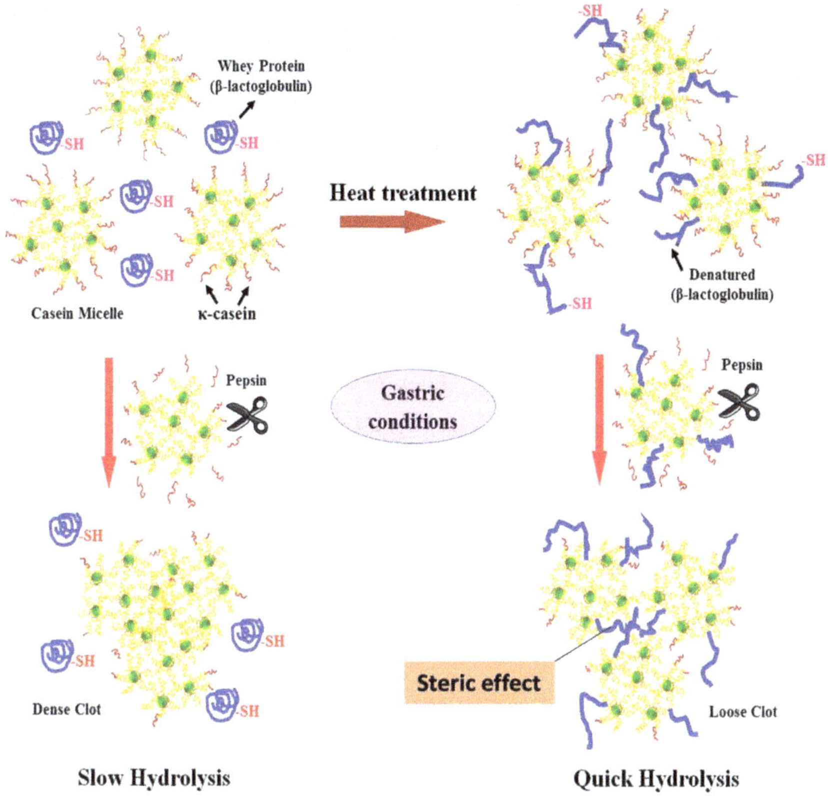


**Figure 3** Schematic diagram of the possible mechanism for the events during the formation of protein curds from unheated milk and heated milk in a gastric environment. From Ye et al. (2019b), reproduced with the permission of Elsevier Inc.

### *3.2.3 Cheese and yoghurt*

Nearly half the milk produced is processed into semi-solid and solid dairy products, like various kinds of cheese and yoghurt. Cheese is a concentrated protein gel containing all the components of milk. The gel structure is formed mainly by aggregation of the casein micelles, which is induced by rennet hydrolysis of the κ-casein in the presence of calcium ions. The disintegration of cheese during gastric digestion was dependent on the initial microstructures, different compositions and a range of different textures (Fang et al., 2016). At the end of the gastric digestion, disintegration of the cheese structure was greater when the initial hardness, cohesiveness and chewiness of the cheese were lower. Pepsin hydrolysis contributed to a decrease in the cohesive forces that held the various cheese matrices together, resulting in an increase in disintegration. For example, Mozzarella cheese, with its higher hardness and cohesiveness values, was broken down slowly and continuously during gastric digestion, but the disintegration of Camembert (a ripened and soft cheese with a high fat content) was relatively fast from the beginning of digestion onward. It has been suggested that some other factors may be involved in the digestion of the structured protein network, as no direct relationship between the disintegration kinetics and the rheological properties of different cheeses has been observed. For example, the release of the fat globules can help to degrade the cheese matrix and accelerate disintegration (Fang et al., 2016). Texture-hardening behaviour was observed in Mozzarella cheese during static soaking in simulated gastric juice (Do and Kong, 2018), which suggested that it may have been due to the aggregation of the hydrophobic regions of the micelles, as a result of pepsin digestion and acidic precipitation (Robinson and Wilbey, 1998; Sousa et al., 2001). This behaviour was similar to that observed in the gastric digestion of milk, in which the milk proteins and the fat globules aggregated into clots, induced by pepsin digestion and acidic precipitation (Ye et al., 2016a,b). In a recent study using visual and confocal microscopy, the restructuring of the gel particles was also observed in the in vitro dynamic gastric digestion of a rennet gel at an early digestion stage (Qazi et al., 2021). Further proteolysis of the remaining κ-casein in the rennet gel was observed along with a change in the structure of the gel particles during digestion before the pH dropped to below 5, which indicated that it was induced by the hydrolysis of the κ-casein in the casein micelles. This restructuring to form larger-sized particles in the stomach led to a reduction in the rate of protein digestion in the stomach and slowed the emptying rate of the rennet gel chyme from the stomach to the next step of digestion (Qazi et al., 2021).

In general, yoghurt is made by the fermentation of heated milk. The milk is heated prior to fermentation to denature the whey proteins and to induce the association of the whey proteins with the casein micelles. A curd is formed during the fermentation because of the acid-induced changes in the casein micelles and the denatured whey proteins. In a gastric simulator chamber during in vitro dynamic gastric digestion, the curd structure disintegrated gradually because of the action of pepsin in the stomach environment (Acevedo-Fani et al., 2021). Unlike milk and other dairy products with neutral pH, when yoghurt or acid gel particles were ingested into the stomach after oral processing, the initial pH was below 5 and reduced to around pH 2 at a rapid rate. The proteolytic action of pepsin was promoted from pH 5 to pH 2 and reaches its optimum at pH 2. The restructuring of the proteins because of aggregation of the casein micelles, induced by the hydrolysis of κ-casein, did not occur in the acid gel matrix under gastric conditions because the pH was lower than is suitable for inducing cleavage of the κ-casein for coagulation of the casein micelles. The dissociation of the protein gel matrix from the acid gel was mainly the result of the hydrolysis of all the protein in the matrix by pepsin at low pH. The amounts of β-casein, $\alpha_s$-casein and κ-casein and even of β-lg and α-la decreased at the same rate during digestion of the acid gel (Qazi et al., 2021). This may have resulted in the greater rate and extent of disintegration of the acid gel or yoghurt compared with a rennet gel or cheese under gastric conditions. This was also observed in an in vivo study on mini-pigs, in which the concentration of amino acid in the duodenal effluents and the blood plasma was much higher from an acid gel than from a rennet gel (Barbé et al., 2014). The structure of the acid gel rapidly disintegrated in the gastric fluid, showing limited molecular arrangements of the protein network, whereas the rennet gel formed a compact aggregate within the stomach, which might delay gastric emptying. The digestion behaviours of yoghurt and cheese were also confirmed by a recent in vivo study which investigated postprandial aminoacidemia after consumption of different dairy products, including yoghurt and cheese, using a single-blinded cross-over design with a group of ten participants (Horstman et al., 2021). The results show that the higher increase in blood essential amino acid levels occurred after yoghurt consumption, whereas cheese consumption resulted in extended EAA response curves, suggesting that the product matrix could be as important as protein composition in postprandial aminoacidemia.

## 3.3 Digestion of milk proteins in the small intestine

Protein digestion starts in the stomach by the action of pepsins, followed by intestinal digestion by trypsin, chymotrypsin and various carboxy- and aminopeptidases. Trypsin catalyses the hydrolysis of peptide bonds; it mainly cleaves proteins at C-terminal side of the amino acids lysine and arginine into smaller peptides, which are further hydrolysed into amino acids via other proteases. In the small intestine, almost all milk proteins and peptides released from the stomach are hydrolysed into smaller peptides or amino acids that become available for absorption into the bloodstream. The amount of nitrogen in these peptides or amino acids, compared to the total amount of nitrogen in ingested protein, is defined as protein digestibility.

All milk proteins and the milk proteins in dairy products are highly digestible compared with proteins from other sources, as seen from both in vivo and in vitro studies (Gilani and Sepehr, 2003; Lacroix et al., 2006; Rutherfurd and Moughan, 2003, 2005; Rutherfurd and Moughan, 1998). Casein and whey protein concentrates were found to be almost 100% digestible, based on the protein digestibility-corrected amino acid score in rats (Gilani and Sepehr, 2003). This suggests that the different digestion rates of casein and whey protein due to the coagulation behaviour during gastric digestion may not certainly lead to the difference in the digestibility of these proteins in the intestine. Furthermore, the processing treatments applied to the milk and milk products, such as heat treatment-induced enhanced proteolysis of whey protein and increased gastric emptying rate of milk in a gastric model, do not automatically mean an increased overall digestibility. For example, raw milk, pasteurised milk, UHT milk, sterilised milk and boiled milk did not show a difference in the protein digestibility (Efigênia et al., 1997). In vitro studies show a similar intestinal proteolysis between raw, pasteurised and sterilised milk (Lamothe et al., 2017; Tunick et al., 2016). For the milk products with various protein matrices, the overall nitrogen absorption determined by portal labelled nitrogen absorption was similar in growing pigs fed with milk, yoghurt, and heated yoghurt (Mpassi et al., 2001). This indicates that denaturation induced by heat treatment does not influence intestinal hydrolysis, and hence the overall digestibility, although it affects digestion kinetics through a different stomach behaviour.

However, some interactions and complexations among milk proteins or between milk protein and other components like carbohydrates, induced by severe heating, could decrease the digestibility of the milk protein in the intestine. For example, glycation induced by the Maillard reaction decreases protein digestibility. The Maillard reaction modifies the side chain of the protein by forming the Amadori product lactulosyllysine between the ε-amino groups of lysine (Lys) residues and the reducing sugars like lactose. From the initial reaction, a series of advanced reactions leading to a variety of modified structures are promoted (Van Boekel, 1998). Blocking of the ε-amino groups of Lys residues by the binding of reducing sugars could hinder the cleavage by trypsin and other digestive enzymes during intestinal digestion and, as a consequence, reduce protein digestibility (Marvin et al., 2002; Mauron et al., 1955; Rérat et al., 2002). Some other products of cross-linking between Lys and Cys residues, such as lysinoalanine and lanthionine formed during heat treatments, could also reduce protein digestibility by inhibiting enzymatic proteolysis (van Lieshout et al., 2020; Wada and Lönnerdal, 2014).

## 4 Behaviour of milk fat globules during digestion

Bovine milk contains from 3.5% to 5.0% lipids, which are mainly triacylglycerols and some minor components including diacylglycerols, monoacylglycerols, FFAs, phospholipids and cholesterol. The triacylglycerols in milk contain about 70% saturated fatty acids and about 30% unsaturated fatty acids (Lindmark Månsson, 2008). The short-chain acids butyric (4:0) and caproic (6:0) are esterified at position *sn*-3 of the triacylglycerol molecule. Medium-chain fatty acids (8:0–14:0) as well as 16:0 are preferentially esterified at positions *sn*-1 and *sn*-2. Stearic acid (18:0) esterification is selectively located at position *sn*-1, whereas oleic acid (18:1) shows preference for positions *sn*-1 and *sn*-3 (MacGibbon and Taylor, 2006).

Milk fat provides the major source of energy for the neonate and is an important carrier of lipid-soluble constituents, such as lipid-soluble vitamins (A, D, E and K) and several volatile flavour compounds. Milk fat for human health is somewhat controversial (Mozaffarian, 2016); the saturated fatty acids and cholesterol are considered to be connected with the risk of cardiovascular disease, whereas some milk lipids such as conjugated linoleic acid, sphingomyelin and butyric acid are shown to exhibit anti-cancer activities (Parodi, 2001). Milk fat in natural milk is in the form of oil-in-water emulsions, called fat globules. The diameter of the fat globules ranges from 0.1 μm to 10 μm. Milk fat globules are surrounded and stabilised by the MFGM (Mather, 2000; Walstra, 1995), which is a multilayer consisting of phospholipids, various glycoproteins, enzymes, phosphoproteins and cholesterol, and with a thickness of about 8–10 nm (Dewettinck et al., 2008; Mather, 2000). In homogenised milk and recombined milk, caseins constitute more than 90% of the adsorbed proteins at the surface of the fat globules (Oortwijn and Walstra, 1982; Sharma et al., 1996). Small globules acquire much more protein per unit of surface area because of the preponderance of large casein micelles at the surfaces of small fat globules (Oortwijn and Walstra, 1982). As lipid digestion is essentially an interfacial process, the size of the fat globules and the structure of the MFGM or protein surface layer play key roles in the digestion and absorption of milk fat. These alterations influence the digestion and the bioavailability of the milk fat (Michalski and Januel, 2006; Ye et al., 2010, 2011). Berton et al. (2009) demonstrated that the catalytic behaviour and the efficiency of pancreatic lipase differ when the milk fat is organised, as in native human milk, and when it exists as homogenised human milk fat globules. It is considered that the MFGM has profound effects on the accessibility of the triacylglycerols for lipase-catalysed digestion (Mather and Keenan, 1975).

The digestion of lipids, the release of nutrients and their conversion into an absorbable form are performed mainly in the small intestine. Before the fat globules enter the intestine, they must pass through the mouth and the stomach. Milk stays in the mouth and is in contact with saliva for only a short time before being swallowed. There is little information on the behaviour of the milk fat globules in the oral environment. Extensive reversible flocculation is observed in the fat globules in raw milk when mixed with artificial saliva, primarily because of depletion interaction (Smoczynski and Staniewski, 2014). In contrast, the thicker dairy products, such as cream and ice cream, and solid dairy products such as cheese, stay in the mouth and are in contact with saliva for a longer time, because the food needs to be broken down to a size that is small enough for swallowing. Very little information on dairy products in the mouth is available in the literature, although there are some reports on the milk protein emulsion and the emulsion gel formed with milk proteins (Guo et al., 2013; Luo et al., 2019).

## 4.1 Gastric digestion

Less than 30% of the triacylglycerols are digested in the stomach (Golding and Wooster, 2010). The shorter acids are selectively released during the digestion of milk fat because the lipases preferentially hydrolyse *sn*-3-position fatty acids, which results in short-chain fatty acids (from 4:0 to 10:0) that pass through the stomach wall, enter the portal vein and are transported to the liver where they are oxidised (Jensen, 2002).

All the MFGM proteins can be hydrolysed by gastric pepsin into a range of peptides, but the rate of hydrolysis is different for different MFGM proteins, as demonstrated in an SDS-PAGE analysis of the milk fat globules isolated from raw milk, digested in an in vitro simulated gastric system (Gallier et al., 2012; Ye et al., 2011). For instance, xanthine oxidase was hydrolysed much faster than butyrophilin and PAS 6, and PAS 7 was hydrolysed faster than butyrophilin, but PAS 7 was more sensitive than PAS 6 (Fig. 4). The hydrolysis of interfacial MFGM protein was followed by an apparent flocculation of the milk fat globules, as indicated by an increase in their average size ($d_{3,2}$; volume-surface mean) with an increase in the gastric digestion time (Ye et al., 2011). An examination of the gastric digesta by confocal laser scanning microscopy showed that the flocculation of the fat globules started at the early stages of digestion and appeared to be enhanced at long digestion times (≥ 30 min) (Ye et al., 2011). The fat globules were linked through some protein or peptide material between the globules. A closer examination of the microstructure revealed that the fat globules were physically entrapped in the casein aggregate network and that they remained essentially intact during gastric digestion. An in vivo

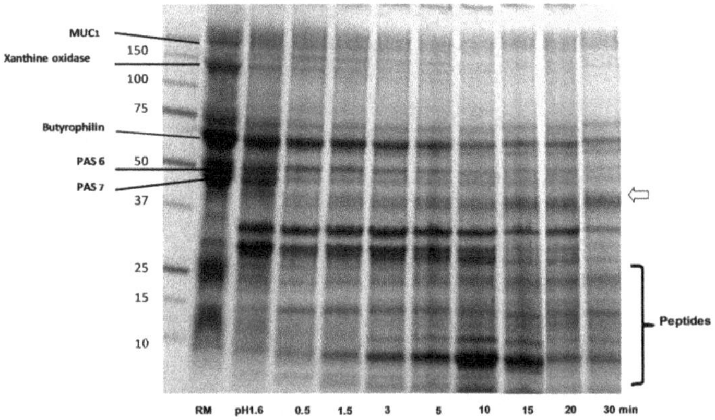


**Figure 4** SDS-PAGE patterns of the MFGM proteins obtained from the cream material of raw milk (RM) samples incubated in simulated gastric fluid containing pepsin at 0.1 mg/mL, as a function of the digestion time. MUC, mucin; PAS, periodic acid Schiff. Reproduced with permission from Ye et al. (2011).

rat experiment using cream rather than milk, to avoid the effect of casein, also showed that the fat globule size did not change during digestion (Gallier et al., 2013). This confirmed that some of the peptides hydrolysed from the MFGM proteins remained at the globule surface and provided sufficient electrostatic repulsion and steric barriers to prevent the coalescence of the fat globules. In addition, the phospholipid in the MFGM would be expected to remain at the surface and prevent the coalescence of the fat globules, even in the absence of MFGM proteins (Brown and London, 2000).

The behaviour and the digestion of milk fat during gastric digestion are also influenced largely by the coagulation of the milk protein under gastric conditions. In whole milk, the structure and the breakdown of the clots during gastric digestion influence the release of fat from the stomach into the small intestine. The rate of release of the fat globules from the clots during gastric digestion showed a linear correlation with the disintegration of the clots, under gastric conditions in fresh milk, homogenised milk and heated homogenised milk (Ye et al., 2017). This result suggests that, perhaps surprisingly, the manner in which the fat globules are incorporated within the clot matrix does not control their release. In untreated milk, the fat globules are trapped within the casein clot matrix, but there are no interactions between proteins and fat globules. In comparison, in homogenised milk, the surfaces of the fat globules are covered with casein, so that the globules themselves become part of the matrix as the clot is formed. In heated homogenised milk, the fat globule surfaces contain both denatured whey proteins and caseins, and again participate in the formation of the clot matrix.

Confocal images showed that some fat globules appeared to coalesce within the matrix of the clot (Fig. 5), as evidenced by the large area of fat seen in the clot; in contrast, other fat globules were within the aqueous phase (Ye et al., 2017). In heated homogenised milk, the structure of the clots was much more open, with blocks of aggregated protein. Within the clot, the fat globules appeared to be clumped in separate areas. In addition, the fat globules in the aqueous phase were apparently larger than the fat globules in the milk before digestion, suggesting that coalescence of the fat globules took place as the MFGM proteins surrounding the fat globules were hydrolysed by pepsin (Ye et al., 2011). These results suggest that differences in the structure of the clot or coagulum formed from milks with various treatments also affect the delivery of the fat globules to the next digestion step.

### *4.2 Lipolysis of milk fat in the small intestine*

Lipid digestion takes place mainly in the small intestine and is catalysed by pancreatic lipases; the digestion of lipids by the gastric lipases is limited (< 30%) (Bauer et al., 2005; Mun et al., 2007; Singh et al., 2009). Pancreatic

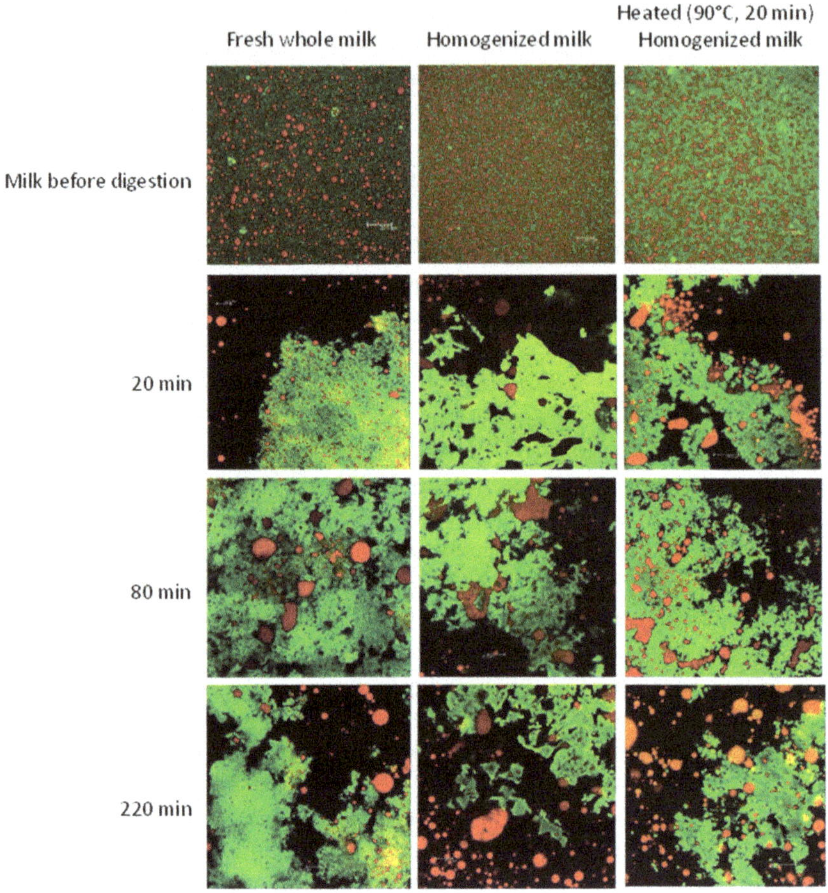


**Figure 5** Confocal microscopy images of untreated whole milk, homogenised milk and heated homogenised milk (90°C for 20 min) at different times (from 0 min to 220 min) during gastric digestion. Red shows the fat and green shows the protein. The scale bar in all images is 20 μm. From Ye et al. (2017), reproduced with the permission of Elsevier Inc.

lipases can work efficiently at about pH 6.5 (Singh and Ye, 2013), although their optimum pH is 8–9 (Patton and Carey, 1981). When the partially digested lipid droplets pass into the small intestine, the pancreatic lipase tends to adsorb at the droplet surface as a complex with co-lipase and/or bile salts (Bauer et al., 2005). Triacylglycerols and diacylglycerols are then broken down to FFAs and 2-monoacylglycerols. It has been reported that pancreatic lipase has no obvious specificity for the chain length of fatty acids, but preferentially cleaves their *sn*-1 and *sn*-3 positions (Mu and Høy, 2004). This complexation reaction of pancreatic lipase at the lipid-droplet interface requires the presence of bile salts, co-lipase and calcium (Hur et al., 2011). Its degree of binding appears to depend on the electrical charge of the interface and competitive adsorption with bile salts, digestion products or other surfactants (McClements et al., 2008). When partially digested food products pass into the small intestine from the stomach, the pH undergoes a rapid increase from the highly acidic environment (pH 1–3) in the stomach to the neutral environment (pH 6–7.5) in the duodenum, because of the secretion of sodium bicarbonate. This neutral pH provides an optimal environment for the action of pancreatic enzymes (Golding and Wooster, 2010; Hur et al., 2011; McClements et al., 2008). The increase in pH may cause some critical changes in the physicochemical properties of the proteins, for example, a reversal of the protein charge.

Under intestinal conditions, the milk fat globules change in size, surface charge and microstructure (Liang et al., 2018; Ye et al., 2010). Using a simple in vitro standardised intestinal model, Ye et al. (2010) compared the behaviour of the fat globules in raw milk, which had an intact MFGM, with that of the fat globules in recombined milk, in which the fat globules were coated by casein micelles and whey proteins (i.e. no MFGM). These samples had similar initial average fat globule sizes; during the in vitro intestinal digestion, the average size increased dramatically during the early stages, followed by a rapid decrease and then a gradual increase with an increase in the digestion time. The zeta-potential was found to increase (in negative value) dramatically for both samples, but there were no differences in the rate of change in the zeta-potential between the raw milk sample and the recombined milk sample. For the lipolysis, it was shown that the rate of release of FFAs during incubation in a simulated intestinal fluid in the absence of bile extract was slower in the raw milk than in the recombined milk, but that the final extent of lipid digestion after 290 min was similar in both samples (Fig. 6). In the presence of bile extract (5.0 mg/mL), the rate of lipid digestion by pancreatic lipase was much faster and almost similar in both milks. However, the total amount of fatty acids released was slightly lower in the raw milk sample than in the recombined milk sample. It has also been reported that homogenised human milk was more rapidly hydrolysed by pancreatic lipase than native human milk, which showed a lag phase in the fatty acid release profile (Berton et al., 2009). This can probably be attributed to the inhibitory effect of components in the MFGM on the activity of lipase. The dense MFGM contains mainly phospholipids, which may slow down pancreatic lipase penetrating through the membrane and reaching the core fat. In contrast, pancreatic lipase may have higher affinity for the micellar casein surface of the fat globules in recombined milk, leading to more lipase binding with the interface of the fat globules. It has been reported that more lipase associates with casein on the surface of fat globules in milk that is agitated or homogenised (Cartier and Chilliard, 1989).

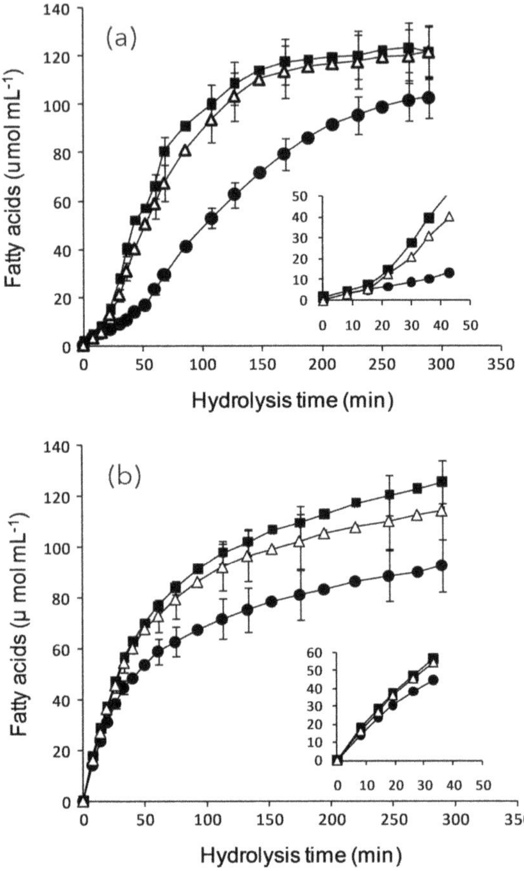


**Figure 6** Amount of fatty acids released from the fat globules in raw milk (●), recombined milk A (■) and recombined milk B (Δ) after digestion with pancreatic lipase, in the absence (a) and in the presence (b) of bile extract, as a function of time. Inset plots present the amount of fatty acids released from the samples at an early stage. Each data point is the average ± SE of two determinations on three separate emulsions. From Ye et al. (2010), reproduced with the permission of Elsevier Inc.

## 5 Digestion of milk lactose

The carbohydrate in mammalian milk includes lactose, oligosaccharides (OSs), very small amounts of monosaccharides and some glycosylated glycoproteins in κ-casein and milk fat globule membrane protein. Lactose is a major source of carbohydrate for all mammalian neonates. Lactose is synthesised only in the mammary gland, with varying concentrations according to species, from 0% to 7%. Cow milk contains 4-5% lactose; human milk contains the highest content of lactose (7%) (Palmiter, 1969). Lactose is a disaccharide consisting of glucose and galactose, linked together via a β-1→4 bond. Hydrolysis of this bond requires a specific enzyme called lactase-phlorizin hydrolase (LPH), or simply lactase, which belongs to a group of intestinal disaccharidases located on the brush border of the small intestine. Lactose is soluble in water and stays in serum phase of milk. When protein and fat globules in milk or dairy products form the coagulum due to the coagulation of milk protein under gastric conditions, lactose is quickly released to the intestine with milk serum. Lactose is hydrolysed by lactase to its components in the intestine, where the glucose and galactose are absorbed into the bloodstream through the intestinal lining (Forsgård, 2019; Montgomery et al., 2007).

Lactase-phlorizin hydrolase (LPH), or lactase, belongs to a group of intestinal disaccharidases located on the brush border of the small intestine. The abundance of lactase is highest in the proximal part of the jejunum and progressively declines towards the ileum. In most mammals, intestinal lactase activity is high at birth but starts to progressively decline after weaning, eventually curtailing the ability to digest dietary lactose (Montgomery et al., 2007; Ugidos-Rodríguez et al., 2018). However, in some humans, a genetic trait enables intestinal lactase activity to persist into adulthood. If too little quantity of lactase is produced in the small intestine, lactose moves into the colon instead of being processed and absorbed. In the colon, colonic microbes, such as the lactic acid bacteria Lactobacillus and Bifidobacterium, possess β-galactosidase activity (i.e. bacterial lactase) and can hydrolyse lactose to glucose and galactose and subsequently ferment them to lactate, SCFAs, and gases, such as $H_2$, $CO_2$, and $CH_4$ (He et al., 2008). The accumulation of fermentation products increases the osmotic load in the colonic lumen, leading to the various gastrointestinal (GI) symptoms of lactose intolerance (He et al., 2008). People with lactose intolerance, which is also called lactose malabsorption, are unable to fully digest the lactose in milk, who have diarrhoea, gas and bloating after eating or drinking dairy products containing lactose. Globally, ~70% of adults are deficient in intestinal lactase, which varies between populations and ethnicities. For example, in Southeast Asia > 90% are lactase-deficient as adults, whereas it is only 5% of the population in Northwest Europe (Bayless et al., 2017).

## 6 Conclusions

The digestion and absorption of milk components are not only based on the high quality and the content of individual nutrients in the diet, but are also largely determined by the structure and matrix of the food. Extensive studies have demonstrated that the digestion kinetics of the protein and lipid in milk and dairy products are strongly influenced by the structure of the original food and the restructuring of the milk proteins during digestion. In general, the influence of the structure on the digestion and absorption of nutrients is mainly through the kinetics of digestion, by affecting the disintegration of the structure during digestion and the rate of transmission of nutrient particles from the stomach to the next digestion step in the gastrointestinal tract. Most of the structures among dairy products are formed by protein interactions during processing. Nevertheless, the restructuring induced by the interactions between milk proteins and biophysical fluids during digestion plays an important role in the digestion, in which the milk proteins not only are hydrolysed to provide peptides and amino acids for nutrition, but also provide the structural aspects for manipulating the digestion rates of nutrients in milk and dairy products. More investigation on the restructuring interactions and the influence of pre-treatment and processing on the formation of structure and on potential applications in the development of healthy dairy products is required.

## 7 Where to look for further information

In recent years, tremendous research effort has gone into understanding the digestion behaviour and digestibility of the proteins in milk and dairy products. Several comprehensive review articles have been published for summarising the studies in the area:

- Huppertz, T. and Chia, L. W. (2020). Milk protein coagulation under gastric conditions: a review. *International Dairy Journal*, 113, 104882.
- Mulet-Cabero, A. I., Mackie, A. R., Brodkorb, A. and Wilde, P. J. (2020). Dairy structures and physiological responses: a matter of gastric digestion. *Critical Reviews in Food Science and Nutrition*, 60(22), 3737–3752. https://doi.org/10.1080/10408398.2019.1707159.
- Van Lieshout, G. A. A., Lambers, T. T., Bragt, M. C. E. and Hettinga, K. A. (2020). How processing may affect milk protein digestion and overall physiological outcomes: a systematic review. *Critical Reviews in Food Science and Nutrition*, 60(14), 2422-2445. https://doi.org/10.1080/10408398.2019.1646703.
- Ye, A. (2021). Gastric colloidal behaviour of milk protein as a tool for manipulating nutrient digestion in dairy products and protein emulsions. *Food Hydrocolloids*, 115, 106599. https://doi.org/10.1016/j.foodhyd.2021.106599.

## 8 References

Acevedo-Fani, A., Ochoa-Grimaldo, A., Loveday, S. M. and Singh, H. (2021). Digestive dynamics of yoghurt structure impacting the release and bioaccessibility of the flavonoid rutin. *Food Hydrocolloids* 111, 106215.

Anema, S. G. and Li, Y. (2003a). Association of denatured whey proteins with casein micelles in heated reconstituted skim milk and its effect on casein micelle size. *The Journal of Dairy Research* 70(1), 73–83.

Anema, S. G. and Li, Y. (2003b). Effect of pH on the association of denatured whey proteins with casein micelles in heated reconstituted skim milk. *Journal of Agricultural and Food Chemistry* 51(6), 1640–1646.

Azpiroz, F. and Malagelada, J. R. (1990). Perception and reflex relaxation of the stomach in response to gut distention. *Gastroenterology* 98(5 Pt 1), 1193–1198.

Barbé, F., Ménard, O., Le Gouar, Y., Buffière, C., Famelart, M. H., Laroche, B., Le Feunteun, S., Dupont, D. and Rémond, D. (2013). The heat treatment and the gelation are strong determinants of the kinetics of milk proteins digestion and of the peripheral availability of amino acids. *Food Chemistry* 136(3–4), 1203–1212.

Barbé, F., Ménard, O., Le Gouar, Y., Buffière, C., Famelart, M. H., Laroche, B., Feunteun, S. L., Rémond, D. and Dupont, D. (2014). Acid and rennet gels exhibit strong differences in the kinetics of milk protein digestion and amino acid bioavailability. *Food Chemistry* 143, 1–8.

Bateman, L., Ye, A. and Singh, H. (2010). In vitro digestion of β-lactoglobulin fibrils formed by heat treatment at low pH. *Journal of Agricultural and Food Chemistry* 58(17), 9800–9808.

Bauer, E., Jakob, S. and Mosenthin, R. (2005). Principles of physiology of lipid digestion. *Asian-Australasian Journal of Animal Sciences* 18(2), 282–295.

Bayless, T. M., Brown, E. and Paige, D. M. (2017). Lactase non-persistence and lactose intolerance. *Current Gastroenterology Reports* 19(5), 23.

Berton, A., Sebban-Kreuzer, C., Rouvellac, S., Lopez, C. and Crenon, I. (2009). Individual and combined action of pancreatic lipase and pancreatic lipase-related proteins 1 and 2 on native versus homogenized milk fat globules. *Molecular Nutrition and Food Research* 53(12), 1592–1602.

Boirie, Y., Dangin, M., Gachon, P., Vasson, M. P., Maubois, J. L. and Beaufrère, B. (1997). Slow and fast dietary proteins differently modulate postprandial protein accretion. *Proceedings of the National Academy of Sciences of the United States of America* 94(26), 14930–14935.

Brown, D. A. and London, E. (2000). Structure and function of sphingolipid-and cholesterol-rich membrane rafts. *Journal of Biological Chemistry* 275(23), 17221–17224.

Carlson, A., Hill Jr., C. G. and Olson, N. F. (1987). Kinetics of milk coagulation: I. The kinetics of kappa casein hydrolysis in the presence of enzyme deactivation. *Biotechnology and Bioengineering* 29(5), 582–589.

Cartier, P. and Chilliard, Y. (1989). Lipase redistribution in cows' milk during induced lipolysis: I. Activation by agitation, temperature change, blood serum and heparin. *Journal of Dairy Research* 56(5), 699–709.

Chobert, J. M., Briand, L., Grinberg, V. and Haertlé, T. (1995). Impact of esterification on the folding and the susceptibility to peptic proteolysis of β-lactoglobulin. *Biochimica et Biophysica Acta* 1248(2), 170–176.

Dalgalarrondo, M., Dufour, E., Chobert, J.-M., Bertrand-Harb, C. and Haertlé, T. (1995). Proteolysis of β-lactoglobulin and β-casein by pepsin in ethanolic media. *International Dairy Journal* 5(1), 1–14.

Dalgleish, D. G. and Corredig, M. (2012). The structure of the casein micelle of milk and its changes during processing. *Annual Review of Food Science and Technology* 3, 449–467.

Dewettinck, K., Rombaut, R., Thienpont, N., Le, T. T., Messens, K. and Van Camp, J. (2008). Nutritional and technological aspects of milk fat globule membrane material. *International Dairy Journal* 18(5), 436–457.

Do, D. H. T. and Kong, F. (2018). Texture changes and protein hydrolysis in different cheeses under simulated gastric environment. *LWT* 93, 197–203.

Donato, L., Guyomarc'h, F., Amiot, S. and Dalgleish, D. G. (2007). Formation of whey protein/κ-casein complexes in heated milk: preferential reaction of whey protein with κ-casein in the casein micelles. *International Dairy Journal* 17(10), 1161–1167.

Dupont, D., Mandalari, G., Mollé, D., Jardin, J., Rolet-Répécaud, O., Duboz, G., Léonil, J., Mills, C. E. and Mackie, A. R. (2010). Food processing increases casein resistance to simulated infant digestion. *Molecular Nutrition and Food Research* 54(11), 1677–1689.

Efigênia, M., Povoa, B. and Moraes-Santos, T. (1997). Effect of heat treatment on the nutritional quality of milk proteins. *International Dairy Journal* 7(8–9), 609–612.

Fang, X., Rioux, L.-E., Labrie, S. and Turgeon, S. L. (2016). Commercial cheeses with different texture have different disintegration and protein/peptide release rates during simulated in vitro digestion. *International Dairy Journal* 56, 169–178.

Fardet, A., Dupont, D., Rioux, L. E. and Turgeon, S. L. (2019). Influence of food structure on dairy protein, lipid and calcium bioavailability: a narrative review of evidence. *Critical Reviews in Food Science and Nutrition* 59(13), 1987–2010.

Forsgård, R. A. (2019). Lactose digestion in humans: intestinal lactase appears to be constitutive whereas the colonic microbiome is adaptable. *The American Journal of Clinical Nutrition* 110(2), 273–279.

Fox, P. F. and Walley, B. F. (1971). Influence of sodium chloride on the proteolysis of casein by rennet and by pepsin. *Journal of Dairy Research* 38(2), 165–170.

Furlund, C. B., Ulleberg, E. K., Devold, T. G., Flengsrud, R., Jacobsen, M., Sekse, C., Holm, H. and Vegarud, G. E. (2013). Identification of lactoferrin peptides generated by digestion with human gastrointestinal enzymes. *Journal of Dairy Science* 96(1), 75–88.

Gallier, S., Cui, J., Olson, T. D., Rutherfurd, S. M., Ye, A., Moughan, P. J. and Singh, H. (2013). In vivo digestion of bovine milk fat globules: effect of processing and interfacial structural changes. I. Gastric digestion. *Food Chemistry* 141(3), 3273–3281.

Gallier, S., Ye, A. and Singh, H. (2012). Structural changes of bovine milk fat globules during in vitro digestion. *Journal of Dairy Science* 95(7), 3579–3592.

Gilani, G. S. and Sepehr, E. (2003). Protein digestibility and quality in products containing antinutritional factors are adversely affected by old age in rats. *The Journal of Nutrition* 133(1), 220–225.

Golding, M. and Wooster, T. J. (2010). The influence of emulsion structure and stability on lipid digestion. *Current Opinion in Colloid and Interface Science* 15(1–2), 90–101.

Guo, M. R., Fox, P. F., Flynn, A. and Kindstedt, P. S. (1995). Susceptibility of β-lactoglobulin and sodium caseinate to proteolysis by pepsin and trypsin. *Journal of Dairy Science* 78(11), 2336–2344.

Guo, Q., Ye, A., Lad, M., Dalgleish, D. and Singh, H. (2013). The breakdown properties of heat-set whey protein emulsion gels in the human mouth. *Food Hydrocolloids* 33(2), 215–224.

He, T., Venema, K., Priebe, M. G., Welling, G. W., Brummer, R. J. and Vonk, R. J. (2008). The role of colonic metabolism in lactose intolerance. *European Journal of Clinical Investigation* 38(8), 541–547.

Horne, D. S. (2020). Casein micelle structure and stability. In: Thompson, A. Boland, M. and Singh, H. (Eds.), *Milk Proteins* (3rd edn, pp. 213–250). Amsterdam: Elsevier.

Horstman, A. M. H., Ganzevles, R. A., Kudla, U., Kardinaal, A. F. M., van den Borne, J. J. G. C. and Huppertz, T. (2021). Postprandial blood amino acid concentrations in older adults after consumption of dairy products: the role of the dairy matrix. *International Dairy Journal* 113, 104890.

Hunt, J. N. and Stubbs, D. F. (1975). The volume and energy content of meals as determinants of gastric emptying. *The Journal of Physiology* 245(1), 209 –225.

Huppertz, T. and Chia, L. W. (2020). Milk protein coagulation under gastric conditions: a review. *International Dairy Journal* 113, 104882.

Hur, S. J., Lim, B. O., Decker, E. A. and McClements, D. J. (2011). In vitro human digestion models for food applications. *Food Chemistry* 125(1), 1–12.

Jensen, R. G. (2002). The composition of bovine milk lipids: January 1995 to December 2000. *Journal of Dairy Science* 85(2), 295–350.

Kethireddipalli, P. and Hill, A. R. (2015). Rennet coagulation and cheesemaking properties of thermally processed milk: overview and recent developments. *Journal of Agricultural and Food Chemistry* 63(43), 9389–9403.

Kethireddipalli, P., Hill, A. R. and Dalgleish, D. G. (2011). Interaction between casein micelles and whey protein/κ-casein complexes during renneting of heat-treated reconstituted skim milk powder and casein micelle/serum mixtures. *Journal of Agricultural and Food Chemistry* 59(4), 1442–1448.

Lacroix, M., Bon, C., Bos, C., Léonil, J., Benamouzig, R., Luengo, C., Fauquant, J., Tomé, D. and Gaudichon, C. (2008). Ultra high temperature treatment, but not pasteurization, affects the postprandial kinetics of milk proteins in humans. *The Journal of Nutrition* 138(12), 2342–2347.

Lacroix, M., Bos, C., Léonil, J., Airinei, G., Luengo, C., Daré, S., Benamouzig, R., Fouillet, H., Fauquant, J., Tomé, D. and Gaudichon, C. (2006). Compared with casein or total milk protein, digestion of milk soluble proteins is too rapid to sustain the anabolic postprandial amino acid requirement. *The American Journal of Clinical Nutrition* 84(5), 1070–1079.

Lamothe, S., Rémillard, N., Tremblay, J. and Britten, M. (2017). Influence of dairy matrices on nutrient release in a simulated gastrointestinal environment. *Food Research International* 92, 138–146.

Liang, L., Zhang, X., Wang, X., Jin, Q. and McClements, D. J. (2018). Influence of dairy emulsifier type and lipid droplet size on gastrointestinal fate of model emulsions: in vitro digestion study. *Journal of Agricultural and Food Chemistry* 66(37), 9761–9769.

Lindahl, A., Ungell, A. L., Knutson, L. and Lennernäs, H. (1997). Characterization of fluids from the stomach and proximal jejunum in men and women. *Pharmaceutical Research* 14(4), 497–502.

Lindmark Månsson, H. (2008). Fatty acids in bovine milk fat. *Food and Nutrition Research* 52(1), 1821.

Luo, N., Ye, A., Wolber, F. M. and Singh, H. (2019). Structure of whey protein emulsion gels containing capsaicinoids: impact on in-mouth breakdown behaviour and sensory perception. *Food Hydrocolloids* 92, 19-29.

MacGibbon, A. and Taylor, M. (2006). Composition and structure of bovine milk lipids. In: Fox, P. F. and McSweeney, P. L. H. (Eds.), *Advanced Dairy Chemistry 2 Lipids* (pp. 1-42). Boston, USA: Springer.

Mahe, S., Roos, N., Benamouzig, R., Davin, L., Luengo, C., Gagnon, L., Gaussergès, N., Rautureau, J. and Tomé, D. (1996). Gastrojejunal kinetics and the digestion of [15N] beta-lactoglobulin and casein in humans: the influence of the nature and quantity of the protein. *The American Journal of Clinical Nutrition* 63(4), 546-552.

Marvin, L. F., Parisod, V., Fay, L. B. and Guy, P. A. (2002). Characterization of lactosylated proteins of infant formula powders using two-dimensional gel electrophoresis and nanoelectrospray mass spectrometry. *Electrophoresis* 23(15), 2505-2512.

Mather, I. H. (2000). A review and proposed nomenclature for major proteins of the milk-fat globule membrane. *Journal of Dairy Science* 83(2), 203-247.

Mather, I. H. and Keenan, T. W. (1975). Studies on the structure of milk fat globule membrane. *The Journal of Membrane Biology* 21(1-2), 65-85.

Mauron, J., Mottu, F., Bujard, E. and Egli, R. H. (1955). The availability of lysine, methionine and tryptophan in condensed milk and milk powder; in vitro digestion studies. *Archives of Biochemistry and Biophysics* 59(2), 433-451.

McClements, D. J., Decker, E. A., Park, Y. and Weiss, J. (2008). Designing food structure to control stability, digestion, release and absorption of lipophilic food components. *Food Biophysics* 3(2), 219-228.

Michalski, M.-C. and Januel, C. (2006). Does homogenization affect the human health properties of cow's milk? *Trends in Food Science and Technology* 17(8), 423-437.

Miranda, G. and Pelissier, J.-P. (1983). Kinetic studies of in vivo digestion of bovine unheated skim-milk proteins in the rat stomach. *Journal of Dairy Research* 50(1), 27-36.

Miranda, G. and Pelissier, J.-P. (1987). Influence of heat treatment of bovine skim-milk on in vivo digestion in rat stomach. *Le Lait* 67(3), 365-377.

Montgomery, R. K., Krasinski, S. D., Hirschhorn, J. N. and Grand, R. J. (2007). Lactose and lactase–who is lactose intolerant and why? *Journal of Pediatric Gastroenterology and Nutrition* 45(Suppl. 2), S131-S137.

Mozaffarian, D. (2016). Dietary and policy priorities for cardiovascular disease, diabetes, and obesity: a comprehensive review. *Circulation* 133(2), 187-225.

Mpassi, D., Rychen, G., Mertes, M. and Laurent, F. (2001). Portal absorption of 45Ca from labelled milk, yoghurt or heat treated yoghurt in the growing pig. *International Dairy Journal* 11(10), 809-815.

Mu, H. and Høy, C. E. (2004). The digestion of dietary triacylglycerols. *Progress in Lipid Research* 43(2), 105-133.

Mulet-Cabero, A.-I., Mackie, A. R., Brodkorb, A. and Wilde, P. J. (2020). Dairy structures and physiological responses: a matter of gastric digestion. *Critical Reviews in Food Science and Nutrition* 60(22), 3737-3752.

Mulet-Cabero, A.-I., Mackie, A. R., Wilde, P. J., Fenelon, M. A. and Brodkorb, A. (2019). Structural mechanism and kinetics of in vitro gastric digestion are affected by process-induced changes in bovine milk. *Food Hydrocolloids* 86, 172-183.

Mun, S., Decker, E. A. and McClements, D. J. (2007). Influence of emulsifier type on in vitro digestibility of lipid droplets by pancreatic lipase. *Food Research International* 40(6), 770–781.

N'Goma, J.-C. B., Amara, S., Dridi, K., Jannin, V. and Carrière, F. (2012). Understanding the lipid-digestion processes in the GI tract before designing lipid-based drug-delivery systems. *Therapeutic Delivery*, 3(1), 105–124.

Nik, A. M., Wright, A. J. and Corredig, M. (2010). Interfacial design of protein-stabilized emulsions for optimal delivery of nutrients. *Food and Function* 1(2), 141–148.

Oberg, C. J., Merrill, R. K., Brown, R. J. and Richardson, G. H. (1992). Effects of milk-clotting enzymes on physical properties of Mozzarella cheese. *Journal of Dairy Science* 75(3), 669–675.

O'Mahony, J. and Fox, P. (2013). Milk proteins: introduction and historical aspects. In: Fox, P. F. and McSweeney, P. L. H. (Eds.), *Advanced Dairy Chemistry* (pp. 43–85). New York, NY: Springer.

O'Mahony, J. and Fox, P. (2014). Milk: an overview. In: *Milk Proteins* (pp. 19–73). Elsevier.

Oortwijn, H. and Walstra, P. (1982). The membranes of recombined fat globules. 3. mode of formation. *Netherlands Milk and Dairy Journal* 36, 103–113.

Palmiter, R. D. (1969). What regulates lactose content in milk? *Nature* 221(5184), 912–914.

Parodi, P. W. (2001). Cow's milk components with anti-cancer potential. *Australian Journal of Dairy Technology* 56(2), 65.

Patton, J. S. and Carey, M. C. (1981). Inhibition of human pancreatic lipase-colipase activity by mixed bile salt-phospholipid micelles. *American Journal of Physiology* 241(4), G328–G336.

Patton, S. (2017). *Milk: Its Remarkable Contribution to Human Health and Well-Being*. London, UK: Routledge.

Peram, M. R., Loveday, S. M., Ye, A. and Singh, H. (2013). In vitro gastric digestion of heat-induced aggregates of β-lactoglobulin. *Journal of Dairy Science* 96(1), 63–74.

Qazi, H. J., Ye, A., Acevedo-Fani, A. and Singh, H. (2021). In vitro digestion of curcumin-nanoemulsion-enriched dairy protein matrices: Impact of the type of gel structure on the bioaccessibility of curcumin. *Food Hydrocolloids* 117, 106692.

Reddy, I. M., Kella, N. K. D. and Kinsella, J. E. (1988). Structural and conformational basis of the resistance of. Beta.-lactoglobulin to peptic and chymotryptic digestion. *Journal of Agricultural and Food Chemistry* 36(4), 737–741.

Rérat, A., Calmes, R., Vaissade, P. and Finot, P. A. (2002). Nutritional and metabolic consequences of the early Maillard reaction of heat treated milk in the pig. Significance for man. *European Journal of Nutrition* 41(1), 1–11.

Robinson, R. and Wilbey, R. (1998). Cheese maturation. In: Scott, R, Robinson, R. K. and Wilbey, R. A. (Eds.), *Cheesemaking Practice* (pp. 271–287). Boston, MA, USA: Springer.

Roy, D., Ye, A., Moughan, P. J. and Singh, H. (2020a). Composition, structure, and digestive dynamics of milk from different species–a review. *Frontiers in Nutrition* 7, 577759.

Roy, D., Ye, A., Moughan, P. J. and Singh, H. (2020b). Gelation of milks of different species (dairy cattle, goat, sheep, red deer, and water buffalo) using glucono-δ-lactone and pepsin. *Journal of Dairy Science* 103(7), 5844–5862.

Rutherfurd, S. M. and Moughan, P. J. (1998). The digestible amino acid composition of several milk proteins: application of a new bioassay. *Journal of Dairy Science* 81(4), 909–917.

Rutherfurd, S. M. and Moughan, P. J. (2003). The rat as a model animal for the growing pig in determining ileal amino acid digestibility in soya and milk proteins. *Journal of Animal Physiology and Animal Nutrition* 87(7-8), 292-300.

Rutherfurd, S. M. and Moughan, P. J. (2005). Digestible reactive lysine in selected milk-based products. *Journal of Dairy Science* 88(1), 40-48.

Sah, B. N. P., McAinch, A. J. and Vasiljevic, T. (2016). Modulation of bovine whey protein digestion in gastrointestinal tract: a comprehensive review. *International Dairy Journal* 62, 10-18.

Sharma, R., Singh, H. and Taylor, M. W. (1996). Composition and structure of fat globule surface layers in recombined milk. *Journal of Food Science* 61(1), 28-32.

Singh, H. and Ye, A. (2013). Structural and biochemical factors affecting the digestion of protein-stabilized emulsions. *Current Opinion in Colloid and Interface Science* 18(4), 360-370.

Singh, H., Ye, A. and Horne, D. (2009). Structuring food emulsions in the gastrointestinal tract to modify lipid digestion. *Progress in Lipid Research* 48(2), 92-100.

Smoczynski, M. and Staniewski, B. (2014). Saliva-induced flocculation of milk fat globules studied by image analysis. *Journal of International Scientific Publications: Agriculture & Food* 2, 542-548.

Sousa, M. J., Ardö, Y. and McSweeney, P. L. H. (2001). Advances in the study of proteolysis during cheese ripening. *International Dairy Journal* 11(4-7), 327-345.

Thorning, T. K., Bertram, H. C., Bonjour, J. P., De Groot, L., Dupont, D., Feeney, E., Ipsen, R., Lecerf, J. M., Mackie, A., McKinley, M. C., Michalski, M. C., Rémond, D., Risérus, U., Soedamah-Muthu, S. S., Tholstrup, T., Weaver, C., Astrup, A. and Givens, I. (2017). Whole dairy matrix or single nutrients in assessment of health effects: current evidence and knowledge gaps. *The American Journal of Clinical Nutrition* 105(5), 1033-1045.

Tunick, M. H., Ren, D. X., Van Hekken, D. L., Bonnaillie, L., Paul, M., Kwoczak, R. and Tomasula, P. M. (2016). Effect of heat and homogenization on in vitro digestion of milk. *Journal of Dairy Science* 99(6), 4124-4139.

Ugidos-Rodríguez, S., Matallana-González, M. C. and Sánchez-Mata, M. C. (2018). Lactose malabsorption and intolerance: a review. *Food and Function* 9(8), 4056-4068.

Van Boekel, M. A. J. S. (1998). Effect of heating on Maillard reactions in milk. *Food Chemistry* 62(4), 403-414.

van Lieshout, G. A. A., Lambers, T. T., Bragt, M. C. E. and Hettinga, K. A. (2020). How processing may affect milk protein digestion and overall physiological outcomes: a systematic review. *Critical Reviews in Food Science and Nutrition* 60(14), 2422-2445.

Vasbinder, A. J., Alting, A. C. and de Kruif, K. G. (2003). Quantification of heat-induced casein-whey protein interactions in milk and its relation to gelation kinetics. *Colloids and Surfaces, Part B: Biointerfaces* 31(1-4), 115-123.

Vist, G. E. and Maughan, R. J. (1995). The effect of osmolality and carbohydrate content on the rate of gastric emptying of liquids in man. *The Journal of Physiology* 486(2), 523-531.

Wada, Y. and Lönnerdal, B. (2014). Effects of different industrial heating processes of milk on site-specific protein modifications and their relationship to in vitro and in vivo digestibility. *Journal of Agricultural and Food Chemistry* 62(18), 4175-4185.

Walstra, P. (1995). Physical chemistry of milk fat globules. *Advanced Dairy Chemistry* 2, 131-178.

Wang, X., Ye, A., Lin, Q., Han, J. and Singh, H. (2018). Gastric digestion of milk protein ingredients: study using an in vitro dynamic model. *Journal of Dairy Science* 101(8), 6842-6852.

Ye, A., Anema, S. G. and Singh, H. (2008). Changes in the surface protein of the fat globules during homogenization and heat treatment of concentrated milk. *The Journal of Dairy Research* 75(3), 347-353.

Ye, A., Cui, J., Dalgleish, D. and Singh, H. (2016a). The formation and breakdown of structured clots from whole milk during gastric digestion. *Food and Function* 7(10), 4259-4266.

Ye, A., Cui, J., Dalgleish, D. and Singh, H. (2016b). Formation of a structured clot during the gastric digestion of milk: impact on the rate of protein hydrolysis. *Food Hydrocolloids* 52, 478-486.

Ye, A., Cui, J., Dalgleish, D. and Singh, H. (2017). Effect of homogenization and heat treatment on the behavior of protein and fat globules during gastric digestion of milk. *Journal of Dairy Science* 100(1), 36-47.

Ye, A., Cui, J. and Singh, H. (2010). Effect of the fat globule membrane on in vitro digestion of milk fat globules with pancreatic lipase. *International Dairy Journal* 20(12), 822-829.

Ye, A., Cui, J. and Singh, H. (2011). Proteolysis of milk fat globule membrane proteins during in vitro gastric digestion of milk. *Journal of Dairy Science* 94(6), 2762-2770.

Ye, A., Liu, W., Cui, J., Kong, X., Roy, D., Kong, Y., Han, J. and Singh, H. (2019). Coagulation behaviour of milk under gastric digestion: effect of pasteurization and ultra-high temperature treatment. *Food Chemistry* 286, 216-225.

Ye, A., Roy, D. and Singh, H. (2020). Structural changes to milk protein products during gastrointestinal digestion. In: Thompson, A., Boland, M. and Singh, H. (Eds.), *Milk Proteins* (3rd edn, pp. 671-700). Amsterdam: Elsevier.

Ye, A., Singh, H., Taylor, M. W. and Anema, S. (2004). Interactions of whey proteins with milk fat globule membrane proteins during heat treatment of whole milk. *Le Lait* 84(3), 269-283.

Zeece, M., Huppertz, T. and Kelly, A. (2008). Effect of high-pressure treatment on in-vitro digestibility of β-lactoglobulin. *Innovative Food Science and Emerging Technologies* 9(1), 62-69.

# Chapter 3

## Milk composition and properties: interspecies comparison

*Golfo Moatsou, Agricultural University of Athens, Greece*



## 1 Introduction

Milk and milk constituents are valuable natural products produced from various mammals, some of which can be farm animals. Research on cow milk has been carried out for more than a century, and despite the great amount of scientific and technological information that has been obtained, new findings are constantly emerging. Cow milk accounts for the largest part of global milk production, which is due to cow's ability to produce and store large quantities of milk and the convenient lactation period. Milk from other ruminants such as buffaloes, goats, sheep, camels and yaks is produced in various regions of the world often from indigenous breeds (http://www.fao.org/dairy-production-products/production/dairy-animals).

Cow milk constituents and treatments are the objectives of the vast part of research and manufacturing activities in the dairy field. Thus, this milk kind can be considered as a 'reference' for the literature. Albeit fewer, there is a considerable number of publications on buffaloes and small ruminants' milk, and recently, publications on camel and yak milk have been presented. The traditional use of milk from equids, that is donkey and horse, as a nutraceutical and its compositional similarities to human milk has attracted the interest of consumers and academia. In particular, a great part of recent research on milk constituents and non-cow milk is driven by the constant efforts to formulate

http://dx.doi.org/10.19103/AS.2022.0099.03

**Table 1** Range of [a] and average [b] gross composition (g/100 mL) of ruminants and equid milk according to selected reviews

| Parameters | Cow | Buffalo | Goat | Sheep | Yak | Camel | Donkey | Horse |
|---|---|---|---|---|---|---|---|---|
| Total solids | 11.8–13 [a]<br>12.7 [b] | 15.7–17.2 [a]<br>16.9 [b] | 11.9–16.3 [a]<br>13.2 [b] | 18.1–20 [a]<br>17.8 [b] | 13.5–18.4 [a]<br>16.7 [b] | 11.9–15 [a]<br>13 [b] | 8.8–11.7 [a]<br>9.5 [b] | 9.3–11.6 [a]<br>10.4 [b] |
| Fat | 3.3–5.4 [a]<br>3.8 [b] | 5.3–9.0 [a]<br>7.2 [b] | 3–7.2 [a]<br>4.3 [b] | 5–9 [a]<br>7.4 [b] | 5.3–9.5 [a]<br>6.4 [b] | 2–6 [a]<br>4.3 [b] | 0.3–1.8 [a]<br>1.4 [b] | 0.3–4.2 [a]<br>1.4 [b] |
| Protein | 3–3.9 [a]<br>3.4 [b] | 2.7–4.7 [a]<br>4.2 [b] | 3–5.2 [a]<br>3.6 [b] | 4.5–7 [a]<br>5.7 [b] | 4.2–5.9 [a]<br>4.9 [b] | 2.4–4.2 [a]<br>3.6 [b] | 1.4–2 [a]<br>1.7 [b] | 1.4–3.2 [a]<br>2 [b] |
| Lactose | 4.4–5.6 [a]<br>4.8 [b] | 3.2–4.9 [a]<br>4.8 [b] | 3.2–5 [a]<br>4.4 [b] | 4.1–5.9 [a]<br>4.8 [b] | 3.3–6.2 [a]<br>5 [b] | 3.5–5.1 [a]<br>4.9 [b] | 5.8–7.4 [a]<br>6.6 [b] | 5.6–7.2 [a]<br>6.6 [b] |
| Ash | 0.7–0.8 [a]<br>0.7 | 0.8–0.9 [a]<br>0.8 [b] | 0.7–0.9 [a]<br>0.8 [b] | 0.8–1 [a]<br>0.9 [b] | 0.4–1 [a]<br>0.8 [b] | 0.7–0.9 [a]<br>0.8 [b] | 0.3–0.5 [a]<br>0.4 [b] | 0.3–0.5 [a]<br>0.4 [b] |

[a] Claeys et al., 2014; [b] Alichanidis et al., 2016.

human milk substitutes. In this respect, emphasis is given to the detailed characterization of minor compounds of small ruminants' and equid milk. Over the last years, review articles that compare the milk characteristics of various mammals have appeared in the literature (e.g. Barlowska et al., 2011; Martemucci and D'Alessandro, 2013; Claeys et al., 2014; Gantner et al., 2015; Faccia et al., 2020; Roy et al., 2020). Moreover, two books have been published entitled *Non-Bovine Milk and Milk Products* (Academic Press, 2016) and *Handbook of Milk of Non-Bovine Mammals, 2nd edition* (Wiley, 2017).

This chapter is a comparative presentation of the characteristics of milk produced from ruminants and equids. Emphasis is on topics associated with technological and nutritional/biological performance of milk, that is, milk fat globule, casein micelle, enzymes and serum minor compounds. Table 1 is an outline of the gross composition of the milk of farm animals, which is the objective of the present review.

## 2 Lipids: neutral, polar and unsaponifiable lipids

### *2.1 Composition of the lipid fraction: fatty acids' patterns*

Lipids are esters of fatty acids and related components that are soluble in nonpolar organic solvents and almost insoluble in water. The term milk fat corresponding to a mixture of triglycerides or triacylglycerols (TAG) is used as more than 98% of the lipid fraction of ruminants milk consists of TAG located in the milk fat globule (MFG) core. In brief, the components of milk lipid fraction are classified as follows: i. neutral lipids, mainly triglycerides and traces of di- and monoglycerides, ii. free fatty acids (FFA), existed in traces, iii. compound amphipolar lipids, mostly phospholipids (PL) and to a lesser extend glycolipids, mainly located in the milk fat globule membrane (MFGM), and iv. unsaponifiable lipids, such as cholesterol, carotenoids and fat-soluble vitamins located in both the MFGM and the MFG core (Walstra et al., 2006).

Table 2 is a comparative synopsis of specific parameters of the complex lipid fraction of milk from ruminants (RS) – cow, buffalo, goat, sheep, yak and camel – and equids (ES) – donkey and mare, which appeared in selected publications that are for the most part comparative studies. Variability in the values reported by various research groups can be attributed to the well-documented natural variation of milk, animal diet, milking conditions or analytical methodology. The polar and unsaponifiable lipids' profiles are presented in Tables 3 and 4, respectively.

Non-ruminant donkey and mare milk contain low quantities of fat, and among ruminants, buffalo, yak and sheep milk exhibit the highest fat content. TAG constitute the most part – 97–98% – of lipids in RS milk similarly to the human counterpart, whereas polar lipids are 0.5–1%. In the lipid fraction of

**Table 2** Specific parameters of the composition of milk lipid fraction – interspecies comparison, based on selected publications. TAG, most abundant triglycerides; SFA, saturated fatty acids; TPL, total polar lipids; MUFA, monounsaturated fatty acids; PUFA, polyunsaturated fatty acid; MC-SFA, medium chain length saturated fatty acids C6:0-C12:0; C4:0, butyric (butanoic) acid; C16:0, palmitic acid; CLA, conjugated linoleic acid C18:2*c*9*t*11; n-6/n-3, omega-6 to omega-3 PUFA ratio; AI, atherogenicity index = [C12:0 + (4 × C14:0)+C16:0]/∑UFA

| Parameters | Cow | Buffalo | Goat | Sheep | Yak | Camel | Donkey | Horse |
|---|---|---|---|---|---|---|---|---|
| Fat[a] | 3.71 ± 0.61 [b]<br>3.8 ± 0.26 [o]<br>3.7 [p]<br>3.89 ± 0.19 [qq] | 7.3 ± 1 [h]<br>7-9.2 [m]<br>6.45 ± 1.88 [qq] | 3-7.2 [pp]<br>3.58 ± 0.30 [qq] | 5.73-7.4 [nn]<br>5.67 ± 0.55 [qq] | 7.1 ± 0.4 [o]<br>5.6 [p]<br>7.22 ± 0.04 [dd] | 4.56 ± 1.87 [b]<br>3.8 [p]<br>5-6.5 [cc]<br>6.40 ± 2.69 [gg]<br>2.19 ± 1.24 [qq] | 0.4 [p]<br>0.76 [hh]<br>0.15-0.19 [ii]<br>0.1-3.8 [kk]<br>0.34 ± 0.27 [qq]<br>0.94 [zz] | 0.68 ± 0.2 [b]<br>1.2 [p]<br>0.42 ± 0.18 [y]<br>0.41-1.2 [z]<br>0.72 ± 0.82 [qq] |
| TAG | C38, C50 [qq]<br>C36, C38, C50 [rr]<br>C36, C38, C40 [zz] | C38, C50 [qq] | C38, C40, C42 [qq]<br>C42, c50 [rr] | C38, C40, C52 [qq] | | C48, C50, C52 [qq] | C50, C52, C54 [qq]<br>C52 [zz] | C44, C50, C52 [qq] |
| TPL[b] | 63-483 [b]<br>195-413 [f]<br>140 ± 20 [h]<br>311 ± 2 [dd] | 189 ± 9 [h] | 276 ± 20 [e]<br>196-202 [f]<br>178-285 [g] | 298 ± 13 [e] | 371 ± 11 [dd] | 257-660 [b] | 33-39 [f]<br>42.8-97.1 [ll] | 52.6-87.9 [b]<br>47.2 ± 8.7 [aa] |
| *Fatty acids profile, % of fat* | | | | | | | | |
| SFA[c] | 75.5 ± 3.8 [c]<br>70.5 ± 2.9 [n]<br>68 ± 2.6 [r]<br>60.5 [u]<br>66.6 ± 3.2 [w] | 82.2-68.5 [m]<br>74 ± 4.5 [c]<br>73.5 ± 3 [n] | 70.2 ± 2.6 [n]<br>69.4 ± 2.6 [r]<br>66.5 [u] | 71 ± 5.2 [c]<br>69.4 ± 2.9 [r]<br>64.1 [u] | 72.5 ± 1.9 [n]<br>64.7 ± 1.5 [w]<br>64.6 ± 2.4 [ff] | 53.7 ± 4.9 [c]<br>51.9 ± 2.2 [n] | 58 ± 2.2 [c]<br>46.7-67.7 [hh]<br>41-63 [k]<br>42.9-62.6 [mm]<br>58.3 [zz] | 49.4 ± 3.8 [y]<br>55.8 [x]<br>45.7 [aa] |

| | | | | | | | | |
|---|---|---|---|---|---|---|---|---|
| MUFA[c] | 21.6 ± 1.7 [c]<br>25.1 ± 2.6 [n]<br>28.4 ± 2.3 [r]<br>32.9 [u]<br>32.6 ± 1.8 [w] | 16.6–27.8 [m]<br>23.6 ± 0.4 [c]<br>24.2 ± 2.8 [n] | 25.9 ± 2.6 [n]<br>26.1 ± 2.4 [r]<br>27.4 [u] | 24.4 ± 0.8 [c]<br>26.9 ± 2.5 [r]<br>28.7 [u] | 25.3 ± 1.8 [n]<br>32.9 ± 2.2 [w]<br>31.5 ± 1.5 [ff] | 41 ± 1.2 [c]<br>39.6 ± 2.4 [n] | 35.1 ± 1.5 [c]<br>15.3–35 [hh]<br>11–28 [kk]<br>24.2–35.5 [mm]<br>20.9 [zz] | 33.6 ± 1.5 [y]<br>25.2 [x]<br>32.7 [aa] |
| PUFA[c] | 1.9 ± 0.2 [c]<br>4.5 ± 0.6 [n]<br>3.6 ± 1 [r]<br>5.7 [u]<br>1.8 ± 0.11 [w] | 1.8–3.7 [m]<br>2.3 ± 0.4 [c]<br>2.3 ± 0.5 [n] | 3.9 ± 0.4 [n]<br>4.5 ± 0.8 [r]<br>4.1 [u] | 4.2 ± 0.7 [c]<br>3.8 ± 0.8 [r]<br>5.8 [u] | 2.3 ± 0.3 [n]<br>2.41 ± 0.31 [w]<br>3.9 ± 0.14 [ff] | 5.2 ± 0.3 [c]<br>8.5 ± 0.9 [n] | 6.1 ± 1.3 [c]<br>15.2–30.5 [hh]<br>24–28 [kk]<br>13.2–22 [mm]<br>18.9 [zz] | 21.3 ± 2.4 [y]<br>19 [x]<br>19.8 [aa] |
| C4:0[c] | 8.7 ± 0.8 [c]<br>3.5 ± 0.3 [n]<br>3.2 ± 0.23 [r]<br>3.21 [u]<br>1.54 ± 0.12 [w] | 8.5 ± 0.7 [c]<br>3.9–4.1 [m]<br>6.4 ± 0.7 [n] | 1.9 ± 0.2 [n]<br>2 ± 0.4 [r]<br>2.26 [u] | 6.1 ± 0.7 [c]<br>2.47 ± 0.3 [r]<br>3.60 [u] | 4.4 ± 0.7 [n]<br>2.28 ± 0.07 [w] | ND [c]<br>ND [n] | 1.06 ± 0.21 [c]<br>0.32–0.6 [hh]<br>0.57 [zz] | 0.2 [x]<br>0.3–0.9 [hh] |
| MC-SFA[c] | 16.2 ± 1.3 [c]<br>17 ± 2.7 [w]<br>9.8 [n]<br>9.2 [r]<br>8.9–9.8 [pp] | 11.6 ± 1.7 [c] | 20.8 [n]<br>18.9 [r]<br>16.8–19.6 [pp] | 20.2 ± 2.6 [c]<br>13 [r]<br>13.0–16.3 [pp] | 13.0 ± 1.2 [w] | 1.05 ± 0.07 [c] | 8.9 ± 1.6 [c]<br>17.06 [zz] | |
| C16:0[c] | 30.4 ± 0.8 [c]<br>32.0 ± 2.4 [n]<br>32.2 ± 1.6 [r]<br>27.1 [u]<br>33.6 ± 2 [w] | 34.6 ± 1.9 [c]<br>27–29 [m]<br>33.6 ± 2.9 [n] | 26.2 ± 1.9 [n]<br>27.7 ± 1.8 [r]<br>26 [u] | 25.4 ± 1.5 [c]<br>29.3 ± 2 [r]<br>20.9 [u] | 34.3 ± 2.1 [n]<br>31.4 ± 0.7 [w]<br>26.6 ± 2.6 [ff] | 27.1 ± 0.7 [c]<br>18.8 ± 1.7 [n]<br>35 [cc]<br>31.3 [gg] | 31.2 ± 1.7 [c]<br>11.5–29.2 [hh]<br>20–22 [kk]<br>26.3 [zz] | 23.8 [x]<br>35 [aa]<br>12.4–28.5 [hh] |

*(Continued)*

**Table 2** (*Continued*)

| Parameters | Cow | Buffalo | Goat | Sheep | Yak | Camel | Donkey | Horse |
|---|---|---|---|---|---|---|---|---|
| CLA[c] | 0.70 ± 0.02 [h]<br>0.2 ± 2.4 [k]<br>0.62 ± 0.11 [n]<br>0.38 ± 0.18 [r]<br>0.46 [u] | 0.90 ± 0.06 [h]<br>0.4–1 [k]<br>0.53–0.72 [m]<br>0.28 ± 0.06 [n] | 0.3–1.2 [k]<br>0.85 ± 0.15 [n]<br>0.51 ± 0.13 [r]<br>0.63 [u] | 0.6–1.1 [k]<br>0.52 ± 0.17 [r]<br>0.87 [u]<br>0.65 ± 0.31 [oo] | 0.28 ± 0.06 [n]<br>0.92 ± 0.10 [ff] | 1.85 ± 0.30 [n] | Traces [k] | 0.09 [x] |
| AI | 2.7 ± 0.5 [n] | 3.1 ± 0.6 [n] | 2.7 ± 0.4 [n] | 1.8–2.7 [t] | 2.5 ± 0.24 [n] | 1 ± 0.2 [n]<br>2.30–3.72 [gg] | 0.60–1.53 [mm] | 0.85 ± 0.13 [y] |
| n-6/n-3 | 3.7 ± 0.6 [c]<br>2.2 ± 0.2 [h]<br>10.3 ± 0.8 [n] | 4.2 ± 1.2 [c]<br>1.3 ± 1 [h]<br>10.3 ± 1.3 [n] | 10.1 ± 1.1 [n] | 2.4 ± 0.3 [c]<br>4.46, 4.13 [oo] | 1.8 ± 0.3 [n] | 14.2 ± 1.3 [c]<br>2.6 ± 0.3 [n] | 16.5 ± 0.9 [c]<br>1.7–0.29 [kk]<br>0.79–2.31 [mm]<br>1.26 [zz] | 1.02 [x] |

[a] Et-Thakafy et al., 2017; [b] Garcia et al., 2012; [c] Zou et al., 2013; [d] Ali et al., 2017 (milk powders); [e] Zancada et al., 2013; [f] Russo et al., 2013; [g] Argov-Argaman et al., 2016; [h] Menard et al., 2010; Abd El-Salam and El-Shibiny, 2011; [k] Gantner et al., 2015; [l] Nguyen et al., 2015; [m] Zotos and Bampidis, 2014; [n] Yang et al., 2018 (Bactrian camel); [o] Ji et al., 2017; [p] Faccia et al., 2020; [q] Cebo et al., 2012a (depending on αs1-casein genotype) [r] Castro-Gomez et al., 2014; [s] Bonczar et al., 2016; [t] Sinanoglou et al., 2015; [u] Rodríguez-Alcala et al., 2009; [v] Rodríguez-Alcala and Fontecha, 2010; [w] Luo et al., 2020; [x] Malacarne et al., 2002; [y] Markiewicz-Keszycka et al., 2014; [z] Markiewicz-Keszycka et al., 2015; [aa] Barello et al., 2008; [bb] Meena et al., 2014; [cc] Khalesi et al., 2017; [dd] Luo et al., 2018; [ee] He et al., 2010; [ff] He et al., 2011 (various species); [gg] Konuspayeva et al., 2008; [hh] Salimei and Fantuz, 2012; [ii] Licitra et al., 2019; [kk] Altomonte et al., 2019; [ll] Contarini et al., 2017 (depending on breed and season); [mm] Massouras et al., 2017 (depending on stage of lactation); [nn] Moatsou and Sakkas, 2019; [oo] Revilla et al., 2017; [pp] Bernard et al., 2018; [qq] Smiddy et al., 2012; [rr] Yao et al., 2016; [ss] Kondyli et al., 2012 (depending on season); [tt] Rocchi et al., 2016 (retinoids); [uu] Martini et al., 2018b; [vv] Gentili et al., 2013; [ww] Haddadin et al., 2008 (depending on season); [xx] Raynal-Ljutovac et al., 2008; [yy] Kulyar et al., 2021; [zz] Gastaldi et al., 2010.

[a] g/100 mL milk.

[b] µg/mL milk.

[c] % total lipids.

**Table 3** Lipids and parameters related to MFG and MFGM - interspecies comparison, based on selected publications. SFA, saturated fatty acids; MUFA, monounsaturated fatty acids, PUFA, polyunsaturated fatty acid; PE, phospatidylethanolamine; PC, phosphatidylcholine; PS, phosphatidylserine; PI, phosphatidylinositol; PA, phosphatidic acid; SM, sphingomyelin; MFG, milk fat globule; MFGM, milk fat globule membrane

| Parameter | Cow | Buffalo | Goat | Sheep | Yak | Camel | Donkey | Horse |
|---|---|---|---|---|---|---|---|---|
| ØMFG[a] | 0.22-5.32 [b]<br>3.6 [h]<br>4.15 ± 0.15 [o]<br>3.9 [p]<br>3.03 ± 0.01 [w]<br>1.6-4.9 [bb]<br>4.89 ± 0.17 [rr] | 4.9 [h]<br>5 ± 0.4 [l]<br>0.4-15.8 [l]<br>3.9-7.7 [bb] | 3.1-3.6 [g]<br>3.1 ± 0.24<br>−3.65 ± 0.19 [q]<br>1.1-3.9 [bb]<br>2.8-3.5 [pp]<br>3.64 ± 0.33 [rr] | 3-3.8 [nn]<br>3.3-3.5 [pp] | 4.19 ± 0.14 [o]<br>4.4 [p]<br>4.57 ± 0.09 [w]<br>6.09 ± 0.02 [dd]<br>4.1 ± 4.34 [ee] | 1.05-2.33 [b]<br>3 [p]<br>1.1-2.1 [bb] | 1.9 [p]<br>2 [kk] | 0.91-1.94 [b]<br>2.5 [p]<br>2-3 [x] |
| CHOL/MFGM[b] | 1.7 ± 0.1 [a]<br>1.7 ± 0.05 [dd] | | 1.5 ± 0.8 [a] | 1.8 ± 0.1 [a] | 1.91 ± 0.03 [dd] | | | |
| CHOL/SM[c] | 3.6 [a]<br>2.64 ± 0.12 [dd] | | 2.9 [a] | 2.8 [a] | 2.61 ± 0.15 [dd] | | | |
| TPL, % [d] | 0.18-1.27 [b]<br>0.47 ± 0.02 [c]<br>0.36 ± 0.07 [h]<br>0.36 [v]<br>0.88 ± 0.006 [dd] | 0.32 ± 0.02 [c]<br>0.26 ± 0.05 [h]<br>0.20-0.31 [i] | 0.43 ± 0.02 [c]<br>0.65 [v] | 0.38 [v] | 0.51 ± 0.015 [dd] | 0.49-1.79 [b]<br>0.47 ± 0.03 [c] | 1.14 ± 2.11 [ll] | 0.76-1.73 [b]<br>5 [x] |
| *MFGM fatty acids profile, % of TPL* | | | | | | | | |
| SFA | 58.5 ± 0.23 [d]<br>56.7 ± 3.6 [c] | 55.6 ± 0.26 [d]<br>60.6 ± 5 [c] | 55.3 ± 0.22 [d]<br>65 [e] | 51.2 ± 0.28 [d]<br>47 ± 3.5 [c]<br>51.7 [e] | | 56.6 ± 0.29 [d]<br>44.6 ± 3.3 [c] | 50.4 ± 3.5 [c] | 59.1 [aa] |
| MUFA | 31.7 ± 0.18 [d]<br>34 ± 2.1 [c] | 33.1 ± 0.08 [d]<br>26.3 ± 2 [c] | 33.8 ± 0.16 [d]<br>27.6 [e] | 41 ± 0.03 [d]<br>38.9 ± 2.7 [c]<br>38.7 [e] | | 29.4 ± 0.07 [d]<br>33 ± 1.4 [c] | 30 ± 2.3 [c] | 18.3 [aa] |

(Continued)

**Table 3** *(Continued)*

| Parameter | Cow | Buffalo | Goat | Sheep | Yak | Camel | Donkey | Horse |
|---|---|---|---|---|---|---|---|---|
| PUFA | 9.8 ± 0.01 [d]<br>9 ± 0.5 [c] | 11.3 ± 0.15 [d]<br>13.1 ± 1 [c] | 11 ± 0.01 [d]<br>7.5 [e] | 7.9 ± 0.03 [d]<br>14.1 ± 1.2 [c]<br>9.6 [e] | | 14.1 ± 0.06 [d]<br>22.4 ± 1.1 [c] | 20 ± 1.1 [c] | 17.9 [aa] |
| *Phospholipid class, % of TPL* | | | | | | | | |
| PE | 31.4 [b]<br>30.2 ± 2.7 [c]<br>29.1 ± 1.6 [h]<br>46.7 ± 1.7 [r]<br>36.6 [v]<br>23.4 ± 0.1 [rr] | 31.1 ± 1.4 [c]<br>29.4 ± 1.4 [h] | 26.9 ± 1.1 [e]<br>46.1 ± 2.7 [r]<br>29.2 [v]<br>19.9 ± 0.1 [rr] | 35.9 ± 1.8 [c]<br>26.1 ± 2.3 [e]<br>43.2 ± 2 [r]<br>32.7 [v] | 28.2 [dd] | 30 [b]<br>35.5 ± 1.4 [c] | 30.6 ± 2.1 [c]<br>21-23 [f]<br>34.6 ± 11.2 [ll] | 18.6 [b]<br>31 [x]<br>7.8 ± 0.6 [aa] |
| PC | 28.7 [b]<br>25.2 ± 1.9 [c]<br>21.8 ± 1.7 [h]<br>32.2 ± 1.9 [r]<br>24.6 [v]<br>33.1 ± 0.2 [rr] | 29.8 ± 1.3 [c]<br>24.0 ± 1.7 [h] | 31.9 ± 1.2 [e]<br>31.5 ± 1 [r]<br>26.3 [v]<br>31.6 ± 0.2 [rr] | 29.6 ± 1.6 [c]<br>27.0 ± 0.9 [e]<br>30.5 ± 2.5 [r]<br>27.2 [v] | 23.12 [dd] | 19.3 [b]<br>25.6 ± 1.9 [c] | 25.3 ± 1.5 [c]<br>11.3-0 [f]<br>24.9 ± 10 [ll] | 21.3 [b]<br>19 [x]<br>46.7 ± 4.5 [aa] |
| PS | 11.2 [b]<br>7.3 ± 1 [c,v]<br>11.8 ± 2 [h]<br>0.12 ± 0.34 [r]<br>9.1 ± 0.1 [rr] | 3.6 ± 0.4 [c]<br>11.3 ± 1.3 [h] | 13.9 ± 0.7 (PS+PI) [e]<br>2.41 ± 0.78 [r]<br>7.7 [v]<br>14 ± 0.05 [rr] | 3.4 ± 0.5 [c]<br>10.7 ± 0.5 [e]<br>1.6 ± 0.76 [r]<br>5 [v] | 15.6 ± 0.21 (PS+PI) [dd] | 9.2 [b]<br>4.8 ± 0.2 [c] | 4 ± 0.4 [c]<br>50-63 [f]<br>10.2 ± 3.7 [ll] | 8.1 [b]<br>16 [x]<br>5.5 ± 1.5 [aa] |
| PI | 3.6 [b]<br>9.9 ± 0.9 [c]<br>10.4 ± 0.6 [h]<br>0.10 ± 0.28 [r]<br>6.2 [v]<br>9 ± 0.1 [rr] | 4 ± 0.4 [c]<br>10.6 ± 0.5 [h] | 2.21 ± 0.43 [r]<br>5.8 [v]<br>9.4 ± 0.1 [rr] | 3.5 ± 0.6 [c]<br>6.4 ± 0.7 [e]<br>1.53 ± 0.69 [r]<br>4.2 [v] | | 4.3 [b]<br>6.1 ± 0.9 [c] | 4.2 ± 0.6 [c]<br>ND [f]<br>13.7 ± 6.2 [ll] | 6.5 [b]<br>traces [x]<br>2.8 ± 0.7 [aa] |

| | | | | | | | | |
|---|---|---|---|---|---|---|---|---|
| PA | ND [b]<br><0.01 [r] | | ND [r] | ND [r] | | 0.9 [b] | | 3.1 [b] |
| SM | 19.9 [b]<br>27.4 ± 1.1 [c]<br>26.9 ± 1.6 [h]<br>19.8 ± 1.1 [r]<br>20.3 [v]<br>25.4 ± 0.2 [rr] | 31.6 ± 2 [c]<br>24.8 ± 0.7 [h] | 27.3 ± 0.4 [e]<br>17.8 ± 2.7 [r]<br>23.2 [v]<br>25 ± 0.2 [rr] | 27.8 ± 0.9 [c]<br>29.7 ± 2.1 [e]<br>22.3 ± 1.4 [r]<br>26.1 [v] | 33.1 ± 0.23 [dd] | 24.6 [b]<br>36 ± 2.6 [c] | 36.0 ± 2.6 [c]<br>14–18 [f]<br>16.7 ± 7 [ll] | 22.2 [b]<br>34 [x]<br>37.1 ± 2.2 [aa] |

[a] Et-Thakafy et al., 2017; [b] Garcia et al., 2012; [c] Zou et al., 2013; [d] Ali et al., 2017 (milk powders); [e] Zancada et al., 2013; [f] Russo et al., 2013; [g] Argov-Argaman et al., 2016; [h] Menard et al., 2010; Abd El-Salam and El-Shibiny, 2011; [k] Gantner et al., 2015; [l] Nguyen et al., 2015; [m] Zotos and Bampidis, 2014; [n] Yang et al., 2018 (Bactrian camel); [o] Ji et al., 2017; [p] Faccia et al., 2020; [q] Cebo et al., 2012a (depending on αs1-casein genotype) [r] Castro-Gomez et al., 2014; [s] Bonczar et al., 2016; [t] Sinanoglou et al., 2015; [u] Rodríguez-Alcala et al., 2009; [v] Rodríguez-Alcala and Fontecha, 2010; [w] Luo et al., 2020; [x] Malacarne et al., 2002; [y] Markiewicz-Keszycka et al., 2014; [z] Markiewicz-Keszycka et al., 2015; [aa] Barello et al., 2008; [bb] Meena et al., 2014; [cc] Khalesi et al., 2017; [dd] Luo et al., 2018; [ee] He et al., 2010; [ff] He et al., 2011 (various species); [gg] Konuspayeva et al., 2008; [hh] Salimei and Fantuz, 2012; [ii] Licitra et al., 2019; [kk] Altomonte et al., 2019; [ll] Contarini et al., 2017 (depending on breed and season); [mm] Massouras et al., 2017 (depending on stage of lactation); [nn] Moatsou and Sakkas, 2019; [oo] Revilla et al., 2017; [pp] Bernard et al., 2018; [qq] Smiddy et al., 2012; [rr] Yao et al., 2016; [ss] Kondyli et al., 2012 (depending on season); [tt] Rocchi et al., 2016 (retinoids); [uu] Martini et al., 2018b; [vv] Gentili et al., 2013; [ww] Haddadin et al., 2008 (depending on season); [xx] Raynal-Ljutovac et al., 2008; [yy] Kulyar et al., 2021; [zz] Gastaldi et al., 2010.

[a] ØMFG, milk fat globule diameter in μm.

[b] Cholesterol expressed in mg/$m^2$ MFGM.

[c] Ratio of cholesterol to sphingolipids expressed in mol/mL.

[d] Total polar lipids expressed as percentage of total lipids.

**Table 4** Unsaponifiable lipids of milk – interspecies comparison, based on selected publications

| Lipids | Cow | Buffalo | Goat | Sheep | Yak | Camel | Donkey | Horse |
|---|---|---|---|---|---|---|---|---|
| Cholesterol<br>mg/g fat | 3.3 ± 0.2 [a]<br>4.3 ± 0.5 [r]<br>3.5 ± 0.08 [s]<br>3.12 ± 0.1 [dd]<br>4.2 ± 0.04 [qq]<br>2.9 [rr] | 2.62–3.54 [i]<br>0.76–1.13 [m]<br>4.7 ± 0.1 [qq] | 3.3 ± 0.2 [a]<br>3.9 ± 0.2 [r]<br>2.7 ± 0.03 [s]<br>5.4 ± 0.06 [qq]<br>2.5 [rr] | 2.7 ± 0.1 [a]<br>3.7 ± 0.4 [r]<br>2.3 ± 0.06 [s]<br>4.7 ± 0.1 [qq] | 2.35 ± 0.06 [dd]<br>1.8–2.9 [ff] | 5.8 ± 1.2 [gg]<br>5.9 ± 0.25 [qq] | 3.3-5.6 [ll]<br>8.2 ± 0.33 [qq] | 5.04–7.3 [z]<br>8.8 ± 0.35 [qq] |
| Vitamin A (retinol)<br>μg/100 mL milk | 57.86 ± 24.48 [oo]<br>66 [tt]<br>210 [vv]<br>40 [xx] | 35.5–190.1 [i]<br>86 [tt]<br>287 [vv]<br>80 [xx] | 23–28 [ss]<br>306 [tt]<br>432 [vv]<br>40 [xx] | 38–53 [ss]<br>102 [tt]<br>432 [vv] | 52 [yy]<br>13.9–89.8 [aaa] | 26.7 ± 8 [ww] | 1.7–58.6 [kk]<br>58.6 [vv] | 40.3 ± 8.1 [y]<br>9.3-34 [hh] |
| β-carotene<br>μg/100 mL milk | 24.3 [vv]<br>20 [xx] | ND [i]<br>ND [vv] | ND [vv]<br>0 [xx] | ND [vv] | | | ND [vv] | 38.8 ± 10.2 [y] |
| Vitamin D (cholecalciferol)<br>μg/100 mL milk | ND [vv]<br>0.08 [xx] | ND [vv] | ND [vv]<br>0.06 [xx] | ND [vv]<br>0.18 [xx] | 2.05 [yy]<br>0.15–3.95 [aaa] | 0.3 ± 0.02 [ww] | 0–2.3 [kk]<br>2.23 ± 0.24 [vv] | 0.49 ± 0.16 [y]<br>0.3 [hh] |
| Vitamin E (α-tocopherol)<br>μg/100 mL milk | 99.6 ± 93.2 [oo]<br>378 [vv]<br>110 [xx] | 38–136 [i]<br>121 [vv] | 159–162 [ss]<br>655 [vv] | 197–267 [ss]<br>305 [vv] | 61 [yy]<br>30.2–91.9 [aaa] | 1.78 ± 0.58 [ww] | 80.7 [vv] | 26–113 [hh] |
| Vitamin K<br>(phyllo-menaquinone)<br>μg/100 mL | 0.88 [vv] | 0.55 [vv] | Traces [vv] | 11.5 [vv] | | | ND [vv] | 17.4 ± 7.5 [y]<br>2.9 [hh] |

[i] Abd El-Salam and El-Shibiny, 2011; [y] Markiewicz-Keszycka et al., 2014; [kk] Altomonte et al., 2019; [oo] Revilla et al., 2017; [qq] Smiddy et al., 2012; [ss] Kondyli et al., 2012 (depending on season); [tt] Rocchi et al., 2016 (retinoids); [uu] Martini et al., 2018a; [vv] Gentili et al., 2013; [ww] Haddadin et al., 2008 (depending on season); [xx] Raynal-Ljutovac et al., 2008; [yy] Kulyar et al., 2021; [aaa] Ma et al., 2017.

horse and donkey, the TAG account for 80–85%, polar lipids 5–10% and FFA 9.5% (Claeys et al., 2014). The TAG composition differs markedly between RS and ES in terms of structure abundancy. Among RS milk, the TAG profile of camel milk is strongly differentiated because the concentration of TAG at >C48 is more than 70%, opposite to lower than 35% observed in other RS milk. TAG at >C48 are also a characteristic feature of equid milk fat consisting of approximately 50% of the TAG fraction (Smiddy et al., 2012).

As depicted in Table 2, approximately two-thirds of the fatty acid profiles in RS milk consist of saturated fatty acids (SFA) and less than 5% is polyunsaturated fatty acids (PUFA), opposite to the lipid composition of RS feed. The fatty acid profile of fat in RS milk is due to a variety of synthetic pathways and is strongly influenced by the feed compounds. Biosynthesis of milk fat in the mammary gland of ruminants results in TAG with a low melting range, with a high relative percentage of short-chain butyric acid (C4:0) and medium-length SFA (C6:0-C14:0). Their synthesis starts from β-hydroxybutyrate and acetate produced in the rumen while C16 and C18 acids come from blood triglycerides (Walstra et al., 2006).

The change of structure of dietary unsaturated lipids by isomerization and hydrogenation takes place in the rumen mainly by bacteria. During ruminal biohydrogenation, *cis* and *trans* isomers of C18:1, conjugated linoleic acids (CLA) and conjugated linolenic acids, are formed (Ferlay et al., 2017). CLA is present only in the RS milk and dairy products; it is 15–20% of milk PUFA and corresponds to a group of isomers of C18:2 with a conjugated double bond. Rumenic acid, that is, C18:2 *c*9 *t*11 isomer, is 76–90% of RS milk CLA (Rodríguez-Alcala et al., 2009; Alichanidis et al., 2016). It exhibits various types of biofunctionality such as immunomodulation, reduction of blood cholesterol, triglycerides and glucose and inhibition of osteoporosis. Variability of the regulation of biohydrogenation among RS species and of diet configure the fine composition of milk lipids, in particular, the CLA and PUFA contents (Martemucci and D'Alessandro, 2013; Ferlay et al., 2017; Bernard et al., 2018).

Palmitic acid (C16:0) is a great part of RS milk fatty acids. The same holds true for the group of low and medium-chain fatty acids, which is a typical feature of RS milk fat – especially of goat and sheep – with the exception of camel milk (Table 2). The SFA relative content of the latter is lower, and the monounsaturated fatty acid (MUFA) fraction – consisted of palmitoleic (C16:1) and oleic (C18:1) fatty acids – is more abundant compared to other ruminants (Zou et al., 2013). Particularities of the fatty acid profile are connected with the characteristic flavour of goat milk. The enzymatic release of butyric, capric, caproic and caprylic acids is related to the development of rancid taste. However, in fresh milk with low lipolysis levels, the intense flavour comes from C6:0 to C9:0 FFA and more specifically from volatile branched-chain C9 and C10 as the 4-methyl and 4-ethyl-C8, which are 0.3–0.4 g/kg of fat in goat milk (Chilliard et al., 2003).

The absence of C4-C8 SFA in camel milk fat has been attributed either to the rapid metabolism by camel tissues before being excreted in the milk or to the camel diet (Haddad et al., 2010).

The high content of PUFA and the lower percentage of SFA in equid milk are due to the lack of biohydrogenation of fatty acids before absorption in monogastric animals; thus, there is a direct effect of feed on the fatty acid composition. Consequently, ES milk contains considerable amounts of the essential C18:2n6 (linoleic acid) and C18:3n3 (α-linolenic acid), while low quantities of C20:4n6 (arachidonic acid, AA), C20:5n3 (eicosapentaenoic acid) and C22:6n3 (docosahexaenoic acid) have been detected (Contarini et al., 2017; Gastaldi et al., 2010; Salimei and Fantuz, 2012). As summarized by Verduci et al. (2019), donkey milk does not contain *trans* fatty acids and CLA, while horse milk contains *trans* fatty acids and CLA in negligible quantities, as indicated in Table 2.

Indices related to the dietary role of lipids vary between various milk kinds (Table 2). The atherogenicity index (AI) that relates the pro-atherogenic C12:0, C14:0 and C16:0 with the anti-atherogenic UFAs is estimated by the formula $[C12{:}0 + (4 \times C14{:}0) + C16{:}0]/\sum UFA]$ and is influenced by the factors that affect the fatty acid composition of milk. Apparently, the low SFA and high UFA content of equid milk explain their low AI and the lower AI of camel milk compared to other RS milk. There are great differences among the values presented in the literature for the ratio of omega-6 to omega-3 (n-6:n-3) PUFA; however, it is evident that ES milk is advantageous. The n-3 fatty acids protect cardiovascular health, while n-6 FAs are potentially prone to low-density lipoprotein oxidation. The optimum n-6/n-3 ratio is equal to 1, and values higher than 5 are related to the pathogenesis of cardiovascular diseases, cancer, inflammatory and autoimmune diseases (Russo, 2009).

TAG composition and structure affect milk fat digestibility. As summarized by Gallier et al. (2020), short chain-SFA and medium chain-SFA at the sn-1,3 positions are quickly released in the stomach, absorbed as fatty acid in the intestine and rapidly degraded in the liver. The high content of goat milk in this group of SFA increases its digestibility. Fatty acids at the sn-2 position are released in the intestine as sn-2 monoglycerides, which are more polar than FFA and more easily absorbed. Long-chain fatty acids in the sn-1,3 positions are released as FFA, solubilized into bile salt micelles and absorbed in the small intestine. Long-chain SFA and in particular palmitic acid (C16:0) in the sn-1,3 positions released as FFA can form insoluble crystalline acid soaps that inhibit the absorption of palmitic acid and calcium. In human milk, the sn-2 palmitic acid is 51–88%, higher than 37–45% and 35% in cow and goat milk, respectively (Gallier et al., 2020). In these terms, donkey milk is advantageous because sn-2 palmitic acid has been found 72.9% of the total, similar to 74.8% of human milk and higher than the 47.1% in human milk (Gastaldi et al., 2010).

## 2.2 Organization of the lipid fraction: milk fat globule

Milk is an oil-in-water emulsion with nearly all fat existing in separate MFGs with a variable diameter ranging from about 0.1 μm to 15 μm, which are surrounded by a surface layer or membrane that prevents them from coalescence. In fresh milk, the ζ-potential of MFG is −12 mV. The composition of MFGM is completely different from either milk fat or milk plasma, and it is similar to a cell membrane (Walstra et al., 2006). In brief, the MFGM is a rather thin complex interfacial tri-layer of 8–20 nm, which in cow milk contains 69–73% lipids, 22–24% proteins – mostly glycoproteins – and enzymes. Lipids of MFGM are mainly polar PL and glycolipids and unsaponifiable lipids, that is, cholesterol, carotenoids and fat-soluble vitamins. Proteins are embedded in the phospholipid tri-layer. Detailed descriptions of the composition and structure of MFGM are presented in several comprehensive reviews, for example, Bernard et al. (2018) and Lopez (2011). The profile of the polar lipids and parameters related to the MFG of various milk kinds are summarized in Table 3.

Although the size of MFG of a particular milk kind is variable, a low mean diameter coincides with a high proportion of small-size MFG. Equid milk contains smaller MFG compared to ruminants milk. Among ruminants, camel milk has the smallest MFG, followed by goat, sheep and cow milk, whereas buffalo and yak milk contain large MFG (Table 3). Specific surface area ($m^2/g$) is higher in smaller MFG. The ζ-potential is related to MFG size as it expresses its charge and can be influenced by the glycoprotein and glycolipids of the MFGM and the mineral composition of milk (Cebo et al., 2012a).

According to the review of Argov-Argaman (2019), the MFG size is strongly influenced by the lactation stage and depends on the size of the intracellular lipid droplets (LD). They propose the following mechanisms involved in the regulation of LD size which can explain the changes of MFG size under various metabolic conditions: i. co-regulation with intracellular triglyceride levels, ii. membrane material availability and iii. fusion, controlled by the polar lipid composition of the membrane involving mitochondrial number and activity. The MFG size is affected by genetic, physiological and environmental factors. For example, the MFG size of goat milk is affected by the energy and neutral detergent fibre of animal feed, the milk fat content, the breed and the lactation stage (Argov-Argaman et al., 2016). An effect of genetic factors on MFG size is documented by Cebo et al. (2012a), who report that strong αs1-casein genotypes coincide with a higher fat content of goat milk, a larger MFG diameter, a lower concentration of PL and a lower absolute value of ζ-potential compared to counterparts with a null αs1-casein variant.

The size of MFG affects the technological behaviour and digestibility of milk. The catalytic efficiency of human pancreatic lipase has been found 4.6-fold higher on small (1.8 μm) than large (6.7 μm) native MFGs, related to a 3.6-fold larger available surface (Berton et al., 2012). Accordingly, the increased *in vitro* digestibility estimated as the release of FFA has been found in the order camel > goat > cow > buffalo opposite to the order of MFG size (Meena et al., 2014).

Most part of milk PL is in the MFGM and a minor part is present in the membrane compounds of defatted milk. Each PL consists of various molecular species in terms of the esterified fatty acids (Walstra et al., 2006). The variability in the quantity and the composition of PL fraction shown in Table 3 is attributed to the diversity of the applied analytical methods and to the effects of fat content and MFG size, diet, breed and other genetic factors (Lopez et al., 2014; Bernard et al., 2018; Argov-Argaman et al., 2013, 2016). Several positive implications of polar lipids on human health have been suggested such as anticancer activity related to colon cancer, positive effects on brain development, neurological and neurocognitive diseases, lipid metabolism, cardiovascular diseases, metabolic syndrome and skin (Bernard et al., 2018; Verardo et al., 2017).

The most abundant PL in milk are glycerolphosphatides, that is, mainly phospatidylcholine (PC), phospatidylethanolamine (PE), phospadtidylinositol (PI), phosphatidylserine (PS) and sphingolipids, that is, mostly sphingomyelin (SM). Due to their amphiphilic character, they exhibit emulsifying properties and delivering capacity for lipophilic substances. Short and medium-chain fatty acids are scarce, and the concentration of MUFA and PUFA is higher in PL compared to milk TAG. PE is highly unsaturated, PC is more saturated compared to other PL and SM is highly saturated because it contains mainly C16:0, C18:0, C18:1, C22:0, C23:0 and C24:0 (Contarini and Povolo, 2013).

According to the comparative studies of Ali et al. (2017) and Zou et al. (2013), the SFA content of PL of cow milk was higher than that of buffalo, camel, goat and sheep counterparts. C14:0, C16:0, C18:0, C18:1 and C18:2 have been reported as the major fatty acids of ruminants milk PL, whereas in donkey milk PL, the highest concentration of PUFA was observed. Among ruminants, camel PL exhibited the highest PUFA content (Ali et al., 2017; Zou et al., 2013). PL fatty acids profile similar to that of donkey milk has been reported for horse milk (Barello et al., 2008). The most abundant fatty acids in the total PL fraction of sheep and goat milk are C18:1, C16:0 and C18:0. Sheep milk has been found to contain more very long-chain fatty acids C22-C24 (9.5 vs 2.2) and less medium-chain fatty acids C10-C15 (3.4 vs 19.6) than goat milk (Zancada et al., 2013). Zou et al. (2013) estimated the similarity of the lipid fractions of various milk kinds to human milk considering a combination of parameters related to the profile of fatty acids, triglyceride and polar lipids. The similarity for cow, buffalo, sheep, camel and donkey milk was 79.6, 74, 96.4, 80.2 and 84.2, respectively. The composition of polar lipids is affected by the MFG size. Total UFA, MUFA, cholesterol and CLA content is higher in small (Ø $3.3 \pm 1.2$ μm) compared to large MFG (Ø $7.6 \pm 0.9$ μm). Moreover, large MFGs exhibit higher PI and lower PE concentrations than small counterparts (Lu et al., 2016a).

Comparative studies of the complex protein fraction of MFGM of various milk kinds have been performed. Detailed protein composition of MFGM of cow, goat, yak and human milk has been presented by Lu et al. (2016b). Fifty conserved proteins across species mainly involved in the process of protein-/vesicle-mediated transport and secretion of lipids to milk were identified. The abundance of major proteins conserved across species was differentiated. Xanthine dehydrogenase and stomatin were 25% and 3% in goat MFGM protein fraction, whereas CD36 was 15–25 fold higher in yak milk. Although lipoprotein lipase (LPL) activity was expressed in the MFGM of cow, goat and human milk, its abundance in the latter was 100-fold higher; also, carboxy ester lipase was expressed in human MFGM. Moreover, variation in host defence proteins across species is reported. This group was more enhanced in human MFGM, whereas the antibacterial cathelicidins – that come from neutrophils – were uniquely present in cow MFGM.

Yang et al. (2015) presented a classification of MFGM proteins of various milk kinds. Cellular processes followed by localization were the most abundant biological processes. The predominant cellular components were related to the membrane. Binding followed by catalytic activity were the most abundant molecular functions. Among 520 identified proteins, only 18 were estimated in similar quantities in all milk kinds. Xanthine dehydrogenase/oxidase, β-casein, lactadherin-like protein, bile salt-dependent lipase, αs2-casein, butyrophilin subfamily 1 and fatty acid synthetase were the most variable proteins among species. Analysis of MFGM proteome revealed four major clusters: cow/yak milk, buffalo/goat milk, cow/buffalo/yak/goat milk and camel/horse/human milk. The same group (Yang et al., 2016) investigated glycoproteins and N-linked glycosylation sites of MFGM of the aforementioned milk kinds. They reported that their predominant function was the 'response to stimulus' category of biological processes and that their glycosylation profile classified broadly milk from ruminant and non-ruminant groups. Ji et al. (2017) ascertained different quantitative profiles for cow and yak MFGM. Proteins with higher concentration in yak MFGM are related to immune function; in particular, glycosylation-dependent cell adhesion molecule 1 (GlyCAM-1) was 10-fold more abundant in yak milk compared to cow milk, whereas CD59 and lactadherin levels were approximately 4.5-fold more.

In their review, Bernard et al. (2018) present an atlas of 202 MFGM proteins common for cow, goat and sheep milk. They reported that 8 of the 30 biological process terms were related to lipids or cholesterol metabolism and 4 to hormone responses. Among the proteins specifically identified in each milk kind, 29 in sheep, 20 in goat and 5 in cow MFGM are related to lipids or fatty acids. Major MFGM proteins butyrophilin, lactadherin, xanthine oxidase, adipophilin have been identified in horse and camel milk (Barello et al., 2008; Saadaoui et al., 2013). Species-specific peculiarities related to the extended glycosylation of butyrophilin and the complexity of lactadherin in horse MFGM in terms of four isoforms have been reported (Cebo et al., 2012b).

As mentioned above, the size and ζ-potential of goat MFG are affected by the αs1-casein genotype (Cebo et al., 2012a). In addition, the expression of lactadherin and stomatin was enhanced in the MFGM from goats with null αs1-casein alleles, whereas the expression of the major proteins MUC-1, butyrophilin and adipophilin was not influenced by the αs1-casein genotype (Cebo et al., 2012a). Differentiation in the concentration of MFGM proteins in relation to MFG size has been observed by Lu et al. (2016a). Proteins mainly associated with protein transport/stabilization, host defence and cytoskeleton organization were in higher concentrations in the MFGM of large MFG. The concentration of proteins associated with lipid metabolic processes and cholesterol synthesis was higher in small MFG.

### *2.3 Unsaponifiable lipids: cholesterol, carotenoids and fat-soluble vitamins*

Cholesterol and tocopherol (vitamin E) are components of MFGM, and retinol (vitamin A), calciferol (vitamin D) and menaquinone (Vitamin K) are located in the fat core. Cholesterol content of cow milk is 40 mg/100 g MFG, 0.2 mg/m$^2$ fat surface or 2% of dry membrane material (Walstra et al., 2006). According to most of the publications, equid milk contains more cholesterol and small ruminants' milk fat contains less cholesterol per gram fat than cow milk fat (Table 4). Considering the fat content of various milk kinds, cholesterol per unit milk is in the following order: buffalo > sheep > goat > cow > camel > horse > donkey (Smiddy et al., 2012).

Fat-soluble vitamins are constituents of the MFG, and it is expected that they will be present at high concentrations in milk kinds with high fat content. This hypothesis is not confirmed by the highly variable values presented in Table 4. The vitamin content of milk depends on the maternal status and diet (Alichanidis et al., 2016), and thus, a seasonal effect has been observed (e.g., Haddadin et al., 2008; Kondyli et al., 2012; Valdivielso et al., 2015). The lack of carotenoids in non-cow milk kinds coincides with their whiter colour. In contrast to other ruminants milk, considerable amounts of vitamin D have been determined in yak and to a lesser extent in camel milk; however, the tocopherol content of camel milk appears low (Table 4).

## 3 Nitrogenous components: proteins, enzymes and non-protein nitrogen compounds

### *3.1 Proteins: caseins, casein micelle and whey (serum) proteins*

The group of nitrogenous components of milk consists of proteins and non-protein nitrogen (NPN) compounds, which account for up to 5% of total nitrogen.

Milk proteins are a very complex group of components, which are categorized into caseins precipitated at pH 4.6 and serum proteins, both fractions being affected by genetic polymorphisms. Casein is a mixture of several proteins and is present in milk in the form of non-heat-denatured casein micelles that are large aggregates containing also colloidal calcium phosphate. On the other hand, the fraction of serum or whey proteins consists of various mostly globular heat-sensitive proteins present in a dissolved form in the milk serum (Walstra et al., 2006; Holt et al., 2013).

Although milk and mammary genes are more highly conserved than other genes of mammals, there is protein variability between species in regard to amino acid sequence, post-translational modifications and quantities. The pattern of post-translational modifications such as phosphorylation and glycosylation affects the properties of casein and varies between species. Human and equid milk exhibit more complex patterns with variable phosphorylation levels compared to ruminants, while camel caseins are the least phosphorylated among ruminants. The complex mechanisms involved in these phenomena are summarized by Martin et al. (2011b). Proteomics and advanced microscopy methods are utilized for the detailed, in-depth study of the complex milk proteome and casein micelle structure.

The quantitative profile of the nitrogenous fractions of various milk kinds reported in selected publications is presented in Table 5. Since protein biosynthesis is genetically controlled, the variability shown in Table 5 is expected. For example, the effect of the breed of small ruminants on the gross composition of milk has been concisely reviewed by Tamime et al. (2011). Particular qualitative aspects related to the proteins and casein micelle of milk kinds other than those of cows' are briefly discussed in the following sections. The mineral profile that is related to the casein micelle structure is reviewed in Section 3.4.

Differences between cow and buffalo milk in regard to caseins and major whey proteins are limited. The phosphorylation level of αs2-casein and sialic acid content of κ-casein is lower while lactoferrin - 0.320 mg/mL - and proteose peptone content of buffalo milk are higher, that is, 330 mg/100 mL (Murtaza et al., 2017). In fact, lactoferrin of buffalo milk varies from 0.05 mg/mL in Egyptian buffaloes to 3.40 mg/mL in Indian Murrah buffaloes (Abd-El-Salam and El-Shibiny, 2011). β-Lactoglobulin is the main serum protein of buffalo milk (Ahmad et al., 2009; Abd El-Salam and El-Shibiny, 2011). Buffalo casein is less hydrated than that of cow milk, that is, 2.6–2.9 g water/g casein and contains lower amounts of sialic acid, hexose, hexosamine but higher amounts of Ca, that is, 11.1 g of Ca atoms/$10^5$ g of casein (Murtaza et al., 2017). Ahmad et al. (2009) in their comparative study estimated the micellar hydration for buffalo and cow milk as 2 and 2.3 g water/g dry casein, respectively. The casein micelles of buffalo milk are more mineralized than those of cow milk containing 2.85–3.58 g/100 g Ca,

**Table 5** Nitrogenous fractions of milk – interspecies comparison, based on selected publications. CN, casein; NPN, non-protein nitrogen components; TN, total nitrogen; WP, whey proteins; β-lg, β-lactoglobulin; α-la, α-lactalbumin; SA, serum albumin; PP, proteose peptone; LYZ, lysozyme; IG, immunoglobulins; LF, la ctoferrin

| Fractions | Cow | Buffalo | Goat | Sheep | Yak | Camel | Donkey | Horse |
|---|---|---|---|---|---|---|---|---|
| Casein g/100 g | 2.6 [a]<br>2.4–2.8 [m]<br>2.93 [q] | 2.94 [b]<br>3.46 ± 1.1 [d] | 1.6–2.6 [i,q]<br>2.3–3.8 [l]<br>2.3–4.6 [m,n] | 4.1–6.6 [l]<br>4.44 [p]<br>4.9 [q] | 3.06–4.21 [w]<br>4.16 ± 0.35 [x] | 2.3 [q,aa]<br>1.9 [y]<br>1.63–2.76 [z] | 0.78 [cc]<br>0.12 [dd]<br>0.56 ± 0.16 [ff] | 1.4 [cc]<br>0.17 [dd]<br>0.73–9.1 [ee] |
| αs1-CN % of CN | 41.1 [a]<br>38 [o]<br>42.9–53.6 [m] | 40 [b]<br>30.2–31 [c]<br>27.5–28.5 [e]<br>41.2 [f] | 4.5–34 [l]<br>0–28 [m,n]<br>5 [o] | 16 [o]<br>26 [l] | 43 [v]<br>28.8–30.4 [w]<br>28.8 [x] | 15 [y]<br>21 [z]<br>22 [aa] | 25 [s] | 17.8 [cc]<br>17.9 [dd]<br>40.2–59 [gg] |
| αs2-CN % of CN | 10.8 [a]<br>10 [o]<br>10.7–14.3 [m] | 6–9 [b]<br>13–17.6 [c]<br>12.1–12.7 [e]<br>6.3 [f] | 9–25 [l]<br>10–25 [m,n]<br>25 [o] | 15 [o]<br>14 [l] | 12 [v]<br>11.1–12.4 [w] | 9.8 [y]<br>9.6 [aa] | 2 [s] | 1.4 [cc,dd] |
| β-CN % of CN | 33.1 [a]<br>33–39 [o]<br>32.1–39.3 [m] | 35 [b]<br>28–33.9 [c]<br>32.5–33.5 [e]<br>36.1 [f] | 34–64 [l]<br>0–64 [m,n]<br>50–64 [o] | 39–47 [o]<br>42 [l] | 32 [v]<br>44.9–47.3 [w]<br>32.5 ± 7.5 [x] | 66.3 [y]<br>65 [z,aa] | 70 [s] | 78.6 [cc,dd]<br>40.1–51.4 [gg] |
| κ-CN % of CN | 11.9 [a]<br>11–13 [o]<br>10.7–14.3 [m] | 12 [b,f]<br>7–15.4 [c]<br>9.9–11.1 [e] | 10–19 [l]<br>15–29 [m,n]<br>10–20 [o] | 7–10 [o]<br>10 [l] | 13 [v]<br>12.3–14 [w]<br>9.4 ± 0.24 [x] | 1.7 [y]<br>3.47 [z]<br>3.3 [aa] | <1 [s] | 1.8 [cc,dd]<br>7.71 [gg] |

| | | | | | | | | |
|---|---|---|---|---|---|---|---|---|
| γ-CN<br>% of CN | 3.1 [a] | 4.1 [f] | 5-6 [l] | | | | | |
| NPN<br>% of TN | 5 [a]<br>5.4 [b]<br>5.2 [q] | 5.7 [b]<br>2.3 [d] | 3-5 [i]<br>7-12 [l]<br>3.6 [q] | 6-9 [l]<br>5.2 [p]<br>4.6 [q] | 5.5 ± 0.32 [t]<br>5.21-5.64 [w] | 7.5 [q]<br>9 [kk] | 17.6 [cc]<br>12.7 [ff] | 9.7 [cc]<br>11.2 [ee] |
| Ø CN, nm | 180 [h,s]<br>150 [q]<br>171-175 [u]<br>178.5 [v] | 190 [b]<br>180 [d]<br>110-160 [f] | 260 [h]<br>206-260 [l]<br>230-270 [q] | 193 [i]<br>170-220 [q] | 220 [s]<br>206 [u]<br>218.6 [v] | 280-550 [q]<br>380 [s] | 298 [s] | 255 [s,cc]<br>255-311 [ee] |
| WP<br>g/100 g | 0.63 [a]<br>0.5-0.7 [m] | 0.72 [b]<br>0.79 [d] | 0.37-0.7 [h]<br>0.3-1.2 [l]<br>0.37-0.7 [m,n] | 0.8-1.6 [l]<br>0.98 [o] | 0.96 [t]<br>0.96-1.08 [w] | 0.63-0.8 [z]<br>0.85 [aa] | 0.58 [cc]<br>0.107 [dd]<br>0.78 ± 0.09 [ff]<br>0.83 [gg] | 0.83 [cc]<br>0.13 [dd]<br>0.71-1.39 [cc] |
| β-lg<br>% of WP | 51 [a]<br>28.6-57.1 [m]<br>50 [o] | 42.7 [b]<br>50.3 [c] | 34-77 [l]<br>39.2-72.1 [m,n] | 51 [l,o] | 61.5 ± 5.2 [t]<br>54.7-67.1 [w] | - | 59.6 [cc]<br>29.8 [dd]<br>40.4-40.8 [hh] | 36.1 [cc]<br>30.7 [dd]<br>25.3-36.3 [ee] |
| α-la<br>% of WP | 19 [a]<br>14.3-21.4 [m]<br>25 [o] | 32.7 [b]<br>16.2 [c] | 17-50 [l]<br>40 [o] | 25 [o]<br>25 [l] | 5.2 ± 1.04 [t]<br>5.79 ± 3.13 [w] | 27 [aa] | 32.8 [cc]<br>22.6 [dd]<br>21.6-23.1 [hh] | 39.8 [cc]<br>28.5 [dd]<br>27.5-29.7 [ee] |
| SA<br>% of WP | 6.3 [a]<br>1.4-5.7 [m] | 10 [c] | 5-22 [l]<br>5.1-21.5 [m,n] | 6 [l] | 6.5 ± 13.3 [w] | 26 [aa] | 6.6 [cc]<br>6.2 [dd]<br>2.7-2.8 [hh] | 4.5 [cc,ee]<br>4.4 [dd,ee] |
| PP | 12.7[a] [a]<br>13.0[a] [o] | 27.4[a] [b] | 9[a] [o] | 6[a] [o]<br>1-1.8[2] (r) | 0.137[b] [w] | | 3.5 ± 0.7[b] [ff] | |

(Continued)

**Table 5** *(Continued)*

| Fractions | Cow | Buffalo | Goat | Sheep | Yak | Camel | Donkey | Horse |
|---|---|---|---|---|---|---|---|---|
| LYZ | 0.13$^{c}$ [ii] | | 0.25$^{c}$ [ii] | 0.10$^{c}$ [ii] | | | 1$^{b}$ [cc]<br>0.1$^{b}$ [dd]<br>25–27$^{1}$ [hh] | 1.1$^{b}$ [cc]<br>0.22$^{b}$ [dd]<br>0.77–1$^{2}$ [ee] |
| IG | 12.7$^{a}$ [a]<br>8.6–14.3$^{a}$ [m]<br>12$^{1}$ [m] | 10.7$^{b}$ [c]<br>8.7$^{a}$ [d] | 0.5–1.3$^{b}$ [l]<br>4.6–21.4$^{a}$ [m,n]<br>18$^{a}$ [o] | 12$^{a}$ [o]<br>0.5–0.7$^{b}$ [l] | | 1.25$^{b}$ [y]<br>18$^{a}$ [aa] | 11.5 [dd]<br>1.3-2.6 [hh] | 10.8 [cc]<br>19.6 [dd]<br>18.7–20.9 [ee] |
| LF | 1.6$^{a}$ [a]<br>0.02–0.2$^{b}$ [w] | 4.4$^{a}$ [b]<br>0.1–0.15$^{b}$ [w] | 0.02–0.3$^{b}$ [l]<br>0.02–0.2$^{b}$ [n] | 0.7–0.9$^{b}$ [l] | 0.04–0.67$^{b}$ [w] | 0.02–2.1$^{b}$ [y]<br>0.95$^{b}$ [h] | 0.037$^{b}$ [cc]<br>0.005$^{b}$ [dd]<br>4–4.3$^{a}$ [hh] | 0.08$^{b}$ [cc]<br>0.07$^{b}$ [dd] |
| CN/WP | 4.5 [a]<br>4.7 [g,s] | 5.5 [a]<br>4.1 [b]<br>4.3 [d]<br>4.6 [g] | 5 [a]<br>3.5 [g]<br>4 [l] | 5.6 [a]<br>3.1 [g]<br>4.4 [l] | 4.5 [g,s]<br>3.7 ± 0.24 [t]<br>3.2–4.1 [w] | 3 [a]<br>2.7–3.2 [g]<br>1.7 [s]<br>2.8 [aa] | 1 [a]<br>1.28 [g]<br>1.3 [s,cc]<br>1.12 [dd] | 1.08 [a]<br>1.1 [g,s]<br>1.69 [cc]<br>1.32 [dd] |

[a] Walstra et al., 2006; [b] Murtaza et al., 2017; [c] Abd El-Salam and El-Shibiny, 2011; [d] Ahmad et al., 2008; [e] Li et al., 2016; [f] Sindhu and Arora, 2011; [g] Claeys et al., 2014; [h] Park, 2017; Park, 2007; [k] Raynal-Ljutovac et al., 2008; [l] Alichanidis et al., 2016; [m] Verruck et al., 2019; [n] Amigo and Fontecha, 2011; [o] Tamime et al., 2011; [p] Ramos and Juarez, 2011; [q] Bornaz et al., 2009; [r] Moatsou and Sakkas, 2019; [s] Faccia et al., 2020; [t] Li et al., 2011; [u] Wang et al., 2013; [v] Zhang et al., 2017; [w] Ma et al., 2017; [x] Zhang et al., 2020; [y] Mati et al., 2017; [z] Al haj and Al Kanhal, 2010; [aa] Hailu et al., 2016; [bb] Attia et al., 2000; [cc] Uniacke-Lowe and Fox, 2011; [dd] Salimei and Fantuz, 2012; [ee] Salimei and Park, 2017; [ff] Malacarne et al., 2019; [gg] Malacarne et al., 2002; [hh] Guo et al., 2007 [ii] Chandan et al., 1968; [kk] El-Agamy, 2017.

$^{a}$ % of WP.

$^{b}$ mg/g.

$^{c}$ µg/mL.

1.42–1.85 g/100 g Pi, 0.12–0.19 g/100 g Mg and 0.61–0.77 g/100 g citrate (Abd El-Salam and El-Shibiny, 2011), while the 88.8% of Ca and the 66.7% of Pi are in the colloidal phase (Ahmad et al., 2009).

Goat milk contains considerably more β-casein and less αs1-casein than cow milk. The unusual and very extended genetic polymorphism of casein in goat milk – with emphasis to αs1-casein – also includes null alleles (e.g. Martin et al., 1999; Moatsou et al., 2008; Amigo and Fontecha, 2011), affects milk composition and properties and explains the wide range of casein content appearing in Table 5. Non-centrifugal casein is 8.7% of total casein in goat milk higher than the 5.7% in cow counterparts, and the same holds true for the mineralization of the micelle. The latter along with high ionic calcium content has been related to the heat instability of goat milk (Park, 2017). The substantial effect of αs1-casein polymorphism on the characteristics of the goat casein micelles has been shown in the study of Remeuf (1993). Goat milk with strong αs1-casein AA alleles producing 7 g/L has casein micelles with lower mean diameter than milk with weak FF alleles producing 0.9 g/L, that is, 221 vs 268 nm. The same holds true for the mineralization level, that is, 31.6 vs 34.3 mg Ca/g casein; however, the hydration level is similar −1.71 vs 1.74 g water/g dry micelle. Also, the firmness of rennet curds from 'strong' milk was much higher compared to 'weak' counterparts.

Opposite to cow milk, the most abundant sheep milk casein is β-casein, – 42% of total casein on average ranging from 31 to 58% – while sheep milk micelles are larger, more mineralized, less hydrated and have lower colloidal stability (Alichanidis et al., 2016; Park, 2007; Raynal-Ljutovac et al., 2007). According to Pellegrini et al. (1994), the characteristics of sheep casein micelles are, on average, as follows: level of hydration 1.82 ± 0.04 g water/g dry micelle, mineralization 33.4 ± 1.78 mg Ca/g casein, mean diameter 193 ± 7.3 nm, αs1-casein 34.5 ± 1.1%, αs2-casein 13.1 ± 1.1%, β-casein 42 ± 0.9% and κ-casein 10.4 ± 0.7%. Pirisi et al. (1999) reported that the diameter of sheep milk micelles is affected significantly by the αs1-casein genotype and is inversely correlated with αs1-casein concentration ranging from 194.9 ± 0.33 to 220.5 ± 16.3 nm at pH 5.5. The comparative study of Bornaz et al. (2009) showed that the mean diameter of sheep casein micelles is 180 nm, ranging from 170 to 220 nm, while that of cow milk micelles is 150 nm, ranging from 90 to 210 nm. Acid wheys from sheep milk of various breeds contain a higher β-lactoglobulin to α-lactalbumin ratio – from 3.91 to 6.65 – than the ratio of 3.09–3.37 estimated in cow milk (Moatsou et al., 2005). Interestingly, sheep milk lactoferrin has been found tolerant at 63°C for 30 min in contrast to cow and human counterparts, which lost 60% and 44% of their antiretroviral activity, respectively (Parrón et al., 2018).

Yak milk casein micelles are more hydrated than cow counterparts, that is, 2.16–2.31 vs 1.92–2.01 g water/g dry micelle (Wang et al., 2013). They have higher micellar calcium and phosphorus concentration than cow milk, that is, 26 mM and 19.9 mM vs 23.13 mM and 11.93 mM. In accordance with the higher diameter of yak casein micelles, the percentage of κ-casein on total casein is lower compared to cow milk (Zhang et al., 2017). Yak milk contains double the quantity of β-lactoglobulin – 15% of total protein – as compared to cow milk and two to six times higher lactoferrin (Verduci et al., 2019; Wiener, 2011). According to Ma et al. (2017), the IgG content of mature yak milk ranges from 0.5 mg/mL to 1.07 mg/mL that is higher compared to human milk. The comparison of the rennet clotting behaviour of yak and cow milk showed that the rennet clotting time of the former was longer as opposed to the curd firming rate and gel firmness. The higher calcium content and yak casein micelle size resulted in paracasein micelles with a hydrodynamic radius of 1620 nm vs 789 nm of cow counterparts (Zhang et al., 2020).

Camel milk contains much less κ-casein, which is 3.5% of total casein opposite to approximately 13% in cow milk. β-lactoglobulin cannot be estimated in camel milk, the most common serum protein is α-lactalbumin, and the main casein is β-casein similar to human milk (Mati et al., 2017). The abundance of β-casein, which is more sensitive to peptic hydrolysis than α-casein, potentially improves digestibility and limits the incidence of allergies. The lack of β-lactoglobulin coincides with higher heat stability than other ruminants milk. Lactoferrin of camel milk does not retain iron at pH 3–4. Moreover, whey basic protein that has not been found in other milk kinds is present at 1.7 g/L (Mati et al., 2017). GlyCAM-1, PGRP-S and proteose-peptone 3 are found in higher concentrations in camel milk compared to cow counterparts, and the variants IgG2 and IgG3 not present in cow milk have been found in camel milk (Hailu et al., 2016). Khalesi et al. (2017) report that protective proteins and enzymes related to the immunological system such as lactoferrin, lysozyme, lactoperoxidase, N-acetylglucosaminidase, IgA and IgM, serum albumin and insulin-like protein, which mimic the interaction of insulin with its receptor, are present in higher amounts in camel compared to cow milk. It is suggested that these protective factors cause a delay in the thermophilic fermentation of lactose (Al haj and Al Kanhal, 2010).

The gel-forming properties of camel milk differ from others ruminants milk due to particularities of the colloidal phase. In particular, camel casein micelle is larger (Table 3) and contains relatively low levels of casein and high content of minerals and citrate; the latter is 30 vs 4 mg/g casein in camel and cow milk, respectively. In contrast to high mineralization – 97.9 mg total mineral or 42.6 mg Ca/g casein – dromedary casein micelles exhibit high hydration level also, that is, 3.32 g water/g protein. In camel milk, 15.4% of total caseins, that is, two times that in cow milk, are soluble (Attia et al., 2000). The abundance of insoluble salts results in the solubilisation of P and citrate at pH 4.8 and not at pH 5.2. Moreover, the lack of β-lactoglobulin that is strongly involved in the heat-induced aggregation of other ruminants milk results in non-typical protein gel networks (Attia et al., 2000; Hailu et al., 2016). The poor rennet clotting properties are due to the incomplete secondary phase (Al haj and Al Kanhal, 2010).

Equid milk nitrogenous fractions exhibit profound differences from that of ruminants milk. In general, the casein to whey protein ratio is less than one-third of that of cow milk, the casein micelle is larger and contains mainly β-casein while κ-casein is present in trace amounts. αs1-, β- and γ-caseins and proteose peptone have been identified in horse milk, while the levels of κ-casein are very low. Horse β-casein has three to seven phosphorylation sites. αs1-casein is considered similar to cow counterparts and has six potential phosphorylation sites and one phosphorylation centre vs eight/nine and two centres of cow milk. Horse milk contains approximately 10.1 mM micellar calcium and approximately 2.6 mM micellar inorganic phosphorus, that is, the micellar calcium to casein ratio is >20:1. At neutral pH, the rennet action on horse casein does not result in gel formation, and the same holds true for acidification at pH 4.2. High levels of ionic calcium approximately 2.6–3 mmol/L compared to 1.8 mmol/L of cow milk and very low levels of κ-casein are related to low heat stability at pH < 6.8 of this milk kind (Uniacke-Lowe and Fox, 2011). Opposite to cow milk, κ-casein of horse milk is positively charged and does not include any non-glycosylated form, which decreases its susceptibility to chymosin; moreover, it is cleaved by chymosin at Phe97-Ile98. These particularities imply a different chymosin-induced clotting mechanism for horse milk (Uniacke-Lowe et al., 2010).

Equid milk contains β-lactoglobulin as opposed to human and camel milk, and the lactoferrin content of horse milk is similar to human milk and higher than that of donkey and cow counterparts. Although the iron-binding capacity of lactoferrin is similar for all milk kinds, its pH dependence differs in equid milk. Horse β-lactoglobulin that lacks sulfhydryl group and α-lactalbumin is much more thermostable than cow counterparts as it does not undergo significant denaturation at temperatures lower than 100°C (Uniacke-Lowe and Fox, 2011). The low ζ-potential of donkey casein micelles has been associated with its poor thermal stability. Although the colloidal stability is low, weak gel formation during acidic and enzyme coagulation of donkey milk takes place. The poor rennet clotting behaviour is attributed to the aggregation mode of the hydrolysed micelles during the secondary phase of coagulation (Faccia et al., 2020).

Non-cow milk has been suggested as a means to overcome the well-known cow milk protein allergy, which is more frequent in infants. Due to the lower β-lactoglobulin to α-lactalbumin ratio compared to cow milk and the low αs1-casein, goat milk has been proposed for consumption for those who are allergic to cow milk; nevertheless, there is cross-reactivity between the proteins of all ruminants milk (El-Agamy, 2007). Data reviewed by Uniacke-Lowe et al.

(2010) showed that the cross-reaction of horse milk with cow milk would be much lower compared to ruminants milk mainly due to the low quantities of αs2-casein and the greater similarity (44%) of equine αs1-casein to the human counterpart. The similarity is lower between human and camel milk (36%) and much lower between human and cow, goat and sheep milk (28–29%). Salimei and Fantuz (2012) and Salimei and Park (2017) presented clinical results, which suggest that both donkey and horse milk can be used in the feeding of infants and children allergic to cow milk. There is evidence that camel milk can be also used as a non-allergenic alternative to cow milk due to the lack of β-lactoglobulin, high β-casein and low αs-casein contents (Al haj and Al Kanhal, 2010; Khalesi et al., 2017; El-Agamy, 2017).

The protein profile and the structural elements of milk are very important factors for milk digestibility. The use of non-cow milk in infant or elderly nutrition has been suggested to overcome the firm curds of cow milk formed in the stomach. In this respect, the fragile curds/flocks of acidified horse and donkey milk and the soft acid or rennet curds of camel and goat milk are considered advantageous. Inglingstad et al. (2010) compared the gastrointestinal digestion of proteins of various kinds of milk. Caseins from all species were digested very fast, and equine and human caseins were rapidly degraded even at the gastric juice pH 2.5. Horse β-lactoglobulin was highly degraded, but 60% of the cow and goat counterparts remained intact; α-lactalbumin of all milk kinds was very resistant. Interestingly, 60% of horse milk lysozyme was not degraded. In the study of Tagliazucchi et al. (2018) that compares ruminants milk, the goat milk exhibited the highest digestibility expressed as released amino groups followed by camel milk. The peptides produced after digestion of goat and sheep milk exhibited the highest similarity with 151 common sequences. Mechanisms and phenomena related to milk digestion have been recently reviewed by Huppertz and Chia (2021) and Roy et al. (2020). Roy et al. (2020) concluded that the research findings on milk digestion of various milk kinds are not solid and proposed further comparative studies that involve also the necessary treatments applied to milk before consumption.

## *3.2 Important enzymes*

Endogenous enzymes of cow milk in terms of action, activity and distribution have been the objective of numerous studies. Literature on the endogenous enzymes of non-cow milk kinds is limited or even scarce and has been recently reviewed by Albenzio et al. (2021). A concise comparative presentation of particular topics related to endogenous enzymes is given in the subsequent paragraphs.

Plasmin, which is part of a complex system consisting of active and inactive forms of the enzyme, activators and inhibitors, is the most important milk protease (Ismail and Nielsen, 2010). The increase in somatic cell count (SCC) during mastitis coincides with an increase in plasminogen activators and an increase in plasmin activity. An increase in the rate of plasminogen activation to plasmin is observed in late-lactation milk and in milk from older cows, assigned to increased plasminogen activators (France et al., 2021).

Buffalo milk contains all parts of the plasmin system, that is, plasmin, plasminogen, plasminogen activators and inhibitors of the latter. Plasmin activity and most of the plasminogen activators are associated with the casein micelles, similar to cow counterparts (Abd El-Salam and El-Shibiny, 2011; Fantuz et al., 1998). A γ-casein, that is, β-casein f(69–209) not equivalent to any γ-casein of cow milk along with its complementary proteose peptone β-casein f(1–68) has been found in river buffalo milk (Di Luccia et al., 2009).

The plasminogen to plasmin ratio and the sum of plasmin plus plasminogen-derived activities are lower in goat milk than those observed in cow milk (Moatsou, 2010). Leitner et al. (2004b) found that PL activity from infected udders –$1.75 \times 10^6$ SCC – was almost double that in milk with low SCC. They suggest that the plasmin system is 'slower' in goats than in sheep and less sensitive to external factors such as infection. The advancement of lactation increases plasmin activity and decreases the plasminogen to plasmin ratio in goat milk (Cortellino et al., 2006).

Plasmin is the predominant proteolytic enzyme of sheep milk, and its activity is affected by subclinical mastitis (Albenzio et al., 2005; Caroprese et al., 2007; Leitner et al., 2004a). In particular, the highest activity has been observed in sheep milk with SCC higher than $2 \times 10^6$ cells/mL and the lowest in milk with SCC lower than $3 \times 10^5$ cells/mL (Albenzio and Santillo, 2011). The findings for the effect of the stage of lactation on the plasmin system of sheep milk are not consistent (Moatsou, 2010). It has been suggested that the plasmin and plasminogen activity and their ratio do not vary significantly throughout the lactation period of sheep when SCC is lower than $600 \times 10^3$ cells/mL (Albenzio et al., 2009).

Plasmin and plasminogen activities have been detected in individual camel milk samples with proteolytic activity on camel casein similar to that of cow milk plasmin (Baer et al., 1994). Fragments of β-casein missing 946 Da that could be the first seven N-terminal residues have been considered as a consequence of plasmin activity in camel milk (Ryskaliyeva et al., 2018). Very low but detectable plasmin activity in horse milk sodium caseinate is reported (Egito et al., 2003) opposite to the high activity and the intense appearance of γ-caseins in fresh horse milk (Humbert et al., 2005); however, plasminogen was not detected by means of the usual activation protocols.

Lipoprotein lipase (LPL) is a glycoprotein electrostatically and hydrophobically associated with the casein micelle in cow milk active at the oil-water interface, which accounts for almost all the lipolytic activity in raw milk. The characterization of the enzyme and lipolysis in milk are presented in detail by Deeth (2006, 2021). Much lower LPL activity has been reported for small ruminants' milk, that is, about one-third and one-tenth that of cow milk for goat and sheep milk, respectively. In contrast to cow milk, Chilliard et al. (1984) estimated that only 8% of goat milk LPL is associated with casein micelles and approximately half of its activity is in the cream phase of goat milk. Spontaneous lipolysis, LPL activity and flavour are higher in goat milk with weak as1-casein genotype FF than in counterparts with strong genotype AA, while the fat content is lower (Delacroix-Buchet et al., 1996).

The tendency of goat milk to spontaneous lipolysis - despite the lower activity of the enzyme - and its relation with nutritional and physiological factors is explained in the review of Chilliard et al. (2003). They suggest that the lower fat secretion in the secretory cells in animals with weak alleles decreases the need for LPL at the basal membrane and increases the availability of LPL for secretion into the milk. Moreover, they found that the proteose peptone fraction inhibits both spontaneous and induced lipolysis of goat milk similar to cow milk.

The above-mentioned low activity of LPL of sheep milk has been confirmed in the mammary tissue; it has been estimated as 13 times lower than that in cow mammary tissue although the transfer of blood fat to milk has been found to be about the same in both species (Yousef et al., 1969). According to the findings of Chavarri et al. (1998), the hydrolysis pattern of sheep milk fat is consistent with that reported for cow LPL, pasteurization causes 73–95% inactivation and a slight downward trend in activity appears as lactation advances.

Information on endogenous lipase in buffalo milk is scarce; its activity increases during lactation and is affected by the type of feeding (Albenzio et al., 2021). Purification of LPL from yak milk is reported by He et al. (2012) who found that a PP3-containing fraction of proteose peptone can enhance its activity, while other fractions suppress it. In the proteomic analysis of Felfoul et al. (2017), LPL isoform 3 was detected in raw camel milk and in that of Ryskaliyeva et al. (2018) LPL of 46.5 kDa was found in skimmed camel milk. High residual lipase activity close to 20% of the initial activity is reported by Lorenzen et al. (2011) after pasteurization of camel milk. Lipase activity has been detected in horse milk, which has different characteristics from LPL, that is, it was activated with 2–8 volumes of blood serum, inactivated by 95% when treated at 60°C for 30 min, being sensitive to sodium chloride and not significantly affected by heparin (Chilliard and Doreau, 1985).

The use of alkaline phosphatase (ALP) as a legislative indicator for the adequate pasteurization of milk has been a motivation for the study of its behaviour in various milk kinds. ALP is a glycoprotein containing sialic acid, occurring in all mammalian milk in considerably variable quantities. In cow milk, it is distributed between skim milk (~55%) and cream (~45%), in particular, in MFGM (Farkye, 2021).

Variable values of ALP activity in the milk of various species have been observed. Raynal-Ljutovac et al. (2007) suggest that ALP is not related to the fat content of a particular milk kind but rather to species uniqueness. Higher ALP activity in buffalo milk but with a heat inactivation pattern similar to cow milk was reported by different sources (Abd El-Salam and El-Shibiny, 2011; Lombardi et al., 2000; Sharma et al., 2009). As summarized by Moatsou (2010) and Alichanidis et al. (2016), the ALP activity in bulk raw sheep milk is much higher than that in cow and goat milk - 8000 to 17 000, 1800 to 4800 and 120 to 1300 μg phenol/mL, respectively. It is negatively correlated to daily milk production and positively correlated to the sheep milk fat content. Mastitis increases the activity of this enzyme whereas there are contradictory findings for the effect of lactation. A substantial increase of ALP activity within 45–120 days of sheep lactation was reported (Martini et al., 2013). Pasteurization at 72.5°C for 15 s decreased the activity from 61 to 0.223 U/L in goat milk, from 1459 to 0.181 in sheep milk and from 744 to 0.085 U/L in cow milk (Klotz et al., 2008). The findings of Lorenzen et al. (2010) confirmed that ALP is more heat stable in small ruminants' milk compared to cow milk.

The ALP activity in camel milk has been found to be 21.31 ± 1.26 μg/mL, which is lower than those in cow and buffalo milk, that is, 575.00 ± 18.42 μg/mL and 253.75 ± 8.22 μg/mL, respectively (Yoganandi et al., 2014). ALP is not been proposed as a pasteurization indicator for this milk kind due to high residual activity (Wernery et al., 2008; Lorenzen et al., 2011).

Marchand et al. (2009) did not detect any specific association of ALP with MFGM in horse milk and report 35–350 times lower activity than in cow counterparts. They concluded that this enzyme could not be a useful pasteurization indicator for horse milk due to the low D values in the range of pasteurization temperatures and the low initial activity. Giacometti et al. (2016) proposed ALP as an indicator for proper pasteurization of donkey milk; the 36 047 ± 1009 mU/L ALP activity in raw donkey milk was decreased to 106 ± 10.5 mU/L after pasteurization at 65°C for 30 min.

Lactoperoxidase (LPO) is the predominant antimicrobial factor of cow milk, which acts in the presence of sufficient concentrations of hydrogen peroxide and thiocyanate ion (SCN) from the animal feed. It is the second most abundant enzyme in cow milk after xanthine oxidoreductase with a concentration of 10–30 μg/mL making up about 0.5% of the whey proteins. Its activity can vary throughout lactation between different breeds and individuals. Xanthine oxidoreductase (XOR) and superoxide dismutase can enhance the activity of lactoperoxidase by forming hydrogen peroxide from various sources. The principal function of the enzyme is the protection of the udder and the nursing calf through the potent antibacterial properties of the oxidized forms of thiocyanate. LPO activity is used as an indicator for heat treatments of milk more severe than pasteurization as it is completely inactivated at temperatures >78°C for 15 s (Kussendrager and Van Hooijdonk, 2000; Seifu et al., 2005).

LPO is the most abundant enzyme in buffalo milk, appears in variable concentrations and it exhibits higher activity than that in cow milk (Abd El-Salam and El-Shibiny, 2011; Ahmad et al., 2013). According to many publications, the LPO activity in goat is higher than that in cow and sheep milk, and both the enzyme activity and the SCN concentration in small ruminants' milk are affected by the animal individuality, breed and stage of lactation (Moatsou, 2010). However, Dumitraşcu et al. (2012) reported that the respective LPO activity for cow, sheep and goat milk is 970 ± 12, 1720 ± 50 and 810 ± 50 U/L. Very recently, Zou et al. (2021) using a different analytical method reported 4542 ± 1006, 1200 ± 140, 20.9 ± 13.4 and 167 ± 33 U/L LPO activity for cow, goat, human and camel milk, respectively.

High LPO activity in raw camel milk (869–1172 U/L) has been estimated, which is not detectable after pasteurization being less heat stable than that of cow milk LPO, exhibiting much higher D values at 71°C and 67°C (Tayefi-Nasrabadi et al., 2011). As it is destroyed at 75°C for 15 s, LPO has been suggested as an indicator for pasteurization of camel milk (Wernery et al., 2013). The SCN content of camel milk is higher than that of other ruminants milk and varies considerably from 2.4 to 33 ppm (Kamau et al., 2010). The concentration of LPO in donkey milk has been found to be 0.11 mg/L that is seven times lower than that in human milk and very much lower than the 30–100 mg/L found in cow milk (Vincenzetti et al., 2017).

The concentration of antimicrobial lysozyme (muramidase) is very low in ruminants milk, but it contributes substantially to the whey protein fraction of equid milk (Table 3). Chandan et al. (1968) report that cow and sheep milk contain similar quantities of lysozyme, that is, 13 and 10 μg/100 mL, respectively, while the content in goat milk is much higher being 25 μg/100 mL. Findings for the concentration of lysozyme in buffalo milk are inconsistent. According to the review of Ahmad et al. (2013), buffalo milk contains on average 20 μg/100 mL lysozyme with activity higher than that in cow milk, which could contribute to the lower mastitis incidence in buffalos. Two to ten times higher lysozyme activity has been estimated in buffalo milk not affected by treatment at 74°C for 1 min (Abd El-Salam and El-Shibiny, 2011). Considerably variable levels of lysozyme are reported for camel milk lysozyme, that is, from 500 μg/100 mL (Duhaiman, 1988) to 15 μg/100 mL, with lytic activity higher than that in cow milk but lower than that in egg white counterparts (Elagamy et al., 1996). However, Kappeler (1998) did not detect lysozyme activity in camel milk, or a cDNA corresponding to C-type lysozymes.

The well-established antibacterial activity in equid milk is associated mainly with the high concentration of lysozyme. The concentration of lysozyme in horse milk varies from 0.77 g/kg to 1 g/kg (Salimei and Park, 2017), in donkey milk from 0.9 g/L to 3 g/L and in human milk from 0.8 g/L to 1.1 g/L (Altomonte et al., 2019). Moreover, the activity of lysozyme in donkey milk ranges from 4000 U/mL to 5000 U/mL that is much higher than the 0.0292 U/mL found in cow milk and very much lower than the 39 000 U/mL estimated in human milk (Vincenzetti et al., 2008; Martini et al., 2018b). Eighty-five per cent of the initial activity of lysozyme has been estimated in horse milk treated at 75°C for 10 min, that is, under conditions appropriate for the treatment of infant formulas (Civardi et al., 2007). Lysozyme in donkey milk survives heating at 90°C for 1 min and 100°C for 5 min but is destroyed at 121°C for 10 min (Coppola et al., 2002).

Xanthine oxidoreductase (XOR) is a complex dimeric molybdenoflavoenzyme that contains two Mo centres, comes from blood and its activity depends on the molybdenum (Mo) content of cow feed. It is very abundant in cow milk – approximately 140 mg/L – and constitutes 20% of the protein content of MFGM. The most important role of XOR in milk is in the secretion of milk fat, in which it does not act as an enzyme but as a protein. There are two forms of this enzyme in cow milk, that is, xanthine oxidase (XO) and xanthine dehydrogenase (XDH) (Moatsou, 2010). Due to Mo deficiency, XOR activity in small ruminants' milk is lower compared to cow milk. In particular, the specific XO activity of sheep milk enzyme has been found 2.5 times lower than that of cow counterparts in accordance with the lower Mo concentration of the former, that is, 0.18 atoms Mo vs 0.55 atoms Mo/enzyme subunit in cow milk (Benboubetra et al., 2004). The lower XO activity in goat milk compared to cow and sheep counterparts is attributed to the very low molybdenum content of 0.09 atoms Mo/enzyme subunit (Atmani et al., 2004). However, no differences were observed in the dehydrogenase form of the enzyme between the three milk kinds (Atmani et al., 2005).

Less XOR activity has been found in buffalo milk, while only one-fourth of it is reported to be in the cream phase (Abd El-Salam and El-Shibiny, 2011). XOR concentration up to 22.2 mg/L has been found in camel milk, but it did not exhibit detectable activity towards xanthine or nicotinamide adenine dinucleotide. The mean molybdenum content per subunit was 0.02 ± 0.002, much lower than the value of 0.56 ± 0.051 atoms of Mo estimated for cow milk counterparts (Baghiani et al., 2003). XDH/XO with molecular mass and electrophoretic profile similar to cow and goat milk counterparts has been identified in the MFG protein fraction of pony horse milk (Cebo et al., 2012b).

### 3.3 Non-protein nitrogen compounds

Approximately 5% of total nitrogen in cow milk is NPN. NPN consists of intermediate products of the protein metabolism by the animal such as ammonia (~1 mg/100 mL milk, ~3 g N/100 g NPN), urea (~25 mg/100 mL milk, ~67 g N/100 g NPN), creatine, creatinine and uric acid. Trace amounts of small peptides, free amino acids (FFA, ~28 mg/100 mL milk, ~15 g N/100 g NPN) and their derivatives such as amines are included in NPN. Orotic acid typically occurring in ruminant milk is found in this fraction (~7 mg/100 mL milk, ~6 g N/100 g NPN). Ribonucleic acids and their degradation products such as phosphate esters and organic bases are also part of the NPN fraction milk (Walstra et al., 2006). According to the findings of Indyk and Woollard (2004), the orotic acid, uric acid and creatinine concentrations in cow milk are 3-7 mg/100 mL, 0.9-2.4 mg/100 mL and 0.6-1.2 mg/100 mL, respectively.

NPN contribution on total N in goat milk is variable due to the high variability of its protein fraction, but it is higher than that in cow milk. Goat milk contains 0.4% NPN, while cow, sheep and human milk contain 0.2%, 0.8% and 0.5% NPN, respectively (Park, 2007). A limited concentration range throughout lactation - from 41.2 mg/100 mL to 43.8 mg/100 mL - is reported by Prosser et al. (2008) for the NPN of reconstituted whole milk powders. Urea, which is the major component of the NPN fraction, has been found highly variable - from 4 mg/100 mL to 54 mg/100 mL - in the milk of six different goat breeds (Mayer and Fiechter, 2012b). In the same study, orotic acid ranges from 0.85 mg/100 g to 2.45 mg/100 g. Avondo et al. (2015) found that urea is significantly lower in milk with 'strong' αs1-casein AA genotype than in counterparts with 'weak' FF alleles, that is, 56.6 mg/100 mL vs 69.2 mg/100 mL; interestingly, no diet effect was observed. Indyk and Woollard (2004) estimated 2.2 mg orotic acid, 1.6 mg uric acid and 0.5 mg creatinine/100 mL for goat milk.

NPN accounts for 5-6.8% of total nitrogen in sheep milk; NPN compounds are urea (45%), FFA (16%), creatine (2.4%), creatinine (1.7%), ammonium nitrogen (1%) and uric acid (2.1%) (Ramos and Juarez, 2011). Mayer and Fiechter (2012a) estimated 43.2 ± 7 mg urea and 1.7 ± 0.31 mg orotic acid per 100 g sheep milk. Contents of 3 mg/100 mL, 2.6 mg/100 mL and 2 mg/100 mL for orotic acid, uric acid and creatinine, respectively, have been also reported for sheep milk (Indyk and Woollard, 2004).

Similarly to cow milk, the NPN fraction of buffalo milk is approximately 5% of the total nitrogen. The profile of NPN compounds is also similar to cow counterparts except for lower urea content that is within 17-22 mg/100 mL (Sindhu and Arora, 2011). Moreover, average orotic acid content of 2.8 mg/100 mL has been reported (Abd El-Salam and El-Shibiny, 2011). The NPN content of yak milk expressed on total N varies widely from 3.7% to 10.3%. More than 50% consists of urea that ranges from 34 mg/100 g to 59 mg/100 g (Ma et al., 2017).

According to the review of El-Agamy (2017), the average NPN content of camel milk is 185 ± 0.11 mg/100 mL or 9% of total nitrogen; nevertheless, an increase from 11% to 14% with lactation progress is reported. NPN content of 7.1-7.9% on total nitrogen has been estimated by Bornaz et al. (2009). A very wide range of urea concentrations affected by the season and correlated to the total protein concentration have been found in milk from different camel species

(Faye et al., 2010). The mean value of urea concentration was 8.16 ± 6.04 mg/100 mL - that is lower than that in cow milk - and ranged from 0 to 29 mg/100 mL in spring; accordingly, very low NPN < 2% of the nitrogen content is reported.

The NPN fraction of donkey milk varies from 0.15 g/100 g to 0.23 g/100 g or 10–12% of total N (Guo et al., 2007). Malacarne et al. (2019) report that NPN is 7.52–13.8% of total N throughout lactation. According to the review of Salimei and Fantuz (2012), the urea content in equid milk is 20–35 mg/100 g corresponding to 40% of NPN. Altomonte et al. (2019) suggest that the donkey milk NPN fraction for the most part consists of peptides derived from β- and αs2-caseins. NPN in horse milk increases from <2% of total N at parturition to >10% after 2 weeks. In mature horse milk, NPN is on average 38.1 mg/100 mL or 10–15% of the total N that is intermediate between the values of 25% in human milk and 5% in ruminant milk (Uniacke-Lowe et al., 2010). Malacarne et al. (2002) report that 11–21% of total N in horse milk is NPN, that is 26–55 mg N/100 g. In contrast to ruminants milk, traces of orotic acid are present in horse milk NPN, that is, <0.1 mg/100 mL, and the uric acid and creatinine have been found to be 0.4 mg/100 mL and 0.3 mg/100 mL, respectively (Indyk and Woollard, 2004).

The amino acid taurine, nucleosides and polyamines are components of the NPN fraction of milk, which are of great biological importance for the early stages of life. Their qualitative and quantitative variability in different milk kinds is taken into consideration for the development of infant and follow-up formulae. The 10–20% of the NPN fraction of milk consists of FAA that are more easily absorbed, and therefore, they are very significant in nutritional terms. The total concentration of free amino acids (FAA) in various kinds of milk has been reviewed by Alichanidis et al. (2016), as follows (mg/kg): cow 33–51, buffalo 15, goat 153–243, sheep 95, yak 47, camel 49–50, donkey 50, horse 232 and human 367–670 mg/kg. It is evident that goat and horse milk contains the highest amount of FAA, which however is lower than that in human milk. In general, the most part of FAA fraction is the nonessential amino acids glutamic acid, glycine, aspartic acid and alanine, while the other amino acids are present in very low concentrations. Among the FAA, the biofunctional taurine, which does not occur as protein-bound amino acid, is of specific interest especially for the newborn (Park, 2007; Alichanidis et al., 2016; Prosser, 2021). Taurine is the most abundant FAA in goat milk (45–70 mg/kg) with quantities similar to that of human milk (35–66 mg/kg). Also, it contributes substantially to the FAA fraction of sheep milk (17 mg/kg) while its concentration in cow milk is low (1.6–6.4 mg/kg). Camel, buffalo and equid milk have low contents of taurine <4 mg/kg (Alichanidis et al., 2016).

Nucleosides (NS, i.e. N-glycosides of pyrimidines and purines), nucleotides (NT, i.e. nucleosides with a phosphate group) and nucleobases are included in the NPN fraction of milk in micromolar concentrations. They are important for cellular functions, acting as precursors of nucleic acids and their concentration is much higher in the colostrum than in milk. They act beneficially for the immune system, gastrointestinal function and cell proliferation and apoptosis. Although they are synthesized de novo, the dietary sources of NTs are conditionally essential during periods of rapid growth of tissues; in this respect, NTs are permitted supplements for infant and follow-up formulae (Gill and Indyk, 2007; Martin et al., 2011a). The composition and quantitative characteristics of this group of compounds are specific for each milk kind. The findings of comparative studies in the mature milk of various animals along with the NS/NT profile reported for human milk are summarized in Table 6.

Cytidine and cytidine 5′-monophosphate are the most abundant forms in cow milk, while uridine and uridine 5′-monophosphate are the dominant forms in goat and sheep milk but not always detected in cow milk (Table 6). The physiological capacity of NS/NT in milk is expressed as 'total potentially available nucleosides' (TPAN). TPAN includes the free nucleotides and also other milk compounds such as nucleoproteins, nucleic acids and nucleos(t)ide derivatives, which are digested in the infant's gastrointestinal tract to yield physiologically available nucleosides (Gill and Indyk, 2007; Martin et al., 2011a).

TPAN has been found higher in ruminants' milk than in human milk, and it was the highest in sheep milk, i.e. 79 ± 5, 974 ± 28, 3741 ± 98 μmol/L in cow, goat and sheep milk. The contribution of NS to the TPAN of cow milk was 30% in contrast to minor contribution in goat and sheep milk; in turn, >30% of TPAN of the latter consisted of NTs and NT adducts (Gill et al., 2012). Orotic acid is the most abundant NT-related compound in cow milk, but it is essentially absent in human milk (Gill and Indyk, 2007). Gil and Sanchez-Medina (1981) observed a strong effect of lactation stage on the NT content of ruminants milk and qualitative and quantitative differences among mature cow, goat and sheep milk.

Polyamines (PAs), that is, putrescine (PUT), spermidine (SPD) and spermine (SPM), are aliphatic molecules with amine groups distributed along their structure occurring as free cations or conjugates bound to proteins and nucleic acids. PAs play an important role in the regulation of cell growth and proliferation, stabilization of negative charges of DNA, RNA transcription, protein synthesis, apoptosis, regulation of the immune response and in the maturation of the neonatal gut. Spermine is the most active, and putrescine is the least active in the control of various biological processes (Larqué et al., 2007). The NPN fraction of human milk contains significant amounts of PAs, mainly spermine and spermidine, which are higher in colostrum and transition milk compared to mature milk. The variable concentrations reported for human milk can be assigned to the post-partum period of time, time of delivery or nutritional mother status. For example, at 7 days post-partum, the range of the reported concentrations is 0.24–0.33, 2.20–7.11 and 2.76–3.13 μmol/L for PUT, SPD and SPM, respectively (Galitsopoulou et al., 2015).

**Table 6** Free nucleosides and nucleotides content (μmol/L) of mature milk (>15 days) - interspecies comparison, based on selected publications. CMP, cytidine 5′-monophosphate; UMP, uridine 5′-monophosphate; IMP, inosine 5′-monophosphate, GMP, guanosine 5′-monophosphate; AMP, adenosine 5′-monophosphate

| Component | Cow | Goat | Sheep | Camel | Donkey | Human |
|---|---|---|---|---|---|---|
| *Nucleosides* | | | | | | |
| Cytidine | 9 ± 1 [b]<br>7 ± 2-7 ± 0 [i] | 1.6 ± 1 [b]<br>8.8 ± 0.5-6.3 ± 0.1 [d]<br>2.26 [e] | 23 ± 0.1 [b]<br>6.7 ± 0.5-4.9 ± 0.2 [d]<br>7.4 [e] | 2.93 ± 0.09 [d] | 6.6 ± 1.7 [f] | 5 [k] |
| Uridine | 19 ± 1 [b]<br>13 ± 1-15 ± 5 [i] | 113 ± 4 [b]<br>76.3 ± 0.6-77.6 ± 0.2 [d]<br>39.01 [e] | 148 ± 11 [b]<br>67.8 ± 0.8-60.4 ± 0.3 [d]<br>115.3 [e] | 10.5 ± 0.19 [d] | 44.5 ± 1.8 [f] | 4 [k] |
| Inosine | | 60.6 ± 0.3-56.1 ± 0.2 [d]<br>3.87 [e] | 41.2 ± 0.3-34.4 ± 0.2 [d]<br>6.6 [e] | 9.3 ± 0.13 [d] | Trace [f] | |
| Guanosine | ND [b]<br>ND [i] | ND [b]<br>2.9 ± 0.01-3.1 ± 0.09 [d]<br>0.63 [e] | 6 ± 0 [b]<br>2.1 ± 0-2.1 ± 0.1 [d]<br>1.2 [e] | 2.65 ± 0.04 [d] | 20.7 ± 1.6 [f] | 1 [k] |
| Adenosine | ND [b]<br>ND-1 ± 0 [i] | ND [b]<br>2.4 ± 0.04-3.7 ± 0.07 [d]<br>8.56 [e] | ND [b]<br>8.8 ± 0.1-7.1 ± 0.1 [d]<br>1.9 [e] | 0.15 ± 0.01 [d] | 1.3 ± 0.08 [f] | 3 [k] |
| *Nucleotides* | | | | | | |
| Orotic acid | 163.8 [a] | 24.8 [a]<br>32.6 [e] | 37.5 [a]<br>55.2 [e] | | 0.65 ± 0.13 | |

(*Continued*)

**Table 6** *(Continued)*

| Component | Cow | Goat | Sheep | Camel | Donkey | Human |
|---|---|---|---|---|---|---|
| CMP | 45.4 ± 2.6 [a]<br>33 ± 1 [b]<br>30.3 ± 1.9 [c]<br>39 ± 0–30 ± 2 [i] | 49.9 ± 4.7 [a]<br>36 ± 2 [b]<br>82.6 ± 1.5 [c]<br>8.68 [e]<br>21.7 [g] | 90.2 ± 7.7 [a]<br>57 ± 3 [b]<br>55.4 ± 1.24 [c]<br>21.7 [e] | | 8.57 ± 3.76 [f] | ND–106 [h]<br>19 [k] |
| UMP | 29.2 ± 6.5 [a]<br>ND [b, c]<br>1 ± 1-1 ± 0 [i] | 123.3 ± 15.4 [a]<br>372 ± 8 [b]<br>259 ± 6.5 [c]<br>145.4 [e]<br>30.8 [g] | 258.5 ± 22 [a]<br>1874 ± 44 [b]<br>126.2 ± 7.1 [c]<br>250.3 [e] | | 23 ± 1.1 [f] | ND–21 [h]<br>13 [k] |
| IMP | | | | | | ND–8 [h] |
| GMP | ND [a,b,c,i] | ND [a]<br>94 ± 5 [b]<br>ND [c]<br>6.6 [e]<br>44.1 [g] | 3.4 ± 2.2 [a]<br>63 ± 0 [b]<br>ND [c]<br>4.1 [e] | | Trace [f] | ND–6 [h]<br>3 [k] |
| AMP | 34 ± 2.6 [a]<br>ND [b]<br>trace [c]<br>8.1 ± 1–4 ± 0 [i] | 50.8 ± 11.6 [a]<br>24 ± 2 [b]<br>96.2 ± 2.1 [c]<br>5.55 [e]<br>2.9 [g] | 120.8 ± 8.7 [a]<br>121 ± 0 [b]<br>60.8 ± 2.3 [c]<br>18.3 [e] | | ND [f] | ND–26 [h]<br>13 [k] |

[a] Gil and Sanchez-Medina, 1981 (enzymic methods); [b] Gill et al., 2012; [c] Ferreira et al., 2001; [d] Martin et al., 2016 ; [e] Plakantara et al., 2010; [f] Vincenzetti et al., 2014; [g] Prosser et al., 2008 (reconstituted whole goat milk powder); [h] Gill et al., 2011; Gill et al., 2011; [k] Martin et al., 2011a.

Due to their importance, PAs from the milk of ruminant or non-ruminant animals are suggested for the supplementation of infant and follow-up formulae. According to the limited number of relevant studies, their quantity and profile are differentiated among animal species. As cited in the reviews of Michaelidou (2008) and Giorgio et al. (2018), mature cow milk contains 0–1, 1–4.7 and 1–4.0 μM PUT, SPD and SPM respectively.

Płoszaj et al. (1997) report that the polyamine pattern of goat milk is related to breed, age, offspring number, lactation period, milking time and individual animal. They found significant differences in the average level of PUT and SPD in the milk of different goat breeds throughout lactation. The estimated concentrations were 6.00 ± 0.27 and 5.12 ± 0.28 for PUT, 39.67 ± 0.82 and 26.00 ± 0.85 μmol/L for SPD and 3.18 ± 0.14 and 3.80 ± 0.17 μmol/L for SPM. Galitsopoulou et al. (2015) reported total PAs content as 1.85 μmol/L for mature goat milk and as 0, 1.04 and 0.81 μmol/L for PUT, SPD and SPM, respectively. PAs in reconstituted whole goat milk powders ranged from 500 μg/L to 700 μg/L within the stage of lactation, while in the cow counterparts, their concentration was 0.041 mg/100 mL (Prosser et al., 2008). Very recently, Tolenaars et al. (2021) estimated the total PAs in goat milk as 664 ± 33 μg/L, consisting of 29 ± 4, 331 ± 26 and 304 ± 17 μg/L PUT, SPD and SPM, respectively. Galitsopoulou et al. (2015) estimated that mature sheep milk at 15 days post-partum contained 0.40, 2.05 and 2.39 μmol/L PUT, SPD and SPM, respectively, that is 4.85 μmol/L for total PAs, much higher than their findings for goat milk. Finally, extremely variable concentrations for PAs in donkey milk from animals of different ages were found by La Torre et al. (2010), that is, 5.79–32.78, 5.20–51.06 and 2.94–374.79 μg/L for SPD, SPM and PUT, respectively.

## 4 Mineral fraction: concentrations and distribution

The mineral content and profile of milk is correlated with the protein content and is affected by the animal breed, diet, stage of lactation, health condition of the udder and by the analytical procedure utilized for the determination. These factors explain the variability in the concentration of major and minor minerals and some trace elements of ruminants and equid milk presented in Table 7.

In brief, minerals or milk salts is a complex dynamic fraction of milk accounting for 8–9 g/L of cow milk (Gaucheron, 2005). It consists of cations and ions that are associated between themselves or with proteins. Sodium, potassium and chloride are diffusible and part of calcium, magnesium, phosphate and citrate are associated with casein. Approximately one-third of calcium, two-thirds of magnesium, half the inorganic phosphate and more than 90% of citrate are in the serum phase of cow milk. Minerals in milk serum exist as free ions or are associated with other ions. Limited changes in the physicochemical conditions

**Table 7** Milk minerals – interspecies comparison based on selected publications

| Minerals | Cow | Buffalo | Goat | Sheep | Yak | Camel | Donkey | Horse |
|---|---|---|---|---|---|---|---|---|
| *mg/100 g* | | | | | | | | |
| Ca | 120 [e]<br>104–128 [t] | 191 [k]<br>112–195 [f]<br>182.6 [gg] | 126 [e]<br>132 ± 1.9 [w]<br>134 [x]<br>128.8 ± 10.2 [z] | 195–200 [e]<br>197.5 ± 2.5 [y]<br>184.6 ± 17.8 [z] | 129 [k]<br>154.5 ± 14.6 [s]<br>152.5–154.5 [i]<br>208 ± 39 [cc]<br>148 [dd] | 137 ± 13.3 [b]<br>114–154 [k]<br>114 ± 13 [r]<br>106–157 [bb]<br>123 ± 4 [ii] | 67.8 [g]<br>91 [k]<br>54.4 ± 21.1 [l]<br>80.7 ± 13.8 [m]<br>53.5–94.7 [n] | 95 [k]<br>77.5 ± 12 [p]<br>50–130 [q] |
| P | 92 [e]<br>93–99 [t]<br>93 [ee] | 85.1–145 [f]<br>185 [k]<br>82.4 [gg] | 97 [e]<br>97.7 ± 1.3 [w]<br>121 [x]<br>100.9 ± 7.3 [z] | 124–158 [e]<br>141 ± 1.7 [y]<br>145.4 ± 9.3 [z] | 92.2–102.3 [i]<br>106 [k]<br>164 ± 0.8 [cc]<br>78 [dd] | 83 ± 15.5 [b]<br>86–131 [k]<br>92.2 ± 7.01 [s]<br>58–104 [bb]<br>102 ± 2 [ii] | 38.6 [g]<br>61 [k]<br>43.4 ± 9.5 [l]<br>63.8 ± 7 [m]<br>36.9–58.9 [n] | 58 [k]<br>21.1 ± 0.5 [p]<br>20–120 [q] |
| K | 150 [e]<br>135 [g]<br>121–168 [t]<br>155 [ee] | 112 [k]<br>91–181.5 [f]<br>106.6 [gg] | 190 [e]<br>181 [g]<br>152 ± 3.5 [w]<br>181 [x] | 136–140 [e]<br>124 [g]<br>138 ± 2 [y]<br>124.8 ± 15 [z] | 95 [k]<br>151 ± 2.8 [cc]<br>107 [dd] | 108.5 ± 31.1 [b]<br>151–186 [k]<br>156 ± 38 [r]<br>172 ± 6 [ii] | 50.8 [g]<br>50 [k]<br>110.3 ± 17.4 [l]<br>74.6 ± 8 [m]<br>40.5–58 [n] | 46.5 [g]<br>51 [k]<br>30–80 [q] |
| Mg | 11 [e]<br>9.7–14.6 [t]<br>11 [ee] | 8–39 [f]<br>12 [k]<br>18 [gg] | 13 [e]<br>15.9 ± 0.4 [w]<br>16 [x]<br>13.8 ± 1.7 [z] | 18–21 [e]<br>19.5 ± 3 [y]<br>19.2 ± 2.5 [z] | 10 [k]<br>15.1–15.4 [i]<br>8.5 ± 0.4 [cc]<br>6 [dd] | 9 ± 0.86 [b]<br>8.1–13 [k]<br>10.5 ± 1.8 [r]<br>15.4 ± 1.32 [s]<br>7.5–16 [bb]<br>9 ± 3 [ii] | 3.6 [g]<br>4 [k]<br>6.1 ± 1.1 [l]<br>8.2 ± 1.3 [m]<br>4.2–7.9 [n] | 3.9 [g]<br>7 [k]<br>2.75 ± 0.1 [p]<br>1.1–4 [q] |

| | | | | | | | | |
|---|---|---|---|---|---|---|---|---|
| Na | 45 [e]<br>55.7 [g]<br>39.1-64.4 [t]<br>43 [ee] | 35-95.5 [f]<br>47 [k]<br>43.9 [gg] | 38 [e]<br>47.2 [g]<br>59.4 ± 1.6 [w]<br>41 [x]<br>31.7 ± 3.5 [z] | 44-58 [e]<br>47.2 [g]<br>39 ± 7 [y]<br>44.2 ± 11.3 [z] | 29 [k]<br>28.5 ± 2.2 [cc]<br>48 [dd] | 57.5 ± 11.8 [b]<br>66-66.4 [k]<br>59 ± 16 [r]<br>36-62 [bb]<br>66 ± 2 [ii] | 21.9 [g]<br>22 [k]<br>43.8 ± 14.6 [l]<br>14 ± 1 [m]<br>15.7-24.4 [n] | 13.1 [g]<br>16 [k]<br>7-20 [q] |
| Cl | 107 [g]<br>110 [e]<br>77-121 [t]<br>89 [ee] | 57 [k]<br>58-75 [f]<br>57.7 [gg] | 160 [e]<br>121 [g]<br>150 [x]<br>175.5 ± 10.8 [z] | 110-112 [e]<br>60.3 [g]<br>119.6 ± 10.2 [z] | 200-280 [hh] | 142 [aa] | 33.7 [g]<br>34 [k]<br>36.6 ± 14.4 [l] | 23.4 [g]<br>19 [k]<br>20-60 [q] |
| Cit | 132-208 [t]<br>177 [g]<br>184 [hh] | 144-159 [f]<br>159.2 [gg] | 101.8 ± 11.3 [z]<br>103.7 [g] | 203.4 ± 153.5 [z]<br>94.1 [g] | 207-264 [hh] | 128 [aa]<br>182 ± 4 [ii] | | 44.2 ± 1.5 [p] |
| *μg/100 g* | | | | | | | | |
| Fe | 46 [e]<br>35.2 [u]<br>145 ± 52 [v]<br>30 [ff] | 170 [k]<br>42.1 [u]<br>101 ± 24 [v] | 55 [e]<br>46.3 [u]<br>108 ± 38 [v]<br>60 ± 3 [w] | 72-122 [e]<br>100 ± 0 [y] | 570 [k]<br>570 ± 40 [s]<br>570-1250 [i]<br>254 ± 64 [v] | 44 ± 0.8 [b]<br>210 [k]<br>29 ± 9 [r]<br>129 ± 73 [v] | 229 ± 17.2 [l]<br>43-188 [m]<br>10.1 ± 4.7 [o] | 22.4 [g]<br>100 [k]<br>22-146 [q] |
| Zn | 380 [e]<br>323 [u]<br>436 ± 127 [v]<br>400 [ff] | 500 [k]<br>463 [u]<br>400 ± 84 [v] | 340 [e]<br>295 [u]<br>311 ± 81 [v]<br>370 ± 10 [w]<br>560 [x] | 520-747 [e]<br>600 ± 100 [y] | 176-731 [i]<br>900 [k]<br>700 ± 44 [s]<br>476 ± 149 [v] | 58 ± 20 [b]<br>600-700 [k]<br>581 ± 114 [v] | 224 ± 70 [l]<br>199 ± 51 [m]<br>225 ± 27 [o] | 183.5 [g]<br>200 [k]<br>183.5 [p]<br>90-640 [q] |

(Continued)

**Table 7** (*Continued*)

| Minerals | Cow | Buffalo | Goat | Sheep | Yak | Camel | Donkey | Horse |
|---|---|---|---|---|---|---|---|---|
| Se | 3 [e]<br>2.1 [u]<br>3.7 ± 1.4 [v]<br>1 [ff] | 2 [u]<br>3.2 ± 1.1 [v] | 2 [e]<br>1.7 [u]<br>2.8 ± 1 [v]<br>1.33 [x] | 3.1 [e]<br>1.7 ± 1 [y] | 1.4 ± 0.53 [v] | 2.9 ± 1.8 [v] | 0.45 ± 0.14 [o] | |
| Cu | 22 [e]<br>2.1 [u]<br>16.5 ± 5.8 [v]<br>9 [ff] | 20 [k]<br>4.8 [u]<br>20.9 ± 9.3 [v] | 30 [e]<br>8.5 [u]<br>20.8 ± 9.8 [v]<br>50 [x] | 40–68 [e]<br>100 ± 0 [y] | 107 ± 8 [s]<br>45–65 [i]<br>52.2 ± 11.5 [v] | 24.8 ± 5.5 [v] | 16 ± 6 [m]<br>9.8 ± 2.9 [o] | 15.5 [g]<br>50 [k]<br>15.5 [p]<br>20–100 [q] |
| Mn | 6 [e]<br>2.3 [u]<br>18.7 ± 12.5 [v]<br>3 [ff] | 5.1 [u]<br>16.9 ± 4.9 [v] | 8 [e]<br>3.6 [u]<br>15.6 ± 3.1 [v]<br>6.5 ± 0.06 [w]<br>32 [x] | 5.3–9 [e]<br>7.15 ± 1.8 [y] | 6 ± 2 [i]<br>6 ± 1 [s]<br>25.6 ± 6 [v] | 5 ± 3 [r]<br>18.8 ± 10.9 [v] | 0.36 ± 0.13 [o] | 1.4 [g,p] |
| I | 7 [e]<br>10.77 [ff] | 4 [k] | 8 [e] | 10.4 [e] | | | | 0.4 [q] |
| Co | 0.03 [u]<br>0.05 [ff] | 0.04 [u] | 0.014 [u] | | | | | |

[a] Markiewicz-Keszycka et al., 2015 (depending on stage of lactation), [b] Haddadin et al., 2008 (depending on season), [c] Salimei and Fantuz, 2012, [d] Aspri et al., 2016, [e] Raynal-Ljutovac et al., 2008, [f] Abd-El-Salam and El-Shibiny, 2011, [g] Uniacke-Lowe and Fox, 2011, [h] Kulyar et al., 2021, Ma et al., 2017, [k] Medhammar et al., 2012, [l] Malacarne et al., 2019, [m] Fantuz et al., 2012, [n] Massouras et al., 2017, [o] Fantuz et al., 2009, [p] Uniacke-Lowe, 2011, [q] Doreau and Martin-Rosset, 2011; [r] Al haj and Al Kanhal, 2010; [s] Li et al., 2011; [t] Gaucheron, 2005; [u] Zhou et al., 2017; [v] Chen et al., 2020; [w] Kondyli et al., 2007 (depending on season); [x] Park, 2017; [y] Balthazar et al., 2017; [z] Mayer and Fiechter, 2012a; [aa] El-Agamy, 2017; [bb] Alhadrami, 2011; [cc] Zhang et al., 2020; [dd] Wiener, 2011; [ee] Cashman et al., 2011a; [ff] Cashman et al., 2011b; [gg] Sindhu and Arora, 2011; [hh] Wang et al., 2013; [ii] Attia et al., 2000.

such as temperature or pH modify their composition and partition between colloidal and serum phase that in turn affect the stability of casein micelles (Gaucheron, 2005). The nutritional significance of the inorganic fraction of milk is concisely presented by Cashman (2011a,b).

Among ruminants' milk presented in Table 7, buffalo and sheep milk exhibit the highest Ca, P and Mg contents in accordance with their high casein content. Potassium is the most abundant mineral in cow, goat and camel milk, while goat and camel milk contains the highest amount of chloride. Also, differences are observed in the Ca to P or K to Na ratios among various milk kinds. Differences have been also reported for the minor and trace elements. Cu, Fe, Cr and Ni were found in higher concentrations in cow milk than in sheep and goat milk and the opposite holds true for Ba and Sr (Coni et al., 1996). Ruminants' milk are not Fe sources, but among them, yak milk is the richest in Fe. It has been found that Fe, Mn and Se contents of various ruminants' milk are related to those in drinking water while Cu and Se were related to those in feed (Zhou et al., 2017).

The results of Holt and Jenness (1984) showed that common general principles determine the partition of salts in various milk kinds and that calcium phosphate in milk is proportional to the casein content of each milk kind. However, differences in the partition of minerals of various milk kinds have been reported, for example, colloidal Ca in yak, camel, donkey and horse milk is 58%, 70%, 45% and 60% of total Ca, respectively (Faccia et al., 2020). Aspects related to casein micelle mineralization of different milk kinds are presented in Section 3.3.1.

The minerals of buffalo milk are on average 50% more compared to cow milk and their distribution is different. According to the comparative study of Ahmad et al. (2009), total Ca in buffalo milk is twice that in cow milk, that is, 65.2 and 33.1 mM, respectively; moreover, buffalo milk contains 32.7 mM Pi, whereas cow milk has 20.8 mM. Different distributions between aqueous and colloidal phase were observed; 11.2% of Ca and 33.3% of Pi are in the aqueous phase of buffalo milk, which is much lower than the respective 26.2% and 52.4% in cow counterparts. Moreover, they report that the molar Ca to Pi ratio in buffalo milk is 1.99 higher than the 1.59 in cow milk. The higher casein and Pi content results in a higher buffering capacity of buffalo milk (Ahmad et al., 2008). A large proportion 67.6–82.6% of total Ca is in the insoluble form apparently due to high casein content; approximately 60% of P is colloidal P, soluble Mg is ~50%, while Na, K, Cl are completely present in the soluble phase (Abd-El-Salam and El-Shibiny, 2011). Accordingly, 21.8% of total Ca and 31.8% of total P are soluble in buffalo milk, whereas 72% of the citrate was soluble that is lower than the 91.5% soluble citrate of cow milk (Sindhu and Arora, 2011).

In general, goat milk contains more calcium, phosphorus, potassium, magnesium and chloride and less sodium than cow milk (Park, 2017). The soluble calcium in sheep milk has been estimated as 27.5% of the total that is lower than the 37.3% and 34.8% estimated for cow and goat milk, respectively (Raynal and Remeuf, 1998) in accordance with the findings of Holt and Jenness (1984). However, a lower portion of soluble calcium has been reported for heat-stable goat milk by Morgan et al. (2000); on average, 28% of total calcium was soluble against 31% in heat-unstable counterparts.

Yaks are natural grazers on the Qinghai-Tibetan plateau, where the intense monthly differences in feeding contribute to the variability of the mineral concentration of their milk that ranges from 0.57 mg/kg to 1.25 mg/kg (Ma et al., 2017). Li et al. (2011) reported calcium content of 1545.45 ± 145.61 mg/kg for yak milk which is higher compared to cow counterparts, but they found that total phosphorus content, that is, 90.3–94.5 mg/100 g, was similar for these two milk kinds. According to Zhang et al. (2017), yak and cow milk have similar ash content that is close to 0.8%, but the mean content of micellar calcium and phosphorus is 104 mg/100 mL and 62 mg/100 mL, which is higher than the respective 93 mg/100 mL and 37 mg/100 mL estimated for cow milk. On average, 57.7% of calcium in yak milk has been found in the micellar form that is much lower than the 71.2% estimated for cow milk in the comparative study of Wang et al. (2013). Higher mean percentages of colloidal magnesium than in cow milk - 35.2% vs 25% - and of colloidal phosphorus - 56.5% vs 51.3% - were also observed in this study.

Ca, P and Mg contents close to and Na, K, Fe, Cu and Mn contents higher than those in cow milk are reported for camel milk (Al haj and Al Kanhal, 2010). El-Agamy (2017) suggest that the low citrate content of camel milk enhances lactoferrin activity, and reported a Ca to P ratio equal to 1.5 and 1.29 for camel and cow milk, respectively. The 65% of calcium is in the micellar phase of camel milk similarly to cow milk, but 63.3%, 65.6% and 31.8% of the Mg, P and citrate are in the colloidal phase of camel milk (Attia et al., 2000).

The findings of Fantuz et al. (2012) show that the concentration of macrominerals in donkey milk is lower than that in ruminants milk but higher than that in human milk. Moreover, the Ca/P ratio is lower than that in human milk and similar to that in cow and goat milk, while the order of their concentration is Ca > K > P > Na > Mg. A significant effect of the stage of lactation was observed. Donkey contains a lower concentration of Zn, Fe, Mn and Se compared to cow and goat milk (Fantuz et al., 2013). On the other hand, Malacarne et al. (2019) report that the most abundant mineral is K followed by Ca, Na and P that are significantly affected by the lactation stage. In addition, they found low concentrations of microelements Fe, Zn and Cu at <2.5 mg/kg. The highest concentration for K and similar concentrations for Na and Ca have been also reported by Massouras et al. (2017) who observed significant variation of the Ca/P and Na/K ratios during the lactation period. The 62.9% of Ca, 53.1% of P and 32.6% of Mg in donkey milk are associated with casein. Colloidal Ca and P in mmol/g casein is twice that in cow milk while pH reduction caused an increase of P in the ultrafiltrate (Fantuz et al., 2020). The findings of Malacarne et al. (2019) are similar for Ca and Mg distribution, but 64% of P has been found to be associated with casein; the colloidal Ca to colloidal Pi molar ratio was 1.51, and a significant decrease of mineralization has been observed with the progress of lactation.

## 5 Serum compounds: lactose, oligosaccharides and vitamins

The major carbohydrate of milk is lactose, or 0-4-D-galactopyranosyl-(1,4)-glucopyranose, which is a reducing sugar unique to the milk of mammals. The osmotic pressure of milk that is isotonic to blood is mainly determined by lactose and dissolved salts. Cow milk contains traces of glucose and galactose but no polysaccharides. Lactose is synthesized exclusively in the Golgi vesicles of the lactating cells due to the presence of α-lactalbumin, which is also unique to milk (Walstra et al., 2006). The synthesis of lactose affects the volume of milk and the concentration of casein by drawing water osmotically into the Golgi vesicles (Fox, 2011). Nutritional and technological significance of lactose is concisely presented by Schaafsma (2008) and Gänzle et al. (2008).

Lactose is the major solid constituent of most milk kinds. However, the most abundant solid compound in buffalo and yak milk is fat followed by lactose, while fat and protein contents are higher than that of lactose in sheep milk (Table 1). The highest lactose content in Table 1 is that of donkey and horse milk that is similar to the 6.3–7% fat content of human milk (Claeys et al., 2014). The concentration of lactose increases slightly during the early stages of lactation, but at the end of lactation, it is about 70% of the maximum value. On the contrary, the lactose content of the equid milk (horse and donkey) increases during lactation, reaching values in the range of 6.0–7.4% (Fox, 2011).

Lactose occupies the reducing end of the milk oligosaccharides (OSs) that are formed by 3–10 linear or branched monosaccharides and usually contain fucose and/or N-acetylneuraminic acid (Fox, 2011). OSs are minor compounds of ruminants' milk, but they consist of a great part of human milk components, that is, 12–13 g/L and 22–24 g/L for mature milk and colostrum, respectively. They are considered as prebiotics and nutrients for brain development and can act as antiadhesive antimicrobials and immune modulators. Therefore, they are of great importance for the development of newborn, and their concentration is higher in the colostrum than in mature milk. There are excellent comprehensive (e.g. Bode, 2012; Urashima et al., 2011, 2013; Hobbs et al., 2021) or brief reviews (e.g. Giorgio et al., 2018 or Sousa et al., 2019b) on the biofunctionality of human milk OS fraction. Milk OSs are either neutral or acidic, which contain sialic acid, that is, N-acetylneuraminic acid often present in the milk of all mammals and N-glycolylneuraminic acid present in that of non-human mammals (Yan et al., 2018). The human OSs are produced by the addition of Neu5Ac and/or L-fucose (Fuc) residues to type I unit (Gal(1–3)GlcNAc) and type II unit (Gal(β1–4)GlcNAc), while OSs from other milk kinds mostly have type II unit (Urashima et al., 2011). OS concentrations in milk of cow, goat and sheep are 100–1000-fold lower than that in human milk, with less structure diversity and more sialylated and less fucosylated OSs (Bode, 2012).

The suitability of OS fraction of milk of non-human mammals is constantly studied during the last years - often comparatively - in relation to the development of infant formulae. The findings are not consistent as many factors affect the results such as lactation stage, genetic variability or analysis method.

Abd El-Salam and El-Shibiny (2011) report higher OS content of buffalo milk colostrum and milk during lactation than that of cow milk, that is, 26 mg/100 mL and 17.2–10.2 mg/100 mL for buffalo colostrum and milk vs 20.3 mg/100 mL and 14.7–8.7 mg/100 mL for cow counterparts. The findings of Wang et al. (2020) indicate that the OSs in yak milk are very close to those in cow milk in terms of structure and content.

The difference in sialylated OSs between human and animal milk is the lack of mono- and disialyated lacto-N-tetraose (LNT and DSLNT) in the latter, along with the lower diversity, the lack of fucosylation and the short chain lengths observed in animal milk (Yan et al., 2018); however, Wang et al. (2020) detected DSLNT in goat milk. The fine composition of sialic acid is differentiated among various milk kinds, that is, the ratio of Neu5Ac/Neu5Gc in goat milk is 36/64, in sheep milk is 6/94, in cow milk is 97/3, while in camel and horse milk are 99.7/0.3 and 99/1, respectively (Albrecht et al., 2014).

Of particular interest is the OS fraction of goat milk. The concentration of OSs in goat milk is reported to be 250–300 mg/L that is much higher than that in cow - 30–60 mg/L - or in sheep - 20–40 mg/L - milk (Urashima et al., 2011). According to several publications recently reviewed in detail by van Leeuwen et al. (2020), the goat milk OS fraction ranges from 60 mg/L to 350 mg/L in mature milk and from 200 mg/L to 650 mg/L in colostrum and contains structures that are in common with the human counterpart. Reported common neutral structures are the 2′-fucosyllactose (2′-FL), 3′-galactosylactose (3′-GL), 6′ galactosylactose (6′-GL), lacto-N-triose, lacto-N-neotraose and lacto-N-neohexanose (LNnH), and common acidic structures are the 3′-sialyllactose (3′-SL), 6′-sialyllactose (6′-SL) and sialyl-lacto-N-neohexanose (S-LNnH). The findings of Wang et al. (2020) show that the neutral OS 2′-FL being abundant in human milk was found only in goat milk and not in cow, donkey or yak counterparts.

Four times as much sialic acid has been estimated in goat milk as in cow milk, that is, ~230 vs 60 mg sialic acid/kg fresh milk (Amigo and Fontecha, 2011). Recently, total sialic acid content of goat milk has been reported as 19.4 ± 0.5 mg/100 mL, consisted of 5.6 ± 0.3 mg/100 mL Neu5Ac and 13.8 ± 0.5 mg/100 mL Neu5Gc (Tolenaars et al., 2021). Considering the pattern and quantity of

OS fraction along with the qualitative characteristics of fat, protein and NPN fraction, goat milk is suggested as an adequate base for the infant formulae (Prosser, 2021).

Two neutral OSs and five acidic OSs have been characterized in market dromedary camel milk (Alhaj et al., 2013). More neutral OSs have been identified in camel milk (16%) compared to horse (10%), sheep (14%), goat (5%) and cow (9%) milk, while the respective content in human milk was 70% (Albrecht et al., 2014). Yan et al. (2018) concluded that there are similarities in the sialyllactose pattern of camel and buffalo milk.

Srivastava et al. (2014) isolated 0.68, 0.51 and 0.49 g/L OSs, from horse colostrum, transition and mature milk, respectively. Monti et al. (2015) reported high variability in the content (intra- and inter-species) and the pattern (intra-species) of the sialylated OSs 3′-SL, 6′-SL and disialyl-lacto-N-tetraose (DSLNT) among human, equids and ruminants milk. On average, horse milk contained 68.5, donkey 14.5, goat 67.4, cow 27.9 and human >196 μg/mL 3′-SL. The 6′-SL being >46 μg/mL in human milk, 18.7 in donkey milk and 2 in mare milk was not detected in ruminants milk. The DSLNT being >450 μg/mL in human milk was found in very low and non-detectable amounts in equids and ruminants milk, respectively. The same trend in the profiles of horse and donkey milk was observed later by Orlandi et al. (2016). In a more detailed study, Licitra et al. (2019) studied five donkey milk sialylated OSs and report that their total concentration decreased from 39 mg/L at 15 days to 24.5 mg/L at 60 days of lactation, and they suggest donkey milk as a promising source of sialytated OSs, in particular of 6′-SL. Also, the findings of Yan et al. (2018) show that only donkey milk among the milk from eight farm animals show 3'-SL/6'-SL pattern similar to human milk, with more 6'-SL than 3'-SL.

It is well-known that milk is a good source of vitamins. Fat-soluble vitamins belonging to milk lipids have been presented in Section 3.2.3 (Table 4). Various reports for the content of different milk kinds in water-soluble vitamins are presented in Table 8. Of particular importance is the existence in milk of specific vitamin-binding proteins connected to folate, vitamin D, vitamin B12 and riboflavin. Their aim is to inhibit the microbial uptake of vitamins and to facilitate their cellular uptake and absorption in various parts of the gastrointestinal tract (Wynn et al., 2011). Folate from blood plasma is sequestered in milk by the soluble folate-binding protein, which is a soluble whey protein (Nygren-Babol and Jägerstad, 2012). Milk is an important source of vitamin B12, which is obtained from food of animal origin. B12 vitaminers in milk are bound to specific mammalian B12-binding proteins for transport, for example, transcobalamin. Although cobalamin concentration in milk is lower compared to meat, it exhibits higher bioavailability (Li et al., 2019).

Despite the great variability observed in the findings of various studies, some conclusions can be drawn from Table 8. Among ruminants, the reference

**Table 8** Water-soluble vitamins – interspecies comparison, based on selected publications

| Vitamins | Cow | Buffalo | Goat | Sheep | Yak | Camel | Donkey | Horse |
|---|---|---|---|---|---|---|---|---|
| Ascorbic acid, mg/100 g (vitamin C) | 1 [e]<br>0.94 [x] | 2.5 [k]<br>0.86-2.03 [f]<br>2.2 [ff] | 1.3 [e]<br>5.48 ± 0.09 [w]<br>1.29 [x] | 5 [e]<br>4.6 ± 0.4 [y] | | 3.3 ± 0.17 [b]<br>6.7 [k]<br>2.4-3.6 [bb] | 0.35-0.5 [d] | 5-15 [a]<br>1.7-14.7 [c]<br>15 [g]<br>4.3 [k] |
| Thiamin, µg/100 g (vitamin B1) | 40 [e]<br>45 [x] | 38.7-53 [f]<br>50 [k]<br>20-50 [ff] | 50 [e]<br>260 ± 16 [w]<br>68 [x] | 80 [e]<br>70 [y] | 36 [h]<br>23.6-48.5 [i] | 48 ± 7 [b]<br>33-60 [bb] | 41 [d]<br>60 [k]<br>41 [g] | 30 [g, k] |
| Riboflavin, µg/100 g (vitamin B2) | 170 [e]<br>160 [x] | 132-183 [f]<br>110 [k]<br>159 [ff] | 140 [e]<br>112 ± 3 [w]<br>210 [x] | 350 [e]<br>300 ± 20 [y] | 50 [h]<br>79.5-20.8 [i] | 168 ± 14 [b]<br>60-120 [k]<br>42-80 [bb] | 64 [d]<br>30 [k]<br>64 [g] | 37 [c]<br>30 [g]<br>20 [k] |
| Niacin, µg/100 g (vitamin B3) | 90 [e]<br>80 [x] | 170 [k] | 200 [e]<br>270 [x] | 420 [e]<br>410 ± 50 [y] | 2.5 [h]<br>2.6-1.6 [i] | 78 ± 10 [b]<br>460 [bb] | 74 [d]<br>90 [k] | 70 [k]<br>140 [g] |
| Pantothenic acid, µg/100 g (vitamin B5) | 340 [e]<br>320 [x] | | 310 [e,x]<br>150 [k] | 410 [e]<br>430 ± 20 [y] | | 368 ± 21.3 [b]<br>88 [bb] | | |
| Pyridoxin, µg/100 g (vitamin B6) | 40 [e]<br>42 [x] | 330 [k]<br>325 [ff] | 50 [e]<br>46 [x] | 80 [e]<br>70 ± 10 [y] | 20.7 [h]<br>0.4-40.8 [i] | 55 ± 8.1 [b]<br>52 [bb] | | 30 [g] |
| Biotin, µg/100 g (vitamin B7) | 2 [e,x] | | 2 [e]<br>1.5 [x] | ND [e]<br>2.5 ± 0 [y] | | | | |
| Cobalamin, µg/100 g (vitamin B12) | 0.35 [e,x] | 0.11-0.5 [f]<br>0.40 [k] | 0.06 [e,x] | 0.71 [e]<br>0.66 ± 0.05 [y] | | 0.85 ± 0.024 [b]<br>0.2 [bb] | 110 [d,g] | 0.3 [g] |
| Folic acid, µg/100 g | 5.3 [e]<br>5 [x] | 1.82-3.78 [f]<br>0.6 [k] | 1 [e,x] | 5 [e]<br>6 ± 0.06 [y] | | 8.7 ± 1.5 [b]<br>0.4 [bb] | | |

[a] Markiewicz-Keszycka et al., 2015 (depending on stage of lactation), [b] Haddadin et al., 2008; [c] Salimei and Fantuz, 2012, [d] Aspri et al., 2016, [e] Raynal-Ljutovac et al., 2008, [f] Abd-El-Salam and El-Shibiny, 2011, [g] Uniacke-Lowe and Fox, 2011, [h] Kulyar et al., 2021, Ma et al., 2017, [k] Medhammar et al., 2012, [l] Malacarne et al., 2019, [m] Fantuz et al., 2012, [n] Massouras et al., 2017, [o] Fantuz et al., 2009, [p] Uniacke-Lowe, 2011, [q] Doreau and Martin-Rosset, 2011; [r] Al haj and Al Kanhal, 2010; [s] Li et al., 2011; [t] Gaucheron, 2005; [u] Zhou et al., 2017; [v] Chen et al., 2020; [w] Kondyli et al., 2007 (depending on season); [x] Park, 2017; [y] Balthazar et al., 2017; [z] Mayer and Fiechter, 2012a; [aa] El-Agamy, 2017; [bb] Alhadrami, 2011; [cc] Zhang et al., 2020; [dd] Wiener, 2011; [ee] Cashman et al., 2011a; [ff] Cashman et al., 2011b; [gg] Sindhu and Arora, 2011; [hh] Wang et al., 2013; [ii] Attia et al., 2000.

cow milk has the lowest quantity of vitamin C, whereas camel milk contains considerable amounts of this vitamin. The folic acid content of goat milk is the lowest, and buffalo milk is superior in pyridoxin. Sheep milk has the highest cobalamin content among other ruminants milk, which however is much lower than that reported for donkey milk.

## 6 Conclusion and future trends

The developed research methodology for the study of cow milk has been utilized in recent years for the study of non-cow milk. The findings reviewed in this chapter - although limited for the most milk kinds - highlight important peculiarities regarding the fine composition, fat emulsion, colloidal phase, technological behaviour, nutritional quality and biological value.

Future perspectives arising from the interspecies comparison of milk are related to the utilization in human nutrition of milk from various farm animals and isolated fractions and constituents As a result, new activities will emerge for food and nutrition science and for all types and sizes of a sustainable dairy sector. Moreover, there will be a positive effect on biodiversity and climate change issues. In particular, milk from ruminants other than cow and from non-ruminants can be excellent bases for specific types of dairy products - apart from beverage milk - with great importance for niche or local markets. At the same time, they can be sources of valuable biofunctional ingredients for the globalized dairy and nutraceutical industry. Thorough knowledge of the composition, structure and nutritional potential and of the influence of genetic, physiological and environmental factors on them is a prerequisite. The same holds true for the dairy streams formed during product manufacture from non-cow milk kinds. The application of powerful omics analytical tools and strategies (e.g. Scano et al., 2019; Manoni et al., 2020; Verma et al., 2020) can provide more and detailed information about these aspects.

On the other hand, technological experimental studies to investigate the ability of non-cow milk kinds to be bases for various categories of dairy products are needed. In this respect, the potential of non-thermal treatments (e.g. Huppertz, 2016; Nunes and Tavares, 2019) to ensure hygiene conditions without affecting the enhanced biofunctionality of particular milk kinds should be considered. Studies on the digestion pattern of various kinds of milk and products made therefrom at various stages of life (e.g. Hernández-Olivas et al., 2020; Aalaei et al., 2021; Phosanam et al., 2021) are also necessary. The development of separation and isolation protocols yielding efficiently minor fractions of milk, which can be incorporated in infant or athlete or senior formulae, without loss of their biological value is a further topic of interest, for example, OS fraction (e.g. Martín-Ortiz et al., 2019; Sousa et al., 2019a) or MFGM material (e.g. Fontecha et al., 2020; Mohan et al., 2020; Silva et al., 2021).

It is evident that much more and multidisciplinary research is needed to assess the characteristics and potential of each milk kind. The outcomes are expected to contribute to scientific and sustainable social and economic development worldwide.

## 7 References

Aalaei, K., Khakimov, B., De Gobba, C. and Ahrné, L. (2021). Digestion patterns of proteins in pasteurized and ultra-high temperature milk using in vitro gastric models of adult and elderly, *J. Food Eng.* 292, 110305.

Abd El-Salam, M. H. and El-Shibiny, S. (2011). A comprehensive review on the composition and properties of buffalo milk, *Dairy Science & Technol.* 91(6), 663–699.

Ahmad, S., Anjum, F. M., Huma, N., Sameen, A. and Zahoor, T. (2013). Composition and physico-chemical characteristics of buffalo milk with particular emphasis on lipids, proteins, minerals, enzymes and vitamins, *J. Anim. Plant Sci.* 23(1), 62–74.

Ahmad, S., Gaucher, I., Rousseau, F., Beaucher, E., Piot, M., Grongnet, J. F. and Gaucheron, F. (2008). Effects of acidification on physico-chemical characteristics of buffalo milk: a comparison with cow's milk, *Food Chem.* 106(1), 11–17.

Ahmad, S., Piot, M., Rousseau, F., Grongnet, J. F. and Gaucheron, F. (2009). Physico-chemical changes in casein micelles of buffalo and cow milks as a function of alkalinisation, *Dairy Sci. Technol.* 89(3–4), 387–403.

Al haj, O. A. and Al Kanhal, H. A. (2010). Compositional, technological and nutritional aspects of dromedary camel milk, *Int. Dairy J.* 20(12), 811–821.

Albenzio, M., Caroprese, M., Santillo, A., Marino, R., Muscio, A. and Sevi, A. (2005). Proteolytic patterns and plasmin activity in ewes' milk as affected by somatic cell count and stage of lactation, *J. Dairy Res.* 72(1), 86–92.

Albenzio, M. and Santillo, A. (2011). Biochemical characteristics of ewe and goat milk: effect on the quality of dairy products, *Small Rumin. Res.* 101(1–3), 33–40.

Albenzio, M., Santillo, A., Caroprese, M., d'Angelo, F., Marino, R. and Sevi, A. (2009). Role of endogenous enzymes in proteolysis of sheep milk, *J. Dairy Sci.* 92(1), 79–86.

Albenzio, M., Santillo, A. and Moatsou, G. (2021). The enzymology of non-bovine milks. In: Kelly, A. and Bach Larsen, L. (Eds), *Agents of Change: Enzymes in Milk and Dairy Products*, Springer Nature, Switzerland, pp. 181–208.

Albrecht, S., Lane, J. A., Marino, K., Al Busadah, K. A., Carrington, S. D., Hickey, R. M. and Rudd, P. M. (2014). A comparative study of free oligosaccharides in the milk of domestic animals, *Br. J. Nutr.* 111(7), 1313–1328.

Alhadrami, G. A. (2011). Dairy animals: camel. In: Fuquay, J. W., Fox, P. F. and McSweeney, P. L. H. (Eds), *Encyclopedia of Dairy Sciences*, Academic Press, San Diego, CA, pp. 1:351–357.

Alhaj, O. A., Taufik, E., Handa, Y., Fukuda, K., Saito, T. and Urashima, T. (2013). Chemical characterisation of oligosaccharides in commercially pasteurised dromedary camel (Camelus dromedarius) milk, *Int. Dairy J.* 28(2), 70–75.

Ali, A. H., Zou, X., Huang, J., Abed, S. M., Tao, G., Jin, Q. and Wang, X. (2017). Profiling of phospholipids molecular species from different mammalian milk powders by using ultra-performance liquid chromatography-electrospray ionization-quadrupole-time of flight-mass spectrometry, *J. Food Comp. Anal.* 62, 143–154.

Alichanidis, E., Moatsou, G. and Polychroniadou, A. (2016). The composition and the properties of non-cow milks and products. In: Tsakalidou, E. and Papadimitriou, K. (Eds), *Non-Bovine Milk and Milk Products.*, Academic Press, UK, pp. 81–116.

Altomonte, I., Salari, F., Licitra, R. and Martini, M. (2019). Donkey and human milk: insights into their compositional similarities, *Int. Dairy J.* 89(2), 111–118.

Amigo, L. and Fontecha, J. (2011). Goat milk. In: Fuquay, J. W., Fox, P. F. and McSweeney, P. L. H. (Eds), *Encyclopedia of Dairy Sciences*, Academic Press, San Diego, CA, pp. 3:484–493.

Argov-Argaman, N. (2019). Symposium review: milk fat globule size: practical implications and metabolic regulation, *J. Dairy Sci.* 102(3), 2783–2795.

Argov-Argaman, N., Hadaya, O., Glasser, T., Muklada, H., Dvash, L., Mesilati-Stahy, R. and Yan Landau, S. Y. (2016). Milk fat globule size, phospholipid contents and composition of milk from purebred and Alpine-crossbred Mid-Eastern goats under confinement or grazing condition, *Int. Dairy J.* 58, 2–8.

Argov-Argaman, N., Mida, K., Cohen, B. C., Visker, M. and Hettinga, K. (2013). Milk fat content and DGAT1 genotype determine lipid composition of the milk fat globule membrane, *PLoS ONE* 8(7), e68707.

Aspri, M., Economou, N. and Papademas, P. (2016). Donkey milk: an overview on functionality, technology, and future prospects, *Food Rev. Int.* 33(3), 316–333.

Atmani, D., Banghiani, A., Harrison, R. and Benboubetra, M. (2005). NADH oxidation and superoxide production by caprine milk oxidoreductase, *Int. Dairy J.* 15(11), 1113–1121.

Atmani, D., Benboubetra, M. and Harrison, R. (2004). Goat's milk xanthine oxidoreductase is grossly deficient in molybdenum, *J. Dairy Res.* 71(1), 7–13.

Attia, H., Kherouatou, N., Nasri, M. and Khorchani, T. (2000). Characterization of the dromedary milk casein micelle and study of its changes during acidification, *Lait* 80(5), 503–515.

Avondo, M., Pennisi, P., Lanza, M., Pagano, R. I., Valenti, B., Di Gregorio, P., De Angelis, A., Giorgio, D. and Di Trana, A. (2015). Effect of the αs1-casein genotype and its interaction with diet degradability on milk production, milk quality, metabolic and endocrinal response of Girgentana goats, *Small Rumin. Res.* 123(1), 136–141.

Baer, A., Ryba, I. and Farah, Z. (1994). Plasmin activity in camel milk, *LWT Food Sci. Technol.* 27(6), 595–598.

Baghiani, A., Harrison, R. and Benboubetra, M. (2003). Purification and partial characterisation of camel milk xanthine oxidoreductase, *Arch. Physiol. Biochem.* 111(5), 407–414.

Balthazar, C. F., Pimentel, T. C., Ferrão, L. L., Almada, C. N., Santillo, A., Albenzio, M., Mollakhalili, N., Mortazavian, A. M., Nascimento, J. S., Silva, M. C., Freitas, M. Q., Sant'Ana, A. S., Granato, D. and Cruz, A. G. (2017). Sheep milk: physicochemical characteristics and relevance for functional food development, *Compr. Rev. Food Sci. Food Saf.* 16(2), 247–262.

Barello, C., Garoffo, L. P., Montorfano, G., Zava, S., Berra, B., Conti, A. and Giuffrida, M. G. (2008). Analysis of major proteins and fat fractions associated with mare's milk fat globules, *Mol. Nutr. Food Res.* 52(12), 1448–1456.

Barlowska, J., Szwajkowska, M., Litwinczuk, Z. and Krol, J. (2011). Nutritional value and technological suitability of milk from various animal species used for dairy production, *Compr. Rev. Food Sci. Food Saf.* 10(6), 291–302.

Benboubetra, M., Baghiani, A., Atmani, D. and Harrison, R. (2004). Physicochemical and kinetic properties of purified sheep's milk xanthine oxidoreductase, *J. Dairy Sci.* 87(6), 1580–1584.

Bernard, L., Bonnet, M., Delavaud, C., Delosière, M., Ferlay, A., Fougère, H. and Graulet, B. (2018). Milk fat globule in ruminant: major and minor compounds, nutritional regulation and differences among species, *Eur. J. Lipid Sci. Technol.* 120(5), 1700039.

Berton, A., Rouvellac, S., Robert, B., Rousseau, F., Lopez, C. and Crenon, I. (2012). Effect of the size and interface composition of milk fat globules on their in vitro digestion by the human pancreatic lipase: native versus homogenized milk fat globules, *Food Hydrocoll.* 29(1), 123–134.

Bode, L. (2012). Human milk oligosaccharides: every baby needs a sugar mama, *Glycobiology* 22(9), 1147–1162.

Bonczar, G., Walczycka, M. B., Domagała, J., Maciejowski, K., Najgebauer-Lejko, D., Sady,M. and Wszołek, M. (2016). Effect of dairy animal species and of the type of starter cultures on the cholesterol content of manufactured fermented milks, *Small Rumin. Res.* 136(1), 22–26.

Bornaz, S., Sahli, A., Attalah, A. and Attia, H. (2009). Physicochemical characteristics and renneting properties of camels' milk: a comparison with goats', ewes' and cows' milks, *Int. J. Dairy Technol.* 62(4), 505–513.

Caroprese, M., Marzano, A., Schena, L., Marino, R., Santillo, A. and Albenzio, M. (2007). Contribution of macrophages to proteolysis and plasmin activity in ewe bulk milk, *J. Dairy Sci.* 90(6), 2767–2772.

Cashman, K. D. (2011a). Milk salts: macroelements, nutritional significance. In: Fuquay, J. W., Fox, P. F. and McSweeney, P. L. H. (Eds), *Encyclopedia of Dairy Sciences*, Academic Press, San Diego, CA, pp. 3:925–932.

Cashman, K. D. (2011b). Milk salts: trace elements, nutritional significance. In: Fuquay, J. W., Fox, P. F. and McSweeney, P. L. H. (Eds), *Encyclopedia of Dairy Sciences*, Academic Press, San Diego, CA, pp. 933–940.

Castro-Gomez, M. P., Rodriguez-Alcala, L. M., Calvo, M. V., Romero, J., Mendiola, J. A., Ibanez, E. and Fontecha, J. (2014). Total milk fat extraction and quantification of polar and neutral lipids of cow, goat, and ewe milk by using a pressurized liquid system and chromatographic techniques, *J. Dairy Sci.* 97(11), 6719–6728.

Cebo, C., Lopez, C., Henry, C., Beauvallet, C., Ménard, O., Bevilacqua, C., Bouvier, F., Caillat, H. and Martin, P. (2012a). Goat αs1-casein genotype affects milk fat globule physicochemical properties and the composition of the milk fat globule membrane, *J. Dairy Sci.* 95(11), 6215–6229.

Cebo, C., Rebours, E., Henry, C., Makhzami, S., Cosette, P. and Martin, P. (2012b). Identification of major milk fat globule membrane proteins from pony mare milk highlights the molecular diversity of lactadherin across species, *J. Dairy Sci.* 95(3), 1085–1098.

Chandan, R. C., Parry, R. M. Jr. and Shahani, K. M. (1968). Lysozyme, lipase and ribonuclease in milk of various species, *J. Dairy Sci.* 51(4), 606–607.

Chavarri, F., Santistebam, A., Virto, M. and de Renobales, M. (1998). Alkaline phosphatase, acid phosphatase, lactoperoxidase and lipoprotein lipase activities in industrial ewe's milk and cheese, *J. Agric. Food Chem.* 46(8), 2926–2932.

Chen, L., Li, X., Li, Z. and Deng, L. (2020). Analysis of 17 elements in cow, goat, buffalo, yak, and camel milk by inductively coupled plasma mass spectrometry (ICP-MS), *RSC Adv.* 10(12), 6736–6742.

Chilliard,Y. and Doreau, M. (1985). Characterization of lipase in mare milk, *J. Dairy Sci.* 68(1), 37–39.

Chilliard, Y., Ferlay, A., Rouel, J. and Lamberet, G. (2003). A review of nutritional and physiological factors affecting goat milk synthesis and lipolysis, *J. Dairy Sci.* 86(5), 1751–1770.

Chilliard, Y., Selselet-Attou, P. B. and Morand-Fehr, P. (1984). Characteristics of the lipolytic system in goat milk, *J. Dairy Sci.* 67(10), 2216–2223.

Civardi, G., Curadi, M. C., Orlandi, M., Cattaneo, T. M. P. and Giangiacomo, R. (2007). Mare's milk: monitoring the effect of thermal treatments on whey proteins stability by SDS-capillary electrophoresis (CE-SDS), *Milchwissenschaft* 62(1), 32–33.

Claeys, W. L., Verraes, C., Cardoen, S., De Block, J., Huyghebaert, A., Raes, K., Dewettinck, K. and Herman, L. (2014). Consumption of raw or heated milk from different species: an evaluation of the nutritional and potential health benefits', *Food Control* 42(1), 188–201.

Coni, E., Bocca, A., Coppolelli, P., Caroli, S., Cavallucci, C. and Marinucci, M. T. (1996). Minor and trace element content in sheep and goat milk and dairy products, *Food Chem.* 57(2), 253–260.

Contarini, G., Pelizzola, V., Scurati, S. and Povolo, M. (2017). Polar lipid of donkey milk fat: phospholipid, ceramide and cholesterol composition, *J. Food Comp. Anal.* 57, 16–23.

Contarini, G. and Povolo, M. (2013). Phospholipids in milk fat: composition, biological and technological significance, and analytical strategies, *Int. J. Mol. Sci.* 14(2), 2808–2831.

Coppola, R., Salimei, E., Succi, M., Sorrentino, E., Nanni, M., Ranieri, P., Belli Blanes, R. and Grazia, L. (2002). Behaviour of *Lactobacillus rhamnosus* strains in ass's milk, *Ann. Microbiol.* 52, 55–60.

Cortellino, G., Locci, F. and Rampilli, M. (2006). An investigation of the plasmin-plasminogen system in caprine milk and cheese, *Int. Dairy J.* 16(6), 619–622.

Deeth, H. C. (2006). Lipoprotein lipase and lipolysis in milk, *Int. Dairy J.* 16(6), 555–562.

Deeth, H. C. (2021). Lipases from milk and other sources. In: Kelly, A. and Bach Larsen, L. (Eds), *Agents of Change: Enzymes in Milk and Dairy Products*, Springer Nature, Switzerland, pp. 245–267.

Delacroix-Buchet, A., Degas, C., Lamembert, G. and Vassal, L. (1996). Effect of AA and FF caprine αs1-casein variants on cheesemaking, *Lait* 76(3), 217–241.

Di Luccia, A., Picariello, G., Trani, A., Alviti, G., Loizzo, P., Faccia, M. and Addeo, F. (2009). Occurrence of β-casein fragments in cold-stored and curdled river buffalo (Bubalus bubalis L.) milk, *J. Dairy Sci.* 92(4), 1319–1329.

Doreau, M. and Martin-Rosset, W. (2011). Horse. In: Fuquay, J. W., Fox, P. F. and McSweeney, P. L. H. (Eds), *Encyclopedia of Dairy Sciences*, Academic Press, San Diego, CA, pp. 1:358–364.

Duhaiman, A. S. (1988). Purification of camel milk lysozyme and its lytic effect on Escherichia coli and Micrococcus lysodeikticus, *Comp. Biochem. Physiol. B* 91(4), 793–796.

Dumitraşcu, L., Stănciuc, N., Stanciu, S. and Râpeanu, G. (2012). Thermal inactivation of lactoperoxidase in goat, sheep and bovine milk–a comparative kinetic and thermodynamic study, *J. Food Eng.* 113(1), 47–52.

Egito, A. S., Girardet, J. -M., Poirson, C., Mollé, D., Humbert, G., Miclo, L. and Gaillard, J. -L. (2003). Action of plasmin on equine β-casein, *Int. Dairy J.* 13(10), 813–820.

El-Agamy, E. I. (2007). The challenge of cow milk protein allergy, *Small Rumin. Res.* 68(1-2), 64–72.

El-Agamy, E. I. (2017). Camel milk. In: Park, Y. W., Haenlein, G. F. W. and Wendorff, W. L. (Eds), *Handbook of Milk of Non-Bovine Animals* (2nd edn.), John Wiley and Sons, Hoboken, NJ, pp. 409–480.

Elagamy, E. I., Ruppanner, R., Ismail, A., Champagne, C. P. and Assaf, R. (1996). Purification and characterization of lactoferrin, lactoperoxidase, lysozyme and immunoglobulins from camel's milk, *Int. Dairy J.* 6(2), 129–145.

Et-Thakafy, O., Guyomarc'h, F. and Lopez, C. (2017) Lipid domains in the milk fat globule membrane: dynamics investigated in situ in milk in relation to temperature and time, *Food Chem.* 220, 352–361.

Faccia, M., D'Alessandro, A. G., Summer, A. and Hailu, Y. (2020). Milk products from minor dairy species: a review, *Animals (Basel)* 10(8), 1260.

Fantuz, F., Baldi, A., Dellorto, V., Polidori, F., Rossi, C. S., Politis, I. and Heegaard, C. W. (1998). Distribution of plasminogen activator forms in different fractions of buffalo milk, *J. Dairy Res.* 65(3), 521–527.

Fantuz, F., Ferraro, S., Todini, L., Cimarelli, L., Fatica, A., Marcantoni, F. and Salimei, E. (2020). Distribution of calcium, phosphorus, sulfur, magnesium, potassium, and sodium in major fractions of donkey milk, *J. Dairy Sci.* 103(10), 8741–8749.

Fantuz, F., Ferraro, S., Todini, L., Mariani, P., Piloni, R. and Salimei, E. (2013). Essential trace elements in milk and blood serum of lactating donkeys as affected by lactation stage and dietary supplementation with trace elements, *Animal* 7(11), 1893–1899.

Fantuz, F., Ferraro, S., Todini, L., Piloni, R., Mariani, P. and Salimei, E. (2012). Donkey milk concentration of calcium, phosphorus, potassium, sodium and magnesium, *Int. Dairy J.* 24(2), 143–145.

Fantuz, F., Maglieri, C., Lebboroni, G. and Salimei, E. (2009). Ca, Mg, Zn, Fe, Cu and Mn content of ass's milk, *Ital. J. Anim. Sci.* 8(Suppl.2), 703-705.

Farkye, N. Y. (2021). Phosphatases in milk. Alkaline phosphatase. In: Kelly, A. and Bach Larsen, L. (Eds), *Agents of Change: Enzymes in Milk and Dairy Products*, Springer Nature, Switzerland, pp. 86–94.

Faye, B., Konuspayeva, G. and Loiseau, G. (2010). Variability of urea concentration in camel milk in Kazakhstan, *Dairy Sci. Technol.* 90(6), 707–713.

Felfoul, I., Jardin, J., Gaucheron, F., Attia, H. and Ayadi, M. A. (2017). Proteomic profiling of camel and cow milk proteins under heat treatment, *Food Chem.* 216, 161–169.

Ferlay, A., Bernard, L., Meynadier, A. and Malpuech-Brugère, C. (2017). Production of trans and conjugated fatty acids in dairy ruminants and their putative effects on human health: a review, *Biochimie* 141, 107–120.

Ferreira, I. M. P. L. V. O., Mendes, E., Gomes, A. M. P., Faria, M. A. and Ferreira, M. A. (2001). The determination and distribution of nucleotides in dairy products using HPLC and diode array detection, *Food Chem.* 74(2), 239–244.

Fontecha, J., Brink, L., Wu, S., Pouliot, Y., Visioli, F. and Jiménez-Flores, R. (2020). Sources, production, and clinical treatments of milk fat globule membrane for infant nutrition and well-being, *Nutrients* 12(6), 1607.

Fox, P. F. (2011). Lactose: chemistry, properties. In: Fuquay, J. W., Fox, P. F. and McSweeney, P. L. H. (Eds), *Encyclopedia of Dairy Sciences*, Academic Press, San Diego, CA, pp. 3:173–181.

France, T. C., O'Mahony, J. A. and Kelly, A. L. (2021). The plasmin system in milk and dairy products. In: Kelly, A. and Bach Larsen, L. (Eds), *Agents of Change: Enzymes in Milk and Dairy Products*, Springer Nature, Switzerland, pp. 11–55.

Galitsopoulou, A., Michaelidou, A. M., Menexes, G. and Alichanidis, E. (2015). Polyamine profile in ovine and caprine colostrum and milk, *Food Chem.* 173(1), 80–85.

Gallier, S., Tolenaars, L. and Prosser, C. (2020). Whole goat milk as a source of fat and milk fat globule membrane in infant formula, *Nutrients* 12(11), 3486.

Gantner, V., Mijić, P., Baban, M., Škrtić, Z. and Turalija, A. (2015). The overall and fat composition of milk of various species, *Mljekarstvo* 65(4), 223–231.

Gänzle, M. G., Haase, G. and Jelen, P. (2008). Lactose: crystallization, hydrolysis and value added derivatives, *Int. Dairy J.* 18(7), 685–694.

Garcia, C., Lutz, N. W., Confort-Gouny, S., Cozzone, P. J., Armand, M. and Bernard, M. (2012). Phospholipid fingerprints of milk from different mammalians determined by 31P NMR: towards specific interest in human health, *Food Chem.* 135(3), 1777–1783.

Gastaldi, D., Bertino, E., Monti, G., Baro, C., Fabris, C., Lezo, A., Medana, C., Baiocchi, C., Mussap, M., Galvano, F. and Conti, A. (2010). Donkey's milk detailed lipid composition, *Front. Biosci. (Elite Ed.)* 2(2), 537–546.

Gaucheron, F. (2005). The minerals of milk, *Reprod. Nutr. Dev.* 45(4), 473–483.

Gentili, A., Caretti, F., Bellante, S., Ventura, S., Canepari, S. and Curini, R. (2013). Comprehensive profiling of carotenoids and fat-soluble vitamins in milk from different animal species by LC-DAD-MS/MS hyphenation, *J. Agric. Food Chem.* 61(8), 1628–1639.

Giacometti, F., Bardasi, L., Merialdi, G., Morbarigazzi, M., Federici, S., Piva, S. and Serraino, A. (2016). Shelf life of donkey milk subjected to different treatment and storage conditions, *J. Dairy Sci.* 99(6), 4291–4299.

Gil, A. and Sanchez-Medina, F. (1981). Acid-soluble nucleotides of cow's, goat's and sheep's milks, at different stages of lactation, *J. Dairy Res.* 48(1), 35–44.

Gill, B. D. and Indyk, H. E. (2007). Determination of nucleotides and nucleosides in milks and pediatric formulas: a review, *J. AOAC Int.* 90(5), 1354–1364.

Gill, B. D., Indyk, H. E. and Manley-Harris, M. (2011). Determination of totally potentially available nucleosides in bovine milk, *Int. Dairy J.* 21(1), 34–41.

Gill, B. D., Indyk, H. E. and Manley-Harris, M. (2012). Determination of totally potentially available nucleosides in bovine caprine, and ovine milk, *Int. Dairy J.* 24(1), 40–43.

Giorgio, D., Di Trana, A. and Claps, S. (2018). Oligosaccharides, polyamines and sphingolipids in ruminant milk, *Small Rumin. Res.* 160(1), 23–30.

Guo, H. Y., Pang, K., Zhang, X. Y., Zhao, L., Chen, S. W., Dong, M. L. and Ren, F. Z. (2007). Composition, physicochemical properties, nitrogen fraction distribution and amino acid profile of donkey milk, *J. Dairy Sci.* 90(4), 1635–1643.

Haddad, I., Mozzon, M., Strabbioli, R. and Frega, N. G. (2010). Stereospecific analysis of triacylglycerols in camel (Camelus dromedarius) milk fat, *Int. Dairy J.* 20(12), 863–867.

Haddadin, M. S. Y., Gammoh, S. I. and Robinson, R. K. (2008). Seasonal variations in the chemical composition of camel milk in Jordan, *J. Dairy Res.* 75(1), 8–12.

Hailu, Y., Hansen, E. B., Seifu, E., Eshetu, M., Ipsen, R. and Kappeler, S. (2016). Functional and technological properties of camel milk proteins: a review, *J. Dairy Res.* 83(4), 422–429.

He, S., Ma, Y., Wang, J., Li, Q., Tang, S. and Li, H. (2012). Effects of proteose-peptone fractions from yak milk on lipoprotein lipase lipolysis, *Int. J. Dairy Technol.* 65(1), 32–37.

He, S., Ma, Y., Wang, J., Li, Q., Tang, S., Zhao, C., Li, H. and Maubois, J.-L. (2010). Characterization of fat globules and milk fat globule membrane proteins in milk of different yak breeds, *Dairy Sci. Technol.* 90(5), 601–609.

He, S., Ma, Y., Wang, J., Li, Q., Yang, X., Tang, S. and Li, H. (2011). Milk fat chemical composition of yak breeds in China, *Journal of Food Composition and Analysis* 24(2), 223-230, Food Comp, J. Anal 24(2), 223-230.

Hernández-Olivas, E., Muñoz-Pina, S., Sánchez-García, J., Andrés, A. and Heredia, A. (2020). Understanding the role of food matrix on the digestibility of dairy products under elderly gastrointestinal conditions, *Food Res. Int.* 137, 109454.

Hobbs, M., Jahan, M., Ghorashi, S. A. and Wang, B. (2021). Current perspective of sialylated milk oligosaccharides in mammalian milk: implications for brain and gut health of newborns, *Foods* 10(2), 473.

Holt, C., Carver, J. A., Ecroyd, H. and Thorn, D. C. (2013). Invited review: Caseins and the casein micelle: their biological functions, structures, and behavior in foods, *J. Dairy Sci.* 96(10), 6127-6146.

Holt, C. and Jenness, R. (1984). Interrelationships of constituents and partition of salts in milk samples from 8 species, *Comp. Biochem. Physiol. A Comp. Physiol.* 77(2), 275-282.

Humbert, G., Chang, O. and Gaillard, J. L. (2005). Plasmin activity and plasminogen activation in equine milk, *Milchwissenschaft* 60(2), 134-137.

Huppertz, T. (2016). Novel technologies for dairy processing: applications for non-cow milk. In: Tsakalidou, E. and Papadimitriou, K. (Eds), *Non-Bovine Milk and Milk Products*, Academic Press, UK, pp. 187-208.

Huppertz, T. and Chia, L. W. (2021). Milk protein coagulation under gastric conditions: a review, *Int. Dairy J.* 113, 104882.

Indyk, H. E. and Woollard, D. C. (2004). Determination of orotic acid, uric acid, and creatinine in milk by liquid chromatography, *J. AOAC Int.* 87(1), 116-122.

Inglingstad, R. A., Devold, T. G., Eriksen, E. K., Holm, H., Jacobsen, M., Liland, K. H., Rucke, E. O. and Vegarud, G. E. (2010). Comparison of the digestion of caseins and whey proteins in equine, bovine, caprine and human milks by human gastrointestinal enzymes, *Dairy Sci. Technol.* 90(5), 549-563.

Ismail, B. and Nielsen, S. S. (2010). Invited review: plasmin protease in milk: current knowledge and relevance to dairy industry, *J. Dairy Sci.* 93(11), 4999-5009.

Ji, X., Li, X., Ma, Y. and Li, D. (2017). Differences in proteomic profiles of milk fat globule membrane in yak and cow milk, *Food Chem.* 221, 1822-1827.

Kamau, P. M. N., Lamuka, P. O. and Wangoh, J. (2010). Effect of lactoperoxidase-thiocyanate-hydrogen peroxide system and storage temperature on keeping quality of raw camel milk, *Afr. J. Food Agric. Nutr. Dev.* 10(10), 4186-4201.

Kappeler, S. (1998). Compositional and structural analysis of camel milk proteins with emphasis on protective proteins, Dissertation for the degree of Doctor of Technical Sciences, Swiss Federal Institute of Technology, Zurich, Diss. ETH No. 12947.

Khalesi, M., Salami, M., Moslehishad, M., Winterburn, J. and Moosavi-Movahedi, A. A. (2017). Biomolecular content of camel milk: a traditional superfood towards future healthcare industry, *Trends Food Sci. Technol.* 62, 49-58.

Klotz, V., Hill, A., Warriner, K., Griffiths, M. and Odumeru, J. (2008). Assessment of the colorimetric and fluorometric assays for alkaline phosphatase activity in cow's, goat's, and sheep's milk, *J. Food Prot.* 71(9), 1884-1888.

Kondyli, E., Katsiari, M. C. and Voutsinas, L. P. (2007). Variations of vitamin and mineral contents in raw goat milk of the indigenous Greek breed during lactation, *Food Chem.* 100(1), 226-230.

Kondyli, E., Svarnas, C., Samelis, J. and Katsiari, M. C. (2012). Chemical composition and microbiological quality of ewe and goat milk of native Greek breeds, *Small Rumin. Res.* 103(2-3), 194-199.

Konuspayeva, G., Lemarie, É, Faye, B., Loiseau, G. and Montet, D. (2008). Fatty acidand cholesterol composition of camel's (Camelus bactrianus, Camelus dromedarius and hybrids) milk in Kazakhstan, *Dairy Sci. Technol.* 88(3), 327-340.

Kulyar, M. F.-E.-A., Yao, W., Ding, Y., Li, K., Zhang, L., Li, A., Waqas, M., Huachun, P., Quan, M., Zeng, Z., Mehmood, K., Sizhu, S. and Li, J. (2021). Bioactive potential of yak's milk and its products; pathophysiological and molecular role as an immune booster in antibiotic resistance, *Food Biosci.* 39, 100838.

Kussendrager, K. D. and Van Hooijdonk, A. C. M. (2000). Lactoperoxidase: physicochemical properties, occurrence, mechanism of action and applications, *Br. J. Nutr.* 84(Suppl.1), 19-25.

La Torre, G. L., Saitta, M., Giorgia Potortì, A., Di Bella, G. and Dugo, G. (2010). High performance liquid chromatography coupled with atmospheric pressure chemical ionization mass spectrometry for sensitive determination of bioactive amines in donkey milk, *J. Chromatogr. A* 1217(32), 5215-5224.

Larqué, E., Sabater, M. and Zamora, S. (2007). Biological significance of dietary polyamines, *Nutrition* 23(1), 87-95.

Leitner, G., Chaffer, M., Shamay, A., Shapiro, F., Merin, U., Erza, E., Saran, A. and Silanikove, N. (2004a). Changes in milk composition as affected by subclinical mastitis in sheep, *J. Dairy Sci.* 87(1), 46-52.

Leitner, G., Merin, U. and Silanikove, N. (2004b). Changes in milk composition as affected by subclinical mastitis in goats, *J. Dairy Sci.* 87(6), 1719-1726.

Li, H., Ma, Y., Li, Q., Wang, J., Cheng, J., Xue, J. and Shi, J. (2011). The chemical composition and nitrogen distribution of Chinese yak (Maiwa) milk, *Int. J. Mol. Sci.* 12(8), 4885-4895.

Li, S., Li, L., Zeng, Q., Lu, J. and Ren, D. (2016). Separation and quantification of milk casein from different buffalo breeds, *J. Dairy Res.* 83(3), 317-325.

Li, Y., Gill, B. D., Grainger, M. N. C. and Manley-Harris, M. (2019). The analysis of vitamin B12 in milk and infant formula: a review, *Int. Dairy J.* 99, 104543.

Licitra, R., Li, J., Liang, X., Altomonte, I., Salaria, F., Yan, J. and Martini, M. (2019). Profile and content of sialylated oligosaccharides in donkey milk at early lactation, *LWT Food Sci. Technol.* 115, 108437.

Lombardi, P., Avallone, L., D'Angelo, A., Mor, T. and Bogin, E. (2000). Buffalo-milk enzyme levels, their sensitivity to heat inactivation, and their possible use as markers for pasteurization, *J. Food Prot.* 63(7), 970-973.

Lopez, C. (2011). Milk fat globules enveloped by their biological membrane: unique colloidal assemblies with a specific composition and structure, *Curr. Opin. Colloid Interface Sci.* 16(5), 391-404.

Lopez, C., Briard-Bion, V.. and Ménard, O. (2014). Polar lipids, sphingomyelin and long-chain unsaturated fatty acids from the milk fat globule membrane are increased in milks produced by cows fed fresh pasture based diet during spring, *Food Res. Int.* 58, 59-68.

Lorenzen, P. C., Martin, D., Clawin-Rädecker, I., Barth, K. and Knappstein, K. (2010). Activities of alkaline phosphatase, glutamyltransferase and lactoperoxidase in cow, sheep and goat's milk in relation to heat treatment, *Small Rumin. Res.* 89(1), 18-23.

Lorenzen, P. C., Wernery, R., Johnson, B., Jose, S. and Wernery, U. (2011). Evaluation of indigenous enzyme activities in raw and pasteurised camel milk, *Small Rumin. Res.* 97(1-3), 79-82.

Lu, J., Argov-Argaman, N., Anggrek, J., Boeren, S., van Hooijdonk, T., Vervoort, J. and Hettinga, K. A. (2016a). The protein and lipid composition of the membrane of milk fat globules depends on their size, *J. Dairy Sci.* 99(6), 4726-4738.

Lu, J., Wang, X., Zhang, W., Liu, L., Pang, X., Zhang, S. and Lv, J. (2016b). Comparative proteomics of milk fat globule membrane in different species reveals variations in lactation and nutrition, *Food Chem.* 196, 665-672.

Luo, J., Huang, Z., Liu, H., Zhang, Y. and Ren, F. (2018). Yak milk fat globules from the Qinghai-Tibetan Plateau: membrane lipid composition and morphological properties, *Food Chem.* 245, 731-737.

Luo, J., Liu, L., Liu, T., Shen, Q., Liu, C., Zhou, H. and Ren, F. (2020). Simulated in vitro infant gastrointestinal digestion of yak milk fat globules: a comparison with cow milk fat globules, *Food Chem.* 314, 126160.

Ma, Y., He, S. and Park, Y. W. (2017). Yak milk. In: Park, Y. W., Haenlein, G. F. W. and Wendorff, W. L. (Eds), *Handbook of Milk of Non-Bovine Animals* (2nd edn.), John Wiley and Sons, Hoboken, NJ, pp. 481-513.

Malacarne, M., Criscione, A., Franceschi, P., Bordonaro, S., Formaggioni, P., Marletta, D. and Summer, A. (2019). New insights into chemical and mineral composition of donkey milk throughout nine months of Lactation, *Animals (Basel)* 9(12), 1161.

Malacarne, M., Martuzzi, F., Summer, A. and Mariani, P. (2002). 'Protein and fat composition of mares' milk: some nutritional remarks with reference to human and cows' milk, *Int. Dairy J.* 12(11), 869-877.

Manoni, M., Di Lorenzo, C., Ottoboni, M., Tretola, M. and Pinotti, L. (2020). Comparative proteomics of milk fat globule membrane (MFGM) proteome across species and lactation stages and the potentials of MFGM fractions in infant formula preparation, *Foods* 9(9), 9091251.

Marchand, S., Merchiers, M., Messens, W., Coudijzer, K. and De Block, J. (2009). Thermal inactivation kinetics of alkaline phosphatase in equine milk, *Int. Dairy J.* 19(12), 763-767.

Markiewicz-Kęszycka, M., Czyżak-Runowska, G., Wójtowski, J., Jóźwik, A., Pankiewicz, R., Łęska, B., Krzyżewski, J., Strzałkowska, N., Marchewka, J. and Bagnicka, E. (2015). Influence of stage of lactation and year season on composition of mares' colostrum and milk and method and time of storage on vitamin C content in mares' milk, *J. Sci. Food Agric.* 95(11), 2279-2286.

Markiewicz-Keszycka, M., Wojtowski, J., Czyzak-Runowska, G., Kuczyńska, B., Puppel, K., Krzyżewski, J., Strzałkowska, N., Jóźwik, A. and Bagnicka, E. (2014). Concentration of selected fatty acids, fat-soluble vitamins and β-carotene in late lactation mares' milk, *Int. Dairy J.* 38(1), 31-36.

Martemucci, G. and D'Alessandro, A. G. (2013). Progress in nutritional and health profile of milk and dairy products: a novel drug target, *Endocr. Metab. Immune Disord. Drug Targets* 13(3), 209-233.

Martin, D., Walte, H.-G. and Lorenzen, P. Chr. (2016). Ribonucleosides in raw milk and heat treated milk samples from cows, sheep, goats and camels and their potential biotechnological application, *Small Rumin. Res.* 137, 162-168.

Martin, D., Schlimme, E. and Tait, D. (2011a). Nucleosides and nucleotides in milk. In: Fuquay, J. W., Fox, P. F. and McSweeney, P. L. H. (Eds), *Encyclopedia of Dairy Sciences*, Academic Press, San Diego, CA, pp. 3:971-979.

Martin, P., Cebo, C. and Miranda, G. (2011b). Inter-Species Comparison of Milk proteinS: quantitative variability and molecular diversity. In: Fuquay, J. W., Fox, P. F. and McSweeney, P. L. H. (Eds), *Encyclopedia of Dairy Sciences*, Academic Press, San Diego, CA, pp. 821-842.

Martin, P., Ollivier-Bousquet, M. and Grosclaude, F. (1999). Genetic polymorphism of caseins: a tool to investigate casein micelle organization, *Int. Dairy J.* 9(3-6), 163-171.

Martín-Ortiz, A., Moreno, F. J., Ruiz-Matute, A. I. and Sanz, M. L. (2019). Selective biotechnological fractionation of goat milk carbohydrates, *Int. Dairy J.* 94, 38-45.

Martini, M., Altomonte, I., Pesi, R., Tozzi, M. G. and Salari, F. (2013). 'Fat globule membranes in ewes' milk: the main enzyme activities during lactation, *Int. Dairy J.* 28(1), 36-39.

Martini, M., Altomonte, I., Licitra, R. and Salari, F. (2018a). Short communication: technological and seasonal variations of vitamin D and other nutritional components in donkey milk, *J. Dairy Sci.* 101(10), 8721-8725.

Martini, M., Salari, F., Altomonte, I., Ragona, G., Piazza, A., Gori, R., Casati, D. and Brajon, G. (2018b). Effects of pasteurization and storage conditions on donkey milk nutritional and hygienic characteristics, *J. Dairy Res.* 85(4), 445-448.

Massouras, T., Triantaphyllopoulos, K. A. and Theodossiou, I. (2017). Chemical composition, protein fraction and fatty acid profile of donkey milk during lactation, *Int. Dairy J.* 75, 83-90.

Mati, A., Senoussi-Ghezali, C., Si Ahmed Zennia, S., Almi-Sebbane, D., El-Hatmi, H. and Girardet, J.-M. (2017). Dromedary camel milk proteins, a source of peptides having biological activities–a review, *Int. Dairy J.* 73, 25-37.

Mayer, H. K. and Fiechter, G. (2012a). Physical and chemical characteristics of sheep and goat milk in Austria, *Int. Dairy J.* 24(2), 57-63.

Mayer, H. K. and Fiechter, G. (2012b). Physicochemical characteristics of goat's milk in Austria-seasonal variations and differences between six breeds, *Dairy Science & Technol.* 92(2), 167-177.

Medhammar, E., Wijesinha-Bettoni, R., Stadlmayr, B., Nilsson, E., Charrondiere, U. R. and Burlingame, B. (2012). Composition of milk from minor dairy animals and buffalo breeds: a biodiversity perspective, *J. Sci. Food Agric.* 92(3), 445-474.

Meena, S., Rajput, Y. S. and Sharma, R. (2014). Comparative fat digestibility of goat, camel, cow and buffalo milk, *Int. Dairy J.* 35(2), 153-156.

Menard, O., Ahmad, S., Rousseau, F., Briard-Bion, V., Gaucheron, F. and Lopez, C. (2010). Buffalo vs. cow milk fat globules: size distribution, zeta-potential, compositions in total fatty acids and in polar lipids from the milk fat globule membrane, *Food Chem.* 120(2), 544-551.

Michaelidou, A. M. (2008). Factors influencing nutritional and health profile of milk and milk products, *Small Rumin. Res.* 79(1), 42-50.

Moatsou, G. (2010). Indigenous enzymatic activities in ovine and caprine milks, *Int. J. Dairy Technol.* 63(1), 16-31.

Moatsou, G. and Sakkas, L. (2019). Sheep milk components: focus on nutritional advantages and biofunctional potential, *Small Rumin. Res.* 180, 86-99.

Moatsou, G., Hatzinaki, A., Samolada, M. and Anifantakis, E. (2005). Major whey proteins in ovine and caprine acid wheys from indigenous Greek breeds, *Int. Dairy J.* 15(2), 123-131.

Moatsou, G., Moschopoulou, E., Molle, D., Gagnaire, V., Kandarakis, I. and Leonil, J. (2008). Comparative study of the protein fraction of goat milk from the indigenous Greek breed and from international breeds, *Food Chem.* 106(2), 509-520.

Mohan, M. S., O'Callaghan, T. F., Kelly, P. and Hogan, S. A. (2020). Milk fat: opportunities, challenges and innovation, *Crit. Rev. Food Sci. Nutr.* 61(14), 2411-2443.

Monti, L., Cattaneo, T. M. P., Orlandi, M. and Curadi, M. C. (2015). Capillary electrophoresis of sialylated oligosaccharides in milk from different species, *J. Chromatogr. A* 1409, 288-291.

Morgan, F., Jacquet, F., Micault, S., Bonnin, V. and Jaubert, A. (2000). Study on the compositional factors involved in the variable sensitivity of caprine milk to high temperature processing, *Int. Dairy J.* 10(1-2), 113-117.

Murtaza, M. A., Pandya, A. J. and Khan, M. M. H. (2017). Buffalo milk utilization for dairy products. In: Park, Y. W., Haenlein, G. F. W. and Wendorff, W. L. (Eds), *Handbook of Milk of Non-Bovine Animals* (2nd edn.), JohnWiley and Sons, Hoboken, NJ, pp. 284-342.

Nguyen, H. T. H., Ong, L., Beaucher, E., Madec, M. N., Kentish, S. E., Gras, S. L. and Lopez, C. (2015). Buffalo milk fat globules and their biological membrane: in situ structural investigations, *Food Res. Int.* 67, 35-43.

Nunes, L. and Tavares, G. M. (2019). Thermal treatments and emerging technologies: impacts on the structure and techno-functional properties of milk proteins', *Trends Food Sci. Technol.* 90, 88-99.

Nygren-Babol, L. and Jägerstad, M. (2012). Folate-binding protein in milk: a review of biochemistry, physiology, and analytical methods, *Crit. Rev. Food Sci. Nutr.* 52(5), 410-425.

Orlandi, M., Curadi, M. C., Monti, L. and Cattaneo, T. M. P. (2016). Sialylated oligosaccharides in mare and ass milk: preliminary results, *Progr. Nutr.* 18(3), 283-287.

Park, Y. W. (2007). Rheological characteristics of goat and sheep milk, *Small Rumin. Res.* 68(1-2), 73-87.

Park, Y. W. (2017). Goat milk - chemistry and nutrition. In: Park, Y. W., Haenlein, G. F. W. and Wendorff, W. L. (Eds), *Handbook of Milk of Non-Bovine Animals* (2nd edn.), JohnWiley and Sons, Hoboken, NJ, pp. 42-83.

Parrón, J. A., Ripollés, D., Ramos, S. J., Pérez, M. D., Semen, Z., Rubio, P., Calvo, M. and Sánchez, L. (2018). Antirotaviral potential of lactoferrin from different origin: effect of thermal and high pressure treatments, *Biometals* 31(3), 343-355.

Pellegrini, O., Remeuf, F. and Rivemale, M. (1994). Évolution des caractéristiques physico-chimiques et des paramètres de coagulation du lait de brebis collecté dans la région de Roquefort, *Lait* 74(6), 425-442.

Phosanam, A., Chandrapala, J., Huppertz, T., Adhikari, B. and Zisu, B. (2021). In vitro digestion of infant formula model systems: influence of casein to whey protein ratio, *Int. Dairy J.* 117, 105008.

Pirisi, A., Piredda, G., Papoff, C. M., Di Salvo, R., Pintus, S., Garro, G., Ferranti, P. and Chianese, L. (1999). Effects of sheep alpha s1-casein CC, CD and DD genotypes on milk composition and cheesemaking properties, *J. Dairy Res.* 66(3), 409-419.

Plakantara, S., Michaelidou, A. M., Polychroniadou, A., Menexes, G. and Alichanidis, E. (2010). Nucleotides and nucleosides in ovine and caprine milk during lactation, *J. Dairy Sci.* 93(6), 2330-2337.

Płoszaj, T., Ryniewicz, Z. and Motyl, T. (1997). Polyamines in goat's colostrum and milk, *Comp. Biochem. Physiol., B Biochem. Mol. Biol.* 118(1), 45-52.

Prosser, C. G. (2021). Compositional and functional characteristics of goat milk and relevance as a base for infant formula, *J. Food Sci.* 86(2), 257-265.

Prosser, C. G., McLaren, R. D., Frost, D., Agnew, M. and Lowry, D. J. (2008). Composition of the non-protein nitrogen fraction of goat whole milk powder and goat milk-based infant and follow-on formulae, *Int. J. Food Sci. Nutr.* 59(2), 123–133.

Ramos, M. and Juarez, M. (2011). Sheep milk. In: Fuquay, J. W., Fox, P. F. and McSweeney, P. L. H. (Eds), *Encyclopedia of Dairy Sciences*, Academic Press, San Diego, CA, pp. 494–502.

Raynal, K. and Remeuf, F. (1998). The effect of heating on physicochemical and renneting properties of milk: a comparison between caprine, ovine and bovine milk, *Int. Dairy J.* 8(8), 695–706.

Raynal-Ljutovac, K., Lagriffoul, G., Paccard, P., Guillet, I. and Chilliard, Y. (2008). Composition of goat and sheep milk products: an update, *Small Rumin. Res.* 79(1), 57–72.

Raynal-Ljutovac, K., Park, Y. W., Gaucheron, F. and Bouhallab, S. (2007). Heat stability and enzymatic modifications of goat and sheep milk, *Small Rumin. Res.* 68(1–2), 207–220.

Remeuf, F. (1993). Infuence du polymorphisme genetique de la caseine as1 caprine sur les caracteristiques physico-chimiques et technologiques du lait, *Lait* 73(5–6), 549–557.

Revilla, I., Escuredo, O., González-Martín, M. I.. and Palacios, C. (2017). Fatty acids and fat-soluble vitamins in ewe's milk predicted by near infrared reflectance spectroscopy. Determination of seasonality, *Food Chem.* 214, 468–477.

Rocchi, S., Caretti, F., Gentili, A., Curini, R., Perret, D. and Pérez-Fernández, V. (2016). Quantitative profiling of retinyl esters in milk from different ruminant species by using high performance liquid chromatography-diode array detection-tandem mass spectrometry, *Food Chem.* 211, 455–464.

Rodríguez-Alcalá, L. M. and Fontecha, J. (2010). Major lipid classes separation of buttermilk, and cows, goats and ewes milk by high performance liquid chromatography with an evaporative light scattering detector focused on the phospholipid fraction, *J. Chromatogr. A* 1217(18), 3063–3066.

Rodríguez-Alcalá, L. M., Harte, F. and Fontecha, J. (2009). Fatty acid profile and CLA isomers content of cow, ewe and goat milks processed by high pressure homogenization, *Innov. Food Sci. Emerg. Technol.* 10(1), 32–36.

Roy, D., Ye, A., Moughan, P. J. and Singh, H. (2020). Composition, structure, and digestive dynamics of milk From different species–a review, *Front. Nutr.* 7, 577759.

Russo, G. L. (2009). Dietary n-6 and n-3 polyunsaturated fatty acids: from biochemistry to clinical implications in cardiovascular prevention, *Biochem. Pharmacol.* 77(6), 937–946.

Russo, M., Cichello, F., Ragonese, C., Donato, P., Cacciola, F., Dugo, P. and Mondello, L. (2013). Profiling and quantifying polar lipids in milk by hydrophilic interaction liquid chromatography coupled with evaporative light-scattering and mass spectrometry detection, *Anal. Bioanal. Chem.* 405(13), 4617–4626.

Ryskaliyeva, A., Henry, C., Miranda, G., Faye, B., Konuspayeva, G. and Martin, P. (2018). Combining different proteomic approaches to resolve complexity of the milk protein fraction of dromedary, Bactrian camels and hybrids, from different regions of Kazakhstan, *PLoS ONE* 13(5), e0197026.

Saadaoui, B., Henry, C., Khorchani, T., Mars, M., Martin, P. and Cebo, C. (2013). Proteomics of the milk fat globule membrane from Camelus dromedarius, *Proteomics* 13(7), 1180–1184. DOI:10.1002/pmic.201200113.

Salimei, E. and Fantuz, F. (2012). Equid milk for human consumption, *Int. Dairy J.* 24(2), 130–142.
Salimei, E. and Park, Y. W. (2017). Mare milk. In: Park, Y. W., Haenlein, G. F. W. and Wendorff, W. L. (Eds), *Handbook of Milk of Non-Bovine Animals* (2nd edn.), John Wiley and Sons, Hoboken, NJ, pp. 369–408.
Scano, P., Cusano, E., Caboni, P. and Consonni, R. (2019). NMR metabolite profiles of dairy: a review, *Int. Dairy J.* 90, 56–67.
Schaafsma, G. (2008). Lactose and lactose derivatives as bioactive ingredients in human nutrition, *Int. Dairy J.* 18(5), 458–465.
Seifu, E., Buys, E. M. and Donkin, E. F. (2005). Significance of the lactoperoxidase system in dairy industry and its potential application: a review, *Trends Food Sci. Technol.* 16(4), 137–154.
Sharma, R., Kaur, S., Rajput, Y. S. and Kumar, R. (2009). Activity and thermal stability of indigenous enzymes in cow, buffalo and goat milk, *Milchwissenschaft* 64(2), 173–175.
Silva, R. C. D., Colleran, H. L. and Ibrahim, S. A. (2021). Milk fat globule membrane in infant nutrition: A dairy industry perspective, *J. Dairy Res.* 88(1), 105–116.
Sinanoglou, V. J., Koutsouli, P., Fotakis, C., Sotiropoulou, G., Cavouras, D. and Bizelis, I. (2015). Assessment of lactation stage and breed effect on sheep milk fatty acid profile and lipid quality indices, *Dairy Sci. Technol.* 95(4), 509–531.
Sindhu, J. S. and Arora, S. (2011). Buffalo milk. In: Park, Y. W., Haenlein, G. F. W. and Wendorff, W. L. (Eds), *Handbook of Milk of Non-Bovine Animals* (2nd edn.), John Wiley and Sons, Hoboken, NJ, pp. 3:503–511.
Smiddy, M. A., Huppertz, T. and van Ruth, S. M. (2012). Triacylglycerol and melting profiles of milk fat from several species, *Int. Dairy J.* 24(2), 64–69.
Sousa, Y. R. F., Araújo, D. F. S., Pulido, J. O., Pintado, M. M. E., Martínez-Férez, A. and Queiroga, R. C. R. E. (2019a). Composition and isolation of goat cheese whey oligosaccharides by membrane technology, *Int. J. Biol. Macromol.* 139, 57–62.
Sousa, Y. R. F., Medeiros, L. B., Pintado, M. M. E. and Queiroga, R. C. R. E. (2019b). Goat milk oligosaccharides: composition, analytical methods and bioactive and nutritional properties, *Trends Food Sci. Technol.* 92, 152–161.
Srivastava, A., Tripathi, R., Soni, V. K., Misra-Bhattacharya, S. and Deepak, D. (2014). Isolation of mare's milk oligosaccharide fraction of colostrum, transitional, and mature phases promotes in vitro oxidative burst in murine macrophages, *J. Equine Vet. Sci.* 34(8), 1009–1015.
Tagliazucchi, D., Martini, S., Shamsia, S., Helal, A. and Conte, A. (2018). Biological activities and peptidomic profile of in vitro-digested cow, camel, goat and sheep milk, *Int. Dairy J.* 81, 19–27.
Tamime, A. Y., Wszolek, M., Božanić, R. and Özer, B. (2011). Popular ovine and caprine fermented milks, *Small Rumin. Res.* 101(1–3), 2–16.
Tayefi-Nasrabadi, H., Hoseinpour-fayzi, M. A. and Mohasseli, M. (2011). Effect of heat treatment on lactoperoxidase activity in camel milk: a comparison with bovine lactoperoxidase, *Small Rumin. Res.* 99(2–3), 187–190.
Tolenaars, L., Romanazzi, D., Carpenter, E., Gallier, S. and Prosser, C. G. (2021). Minor dietary components intrinsic to goat milk and goat milk formulas, *Int. Dairy J.* 117, 105012.
Uniacke-Lowe, T. (2011). Studies on equine milk and comparative studies on equine and bovine milk systems, PhD Thesis, University College Cork.

Uniacke-Lowe, T. and Fox, P. F. (2011). Equid milk. In: Fuquay, J. W., Fox, P. F. and McSweeney, P. L. H. (Eds), *Encyclopedia of Dairy Sciences*, Academic Press, San Diego, CA, pp. 3:518–529.

Uniacke-Lowe, T., Huppertz, T. and Fox, P. F. (2010). Equine milk proteins: chemistry, structure and nutritional significance, *Int. Dairy J.* 20(9), 609–629.

Urashima, T., Asakuma, S., Kitaoka, M. and Messer (2011). Indigenous oligosaccharides in milk. In: Fuquay, J. W., Fox, P. F. and McSweeney, P. L. H. (Eds), *Encyclopedia of Dairy Sciences*, Academic Press, San Diego, CA, pp. 3:241–273.

Urashima, T., Taufik, E., Fukuda, K. and Asakuma, S. (2013). Recent advances in studies on milk oligosaccharides of cows and other domestic farm animals, *Biosci. Biotechnol. Biochem.* 77(3), 455–466.

Valdivielso, I., Bustamante, M. Á, Buccioni, A., Franci, O., Ruiz de Gordoa, J. C., De Renobales, M. and Barron, L. J. (2015). Commercial sheep flocks – fatty acid and fat-soluble antioxidant composition of milk and cheese related to changes in feeding management throughout lactation, *J. Dairy Res.* 82(3), 334–343.

van Leeuwen, S. S., Te Poele, E. M., Chatziioannou, A. C., Benjamins, E., Haandrikman, A. and Dijkhuizen, L. (2020). Goat milk oligosaccharides: their diversity, quantity, and functional properties in comparison to human milk oligosaccharides, *J. Agric. Food Chem.* 68(47), 13469–13485.

Verardo, V., Gómez-Caravaca, A. M., Arráez-Román, D. and Hettinga, K. (2017). Recent advances in phospholipids from colostrum, milk and dairy by-products, *Int. J. Mol. Sci.* 18(1), 173.

Verduci, E., D'Elios, S., Cerrato, L., Comberiati, P., Calvani, M., Palazzo, S., Martelli, A., Landi, M., Trikamjee, T. and Peroni, D. G. (2019). Cow's milk substitutes for children: nutritional aspects of milk from different mammalian species, special formula and plant-based beverages, *Nutrients* 11(8), 1739.

Verma, M., Dige, M. S., Gautam, D., De, S. and Rout, P. K. (2020). Functional milk proteome analysis of genetically diverse goats from different agro-climatic regions', *J. Proteomics* 227, 103916.

Verruck, S., Dantas, A. and Prudencio, E. S. (2019). Functionality of the components from goat's milk, recent advances for functional dairy products development and its implications on human health, *J. Funct. Foods* 52, 243–257.

Vincenzetti, S., Polidori, P., Mariani, P., Cammertoni, N., Fantuz, F. and Vita, A. (2008). Donkey's milk protein fractions characterization, *Food Chem.* 106(2), 640–649.

Vincenzetti, S., Pucciarelli, S., Nucci, C., Polzonetti, V., Cammertoni, N. and Polidori, P. (2014). Profile of nucleosides and nucleotides in donkey's milk, *Nucleosides Nucleotides Nucleic Acids* 33(10), 656–667.

Vincenzetti, S., Pucciarelli, S., Polzonetti, V. and Polidori, P. (2017). Role of proteins and of some bioactive peptides on the nutritional quality of donkey milk and their impact on human health, *Beverages* 3(4), 34.

Walstra, P., Wouters, J. T. M. and Geurts, T. J. (2006). Chapter 2. 'Milk components' and Chapter. In: *Colloidal Particles of Milk* (vol. 3), Chapter 2.3 'Lipids'. In *Dairy Science and Technology* (2nd edn.), CRC Press Press Taylor and Francis Group, Boca Raton, FL, pp. 17–97 and 109–157.

Wang, P., Liu, H., Wen, P., Zhang, H., Guo, H. and Ren, F. (2013). The composition, size and hydration of yak casein micelles, *Int. Dairy J.* 31(2), 107–110.

Wang, Y., Zhou, X., Gong, P., Chen, Y., Feng, Z., Liu, P., Zhang, P., Wang, X., Zhang, L. and Song, L. (2020). Comparative major oligosaccharides and lactose between Chinese human and animal milk, *Int. Dairy J.* 108, 104727.

Wernery, U., Fischbach, S., Johnson, B. and Jose, S. (2008). Evaluation of alkaline phosphatase (ALP). *γ-Glutamyl Transferase (GGT) and Lactoperoxidase (LPO) Activities for Their Suitability as Markers of Camel Milk Heat Inactivation, Milchwissenschaft* 63(3), 265–267.

Wernery, U., Wernery, R., Masko, O., Johnson, B., Gnanaraj, B., Jose, S., Nagy, P. and Lorenzen, P. C. (2013). Lactoperoxidase: a suitable enzymatic marker of camel milk pasteurization, *J. Camel Pract. Res.* 20(1), 35–38.

Wiener, G. (2011). Dairy animals: yak. In: In, J., Fuquay, J. W., Fox, P. F. and McSweeney, P. L. H. (Eds), *Encyclopedia of Dairy Sciences*, Academic Press, San Diego, CA, pp. 1:343–350.

Wynn, P. C., Morgan, A. D. and Sheehy, P. A. (2011). Minor proteins, bovine serum albumin, vitamin-binding proteins. In: Fuquay, J. W., Fox, P. F. and McSweeney, P. L. H. (Eds), *Encyclopedia of Dairy Sciences*, Academic Press, San Diego, CA, pp. 3:795–800.

Yan, J., Ding, J., Jin, G., Yu, D., Yu, L., Long, Z., Guo, Z., Chai, W.. and Liang, X. (2018). Profiling of sialylated oligosaccharides in mammalian milk using online solid phase extraction hydrophilic interaction chromatography coupled with negative-ion electrospray mass spectrometry, *Anal. Chem.* 90(5), 3174–3182.

Yang, J., Zheng, N., Wang, J. and Yang, Y. (2018). Comparative milk fatty acid analysis of different dairy species, *Int. J. Dairy Technol.* 71(2), 544–550.

Yang, Y., Zheng, N., Wang, W., Zhao, X., Zhang, Y., Han, R., Ma, L., Zhao, S., Li, S., Guo, T., Zang, C. and Wang, J. (2016). N-glycosylation proteomic characterization and cross-species comparison of milk fat globule membrane proteins from mammals, *Proteomics* 16(21), 2792–2800.

Yang, Y., Zheng, N., Zhao, X., Zhang, Y., Han, R., Ma, L., Zhao, S., Li, S., Guo, T. and Wang, J. (2015). Proteomic characterization and comparison of mammalian milk fat globule proteomes by iTRAQ analysis, *J. Proteomics* 116, 34–43.

Yao, Y., Zhao, G., Xiang, J., Zou, X., Jin, Q. and Wang, X. (2016). Lipid composition and structural characteristics of bovine, caprine and human milk fat globules, *Int. Dairy J.* 56, 64–73.

Yoganandi, J., Jain, A. K., Mehta, B. M., Wadhwani, K. N. and Aparnathi, K. D. (2014). Comparative studies on selected enzymes activities of camel, cow and buffalo milk, *J. Camel Pract. Res.* 21(2), 249–252.

Yousef, I. M., Emery, R. S., Askew, E. W., Benson, J. D., Thomas, J. W. and Huber, J. T. (1969). Comparison of several enzymes and components of milk, blood and tissues between cows and ewes, *J. Dairy Sci.* 52(10), 1577–1581.

Zancada, L., Pérez-Díez, F., Sánchez-Juanes, F., Alonso, J. M., García-Pardo, L. A. and Hueso, P. (2013). Phospholipid classes and fatty acid composition of ewe's and goat's milk, *Grasas Aceites* 64(3), 304–310.

Zhang, J., Yang, M., Cai, D., Hao, Y., Zhao, X., Zhu, Y., Zhu, H. and Yang, Z. (2020). Composition, coagulation characteristics, and cheese making capacity of yak milk, *J. Dairy Sci.* 103(2), 1276–1288.

Zhang, Y., Li, Y., Wang, P., Tian, Y., Liang, Q. and Ren, F. (2017). Rennet-induced coagulation properties of yak casein micelles: a comparison with cow casein micelles, *Food Res. Int.* 102, 25–31.

Zhou, X., Qu, X., Zhao, S., Wang, J., Li, S. and Zheng, N. (2017). Analysis of 22 elements in milk, feed, and water of dairy cow, goat, and buffalo from different regions of China, *Biol. Trace Elem. Res.* 176(1), 120–129.

Zotos, A. and Bampidis, V. A. (2014). Milk fat quality of Greek buffalo (*Bubalus bubalis*), *J. Food Compos. Anal.* 33(2), 181–186.

Zou, X., Huang, J., Jin, Q., Guo, Z., Liu, Y., Cheong, L., Xu, X. and Wang, X. (2013). Lipid composition analysis of milk fats from different mammalian species: potential for use as human milk fat substitutes, *J. Agric. Food Chem.* 61(29), 7070–7080.

Zou, Z., Bauland, J., Hewavitharana, A. K., Al-Shehri, S. S., Duley, J. A., Cowley, D. M., Koorts, P., Shaw, P. N. and Bansal, N. (2021). A sensitive, high-throughput fluorescent method for the determination of lactoperoxidase activities in milk and comparison in human, bovine, goat and camel milk, *Food Chem.* 339, 128090.

# Part 2

## Proteins and lipids

# Chapter 4

## Caseins and casein micelles

*Thom Huppertz, FrieslandCampina and Wageningen University & Research, The Netherlands and Victoria University, Australia; and Inge Gazi, Utrecht University and Netherlands Proteomics Center, The Netherlands*



### 1 Introduction

In this chapter, we will focus on the major class of proteins in milk - the caseins - with particular emphasis on those in the milk of domesticated cattle, *Bos taurus*. With more than 700 billion kilograms of bovine milk produced annually around the world (IDF, 2020) and taking an average casein content of the milk of ~2.5%, ~17 billion kilograms of bovine casein is produced globally on an annual basis, most of which is used for human nutrition. This makes casein the third-largest protein class available for human consumption, after wheat and rice proteins. In addition to being a high-quality source of dietary protein based on amino acid composition and digestibility (Mathai et al., 2017), the caseins have played a central role in the development of the dairy industry. The controlled coagulation of casein by enzymes or acid forms the basis of many dairy products (e.g. yoghurt, cheese and other fermented dairy products) and dairy ingredients (Walstra et al., 2005). This has allowed not only a great diversification into product categories beyond liquid dairy but also the creation of cheese whey or acid whey as co-products and base materials for the manufacture of many nutritional ingredients (Smithers, 2008, 2015).

The importance of the coagulation properties of caseins is also highlighted by the fact that the first designation toward the name casein, that is, Käse (Berzelius, 1814) or Käsestoff (Braconnot, 1830) was already reported before

http://dx.doi.org/10.19103/AS.2022.0099.04

the term 'protein' was even used. However, considering the utilization of casein from the perspective of the dairy sector, it is important to realize that this is, in essence, made possible by the fact that caseins encounter similar conditions in the gastrointestinal tract of the neonate calf (Longenbach and Heinrichs, 1998). Also there, enzymatic and acid-induced coagulation and salt solubilization are key physiochemical changes encountered by the caseins, allowing controlled release of both protein and salts throughout the digestive process (Longenbach and Heinrichs, 1998). Similar processes occur when milk is ingested by humans (Huppertz and Chia, 2021). Hence, one may consider that mimicking some key aspects of the digestive process actually forms the basis for casein conversion in the dairy industry and that the structuring of caseins to fulfill their natural nutritional role enables this.

When considering the caseins in bovine milk and in the milk of most other species, their natural function entails various aspects. As proteins, the caseins are an excellent and highly digestible source of essential amino acids and nitrogen (Mathai et al., 2017; Huppertz and Chia, 2021). Next, the casein sequences also include several shorter peptides, which, when released, can have beneficial biological activity (Vargas-Bello-Perez et al., 2019). Furthermore, the caseins are an essential carrier of salts and thereby allow the transport of large amounts of calcium and phosphate from the mother to the neonate at concentrations far above their solubility limits (Lenton et al., 2015, 2020). The ability of caseins to act as carriers of salts is due to the presence of clusters of phosphorylation on three out of the four casein variants ($\alpha_{s1}$-casein, $\alpha_{s2}$-casein and β-casein) which, combined with calcium and inorganic phosphate, can form a matrix of protein-stabilized calcium phosphate nanoclusters (Holt, 1992; De Kruif and Holt, 2003; Horne, 2020). The inclusion of a fourth casein - κ-casein - ensures that this casein–calcium phosphate matrix becomes not just one big slab of material but that colloidal particles, that is, casein micelles, are formed (Holt, 1992; De Kruif and Holt, 2003; Horne, 2020).

In this chapter, we will review our current understanding of caseins and casein micelles, with particular emphasis on recent developments. For historical overviews on specific topics, we will refer to other publications in the field.

## 2 Chemistry of caseins

The chemistry of caseins is described extensively elsewhere (Swaisgood, 2003; Huppertz, 2013). Here, we will provide a brief overview with particular emphasis on those topics that we believe are most important in relation to casein interactions and the formation and stability of casein micelles. As outlined previously, the casein class in bovine milk consists of four gene products, that is, $\alpha_{s1}$-casein, $\alpha_{s2}$-casein, β-casein and κ-casein. The amino acid sequence of each of the caseins is shown in Fig. 1. For each of the caseins, various genetic variants have been identified across all bovine species (Swaisgood, 2003; Huppertz, 2013). The milk of *Bos taurus* is dominated by a single $\alpha_{s1}$-casein and a single $\alpha_{s2}$-casein genetic variant, respectively, B and A. The most common *Bos taurus* β- and κ-casein genetic variants are shown in Table 1. In addition, all caseins undergo some post-translational modifications (Swaisgood, 2003; Huppertz, 2013). For $\alpha_{s1}$-, $\alpha_{s2}$- and β-casein, post-translational modification is in the form of phosphorylation of Ser, and to a smaller extent of Thr residues, to form SerP and ThrP residues, respectively. In the case of κ-casein, both phosphorylation and glycosylation (of Thr residues *via* O-glycosylation) can occur. In addition,

(a)

| | | | | | | | | | | | | | | | | | | | | | | |
|---|---|---|---|---|---|---|---|---|---|---|---|---|---|---|---|---|---|---|---|---|---|---|
| **1** | Arg | Pro | Lys | His | Pro | Ile | Lys | His | Gln | Gly | **10** | Leu | Pro | Gln | Glu | Val | Leu | Asn | Glu | Asn | Leu | **20** |
| **21** | Leu | Arg | Phe | Phe | Val | Ala | Pro | Phe | Pro | Glu | **30** | Val | Phe | Gly | Lys | Glu | Lys | Val | Asn | Glu | Leu | **40** |
| **41** | Ser | Lys | Asp | Ile | Gly | SerP | Glu | SerP | Thr | Glu | **50** | Asp | Gln | Ala | Met | Glu | Asp | Ile | Lys | Gln | Met | **60** |
| **61** | Glu | Ala | Glu | SerP | Ile | SerP | SerP | SerP | Glu | Glu | **70** | Ile | Val | Pro | Asn | SerP | Val | Glu | Gln | Lys | His | **80** |
| **81** | Ile | Gln | Lys | Glu | Asp | Val | Pro | Ser | Glu | Arg | **90** | Tyr | Leu | Gly | Tyr | Leu | Glu | Gln | Leu | Leu | Arg | **100** |
| **101** | Leu | Lys | Lys | Tyr | Lys | Val | Pro | Gln | Leu | Glu | **110** | Ile | Val | Pro | Asn | SerP | Ala | Glu | Glu | Arg | Leu | **120** |
| **121** | His | Ser | Met | Lys | Glu | Gly | Ile | His | Ala | Gln | **130** | Gln | Lys | Glu | Pro | Met | Ile | Gly | Val | Asn | Gln | **140** |
| **141** | Glu | Leu | Ala | Tyr | Phe | Tyr | Pro | Glu | Leu | Phe | **150** | Arg | Gln | Phe | Tyr | Gln | Leu | Asp | Ala | Tyr | Pro | **160** |
| **161** | Ser | Gly | Ala | Trp | Tyr | Tyr | Val | Pro | Leu | Gly | **170** | Thr | Gln | Tyr | Thr | Asp | Ala | Pro | Ser | Phe | Ser | **180** |
| **181** | Asp | Ile | Pro | Asn | Pro | Ile | Gly | Ser | Glu | Asn | **190** | Ser | Glu | Lys | Thr | Thr | Met | Pro | Leu | Trp | | **199** |

(b)

| | | | | | | | | | | | | | | | | | | | | | | |
|---|---|---|---|---|---|---|---|---|---|---|---|---|---|---|---|---|---|---|---|---|---|---|
| **1** | Lys | Asn | Thr | Met | Glu | His | Val | SerP | SerP | SerP | **10** | Glu | Glu | Ser | Ile | Ile | SerP | Gln | Glu | Thr | Tyr | **20** |
| **21** | Lys | Gln | Glu | Lys | Asn | Met | Ala | Ile | Asn | Pro | **30** | Ser | Lys | Glu | Asn | Leu | Cys | Ser | Thr | Phe | Cys | **40** |
| **41** | Lys | Glu | Val | Val | Arg | Asn | Ala | Asn | Glu | Glu | **50** | Glu | Tyr | Ser | Ile | Gly | SerP | SerP | SerP | Glu | Glu | **60** |
| **61** | SerP | Ala | Glu | Val | Ala | Thr | Glu | Glu | Val | Lys | **70** | Ile | Thr | Val | Asp | Asp | Lys | His | Tyr | Gln | Lys | **80** |
| **81** | Ala | Leu | Asn | Glu | Ile | Asn | Gln | Phe | Tyr | Gln | **90** | Lys | Phe | Pro | Gln | Tyr | Leu | Gln | Tyr | Leu | Tyr | **100** |
| **101** | Gln | Gly | Pro | Ile | Val | Leu | Asn | Pro | Trp | Asp | **110** | Gln | Val | Lys | Arg | Asn | Ala | Val | Pro | Ile | Thr | **120** |
| **121** | Pro | Thr | Leu | Asn | Arg | Glu | Gln | Leu | SerP | Thr | **130** | SerP | Glu | Glu | Asn | Ser | Lys | Lys | Thr | Val | Asp | **140** |
| **141** | Met | Glu | SerP | Thr | Glu | Val | Phe | Thr | Lys | Lys | **150** | Thr | Lys | Leu | Thr | Glu | Glu | Glu | Lys | Asn | Arg | **160** |
| **161** | Leu | Asn | Phe | Leu | Lys | Lys | Ile | Ser | Gln | Arg | **170** | Tyr | Gln | Lys | Phe | Ala | Leu | Pro | Gln | Tyr | Leu | **180** |
| **181** | Lys | Thr | Val | Tyr | Gln | His | Gln | Lys | Ala | Met | **190** | Lys | Pro | Trp | Ile | Gln | Pro | Lys | Thr | Lys | Val | **199** |
| **201** | Ile | Pro | Tyr | Val | Arg | Tyr | Leu | | | | **207** | | | | | | | | | | | |

(c)

| | | | | | | | | | | | | | | | | | | | | | | |
|---|---|---|---|---|---|---|---|---|---|---|---|---|---|---|---|---|---|---|---|---|---|---|
| **1** | Arg | Glu | Leu | Glu | Glu | Leu | Asn | Val | Pro | Gly | **10** | Glu | Ile | Val | Glu | SerP | Leu | SerP | SerP | SerP | Glu | **20** |
| **21** | Glu | Ser | Ile | Thr | Arg | Ile | Asn | Lys | Lys | Ile | **30** | Glu | Lys | Phe | Gln | SerP | Glu | Glu | Gln | Gln | Gln | **40** |
| **41** | Thr | Glu | Asp | Glu | Leu | Gln | Asp | Lys | Ile | His | **50** | Pro | Phe | Ala | Gln | Thr | Gln | Ser | Leu | Val | Tyr | **60** |
| **61** | Pro | Phe | Pro | Gly | Pro | Ile | Pro | Asn | Ser | Leu | **70** | Pro | Gln | Asn | Ile | Pro | Pro | Leu | Thr | Gln | Thr | **80** |
| **81** | Pro | Val | Val | Val | Pro | Pro | Phe | Leu | Gln | Pro | **90** | Glu | Val | Met | Gly | Val | Ser | Lys | Val | Lys | Glu | **100** |
| **101** | Ala | Met | Ala | Pro | Lys | His | Lys | Glu | Met | Pro | **110** | Phe | Pro | Lys | Tyr | Pro | Val | Glu | Pro | Phe | Thr | **120** |
| **121** | Glu | Ser | Gln | Ser | Leu | Thr | Leu | Thr | Asp | Val | **130** | Glu | Asn | Leu | His | Leu | Pro | Leu | Pro | Leu | Leu | **140** |
| **141** | Gln | Ser | Trp | Met | His | Gln | Pro | His | Gln | Pro | **150** | Leu | Pro | Pro | Thr | Val | Met | Phe | Pro | Pro | Gln | **160** |
| **161** | Ser | Val | Leu | Ser | Leu | Ser | Gln | Ser | Lys | Val | **170** | Leu | Pro | Val | Pro | Gln | Lys | Ala | Val | Pro | Tyr | **180** |
| **181** | Pro | Gln | Arg | Asp | Met | Pro | Ile | Gln | Ala | Phe | **190** | Leu | Leu | Tyr | Gln | Glu | Pro | Val | Leu | Gly | Pro | **199** |
| **201** | Val | Arg | Gly | Pro | Phe | Pro | Ile | Ile | Val | | **209** | | | | | | | | | | | |

(d)

| | | | | | | | | | | | | | | | | | | | | | | |
|---|---|---|---|---|---|---|---|---|---|---|---|---|---|---|---|---|---|---|---|---|---|---|
| **1** | Pyr | Glu | Gln | Asn | Gln | Glu | Gln | Pro | Ile | Arg | **10** | Cys | Glu | Lys | Asp | Glu | Arg | Phe | Phe | Ser | Asp | **20** |
| **21** | Lys | Ile | Ala | Lys | Tyr | Ile | Pro | Ile | Gln | Tyr | **30** | Val | Leu | Ser | Arg | Tyr | Pro | Ser | Tyr | Gly | Leu | **40** |
| **41** | Asn | Tyr | Tyr | Gln | Gln | Lys | Pro | Val | Ala | Leu | **50** | Ile | Asn | Asn | Gln | Phe | Leu | Pro | Tyr | Pro | Tyr | **60** |
| **61** | Tyr | Ala | Lys | Pro | Ala | Ala | Val | Arg | Ser | Pro | **70** | Ala | Gln | Ile | Leu | Gln | Trp | Gln | Val | Leu | Ser | **80** |
| **81** | Asn | Thr | Val | Pro | Ala | Lys | Ser | Cys | Gln | Ala | **90** | Gln | Pro | Thr | Thr | Met | Ala | Arg | His | Pro | His | **100** |
| **101** | Pro | His | Leu | Ser | Phe | Met | Ala | Ile | Pro | Pro | **110** | Lys | Lys | Asn | Gln | Asp | Lys | Thr | Glu | Ile | Pro | **120** |
| **121** | Thr | Ile | Asn | Thr | Ile | Ala | Ser | Gly | Glu | Pro | **130** | Thr | Ser | Thr | Pro | Thr | Thr | Glu | Ala | Val | Glu | **140** |
| **141** | Ser | Thr | Val | Ala | Thr | Leu | Glu | Asp | SerP | Pro | **150** | Glu | Val | Ile | Glu | Ser | Pro | Pro | Glu | Ile | Asn | **160** |
| **161** | Thr | Val | Gln | Val | Thr | Ser | Thr | Ala | Val | | **169** | | | | | | | | | | | |

**Figure 1** Amino acid sequence of the mature casein proteins, excluding signal peptides. (a) $\alpha_{s1}$-casein B-8P; (b) $\alpha_{s2}$-casein A-11P; (c) β-casein A2-5P; (d) κ-casein A-1P; highlighted are the following post-translational modifications: **Cys** = cysteine; **SerP** = phosphorylated serine; **Pyr** = cyclization of protein N-terminal glutamine to pyroglutamic acid.

**Table 1** Differences in amino acid sequence between the most common genetic variants of β- and κ-casein shown as a function of the reference proteoforms β-CN A2-5P and κ-CN A-1P, respectively

| Protein | Genetic variant | Amino acid position in mature protein | | | | | Proteoform | Mass (Da) | |
|---|---|---|---|---|---|---|---|---|---|
| | | | | | | | | Monoisotopic | Average |
| | | **67** | **93** | **106** | **122** | **152** | | | |
| β-casein | A2 | Pro | Met | His | Ser | Pro | β-CN A2-5P | 23 968.15 | 23 982.89 |
| | A1 | His | | | | | β-CN A1-5P | 24 008.16 | 24 022.91 |
| | A3 | | | Gln | | | β-CN A3-5P | 23 959.15 | 23 973.88 |
| | B | His | | | Arg | | β-CN B-5P | 24 077.23 | 24 092.02 |
| | F | His | | | | Leu | β-CN F-5P | 24 024.19 | 24 038.96 |
| | I | | Leu | | | | β-CN I-5P | 23 950.19 | 23 964.85 |
| | | **136** | | **148** | **155** | | | | |
| κ-casein | A | Thr | | Asp | Ser | | κ-CN A-1P | 19 025.53 | 19 037.14 |
| | B | Ile | | Ala | | | κ-CN B-1P | 18 993.58 | 19 005.18 |
| | E | | | | Gly | | κ-CN E-1P | 18 995.52 | 19 007.11 |

both $\alpha_{s2}$-casein and κ-casein contain two Cys residues each, which form intra- and intermolecular disulfide bonds (Swaisgood, 2003; Huppertz, 2013).

The degree of phosphorylation varies between the caseins. For $\alpha_{s1}$-casein, the 8P and 9P variants are most commonly observed, but 7P variants have also been found (Nilsson et al., 2020). For $\alpha_{s2}$-casein, next to the most commonly observed 10P, 11P and 12P variants, variants with 9, 13, 14 or 15 SerP residues have also been reported (Fang et al, 2016). For β-casein, only a 4P variant has been observed, next to the most commonly found 5P variant. For κ-casein, next to the 1P variant which is found most often, 0P, 2P and 3P variants have also been reported (Swaisgood, 2003; Huppertz, 2013; Bijl et al., 2020). The Golgi casein kinase (GC-K) responsible for the phosphorylation of all caseins has been identified as the FAM20C kinase (Ishikawa et al., 2012; Tagliabracci et al., 2012). The kinase is highly specific for the Ser–Xxx–Glu/SerP motif. The phosphorylation clusters in the caseins occur in sequence regions that are particularly rich in both Ser and Glu residues. This ensures the formation of a phosphorylation train starting at the C-terminal Ser that is found in the Ser–Xxx–Glu motif and continuing toward the proximal N-terminal Ser residues that fit the motif.

Next to the number of SerP residues, their positioning within the casein molecules is also of particular relevance. In $\alpha_{s1}$-, $\alpha_{s2}$- and β-casein, so-called clusters of phosphorylation are found, that is, three adjacent SerP residues. It is worthwhile noting that all these caseins have SerP–SerP–SerP–Glu–Glu sequences, that is, residues 66–70 of $\alpha_{s1}$-casein, residues 17–21 of β-casein and residues 8–12 and 56–60 of $\alpha_{s2}$-casein (Fig. 1). For $\alpha_{s2}$-casein, an additional SerP is found just after this sequence, at position 61, whereas for β-casein, an additional SerP is found at position 15, that is, also in close proximity to the $(SerP)_3$–$Glu_2$ sequence (Fig. 1). It has been established that these clusters of phosphorylation are essential in stabilizing calcium phosphate nanoclusters, as outlined in further detail in Section 3.2.1. For both $\alpha_{s1}$- and $\alpha_{s2}$-casein, a SerP–X–SerP sequence is also present, at respective positions 46–48 and 129–131, with X being Glu and Thr, respectively (Fig. 1).

O-glycosylation occurs in the C-terminal part of κ-casein, with seven Thr having been identified as glycosylation sites, that is, $Thr^{121}$, $Thr^{131}$, $Thr^{133}$, $Thr^{136}$ (not in variant B in light of the $Thr^{136} \rightarrow Ile^{136}$ mutation), $Thr^{142}$, $Thr^{145}$ and $Thr^{165}$ (Sunds et al., 2019). Although κ-casein molecules with up to six glycosylated Thr residues have been found (Holland et al., 2006), the non-, mono- and diglycosylated variants of κ-casein are by far the most abundant (Holland et al., 2004, 2005). All milk samples investigated to date, including those from individual cows, contain a mixture of glycosylated and non-glycosylated κ-casein, with variation observed in the ratio between these forms of κ-casein (Bijl et al., 2014; Day et al., 2015; Poulsen et al., 2016; Sunds et al., 2019). The exact reason for the variability in κ-casein glycosylation is not known at present, but it is interesting to note that a relation between κ-casein glycosylation and κ-casein polymerization has been established; that is, higher-order polymers of κ-casein have been found to contain less glycosylated κ-casein than lower-order aggregates, for example, dimers or trimers (Groves et al., 1992). However, κ-casein dimers and trimers in milk also contain non-glycosylated κ-casein (Groves et al., 1992). This does draw the question of whether glycosylation affects the polymerization of κ-casein or vice versa. Conceptually, κ-casein polymerization affecting subsequent glycosylation may be more appealing, considering the fact that intermolecular disulfide bonds between κ-casein form in the N-terminal part ($Cys^{11}$ and $Cys^{88}$) whereas glycosylation occurs in the C-terminus. Dimers and higher-order polymers of (106–169) of κ-casein (i.e., the casein macropeptide; CMP) have been shown to occur and even remain intact under reducing conditions during SDS-PAGE electrophoresis (Farías et al., 2010; Mikkelsen et al., 2005). It would appear, though, that these form after the release of the peptide from κ-casein because intact κ-casein is readily reduced to monomers under the same conditions.

Next to κ-casein, $\alpha_{s2}$-casein also forms disulfide bonds *via* its two Cys residues, that is, $Cys^{36}$ and $Cys^{40}$. Unlike κ-casein, for which disulfide bonding appears to be (virtually) exclusively intermolecular, both inter- and intramolecular disulfide bonds have been observed for $\alpha_{s2}$-casein (Rasmussen et al., 1992, 1994; Bouguyon et al., 2006). Disulfide bonds in $\alpha_{s2}$-casein dimers may be in both parallel and anti-parallel orientations (Rasmussen et al., 1992, 1994). Higher orders of polymerization than dimers have not been observed for $\alpha_{s2}$-casein, nor have disulfide-linked polymers between $\alpha_{s2}$-casein and κ-casein been observed. The position of the Cys residues in $\alpha_{s2}$-casein is between the two aforementioned SerP–SerP–SerP–Glu–Glu at positions 81–12 and 56–60 (Fig. 1). The mode of disulfide bonding of $\alpha_{s2}$-casein (intra- vs. intermolecular, parallel vs. anti-parallel orientation) could have implications for the role of $\alpha_{s2}$-casein in the stabilization of calcium phosphate nanoclusters in casein micelles. This topic, however, has not been studied to date.

# 3 Casein micelles

## 3.1 Compositional and physicochemical properties of casein micelles

In bovine milk, casein occurs predominantly in the form of casein micelles. In addition, a smaller proportion of casein is also found in what is commonly referred to as the fraction of non-micellar casein. Separation of micellar and non-micellar casein is typically done on the basis of (ultra)centrifugation or (micro)filtration, and as such, the classification of micellar or non-micellar is typically a size-based classification. There is, however, no unequivocal definition of the non-micellar casein fraction. Particle size analysis reveals that the casein micelles in milk have an average radius of 70–120 nm, with differences observed between the milk from different cows (Glantz et al., 2010; De Kruif and Huppertz, 2012; Bijl et al., 2014; Gustavsson et al., 2014; Logan et al., 2015; Ketto et al., 2017) and all individual cows showing a relatively narrow size distribution (De Kruif and Huppertz, 2012). Bulk milk narrows the variation in average particle size but broadens the size distribution (De Kruif and Huppertz, 2012). Electron microscopic analysis reveals that the casein micelles have a near-spherical shape (McMahon and McManus, 1998; Dalgleish et al., 2004; Trejo et al., 2011; Hettiarachchi et al., 2020). Compositionally, the dry matter of the micelles has been shown to consist primarily of protein (93–95%), with the remainder being salts, commonly referred to as micellar calcium phosphate (MCP) (De Kruif et al., 2012; Huppertz et al., 2017). Next to calcium and phosphate, this fraction also contains some magnesium and citrate. The levels of MCP in the casein micelles, also referred to as casein mineralization, vary notably between the milk from individual cows, with more than a 1.5-fold difference observed between milk with high and low levels of mineralization (Bijl et al., 2019; Huppertz et al., 2021). On a wet basis, casein micelles have been estimated to contain approximately 75% water, thus making them highly hydrated (Dalgleish et al., 1989; Huppertz et al., 2017).

### 3.2 Structural aspects of casein micelles

While the composition and physicochemical properties of casein micelles in bovine milk are widely agreed on, their structure has been a widely debated topic for many decades. Part of this, arguably, is due to the fact that the structure of a casein micelle cannot be 'simply visualized'. For a particle of defined size, mass and composition, the structure thus is essentially related to the distribution of material throughout the particle. Within this context, three aspects can be distinguished in relation to casein micelle structure, which will be covered in this section, which are:

1 salts and their distribution and interactions;
2 proteins and their distribution and interactions; and
3 water and its distribution.

#### 3.2.1 Salt distribution and interactions

As outlined in the Introduction and in Section 2, bovine milk is strongly supersaturated with respect to most calcium phosphate salts. Because this would entail the risk of pathological calcification of the mammary gland, a sizeable proportion of the calcium and phosphate is encapsulated in the micelles in the form of MCP. For bulk milk, on average, half of the inorganic phosphate and two-thirds of the calcium are in the form of MCP, whereas the remainder is in soluble form, either as free ions or as complexes (White and Davies, 1958; Walstra et al., 2005). While the MCP fraction only represents a small part of the dry matter of a casein micelle, it is crucial in maintaining the structural integrity of the micelle. Chelation or removal of the micellar calcium by, for example, citrate, EDTA or cation exchange resins, leads to the disruption of casein micelles (Ward et al., 1997; Udabage et al., 2000; Smiddy et al., 2006; Grimley et al., 2010; Xu et al., 2016). Based on electron-dense regions in electron micrographs of casein micelles, Knoop et al. (1973) suggested an inhomogeneous distribution of calcium phosphate in the casein micelles. A decade later, McGann et al. (1983a) showed that after deproteination of casein micelles by hydrazine, granular amorphous calcium phosphate particles phosphate ranging several nanometers in diameter remained. Similar calcium phosphate granules were also observed after hydrolysis of caseins in the casein micelles by Pronase, a mixture of endo- and exopeptidases (McGann et al., 1983b). Further analysis of such granules after tryptic digestion of casein micelles has revealed that the phosphorylation clusters remain attached to the surface of these calcium phosphate granules despite extensive protein hydrolysis by the enzyme (Ono et al., 1994). Subsequent research showed that these nanoclusters of calcium phosphate, with a composition and size very similar to those found in the casein micelles, could be produced from supersaturated solutions of salts if sufficient concentrations of one of more phosphopeptides with the aforementioned SerP-SerP-SerP-Glu-Glu motif were present, for example, f(1–25) of β-casein (Holt et al., 1996, 1998). These phosphopeptides were shown to associate on the surface of the calcium phosphate nanoclusters, thereby forming a shell which limits further growth (Holt et al., 1998). The size of the nanoclusters appears to be dependent on the phosphorylation motif of the phosphopeptide(s) used. When nanoclusters are prepared with casein phosphopeptides containing the aforementioned SerP-SerP-SerP-Glu-Glu motif, diameters in the range 4.5-5.0 nm are invariably found (Holt et al., 1998). However, if phosphopeptides from the much more highly phosphorylated protein osteopontin are used, much larger calcium phosphate nanoclusters can be prepared (Holt et al., 2009). In casein micelles, the presence of calcium phosphate nanoclusters has been verified using small-angle X-ray and neutron scattering experiments, a broad overview of which was presented by De Kruif (2014), supplemented by more recent studies (Bouchoux et al., 2015; Ingham et al., 2015, 2016; Day et al., 2017; Nogueira et al., 2021). These studies all strongly support the presence of calcium phosphate nanoclusters with a diameter of 4.5–5.0 nm in casein micelles. A typical casein micelle will contain several hundred of these nanoclusters, at an average spacing of ~18 nm (De Kruif et al., 2012). The ratio of MCP:casein can vary between the milk from individual cows, but the ratio of Ca:P is extremely constant (White and Davies, 1958; Huppertz et al., 2021), indicating a constant nanocluster composition as well. This composition was proposed to be $[Ca_{13.2}(P_i)_{6.5}Mg_{1.0}Cit_{1.3}Casein_{1.0}(SerP)_4]_{49}$ by De Kruif and Holt (2003).

### 3.2.2 Casein interactions

Although both $\alpha_{s2}$-casein and κ-casein are able to form intermolecular disulfide bonds, most interactions between casein molecules are non-covalent. The nature of these interactions has been debated extensively, with some highlighting the essential role of hydrophobic interactions (Horne, 2017, 2020), and others arguing against this and highlighting that backbone interactions are the most crucial (Holt et al., 2013; Thorn et al., 2015; Carver and Holt, 2019). De Kruif et al. (2012) explicitly used the term 'weak interactions' as a collective term for hydrophobic bonding, hydrogen bonding, electrostatic interactions and van der Waals interactions that in a cooperative fashion can result in interactions between casein molecules. The weak interactions specifically excluded the ionic bonds forming between the SerP residues and the calcium phosphate nanoclusters. The cooperative fashion of intermolecular casein interactions is also apparent from the different means by which they can be disturbed, leading to dissociation. This is observed, for example, as a result of increasing net-negative charge on the caseins, either through increases in pH (Vaia et al., 2006) or due to enzymatic deamidation of glutamine residues to glutamic acid residues (Miwa et al., 2010). Furthermore, the addition of detergents like sodium dodecyl sulfate (SDS), whose hydrophobic tail can interact with proteins, results in the disruption of casein interactions (Lefebvre-Cases et al., 1998; Smiddy et al., 2006), as does the addition of chaotropic agents like urea or guanidine hydrochloride (McGann and Fox, 1974; Aoki et al., 1986; Holt, 1998; Smiddy et al., 2006), which disrupt the hydrogen bonding network between water molecules, thereby affecting the hydration shells around protein molecules and consequently protein folding. Because urea does not affect the strong interactions between caseins and calcium phosphate nanoclusters, size-exclusion chromatography in the presence of urea can be applied to separate the caseins which interact with the calcium phosphate nanoclusters and those that do not. Using this approach, Aoki et al. (1986, 1990, 1991, 1992) found that virtually all $\alpha_{s1}$- and $\alpha_{s2}$-casein, part of the β-casein and none of the κ-casein is associated with the nanoclusters in the casein micelles.

Changes in temperature also affect casein interactions. Most notably, cooling can result in dissociation of β-casein from the micelle (Atamer et al., 2017), whereas heating at temperatures greater than 60°C can result in dissociation of κ-casein from the casein micelle (Huppertz, 2016; Dumpler et al., 2017). Dissociation of β-casein on cooling has been attributed to weakened hydrophobic interactions at a lower temperature (Huppertz et al., 2006; Atamer et al., 2017) and is in line with reduced micellization of β-casein observed at the lower temperature in self-association studies (De Kruif and Grinberg, 2002). It is worthwhile noting that only a part of the β-casein in the casein micelle will dissociate on cooling, with maximum values of dissociated β-casein typically not exceeding 30% (Huppertz et al., 2006; Atamer et al., 2017). Presumably, it is the fraction of β-casein not interacting with the calcium phosphate nanoclusters, which can dissociate on weakening of interactions between caseins (Aoki et al., 1990). For κ-casein, the dissociation of micellar κ-casein that occurs at temperatures above 60°C cannot be linked directly to studies on self-association. While it is apparent that the heat-induced dissociation results in a weakening of the aforementioned collective weak interactions, the exact nature has not been established to date. The pH-dependence of the heat-induced dissociation of κ-casein, with a higher extent of dissociation occurring at a higher pH, and dependence on the addition of calcium and phosphate, for example (Huppertz, 2016; Anema, 2020), suggests an involvement of charged residues. In this perspective, it is also interesting to consider the potential role of histidine. With a pKa of the imidazole side group of 6.4, which is temperature-dependent (pKa decrease of ~1 unit between 0°C and 60°C), both increases in pH and in temperature favor the deprotonation of the imidazole group of His. A direct link, however, between His protonation and κ-casein remains to be established.

An area of casein interactions that has received notable research attention in recent years is their ability to form fibrils under certain conditions which, based on thioflavin-T (THT)-staining, have been designated to be amyloid-type fibrils. It has been shown that the two caseins capable of forming such fibrils are κ-casein (Ecroyd et al., 2008; Carver et al., 2010) and $\alpha_{s2}$-casein (Thorn et al., 2008, 2021), that is, the two caseins containing Cys residues, but only when they are in reduced form. When intra- and intermolecular disulfide bonds are present, as they would be in milk, κ-casein fibril formation is suppressed (Ecroyd et al., 2010). The oxidizing environment of the endoplasmic reticulum in this respect is important to ensure the rapid formation of disulfide bonds. Furthermore, the presence of other caseins, namely $\alpha_{s1}$- and β-casein, also suppresses fibril formation by κ- and $\alpha_{s2}$-casein (Thorn et al., 2005). Hence, while κ-casein and $\alpha_{s2}$-casein do have the ability to form amyloid-type fibrils, the process occurs primarily under conditions that are not encountered naturally. It is thus unlikely that these fibrils affect casein structure and casein micelle structure in milk.

### 3.2.3 Hydration of casein micelles

As outlined in Section 3.1, by mass, most of the casein micelle actually consists of water, with ~3–3.5 g of water per gram of dry matter typically reported for casein micelles at natural pH at 20–25°C, based on either water content of ultracentrifugal pellets or (intrinsic) viscosity measurements (assuming a spherical shape of the micelles). However, like for protein and salts, specific distributions of water are also observed in casein micelles. Some water is bound by the protein; this so-called non-freezable water has been estimated to be ~0.5 g per gram of protein (Huppertz et al., 2017). Furthermore, the surface of the casein micelle appears to be highly hydrated, because hydrolysis of κ-casein by chymosin reduces micellar hydration by ~1.0 g water per gram of protein. The remaining water (close to 1.5–2.0 g of water per gram of protein) is entrapped in the micelle structure but freely diffusible (Huppertz et al., 2017). Small-angle neutron scattering (SANS) measurements show these areas as solvent pockets (De Kruif et al., 2012).

## 3.3 Colloidal stability of casein micelles in relation to surface composition

As outlined in the Introduction, the controlled destabilization of casein micelles is the basis of not only the manufacture of cheese, yogurt and many other dairy products but also the coagulation of casein in the stomach. Given the importance of this process, it is not surprising that a considerable amount of research has been dedicated to the colloidal stability of casein micelles. It has been established that casein micelles can be considered as sterically stabilized association colloids, whose stability is mainly governed by a brush of casein segments protruding from the surface of the micelles (Holt and Horne, 1996; Dalgleish, 1998; De Kruif, 1999; Tuinier and De Kruif, 2002). This brush consists to a large extent of the mobile C-terminal part of κ-casein, including the chymosin-sensitive-site $Phe_{105}$-$Met_{106}$; other caseins are also known to be part of the surface (Rollema et al., 1988; Dalgleish, 1998, 2011). The exact proportion of each casein on the surface is unknown, but it is clear that κ-casein plays a major role in surface stabilization because the specific hydrolysis of κ-casein at $Phe_{105}$-$Met_{106}$ by chymosin, or other enzymes with similar specificity, causes sufficient destabilization to make the micelles susceptible to aggregation (De Kruif, 1999; Horne and Banks, 2004; Corredig and Salvatore, 2016). The chymosin-induced hydrolysis of κ-casein reduces the diameter of the casein micelles by ~10 nm. Although the CMP part of κ-casein that is split from the micelle surface does have a notable net-negative charge, the stabilizing action of κ-casein on the micelle surface is believed to be primarily steric rather than electrostatic, because even addition of high levels of salt, which reduces the Debye length to very small values, does not cause coagulation of casein micelles (Huppertz and Fox, 2006). Surface coverage by κ-casein is believed to be in patches of κ-casein oligomers (Dalgleish, 2011), which may occur at approximately similar spacings as the calcium phosphate nanoclusters inside the micelles, that is, ~15–20 nm

(Huppertz et al., 2017). Hence, hydrolysis of some of these oligomers creates bare patches on the micelle surface available for aggregation. Aggregation of enzymatically destabilized micelles, that is, para-casein micelles, has typically been considered on the basis of adhesive hard-sphere theory (De Kruif, 1993, 1999). This approach shows good correlations between theoretical and experimental values. However, the adhesive hard-sphere approach merely considers the aggregating particles, whereas it has become apparent that non-aggregating particles can also affect the aggregation of para-casein micelles. It has, for instance, been shown that both the presence of native whey protein (Gamlath et al., 2018) and the addition of sodium caseinate (Gaygadzhiev et al., 2012; Lin et al., 2017) notably reduce the rate of aggregation of para-casein micelles. Hence, the role of other protein particles should also be considered when considering the aggregation behavior of para-casein micelles. These probably act as barriers between para-casein micelles and thus prevent interaction. On a molar basis, it is important to consider in this respect that the number of whey proteins in milk exceeds the number of casein micelles by approximately four orders of magnitude. Likewise, the presence of even small amounts of casein in non-micellar form (e.g. 5% of total casein) readily creates a large molar excess of non-micellar casein particles over casein micelles, which can interfere with coagulation.

In addition to enzymatic coagulation, acid-induced coagulation of casein micelles is another widely-studied phenomenon. This process occurs on acidification of milk, for example, in the manufacture of yogurt or caseinates. For casein micelles in unheated or mildly pasteurized milk, acid-induced coagulation commences at pH ~5.0. This process has also been described by the adhesive hard-sphere theory, but the loss of steric stabilization in this case is not due to the removal but rather the collapse of the stabilizing κ-casein brush, which occurs as a result of protonation of carboxylate groups of the side chains of amino acid residues in the C-terminus of κ-casein (De Kruif and Zhulina, 1996). However, it is worthwhile noting that at pH 5.0, only ~25% of Glu residues (pKa = 4.5 for the carboxylate side group of Glu) would be protonated, that is, two out of eight Glu residues in the C-terminus of κ-casein. In this context, it is worthwhile considering whether other factors also contribute to the pH-induced collapse of the κ-casein brush. One such factor could lie in the Pro-residues. Unlike other peptide bonds, which occur virtually exclusively in trans configuration, X-Pro peptide bonds show notable cis–trans isomerism which is pH-dependent (Ivanova et al., 2010). Given the position of Pro on positions 101, 109, 110 and 120 in κ-casein (Fig. 1), that is, around the chymosin-sensitive region, it is worthwhile considering whether cis–trans isomerism of peptide bonds involving these Pro-residues affects the stability of the κ-casein brush.

The combined effects of acid- and rennet-induced coagulation have also been researched and it has been established that (1) partial removal of CMP by chymosin increases the pH at which acid coagulation occurs and that (2) aggregation of para-casein micelles is enhanced at lower pH. Furthermore, the combined effects of pH and κ-casein brush collapse with ethanol and/or chymosin hydrolysis on coagulation of casein micelles have also been researched. On its own, ethanol can induce coagulation of casein micelles *via* the collapse of the stabilizing κ-casein brush (Horne, 2016). At lower pH, the amount of ethanol required to induce coagulation of casein micelles is reduced (Horne and Parker, 1980), whereas adding ethanol to milk also increases the pH at which acid-induced coagulation of milk commences (Horne and Parker, 1980). Likewise, partial removal of CMP by chymosin reduces the amount of ethanol required to induce visible coagulation of milk (De Kruif, 1999) and adding ethanol can also enhance rennet-induced coagulation of milk at conditions where the added ethanol does not inhibit the activity of chymosin (O'Connell et al., 2006). It is thus clear that there are various ways of destabilizing casein micelles, which can also work together.

## 3.4 Variation in casein micelles

Over the years, a notable number of studies have been conducted to investigate the variability in the properties of casein micelles. Such studies have mainly focused on casein micelle size and composition, as well as the rennet- and acid-induced coagulation of casein micelles. In terms of casein micelle size, it has been shown that differences in casein micelle size occur between the milk of different cows and that the width of the particle size distribution is notably narrower for milk from individual cows than for bulk milk (De Kruif and Huppertz, 2012). In addition, it has been shown that casein micelle size in the milk from one cow is constant over a milking, over a lactation stage and even over many lactation stages (De Kruif and Huppertz, 2012; Bijl et al., 2014). In other words, casein micelle size appears to be highly defined for each cow. Early studies on centrifugal fractionation of bulk milk concluded that the fractions with smaller casein micelles contained more κ-casein (Creamer et al., 1973; McGann et al., 1980; Dalgleish et al., 1989). However, when relating the κ-casein:total casein ratio to casein micelle size in the milk from individual cows, no such relations are found, and micelle size appears independent of casein composition (De Kruif and Huppertz, 2012; Bijl et al., 2014). However, when genetic variation and post-translational modification of casein are also taken into account, it is clearly noted that milk with κ-casein A has larger casein micelles than milk with κ-casein B (Frederiksen et al., 2011; Bijl et al., 2014; Day et al., 2015; Huppertz et al., 2021) and that a negative correlation is observed between κ-casein B glycosylation and casein micelle size (Bijl et al., 2014). Hence, from studies on milk from individual cows, it appears that not κ-casein content but the genetic variation and post-translational modification of κ-casein strongly affect micelle size. The fact that an inverse correlation between micelle size and κ-casein content has been observed on centrifugal fractionation of (bulk) milk thus appears contrary to these findings. However, it should be kept in mind that, as stated earlier, not all casein in milk is in the micellar fraction and therefore the content and ratios of caseins in the micelles are expected to differ from those of total casein in the milk. Furthermore, Huppertz et al. (2021) recently showed that for milk with high levels of κ-casein glycosylation associated with smaller micelles, the pellets formed on ultracentrifugation were less cohesive and more prone to a disturbance on decanting. Hence, this could lead to the casein that did sediment actually being classified as nonsedimentable.

The exact reason for the relationship between genetic variants, glycosylation of κ-casein and casein micelle size has not been established to date. Explanations may be sought in the area of oligomerization of κ-casein, because, as outlined previously, there appears to be an inverse relationship between κ-casein oligomerization and glycosylation. Hence, a larger number of κ-casein oligomers would be available for casein micelle surface stabilization in milk with κ-casein B and more highly glycosylated κ-casein (Poulsen et al., 2016). Further research is required to establish this relation, as well as to clarify whether the oligomerization of κ-casein precedes glycosylation or vice versa.

In addition to variation in casein micelle size, variation in the mineralization of the casein fraction of milk has also been established in milk from different cows. A recent study by Huppertz et al. (2021) established that in the milk from 48 different Holstein-Friesian cows taken from a single farm in the Netherlands, the degree of mineralization of the casein fraction varied approximately twofold, that is, from 6.7 to 12.9 mmol protein-associated Ca per 10 g of casein. Protein-associated P levels showed a similar degree of variation and the relationship between micellar Ca and micellar P was very constant across the range of samples. Similar trends could be deduced on the re-evaluation of the data of White and Davies (1958) and Bijl et al. (2014), who also observed large variation in casein mineralization in the milk of individual Ayrshire and Holstein-Friesian cows, respectively. Mineral speciation calculations on the data of Bijl et al. (2014) by Bijl et al. (2019) also highlighted such differences, in this case expressed as a saturation parameter, reflecting the degree to which the maximum binding capacity of the casein fraction with respect to micellar salts had been occupied. The correlation of differences in casein mineralization to genetic variants of β-casein, κ-casein and β-lactoglobulin present in the milk of 48 Holstein-Friesian cows did not yield significant differences (Huppertz et al., 2021). No correlation could be established either between casein mineralization and casein micelle size (Huppertz et al., 2021). However, casein mineralization may affect several properties of milk, such as rennet coagulation (Malacarne et al., 2014), *in vitro* gastric coagulation and digestion (Huppertz and Lambers, 2020), heat-induced coagulation (Fox and Hoynes, 1975) and acid-induced coagulation (Anema, 2009).

## 4 Formation of casein micelles

### 4.1 Learnings from in vitro studies

In addition to characterization of casein micelles as found in bovine milk, as well as in milk from other species, a considerable body of research has also been dedicated to *in vitro* studies wherein the formation of casein micelles from various casein sources was studied. Recently, Huppertz et al. (2017) reported that when stock solutions of (1) sodium caseinate, (2) calcium chloride and magnesium chloride and (3) sodium phosphate and trisodium citrate were titrated into a milk permeate solution over 50 min at 37°C, a suspension of casein particles very closely resembling casein micelles was obtained. In addition to similarities in size and turbidity, further similarities were observed in the distribution of caseins between the sedimentable and nonsedimentable phase, as well as the distribution of calcium, magnesium and phosphorus between the soluble phase, the micellar phase and the nanoclusters within the micellar phase. These findings highlight that it is possible to (re)produce casein micelles from sodium caseinate under appropriate conditions. The conditions chosen by Huppertz et al. (2017) were chosen based on several decades of previous research by others, which highlighted some key observations.

When only calcium and phosphate are added to the casein solution, needle-like structures, presumably of calcium phosphate, are observed, but when citrate is also included, micelles are formed (Knoop et al., 1979). This occurs due to the fact that in the absence of citrate, local high concentrations of calcium and phosphate can lead to precipitation if stabilization by caseins occurs at an insufficiently quick rate. Citrate, in this respect, can aid in preventing excessive local supersaturation and hence facilitate the gradual formation of casein micelles.

Variations in casein composition do not affect the distribution of calcium and inorganic phosphate between the sedimentable and nonsedimentable phases of the micelle dispersions but do affect the protein distribution and nitrogen between the sedimentable and nonsedimentable phases (Schmidt et al., 1977). At a given ratio of $\alpha_{s1}$-casein:β-casein, the level of nonsedimentable nitrogen increased with an increasing proportion of κ-casein. In addition, nonsedimentable nitrogen also increased with a decreasing proportion of $\alpha_{s1}$-casein and an increasing proportion of β-casein in the samples (Schmidt et al., 1979).

When keeping casein content and composition constant but varying total Ca and inorganic phosphate, Schmidt et al. (1979) found an inverse correlation between total calcium and serum nitrogen; that is, at a given concentration of inorganic phosphorus, serum nitrogen decreased with an increasing total calcium content. In addition, serum nitrogen content decreased slightly with decreasing inorganic phosphorus content of the samples. Citrate concentration also affects the amount of protein in the supernatant of artificial casein micelle systems; with increasing citrate concentration, the concentration of protein in the ultracentrifugal supernatant was also found to increase (Schmidt et al., 1979). Aoki (1989) also showed that the proportion of caseins that could not be dissociated to monomers under reducing conditions in the presence of 6 M urea, increased with increasing concentration of calcium and phosphate added (at a 1.5:1 ratio) on top of base levels of 10 mmol $L^{-1}$ and 12 mmol $L^{-1}$, respectively, of calcium and phosphate already present. In this study, as for casein micelles from bovine milk, the order of linking of caseins to calcium phosphate was $\alpha_{s1}$-casein > $\alpha_{s2}$-casein > β-casein, whereas no κ-casein was found to associate. This again confirms that at the appropriate ratio of casein:calcium:inorganic phosphate structural elements strongly resembling those in casein micelles can be formed from casein.

## 4.2 Casein micelle biosynthesis

While *in vivo* biosynthesis of casein micelles has been difficult to study, some insights have emerged. As outlined earlier, the synthesis of casein micelles requires some key elements, that is, the caseins and their post-translational modification, as well as the salts. The pathway of biosynthesis should lead to casein micelles with physicochemical properties as described above, that is, a near-spherical shape, of a defined size, and a composition with substructure elements. Farrell et al. (2006) described this process and separated the following key aspects:

1. Synthesis and secretion of caseins;
2. Post-translational modification of caseins;
3. Association of caseins; and
4. Interactions of caseins with salts.

The synthesis of caseins occurs on the ribosomes of the highly developed rough endoplasmic reticulum of the secretory epithelial cells, which is followed by the insertion of the caseins into the lumen of the endoplasmic reticulum and their transport through the Golgi apparatus. At this point, the caseins appear to occur in complexes of about 10 nm in diameter (Farrell et al., 2006). This is notably larger than the size of a casein monomer, thus indicating that some association of caseins has already occurred at this point. Although all caseins show self-association, the association occurring is likely to be mixed association given that the presence of $\alpha_{s1}$-casein is known to reduce self-association of β-casein and κ-casein and in fact favor mixed association (Farrell et al., 2003, 2006). At this stage, the caseins also become phosphorylated by a kinase. The high calcium ion concentration required by this kinase results in the virtually immediate association of calcium ions with phosphoserine residues on the caseins (Bingham and Farrell, 1974). The high local calcium concentrations in the Golgi vesicles are achieved by an ATPase-driven calcium pump. For micelle formation, not only calcium, but also inorganic phosphate is required. It has been suggested that this inorganic phosphate is derived from ADP, as a byproduct of the phosphorylation reaction utilizing ATP. The conversion of ADP to AMP would release inorganic phosphate and thus facilitate micelle formation (Farrell et al., 2006). Considering the 6.5:4 ratio for inorganic to organic phosphate for nanoclusters suggested by De Kruif and Holt (2003), the fact that not all SerP residues may associate with nanoclusters and that some basal levels of phosphate will be present in the vesicles, ADP could indeed be the main source of inorganic phosphate used in nanocluster formation, and hence the formation of casein micelles.

The above process thus suggests that caseins first undergo association reactions before they become phosphorylated and the association of calcium with SerP occurs. Subsequently, increases in concentrations of inorganic phosphate from the conversion of ADP to AMP initiate the formation of nanoclusters. The spatial organization of SerP clusters may function as a template for nanocluster formation in this respect. Huppertz et al. (2017) suggested that the casein micelle consists of a network of subunits, called primary casein particles, which are connected through calcium phosphate nanoclusters, as shown in Fig. 2. In this model, each primary casein particle connects to two nanoclusters and each nanocluster to six primary casein particles (Fig. 2). The structure of the primary casein particles is based on that proposed by Kumosinki et al. (1994a,b) and Farrell et al. (2003), consisting of a framework of four $\alpha_{s1}$-casein molecules to which four β-casein molecules associate, as well as one $\alpha_{s2}$-casein molecule or one or more κ-casein molecules. The $\alpha_{s2}$-casein molecules containing primary casein particles are found in the core of the micelles, whereas those containing κ-casein are found on the surface of the micelles. This subunit model differs from previous submicelle models (Schmidt, 1982; Walstra, 1990; Walstra et al., 2005) by the shape of the primary casein particles (i.e. elongated rather than spherical) and the limited number of nanoclusters (i.e. two) that a primary casein particle can associate with. This model also provides a basis for the fact that casein micelles are porous and that proteins and enzymes can diffuse into and

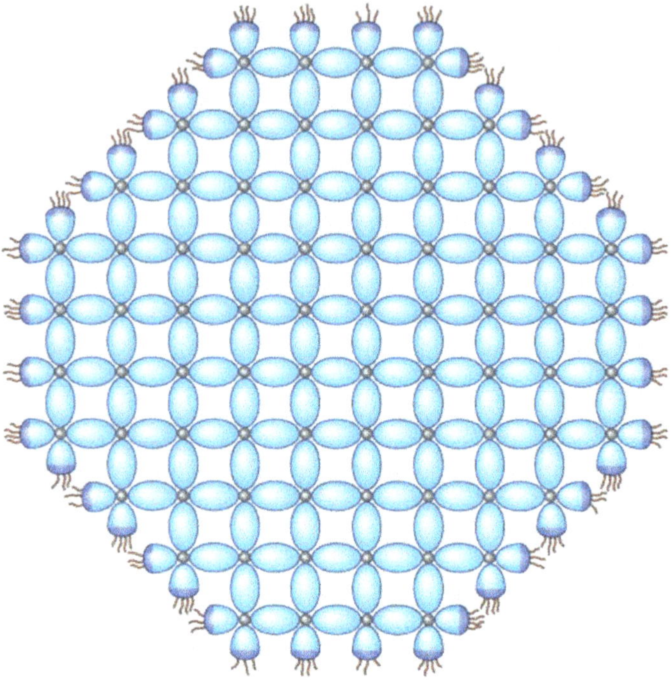

**Figure** 2 Illustration of a transversal section through a casein micelle with a diameter of ~200 nm. Red - colloidal calcium phosphate nanoclusters; blue - primary casein particle, with areas consisting of $\alpha_{S1}$-, $\alpha_{S2}$- and β-casein in light blue and κ-casein in dark blue/brown. Reproduced with permission from: https://www.frieslandcampina.com/en/innovation/casein-micelles/.

out of a casein micelle; it is also in line with SANS measurements, indicating that defined spacings not only of calcium phosphate nanoclusters but also of protein-rich and solvent-rich domains exist in the structure. The rather porous structure is in line with previous suggestions by Dalgleish (2011) and cryo-TEM tomography studies by Trejo et al. (2011) and Hettiarachchi et al. (2020), but deviates from rather uniform other envisaged structural models of casein micelles (De Kruif and Holt, 2003; De Kruif et al., 2012) which envision more homogenous protein matrices. Undoubtedly, further refinements of casein micelle structure will be proposed, discussed, debated and challenged in the years to come.

## 5 Conclusion

Successful discoveries in any field of science can often be measured not only by their academic credentials but also on their potential application to improve the quality of life. For research on caseins and casein micelles, bridging this gap has been a key driver for more than a century of high-quality scientific research that has been hugely important for the development of the dairy sector to its current status. With casein being one of the main protein sources in the human diet, understanding its structure, stability and conversion into different safe and stable products for human consumption has been crucial. With the level of understanding available at present, many aspects can be well understood and applied. However, some frontiers in casein research remain, including further quantitative aspects of casein interactions, variations in casein composition and casein mineralization between individual animals and providing full linkage between properties of casein micelles and their behavior during and post-processing into dairy products. Undoubtedly, advances in analytical techniques and computing power in years to come - combined with new generations of curious minds willing to tackle these topics - will continue to broaden and deepen our understanding and form the basis of further utilization of this unique class of proteins.

## 6 Where to look for further information

Caseins and casein micelles have intrigued scientists for more than two centuries and will undoubtedly continue to do so for the foreseeable future. While the area of food science is the central area of application, the research on caseins and casein micelles has attracted researchers from many other areas, including biochemistry, physical chemistry, colloid science, proteomics, biology, and physiology. It is this combination that has led to the current state-of-the-art, and this would certainly not have been achieved without such multidisciplinary approaches. Key contributions over the years have come from many research groups, including those at the Hannah Research Institute (UK), Wageningen University (NL), INRA (FR), NIZO (NL), CSIRO (AU), Massey University (NZ), University College Cork (IE), University of Wisconsin-Madison (USA), University of Guelph (CA), University of Glasgow (UK), and the US Department of Agriculture (USA). However, this list is by no means complete, and could easily be extended by order of magnitude. Every new piece of research and every fresh look at previous work has the potential to provide valuable new insights into the area, which will increase not only the understanding of the area, but also the utilization potential by the dairy sector.

Publications in the field regularly feature in key dairy journals like International Dairy Journal and Journal of Dairy Science, but also in many other journals. The topic also finds extensive coverage in books, for example, the Advanced Dairy Chemistry series and the various editions of Milk Proteins: from Expression to Food, and is a regularly topic at many conference, including those organized by International Dairy Federation, the NIZO Dairy Conference, and Food Hydrocolloids.

## 7 References

Anema, S. G. (2009). Role of colloidal calcium phosphate in the acid gelation properties of heated skim milk. *Food Chemistry* 114(1), 161-167.

Anema, S. G. (2020). The whey proteins in milk: thermal denaturation, physical interactions, and effects on the functional properties of milk. In: Boland, M. and Singh, H. (Eds) *Milk Proteins* (pp. 325-384). Academic Press, London, UK.

Aoki, T. (1989). Incorporation of individual casein constituents into casein aggregates cross-linked by colloidal calcium phosphate in artificial casein micelles. *Journal of Dairy Research* 56(4), 613-618.

Aoki, T., Kako, Y. and Imamura, T. (1986). Separation of casein aggregates cross-linked by colloidal calcium phosphate from bovine casein micelles by high performance gel chromatography in the presence of urea. *Journal of Dairy Research* 53(1), 53-59.

Aoki, T., Yamada, N. and Kako, Y. (1990). Relation between colloidal calcium phosphate cross-linkage and release of β-casein from bovine casein micelles on cooling. *Agricultural and Biological Chemistry* 54(9), 2287-2292.

Aoki, T., Sakamoto, H. and Kako, Y. (1991). Cross-linking of caseins by colloidal calcium phosphate in the presence of urea. *International Dairy Journal* 1(1), 67-75.

Aoki, T., Umeda, T. and Kako, Y. (1992). The least number of phosphate groups for crosslinking of casein by colloidal calcium phosphate. *Journal of Dairy Science* 75(4), 971-975.

Atamer, Z., Post, A. E., Schubert, T., Holder, A., Boom, R. M. and Hinrichs, J. (2017). Bovine β-casein: isolation, properties and functionality. A review. *International Dairy Journal* 66, 115-125.

Berzelius, J. J. (1814). Über Thierische Chemie. *Schweiggers Journal für Chemie Physik* 11, 261-280.

Bijl, E., de Vries, R., van Valenberg, H., Huppertz, T. and Van Hooijdonk, T. (2014). Factors influencing casein micelle size in milk of individual cows: genetic variants and glycosylation of κ-casein. *International Dairy Journal* 34(1), 135-141.

Bijl, E., Huppertz, T., van Valenberg, H. and Holt, C. (2019). A quantitative model of the bovine casein micelle: ion equilibria and calcium phosphate sequestration by individual caseins in bovine milk. *European Biophysics Journal: EBJ* 48(1), 45-59.

Bijl, E., Holland, J. W. and Boland, M. (2020). Post-translational modifications of caseins. In: Boland, M. and Singh, H. (Eds) *Milk Proteins* (pp. 173-211). Academic Press, London, UK.

Bingham, E. W. and Farrell Jr., H. M. (1974). Casein kinase from the Golgi apparatus of lactating mammary gland. *Journal of Biological Chemistry* 249(11), 3647-3651.

Bouchoux, A., Ventureira, J., Gésan-Guiziou, G., Garnier-Lambrouin, F., Qu, P., Pasquier, C., Pézennec, S., Schweins, R. and Cabane, B. (2015). Correction: structural heterogeneity of milk casein micelles: a SANS contrast variation study. *Soft Matter* 11(4), 806-806.

Bouguyon, E., Beauvallet, C., Huet, J. C. and Chanat, E. (2006). Disulphide bonds in casein micelle from milk. *Biochemical and Biophysical Research Communications* 343(2), 450-458.

Braconnot, H. (1830). Ueber den Käsestoff und die Milch, und deren neue Nutzanwendungen. *Annalen der Physik und Chemie* 95(5), 34-47.

Carver, J. A., Duggan, P. J., Ecroyd, H., Liu, Y., Meyer, A. G. and Tranberg, C. E. (2010). Carboxymethylated-κ-casein: a convenient tool for the identification of polyphenolic

inhibitors of amyloid fibril formation. *Bioorganic and Medicinal Chemistry* 18(1), 222–228.
Carver, J. A. and Holt, C. (2019). Functional and dysfunctional folding, association and aggregation of caseins. *Advances in Protein Chemistry and Structural Biology* 118, 163–216.
Corredig, M. and Salvatore, E. (2016). Enzymatic coagulation of milk. In: McSweeney, P. L. H. and O'Mahony, J. A. (Eds) *Advanced Dairy Chemistry* (pp. 287–307). Springer, New York.
Creamer, L. K., Wheelock, J. V. and Samuel, D. (1973). The distribution of glyco-κ-casein and carbohydrate-free κ-casein between large and small bovine casein micelles, and its implication in micelle structure. *Biochimica et Biophysica Acta* 317(1), 202–206.
Dalgleish, D. G. (1998). Casein micelles as colloids: surface structures and stabilities. *Journal of Dairy Science* 81(11), 3013–3018.
Dalgleish, D. G. (2011). On the structural models of bovine casein micelles—review and possible improvements. *Soft Matter* 7(6), 2265–2272.
Dalgleish, D. G., Horne, D. S. and Law, A. J. R. (1989). Size-related differences in bovine casein micelles. *Biochimica et Biophysica Acta (BBA) - General Subjects* 991(3), 383–387.
Dalgleish, D. G., Spagnuolo, P. A. and Goff, H. D. (2004). A possible structure of the casein micelle based on high-resolution field-emission scanning electron microscopy. *International Dairy Journal* 14(12), 1025–1031.
Day, L., Williams, R. P. W., Otter, D. and Augustin, M. A. (2015). Casein polymorphism heterogeneity influences casein micelle size in milk of individual cows. *Journal of Dairy Science* 98(6), 3633–3644.
Day, L., Raynes, J. K., Leis, A., Liu, L. H. and Williams, R. P. W. (2017). Probing the internal and external micelle structures of differently sized casein micelles from individual cows' milk by dynamic light and small-angle X-ray scattering. *Food Hydrocolloids* 69, 150–163.
De Kruif, C. G. (1993). The turbidity of renneted skim milk. *Journal of Colloid and Interface Science* 156(1), 38–42.
De Kruif, C. G. (1999). Casein micelle interactions. *International Dairy Journal* 9(3–6), 183–188.
De Kruif, C. G. (2014). The structure of casein micelles: a review of small-angle scattering data. *Journal of Applied Crystallography* 47(5), 1479–1489.
De Kruif, C. G. and Zhulina, E. B. (1996). κ-casein as a polyelectrolyte brush on the surface of casein micelles. *Colloids and Surfaces. Part A: Physicochemical and Engineering Aspects* 117(1–2), 151–159.
De Kruif, C. G. and Grinberg, V. Y. (2002). Micellisation of β-casein. *Colloids and Surfaces. Part A: Physicochemical and Engineering Aspects* 210(2–3), 183–190.
De Kruif, C. G. and Holt, C. (2003). Casein micelle structure, functions and interactions. In: Fox, P. F. and McSweeney, P. L. H. (Eds) *Advanced Dairy Chemistry–1 Proteins* (pp. 233–276). Springer, Boston, MA .
De Kruif, C. G. and Huppertz, T. (2012). Casein micelles: size distribution in milks from individual cows. *Journal of Agricultural and Food Chemistry* 60(18), 4649–4655.
De Kruif, C. G., Huppertz, T., Urban, V. S. and Petukhov, A. V. (2012). Casein micelles and their internal structure. *Advances in Colloid and Interface Science* 171–172, 36–52.

Dumpler, J., Wohlschläger, H. and Kulozik, U. (2017). Dissociation and coagulation of caseins and whey proteins in concentrated skim milk heated by direct steam injection. *Dairy Science and Technology* 96(6), 807–826.

Ecroyd, H., Koudelka, T., Thorn, D. C., Williams, D. M., Devlin, G., Hoffmann, P. and Carver, J. A. (2008). Dissociation from the oligomeric state is the rate-limiting step in fibril formation by κ-casein. *Journal of Biological Chemistry* 283(14), 9012–9022.

Ecroyd, H., Thorn, D. C., Liu, Y. and Carver, J. A. (2010). The dissociated form of κ-casein is the precursor to its amyloid fibril formation. *Biochemical Journal* 429(2), 251–260.

Fang, Z. H., Visker, M. H. P. W., Miranda, G., Delacroix-Buchet, A., Bovenhuis, H. and Martin, P. (2016). The relationships among bovine αS-casein phosphorylation isoforms suggest different phosphorylation pathways. *Journal of Dairy Science* 99(10), 8168–8177.

Farías, M. E., Martinez, M. J. and Pilosof, A. M. R. (2010). Casein glycomacropeptide pH-dependent self-assembly and cold gelation. *International Dairy Journal* 20(2), 79–88.

Farrell, H. M., Cooke, P. H., Wickham, E. D., Piotrowski, E. G. and Hoagland, P. D. (2003). Environmental influences on bovine κ-casein: reduction and conversion to fibrillar (amyloid) structures. *Journal of Protein Chemistry* 22(3), 259–273.

Farrell Jr., H. M., Malin, E. L., Brown, E. M. and Qi, P. X. (2006). Casein micelle structure: what can be learned from milk synthesis and structural biology? *Current Opinion in Colloid and Interface Science* 11(2–3), 135–147.

Fox, P. F. and Hoynes, M. C. T. (1975). Heat stability of milk: influence of colloidal calcium phosphate and β-lactoglobulin. *Journal of Dairy Research* 42(3), 427–435.

Frederiksen, P. D., Andersen, K. K., Hammershøj, M., Poulsen, H. D., Sørensen, J., Bakman, M., Qvist, K. B. and Larsen, L. B. (2011). Composition and effect of blending of noncoagulating, poorly coagulating, and well-coagulating bovine milk from individual Danish Holstein cows. *Journal of Dairy Science* 94(10), 4787–4799.

Gamlath, C. J., Leong, T. S. H., Ashokkumar, M. and Martin, G. J. O. (2018). The inhibitory roles of native whey protein on the rennet gelation of bovine milk. *Food Chemistry* 244, 36–43.

Gaygadzhiev, Z., Massel, V., Alexander, M. and Corredig, M. (2012). Addition of sodium caseinate to skim milk inhibits rennet-induced aggregation of casein micelles. *Food Hydrocolloids* 26(2), 405–411.

Glantz, M., Devold, T. G., Vegarud, G. E., Månsson, H. L., Stålhammar, H. and Paulsson, M. (2010). Importance of casein micelle size and milk composition for milk gelation. *Journal of Dairy Science* 93(4), 1444–1451.

Grimley, H. J., Grandison, A. S. and Lewis, M. J. (2010). The effect of calcium removal from milk on casein micelle stability and structure. *Milchwissenschaft* 65(2), 151–154.

Groves, M. L., Dower, H. J. and Farrell, H. M. (1992). Reexamination of the polymeric distributions of κ-casein isolated from bovine milk. *Journal of Protein Chemistry* 11(1), 21–28.

Gustavsson, F., Glantz, M., Buitenhuis, A. J., Lindmark-Månsson, H., Stålhammar, H., Andrén, A. and Paulsson, M. (2014). Factors influencing chymosin-induced gelation of milk from individual dairy cows: major effects of casein micelle size and calcium. *International Dairy Journal* 39(1), 201–208.

Hettiarachchi, C. A., Swulius, M. T. and Harte, F. M. (2020). Assessing constituent volumes and morphology of bovine casein micelles using cryo-electron tomography. *Journal of Dairy Science* 103(5), 3971–3979.
Holland, J. W., Deeth, H. C. and Alewood, P. F. (2004). Proteomic analysis of κ-casein micro-heterogeneity. *Proteomics* 4(3), 743–752.
Holland, J. W., Deeth, H. C. and Alewood, P. F. (2005). Analysis of O-glycosylation site occupancy in bovine κ-casein glycoforms separated by two-dimensional gel electrophoresis. *Proteomics* 5(4), 990–1002.
Holland, J. W., Deeth, H. C. and Alewood, P. F. (2006). Resolution and characterisation of multiple isoforms of bovine κ-casein by 2-DE following a reversible cysteine-tagging enrichment strategy. *Proteomics* 6(10), 3087–3095.
Holt, C. (1992). Structure and stability of bovine casein micelles. *Advances in Protein Chemistry* 43, 63–151.
Holt, C. (1998). Casein micelle substructure and calcium phosphate interactions studied by sephacryl column chromatography. *Journal of Dairy Science* 81(11), 2994–3003.
Holt, C. and Horne, D. S. (1996). The hairy casein micelle: evolution of the concept and its implications for dairy technology. *Netherlands Milk and Dairy Journal* 50, 85–111.
Holt, C., Wahlgren, N. M. and Drakenberg, T. (1996). Ability of a β-casein phosphopeptide to modulate the precipitation of calcium phosphate by forming amorphous dicalcium phosphate nanoclusters. *Biochemical Journal* 314(3), 1035–1039.
Holt, C., Timmins, P. A., Errington, N. and Leaver, J. (1998). A core-shell model of calcium phosphate nanoclusters stabilized by β-casein phosphopeptides, derived from sedimentation equilibrium and small-angle X-ray and neutron-scattering measurements. *European Journal of Biochemistry* 252(1), 73–78.
Holt, C., Sørensen, E. S. and Clegg, R. A. (2009). Role of calcium phosphate nanoclusters in the control of calcification. *The FEBS Journal* 276(8), 2308–2323.
Holt, C., Carver, J. A., Ecroyd, H. and Thorn, D. C. (2013). Invited review: caseins and the casein micelle: their biological functions, structures, and behavior in foods. *Journal of Dairy Science* 96(10), 6127–6146.
Horne, D. S. (2016). Ethanol stability and milk composition. In: McSweeney, P. L. H. and O'Mahony, J. A. (Eds) *Advanced Dairy Chemistry* (pp. 225–246). Springer, New York.
Horne, D. S. (2017). A balanced view of casein interactions. *Current Opinion in Colloid and Interface Science* 28, 74–86.
Horne, D. S. (2020). Casein micelle structure and stability. In: Boland, M. and Singh, H. (Eds) *Milk Proteins* (pp. 213-250). Academic Press, London.
Horne, D. S. and Parker, T. G. (1980). The pH sensitivity of the ethanol stability of individual cow milks. *Netherlands Milk and Dairy Journal* 34, 126–130.
Horne, D. S. and Banks, J. M. (2004). Rennet-induced coagulation of milk. *Cheese: Chemistry, Physics and Microbiology* 1, 47–70.
Huppertz, T. (2013). Chemistry of the caseins. In: McSweeney, P. L. H. and Fox, P. L. (Eds) *Advanced Dairy Chemistry* (pp. 135–160). Springer, Boston, MA.
Huppertz, T. (2016). Heat stability of milk. In: McSweeney, P. L. H. and O'Mahony, S. A. (Eds) *Advanced Dairy Chemistry* (pp. 179–196). Springer, New York.
Huppertz, T. and Fox, P. F. (2006). Effect of NaCl on some physico-chemical properties of concentrated bovine milk. *International Dairy Journal* 16(10), 1142–1148.
Huppertz, T. and Lambers, T. T. (2020). Influence of micellar calcium phosphate on in vitro gastric coagulation and digestion of milk proteins in infant formula model systems. *International Dairy Journal* 107, 104717.

Huppertz, T. and Chia, L. W. (2021). Milk protein coagulation under gastric conditions: a review. *International Dairy Journal* 113, 104882.

Huppertz, T., Hennebel, J., Considine, T., Shakeel-Ur-Rehman, Kelly, A. L. and Fox, P. F. (2006). A method for the large-scale isolation of β-casein. *Food Chemistry* 99(1), 45-50.

Huppertz, T., Gazi, I., Luyten, H., Nieuwenhuijse, H., Alting, A. and Schokker, E. (2017). Hydration of casein micelles and caseinates: implications for casein micelle structure. *International Dairy Journal* 74, 1-11.

Huppertz, T., Heck, J., Bijl, E., Poulsen, N. A. and Larsen, L. B. (2021). Variation in casein distribution and mineralisation in the milk from Holstein-Friesian cows. *International Dairy Journal* 119, 105064.

IDF (2020). *The World Dairy situation 2020*. Bulletin of the IDF 506. International Dairy Federation, Brussels, Belgium.

Ingham, B., Erlangga, G. D., Smialowska, A., Kirby, N. M., Wang, C., Matia-Merino, L., Haverkamp, R. G. and Carr, A. J. (2015). Solving the mystery of the internal structure of casein micelles. *Soft Matter* 11(14), 2723-2725.

Ingham, B., Smialowska, A., Erlangga, G. D., Matia-Merino, L., Kirby, N. M., Wang, C., Haverkamp, R. G. and Carr, A. J. (2016). Revisiting the interpretation of casein micelle SAXS data. *Soft Matter* 12(33), 6937-6953.

Ishikawa, H. O., Xu, A., Ogura, E., Manning, G. and Irvine, K. D. (2012). The Raine syndrome protein FAM20C is a Golgi kinase that phosphorylates bio-mineralization proteins. *PLoS ONE* 7(8), e42988.

Ivanova, G., Yakimova, B., Angelova, S., Stoineva, I. and Enchev, V. (2010). Influence of pH on the cis-trans isomerization of valine-proline dipeptide: an integrated NMR and theoretical investigation. *Journal of Molecular Structure* 975(1-3), 330-334.

Ketto, I. A., Knutsen, T. M., Øyaas, J., Heringstad, B., Ådnøy, T., Devold, T. G. and Skeie, S. B. (2017). Effects of milk protein polymorphism and composition, casein micelle size and salt distribution on the milk coagulation properties in Norwegian Red cattle. *International Dairy Journal* 70, 55-64.

Knoop, A. M., Knoop, E. and Wiechen, A. (1973). Electron microscopical investigations on the structure of the casein micelles. *Netherlands Milk and Dairy Journal* 27, 121-127.

Knoop, A. M., Knoop, E. and Wiechen, A. (1979). Sub-structure of synthetic casein micelles. *Journal of Dairy Research* 46(2), 347-350.

Kumosinski, T. F., King, G. and Farrell, H. M. (1994a). An energy-minimized casein submicelle working model. *Journal of Protein Chemistry* 13(8), 681-700.

Kumosinski, T. F., King, G. and Farrell, H. M. (1994b). Comparison of the three-dimensional molecular models of bovine submicellar caseins with small-angle X-ray scattering. Influence of protein hydration. *Journal of Protein Chemistry* 13(8), 701-714.

Lefebvre-Cases, E., Gastaldi, E. and de la Fuente, B. T. (1998). Influence of chemical agents on interactions in dairy products: effect of SDS on casein micelles. *Colloids and Surfaces, Part B: Biointerfaces* 11(6), 281-285.

Lenton, S., Nylander, T., Teixeira, S. C. M. and Holt, C. (2015). A review of the biology of calcium phosphate sequestration with special reference to milk. *Dairy Science and Technology* 95(1), 3-14.

Lenton, S., Wang, Q., Nylander, T., Teixeira, S. and Holt, C. (2020). Structural biology of calcium phosphate nanoclusters sequestered by phosphoproteins. *Crystals* 10(9), 755.

Lin, Y., Kelly, A. L., O'Mahony, J. A. and Guinee, T. P. (2017). Addition of sodium caseinate to skim milk increases nonsedimentable casein and causes significant changes in rennet-induced gelation, heat stability, and ethanol stability. *Journal of Dairy Science* 100(2), 908-918.

Logan, A., Leis, A., Day, L., Øiseth, S. K., Puvanenthiran, A. and Augustin, M. A. (2015). Rennet gelation properties of milk: influence of natural variation in milk fat globule size and casein micelle size. *International Dairy Journal* 46, 71-77.

Longenbach, J. I. and Heinrichs, A. J. (1998). A review of the importance and physiological role of curd formation in the abomasum of young calves. *Animal Feed Science and Technology* 73(1-2), 85-97.

Malacarne, M., Franceschi, P., Formaggioni, P., Sandri, S., Mariani, P. and Summer, A. (2014). Influence of micellar calcium and phosphorus on rennet coagulation properties of cow's milk. *Journal of Dairy Research* 81(2), 129-136.

Mathai, J. K., Liu, Y. and Stein, H. H. (2017). Values for digestible indispensable amino acid scores (DIAAS) for some dairy and plant proteins may better describe protein quality than values calculated using the concept for protein digestibility-corrected amino acid scores (PDCAAS). *British Journal of Nutrition* 117(4), 490-499.

McGann, T. C. A. and Fox, P. F. (1974). Physico-chemical properties of casein micelles reformed from urea-treated milk. *Journal of Dairy Research* 41(1), 45-53.

McGann, T. C. A., Donnelly, W. J., Kearney, R. D. and Buchhemm, W. (1980). Composition and size distribution of bovine casein micelles. *Biochimica et Biophysica Acta* 630(2), 261-270.

McGann, T. C. A., Kearney, R. D., Buchheim, W., Posner, A. S., Betts, F. and Blumenthal, N. C. (1983a). Amorphous calcium phosphate in casein micelles of bovine milk. *Calcified Tissue International* 35(6), 821-823.

McGann, T. C. A., Buchheim, W., Kearney, R. D. and Richardson, T. (1983b). Composition and ultrastructure of calcium phosphate-citrate complexes in bovine milk systems. *Biochimica et Biophysica Acta* 760(3), 415-420.

McMahon, D. J. and McManus, W. R. (1998). Rethinking casein micelle structure using electron microscopy. *Journal of Dairy Science* 81(11), 2985-2993.

Mikkelsen, T. L., Frøkiær, H., Topp, C., Bonomi, F., Iametti, S., Picariello, G., Ferranti, P. and Barkholt, V. (2005). Caseinomacropeptide self-association is dependent on whether the peptide is free or restricted in κ-casein. *Journal of Dairy Science* 88(12), 4228-4238.

Miwa, N., Yokoyama, K., Wakabayashi, H. and Nio, N. (2010). Effect of deamidation by protein-glutaminase on physicochemical and functional properties of skim milk. *International Dairy Journal* 20(6), 393-399.

Nilsson, K., Buhelt Johansen, L., de Koning, D. J., Duchemin, S. I., Stenholdt Hansen, M., Stålhammar, H., Lindmark-Månsson, H., Paulsson, M., Fikse, W. F. and Glantz, M. (2020). Effects of milk proteins and posttranslational modifications on noncoagulating milk from Swedish Red dairy cattle. *Journal of Dairy Science* 103(8), 6858-6868.

Nogueira, M. H., Humblot, L., Singh, R. P., Dieude-Fauvel, E., Doumert, B., Nasser, S., Lesur, C., Karoui, R., Delaplace, G. and Peixoto, P. P. S. (2021). The heterogeneous substructure of casein micelles evidenced by SAXS and NMR in demineralized samples. *Food Hydrocolloids* 117, 106653.

O'Connell, J. E., Saracino, P., Huppertz, T., Uniake, T., De Kruif, C. G., Kelly, A. L. and Fox, P. F. (2006). Influence of ethanol on the rennet-induced coagulation of milk. *The Journal of Dairy Research* 73(3), 312-317.

Ono, T., Ohotawa, T. and Takagi, Y. (1994). Complexes of casein phosphopeptide and calcium phosphate prepared from casein micelles by tryptic digestion. *Bioscience, Biotechnology, and Biochemistry* 58(8), 1376-1380.

Poulsen, N. A., Jensen, H. B. and Larsen, L. B. (2016). Factors influencing degree of glycosylation and phosphorylation of caseins in individual cow milk samples. *Journal of Dairy Science* 99(5), 3325-3333.

Rasmussen, L. K., Hojrup, P. and Petersen, T. E. (1992). Localization of two interchain disulfide bridges in dimers of bovine αs2-casein: parallel and antiparallel alignments of the polypeptide chains. *European Journal of Biochemistry* 203(3), 381-386.

Rasmussen, L. K., Højrup, P. and Petersen, T. E. (1994). Disulphide arrangement in bovine caseins: localization of intrachain disulphide bridges in monomers of κ-and αs2-casein from bovine milk. *Journal of Dairy Research* 61(4), 485-493.

Rollema, H. S., Brinkhuis, J. A. and Vreeman, H. J. (1988). 1H-NMR studies of bovine κ-casein and casein micelles. *Netherlands Milk and Dairy Journal* 42(2), 233-248.

Schmidt, D. G. (1982). Association of caseins and casein micelle structure. *Developments in Dairy Chemistry* 1, 61-86.

Schmidt, D. G., Koops, J. and Westerbeek, D. (1977). Properties of artificial casein micelles. 1. Preparation, size distribution and composition. *Netherlands Milk and Dairy Journal* 31, 328-341.

Schmidt, D. G., Both, P. and Koops, J. (1979). Properties of artificial casein micelles. 3. Relationship between salt composition, size and stability towards ethanol, dialysis and heat. *Netherlands Milk and Dairy Journal* 33, 40-48.

Smiddy, M. A., Martin, J. E., Kelly, A. L., De Kruif, C. G. and Huppertz, T. (2006). Stability of casein micelles cross-linked by transglutaminase. *Journal of Dairy Science* 89(6), 1906-1914.

Smithers, G. W. (2008). Whey and whey proteins–from 'gutter-to-gold'. *International Dairy Journal* 18(7), 695-704.

Smithers, G. W. (2015). Whey-ing up the options-Yesterday, today and tomorrow. *International Dairy Journal* 48, 2-14.

Sunds, A. V., Poulsen, N. A. and Larsen, L. B. (2019). Short communication: application of proteomics for characterization of caseinomacropeptide isoforms before and after desialidation. *Journal of Dairy Science* 102(10), 8696-8703.

Swaisgood, H. E. (2003). Chemistry of the caseins. In: Fox, P. F. and McSweeney, P. L. H. (Eds) *Advanced Dairy Chemistry–1 Proteins* (pp. 139-201). Springer, Boston, MA.

Tagliabracci, V. S., Engel, J. L., Wen, J., Wiley, S. E., Worby, C. A., Kinch, L. N., Xiao, J., Grishin, N. V. and Dixon, J. E. (2012). Secreted kinase phosphorylates extracellular proteins that regulate biomineralization. *Science* 336(6085), 1150-1153.

Thorn, D. C., Meehan, S., Sunde, M., Rekas, A., Gras, S. L., MacPhee, C. E., Dobson, C. M., Wilson, M. R. and Carver, J. A. (2005). Amyloid fibril formation by bovine milk κ-casein and its inhibition by the molecular chaperones αS-and β-casein. *Biochemistry* 44(51), 17027-17036.

Thorn, D. C., Ecroyd, H., Sunde, M., Poon, S., and Carver, J. A. (2008). Amyloid fibril formation by bovine milk αs2-casein occurs under physiological conditions yet is prevented by its natural counterpart, αs1-casein. *Biochemistry*, *47*(12), 3926-3936.

Thorn, D. C., Ecroyd, H., Carver, J. A. and Holt, C. (2015). Casein structures in the context of unfolded proteins. *International Dairy Journal* 46, 2-11.

Thorn, D. C., Bahraminejad, E., Grosas, A. B., Koudelka, T., Hoffmann, P., Mata, J. P., Devlin, G. L., Sunde, M., Ecroyd, H., Holt, C. and Carver, J. A. (2021). Native disulphide-linked dimers facilitate amyloid fibril formation by bovine milk αS2-casein. *Biophysical Chemistry* 270, 106530.

Trejo, R., Dokland, T., Jurat-Fuentes, J. and Harte, F. (2011). Cryo-transmission electron tomography of native casein micelles from bovine milk. *Journal of Dairy Science* 94(12), 5770–5775.

Tuinier, R. and De Kruif, C. G. (2002). Stability of casein micelles in milk. *The Journal of Chemical Physics* 117(3), 1290–1295.

Udabage, P., McKinnon, I. R. and Augustin, M. A. (2000). Mineral and casein equilibria in milk: effects of added salts and calcium-chelating agents. *Journal of Dairy Research* 67(3), 361–370.

Vaia, B., Smiddy, M. A., Kelly, A. L. and Huppertz, T. (2006). Solvent-mediated disruption of bovine casein micelles at alkaline pH. *Journal of Agricultural and Food Chemistry* 54(21), 8288–8293.

Vargas-Bello-Pérez, E., Márquez-Hernández, R. I. and Hernández-Castellano, L. E. (2019). Bioactive peptides from milk: animal determinants and their implications in human health. *Journal of Dairy Research* 86(2), 136–144.

Walstra, P. (1990). On the stability of casein micelles. *Journal of Dairy Science* 73(8), 1965–1979.

Walstra, P., Walstra, P., Wouters, J. T. and Geurts, T. J. (2005). *Dairy Science and Technology*. CRC Press, Boca Raton, FL.

Ward, B. R., Goddard, S. J., Augustin, M. A. and McKinnon, I. R. (1997). EDTA-induced dissociation of casein micelles and its effect on foaming properties of milk. *Journal of Dairy Research* 64(4), 495–504.

White, J. C. D. and Davies, D. T. (1958). 712. The relation between the chemical composition of milk and the stability of the caseinate complex: I. General introduction, description of samples, methods and chemical composition of samples. *Journal of Dairy Research* 25(2), 236–255.

Xu, Y., Liu, D., Yang, H., Zhang, J., Liu, X., Regenstein, J. M., Hemar, Y. and Zhou, P. (2016). Effect of calcium sequestration by ion-exchange treatment on the dissociation of casein micelles in model milk protein concentrates. *Food Hydrocolloids* 60, 59–66.

# Chapter 5

## Advances in dairy protein science: whey proteins

*Dimuthu Bogahawaththa and Todor Vasiljevic, Victoria University, Australia*



### 1 Introduction

Milk is a complex biological suspension that is primarily composed of fat globules, casein micelles (colloidal phase) and milk serum (serum phase). The serum phase is broadly known as 'whey', which is literally the remaining soluble fraction of the milk after coagulation and separation of milk curd during manufacture of caseins, cheeses and yoghurts (Deeth and Bansal, 2019; Guo and Wang, 2019b; Nishanthi et al., 2017c). The whey has also been classified and termed based on the production process involved. The two main types are 'sweet whey' and 'acid whey', which primarily result from acid and rennet coagulation of the caseins, respectively. Application of a high salt concentration during the production of some types of cheese yields 'salty whey' (Nishanthi et al., 2017c). Whey comprises about 80–90% of the total milk volume and over 50% of the total milk solids. Whey is composed mainly of water, while the major proportion of solids is composed of lactose with varying levels of valuable whey proteins and minerals, depending on the processing history (Deeth and Bansal, 2019; Jauregui-Rincón et al., 2018).

http://dx.doi.org/10.19103/AS.2022.0099.06

After decades of being considered only a by-product of cheese production and as waste, whey and its components have been given a significant role in the food industry due to the advancement of the science and technology (Bansal and Bhandari, 2016). The global market for whey protein products has been estimated at US$ 9.7 billion in 2020, and is projected to reach US$ 16.4 billion by 2027, with a compound annual growth rate of 7.8% (Market Research, 2021). As one of the main products, whey protein concentrate is expected to grow at a compound annual growth rate of 8.9% and reach US$ 6.1 billion during the same time. Whey protein powders of varying grades of purity and with different ratios of the major whey proteins are widely used ingredients due to their broad range of functionalities, along with exceptional nutritional and physiological properties. They are used in a number of food applications, including but not limited to protein supplements, baked goods, dairy and meat products.

The key functional properties of whey proteins include water-binding, solubility, viscosity, heat coagulation, gelation, foaming, emulsification, film/coat formation and encapsulation. The main factors that limit the application of whey protein powders in food systems include compositional variations governed by the manufacturing processes, physicochemical modifications during storage, and particularly, poor heat stability. The functional properties of whey proteins predominantly depend on their intrinsic characteristics, which can be modulated by extrinsic or environmental factors including temperature, pH, ionic strength and shear (Nishanthi et al., 2017a,b,c; 2018a,b; Liyanaarachchi, 2017; Singh, 2011). Hence, the aim of this chapter is to discuss heat-induced denaturation and aggregation of whey proteins under different environmental conditions and relate them with the functionality and other methods for modulation of whey protein functionality. In particular, heat-induced denaturation and aggregation of immunoglobulins have been discussed, as they are considered increasingly important as biologically/immunologically active whey proteins.

## 2 Whey proteins: an overview

The soluble proteins present in the serum phase of the milk or in the whey, commonly referred to as 'whey proteins', are discussed in this chapter. It is, however, worthwhile to note that there may be differences between proteins in the whey and the serum proteins in the milk. For instance, caseinomacropeptide, also known as glycomacropeptide, found in sweet whey, is not usually considered a serum protein/peptide in the milk (Córdova-Dávalos et al., 2019; Deeth and Bansal, 2019). Furthermore, the physicochemical and structural properties of the whey proteins may vary according to type of whey (Nishanthi et al., 2017c). Whey proteins can be differentiated from the caseins mainly by their native globular conformation, with well-defined secondary and tertiary structures and solubility at pH 4.6. The native globular conformation of whey proteins is manly stabilised by hydrophobic interactions, van der Waal's interactions, ion-pair interactions, hydrogen bonds and disulphide bridges (Anema, 2014; O'Mahony and Fox, 2014; Singh and Havea, 2003). Whey proteins constitute ~20% of the total protein in bovine milk. β-lactoglobulin is the most abundant whey protein (~50%), and the other major whey proteins are α-lactalbumin (~20%), immunoglobulins (~12%) and bovine serum albumin (~8%). The less abundant proteins, at a concentration < 0.3 g $L^{-1}$ in mature bovine milk (Deeth and Bansal, 2019), are considered the minor whey proteins. These include lactoferrin, lactoperoxidase, growth factors, lysozymes and other peptide components (Deeth and Bansal, 2019; Wijayanti et al., 2019). The major whey proteins are the main focus of this chapter. Their important molecular characteristics and functionality have been briefly discussed below and summarised in Table 1. Caseinomacropeptide is also briefly covered in this section as it constitutes about 20–25% of the total proteins in the whey protein products manufactured from sweet whey (Thomä-Worringer et al., 2006). Immunoglobulins are discussed in Section 6.4.1.

## *2.1 β-Lactoglobulin (β-Lg)*

β-Lg is a typical globular protein which belongs to the lipocalin family. The primary structure comprises 162 amino acid residues with molecular weight ~18.3 kDa. Thirteen genetic variants of bovine β-Lg have been reported, of which variants A and B are the most common. β-Lg occurs in monomeric, dimeric or octameric form, mostly depending on temperature, pH, protein concentration and ionic strength. The secondary structure of the β-Lg monomer is composed of about 8% α-helices, 45% β-sheets and 47% random coils. The native conformation of monomeric β-Lg is stabilised by hydrophobic and ionic interactions, hydrogen bonds and two disulphide bonds located at $Cys^{66}$–$Cys^{160}$ and $Cys^{106}$–$Cys^{119}$. The β-Lg molecule contains a free thiol group at $Cys^{121}$ that is buried in the native conformation (Deeth and Bansal, 2019; Edwards and Jameson, 2020; Gottschalk et al., 2003; Guo and Wang, 2019a; Sawyer, 2013). β-Lg has a great influence on the functional properties of whey protein ingredients, for example, whey protein isolate (WPI) and whey protein concentrate (WPC), due to its heat-induced denaturation, intra- and interprotein aggregation and gelation when subjected to other environmental factors such as pH, ionic strength, shear and protein concentration (Deeth and Bansal, 2019; Nicolai et al., 2011; Quevedo et al., 2021; Quevedo et al., 2020a).

**Table 1** Molecular characteristics and functionality of major whey proteins

| | Type of whey protein | | | |
|---|---|---|---|---|
| Molecular characteristics / functionality | β-Lactoglobulin | α-Lactalbumin | Bovine serum albumin | Immunoglobulin-G |
| Proportion of total whey proteins (%) | 50 | 20 | 8 | 12 |
| Number of amino acid residues | 162 | 123 | 583 | >500 |
| Molecular mass (kDa) | 18.3 | 14.2 | 66.5 | 160.0 |
| Isoelectric point | 5.13 | 4.2–4.5 | 4.7–4.9 | 5.5–8.3 |
| Number of cysteine residues | 5 | 8 | 35 | |
| Number of S-S groups | 2 | 4 | 17 | |
| Number of S-H groups | 1 | 0 | 1 | |
| Number of proline residues | 2 | 8 | 28 | |
| Composition of secondary structure | 8% α-helices<br>45% β-sheets<br>47% random coils | 26% α-helices<br>14% β-sheets<br>60% unordered structures | 67% α-helices<br>10% turns<br>23% extended chains<br>No β-sheets | 3% α-helices<br>64% β-sheets<br>28% turns<br>5% random coils |
| Native conformation | Globular | Globular | Globular | Globular |
| Initial denaturation Temperature (°C) | 78 | 62 | 64 | 72 |
| Enthalpy of denaturation (kJ $mol^{-1}$) | 311 | 253 | 803 | 500 |
| Solubility | High | High | High | High |
| Foamability and foam stability | High | High | Medium | |
| Emulsification and emulsion stability | High | High | Medium | |
| Gelation (heat-induced) | High | Low | High | High |

*Sources*: (Abd El-Salam et al., 2009; Brodkorb et al., 2016; Deeth and Bansal, 2019; Edwards and Jameson, 2020; Gorga et al., 1989; Liyanaarachchi, 2017; Lu et al., 2015).

## 2.2 α-Lactalbumin (α-La)

α-La is a small compact globular protein that is categorised as a calcium metalloprotein. The primary structure of α-La has 123 amino acid residues with molecular weight ~14.2 kDa. Native α-La exists as monomers, and four genetic variants have been reported (A, B, C and D). The secondary structure of α-La is composed of about 26% α-helices, 14% β-sheets and 60% unordered structures. The three-dimensional structure of α-La consists of two lobes, one of which is made up of α-helices and the other composed of β-sheets and unordered structures. The native conformation of α-La is mainly stabilised by four disulphide bridges located at $Cys^{6}$-$Cys^{120}$, $Cys^{28}$-$Cys^{111}$, $Cys^{60}$-$Cys^{77}$ and $Cys^{73}$-$Cys^{90}$ but no free thiol groups. Binding with calcium plays a great role in proper folding of the α-La molecule and maintains its stability (Brew, 2013; Deeth and Bansal, 2019; Edwards and Jameson, 2020; Guo and Wang, 2019a). α-La is known to be the most heat-stable whey protein, with the ability to withstand heat-induced aggregation (Singh and Havea, 2003), as it contains four disulphide bridges and no free thiol groups, as well the ability to bind with calcium. Such characteristics render important the functionality of α-La, including heat stability and foaming properties (Deeth and Bansal, 2019).

## 2.3 Bovine serum albumin (BSA)

BSA is a highly structured but flexible globular protein, having a single polypeptide chain with 583 amino acid residues in the primary structure, with the molecular weight ~66.5 kDa. BSA is synthesised outside the mammary gland and transported into milk from the blood stream by passive leakage. BSA is a non-glycoprotein and contains no phosphorus. BSA contains 35 cysteine residues that form 17 disulphide bonds, while leaving one free thiol group at $Cys^{34}$. The secondary structure of BSA is dominated by α-helices (67%) and the remainder consists of 10% turns and 23% extended chains but no β-sheets. The native conformation of BSA is characterised by a heart shape that consists of three helical domains formed by 9 disulphide loops of 17 disulphide bridges (Deeth and Bansal, 2019; Edwards and Jameson, 2020; Guo and Wang, 2019a; Lu et al., 2015; Wynn and Sheehy, 2013). BSA possesses a great ligand-binding capacity and especially binds with fatty acids, flavour compounds and metal ions (e.g. calcium). BSA bound with fatty acids could be used as a food ingredient due to its improved heat stability, resistance to enzymatic hydrolysis and enhanced emulsification properties (Deeth and Bansal, 2019).

## 2.4 Caseinomacropeptide (CMP)

CMP is a hydrophilic peptide, the C-terminal end of κ-casein, that contains 64 amino acid residues with a molecular weight of ~6.75 kDa (non-glycosylated monomer). The molecular weight of a glycated CMP can vary from 7.50 to 9.63 kDa. CMP is released from κ-casein after cleavage of the $Phe^{105}$-$Met^{106}$ peptide bond by rennet during cheesemaking. CMP has been suggested to processes a random coil-type structure and its three-dimensional conformation may change depending on pH, ionic strength and presence of other molecules. Post-translational modifications, phosphorylation and glycosylation, occur in the C-terminal end of κ-casein and thus are present in CMP (Córdova-Dávalos et al., 2019; Mann et al., 2019; Thomä-Worringer et al., 2006). CMP is a highly soluble acidic peptide, but solubility is limited at pH 1–5 (isoelectric point ≤ 4). CMP also possesses important bioactivities and desirable functionalities including gelation, emulsification and foaming. It shows good emulsifying capacity at alkaline pH (minimum at pH 4.5–5.5), as well as good foamability (but poor stability) in comparison to egg white. The functionality of CMP depends on the level of glycosylation (Deeth and Bansal, 2019; Sharma et al., 2013; Thomä-Worringer et al., 2006).

## 3 Denaturation and aggregation of whey proteins

The heat-induced denaturation and aggregation behaviour of whey proteins appears to be a complex process and is predominantly governed by β-Lg, as it is the most abundant whey protein. Messens et al. (1997) have defined the terms applicable to heat-induced changes in proteins. Denaturation of a protein is any change in the secondary, tertiary or quaternary conformation, excluding the rupture of peptide bonds in its primary structure, that results in an altered conformation with a totally or partially unfolded polypeptide structure. Association of proteins refers to molecular-level modifications, for instance the formation of dimers, that are governed by weak bonds between specific interaction sites. Protein aggregation (or coagulation, polymerisation and precipitation) involves unspecified interprotein interactions resulting in the formation of large protein complexes with higher molecular weights. Protein gelation refers to a process of aggregation of native and/or (partially) denatured proteins, in an orderly manner, resulting in a three-dimensional network that eventually forms a well-ordered matrix through balanced protein-protein and protein-solvent interactions, enabling significant amounts of water to be held. Figure 1 displays the basic pathway for the progression of denaturation and aggregation of major whey proteins with increase in heating. The process usually begins with dissociation of β-Lg dimers into monomers and a partial unfolding of the native conformation of the whey proteins. This step is reversible upon cooling. An increase in heating leads to irreversible unfolding of the native globular conformation, while exposing hydrophobic residues and free sulphydryl groups, which eventually result in the formation of reactive monomers (e.g. β-Lg and BSA). Further heating can result in the formation of large aggregates with higher molecular weight. The aggregation process ends with the development of larger protein networks or gels (for further information, consult Section 6.6.3.1). It appears that non-covalent interactions (e.g. hydrophobic and electrostatic) are mainly involved in protein aggregation at lower temperatures (65–85°C), whereas covalent interactions (e.g. disulphide) dominate the aggregation process at higher temperatures (>85°C), particularly when forming larger protein aggregates (Havea et al., 2001; Roefs and De Kruif, 1994; Wijayanti et al., 2014; Wijayanti et al., 2019). The recent investigations have reported the involvement of non-disulphide covalent bonds in the aggregation of β-Lg in a concentrated dispersion subjected to heat and shear simultaneously (Section 6.5.1) (Quevedo et al., 2020a,b).

The denaturation and aggregation behaviour of whey proteins is principally governed by changes in their molecular properties, importantly the exposure of hydrophobic residues, sulphydryl and disulphide groups (Walstra et al., 2006). For instance, in terms of β-Lg, the exposure of hydrophobic residues determines the degree of unfolding, while the formation of disulphide bonds and electrostatic repulsions govern the extent of aggregation (Wijayanti et al., 2019). The molecular properties of whey proteins are, however, modulated by many factors such as temperature, pH, ionic strength/salt, protein concentration and shear (Huppertz et al., 2019; Nicolai et al., 2011; Quevedo et al., 2021; Wijayanti et al., 2019). Nicolai et al. (2011), Ryan et al. (2013) and Wijayanti et al. (2019) have reviewed the impact of temperature, pH, ionic strength and protein concentration, either alone or in combination, on molecular-level changes, and thereby on the modulation of denaturation and aggregation of whey proteins. β-Lg usually displays a reversible denaturation at low protein concentration and at low temperature (<60°C). β-Lg proceeds to the irreversible aggregation phase at a high protein concentration or at high temperatures (>70°C). Increase in pH (6.25–8.5) has been shown to accelerate the denaturation of β-Lg. pH can also govern the mechanism of β-Lg aggregation and thereby aggregate size. A pH of 6.9 resulted in smaller aggregates (~40 nm) due to increased availability of the free thiol group, whereas a pH < 6.4 induced the formation of larger aggregates (~100 nm) through non-covalent interaction of smaller aggregates. Such large aggregates tend to precipitate at pH 6.2. Lower pH (< 7) may also lead to the formation of partially denatured intermediates through inter- and intramolecular disulphide bonds. High temperature (85°C) and low pH (5.7–5.9) induce the formation of stable aggregates that are stabilised by covalent and non-covalent interactions. The inclusion of calcium appears to accelerate the denaturation of β-Lg, potentially due to cross-linking between β-Lg molecules and calcium ions or due to electrostatic shielding of negatively charged groups on the protein molecule. Ionic strength also pays an important role in modulating the β-Lg aggregation process (see Section 6.6.2.1 for further information) (Wijayanti et al., 2019).

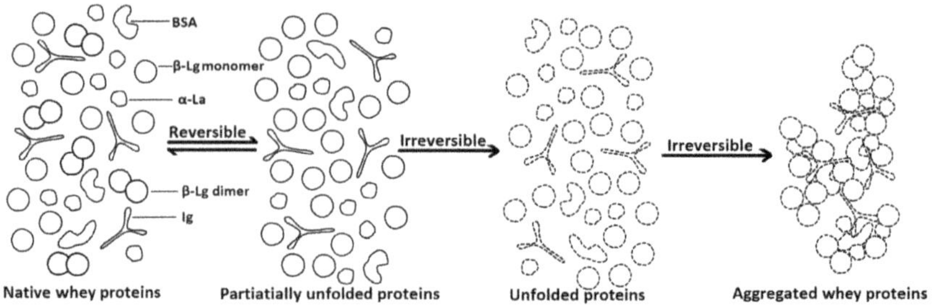


**Figure 1** The basic pathway illustrating heat-induced denaturation and aggregation of major whey proteins (β-Lg - β-lactoglobulin; α-La - α-lactalbumin; BSA - bovine serum albumin; Ig - immunoglobulins).

In addition to β-Lg, the other major whey proteins such α-La, BSA and immunoglobulins influence the heat-induced denaturation and aggregation process greatly (Bogahawaththa et al., 2017b, 2019; Wijayanti et al., 2014). α-La is the most heat-stable (able to withstand heat-induced aggregation) major whey protein, primarily due to the absence of a free thiol group (Calvo et al., 1993). Denaturation of α-La appears to be mostly reversible at temperatures <80°C (Dalgleish et al., 1997; Gezimati et al., 1997; Hines and Foegeding, 1993). Heating at higher temperature (≥95°C) for a longer time (e.g. 15 min) can result in a substantial level of irreversible denaturation (Chaplin and Lyster, 1986). Such heating conditions may induce the rupture of disulphide bonds in the α-La molecule, resulting in reactive free sulphydryl groups, which engage in irreversible disulphide-linked aggregation - similar to the free sulphydryl group of the β-Lg molecule (Chaplin and Lyster, 1986). BSA is considered the most heat-sensitive major whey protein at neutral pH (Takeda et al., 1989; Yamasaki et al., 1991). Heat-induced denaturation and aggregation of BSA appeared mostly similar to β-Lg. However, BSA formed aggregates at a lower temperature than β-Lg (Gezimati et al., 1996b). Heating at > 60°C resulted in irreversible denaturation of BSA and initiated disulphide-linked aggregation. Further heating >70°C led to gelation of BSA through disulphide bonds and hydrophobic interactions (Considine et al., 2007; Murayama and Tomida, 2004; Rüegg et al., 1977; Takeda et al., 1989). The rate of gelation of BSA has been reported to be up to 100 times faster than β-Lg (Hines and Foegeding, 1993). A discussion on heat-induced denaturation and aggregation of immunoglobulins has been provided in Section 6.4.2.

Interprotein interactions also play an important role in modulating the heat-induced denaturation and aggregation of whey proteins (Bogahawaththa, 2018a; Patel, 2007; Wijayanti et al., 2014; Wijayanti et al., 2019). α-La did not aggregate when it was heated alone at 75°C, but in the presence of β-Lg, α-La formed aggregates through hydrophobic and covalent interactions, after being subjected to the same heat treatment (Schokker et al., 2000). In a binary mixture, α-La denatured more rapidly than β-Lg and formed intra- and interprotein aggregates (Calvo et al., 1993; Dalgleish et al., 1997; Gezimati et al., 1997; Havea et al., 2001; Schokker et al., 2000). The covalent aggregates are usually formed through a thiol–disulphide interchange reaction between a free thiol group of a β-Lg monomer and a cysteine residue located on the surface of an α-La molecule (Bogahawaththa et al., 2019; Calvo et al., 1993; Livney et al., 2003). Nevertheless, heat-induced denaturation of β-Lg was not affected notably by α-La in a binary mixture, in comparison to a pure β-Lg dispersion, displaying the leading role of β-Lg in the protein denaturation and aggregation processes (Hong and Creamer, 2002; Schokker et al., 2000). In a binary mixture, BSA was observed to govern the aggregation process over β-Lg at <70°C, because of its lower thermal transition temperature than β-Lg (deWit and Klarenbeek, 1984), whereas both the proteins mostly displayed a similar aggregation rate at <75°C (Gezimati et al., 1996a; Matsudomi et al., 1994). BSA appeared to accelerate the heat-induced aggregation of β-Lg, but the effect of β-Lg on the aggregation of BSA has not been reported (Havea et al., 2001; Hines and Foegeding, 1993; Matsudomi et al., 1994). Heating of a binary mixture of α-La and BSA at 80°C for 30 min formed disulphide-linked interprotein aggregates (Matsudomi et al., 1993). During heating of a BSA and α-La mixture, BSA appeared to self-aggregate via both disulphide and hydrophobic interactions, while α-La did not interact hydrophobically (Havea et al., 2001; Matsudomi et al., 1993). The exposed free thiol groups of BSA can intiate thiol–disulphide interchange interactions with disulphide bonds of α-La leading to form covalent aggregates of α-La itself and with BSA (Havea et al., 2000). BSA was more effective than β-Lg in inducing intra- and/or interprotein aggregation of α-La in respective binary mixtures, due to their characteristic thermal transition temperatures (Havea et al., 2001). When a ternary mixture (β-Lg, α-La and BSA) was heated at 75°C, BSA commenced self-aggregation initially, then disulphide-linked aggregates were formed with the involvement of all three proteins (Havea et al., 2001). Recent studies have reported that in a mixture of β-Lg, α-La, immunoglobulin G (IgG) and BSA, α-La governed the heat-induced denaturation over β-Lg, possibly due to the influence of BSA, and formed disulphide-linked aggregates with the involvement of all four proteins (Bogahawaththa et al., 2017b, 2019). This effect disappeared once the caseins were included into the same whey protein mixture before heating (Bogahawaththa et al., 2019). The heat-induced denaturation of immunoglobulins and its interactions with other whey proteins have been discussed in Section 6.4.2. However, the influence of BSA and immunoglobulins on heat-induced denaturation and aggregation of other whey proteins are not comprehensively understood (Bogahawaththa, 2018a; Wijayanti et al., 2014, 2019). Heat-induced interactions between whey proteins and caseins have been reviewed previously (Anema, 2014; Wang and Guo, 2019a; Wijayanti et al., 2014; Wijayanti et al., 2019). Thus, they are not a focus of this chapter.

## 4 Immunoglobulins: native conformation, denaturation and interactions

In comparison to β-Lg and α-La, the thermal denaturation and aggregation behaviour of immunoglobulins have not been extensively studied or reviewed (Borad and Singh, 2018; Brodkorb et al., 2016; Wijayanti et al., 2014; Wijayanti et al., 2019). This is possibly due to their lower proportion in the whey and thereby relatively low contribution to the nutritional value and functionality of the whey. However, the emerging research findings emphasise the importance of preserving and/ or incorporating biologically active whey components into food systems to advance the quality of the final product (Bogahawaththa, 2018a; Brick et al., 2017; Lagrange and Clark, 2019; Mangano et al., 2019; Sharma, 2019; Xiong et al., 2020). The evolution of applying non-thermal processing methods, instead of traditional thermal treatments, is also focused to retain the biological functions of whey proteins (Bogahawaththa et al., 2018c; Borad and Singh, 2018; Huppertz et al., 2019). Furthermore, the novel concepts of food protein functionality pay attention to some important parameters such as nutritional quality and biological functions, including digestibility and allergenicity of protein ingredients and their modifications during processing (Foegeding, 2015; Foegeding and Davis, 2011). This is in addition to the non-nutritive functional properties of proteins, for instance emulsifying, foaming and gelling, that were traditionally considered valuable.

### 4.1 Native conformation

Immunoglobulins are a group of glycoproteins, antibodies and bioactive proteins (Hurley, 2003; Hurley and Theil, 2013; Marnila and Korhonen, 2011), that constitute about 10% (w/w) of the total whey proteins (Edwards and Jameson, 2020; Wang and Guo, 2019b), and are considered a major component of whey proteins (Deeth and Bansal, 2019; Fox, 2003; Kilara and Vaghela, 2018). Bovine immunoglobulins can be classified into three main classes, namely, IgG, IgM and IgA, based on their structure. IgG consists of two subclasses, termed $IgG_1$ (~80%) and $IgG_2$ (~20%), which are commonly referred to as IgG. IgG constitutes about 70–80% of the total immunoglobulins in milk (Gapper et al., 2007; Hurley and Theil, 2013; Marnila and Korhonen, 2011). The structures of the immunoglobulins have been discussed in detail using schematic representations by many authors (Gapper et al., 2007; Guo and Wang, 2019a; Hurley and Theil, 2013; Marnila and Korhonen, 2011; Saito, 2009; Schroeder Jr and Cavacini, 2010). In brief, the monomeric form of immunoglobulins, for instance IgG, is composed of four subunits; two heavy chains and two light chains that are arranged by disulphide bonds into a Y-shaped molecule (~160 kDa) with two antigen-binding sites (Fig. 2). IgA occurs as either monomeric or dimeric forms and IgM exists in a pentameric structure linked by disulphide bonds.

Immunoglobulins provide a range of protective bioactivities including passing of immune protection to neonates (Hurley and Theil, 2013; Hurley and Theil, 2011; Mehra et al., 2006). The oral ingestion of bovine immunoglobulins has been effective in controlling several infections in human gastrointestinal and upper-respiratory tracts (Mehra et al., 2006; Riera and Álvarez, 2014; Ulfman et al., 2018; Yolken et al., 1985). Immunoglobulins also appear to render a protective effect against development of allergies, including the most prevalent milk protein allergy among children (Abbring et al., 2019; Bogahawaththa et al., 2017a; Bogahawaththa and Vasiljevic, 2020a; Ulfman et al., 2018). The great resistance of immunoglobulins to gastrointestinal digestion, in comparison to the other whey proteins, makes them biologically active throughout the gastrointestinal tract (Jasion and Burnett, 2015; Mahe et al., 1992; Roos et al., 1995). The biological activity of a protein or antibody is associated with its native three-dimensional conformation (Price, 2000). The preservation of this native conformation during food processing and storage is of utmost importance for the expression of biological activity in a food system. Immunoglobulins are highly heat-sensitive, in comparison to β-Lg and α-La, of which-antigen binding sites are more prone to heat-induced changes than the other regions of the molecule (Bogahawaththa et al., 2017b; Gapper et al., 2007; Hurley and Theil, 2013; Oldfield et al., 1998). Processing conditions, for instance application of heating at or above 72°C, may alter their native conformation and the associated beneficial bioactivities (Anema, 2017; Bogahawaththa et al., 2017a; Loss et al., 2011).

## 4.2 Denaturation and interactions

Immunoglobulins are globular proteins. Thus, they can undergo denaturation and a subsequent aggregation process when subjected to heating. In common with other whey proteins, the native structure of immunoglobulins passes through different phases – from reversible unfolding to irreversible unfolding and then aggregation through hydrophobic and/or disulphide interactions (Section 6.3) (Indyk et al., 2008). Anema (2017) recently studied the thermal denaturation kinetics of immunoglobulins in bovine skimmed milk at a wider temperature range (55–80°C) for up to 15 min holding time. Denaturation of immunoglobulins (insoluble at pH 4.6) displayed second-order reaction kinetics. Oldfield et al. (1998) also reported a second-order reaction for denaturation of IgG in bovine skimmed milk (70–85°C). However, several different reaction orders have been reported previously, for instance, IgG-fortified UHT milk (1.2 mg of IgG/ mL of milk) heated at 62–80°C had a reaction order of 1 (Fukumoto et al., 1994) and IgG in milk heated at 70–90°C showed a reaction order of 1.5 (Law, 1995). The lower reaction orders resulted when the denaturation did not proceed to higher levels, preferably over 90% (Anema, 2017; Oldfield et al., 1998). The activation energy ($E_a$) of immunoglobulins in bovine skimmed milk has been reported at 261 (Anema, 2017) and 269 kJ $mol^{-1}$ (Oldfield et al., 1998). The initial denaturation temperature ($T_D$), temperature at the DSC peak maximum ($T_{tr}$) and enthalpy of denaturation ($\Delta H$) of immunoglobulins were 72°C, 89°C and 500 kJ $mol^{-1}$, respectively (Brodkorb et al., 2016; Kinsella and Whitehead, 1989). A number of other studies have investigated the denaturation kinetics of immunoglobulins in the past (Chen and Chang, 1998; Chen et al., 2000; Dominguez et al., 1997; Domínguez et al., 2001; Li-Chan et al., 1995; Luf et al., 1993; Mainer et al., 1997; Resmini et al., 1989; Riera and Álvarez, 2014). The inconsistency of the results reported can be attributed to the variations in medium (buffer solution, whey, milk or colostrum), immunoglobulin concentration, pH range, heating methods and techniques used to determine the degree of denaturation (Bogahawaththa et al., 2017b; Oldfield et al., 1998). IgG gained higher heat stability in colostrum and milk than in whey or phosphate buffer solution (Chen and Chang, 1998; Li-Chan et al., 1995). IgG dissolved in a phosphate buffer (0.2 mg $mL^{-1}$) was stable for several hours at pH 6–7 and at 37°C, but unstable at highly acidic ($pH < 4$) or basic ($pH > 9$) conditions, due to changes in charge on the molecular surface and subsequent denaturation (Chen and Chang, 1998). Lowering the pH from 6.5 to 5.5 and 4.5 increased the denaturation rate of IgG at 70°C, 72°C and 78°C (Domínguez et al., 2001). Devi et al. (2011) reported that when IgG solutions with different concentrations (25–200 mg $mL^{-1}$ at pH 6–7) were

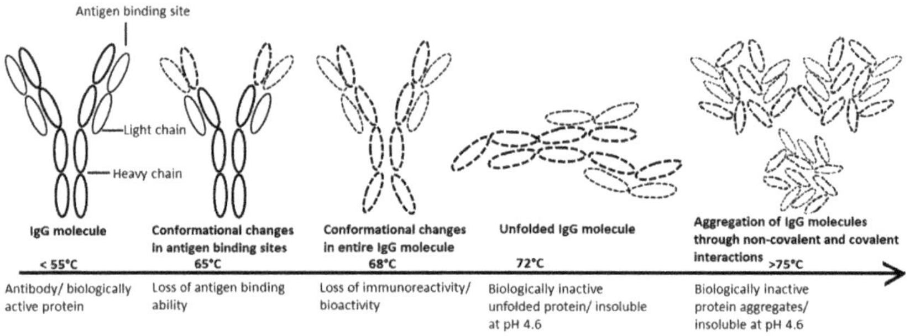


**Figure 2** A simplified pathway of thermal denaturation and aggregation behaviour of bovine immunoglobulin G (IgG) in milk with increase in temperature. Arrow indicates increase in temperature. Conformational modifications begin with changes in antigen-binding sites resulting in loss of antigen-binding ability; proceed to changes in the entire conformation of the molecule causing loss of bioactivity/immunoreactivity; turn into an irreversibly unfolded molecule becoming a biologically inactive protein (insoluble at pH 4.6); and progress to protein aggregation through non-covalent and covalent interactions.

heat treated from 25°C to 95°C, IgG concentration of 25 mg $mL^{-1}$ showed the highest thermal stability.

It is worthwhile to note that, unlike with the other whey proteins, researchers use various methods to determine the degree of denaturation of immunoglobulins. In common with the other whey proteins, irreversibly denatured immunoglobulins are insoluble at pH 4.6 and can be separated from native immunoglobulins and/or other native whey proteins by centrifugation (Anema, 2017). Immunochemical methods are also extensively used to determine the level of denaturation or thermal stability of immunoglobulins. They are mainly two-fold; determination of denaturation of the entire immunoglobulin molecule based on the degree of loss of its immunoreactivity (reactions between immunoglobulins and the antibodies specifically developed against immunoglobulins) or determination of alteration of the antigen-binding sites of the immunoglobulin molecule based on their ability to bind with specific antigens (Domínguez et al., 2001). These methods capture different conformational changes of the immunoglobulin molecule at different stages of the heat-induced denaturation and aggregation process (Fig. 2). Hence, different methods lead to determination of the degree of denaturation or thermal stability of immunoglobulins differently, making it difficult to compare the results (Cao et al., 2007; Indyk et al., 2008).

Anema (2017) reported that denaturation of immunoglobulins in skimmed milk was a slow process at low temperature (~5% denaturation at 55°C for 15 min), but it was rapid at higher temperature (~85–90% denaturation at 80°C for 4 min). This was consistent with Bogahawaththa and Vasiljevic (2020a), that ~12% of immunoglobulins in skimmed milk denatured at 72°C for 15 s (high-temperature short-time pasteurisation – HTST), while the total denaturation occurred at 78°C for 5 min. Borad and Singh (2018) reviewed and summarised that unfolding of immunoglobulins initiates at >65°C, the commercial pasteurisation (72°C for 15 s) can denature about 10–40% of IgG, about 34% of total immunoglobulins denatured under the batch pasteurisation (63°C for 30 min), and almost all the immunoglobulins denatured following the UHT (ultrahigh temperature) treatment. Domínguez et al. (2001) reported that IgG in a phosphate buffer at pH 4 totally lost the antigen-binding ability after heating at 72°C for 5 s. Chen and Chang (1998) and Chen et al. (2000) investigated the application of different food ingredients to improve the thermal stability of IgG. They reported that inclusion of fructose, maltose, sorbitol, glutamic acid, glycine or glycerol at about 5–20% (w/v) into IgG solutions before heating was effective in improving the thermal stability of IgG. Many other studies have investigated the thermal denaturation behaviour and/or thermal stability of immunoglobulins, of which results have been summarised in Table 2.

Heat-induced interactions between the immunoglobulins and with other proteins are rarely studied, in comparison to β-Lg and α-La, as per the literature reported (Wijayanti et al., 2019). Bogahawaththa et al. (2017b) investigated the heat-induced denaturation of bovine IgG and interactions of denatured IgG with β-Lg, α-La and BSA. Native whey proteins, fractionated from the whey of raw skimmed milk using a chromatography method, were used for this study. For all individual protein samples and their mixtures, the final concentration was set at 0.08%, 0.04%, 0.12%, and 0.32% (w/w) for IgG, BSA, α-La, and β-Lg, respectively, to mimic their concentration in the milk. IgG did not appear to change its native conformation, determined by Fourier transform infrared (FTIR) spectroscopy, after heating at 72°C for 15 s. About 55% of the total IgG formed thiol-disulphide-linked aggregates at 100°C for 30 s, determined by the sodium dodecyl sulphate polyacrylamide gel electrophoresis (SDS-PAGE) method, while losing almost all the native conformation. However, a small amount of BSA (~0.016 %) that was present in the IgG sample did not appear to have thiol-disulphide interactions with IgG. After inclusion of β-Lg and α-La into the IgG sample and heating at 100°C for 30 s, all the IgG, ~18% of β-Lg and ~48% of α-La contributed to form aggregates by thiol-disulphide bonds, demonstrating a greater involvement of α-La in protein aggregation over β-Lg. A similar behaviour was observed from α-La in the β-Lg, α-La, IgG and BSA mixture under the same heating conditions. However, in the native whey, β-Lg led the covalent aggregation of proteins over α-La (~74% of β-Lg and ~30% of α-La) after heating at 100°C for 30 s. These results suggested that less abundant whey proteins, for instance IgG, can still modulate the thermal denaturation and aggregation behaviour of mostly available whey proteins (β-Lg and α-La) in different protein mixtures. Further studies have confirmed these results and have also revealed that if the caseins were included into a β-Lg, α-La, IgG and BSA mixture and heated under the same conditions, such a protein mixture can behave in a similar way to skimmed milk, displaying a β-Lg-driven whey protein and casein aggregation through thiol-disulphide interactions (Bogahawaththa et al., 2018b, 2019). The past studies have revealed related evidence; an increase in the whey protein concentration in skimmed milk increased the rate of denaturation of α-La to a greater extent than the β-Lg (Oldfield et al., 2005). Human serum albumin hindered the heat-induced aggregation of human immunoglobulins at 53–56°C for 90 min (Soltis et al., 1979). A few other studies have reported the interaction of immunoglobulins with other whey proteins and caseins. IgG, LF (lactoferrin), BSA, α-La and β-Lg formed large protein aggregates with the caseins via thiol-disulphide bonds when skimmed milk was subjected to temperatures above 100°C (Patel et al., 2006). Heating of IgG alone at 95°C for 20 min resulted in the formation of covalent complexes, but in a skimmed milk system, IgG formed non-covalent aggregates with the caseins (Oh and Richardson, 1991). IgG in a phosphate buffer at pH 6.5 formed covalent aggregates after heating at 72°C for 4 min (Domínguez et al., 2001). Hence, the current evidence emphasises the importance of preserving the native conformation and associated bioactivity of bovine immunoglobulins, as well as the need to conduct further investigations to establish their thermal denaturation and intra- and interprotein interactions, to enable better modulation of thermal denaturation of whey proteins and to incorporate therapeutic and prophylactic properties of immunoglobulins into novel product development.

## 5 Shear as a factor governing modification of whey proteins

Transport phenomena, mainly flow, heat conduction and convection, are integral components of many units of operations of dairy processing. In dairy systems, materials tend to flow, regardless of their physical state (e.g. liquid, semi-solid or solid), if an adequate mechanical force is applied for a sufficient period of time. In rheology, the term 'stress' is used, instead of force, and defined as the force that is acting on a unit area. The stress may result in a deformation in the system, which indicates a change in the distance between two points in the material and is expressed as a 'strain' (Huppertz et al., 2019; Walstra, 2003). Huppertz et al. (2019) have recently explained the general principles applicable for shear-based processing of whey proteins. When milk or a protein dispersion is subjected to a shear stress, for instance during pumping, stirring or homogenisation, it will move in a way of either laminar or turbulent flow. Examples for laminar flows are simple shear, rotational and elongational flow. The type of flow determines the viscosity of a fluid, which is defined as the tendency to resist flow (Huppertz et al., 2019; Rao, 2013). In a Newtonian fluid like milk or a protein dispersion, during a simple shear flow, the viscosity does not depend on shear rate, which is the velocity gradient created in a fluid due to an applied shear stress. Increase in velocity may transform a laminar flow into a turbulent flow resulting in rapid variation of flow velocity, direction and pressure (Rao, 2013).

Huppertz et al. (2019) elaborated the possible mechanisms of deformation of major whey proteins when they are subjected to a simple shear flow at a low concentration (Newtonian fluid). The α-La molecule, as a deformable sphere, tends to elongate while changing its orientation about 45° to the flow direction with liquid spinning inside the molecule. The β-Lg dimer, as a polymer coil, may change its orientation, in a similar way to α-La, and also elongates as a result of simultaneous coil compression and extension towards opposite directions. Although there had been some evidence of deactivation of enzymes after subjecting them to shear (Charm and Lai, 1971; Charm and Wong, 1970; Thomas and Geer, 2011; Tirrell and Middleman, 1975), depending on shear rate and exposure time (Charm and Lai, 1971) and also certain interfacial properties of the protein (e.g. adsorption to solid or air surfaces) (Lee and Choo, 1989; Thomas and Geer, 2011), a model developed by Jaspe and Hagen

**Table 2** A summary of studies that investigated the thermal denaturation and/or thermal stability of bovine immunoglobulins under different conditions using various methods

| Nature of the study | Method used to assess immunoglobulins | Key findings | Reference |
|---|---|---|---|
| Quantification of IgG in differently processed milk in comparison to raw milk | Enzyme-linked immunosorbent assay (ELISA) and radial immunodiffusion (RID) | HTST (high temperature short time) pasteurised milk contained about 65–79% of IgG. IgG was not found in evaporated or UHT (ultrahigh temperature) treated milk. | Kummer et al. (1992) |
| Assessment of changes in secondary structure of IgG and its relationship with immunoreactivity | ELISA and Circular dichroism (CD) for secondary structure changes | >70°C significantly changed the secondary structure of IgG that was related with loss of immunoreactivity. | Li et al. (2005) |
| Changes in IgG content in differently processed milk | ELISA | HTST-pasteurised milk retained about 95% of IgG. | Bogahawaththa et al. (2018c) |
| Comparison of denaturation kinetics of bovine IgG, IgA and IgM | Single radial immunodiffusion (SRID) | D values were reported in the order of IgG > IgA > IgM. Immunoglobulins could withstand up to 70°C for 15 s. | Mainer et al. (1997) |
| Stability of IgG affected by different heat treatments | ELISA | IgG content did not change either after heating at 62.7°C for 30 min or holding at room temperature or 4°C for 24 hours. | Li-Chan et al. (1995) |
| Use of thermal protectants to improve the stability of IgG | ELISA and SRID | Inclusion of 5% fructose/ maltose or 2% glycine preserved over 25% IgG after heating at 95°C for 15 s. | Chen and Chang (1998) |
| Impact of thermal treatment on the quality of colostrum | SRID | Heating at ≥60°C for ≥ 30 min denatured $IgG_1$ significantly, but not $IgG_2$. | Elizondo-Salazar et al. (2010) |
| Effect of heating on antigen-binding activity of IgG in bovine colostrum | ELISA | IgG did not lose antibody-binding activity at 65°C for 60 min. | Dominguez et al. (1997) |
| Assessment of antigen-binding ability of IgG in milk | ELISA | Antigen-binding ability of IgG was >85% after heating at 63°C for 60 min or storing at 37°C for 24 hrs at pH 4–10. | Ohnuki and Otani (2005) |

| | | | |
|---|---|---|---|
| Determination of thermal conditions for inactivation of antigen-binding ability of different immunoglobulins in colostrum | ELISA | Antigen-binding ability of IgG, IgM and IgA was completely inactivated upon heating at 85°C for 30 min, 85°C for 20 min and 80°C for 25 min, respectively. | Ustunol and Sypien (1997) |
| Assessment of the quality of heat-treated colostrum | Turbidimetric immunoassay | Heat treatment at 60°C for 60 min did not reduce the IgG concentration in colostrum. | Johnson et al. (2007) |
| Assessment of the quality of heat-treated colostrum | Turbidimetric immunoassay | Heating colostrum at 63°C for 60 min resulted in 34% reduction of IgG concentration, but treatment at 60°C for 120 min had no effect. | McMartin et al. (2006) |
| Assessment of the quality of heat-treated colostrum | Turbidimetric immunoassay | Heat treatment at 60°C for 60 min can be applied to process the colostrum without affecting IgG concentration. | Donahue et al. (2012) |
| Assessment of the quality of heat-treated colostrum | Turbidimetric immunoassay and microtitration serum neutralisation assay | Heat treatment at 60°C for 120 min did not change IgG concentration and antibody activity. | Godden et al. (2006) |
| Assessment of the quality of processed colostrum | RID | Batch pasteurisation at 60°C for 45 min and freeze-drying reduced immunoglobulins content by 25% in colostral whey powder. | 8Elfstrand et al. (2002) |
| Assessment of antigen-binding capacity of IgG in hyper-immune cheese whey | ELISA | Antigen-binding capacity rapidly decreased with increase in temperature (from 68°C to 80°C) and holding time (from 0 to 60 min). Heating at 72°C for 16 s did not reduce the antigen-binding capacity significantly. | Riera and Álvarez (2014) |
| Assessment of the availability of bovine $IgG_1$ antibodies (anti-rotavirus) in milk and dairy products | Enzyme immunoassay | A detectable level of anti- rotavirus antibodies was found in raw and pasteurised milk but none or little in commercial infant formulas. | Yolken et al. (1985) |

(Continued)

**Table 2** (*Continued*)

| Nature of the study | Method used to assess immunoglobulins | Key findings | Reference |
|---|---|---|---|
| Evaluation of different processing conditions to preserve bioactivity of immunoglobulins in skimmed milk | ELISA | HTST pasteurisation reduced IgG, IgA and IgM concentration by 35%, 55% and 60%, respectively. Micro filtration alone or in combination with ultraviolet treatment did not change immunoglobulin content. | Zhang et al. (2021) |
| Assessment of thermal stability of immunoglobulins in skimmed colostrum under different unit operations | ELISA | Commercial pasteurisation (72°C for 15 s) and batch pasteurisation (63°C for 30 min) resulted in 56% and 25% denaturation of IgG, respectively. Immunoglobulins concentration was not affected by homogenisation and high-shear mixing. | Borad et al. (2019) |
| Evaluation of the impact of simulated industrial processing on bovine milk serum proteins | ELISA | Immunoglobulins were not detected in milk after extended shelf life (125°C for 5 s), UHT sterilisation (135°C for 5 s) and spray-drying (inlet temperature 170°C and outlet temperature 85°C) conditions. Pasteurisation conditions (72°C for 15 s or 63°C for 30 min) denatured 40–70% of immunoglobulins. | Liu et al. (2020) |

(2006) predicted that an extremely high shear rate, about $10^7$ $s^{-1}$, is required to unfold a typical small globular protein (about 100 amino acids) in water. However, unfolding could still occur if the molecular weight of the protein or viscosity of the solution is suffciently high in a laminar flow, or alternatively, if the flow is highly turbulent (Jaspe and Hagen, 2006).

In a flow, the collision frequency of unfolded monomers determines the level of protein aggregation (reversible or irreversible) (Taboada-Serrano et al., 2005), which involves three main mechanisms based on the nature of particle collision, namely Brownian motion (perikinetic aggregation), shear (orthokinetic aggregation) and differential sedimentation (Meyer and Deglon, 2011). According to the aggregation kinetics model suggested by Smoluchowski (1918), the number of collisions predominantly depend on the number of particles and their size. Shear-based aggregation, in particular, is governed by shear rate, the number of particles, as well as viscosity of the medium (Potanin, 1991). Increased frequency and the high energy of collisions create an environment for the formation of intermolecular bonds. However, this mechanism may be disrupted by the high shear rate that lowers contact time between particles and thereby hinders such bond creation, leading to the formation of relatively small aggregates (Zumaeta et al., 2006). While playing a predominant role in aggregation kinetics, the flow mechanisms concurrently generate disruptive forces to break the formed agglomerates (Kim and Kramer, 2006). Furthermore, large aggregates are more susceptible to rupture than the small aggregates, possibly due to the presence of weak spots in their structure. Moreover, the balance between growth of aggregate and shear-controlled breakage largely determines the net growth rate of a protein aggregate and its size (Kim and Kramer, 2006). The mass transfer and reactions among constituents in a solution govern the rate of aggregation. Such a mass transfer-controlled aggregation process, under hydrodynamic shear conditions, is comprised of two distinct phases: diffusion controlled-growth (perikinetic growth) and hydrodynamic shear-controlled growth (orthokinetic growth) (Taylor and Fryer, 1994).

## *5.1 Shear-induced structural changes, denaturation and aggregation of whey proteins*

A number of studies have investigated the structural changes and denaturation behaviour of whey proteins affected by shear alone or in combination with heating. Bekard et al. (2012) studied the shear-induced secondary structural and conformational changes of BSA using an optical quartz Couette cell that allowed them to obtain fluorescence and circular dichroism spectra *in situ*. They reported that when a native BSA dispersion in phosphate buffer at pH 7.4, was sheared at 300 $s^{-1}$ for 60 min, irreversible conformational transition was observed, leading to denaturation. The increase in shear rate resulted in a greater reduction of α-helical content, with an increase in β-sheets and random coils. The degree of conformational changes depended on the exposure time (for instance, unfolding of α-helices occurred after 40 min) but not on the BSA concentration (0.1–1 mg $mL^{-1}$). Sharma and Pandey (2021) subjected a BSA dispersion (5 mg $mL^{-1}$ in Tris-HCL buffer at pH 7.4) to different thermal and shear conditions (55°C, 60°C or 65°C with or without a constant shearing at 300 $s^{-1}$) for up to 120 min using a rheometer. They reported that the shearing at 55°C could not initiate the unfolding of BSA molecules. However, the shearing greatly contributed to modification of the secondary structure at >55°C, for instance, about 30% and 90% of native α-helical content was lost at 65°C without and with shearing, respectively. Simultaneous shearing with heating increased the rates of BSA aggregation by 1.5 fold, in comparison to heating alone. It was concluded that thermal energy appeared to be effective in unfolding the native conformation of BSA, while the shear-generated dissipation energy largely contributed to fibrillation of unfolded molecules, leading to aggregation, even below the melting temperature (63°C).

Ashton et al. (2010) monitored flow-induced conformational changes of a several different proteins, including a β-Lg dispersion (10 mg in 1 mL of distilled deionised water) placed in a Taylor–Couette flow cell using a Raman spectrometer *in situ*. They concluded that the conformational changes observed are subtle and reversible. The degree of β-sheet structures contained in a protein determines its sensitivity to conformational changes induced by shear and the degree to which such changes could occur. β-Lg can withstand flow-induced shear to a greater extent, as its structure consists of substantial amounts of β-sheets (Table 1). Gabovich and Li (2009) observed that a considerably higher mechanical force is required for unfolding a β-protein, in comparison to stretching an α-protein. Ashton et al. (2010) also suggested that smaller protein molecules are more susceptible to shear-induced changes, considering the force required to initiate the unfolding, than the larger proteins. This finding, however, contradicted other studies that assessed complete unfolding of proteins driven by mechanical forces (Alexander-Katz et al., 2006). When raw skimmed milk was subjected to several shear rates at 20°C using a rheometer, mostly reversible secondary structural changes in whey proteins were observed by FTIR spectroscopy (Mediwaththe et al., 2018a). For instance, an increase in shear rate from 0 $s^{-1}$ to 500 $s^{-1}$ resulted in substantial reduction of the native α-helical structure and exposure of β-sheets, which can be primarily linked to disassociation of β-Lg dimers into monomers and then the partial unfolding of the native monomeric structure. Increased β-sheet content and their further modification were detected after increase in shear rate up to 1000 $s^{-1}$ (Mediwaththe et al., 2018a).

Wolz et al. (2016) treated whey protein dispersions of 5–30 % (w/w) in deionised water at pH 6.7–7.0, subjecting them to different shear rates (100–1452 $s^{-1}$) using a rotational rheometer. It appeared that the increase in shear rate, at 80°C for up to 30 min, accelerated the denaturation of β-Lg at the protein concentration of up to 20%, but this effect disappeared at 30% due to rapid denaturation. Application of shear stress alone was not sufficient to induce the denaturation of proteins (Wolz et al., 2016). The authors attributed the greater influence of shear on protein aggregation at low protein concentrations, to the enhanced mobility of molecules and the subsequently increased rate of collisions. Bogahawaththa and Vasiljevic (2020b) reported consistent results in relation to denaturation of β-Lg and α-La in skimmed milk subjected to a range of temperatures (70–100°C for 5 min), alone or in combination with a constant shear rate (1000 $s^{-1}$), using a rheometer. The applied shear was effective in denaturation of β-Lg and α-La even at 70°C due to destabilisation of the native conformation. A greater influence was observed at 85°C resulting in about 25% of extra denaturation than that of the non-sheared milk. The shearing effect rapidly diminished at higher temperatures (≥ 95°C). Moreover, the shear tended to induce aggregation of whey proteins and caseins in skimmed milk.

Quevedo et al. (2020c) investigated the denaturation kinetics of concentrated β-Lg dispersions of 50–70 % (w/w) in deionised water at pH 6.7–7.0, under different temperatures (60–100°C for up to 90 s) and oscillatory shear rates (0.05 $s^{-1}$, 25 $s^{-1}$, 50 $s^{-1}$) using a closed cavity rheometer. An increase in shear rate enhanced the denaturation of β-Lg at every temperature and concentration level tested. As well, it significantly induced the formation of covalent aggregates up to 80°C, but the effect of shear was not apparent after 80°C due to rapid aggregation induced by the higher temperature. Furthermore, they reported that the combination of shear and high temperature induced the formation of non-disulphide covalent interactions at lower protein concentrations apart from other non-covalent and disulphide covalent bonding. In an equivalent experimental setting, shear appeared to modulate the degree of aggregation and aggregation behavior of concentrated β-Lg at different temperature levels (Quevedo et al., 2020a). For instance, increase in shear from 0.05 $s^{-1}$ to 30 $s^{-1}$ increased the degree of aggregation of concentrated β-Lg (50%, 60% or 70% (w/w)) at each temperature level (60°C, 70°C or 80°C for 30 s) and mostly induced the formation of disulphide-linked aggregates. After treatment at 100°C for 30 s, almost all β-Lg aggregated, regardless of the shear rate and the protein concentration, but increase in shear appeared to induce non-disulphide covalent interactions (Quevedo et al., 2020a). When highly concentrated (60–70% (w/w)) binary mixtures of β-Lg and α-La (1:1 or 5:2) were subjected to thermomechanical treatments (60–100°C for up to 90 s and shear rate at 0.06–50 $s^{-1}$) using a closed cavity rheometer, the increase in shear induced the denaturation of proteins, regardless of total protein concentration and the β-Lg:α-La ratio. The effect, however, disappeared at higher temperatures (> 80°C) (Quevedo et al., 2020b). It was also observed that the applied shear was able to alter the aggregation onset temperature and that increase in shear and temperature accelerated the formation of disulphide interactions (Quevedo et al., 2020b). Quevedo et al. (2021) reported that the degree of denaturation of whey proteins increased by 8%, 46% and 14% of a concentrated WPI dispersion (60% (w/w) at pH 6.7), treated in a closed cavity rheometer, when increase in shear rate from 0.06 $s^{-1}$ to 50 $s^{-1}$ at the temperatures of 60°C, 70°C and 80°C for 30 s, respectively. At higher temperature (100°C), the influence of shear on denaturation of whey protein was not noticeable due to fast aggregation of the proteins, induced by heating. Furthermore, this study revealed that the shear could induce denaturation of whey proteins in a wider range of pH (3–9) in combination with heating. However, the influence was weaker, in comparison to the neutral pH (6.7), and it almost disappeared at ≥80°C. Moreover, the increase in shear rate resulted in reduction of the calculated activation energy in relation to denaturation reaction, regardless of the temperature and pH levels tested (Quevedo et al., 2021).

Studies reported the formation, fragmentation and reformation of protein aggregates in skimmed milk under simultaneous application of shear and heating (Mediwaththe et al., 2018a,b). The shear rate at 500 $s^{-1}$ and 1000 $s^{-1}$ appeared to break non-covalently bound protein aggregates and their reformation, respectively, during heating at 72°C. However, at high temperature (140°C), applied shear (500 and 1000 $s^{-1}$) commonly induced a larger amount of non-covalent and a smaller amount of disulphide-linked protein aggregation. In a similar experimental setting, the application of shear at 500 $s^{-1}$ and 1000 $s^{-1}$ contributed to the aggregation of whey proteins and caseins in skimmed milk at 80°C, while shear-induced fragmentation of aggregates was obvious at 120°C (Mediwaththe et al., 2018b). However, at 100°C, shearing at 500 $s^{-1}$ and 1000 $s^{-1}$ induced fragmentation and formation of aggregates, respectively.

Most of the studies discussed above highlighted the potentiality of modulating the denaturation and aggregation behaviour of whey proteins through the simultaneous application of shear and temperature, mostly above 60°C. Some of these studies also indicated the possibility of modifying the functionality of whey proteins through this technique (Quevedo et al., 2020a). However, it is obvious that there are inconsistencies among the results of different studies that have investigated the influence of shear on structural changes and denaturation behaviour of whey proteins. These disparities are mostly ascribed to a several key elements such as physicochemical and structural properties of the protein (e.g. primary structure, native conformation, molecular size, molecular weight, and surface properties), type of shear device used, nature of the shear flow (e.g. laminar or turbulent flow) including flow dynamics and Reynolds number, shear exposure time and other external factors (e.g. temperature, pH, pressure, ionic strength, mineral concentration and other components, and protein concentration), viscosity of the protein dispersion and surface forces between solvent–solid and solvent–air (see Section 6.6.2) (Bekard et al., 2012; Huppertz et al., 2019; Quevedo et al., 2021).

## 6 Modification of whey proteins' functionality

Although whey proteins possess the excellent functional properties, limitations in the application in food systems do exist due to their inherent compositional variability and different processing conditions used in cheese making, the manufacture of WPC/ WPI and the subsequent storage conditions as well (Nishanthi et al., 2017a,b,c; 2018a,b). This may ultimately result in inconsistency in product quality and complexity of the reactivity of whey proteins during processing of foods containing whey proteins (Jovanovic et al., 2005). While different commercial WPCs may contain the same levels of proteins, they still may show substantial variations in physicochemical properties, for example, proximate composition and particle size distribution patterns, which is mostly consistent with the variations, observed in different functional properties (Onwulata et al., 2004). The properties of whey proteins, such as their heat stability and heat-induced gelation and the firmness of created gels of most of the products appear to have a link with calcium content. Another common setback regarding application of the WPC is development of off-flavours during storage (Javidipour and Qian, 2008), mainly due to lipid oxidation. The free radicals and numerous other organic volatile substances released during lipid oxidation may further detrimentally impact the quality of the powders, by catalysing the secondary Maillard browning reaction between proteins and lactose (Morr and Ha, 1993).

The functionality of whey proteins depends on the physicochemical properties of their three-dimensional protein structure including, size, shape, amino acid composition and sequence, net charge, ratio of hydrophobicity to hydrophilicity, molecular flexibility and rigidity, which are, on the other hand, governed by the external environmental factors such as protein concentration, pH, temperature, ionic strength, the type of ions present and the influence of other available food components (Wijayanti et al., 2019). Although whey proteins' functionality mainly depends on the behaviour of β-Lg, the most abundant protein in whey, the overall functionality entirely depends on the combination of the properties of all other components present in the whey protein products. For instance, β-Lg has excellent gelling, foaming and emulsifying properties, while α-La exhibits good emulsifying properties, but its gelling ability is poor (Table 1) (Pearce and Kinsella, 1978). Therefore, in order to stabilise and functionalise whey proteins, it is necessary to modify them by intentionally modulating one or more of the parameters governing their behaviour. In addition to achieving the required functionality, protein modifications may also, for example, lead to improvements in nutritional value through controlled digestibility (Sah et al., 2016) or diminished allergenicity (Golkar et al., 2019).

## 6.1 Approaches to modifications of whey protein functionality

The functionalisation of proteins in general is achieved by chemical, enzymatic or physical means. Chemical modifications involve derivatisation of protein side chains with the main objective of modulating the various non-covalent forces described above. These modifications ultimately lead to structural and thus functional changes. For example, sulphitolysed whey proteins, produced by controlled oxidative breakdown of disulphide bonds (sulphitolysis), had improved surface properties including emulsification and foaming (Klemaszewski and Kinsella, 1991) while digestibility of β-Lg increased with a concomitant decline in antigenicity *in vitro* (Kananen et al., 2000). Enzymatic modifications are usually imparted by application of specific proteases to hydrolyse proteins, yielding a mixture of peptides. The functionality of the hydrolysed proteins is highly dependent on the degree of protein hydrolysis in this case. Changes in the functionality of proteins can also be achieved by inter- and intramolecular cross-linking using specific enzymes, for example, transglutaminase. Transglutaminase (protein-glutamine-γ-glutamyl transferase, EC 2.3.2.13) is a family of enzymes that cross-links proteins between the γ-carboxamide groups of glutamine residues and the ε-amino groups of lysine residues in proteins, peptides and primary amines (Mostafa, 2020). One of the critical problems with the use of proteases is control of the extent of hydrolysis, which can be accomplished with appropriate enzyme selection or process modifications, for example, with the use of immobilised enzymes or process adjustments. Another important issue is the liberation of bitter peptides through hydrolysis that can impart off-flavours during the application. Modification of protein functionality by physical means presents a way to obtain ingredients with unique and desirable functional properties. These changes are usually brought about by changing the system temperature or pressure, the quality of solvent through change of pH, ionic strength or addition of a complexing biopolymer, and/or a combination of these effects. In their native state, whey proteins exist as globular proteins with a well-defined three-dimensional structure. As many functional properties depend on solubility, the high aqueous solubility of whey proteins is an added advantage in this regard. The solubility of a protein depends on a thermodynamic equilibrium between protein-protein and protein-solvent interactions (Zayas, 1997). The intrinsic properties of proteins, such as surface hydrophilicity and hydrophobicity, greatly affect the protein solubility. Due to lack of electrostatic repulsion, proteins in general exhibit minimum solubility around the isoelectric pH. However, whey proteins are highly soluble even at their isoelectric points due to a large ratio of surface hydrophilic to hydrophobic residues. Moreover, they are soluble in the pH range of 2–9, which makes them suitable for incorporation into various food products such as carbonated beverages at acidic pH (Resch et al., 2004).

## 6.2 Improving functionality by controlled aggregation

As outlined in the preceding sections, despite their excellent functional properties, whey proteins change their physicochemical properties during commercial processing, which can hinder their application in the food industry. Processing - especially the mandatory heat application used in the food industry to address safety concerns and prolong the shelf life - promotes various interactions of whey proteins with other food components. These, in addition to changes and adjustments in the surrounding environment (solvent quality, composition, pH, ionic strength), ultimately impact the stability of these proteins, inevitably leading to loss of functionality, including but not limited to solubility and surface properties such as emulsifying and foaming capacities, and consequently macroscopic structural changes of the system. It is therefore imperative for the food industry and consumers alike that the functionality of whey proteins be improved by modifying conventional whey processing approaches by creating inert whey proteins that could beneficially limit heat-induced instabilities. While many approaches have been suggested, they could all be described as a two-step process during which whey proteins are first unfolded and denatured followed by controlled aggregation that would eventually result in different particle sizes and morphologies which could be used in different food applications. The main aim is to 'lock' the structure in, so that it would undergo very minimal or no further changes during later processing. Depending on the environment and molecular architecture, whey proteins can assemble into different structures occupying different dimensions. Aggregation or gelation in a solution takes places by creation of 3D protein networks that may occupy a very small space, as in the case of colloidal dispersions, or extend through the entire space, as in the case of gels. On the other hand, due to their surface activity, they tend to create two-dimensional structures at the oil-water or at the air-water interface. Under certain conditions, proteins can also create one-dimensional self-assembled fibrillar aggregates. Therefore, understanding molecular interactions and structural changes during whey protein aggregation is important, as these clarify the many details of association mechanisms that impact their functionality in food materials.

### 6.2.1 Driving forces of whey protein aggregation

Depending on the sequence of the amino acids, food proteins are described as either globular (folded) or random-coil (unfolded). In globular proteins such as whey proteins, the sequence of amino acids develops into various secondary structures which are closely packed together, giving it a sphere-like shape. Naturally, these proteins are folded into a tertiary structure in which the total free energy is minimised by reducing the total interfacial energy by exposing the more hydrophilic parts of the protein towards water, while burying more hydrophobic sides within the interior. In contrast, random-coil like proteins such as individual caseins exist in an unfolded state. The main reason for the strong tendency of globular food proteins including whey proteins to aggregate is their size. The Derjaguin, Landau, Verwey and Overbeek (DLVO) theory outlines the direct dependence of interaction potential ($F_{DLVO}$) on the colloidal particle size. Therefore, the smaller the particle, the smaller the energy barrier, which needs to be overcome to promote aggregation. While such a description is fairly rudimentary, it supports the propensity of β-Lg to create dimers at intermediate pH at 100 mM (Mercadante et al., 2012). However, the scenario is far more complex than this and the real nature of the attractive forces in native protein systems still remains a matter of debate. A number of studies have indicated that the phase behaviour of protein is governed by the presence of very short-range attractions (Bijsterbosch et al., 1995; Piazza, 2000). The 'short range' is defined as a width of the potential much shorter than the protein hard-core radius (Mezzenga and Fisher, 2013). These attractions have been described by several different models, including the Baxter sticky particle model, in addition to the van der Waals or LJ potentials, with varying degrees of success. In addition to these potentials, other interactions play an important role in the phase behaviour of proteins including hydrophobic interactions, hydration, solvation and hydrogen bonds, hydrophilic interactions, charge residues complementarity, salt-specificity effects and molecular recognition (Bryant and Mcclements, 1998). What complicates all these approaches even more is that they are based on isotropic pair-wise potentials, which do not reflect the heterogeneous surface composition of proteins that induces the presence of non-isotropic potentials, creating a notion of patchy colloidal particles. This naturally follows from the protein interactions being driven by the surface amino acid residues which strongly differ from one protein to the other. These models are further complicated when proteins undergo their transitional change from the folded into unfolded and aggregated states. It is well known that unfolded proteins, which are relevant in dairy processing, are subject to strong aggregation and formation of clusters, with a structural organisation strongly dependent on the environmental conditions. The formation of these aggregates is primarily driven by the presence of hydrophobic patches on the surface of the unfolded proteins resulting from the change in conformation of the proteins during the denaturation process (McSwiney et al., 1994; Qi et al., 1997). Proteins in general lose their native tertiary structure by the presence of denaturants, but most frequently by thermal treatment, elevated pressures or upon adsorption onto surfaces and interfaces. The initially folded sides in the internal part of the protein are suddenly exposed to the aqueous environment, which induces a new hydrophobic component to the attractive potential, as proteins will aggregate to minimise the contacts of the hydrophobic residues with water. These interactions are driven by hydrophobic attractions and electrostatic repulsions and are explicitly dependent on the size of the particles (Mezzenga and Fisher, 2013). This immediately brings up the importance of hydrophobic interactions in protein dispersions, as the competition with the repulsive electrostatic potential can be relatively large - an important factor promoting aggregation in proteins.

The structure of the aggregates formed upon unfolding the globular proteins at high temperatures depends strongly on the pH, ionic strength and protein concentration. Tentatively, three main structural morphologies have been recognised, mainly depending on the pH at heating. Under highly acidic conditions, well below the isoelectric point of whey proteins, fibrils are formed; around the isoelectric point, spherical particles are created with diameters ranging between about 50 nm and a few microns; and at pH closer to neutral, strands are formed with diameters that are most often less than 10 nm and with lengths between tens of nanometres up to tens of microns (Jung et al., 2008). As shown in Figure 3, long fibrillar structures (image on the left) were formed at pH below 3. The created fibrils formed a system that was less thick than the particulate system created around the isoelectric point, shown on the right. Aggregates generated at neutral pH from whey proteins at a concentration below a critical value appear in isolated clusters well dispersed in the solution; thus the fluid remains translucent. These clusters appear to have the same self-similar internal structure with a fractal dimension (df) = 2. The morphology of the whey protein aggregates is usually explained by the fractal gel model, which describes aggregates as an assembly of self-similar primary particles. These discrete fractal aggregates have a mass $M_{Agg}$ with a dimension $\xi$ so that $M_{Agg} \sim \xi^{df}$ with a partially preserved secondary structure of the original globular proteins (Ikeda and Li-Chan, 2004). Secondary aggregation of the strands and the spherical particles occurs at higher protein concentrations, leading to larger fractal clusters. The size of these clusters increases with increasing concentration, until it diverges at a critical concentration above which either a self-supporting gel or a precipitate is created.

For optimal functionality, the aggregation process needs to be controlled and fine-tuned. In principle, it can be described as a two-stage process involving primary and secondary aggregation steps. The primary aggregation leads to the formation of finite-sized aggregates with distinct morphologies which have been termed differently, for example thin elongated aggregates can be called (proto-) fibrils, strands, amyloids, ribbons, nanotubes or polymers. The shape of the strands can be described as rigid, semi-flexible, worm-like, flexible, curly or curvy. The distinction between the rigid and flexible forms depends on the ratio between the length and the thickness of the strand. Semi-flexible would then mean that these two attributes are of the same order of magnitude. Amyloid fibrils describe long semi-flexible fibrils with persistence lengths in the order of microns. On the other hand, spherical aggregates have been termed nano- or microparticles and microgels. Nanoparticles are usually those with a particle size of less than 100 nm while microgels represent small particulate matter composed of a cross-linked network which entraps a large fraction of water.

Stable suspensions of homogeneous spherical microparticles are not easily created by thermal treatment of an aqueous whey protein solution. For example, to create stable microgel suspensions, β-Lg is heated in a narrow pH range around pH 4.6 or around pH 5.8 at a very low ionic strength, yielding particles with a hydrodynamic radius ~200 nm (Schmitt et al., 2009). The transition between the formation of strands and spheres as a function of the pH is sharp and is easily observed visually, as the solutions turn from transparent to become opaque. On the other hand, when microgels are created from WPI, they are dependent on the protein composition and the presence of minerals (Schmitt et al., 2011). Addition of $Ca^{2+}$ can induce the formation of microgels even at neutral pH. At higher $CaCl_2$ concentrations, secondary aggregation takes place, leading to precipitation or gelation. Microgels are only formed when the net charge is reduced below a critical value either by approaching the isoelectric point or by binding of divalent ions. The aggregation process leading to microgel formation starts even when conversion is less than 1%, with well-defined aggregates created with radii of tens of nanometres. Initially, more of the same-sized aggregates are formed, but at a later stage, they grow by fusion and by addition of newly denatured proteins. Secondary aggregation of flexible strands leads to the formation of polydisperse spherical randomly branched aggregates (Fig. 4) and occurs at higher protein concentrations. The average size of aggregates at steady state increases with increasing concentration and diverges at the gel point. At a higher ionic strength (Pouzot et al., 2005) or the pH closer to the isoelectric point (Mehalebi et al., 2008), secondary aggregation starts at lower protein concentrations driven by minimised electrostatic repulsions between the strands. In all cases, the large-scale structure is self-similar with a fractal dimension close to two, which indicates that the protein density of the aggregates decreases approximately linearly with increasing radius. As already mentioned, secondary aggregation of spherical particles usually leads in most situations to gelation or precipitation. From this brief description, it is clear that establishing optimum conditions for

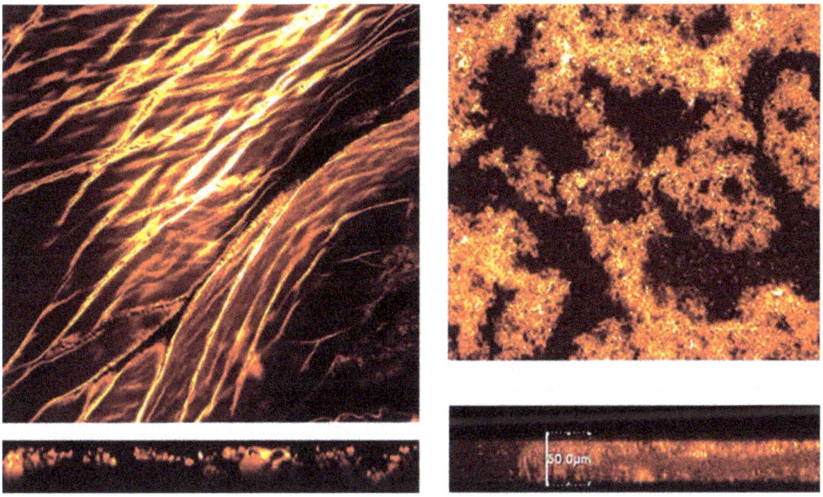

**Figure 3** Different whey protein morphologies created during aggregation at a pH below 3 (on the left) and the pH around the isoelectric point (pH ~ 4.9; on the right) determined by confocal laser scanning microscopy. Cross-sections show differing thickness of generated structures.

creation of particulated whey proteins is needed in order to control the ultimate particle size.

## 6.3 Application of shear for creation of whey proteins microgels

As an additional factor in the creation of particulated whey proteins, shear or cavitation can be used to control the extent of whey protein aggregation. Shear-induced conformational changes propagate weak interactions among the constituent proteins, that may be enhanced or lessened depending on the environment (see Section 6.4). As described already in the preceding sections, during heating of WPC or WPI dispersions, as the temperature is increased, BSA starts denaturing early at a relatively lower temperature, followed by α-La and β-Lg (Considine et al., 2007; Havea et al., 2000). The denaturation pathways of these major whey proteins proceed via initial formation of homo-aggregates followed by the creation of hetero-aggregates (Havea et al., 2001). As pointed out earlier, the initial forces driving the aggregation are hydrophobic attractions (Havea et al., 2004; Havea et al., 2009), and the extent of their involvement is directly proportional to the concentration of the different proteins, especially that of β-Lg (Havea et al., 1998). The choice of process temperature plays a role. In the low temperature range, the unfolding of the proteins' tertiary structure appears to be the rate-limiting step, whereas, at higher temperatures, the aggregation of initial nuclei has been indicated as the governing factor (Anema and McKenna, 1996).

As outlined above, it is possible to create whey protein nano- and/or microparticles by carefully selecting the processing conditions including whey protein concentration, pH, ionic strength and quality of the mineral environment, temperature and shear rate (Erabit et al., 2014; Ipsen, 2017). At low protein concentrations, enlargement of aggregate size appears directly proportional to changes in concentration, without a major impact of shear, while above 10%, shear influences the size of aggregates (Wolz and Kulozik, 2015). Above a critical protein concentration, heating of whey proteins typically results in bulk gelation. Therefore, the traditional approaches applied in microparticulation of whey proteins could be loosely grouped into approaches falling under A or B in Fig. 5. Approach A refers to heating a more concentrated protein solution under appropriate conditions, with simultaneous application of shear under a particular flow regime that would limit the growth of aggregates. Approach B refers to intentional bulk gelation, during which large aggregates are formed in which the primary nuclei particles are held mainly by weak interactions that are readily disrupted by shear (Dissanayake et al., 2012).

These approaches have been trialled commercially with varied degrees of success. One of the first inventions involved manufacturing whey protein microparticles by simultaneous application of a thermal treatment (80–120°C) and high shear (7500–10 000 $s^{-1}$) accompanied by a low pH. These particles were used as a fat replacer under a commercial name, Simplesse[R] (Singer and Dunn, 1990). Several studies reported that the size of aggregates decreased with increasing shear stress (Simmons et al., 2007; Wolz et al., 2016). For shear to be effective, high rates, turbulence and an interactive surface are required (Huppertz et al., 2019). Thus, the impact of shear can be described by two mechanisms: (1) deformation and fragmentation; or (2) erosion of primary particles from the surface of large particles (Taylor and Fryer, 1994; Heffernan et al., 2005). However, whey protein aggregation during simultaneous heating and shearing is far more complex than these two pathways (Simmons et al., 2007; Wolz et al., 2016). Below the denaturation temperature, newly formed aggregates were small and weakly bonded, while above the denaturation temperature, large aggregates were created linked with strong bonds. The heating temperature governed the nature of interactions as weak van der Waals interactions prevailed at low temperature, while disulphide covalent bonds were formed at higher temperatures. Protein denaturation appeared strongly influenced by temperature due to the reaction kinetics. On the other hand, particle aggregation was governed by the applied shear field and magnitude of interactions among the particles. Furthermore, the extent of protein denaturation and particle growth rate of aggregates was directly related to shear rate, which was attributed to the frequency of particle collisions, but the final particle size showed a complex behaviour with a further increase in shear rate (Simmons et al., 2007).

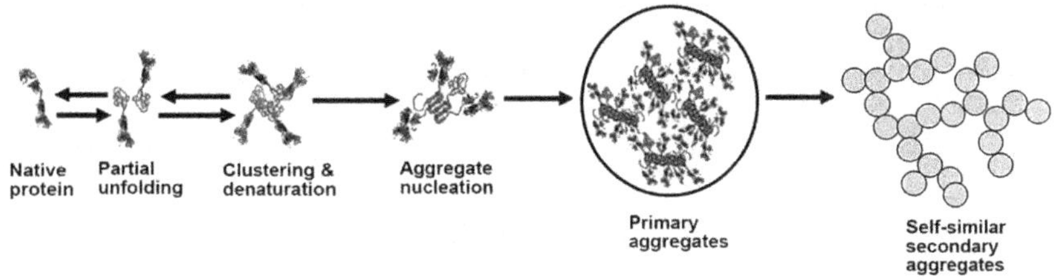


**Figure 4** A simplified pathway illustrating unfolding and aggregation of whey proteins. Unfolding enhances hydrophobic attractions, leading to protein clustering and enabling of strong interactions. At low protein concentration, small primary particles are formed which may grow at higher protein concentration via secondary aggregation into self-similar structures. Double and single arrows denote reversible and irreversible steps, respectively. Nuclei are defined as the smallest net-irreversible aggregates (adapted from Huppertz et al., 2019)

An alternative microparticulation approach to simultaneous heating and shearing is the induction of bulk gelation under defined conditions, to control the rate and extent of aggregation, followed by the application of shear (Fig. 5B). Dissanayake and Vasiljevic (2009) used this method to produce microparticles with an average size of approximately 2–5 μm, applying microfluidisation to microparticulate a 10% WPC dispersion at pH 7 at 140 MPa. The proteins were first denatured by heating at 90°C for 20 min and then microfluidised. The size of the particles was further reduced when the similar approach was applied at low pH, which prevented the creation of covalent bonds through disulphide bonding, allowing microfluidisation to disrupt non-covalent interactions (Dissanayake et al., 2012).

Different types of carbohydrates have been proven to reduce the heat-denaturation/ aggregation of proteins and enhance their colloidal stability. Among them are sugars such as lactose and sucrose (Garrett et al., 1988; Rich and Foegeding, 2000) that could modulate the denaturation of whey proteins and either moderate or delay their aggregation. The influence of these sugars was highly concentration-dependent, where higher concentrations (>10%) resulted in enhancement of aggregation (Kulmyrzaev et al., 2000). The inclusion of similar additives such as ribose and sorbitol also enhanced the heat stability of whey proteins (Rich and Foegeding, 2000; Baier and McClements, 2003; Baier et al., 2004). The protective mechanism of sugars can be described by preferential hydration of proteins (Timasheff, 2002). The presence of stabilising solutes in a protein solution results in increased chemical potentials of both the protein and the additive, creating a thermodynamically unfavourable situation. Such an entropically unfavourable state leads to the preferential exclusion of solutes in the vicinity of proteins and subsequent hydration and stabilisation, which may even be greater in case of denatured proteins (Xie and Timasheff, 1997). However, the inclusion of these compounds, especially reducing sugars,

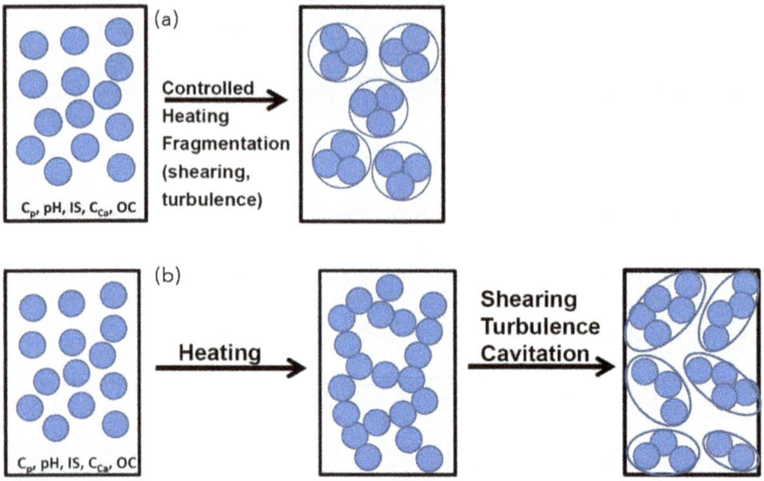


**Figure 5** Two main approaches for production of nano- and microparticles: (a) involves heating with simultaneous shearing/turbulent flow/cavitation of protein dispersions; (b) involves creation of particles by heating of a protein dispersion followed by shearing of created gel. Legend: T - temperature; Cp - protein concentration; IS - ionic strength; $C_{Ca}$ - calcium concentration; OC - other components (polysaccharides, proteins, minerals).

in food systems may result in some undesired interactions (Rich and Foegeding, 2000). Polysaccharides have also been proven effective in enhancing the heat stability of dairy proteins. The well-known candidates are dextran (Nagasawa et al., 1996) and pectin (Zhang et al., 2012). The charges on these polysaccharides interfered with protein-protein interactions during heating and also enhanced the colloidal stability of protein-polysaccharide conjugates. The industrial applications of these methods are very limited, mainly due to the requirement of added non-dairy ingredients in the formulation, and also due to the increased sensitivity to pH and concentration.

Cavitation (sonication) is another method used to obtain smaller whey protein heat aggregates. Instead of preventing denaturation, the application of ultrasound sonication followed by heat denaturation can disperse the aggregates and avoid their reformation upon further heating (Ashokkumar et al., 2009). It was suggested that the either the thiol groups may deactivate due to the combined process of heat treatment and sonication, (Ashokkumar et al., 2009), or the whey proteins may configure into an ultrasonically induced state which is functionally different from the usual heat-denatured state (Gülseren et al., 2010). Successful experiments of ultrasound sonication treatment have been carried out with continuous reactors after preheat treatment where aggregation was controlled (Zisu et al., 2010), which is important in industrial processing. In addition to controlling aggregation, ultrasound treatment is also capable of improving other functional properties of proteins such as emulsification ability, solubility, assisting in processes such as homogenisation, viscosity alteration, extraction, drying, crystallisation, defoaming and ultrasound-assisted extraction (Soria and Villamiel, 2010).

More recently, the research and commercial interests have been focused on the protective influence of the casein micelles against heat aggregation of whey proteins. Such an influence had been suggested as far back as the 1960s (Kenkare et al., 1964). More recently, caseins were indicated as molecular chaperones that would control the aggregation of whey proteins (Kehoe and Foegeding, 2011). The role of the individual caseins, including $\alpha_{S1}$-, $\alpha_{S2}$-, β- and κ-casein as molecular chaperones, has been examined in a number of studies (Kehoe and Foegeding, 2011; Koudelka et al., 2009; Morgan et al., 2005; Treweek et al., 2011; Yong and Foegeding, 2008). Molecular chaperones are presented by a range of cellular proteins with a principal role of governing protein folding and their assembly (Hendrick and Hartl, 1995). For example, heat shock proteins are frequently described as chaperones, due to their role in maintaining a native structure and function of intracellular proteins against a variety of stressors (de Moura et al., 2013). Chaperones are a large group of proteins with similar functionality but they consist of members from a number of protein families that are genetically unrelated (Beissinger and Buchner, 1998; Hendrick and Hartl, 1995; Thorn et al., 2005). The intracellular function of these chaperones has been identified as the protection of cellular proteins against stress conditions. Initially, they were identified to be synthesised in prokaryotic cells when microorganisms were briefly exposed to high temperatures (Beissinger and Buchner, 1998; Hendrick and Hartl, 1995), and were hence called heat shock proteins (e.g. hsp 60 and hsp70). The common function they perform is related to regulating protein folding during synthesis and transportation (Fujiyama et al., 2002; Fujiyama-Nakamura et al., 2009; Suzuki et al., 1998).

Milk proteins, including $\alpha_{S2}$- and κ-casein, have been known to form amyloid fibrils in human mammary tissues where amyloid-like plaques have also been identified within calcified stones in different species (Holt et al., 2013; Thorn et al., 2009). Conversely, the other two caseins would have the capacity to prevent the misfolding of caseins and other milk proteins, as they are known to exhibit molecular chaperone-like properties. It has been reported that $\alpha_{S1}$- and β-casein can protect the heat-, light- and chemically-stressed target proteins against aggregation (Holt et al., 2013). This chaperone-like activity of casein can be industrially utilised to control unfolding and aggregation of milk proteins under stress conditions such as high heat. Individual caseins in general do not have a well-defined structure, and thus appear very dynamic and prone to self-association (Alaimo et al., 1999). This would lead to a micelle formation, as in milk, that would govern their overall behaviour in the presence of environmental stresses (e.g. heat, pH, ionic stress). Proteins would fold in dispersions to shield their hydrophobic zones, and in the process, hydrophobic zones would orient next to each other to reach a thermodynamically favourable configuration (Jöbstl et al., 2006). When proteins unfold under heat or other stress conditions, these hydrophobic zones would expose increasing surface hydrophobicity of the proteins.

The $\alpha_s$-caseins ($\alpha_{S1}$) are considered to interact with partially unfolded whey proteins through hydrophobic surfaces, preventing their normal thiol-disulphide interchange with other whey proteins and eventual aggregation, and their activity appeared to increase with decreasing temperature (Bhattacharyya and Das, 1999). The β-casein also forms complexes with soluble or partially aggregated whey proteins involving electrostatic interactions as well, which is capable of reducing the particle size of whey proteins compared to those in whey protein dispersions with no added β-casein. The smaller complexes created were more soluble and their heat tolerance (up to 145°C) suggests the involvement of hydrophobic interactions, which are enhanced at higher temperatures (Morgan et al., 2005). Even the 10% milk protein concentrate (MPC) dispersions were reported to have a protective influence on whey proteins at retort temperatures (Zhang et al., 2005). The chaperone-like activity of $\alpha_s$-caseins had reduced under molecular crowding conditions, and at slightly alkaline pH values rather than at neutral pH (Morgan et al., 2005). In addition, the β-casein, with a high hydrophobicity, has shown better chaperone-like activity than the $\alpha_{S1}$-casein but when the amphiphilic nature of the proteins was reduced. Nevertheless, although β-casein has consistently been shown to reduce the aggregation of heat-stressed globular proteins, it has not been able to disperse the insoluble aggregates that were formed by heating (Bhattacharyya and Das, 1999; Morgan et al., 2005; Yong and Foegeding, 2008); hence, β-casein can only be considered to display 'chaperone-like activity'. It is concluded that the flexible, dynamic and relatively unstructured and amphipathic nature of $\alpha_s$- and β-caseins are important in their chaperone action (Koudelka et al., 2009). Liyanaarachchi (2017) showed that presence of caseins controlled the aggregation of whey proteins through non-covalent association, which resulted from entrapment of whey proteins (β-Lg) by casein micelles or micellar associations formed with individual (β- and $\alpha_{S1}$) caseins. Elevating the pH to 7.5 resulted in preservation of the secondary conformations of whey proteins in dispersions containing a higher casein fraction during heating. When shearing was introduced, the interactions were further impacted by enhancement of non-covalent bonds that were maximised at pH 7.5. Increase of shear rate during heat processing of whey proteins in the presence of sufficient casein content could successfully enhance the protective influence of caseins on the whey proteins at an increased pH level (Fig. 6). The author attributed such an effect to the enhanced dissociation of caseins from the micelles at higher pH when they were loosely packed, shear-induced dissociation of aggregates and hydrophobic interactions of caseins. Similar observations were obtained when sodium caseinate was used during heat (Liyanaarachchi, 2017) and pressure treatment (He et al., 2019).

## 7 Case studies: Application of whey protein nano- and microparticles

As outlined in previous sections, whey proteins are a highly desirable protein group in the food industry in many applications, owing to their high nutritional and physiological values. Their inherent heat instability hinders their wider applications, as whey protein-containing products usually undergo substantial physical changes during and after heat treatment, which render these products unstable and organoleptically unacceptable. For this reason, their stabilisation through some form modification, preferably physical, may expand their uses. Initially whey protein microparticles, for example, Simplesse[R], were used as fat replacers, which still remains an expanding area of research and commercial interest, while the intentions broadened to creation of smaller whey protein particles (nanoparticles) that would presumably remain kinetically stable in solution. These particles, as depicted in Fig. 5, can be created by either a controlled network growth or controlled fragmentation. Murray (2019) defined these two approaches as 'bottom-up' and 'top-down' formation, respectively. Their obvious application as rheology stabilisers has been long recognised, but more recently, they appear as a 'particle' stabilisers of emulsions and foams. Highly stable emulsions and foams are usually created by application of pressure (homogenisation, air injections) to create small droplets or bubbles that are simultaneously coated by highly active surface agents, which also facilitate size reduction due to reduction in the interfacial free energy. Consequently, these agents create an interfacial barrier, preventing coalescence with neighbouring droplets or bubbles (Yan et al., 2020). When solid particles are used as interfacial agents, this phenomenon is termed 'Pickering stabilisation' (Pickering, 1907). Gong et al. (2014) even suggested using a term - Pickering emulsions - for those stabilised by protein microparticles to distinguish them from true Pickering emulsions. The stabilisation is provided by particles attached to the interface, which remain there indefinitely under quiescent conditions (Dickinson, 2017) by creating a dense mechanical barrier. This jamming of hard particles creates an elastic but highly rigid monolayer. During emulsification, interfacial particle jamming may lead to the creation of non-spherical droplets due to deformation by the applied shear field, which are then prevented from relaxing back to the normal equilibrium spherical shape (Merkel et al., 2015). In their native state, globular proteins such as whey proteins are not considered Pickering stabilisers, as upon adsorption on the interface, they tend to undergo substantial conformational changes which may lead to improved or compromised emulsion stability. For a particle to be considered a Pickering stabiliser, it should remain intact over the life span of an emulsion (Dickinson, 2012).

In a Pickering emulsion, the precise location of each particle at the interface depends on its interactions with neighbouring particles, and on the value of the contact angle θ made at the oil-water interface by the particle's surface. For a solid sphere of a defined radius, the free energy of spontaneous desorption is proportional to $R^2(1-\cos\theta)^2$. This means that even for a relatively small nanoparticle (~10 nm radius) and the contact angle removed from 0° or 180°, this particle can stay effectively permanent at the droplet surface. From the relationship, it is evident that the desorption energy is at its theoretical maximum value when θ = 90°, for example, particle surface equally wetted by oil and water (Fig. 7). However, Finkle's empirical rule (Finkle et al., 1923) states that the preferential wetting of the particle surface by the bulk continuous phase enhances the effectiveness of any mechanical barrier formed on the convex side of the liquid interface. Therefore, proteins and their conjugates may have surfaces with the predominant hydrophilic character and would preferentially stabilise oil-in-water emulsions.

The Pickering stabilisation effect involves the creation of a complete monolayer of closely packed particles on the interface around individual droplets (or bubble), which effectively creates a physical barrier and prevents coalescence. The stabilisation is provided by several factors including (i) an adhesion energy effect which prevents desorption of individual particles from the interface; (ii) a steric hindrance effect by which proximity to other particles restricts lateral movement of particles within the layer; (iii) a capillary pressure

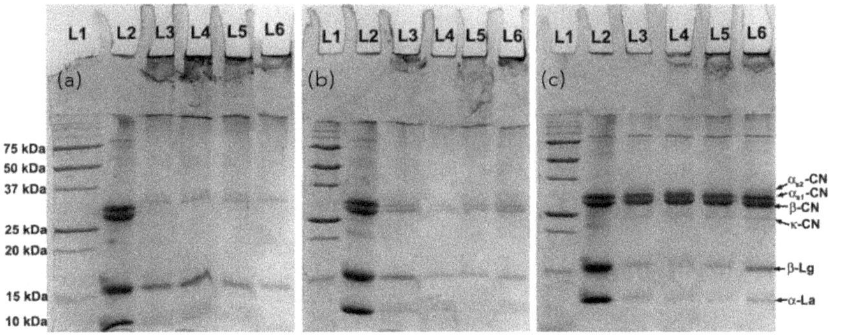


**Figure 6** Non-reducing SDS-PAGE gels after heat treatment of concentrated milk protein dispersions (10% TS) at 100°C for 1 min containing (a) WPI; (b) casein to whey proteins ratio 5:95; and (c) casein to whey proteins ratio 30:70. The samples were adjusted to a pH of 6.7 or 7.5. Lanes L1 to L6 contained the following samples: L1 - molecular marker; L2 - protein standards; L3 - pH 6.7 samples treated at 500 $s^{-1}$ shear rate; L4 - pH 6.7 samples treated at 1500 $s^{-1}$ shear rate; L5 - pH 7.5 samples treated at 500 $s^{-1}$ shear rate; L6 - pH 7.5 samples treated at 1500 $s^{-1}$ shear rate.

effect, which prevents rupturing of the liquid film; and (iv) an interfacial rheological effect, which slows down film drainage (Dickinson, 2017). This layered structure is dependent on the concentration of particles and may transform into a multilayered structure in the excess of particles which can create an additional barrier to coalescence (Wasan et al., 2004).

The ideal situation (Fig. 8a), in which a continuous monolayer is created, is rarely achieved in practice due to limitations associated with the actual concentration of dispersed particles available for adsorption at the freshly formed interface during emulsification or foaming. Such a restricted adsorption results in transient formation of particle-coated droplets, with surface density well below that of the completely saturated monolayer (Dickinson, 2017). This can lead to immediate droplet–droplet coalescence and the creation of larger spherical droplets, or in the case of particle surface density not too far from full saturation coverage, a pair of droplets may form a stable non-spherical droplet through arrested coalescence (Xie et al., 2019) and sharing of a single bridging layer of adsorbed particles (Fig. 8b). Where whey protein nano- and microparticles may be very effective is in addressing the issue of disproportionation when it comes to the stability of foams and aerated systems in general. Basically, the air persistently diffuses from small bubbles into big bubbles through the continuous phase which leads to foam coarsening or loss of air (Dickinson, 2017). The process can be stopped completely only by the formation of a rigid elastic layer at the interface surrounding the bubble, which cannot be achieved even with milk proteins (Dickinson et al., 2002). On the other hand, a close-packed layer of solid particles at the air–water interface can completely inhibit bubble shrinkage due to the ability to sustain sufficient interfacial stress to compensate for the Laplace pressure difference (Du et al., 2003). Often, these solid particles in the bulk system exist in a state of weak aggregation and cannot readily arrange themselves into closely packed monolayers but remain as a thicker, more disordered region of a fractal-type structure consisting of aggregates of interconnected particles held together by attractive forces (Fig. 8c). In addition to greater stability against droplet-bubble coalescence, such systems provide solid-like rheological character-enhancing stability against gravity-induced creaming and serum separation (Hoffmann and Reger, 2014). Another important aspect is the packing density, which appears to be affected by aspect ratio, that is, elongation. Rod-like or fibril-shaped particles provide greater emulsion stability at lower concentration than particles with a lower aspect ratio. This leads to a larger value of free energy of spontaneous desorption for elongated particles, stronger capillary forces between adjacent particles at the liquid interface and greater provision of steric repulsion between the droplets (Fig. 8d) (Dickinson, 2017).

Particulated whey proteins (nanoparticles, microparticles or microgels) provide one advantage over other Pickering stabilisers – their deformability. Since

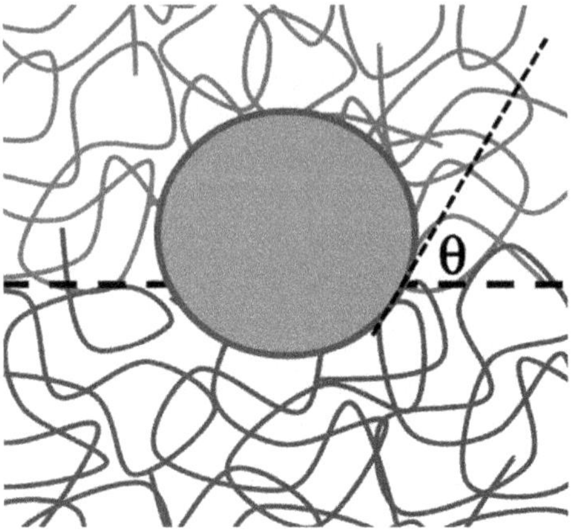


**Figure 7** A particle at the interface between two incompatible polymer solutions with a contact angle (θ) between them.

they are porous, created by controlled denaturation and aggregation mainly via covalent bonding, they are very kinetically stable (Nicolai, 2016). During their creation, they entrap water within their cross-linked polymer network, which gives them an ability to reversibly respond to changes in temperature, pH or ionic strength (Deshmukh et al., 2015). Protein particles with a moderately low degree of internal cross-linking may change shape, from spherical in an aqueous solution to flattened at the oil-water interface, which induces interpenetration of the adsorbed particles (Dickinson, 2017). In addition to their swelling, these particles also function as thickeners and structuring agents in aqueous systems, which makes them increasingly important in further developments of stabilising agents (Dickinson, 2017). The adsorption of whey protein microparticles (β-Lg) at the interface is directly related to their size and bulk concentration (Murphy et al., 2016), with the smaller, faster-adsorbing particles creating more elastic layers. These microparticles provide long-term coalescence stability in oil-in-water emulsions over a wide range of pH levels and ionic strengths (Destribats et al., 2014). Intermolecular disulphide cross-links provide internal structural integrity over a wide range of pH levels, while the particles self-aggregate, especially close to the isoelectric point. This leads to the formation of a structured network of aggregated particles and flocculation of droplets by protein bridging, but prevents droplet coalescence (Destribats et al., 2014). Similarly, a particle layer formation at the air-water interface and aggregated particle network structuring in the continuous phase effectively stabilised the aqueous foams (Schmitt et al., 2014). Another way to prepare these particles is via pre-emulsification of the protein dispersions within a continuous oil phase, followed by heating (Sağlam et al., 2014). Due to the limited availability of water, this approach creates 'perfectly' spherical particles of a uniform surface

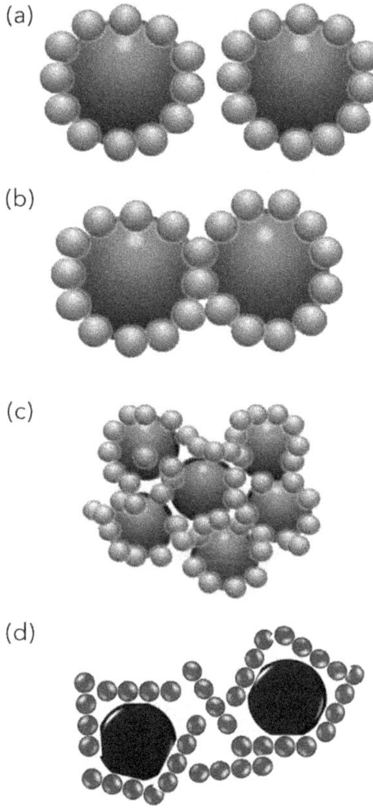

**Figure 8** Possible structural arrangements of droplets (or bubbles) in food systems stabilised by whey proteins particles: (a) bilayer arrangement for two droplets covered by monolayer of dispersed spheres; (b) shared bridging layer stabilising two droplets; (c) partially coated droplets with space-filling aggregated spherical particles; (d) stabilisation with fibrillar structures.

and internal structure at around neutral pH (Sağlam et al., 2014), which are very efficient in forming Pickering oil-in-water emulsions with good stability due to close packing and partial wetting properties (Wu et al., 2015). Maintaining the integrity of the layer at some conditions appears to be an issue as these particles tend to aggregate near the isoelectric point and at high ionic strength, and precipitate in the bulk phase (Wu et al., 2015). Top-down microparticulation of heated whey proteins using high dynamic pressure (microfluidisation) resulted in fully denatured whey protein particles whose average size depended on the pH at heating (Dissanayake et al., 2012; Dissanayake and Vasiljevic, 2009). These particles were twice as efficient as native whey protein dispersions in creating emulsions, and more stable (Dissanayake et al., 2012); however, their foaming ability was diminished completely (Dissanayake and Vasiljevic, 2009). On the other hand, Lazidis et al. (2016) applied moderate shearing conditions during heating of whey protein dispersions at two pH levels, close to and well above the isoelectric point, which resulted in the average particle size of 10 μm and 150 μm, respectively. The overrun achieved with such particles was similar to that created with a standard whey protein dispersion. More importantly, the foam stability was substantially greater, which was attributed to positioning of smaller, more surface-active entities diffusing rapidly to the interface followed by space-filling by larger particles between the bubbles, which prevented liquid drainage (Lazidis et al., 2016). As pointed out in the preceding sections, whey proteins, especially β-Lg, create fibrillar structures at low pH and low ionic strength. These fibrils create space-filling networks at low concentrations which makes them very attractive as thickeners, stabilisers and gelling agents (Kroes-Nijboer et al., 2012). As depicted in Fig. 8d, these protein fibril networks inhibit coalescence by creating viscoelastic layers at liquid interfaces and steric structural barriers between neighbouring liquid droplets or gas bubbles (Dickinson, 2017). It has been demonstrated recently that fibrillisation of whey proteins significantly enhanced foaming capacity and foam stability in comparison to equivalent non-fibrillar heat-denatured proteins (Oboroceanu et al., 2014). Interestingly, these aerated systems prepared with fibrillar whey protein particles possessed textural characteristics similar to those formed by traditional whipping of egg-white protein. In addition to creation of viscoelastic networks around the droplets, the protein fibrils in emulsions trigger depletion flocculation at an appropriate concentration that provides a substantial degree of kinetic stabilisation (Peng et al., 2016).

## 8 Future trends in research

Whey proteins are highly important as food ingredients, both nutritionally and structurally, whose inherent heat-related instabilities could be addressed by physical modifications. Under specific conditions, it is possible to create particles of various sizes and morphologies that can remain stable in aqueous solutions and also serve as very effective functional ingredients for the making and stabilising of emulsions and foams. In general, protein particles are intrinsically surface-active, a characteristic that can be applied to proteins coating either droplets or the bubbles in emulsions or foams. The long-term stability of these systems involves a combination of two separate mechanisms: (i) Pickering stabilisation by adsorbed particles at the interface; and (ii) formation of an aggregated particle network in the bulk phase region between neighbouring oil droplets or bubbles. Through the combined structural and rheological effects, the deleterious physical changes that usually take place during extended emulsion storage, including creaming and coalescence, are prevented.

While the current knowledge has shown the potential of these particles, it is good to keep in mind that neither whey protein preparations nor the food systems in which these proteins are used are pure. Thus, the interactions and interfaces are rather complex. The presence of small quantities of highly active surfactants, usually present in food systems, for example, may have a prominent influence on the rate and extent of lowering of the tension, as well as on the ability of particles to create small droplets or bubbles. Whey protein particles in a real food system will not only compete with other species for the interfaces but also interact and associate with them in the bulk dispersion. For these reasons, it is obviously necessary to gain a fundamental understanding of the mechanisms involved in these interactions using pure and well-defined components. However, the application of these particles needs to be placed in perspective, in terms of their interactions with other food components such as small-molecule surfactants, proteins and hydrocolloids. Long-term stability of droplets and bubbles coated with whey protein particles is also needed, which can be achieved by investigating the time-dependent behaviour of well-defined particle-laden interfaces, preferably with the surface rheology and microstructure probed simultaneously on the same sample, in addition to the computer simulation and theoretical modelling.

More importantly, these nanoparticles and microparticles can serve as encapsulants and delivery vehicles for nutrients, antioxidants, flavour compounds and bioactives, as indicated by various studies. What needs to be addressed is how these particles behave during the gastrointestinal transit and on-site delivery. Questions inevitably arise about what happens to such biopolymer-based particles and the emulsified lipids stabilised by them when they traverse the various stages of the human gastrointestinal tract. As it appears, these particle stabilisers can be viewed as multifunctional ingredients for the development of novel edible colloidal systems with potential health-beneficial applications for controlled release and digestion management.

## 9 Where to look for further information


- Deeth, H. C. and Bansal, N. (2019). *Whey Proteins: From Milk to Medicine*. London, UK: Academic Press.
- Dickinson, E. (2020). Advances in food emulsions and foams: reflections on research in the neo-Pickering era. *Current Opinion in Food Science*, 33, 52–60.
- Dickinson, E. (2019). Strategies to control and inhibit the flocculation of protein-stabilized oil-in-water emulsions. *Food Hydrocolloids*, 96, 209–223.
- Nicolai, T. (2019). Gelation of food protein-protein mixtures. *Advances in Colloid and Interface Science*, 270, 147–164.

- Nicolai, T. and Murray, B. (2017). Particle-stabilized water-in-water emulsions. *Food Hydrocolloids*, 68, 157–163.
- Nicolai, T. (2016). Formation and functionality of self-assembled whey protein microgels. *Colloids and Surfaces B: Biointerfaces*, 137, 32–38.
- Shi, A., Feng, X., Wang, Q. and Adhikari, B. (2020). Pickering and high internal phase Pickering emulsions stabilized by protein-based particles: a review of synthesis, application and prospective. *Food Hydrocolloids*, 109, 106117.
- Yan, X., Ma, C., Cui, F., McClements, D.J., Liu, X. and Liu, F. (2020). Protein-stabilized Pickering emulsions: formation, stability, properties, and applications in foods. *Trends in Food Science and Technology*, 103, 293–303.

## 10 References

Abbring, S., Hols, G., Garssen, J. and van Esch, B. C. A. M. (2019). Raw cow's milk consumption and allergic diseases-the potential role of bioactive whey proteins. *European Journal of Pharmacology* 843, 55–65.

Abd El-Salam, M. H., El-Shibiny, S. and Salem, A. (2009). Factors affecting the functional properties of whey protein products: a review. *Food Reviews International* 25(3), 251–270.

Alaimo, M. H., Wickham, E. D. and Farrell Jr., H. M. (1999). Effect of self-association of α(s1)-casein and its cleavage fractions α(s1)-casein (136–196) and α(s1)-casein (1–197),1 on aromatic circular dichroic spectra: comparison with predicted models. *Biochimica et Biophysica Acta* 1431(2), 395–409. DOI: 10.1016/S0167-4838(99)00062-X.

Alexander-Katz, A., Schneider, M. F., Schneider, S. W., Wixforth, A. and Netz, R. R. (2006). Shear-flow-induced unfolding of polymeric globules. *Physical Review Letters* 97(13), 138101.

Anema, S. G. (2014). Chapter 9 – The whey proteins in milk: thermal denaturation, physical interactions, and effects on the functional properties of milk. In: Singh, H., Boland, M., Thompson, A. (Eds), *Food Science and Technology, Milk Proteins* (2nd edn., pp. 269–318). Elsevier.

Anema, S. G. (2017). A comparison of the kinetics of the thermal denaturation of the immunoglobulins in caprine and bovine skim milk samples. *International Dairy Journal* 65, 1–4.

Anema, S. G. and McKenna, A. B. (1996). Reaction kinetics of thermal denaturation of whey proteins in heated reconstituted whole milk. *Journal of Agricultural and Food Chemistry* 44(2), 422–428.

Ashokkumar, M., Lee, J., Zisu, B., Bhaskarcharya, R., Palmer, M. and Kentish, S. (2009). Hot topic: sonication increases the heat stability of whey proteins. *Journal of Dairy Science* 92(11), 5353–5356. DOI: 10.3168/jds.2009-2561.

Ashton, L., Dusting, J., Imomoh, E., Balabani, S. and Blanch, E. W. (2010). Susceptibility of different proteins to flow-induced conformational changes monitored with Raman spectroscopy. *Biophysical Journal* 98(4), 707–714.

Baier, S. K., Decker, E. A. and McClements, D. J. (2004). Impact of glycerol on thermostability and heat-induced gelation of bovine serum albumin. *Food Hydrocolloids* 18(1), 91–100. DOI: 10.1016/S0268-005X(03)00046-8.

Baier, S. K. and McClements, D. J. (2003). Combined Influence of NaCl and sucrose on heat-induced gelation of bovine serum albumin. *Journal of Agricultural and Food Chemistry* 51(27), 8107–8112. DOI: 10.1021/jf034249m.

Bansal, N. and Bhandari, B. (2016). Functional milk proteins: production and utilization–whey-based ingredients. In: McSweeney, P. and O'Mahony, J. (Eds), *Advanced Dairy Chemistry* (pp. 67-98). New York: Springer.

Beissinger, M. and Buchner, J. (1998). How chaperones fold proteins. *Biological Chemistry* 379(3), 245–259.

Bekard, I. B., Asimakis, P., Teoh, C. L., Ryan, T., Howlett, G. J., Bertolini, J. and Dunstan, D. E. (2012). Bovine serum albumin unfolds in Couette flow. *Soft Matter* 8(2), 385–389.

Bhattacharyya, J. and Das, K. P. (1999). Molecular chaperone-like properties of an unfolded protein, α(s)-casein. *Journal of Biological Chemistry* 274(22), 15505–15509. DOI: 10.1074/jbc.274.22.15505.

Bijsterbosch, B. H., Bos, M. T. A., Dickinson, E., van Opheusden, J. H. J. and Walstra, P. (1995). Brownian dynamics simulation of particle gel formation: from argon to yoghurt. *Faraday Discussions* 101, 51–64.

Bogahawaththa, D. (2018a). Processing-induced modifications of native bovine milk proteins in relation to immunogenicity. PhD thesis. Melbourne, Australia: Victoria University.

Bogahawaththa, D., Ashraf, R., Chandrapala, J., Donkor, O. and Vasiljevic, T. (2018b). In vitro immunogenicity of various native and thermally processed bovine milk proteins and their mixtures. *Journal of Dairy Science* 101(10), 8726–8736.

Bogahawaththa, D., Buckow, R., Chandrapala, J. and Vasiljevic, T. (2018c). Comparison between thermal pasteurization and high pressure processing of bovine skim milk in relation to denaturation and immunogenicity of native milk proteins. *Innovative Food Science and Emerging Technologies* 47, 301–308.

Bogahawaththa, D., Chandrapala, J. and Vasiljevic, T. (2017a). Modulation of milk immunogenicity by thermal processing. *International Dairy Journal* 69, 23–32.

Bogahawaththa, D., Chandrapala, J. and Vasiljevic, T. (2017b). Thermal denaturation of bovine immunoglobulin G and its association with other whey proteins. *Food Hydrocolloids* 72, 350–357.

Bogahawaththa, D., Chandrapala, J. and Vasiljevic, T. (2019). Thermal denaturation of bovine β-lactoglobulin in different protein mixtures in relation to antigenicity. *International Dairy Journal* 91, 89–97.

Bogahawaththa, D. and Vasiljevic, T. (2020a). Denaturation of selected bioactive whey proteins during pasteurization and their ability to modulate milk immunogenicity. *Journal of Dairy Research* 87(4), 484–487.

Bogahawaththa, D. and Vasiljevic, T. (2020b). Shearing accelerates denaturation of β-lactoglobulin and α-lactalbumin in skim milk during heating. *International Dairy Journal* 105, 104674.

Borad, S. G. and Singh, A. K. (2018). Colostrum immunoglobulins: processing, preservation and application aspects. *International Dairy Journal* 85, 201–210.

Borad, S. G., Singh, A. K., Kapila, S., Behare, P., Arora, S. and Sabikhi, L. (2019). Influence of unit operations on immunoglobulins and thermal stability of colostrum fractions. *International Dairy Journal* 93, 85–91.

Brew, K. (2013). α-lactalbumin. In: McSweeney, P. L. H. and Fox, P. F. (Eds), *Advanced Dairy Chemistry (vol. 1A). Proteins: Basic Aspects* (pp. 261–273). New York, NY: Springer Science+Business Media.

Brick, T., Ege, M., Boeren, S., Böck, A., Von Mutius, E., Vervoort, J. and Hettinga, K. (2017). Effect of processing intensity on immunologically active bovine milk serum proteins. *Nutrients* 9(9), 963.

Brodkorb, A., Croguennec, T., Bouhallab, S. and Kehoe, J. J. (2016). Heat-induced denaturation, aggregation and gelation of whey proteins. In: *Advanced Dairy Chemistry* (pp. 155-178). New York, NY: Springer.

Bryant, C. M. and Mcclements, D. J. (1998). Molecular basis of protein functionality with special consideration of cold-set gels derived from heat- denatured whey. *Trends in Food Science and Technology* 9(4), 143-151.

Calvo, M. M., Leaver, J. and Banks, J. M. (1993). Influence of other whey proteins on the heat-induced aggregation of α-lactalbumin. *International Dairy Journal* 3(8), 719-727.

Cao, J., Wang, X. and Zheng, H. (2007). Comparative studies on thermoresistance of protein G-binding region and antigen determinant region of immunoglobulin G in acidic colostral whey. *Food and Agricultural Immunology* 18(1), 17-30.

Chaplin, L. C. and Lyster, R. L. J. (1986). Irreversible heat denaturation of bovine α-lactalbumin. *Journal of Dairy Research* 53(2), 249-258.

Charm, S. E. and Lai, C. J. (1971). Comparison of ultrafiltration systems for concentration of BioLogicals. *Biotechnology and Bioengineering* 13(2), 185-202.

Charm, S. E. and Wong, B. L. (1970). Enzyme inactivation with shearing. *Biotechnology and Bioengineering* 12(6), 1103-1109.

Chen, C.-C. and Chang, H.-M. (1998). Effect of thermal protectants on the stability of bovine milk immunoglobulin G. *Journal of Agricultural and Food Chemistry* 46(9), 3570-3576.

Chen, C.-C., Tu, Y.-Y. and Chang, H.-M. (2000). Thermal stability of bovine milk immunoglobulin G (IgG) and the effect of added thermal protectants on the stability. *Journal of Food Science* 65(2), 188-193.

Considine, T., Patel, H. A., Anema, S. G., Singh, H. and Creamer, L. K. (2007). Interactions of milk proteins during heat and high hydrostatic pressure treatments-A Review. *Innovative Food Science and Emerging Technologies* 8(1), 1-23. DOI: 10.1016/j.ifset.2006.08.003.

Córdova-Dávalos, L. E., Jiménez, M. and Salinas, E. (2019). Glycomacropeptide bioactivity and health: a review highlighting action mechanisms and signaling pathways. *Nutrients* 11(3), 598.

Dalgleish, D. G., Senaratne, V. and Francois, S. (1997). Interactions between α-lactalbumin and β-lactoglobulin in the early stages of heat denaturation. *Journal of Agricultural and Food Chemistry* 45(9), 3459-3464.

Deeth, H. and Bansal, N. (2019). Whey proteins: an overview. In: Deeth, H. and Bansal, N. (Eds), *Whey Proteins: From Milk to Medicine* (pp. 1-50). London, UK: Academic Press.

de Moura, C. S., Lollo, P. C. B., Morato, P. N., Carneiro, E. M. and Amaya-Farfan, J. (2013). Whey protein hydrolysate enhances the exercise-induced heat shock protein (HSP70) response in rats. *Food Chemistry* 136(3-4), 1350-1357. DOI: 10.1016/j.foodchem.2012.09.070.

Deshmukh, O. S., Van Den Ende, D., Stuart, M. C., Mugele, F. and Duits, M. H. G. (2015). Hard and soft colloids at fluid interfaces: adsorption, interactions, assembly & rheology. *Advances in Colloid and Interface Science* 222, 215-227. DOI: 10.1016/j.cis.2014.09.003.

Destribats, M., Rouvet, M., Gehin-Delval, C., Schmitt, C. and Binks, B. P. (2014). Emulsions stabilised by whey protein microgel particles: towards food-grade Pickering emulsions. *Soft Matter* 10(36), 6941–6954. DOI: 10.1039/c4sm00179f.

Devi, V. S., Coleman, D. R. and Truntzer, J. (2011). Thermal unfolding curves of high-concentration bovine IgG measured by FTIR spectroscopy. *The Protein Journal* 30(6), 395–403.

deWit, J. N. and Klarenbeek, G. (1984). Effects of various heat treatments on structure and solubility of whey proteins. *Journal of Dairy Science* 67(11), 2701–2710.

Dickinson, E. (2012). Use of nanoparticles and microparticles in the formation and stabilization of food emulsions. *Trends in Food Science and Technology* 24(1), 4–12. DOI: 10.1016/j.tifs.2011.09.006.

Dickinson, E. (2017). Biopolymer-based particles as stabilizing agents for emulsions and foams. *Food Hydrocolloids* 68, 219–231. DOI: 10.1016/j.foodhyd.2016.06.024.

Dickinson, E., Ettelaie, R., Murray, B. S. and Du, Z. (2002). Kinetics of disproportionation of air bubbles beneath a planar air–water interface stabilized by food proteins. *Journal of Colloid and Interface Science* 252(1), 202–213. DOI: 10.1006/jcis.2002.8405.

Dissanayake, M., Liyanaarachchi, S. and Vasiljevic, T. (2012). Functional properties of whey proteins microparticulated at low pH. *Journal of Dairy Science* 95(4), 1667–1679. DOI: 10.3168/jds.2011-4823.

Dissanayake, M. and Vasiljevic, T. (2009). Functional properties of whey proteins affected by heat treatment and hydrodynamic high-pressure shearing. *Journal of Dairy Science* 92(4), 1387–1397. DOI: 10.3168/jds.2008-1791.

Dominguez, E., Perez, M. D. and Calvo, M. (1997). Effect of heat treatment on the antigen-binding activity of anti-peroxidase immunoglobulins in bovine colostrum. *Journal of Dairy Science* 80(12), 3182–3187.

Domínguez, E., Perez, M. D., Puyol, P., Sánchez, L. and Calvo, M. (2001). Effect of pH on antigen-binding activity of IgG from bovine colostrum upon heating. *Journal of Dairy Research* 68(3), 511–518.

Donahue, M., Godden, S. M., Bey, R., Wells, S., Oakes, J. M., Sreevatsan, S., Stabel, J. and Fetrow, J. (2012). Heat treatment of colostrum on commercial dairy farms decreases colostrum microbial counts while maintaining colostrum immunoglobulin G concentrations. *Journal of Dairy Science* 95(5), 2697–2702.

Du, Z., Bilbao-Montoya, M. P., Binks, B. P., Dickinson, E., Ettelaie, R. and Murray, B. S. (2003). Outstanding stability of particle-stabilized bubbles. *Langmuir* 19(8), 3106–3108. DOI: 10.1021/la034042n.

Edwards, P. J. B. and Jameson, G. B. (2020). Structure and stability of whey proteins. In: Boland, M. and Singh, H. (Eds), *Milk Proteins. From Expression to Food* (3rd edn., pp. 251–291). London, UK: Academic Press.

Elfstrand, L., Lindmark-Månsson, H., Paulsson, M., Nyberg, L. and Åkesson, B. (2002). Immunoglobulins, growth factors and growth hormone in bovine colostrum and the effects of processing. *International Dairy Journal* 12(11), 879–887.

Elizondo-Salazar, J. A., Jayarao, B. M. and Heinrichs, A. J. (2010). Effect of heat treatment of bovine colostrum on bacterial counts, viscosity, and immunoglobulin G concentration. *Journal of Dairy Science* 93(3), 961–967.

Erabit, N., Flick, D. and Alvarez, G. (2014). Formation of β-lactoglobulin aggregates during thermomechanical treatments under controlled shear and temperature conditions. *Journal of Food Engineering* 120, 57–68.

Finkle, P., Draper, H. D. and Hildebrand, J. H. (1923). The theory of emulsification 1. *Journal of the American Chemical Society* 45(12), 2780–2788. DOI: 10.1021/ja01665a002.

Foegeding, E. A. (2015). Food protein functionality–A new model. *Journal of Food Science* 80(12), C2670–C2677.

Foegeding, E. A. and Davis, J. P. (2011). Food protein functionality: a comprehensive approach. *Food Hydrocolloids* 25(8), 1853–1864. DOI: 10.1016/j.foodhyd.2011.05.008.

Fox, P. F. (2003). Milk proteins: general and historical aspects. In: Fox, P. F., McSweeney, P. L. H. (Eds), *Advanced Dairy Chemistry–1 Proteins: Part A/Part B*, (pp. 1–48). Boston, MA: Springer. DOI: 10.1007/978-1-4419-8602-3_1.

Fujiyama, S., Yanagida, M., Hayano, T., Miura, Y., Isobe, T. and Takahashi, N. (2002). Isolation and proteomic characterization of human parvulin associating preribosomal ribonucleo protein complexes. *Journal of Biological Chemistry* 277(44), 42418.

Fujiyama-Nakamura, S., Yoshikawa, H., Homma, K., Hayano, T., Tsujimura-Takahashi, T., Izumikawa, K., Ishikawa, H., Miyazawa, N., Yanagida, M., Miura, Y., Shinkawa, T., Yamauchi, Y., Isobe, T. and Takahashia, N. (2009). Parvulin (Par14), a peptidyl-prolyl cis-trans isomerase, is a novel rRNA processing factor that evolved in the metazoan lineage. *Molecular & Cellular Proteomics* 8(7), 1552–1565.

Fukumoto, L. R., Skura, B. J. and Nakai, S. (1994). Stability of membrane-sterilized bovine immunoglobulins aseptically added to UHT milk. *Journal of Food Science* 59(4), 757–759.

Gabovich, A. M. and Li, M. S. (2009). Mechanical stability of proteins. *The Journal of Chemical Physics* 131(2), 024121.

Gapper, L. W., Copestake, D. E., Otter, D. E. and Indyk, H. E. (2007). Analysis of bovine immunoglobulin G in milk, colostrum and dietary supplements: a review. *Analytical and Bioanalytical Chemistry* 389(1), 93–109.

Garrett, J. M., Stairs, R. A. and Annett, R. G. (1988). Thermal denaturation and coagulation of whey proteins: effect of sugars. *Journal of Dairy Science* 71(1), 10–16. DOI: 10.3168/jds.S0022-0302(88)79518-1.

Gezimati, J., Creamer, L. K. and Singh, H. (1997). Heat-induced interactions and gelation of mixtures of β-lactoglobulin and α-lactalbumin. *Journal of Agricultural and Food Chemistry* 45(4), 1130–1136.

Gezimati, J., Singh, H. and Creamer, L. K. (1996a). Aggregation and gelation of bovine β-lactoglobulin, α-lactalbumin, and serum albumin. In: Parris, N., Kato, A., Creamer, L. K. and Pearce, J. (Eds), *ACS Symposium Series*. Washington, DC: American Chemical Society, 113–123.

Gezimati, J., Singh, H. and Creamer, L. K. (1996b). Heat-induced interactions and gelation of mixtures of bovine β-lactoglobulin and serum albumin. *Journal of Agricultural and Food Chemistry* 44(3), 804–810.

Godden, S., McMartin, S., Feirtag, J., Stabel, J., Bey, R., Goyal, S., Metzger, L., Fetrow, J., Wells, S. and Chester-Jones, H. (2006). Heat-treatment of bovine colostrum. II: effects of heating duration on pathogen viability and immunoglobulin G. *Journal of Dairy Science* 89(9), 3476–3483.

Golkar, A., Milani, J. M. and Vasiljevic, T. (2019). Altering allergenicity of cow's milk by food processing for applications in infant formula. *Critical Reviews in Food Science and Nutrition* 59(1), 159-172. DOI: 10.1080/10408398.2017.1363156.

Gong, X., Hua, L., Wei, J. and Ngai, T. (2014). Tuning the particle-surface interactions in aqueous solutions by soft microgel particles. *Langmuir* 30(44), 13182–13190. DOI: 10.1021/la503573u.

Gorga, J. C., Dong, A., Manning, M. C., Woody, R. W., Caughey, W. S. and Strominger, J. L. (1989). Comparison of the secondary structures of human class I and class II major histocompatibility complex antigens by Fourier transform infrared and circular dichroism spectroscopy. *Proceedings of the National Academy of Sciences* 86(7), 2321–2325.

Gottschalk, M., Nilsson, H., Roos, H. and Halle, B. (2003). Protein self-association in solution: the bovine β-lactoglobulin dimer and octamer. *Protein Science* 12(11), 2404–2411.

Gülseren, I., Alexander, M. and Corredig, M. (2010). Probing the colloidal properties of skimmed milk using acoustic and electroacoustic spectroscopy. Effect of concentration, heating and acidification. *Journal of Colloid and Interface Science* 351(2), 493–500.

Guo, M. and Wang, C. (2019a). Chemistry of whey proteins. In: Guo, M. (Ed), *Whey Protein Production, Chemistry, Functionality, and Applications* (pp. 39–65). Hoboken: John Wiley & Sons, Inc.

Guo, M. and Wang, G. (2019b). History of Whey Production and Whey Protein Manufacturing. In: Guo, M. (Ed), *Whey Protein Production, Chemistry, Functionality, and Applications* (pp. 1–12). Hoboken: John Wiley & Sons, Inc.

Havea, P., Carr, A. J. and Creamer, L. K. (2004). The roles of disulphide and non-covalent bonding in the functional properties of heat-induced whey protein gels. *Journal of Dairy Research* 71(3), 330–339.

Havea, P., Singh, H. and Creamer, L. K. (2000). Formation of new protein structures in heated mixtures of BSA and a-lactalbumin. *Journal of Agricultural and Food Chemistry* 48(5), 1548–1556.

Havea, P., Singh, H. and Creamer, L. K. (2001). Characterization of heat-induced aggregates of ß-lactoglobulin, a-lactalbumin and bovine serum albumin in a whey protein concentrate environment. *Journal of Dairy Research* 68(3), 483–497.

Havea, P., Singh, H., Creamer, L. K. and Campanella, O. H. (1998). Electrophoretic characterization of the protein products formed during heat treatment of whey protein concentrate solutions. *Journal of Dairy Research* 65(1), 79–91.

Havea, P., Watkinson, P. and Kuhn-Sherlock, B. (2009). Heat-induced whey protein gels: protein-protein interactions and functional properties. *Journal of Agricultural and Food Chemistry* 57(4), 1506–1512.

He, J.-S., Gao, Q., Mu, T.-H. and Azuma, N. (2019). Molecular chaperone-like properties of sodium caseinate to suppress the pressure-induced aggregation of β-lactoglobulin. *High Pressure Research* 39(2), 210–217. DOI: 10.1080/08957959.2019.1575967.

Hendrick, J. P. and Hartl, F. U. (1995). The role of molecular chaperones in protein folding. *FASEB Journal* 9(15), 1559–1569. DOI: 10.1096/fasebj.9.15.8529835.

Heffernan, S. P., Zumaeta, N., Cartland-Glover, G. M., Byrne, E. P. and Fitzpatrick, J. J. (2005). Influence of downstream processing on the breakage of whey protein precipitates. *Food and Bioproducts Processing* 83(3), 238–244.

Hines, M. E. and Foegeding, E. A. (1993). Interactions of alpha-lactalbumin and bovine serum albumin with. beta.-lactoglobulin in thermally induced gelation. *Journal of Agricultural and Food Chemistry* 41(3), 341–346.

Hoffmann, H. and Reger, M. (2014). Emulsions with unique properties from proteins as emulsifiers. *Advances in Colloid and Interface Science* 205, 94–104. DOI: 10.1016/j.cis.2013.08.007.
Holt, C., Carver, J. A., Ecroyd, H. and Thorn, D. C. (2013). Invited review: caseins and the casein micelle: their biological functions, structures, and behavior in foods. *Journal of Dairy Science* 96(10), 6127–6146. DOI: 10.3168/jds.2013-6831.
Hong, Y.-H. and Creamer, L. K. (2002). Changed protein structures of bovine β-lactoglobulin B and α-lactalbumin as a consequence of heat treatment. *International Dairy Journal* 12(4), 345-359.
Huppertz, T., Vasiljevic, T., Zisu, B. and Deeth, H. (2019). Deeth, H. and Bansal, N. (Eds), *Novel Processing Technologies, In Whey Proteins: From Milk to Medicine* (pp. 281-334). London, UK: Academic Press.
Hurley, W. L. (2003). Immunoglobulins in mammary secretions. In: Fox, P. F. and McSweeney, P. L. H. (Eds), *Advanced Dairy Chemistry–1 Proteins* (pp. 421–447). Boston, MA: Springer. DOI: 10.1007/978-1-4419-8602-3_9.
Hurley, W. L. and Theil, P. K. (2011). Perspectives on immunoglobulins in colostrum and milk. *Nutrients* 3(4), 442–474.
Hurley, W. L. and Theil, P. K. (2013). Immunoglobulins in mammary secretions. In: Fox, P. F. and McSweeney, P. L. H. (Eds), *Advanced Dairy Chemistry* (pp. 275–294). Boston, MA: Springer.
Ikeda, S. and Li-Chan, E. C. Y. (2004). Raman spectroscopy of heat-induced fine-stranded and particulate beta-lactoglobulin gels. *Food Hydrocolloids* 18(3), 489–498.
Indyk, H. E., Williams, J. W. and Patel, H. A. (2008). Analysis of denaturation of bovine IgG by heat and high pressure using an optical biosensor. *International Dairy Journal* 18(4), 359-366.
Ipsen, R. (2017). Microparticulated whey proteins for improving dairy product texture. *International Dairy Journal* 67, 73–79.
Jasion, V. S. and Burnett, B. P. (2015). Survival and digestibility of orally-administered immunoglobulin preparations containing IgG through the gastrointestinal tract in humans. *Nutrition Journal* 14, 22.
Jaspe, J. and Hagen, S. J. (2006). Do protein molecules unfold in a simple shear flow? *Biophysical Journal* 91(9), 3415-3424.
Jauregui-Rincón, J., Salinas-Miralles, E., Chávez-Vela, N. and Jiménez-Vargas, M. (2018). Glycomacropeptide: biological activities and uses. In: Gigli, I. (Ed), *Whey-Biological Properties and Alternative Uses*. IntechOpen. DOI: 10.5772/intechopen.82144.
Javidipour, I. and Qian, M. C. (2008). Volatile component change in whey protein concentrate during storage investigated by headspace solid-phase microextraction gas chromatography. *Dairy Science and Technology* 88(1), 95–104.
Jöbstl, E., Howse, J. R., Fairclough, J. P. A. and Williamson, M. P. (2006). Noncovalent cross-linking of casein by epigallocatechin gallate characterized by single-molecule force microscopy. *Journal of Agricultural and Food Chemistry* 54(12), 4077-4081. DOI: 10.1021/jf053259f.
Johnson, J. L., Godden, S. M., Molitor, T., Ames, T. and Hagman, D. (2007). Effects of feeding heat-treated colostrum on passive transfer of immune and nutritional parameters in neonatal dairy calves. *Journal of Dairy Science* 90(11), 5189-5198.
Jovanovic, S., Barac, M. and Macej, O. (2005). Whey proteins-Properties and possibility of application. *Mljekarstvo* 55(3), 215-233.

Jung, J. M., Savin, G., Pouzot, M., Schmitt, C. and Mezzenga, R. (2008). Structure of heat-induced beta-lactoglobulin aggregates and their complexes with sodium-dodecyl sulphate. *Biomacromolecules* 9(9), 2477–2486.

Kananen, A., Savolainen, J., Mäkinen, J., Perttilä, U., Myllykoski, L. and Pihlanto-Leppälä, A. (2000). Influence of chemical modification of whey protein conformation on hydrolysis with pepsin and trypsin. *International Dairy Journal* 10(10), 691-697. DOI: 10.1016/S0958-6946(00)00094-7.

Kehoe, J. J. and Foegeding, E. A. (2011). Interaction between β-casein and whey proteins as a function of pH and salt concentration. *Journal of Agricultural and Food Chemistry* 59(1), 349–355.

Kenkare, D. B., Morr, C. V. and Gould, I. A. (1964). Factors affecting the heat aggregation of proteins in selected skimmed milk sera. *Journal of Dairy Science* 47(9), 947–953.

Kilara, A. and Vaghela, M. N. (2018). Whey proteins. In: Yada, R. Y. (Ed), *Proteins in Food Processing* (pp. 93–126). Cambridge, England: Woodhead Publishing Limited.

Kim, J. and Kramer, T. A. (2006). Improved orthokinetic coagulation model for fractal colloids: aggregation and breakup. *Chemical Engineering Science* 61(1), 45–53.

Kinsella, J. E. and Whitehead, D. M. (1989). Proteins in whey: chemical, physical, and functional properties. In: Kinsella, J. E. (Ed), *Advances in Food and Nutrition Research* (Vol. 33, pp. 343–348). Academic Press. DOI: 10.1016/S1043-4526(08)60130-8.

Klemaszewski, J. L. and Kinsella, J. E. (1991). Sulfitolysis of whey proteins: effects on emulsion properties. *Journal of Agricultural and Food Chemistry* 39(6), 1033–1036.

Koudelka, T., Hoffmann, P. and Carver, J. A. (2009). Dephosphorylation of αs- and β-caseins and its effect on chaperone activity: a structural and functional investigation. *Journal of Agricultural and Food Chemistry* 57(13), 5956–5964.

Kroes-Nijboer, A., Venema, P. and Linden, E. V. D. (2012). Fibrillar structures in food. *Food and Function* 3(3), 221–227. DOI: 10.1039/c1fo10163c.

Kulmyrzaev, A., Bryant, C. and McClements, D. J. (2000). Influence of sucrose on the thermal denaturation, gelation, and emulsion stabilization of whey proteins. *Journal of Agricultural and Food Chemistry* 48(5), 1593–1597. DOI: 10.1021/jf9911949.

Kummer, A., Kitts, D. D., Li-Chan, E., Losso, J. N., Skura, B. J. and Nakai, S. (1992). Quantification of bovine IgG in milk using enzyme-linked immunosorbent assay. *Food and Agricultural Immunology* 4(2), 93–102.

Lagrange, V. and Clark, D. C. (2019). Nutritive and therapeutic aspects of whey proteins. In: Deeth, H. and Bansal, N. (Eds), *Whey Proteins: From Milk to Medicine* (pp. 549–577). London, UK: Academic Press.

Law, A. (1995). Heat denaturation of bovine, caprine and ovine whey proteins. *Milchwissenschaft (Germany)* 50, 384–388.

Lazidis, A., Hancocks, R. D., Spyropoulos, F., Kreuß, M., Berrocal, R. and Norton, I. T. (2016). Whey protein fluid gels for the stabilisation of foams. *Food Hydrocolloids* 53, 209–217. DOI: 10.1016/j.foodhyd.2015.02.022.

Lee, Y. K. and Choo, C. L. (1989). The kinetics and mechanism of shear inactivation of lipase from *Candida Cylindracea*. *Biotechnology and Bioengineering* 33(2), 183–190.

Li, S. Q., Bomser, J. A. and Zhang, Q. H. (2005). Effects of pulsed electric fields and heat treatment on stability and secondary structure of bovine immunoglobulin G. *Journal of Agricultural and Food Chemistry* 53(3), 663–670.

Li-Chan, E., Kummer, A., Losso, J. N., Kitts, D. D. and Nakai, S. (1995). Stability of bovine immunoglobulins to thermal treatment and processing. *Food Research International* 28(1), 9–16.

Liu, Y., Zhang, W., Zhang, L., Hettinga, K. and Zhou, P. (2020). Characterizing the changes of bovine milk serum proteins after simulated industrial processing. *LWT* 133, 110101.

Livney, Y. D., Verespej, E. and Dalgleish, D. G. (2003). Steric effects governing disulfide bond interchange during thermal aggregation in solutions of β-lactoglobulin B and α-lactalbumin. *Journal of Agricultural and Food Chemistry* 51(27), 8098–8106.

Liyanaarachchi, W. S. (2017). Microparticulation of dairy proteins. PhD thesis. Melbourne, Australia: Victoria University.

Loss, G., Apprich, S., Waser, M., Kneifel, W., Genuneit, J., Buchele, G., Weber, J., Sozanska, B., Danielewicz, H., Horak, E., van Neerven, R. J., Heederik, D., Lorenzen, P. C., von Mutius, E., Braun-Fahrländer, C. and GABRIELA Study Group (2011). The protective effect of farm milk consumption on childhood asthma and atopy: the GABRIELA study. *Journal of Allergy and Clinical Immunology* 128(4), 766–773.e4. DOI: 10.1016/j.jaci.2011.07.048.

Lu, R., Li, W. W., Katzir, A., Raichlin, Y., Yu, H. Q. and Mizaikoff, B. (2015). Probing the secondary structure of bovine serum albumin during heat-induced denaturation using mid-infrared fiberoptic sensors. *Analyst* 140(3), 765–770.

Luf, W., Feiersinger, G. and Brandl, E. (1993). On kinetics of thermal denaturation of immunoglobulins in milk: application as a heat treatment indicator. In: *Protein and Fat Globule Modification by Heat Treatment, Homogenization and Other* (pp. 147-152). Brussels: Technological Means for High Quality Dairy Products. International Dairy Federation.

Mahe, S., Huneau, J. F., Marteau, P., Thuillier, F. and Tome, D. (1992). Gastroileal nitrogen and electrolyte movements after bovine milk ingestion in humans. *The American Journal of Clinical Nutrition* 56(2), 410–416.

Mainer, G., Sanchez, L., Ena, J. M. and Calvo, M. (1997). Kinetic and thermodynamic parameters for heat denaturation of bovine milk IgG, IgA and IgM. *Journal of Food Science* 62(5), 1034–1038.

Mangano, K. M., Bao, Y. and Zhao, C. (2019). Nutritional properties of whey proteins. In: Guo, M. (Ed), *Whey Protein Production, Chemistry, Functionality, and Applications* (pp. 103–140). Hoboken: John Wiley & Sons, Inc.

Mann, B., Athira, S., Sharma, R., Kumar, R. and Sarkar, P. (2019). Bioactive peptides from whey proteins. In: Deeth, H. and Bansal, N. (Eds), *Whey Proteins: From Milk to Medicine* (pp. 519-547). London, UK: Academic Press.

Market Research (2021). *Global Whey Protein Ingredients Industry*. [online]. Reportlinker .com. Available at: https://www.reportlinker.com/p05442604/Global-Whey-Protein-Ingredients-Industry.html?utm_source=GNW [Accessed 12 January 2021].

Marnila, P. and Korhonen, H. (2011). *Immunoglobulins. Encyclopedia of Dairy Sciences* (3rd edn.). San Diego, CA: John W. Fuquay, Patrick F. Fox and Paul L. H. McSweeney.

Matsudomi, N., Oshita, T. and Kobayashi, K. (1994). Synergistic interaction between β-lactoglobulin and bovine serum albumin in heat-induced gelation. *Journal of Dairy Science* 77(6), 1487–1493.

Matsudomi, N., Oshita, T., Kobayashi, K. and Kinsella, J. E. (1993). alpha-Lactalbumin enhances the gelation properties of bovine serum albumin. *Journal of Agricultural and Food Chemistry* 41(7), 1053–1057.

McMartin, S., Godden, S., Metzger, L., Feirtag, J., Bey, R., Stabel, J., Goyal, S., Fetrow, J., Wells, S. and Chester-Jones, H. (2006). Heat treatment of bovine colostrum. I: effects of temperature on viscosity and immunoglobulin G level. *Journal of Dairy Science* 89(6), 2110–2118.

McSwiney, M., Singh, H. and Campanella, O. H. (1994). Thermal aggregation and gelation of bovine beta-lactoglobulin. *Food Hydrocolloids* 8(5), 441–453.

Mediwaththe, A., Bogahawaththa, D., Grewal, M. K., Chandrapala, J. and Vasiljevic, T. (2018a). Structural changes of native milk proteins subjected to controlled shearing and heating. *Food Research International* 114, 151–158.

Mediwaththe, A., Chandrapala, J. and Vasiljevic, T. (2018b). Shear-induced behaviour of native milk proteins heated at temperatures above 80°C. *International Dairy Journal* 77, 29–37.

Mehalebi, S., Nicolai, T. and Durand, D. (2008). Light scattering study of heat-denatured globular protein aggregates. *International Journal of Biological Macromolecules* 43(2), 129–135.

Mehra, R., Marnila, P. and Korhonen, H. (2006). Milk immunoglobulins for health promotion. *International Dairy Journal* 16(11), 1262–1271.

Mercadante, D., Melton, L. D., Norris, G. E., Loo, T. S., Williams, M. A. K., Dobson, R. C. J. and Jameson, G. B. (2012). Bovine beta-lactoglobulin is dimeric under imitative physiological conditions: dissociation equilibrium and rate constants over the pH range of 2.5–7.5. *Biophysical Journal* 103(2), 303–312.

Merkel, T., Gräf, V., Walz, E. and Schuchmann, H. P. (2015). Production of particle-stabilized non-spherical emulsion drops in simple shear flow. *Chemical Engineering and Technology* 38(8), 1490–1493. DOI: 10.1002/ceat.201500068.

Messens, W., Van Camp, J. and Huyghebaert, A. (1997). The use of high pressure to modify the functionality of food proteins. *Trends in Food Science and Technology* 8(4), 107–112.

Meyer, C. J. and Deglon, D. A. (2011). Particle collision modeling-a review. *Minerals Engineering* 24(8), 719–730.

Mezzenga, R. and Fisher, P. (2013). The self-assembly, aggregation and phase transitions of food protein systems in one, two and three dimensions. *Reports on Progress in Physics. Physical Society* 76(4), 046601.

Morgan, P. E., Treweek, T. M., Lindner, R. A., Price, W. E. and Carver, J. A. (2005). Casein proteins as molecular chaperones. *Journal of Agricultural and Food Chemistry* 53(7), 2670–2683.

Morr, C. V. and Ha, E. Y. W. (1993). Whey protein concentrates and isolates-processing and functional properties. *Critical Reviews in Food Science and Nutrition* 33(6), 431–476.

Mostafa, H. S. (2020). Microbial transglutaminase: an overview of recent applications in food and packaging. *Biocatalysis and Biotransformation* 38(3), 161–177. DOI: 10.1080/10242422.2020.1720660.

Murayama, K. and Tomida, M. (2004). Heat-induced secondary structure and conformation change of bovine serum albumin investigated by Fourier transform infrared spectroscopy. *Biochemistry* 43(36), 11526–11532.

Murphy, R. W., Farkas, B. E. and Jones, O. G. (2016). Dynamic and viscoelastic interfacial behavior of β-lactoglobulin microgels of varying sizes at fluid interfaces. *Journal of Colloid and Interface Science* 466, 12–19. DOI: 10.1016/j.jcis.2015.12.012.

Murray, B. S. (2019). Microgels at fluid–fluid interfaces for food and drinks. *Advances in Colloid and Interface Science* 271, 101990. DOI: 10.1016/j.cis.2019.101990.

Nagasawa, K., Takahashi, K. and Hattori, M. (1996). Improved emulsifying properties of β-lactoglobulin by conjugating with carboxymethyl dextran. *Food Hydrocolloids* 10(1), 63–67. DOI: 10.1016/S0268-005X(96)80055-5.

Nicolai, T. (2016). Formation and functionality of self-assembled whey protein microgels. *Colloids and Surfaces. B, Biointerfaces* 137, 32–38. DOI: 10.1016/j.colsurfb.2015.05.055.

Nicolai, T., Britten, M. and Schmitt, C. (2011). β-Lactoglobulin and WPI aggregates: formation, structure and applications. *Food Hydrocolloids* 25(8), 1945–1962.

Nishanthi, M., Chandrapala, J. and Vasiljevic, T. (2017a). Properties of whey protein concentrate powders obtained by spray drying of sweet, salty and acid whey under varying storage conditions. *Journal of Food Engineering* 214, 137–146. DOI: 10.1016/j.jfoodeng.2017.06.032.

Nishanthi, M., Chandrapala, J. and Vasiljevic, T. (2017b). Compositional and structural properties of whey proteins of sweet, acid and salty whey concentrates and their respective spray-dried powders. *International Dairy Journal* 74, 49–56. DOI: 10.1016/j.idairyj.2017.01.002.

Nishanthi, M., Vasiljevic, T. and Chandrapala, J. (2017c). Properties of whey proteins obtained from different whey streams. *International Dairy Journal* 66, 76–83. DOI: 10.1016/j.idairyj.2016.11.009.

Nishanthi, M., Chandrapala, J. and Vasiljevic, T. (2018a). Impact of storage conditions on solubility, heat stability and emulsifying properties of selected spray dried whey protein concentrates. *LWT* 92, 16–21. DOI: 10.1016/j.lwt.2018.01.068.

Nishanthi, M., Chandrapala, J. and Vasiljevic, T. (2018b). Physical properties of selected spray-dried whey protein concentrate powders during storage. *Journal of Food Engineering* 219, 111–120. DOI: 10.1016/j.jfoodeng.2017.09.021.

Oboroceanu, D., Wang, L., Magner, E. and Auty, M. A. E. (2014). Fibrillization of whey proteins improves foaming capacity and foam stability at low protein concentrations. *Journal of Food Engineering* 121, 102–111. DOI: 10.1016/j.jfoodeng.2013.08.023.

O'Mahony, J. A. and Fox, P. F. (2014). Milk: an overview. In: Singh, H., Boland, M. and Thompson, A. (Eds), *Milk Proteins: From Expression to Food* (pp. 20–61). Waltham: Academic Press.

Oh, S. S. and Richardson, T. (1991). Heat-induced interactions of bovine serum albumin and immunoglobulin. *Journal of Dairy Science* 74(6), 1786–1790.

Ohnuki, H. and Otani, H. (2005). Antigen-binding and protein G-binding abilities of immunoglobulin G in hyperimmunized cow's milk treated under various conditions. *Animal Science Journal* 76(3), 283–290.

Oldfield, D. J., Singh, H. and Taylor, M. W. (2005). Kinetics of heat-induced whey protein denaturation and aggregation in skim milks with adjusted whey protein concentration. *Journal of Dairy Research* 72(3), 369–378.

Oldfield, D. J., Singh, H., Taylor, M. W. and Pearce, K. N. (1998). Kinetics of denaturation and aggregation of whey proteins in skim milk heated in an ultra-high temperature (UHT) pilot plant. *International Dairy Journal* 8(4), 311–318.

Onwulata, C. I., Konstance, R. P. and Thomasula, P. M. (2004). Minimizing variations in functionality of whey protein concentrates from different sources. *Journal of Dairy Science* 87(3), 749–756.

Patel, H. A. (2007). Studies on heat-and pressure-induced interactions of milk proteins. PhD thesis. Palmerston North, New Zealand: Massey University.

Patel, H. A., Singh, H., Anema, S. G. and Creamer, L. K. (2006). Effects of heat and high hydrostatic pressure treatments on disulfide bonding interchanges among the proteins in skim milk. *Journal of Agricultural and Food Chemistry* 54(9), 3409–3420.

Pearce, K. N. and Kinsella, J. E. (1978). Emulsifying properties of proteins: evaluation of a turbidimetric technique. *Journal of Agricultural and Food Chemistry* 26(3), 716–723.

Peng, J., Simon, J. R., Venema, P. and Van Der Linden, E. (2016). Protein fibrils induce emulsion stabilization. *Langmuir* 32(9), 2164–2174. DOI: 10.1021/acs.langmuir.5b04341.

Piazza, R. (2000). Interactions and phase transitions in protein solutions. *Current Opinion in Colloid and Interface Science* 5(1–2), 38–43.

Pickering, S. U. (1907). Emulsions. *Journal of the Chemical Society* 91, 2001–2021.

Potanin, A. A. (1991). On the mechanism of aggregation in the shear flow of suspensions. *Journal of Colloid and Interface Science* 145(1), 140–157.

Pouzot, M., Nicolai, T., Visschers, R. W. and Weijers, M. (2005). X-ray and light-scattering study of the structure of large protein aggregates at neutral pH. *Food Hydrocolloids* 19(2), 231–238.

Price, N. C. (2000). Conformational issues in the characterization of proteins. *Biotechnology and Applied Biochemistry* 31(1), 29–40.

Qi, X. L., Holt, C., McNulty, D., Clarke, D. T., Brownlow, S. and Jones, G. R. (1997). Effect of temperature on the secondary structure of beta-lactoglobulin at pH 6.7, as determined by CD and IR spectroscopy: a test of the molten globule hypothesis. *Biochemical Journal* 324(1), 341–346.

Quevedo, M., Karbstein, H. P. and Emin, M. A. (2021). Influence of thermomechanical treatment and pH on the denaturation kinetics of highly concentrated whey protein isolate. *Journal of Food Engineering* 292, 110294.

Quevedo, M., Kulozik, U., Karbstein, H. P. and Emin, M. A. (2020a). Effect of thermomechanical treatment on the aggregation behaviour and colloidal functionality of β-Lactoglobulin at high concentrations. *International Dairy Journal* 104, 104654.

Quevedo, M., Kulozik, U., Karbstein, H. P. and Emin, M. A. (2020b). Influence of Thermomechanical Treatment and Ratio of β-Lactoglobulin and α-lactalbumin on the Denaturation and Aggregation of Highly Concentrated Whey Protein Systems. *Foods* 9(9), 1196.

Quevedo, M., Kulozik, U., Karbstein, H. P. and Emin, M. A. (2020c). Kinetics of denaturation and aggregation of highly concentrated β-Lactoglobulin under defined thermomechanical treatment. *Journal of Food Engineering* 274, 109825.

Rao, M. A. (2013). *Rheology of Fluid, Semisolid, and Solid Foods: Principles and Applications*. New York: Springer Science & Business Media.

Resch, J. J., Daubert, C. R. and Foegeding, E. A. (2004). The effects of acidulant type on the rheological properties of beta-lactoglobulin gels and powders derived from these gels. *Food Hydrocolloids* 19(5), 851–860.

Resmini, P., Pellegrino, L., Hogenboom, J. and Andreini, R. (1989). Thermal denaturation of whey protein in pasteurized milk. Fast evaluation by HPLC. *Italian Journal of Food Science* 1, 51–62.

Rich, L. M. and Foegeding, E. A. (2000). Effects of sugars on whey protein isolate gelation. *Journal of Agricultural and Food Chemistry* 48(10), 5046–5052. DOI: 10.1021/jf000272l.
Riera, F. and Álvarez, A. (2014). Influence of temperature and pH on the antigen-binding capacity of immunoglobulin G in cheese whey derived from hyper-immune milk. *International Dairy Journal* 37(2), 111–116.
Roefs, S. P. and De Kruif, K. G. (1994). A model for the denaturation and aggregation of β-lactoglobulin. *European Journal of Biochemistry* 226(3), 883–889.
Roos, N., Mahe, S., Benamouzig, R., Sick, H., Rautureau, J. and Tome, D. (1995). 15N-labeled immunoglobulins from bovine colostrum are partially resistant to digestion in human intestine. *The Journal of Nutrition* 125(5), 1238–1244.
Rüegg, M., Moor, U. and Blanc, B. (1977). A calorimetric study of the thermal denaturation of whey proteins in simulated milk ultrafiltrate. *Journal of Dairy Research* 44(3), 509–520.
Ryan, K. N., Zhong, Q. and Foegeding, E. A. (2013). Use of whey protein soluble aggregates for thermal stability–A hypothesis paper. *Journal of Food Science* 78(8), R1105–R1115.
Sağlam, D., Venema, P., van der Linden, E. and de Vries, R. (2014). Design, properties, and applications of protein micro- and nanoparticles. *Current Opinion in Colloid and Interface Science* 19(5), 428–437. DOI: 10.1016/j.cocis.2014.09.004.
Sah, B. N. P., McAinch, A. J. and Vasiljevic, T. (2016). Modulation of bovine whey protein digestion in gastrointestinal tract: a comprehensive review. *International Dairy Journal* 62, 10–18:DOI: 10.1016/j.idairyj.2016.07.003.
Saito, T. (2009). Potential for improving health: immunomodulation by dairy ingredients. *Bioactive Components in Milk and Dairy Products*, 347–362.
Sawyer, L. (2013). β-Lactoglobulin. In: McSweeney, P. L. H. and Fox, P. F. (Eds), *Advanced Dairy Chemistry (vol. 8). Proteins: Basic Aspects* (pp. 211–259). New York, NY: Springer Science+Business Media.
Schmitt, C., Bovay, C. and Rouvet, M. (2014). Bulk self-aggregation drives foam stabilization properties of whey protein microgels. *Food Hydrocolloids* 42, 139–148. DOI: 10.1016/j.foodhyd.2014.03.010.
Schmitt, C., Bovay, C., Vuilliomenet, A. M., Rouvet, M. and Bovetto, L. (2011). Influence of protein and mineral composition on the formation of whey protein heat-induced microgels. *Food Hydrocolloids* 25(4), 558–567.
Schmitt, C., Bovay, C., Vuilliomenet, A. M., Rouvet, M., Bovetto, L., Barbar, R. and Sanchez, C. (2009). Multiscale characterization of individualized beta-lactoglobulin microgels formed upon heat treatment under narrow pH range conditions. *Langmuir* 25(14), 7899–7909. DOI: 10.1021/la900501n.
Schokker, E. P., Singh, H. and Creamer, L. K. (2000). Heat-induced aggregation of β-lactoglobulin A and B with α-lactalbumin. *International Dairy Journal* 10(12), 843–853.
Schroeder Jr., H. W. and Cavacini, L. (2010). Structure and function of immunoglobulins. *Journal of Allergy and Clinical Immunology* 125(2) (Suppl. 2), S41–S52.
Sharma, L. G. and Pandey, L. M. (2021). Thermomechanical process induces unfolding and fibrillation of bovine serum albumin. *Food Hydrocolloids* 112, 106294.
Sharma, R. (2019). Whey proteins in functional foods. In: Deeth, H. and Bansal, N. (Eds), *Whey Proteins: From Milk to Medicine* (pp. 637–663). London, UK: Academic Press.
Sharma, R. N., Rajput, Y. S. and Mann, B. (2013). Chemical and functional properties of glycomacropeptide (GMP) and its role in the detection of cheese whey adulteration in milk: a review. *Dairy Science and Technology* 93(1), 21–43.

Simmons, M. J. H., Jayaraman, P. and Fryer, P. J. (2007). The effect of temperature and shear rate upon the aggregation of whey protein and its implications for milk fouling. *Journal of Food Engineering* 79(2), 517–528.

Singer, N. S. and Dunn, J. M. (1990). Protein microparticulation: the principle and the process. *Journal of the American College of Nutrition* 9(4), 388–397.

Singh, H. (2011). Milk protein products: functional properties of milk proteins. In: Fuquay, J., McSweeney, P., and Fox, P. (Eds), *Encyclopedia of Dairy Sciences* (2nd edn., pp. 887-893). Cambridge, MA: Academic Press.

Singh, H. and Havea, P. (2003). Thermal denaturation, aggregation and gelation of whey proteins. In: Fox, P. F. and McSweeney, P. L. H. (Eds), *Advanced Dairy Chemistry–1 Proteins: Part A/Part B* (pp. 1261–1287). Boston, MA: Springer US.

Smoluchowski, M. (1918). Versuch einer mathematischen Theorie der Koagulationskinetik kolloider Lösungen. *Zeitschrift für physikalische Chemie* 92U, 129–168.

Soltis, R. D., Hasz, D., Morris, M. J. and Wilson, I. D. (1979). The effect of heat inactivation of serum on aggregation of immunoglobulins. *Immunology* 36(1), 37–45.

Soria, A. C. and Villamiel, M. (2010). Effect of ultrasound on the technological properties and bioactivity of food: a review. *Trends in Food Science and Technology* 21(7), 323–331. DOI: 10.1016/j.tifs.2010.04.003.

Suzuki, A., Sugiyama, Y., Hayashi, Y., Nyu-i, N., Yoshida, M., Nonaka, I., Ishiura, S., Arahata, K. and Ohno, S. (1998). MKBP, a novel member of the small heat shock protein family, binds and activates the myotonic dystrophy protein kinase. *Journal of Cell Biology* 140(5), 1113–1124.

Taboada-Serrano, P., Chin, C.-J., Yiacoumi, S. and Tsouris, C. (2005). Modeling aggregation of colloidal particles. *Current Opinion in Colloid and Interface Science* 10(3–4), 123–132.

Takeda, K., Wada, A., Yamamoto, K., Moriyama, Y. and Aoki, K. (1989). Conformational change of bovine serum albumin by heat treatment. *Journal of Protein Chemistry* 8(5), 653–659.

Taylor, S. M. and Fryer, P. J. (1994). The effect of temperature/shear history on the thermal gelation of whey protein concentrates. *Food Hydrocolloids* 8(1), 45–61.

Thomas, C. R. and Geer, D. (2011). Effects of shear on proteins in solution. *Biotechnology Letters* 33(3), 443–456.

Thomä-Worringer, C., Sørensen, J. and López-Fandiño, R. (2006). Health effects and technological features of caseinomacropeptide. *International Dairy Journal* 16(11), 1324–1333.

Thorn, D. C., Ecroyd, H. and Carver, J. A. (2009). The two-faced nature of milk casein proteins: amyloid fibril formation and chaperone-like activity. *Australian Journal of Dairy Technology* 64, 34–40.

Thorn, D. C., Meehan, S., Sunde, M., Rekas, A., Gras, S. L., MacPhee, C. E., Dobson, C. M., Wilson, M. R. and Carver, J. A. (2005). Amyloid fibril formation by bovine milk κ-casein and its inhibition by the molecular chaperones αs- and β-casein. *Biochemistry* 44, 17027–17036.

Timasheff, S. N. (2002). Protein-solvent preferential interactions, protein hydration, and the modulation of biochemical reactions by solvent components. *Proceedings of the National Academy of Sciences of the United States of America* 99(15), 9721-9726. DOI: 10.1073/pnas.122225399.

Tirrell, M. and Middleman, S. (1975). Shear modification of enzyme kinetics. *Biotechnology and Bioengineering* 17(2), 299–303.

Treweek, T. M., Thorn, D. C., Price, W. E. and Carver, J. A. (2011). The chaperone action of bovine milk αs1- and αs2-caseins and their associated form αs- casein. *Archives of Biochemistry and Biophysics* 510(1), 42–52. DOI: 10.1016/j.abb.2011.03.012.

Ulfman, L. H., Leusen, J. H. W., Savelkoul, H. F. J., Warner, J. O. and van Neerven, R. J. J. (2018). Effects of bovine immunoglobulins on immune function, allergy, and infection. *Frontiers in Nutrition* 5, 52.

Ustunol, Z. and Sypien, C. (1997). Heat stability of bovine milk immunoglobulins and their ability to bind Lactococci as determined by an ELISA. *Journal of Food Science* 62(6), 1218–1222.

Walstra, P. (2003). *Physical Chemistry of Foods*. New York: Marcel Dekker.

Walstra, P., Wouters, J. T. M. and Geurts, T. J. (2006). Chapters 3 and 7: *Dairy Science and Technology* (2nd edn.). Boca Raton, FL: CRC Press (pp. 109–125, 225–277),.

Wang, C. and Guo, M. (2019a). Whey protein structure and denaturation and interactions with other food components. In: Guo, M. (Ed), *Whey Protein Production, Chemistry, Functionality, and Applications* (pp. 67–102). Hoboken: John Wiley & Sons, Inc.

Wang, G. and Guo, M. (2019b). Manufacturing technologies of whey protein products. In: Guo, M. (Ed), *Whey Protein Production, Chemistry, Functionality, and Applications* (pp. 13–37). Hoboken: John Wiley & Sons, Inc.

Wasan, D. T., Nikolov, A. D. and Aimetti, F. (2004). Texture and stability of emulsions and suspensions: role of oscillatory structural forces. *Advances in Colloid and Interface Science* 108–109, 187–195, 187–195. DOI: 10.1016/j.cis.2003.10.021.

Wijayanti, H. B., Bansal, N. and Deeth, H. C. (2014). Stability of whey proteins during thermal processing: a review. *Comprehensive Reviews in Food Science and Food Safety* 13(6), 1235–1251. DOI: 10.1111/1541-4337.12105.

Wijayanti, H. B., Brodkorb, A., Hogan, S. A. and Murphy, E. G. (2019) *Thermal Denaturation, Aggregation, and Methods of Prevention, In Whey ProteinS: From Milk to Medicine* (Deeth, H. and Bansal, N., eds), (pp. 185–247). DOI: 10.1016/B978-0-12-812124-5.00006-0.

Wolz, M. and Kulozik, U. (2015). Thermal denaturation kinetics of whey proteins at high protein concentrations. *International Dairy Journal* 49, 95–101.

Wolz, M., Mersch, E. and Kulozik, U. (2016). Thermal aggregation of whey proteins under shear stress. *Food Hydrocolloids* 56, 396–404.

Wu, J., Shi, M., Li, W., Zhao, L., Wang, Z., Yan, X., Norde, W. and Li, Y. (2015). Pickering emulsions stabilized by whey protein nanoparticles prepared by thermal cross-linking. *Colloids and Surfaces. B, Biointerfaces* 127, 96–104. DOI: 10.1016/j.colsurfb.2015.01.029.

Wynn, P. and Sheehy, P. (2013). Minor proteins, including growth factors. In: McSweeney, P. L. H. and Fox, P. F. (Eds), *Advanced Dairy Chemistry (vol. 1A). Proteins: Basic Aspects* (pp. 317–335). New York, NY: Springer Science+Business Media.

Xie, G. and Timasheff, S. N. (1997). Mechanism of the stabilization of ribonuclease A by sorbitol: preferential hydration is greater for the denatured than for the native protein. *Protein Science* 6(1), 211–221. DOI: 10.1002/pro.5560060123.

Xie, Z., Burke, C. J., Mbanga, B., Spicer, P. T. and Atherton, T. J. (2019). Geometry and kinetics determine the microstructure in arrested coalescence of Pickering emulsion droplets. *Soft Matter* 15(46), 9587–9596. DOI: 10.1039/c9sm00435a.

Xiong, L., Li, C., Boeren, S., Vervoort, J. and Hettinga, K. (2020). Effect of heat treatment on bacteriostatic activity and protein profile of bovine whey proteins. *Food Research International* 127, 108688.

Yamasaki, M., Yano, H. and Aoki, K. (1991). Differential scanning calorimetric studies on bovine serum albumin: II. Effects of neutral salts and urea. *International Journal of Biological Macromolecules* 13(6), 322–328.

Yan, X., Ma, C., Cui, F., McClements, D. J., Liu, X. and Liu, F. (2020). Protein-stabilized Pickering emulsions: formation, stability, properties, and applications in foods. *Trends in Food Science and Technology* 103, 293–303. DOI: 10.1016/j.tifs.2020.07.005.

Yolken, R. H., Losonsky, G. A., Vonderfecht, S., Leister, F. and Wee, S. B. (1985). Antibody to human rotavirus in cow's milk. *New England Journal of Medicine* 312(10), 605–610.

Yong, Y. H. and Foegeding, E. A. (2008). Effects of caseins on thermal stability of bovine β-lactoglobulin. *Journal of Agricultural and Food Chemistry* 56(21), 10352–10358. DOI: 10.1021/jf801658u.

Zayas, J. F. (1997). Solubility of proteins. In: *Functionality of Proteins in Food*. Berlin, Heidelberg: Springer. DOI: 10.1007/978-3-642-59116-7_2.

Zhang, S., Zhang, Z., Lin, M. and Vardhanabhuti, B. (2012). Raman spectroscopic characterization of structural changes in heated whey protein isolate upon soluble complex formation with pectin at near neutral pH. *Journal of Agricultural and Food Chemistry* 60(48), 12029–12035. DOI: 10.1021/jf303780c.

Zhang, W., Liu, Y., Li, Z., Xu, S., Hettinga, K. and Zhou, P. (2021). Retaining bioactive proteins and extending shelf life of skim milk by microfiltration combined with ultraviolet-C treatment. *LWT*, 110945.

Zhang, X., Fu, X., Zhang, H., Liu, C., Jiao, W. and Chang, Z. (2005). Chaperone-like activity of β-casein. *The International Journal of Biochemistry & Cell Biology* 37, 1232–1240.

Zisu, B., Bhaskaracharya, R., Kentish, S. and Ashokkumar, M. (2010). Ultrasonic processing of dairy systems in large-scale reactors. *Ultrasonics Sonochemistry* 17(6), 1075–1081. DOI: 10.1016/j.ultsonch.2009.10.014.

Zumaeta, N., Byrne, E. P. and Fitzpatrick, J. J. (2006). Predicting precipitate particle breakage in a pipeline: effect of agitation intensity during precipitate formation. *Chemical Engineering Science* 61(24), 7991–8003.

# Chapter 6

## Understanding nutritional and bioactive properties of whey

*Geoffrey McCarthy, Teagasc Food Research Centre, Ireland; James A. O'Mahony, University College Cork, Ireland; and Mark A. Fenelon and Rita M. Hickey, Teagasc Food Research Centre, Ireland*



### 1 Introduction

Whey is the co-product remaining after milk is coagulated during cheese making (Bansal and Bhandari, 2016). Liquid whey can be concentrated by ultrafiltration (UF) and dried to produce whey protein concentrate (WPC) powders (typically containing 34–80/85% protein). In the manufacture of whey protein isolate (WPI), containing >90% protein on a dry weight basis, UF is used for protein enrichment and is complemented by the use of microfiltration to remove lipid from sweet whey (which is concentrated along with protein on UF of liquid whey). Consequently, WPI has a higher protein concentration and lower fat, ash, and lactose contents than WPC (Hoffman and Falvo, 2004). Demineralized whey powder is a reduced mineral whey powder, with a lower bitter and salty taste (Bansal and Bhandari, 2016). Whey permeate (a deproteinized whey stream) is a further by-product of whey processing and comprises predominantly lactose and minerals (Frankowski et al., 2014). Two different whey products can be obtained from milk processing: sweet whey and acid whey (for macro-composition, see Fig. 1). Sweet whey results from the production of certain cheeses (Cheddar, Swiss,

http://dx.doi.org/10.19103/AS.2022.0099.07

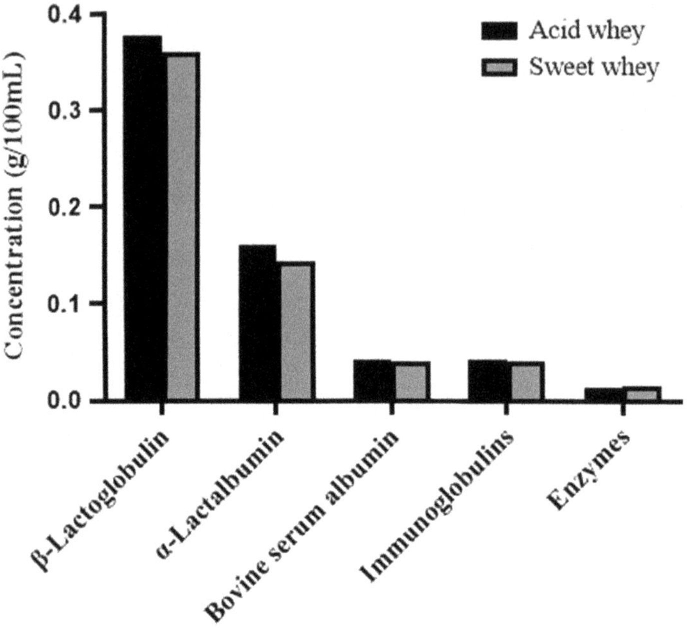


**Figure 1** Concentrations of major proteins found in acid versus sweet whey. Data were compiled from Villarreal (2017) and Sharma (2019).

and Mozzarella), where the controlled precipitation of casein is through enzymatic cleavage. Acid whey may be obtained as a by-product of acid casein production, involving either bacterial fermentation or the addition of organic acids or mineral acids (Ryan and Walsh, 2016). Acid whey may also be obtained as a by-product from the production of acid-coagulated cheeses, for example, cottage cheese and ricotta cheese and also Greek yogurt (Rocha-Mendoza et al., 2021), making it a very diverse dairy process stream. The composition and physicochemical characteristics of both types of whey are similar. In general, whey is composed of approximately 70% of lactose, 14% whey protein, minerals, and residual fat on a dry basis (Kareb and Aïder, 2019). The differences between sweet and acid whey are observed in their acidity and mineral content as well as in some of the individual proteins. Acid whey has a higher content of minerals in comparison to sweet whey, due to the dissolution of colloidal calcium phosphate in casein micelles during the acidification process (Lievore et al., 2015). Sweet whey contains caseinomacropeptide (CMP) produced by the enzymatic hydrolysis of κ-casein. In general, CMP refers to the peptide's non-glycosylated form, and glycomacropeptide (GMP) represents the glycosylated form. The amount of free amino acids (AAs) in sweet whey can vary according to the degree of casein hydrolysis.

Whey proteins are a source of essential amino acids (EAAs), including branched-chain amino acids (BCAAs; leucine, isoleucine, and valine) (Hoffman and Falvo, 2004). Enzymatic hydrolysis of whey proteins in the human gastrointestinal (GI) digestive system, fermentation of milk by lactic acid bacteria (starter cultures), and hydrolysis by microbial or plant proteases result in the release of bioactive peptides. These bioactive peptides act as signaling molecules resulting in various physiological effects on the GI, immune, endocrine, cardiovascular and nervous systems (Brandelli et al., 2015, Barukčić et al., 2019). In industry, the degree of hydrolysis of a particular protein can be controlled; therefore, the hydrolysates can range from whole proteins to peptides to individual AAs. Whey protein hydrolysates (WPH) are more easily digested and absorbed than intact whey protein and are therefore suitable for use in infants, sports and medical nutrition products (Boland et al., 2014). This chapter summarizes the established and emerging biological roles of whey and its protein derivatives.

## 2 Health benefits and applications of whey proteins

Numerous studies have confirmed the valuable impact of whey protein on human health and well-being (summarized in Table 1) such as antioxidant (Athira et al., 2013), antimicrobial (Freedman et al., 1998), antidiabetic (Salehi et al., 2012), antihypertensive (Mullally et al., 1997), and immunomodulatory properties (Badr et al., 2012a). Whey protein may also influence mechanisms involved in reducing body mass (Pal et al., 2010) and reduce or inhibit allergic reactions (Alexander et al., 2010). The following section summarizes the current knowledge on the beneficial effects of whey protein in the human diet.

### *2.1 Modulation of the gut microbiota by whey proteins*

It is increasingly understood that modulating the gut microbiota through diet may influence health outcomes (Singh et al., 2017). Whey proteins can affect the gut microbiota in a variety of ways, such as facilitating the growth of beneficial bacteria, inhibiting the colonization of pathogens, and altering the GI environment. In particular, whey protein may play a vital role in shaping the infant gut microbiota. Bifidobacteria are found in abundance in the breast-fed infant gut and are known to be involved in regulating immunological and inflammatory responses, inhibiting the growth and colonization of pathogens, and modulating mucosal barrier function (Westermann et al., 2016). As an example, Balmer et al. (1989) investigated the fecal microflora of infants fed a whey protein formula (33 infants), a casein formula (29 infants), and breast milk (38 infants) over 15 weeks. Infants receiving the whey protein formula had higher bifidobacteria numbers compared to infants who received the

**Table 1** Clinical studies on whey protein

| Health pillar | Outcome of study | References |
|---|---|---|
| Microbiota | Hydrolyzed whey protein increased bifidobacteria numbers in the stool of infants | Schmelzle et al. (2003) |
| | Whey formula-fed infants displayed higher numbers of bifidobacteria colonizing the gut, replicating a closer microbiota to that of breast-fed infants | Balmer et al. (1989), Hascoët et al. (2011), Schaafsma et al. (2021) |
| Cancer | Whey protein exerts a strong antioxidant activity due to the presence of cysteine, leading to a rise in tissue glutathione. This may be beneficial in cancer prevention due to the associated immune response | Walzem et al. (2002), Bounous and Molson (2003), Cheng et al. (2017) |
| | Whey protein has been shown to potentially deplete tumor cells of glutathione (GSH) which can render them more vulnerable to chemotherapy | Kennedy et al. (1995), Mazzuca et al. (2019), Bumrungpert et al. (2018) |
| | Whey protein hydrolysate treatment resulted in rats with colon cancer developing significantly less macroscopic and microscopic tumors compared to the group fed with untreated whey protein | Attaallah et al. (2012) |
| Osteoporosis | Whey protein intake was shown to increase lumbar spine bone mineral density (BMD) in postmenopausal women | Aoe et al. (2005) |
| | Whey protein was shown to increase lean mass in older women | Thomas et al. (2015) |
| | Whey constituents displayed inhibitory activity against osteoclasts leading to a reduced degradation of bone matrix | Morita et al. (2008), Takada et al. (1997), Toba et al. (2001), Yoneme et al. (2015), Uenishi et al. (2007) |
| Weight management | Whey protein induced a higher feeling of satiety in patients compared to other protein sources | Gilbert et al. (2011), Chungchunlam et al. (2012) |
| | Whey protein administration to obese patients improved some cardiovascular disease risk factors | Wirunsawanya et al. (2018) |
| | Consumption of whey protein resulted in significant reduction in disease risk factors including cholesterol and insulin | Amirani et al. (2020) |
| Immune system | Ingested whey protein stimulated immune cell proliferation and migration to the secondary lymphoid organs in rodents | Badr et al. (2012b) |
| | WPC activated adaptive immunity against antigens in rodents by modulating helper T cells | Ha et al. (2021) |
| | WPI improved skin conditions of psoriasis patients | Prussick et al. (2013) |

casein-predominant formula at 2 weeks. At week 7, the infants receiving the casein formula had a dominant growth of enterococci unlike the breast milk group and the whey formula group. There were very few significant changes in the fecal flora observed at week 15. In a German double-blind randomized study, 102 healthy infants (≤2 weeks of age) received either a standard cows' milk formula or an infant formula (IF) containing hydrolyzed whey protein until 12 weeks of age (Schmelzle et al., 2003). The hydrolyzed whey protein-fed infants were found to have greater numbers of fecal bifidobacteria and produced blood biochemical values similar to that of formula-fed infants. Hascoët et al. (2011) in a study of 190 infants also demonstrated that infants who received a whey-predominant (70% whey, 30% casein) IF, with low protein and phosphate concentrations, displayed a fecal bacterial profile similar in bifidobacteria composition to that of breast-fed infants. The absolute bifidobacterial counts and the relative proportion of bifidobacteria in the stool samples were both significantly higher in infants fed the whey formula compared with those fed the control formula.

Regarding the gut microbiota of adults, Reimer et al. (2017), in a placebo-controlled, double-blind study, randomized 125 overweight/obesity adults to receive isocaloric snack bars of (1) control, (2) inulin-type fructans (ITF), (3) whey protein, and (4) ITF + whey protein. At the end of the intervention (12 weeks), changes in gut microbiota composition were found in groups 2 and 4 with the abundance of *Bifidobacterium* increased, and alpha diversity also decreased in these two groups. In contrast, there were no differences reported in either group 1 or group 3. This study suggests whey protein does not prompt changes to the adult gut microbiota. However, in D-galactose-treated aging mice, changes in intestinal flora structure induced by whey protein were observed using next-generation sequencing analysis (Ma et al., 2021). The relative abundance of *Lactobacillus* and *Stenotrophomonas* was increased and decreased, respectively, after intragastric administration of whey protein. Schaafsma et al. (2021) also recently investigated the effect of a whey protein–galactooligosaccharide-based supplement on sleep quality, stress, and the gut microbiota of healthy Dutch adults. The relative abundance of *Bifidobacterium* increased after 21 days on the supplement compared to the placebo group. Further studies are warranted to gain a better understanding of the broader impacts of whey protein on the gut microbiota.

## 2.2 Whey protein applications in weight management

Whey protein has potential as a functional ingredient that can impact body weight regulation through providing satiety signals that affect both short-term and long-term food intake regulation (Luhovyy et al., 2007). Only some studies have demonstrated a higher satiating effect with whey protein compared to protein from other sources (Gilbert et al., 2011). Chungchunlam et al. (2012) showed in normal-weight adult female subjects that a whey protein-enriched preload drink reduced food intake at a test meal when compared with a maltodextrin carbohydrate-enriched preload drink. Recently, Wirunsawanya et al. (2018) conducted a meta-analysis that evaluated the effects of whey protein on overweight and obese patients. Overall, the analysis concluded that whey protein administration improved body weight, total fat mass, and multiple cardiovascular disease risk factors including total cholesterol. In a further meta-analysis by Amirani et al. (2020), the efficacy of whey protein administration in subjects with metabolic syndrome or related conditions on glycemic and lipid control parameters was reviewed. The authors concluded that whey protein had a beneficial effect on a number of indicators of glycemic control and lipid parameters when compared to the respective controls, for example, a significant reduction in triglyceride levels, total cholesterol, low-density lipoprotein-cholesterol levels, and the total cholesterol/high-density lipoprotein cholesterol ratio. In general, whey protein may be more effective for improving body composition when included as part of a weight loss program. More research is also needed to specifically assess its effects when combined with exercise training.

## 2.3 Whey protein applications associated with immune health

It has been shown that whey proteins boost immunity early in life and as a result may prevent many immune complications (Badr et al., 2011). For instance, ingested whey protein was shown to stimulate immune cell proliferation and migration to the secondary lymphoid organs in a rat model (Badr et al., 2012b). Immunological analyses on the blood of WPC-fed mice showed a reduction in inflammatory cytokines such as interleukin (IL)-1α, IL-1β, IL-10, and tumor necrosis factor-alpha. In contrast, there was an increase in the levels of cytokines IL-2, IL-4, IL-7, IL-8, responsible for maintaining normal immune responses. In addition, the ability of lymphocytes, monocytes, and macrophages to proliferate in response to antigens improved (Badr et al., 2012b). In another immune study, the bioactive effect of WPI was evaluated against psoriasis (a skin condition with thick skin, dry scales, and red patches). The 12-week study found that an intake dose of 20 g/d of the WPI improved the skin conditions of the psoriasis patients (Prussick et al., 2013). To examine the adaptive immune response, Ha et al. (2021) tested the effect of WPC after antigen (hemagglutinin peptide) exposure and re-exposure in rats. The WPC group had decreased IgA and cytotoxic T cells before the antigen injection at day 0 but increased immunoglobulin G (IgG), IL-2, and IL-4 after antigen injection due to increased B cells and T cells. At days 14 and 21, helper T cells were increased, but cytotoxic T cells were not affected by WPC. The authors concluded that WPC may activate adaptive immunity (IgG) against the antigen by modulating helper T cells.

## 2.4 Whey protein applications in cancer therapy

The use of whey protein as a dietary strategy in the field of oncology has recently been reviewed by Teixeira et al. (2019) and Leischner et al. (2021). Glutathione (GSH) is an important antioxidant in cells and is involved in regulating apoptosis. Increased GSH in cancer cells is linked with resistance to chemotherapy because of resistance to apoptosis. Cheng et al. (2017) using rat models confirmed that WPC (0.334 g/kg) supplementation resulted in a 92% increase in liver GSH and a 47% decrease in tumor GSH. These findings suggest that the incorporation of whey protein may be a promising means to overcome treatment resistance in cancer therapy. Studies have suggested that WPHs confer an improved anti-cancer effect when compared to intact whey protein; for instance, rats with colon cancer that were fed WPH developed significantly less macroscopic and microscopic tumors than those fed untreated whey protein (Attaallah et al., 2012).

The effects of whey protein on cancer in human studies have also been summarized by Teixeira et al. (2019). In one such human study on seven patients, Kennedy et al. (1995) investigated five patients with metastatic carcinoma of the breast, one with cancer of the pancreas, and one with cancer of the liver who were fed 30 g of WPC daily for 6 months. In six of the seven patients, the blood lymphocyte GSH levels were substantially above normal initially, suggesting high tumor GSH levels. Two patients exhibited signs of tumor regression and a continued reduction of lymphocyte GSH levels toward normal. Two other patients showed stabilization of the tumor, without normalization of GSH levels. In the three other patients, the disease progressed trending toward higher lymphocyte GSH levels. These results indicate that WPC may deplete tumor cells of GSH and render them less resistant to chemotherapy (Kennedy et al., 1995).

Recently, Bumrungpert et al. (2018) investigated the effects of whey protein supplementation on nutritional status, immunity, GSH levels, and inflammatory markers in cancer patients. A total of 42 cancer patients who received intravenous chemotherapy were randomized in a double-blind controlled trial at the National Cancer Institute in Thailand. Patients received 40 g of WPI plus zinc and selenium (intervention group, $n = 23$) or a maltodextrin oral snack (control group, $n = 19$) every day for 12 weeks. The authors reported a significant increase in albumin (2.9%) and IgG (4.8%), as well as a time-dependent increase in GSH (11.7%) in the whey protein group. This study also suggests that whey protein can improve the immunity of cancer patients undergoing chemotherapy.

Reduced muscle skeletal tissue and lean body mass during chemotherapy correspond with toxicity and poorer survival expectancy. As a result, sarcopenia (the loss of both skeletal muscle mass and function) is becoming a major consideration in chemotherapy response. Mazzuca et al. (2019) performed an exploratory, placebo-controlled, randomized study to evaluate the activity ProLYotin® (which consists of highly purified whey proteins, is casein free, and has a lactose content<2%) in 47 colorectal cancer patients undergoing 5-fluorouracil-based chemotherapy. The patients were assessed physically and nutritionally before chemotherapy and after 3 and 6 months of treatment using a variety of tests. This study suggests that whey protein may prevent the onset of sarcopenia and malnutrition and prevent severe toxicity during chemotherapy.

## 2.5 Whey protein applications in osteoporosis therapy

Osteoporosis is a systemic skeletal disease characterized by decreased bone mineral density (BMD) and deterioration of the microarchitecture of bone tissue with a consequent increase in bone fragility and susceptibility to fracture (WHO, 2004). Several studies have investigated the effects of whey protein on bone health (reviewed by Fardellone et al., 2017). In mice, whey proteins diminished bone turnover (Takada et al., 1997), promoted bone formation and suppressed bone resorption (Toba et al., 2001), and enhanced fracture healing (Yoneme et al., 2015). Whey protein intake was also shown to increase lumbar spine BMD in postmenopausal women (Aoe et al., 2005) and lean mass in older women (Thomas et al., 2015). *In vitro* and *in vivo* studies in mice demonstrated that bovine angiogenin purified from milk basic protein (MBP), which is found in whey, strongly inhibited osteoclast activity which are cells capable of degrading the bone matrix (Morita et al., 2008). Daily doses of MBP 40 mg (equivalent to 400–800 mL of milk) have been deemed sufficient to result in significantly increased BMD and reduced bone resorption (Uenishi et al., 2007).

## 2.6 Whey protein as sports supplements

Resistance and endurance exercise can result in a net negative protein balance. If the body, both pre- and post-exercise, lacks sufficient levels of AAs, side effects including muscle wasting and delayed exercise recovery can result (Wilborn et al., 2013). Therefore, many individuals, particularly athletes, use protein supplements to match the added demands associated with training or competition. Ali et al. (2019) recently reviewed the positive effects of whey protein and resistance exercise on muscle mass, strength performance, and other metabolic changes in trained and untrained humans and animals. The results of many meta-analyses indicate that significantly larger improvements in muscle mass and strength are observed with whey protein ingestion compared to control supplements when used in combination with prolonged resistance training interventions of more than 4 weeks (Cermak et al., 2012, Miller et al., 2014, Naclerio and Larumbe-Zabala, 2016).

## 2.7 Applications of whey protein on infant health

Bovine whey contains many important components similar to those found in human milk and is therefore an important ingredient in a variety of IF, including formula developed for premature infants. While breastfeeding is the gold standard (WHO, 2003), IF-containing whey protein–based ingredients are used as a substitute when breastfeeding is not an option (reviewed by Fenelon et al., 2019). Whey-predominant and whey hydrolysate formulas have been found to aid the infant immune system and influence the infant gut microbiota (see Section 2.1). Whey protein hydrolysate formulas are known to be less allergenic when compared to a standard IF and may decrease the risk of later allergy development in the infant (Szajewska et al., 2004). A meta-analysis review suggests that the frequency of atopic dermatitis was substantially lower among infants who received partially hydrolyzed whey protein-based formula compared to those receiving the intact protein-based formula (Alexander et al., 2010).

Hegar et al. (2021) more recently reviewed the effects of partially hydrolyzed whey protein in preventing allergy, especially atopic disease, and managing functional GI disorders. Several clinical studies showed that partially hydrolyzed whey protein decreases the incidence of eczema. An example is the 15-year follow-up of the German Infant Nutritional Intervention Study (Von Berg et al., 2016). A reduction in the cumulative incidence of eczema and allergic rhinitis was evident with the feeding of partially hydrolyzed whey protein formula compared to the control (cows' milk formula). Partially hydrolyzed whey protein was also found to be useful in the management of functional GI disorders such as regurgitation (Vandenplas et al., 2013), constipation (Vandenplas et al., 2019), and colic (Xinias et al., 2017).

Clinical evidence suggests that whey protein can also increase infant muscle mass (Hoppe et al., 2008), which may support a healthy body composition. Whey protein contains AAs with a very comparable make-up to muscle proteins and a significantly higher proportion of BCAAs as mentioned, which may enhance muscle protein synthesis (Mølgaard et al., 2011). Whey protein also leads to increased whole-body protein synthesis when compared to casein since an important AA in protein metabolism, leucine, is present at higher levels in whey protein when compared to casein (Mølgaard et al., 2011). Studies also suggest that whey protein can support gastric emptying. For instance, Billeaud et al. (1990) performed a study with 201 infants (age: 0–1 year) to examine the impact of milk composition and the influence of gastroesophageal reflux on gastric emptying. Infants were fed various milks (e.g. human milk, whey-predominant formula, casein-predominant formula, follow-up formula, whey hydrolysate formula, or cows' milk). The results demonstrated that gastric emptying took less time with ingestion of whey-predominant formulas when compared with casein-predominant formulas. It should be noted, however, that in a study on preterm infants, gastric emptying rates for casein- and whey-dominant formulas were similar (Thorkelsson et al., 1994). Some studies suggest that partially hydrolyzed formulas may be superior to intact protein formulas for infants with GI issues (Vandenplas et al., 2014, Savino et al., 2003).

## 3 Major whey proteins and peptides

During recent years, whey fractions have gained significant attention due to their associated health benefits. Individual proteins derived from whey have been reported to hold potential biotechnological promise for use in the health sector. Studies that have focused on the health benefits of individual major and minor whey proteins are summarized in Table 2 and described in the following sections.

### *3.1 α-Lactalbumin*

α-Lactalbumin (α-LA) is a small (Mr 14.2 kDa), $Ca^{2+}$-binding, acidic (pI 4.2–4.5) protein, which is a regulatory component of the lactose synthase enzyme system in the lactating mammary gland (Permyakov, 2020). α-LA makes up approximately 22% of total protein and 36% of the whey protein in human milk, while in bovine milk, α-LA constitutes 3.5% of total protein and 17% of whey protein (Layman et al., 2018). Human and bovine α-LA have similar AA compositions, sharing 74% sequence homology and similar biological properties (Kamau et al., 2010). Due to α-LA being a rich source of EAAs, particularly tryptophan, lysine, and cysteine, the food industry has shown an increasing interest in optimizing α-LA isolation methods and developing novel foods intended for use by target consumer groups such as infants and those suffering from various health disorders. Advances in milk separation technology have resulted in processes that yield whey protein fractions with a substantially higher concentration of α-LA than standard bovine whey. These α-LA-enriched WPCs are available commercially and produced through filtration. Highly purified α-LA can be obtained by ion-exchange methods, accounting for more than 93% of total protein (Layman et al., 2018). The role α-LA plays in infant development was recently discussed as part of a review by Almeida et al. (2021) while Layman et al. (2018) discusses its role in infant and adult nutrition. α-LA is important in infant growth and development (Sandström et al., 2008) and

**Table 2** Health-promoting properties of individual whey proteins and peptides

| Protein | Potential benefits to general health | Reference |
|---|---|---|
| α-lactalbumin | Antiproliferative effects in human adenocarcinoma cell lines, anti-cancer/tumoricidal properties | Ho et al. (2012), Lajnaf et al. (2020), Rammer et al. (2010) |
| | Source of branched-chain and essential amino acids, e.g. tryptophan, which stimulates numerous brain responses | Steinberg et al. (1992) |
| | Antioxidant and antibacterial activity | Jaziri et al. (1992), Berthou et al. (1987), Brück et al. (2006), Pellegrini et al. (1999), Berkhout et al. (1997), Oevermann et al. (2003) |
| β-lactoglobulin | Increasing fatty acid absorption | Perez et al. (1992) |
| | Provides immune tolerance when loaded with iron-quercetin complex | Roth-Walter et al. (2021) |
| | Bind ligands and transport small hydrophobic ligands, vitamins A and D, palmitic acid, and other hydrophobic compounds | Pérez and Calvo (1995), Le Maux et al. (2014) |
| Immunoglobulins | Immunomodulatory and anti-inflammatory | Reviewed by Ulfman et al. (2018), Detzel et al. (2015), Adar et al. (2012) |
| | Prebiotic and *in vitro* priming of intestinal surface for colonization by bifidobacteria | Bojsen et al. (2007), Morrin et al. (2020b) |
| | Antiviral/antibacterial | Korhonen et al. (2000), den Hartog et al. (2014), Arenas et al. (2021) |
| Glycomacropeptide | Used as a source of protein in phenylketonuria patients | Etzel (2004), Lim et al. (2007), Ney et al. (2016), Singh et al. (2014) |
| | Human colorectal cancer treatment | Gong et al. (2014) |
| | Antimicrobial and antiviral | Nakajima et al. (2005), Feeney et al. (2017) |
| | Prebiotic effect on beneficial bacteria | O'Riordan et al. (2018), Azuma et al. (1984) |
| | Immunomodulatory | Otani et al. (1996); Sawin et al. (2015) |
| Serum albumin | Source of essential amino acids and binds fatty acids | Gupta et al. (2012), Kassem (2015), van der Vusse (2009) |
| | Applications in delivery of antibacterial treatments | Xue et al. (2021) |

*(Continued)*

**Table 2** (*Continued*)

| Protein | Potential benefits to general health | Reference |
|---|---|---|
| Lactoferrin | Represents a useful vehicle for iron supplementation and possible treatment for anemia and inflammation | Artym et al. (2021) |
| | Prebiotic | Hu et al. (2012) |
| | Iron-binding properties | Fransson et al. (1983), Morel et al. (2022) |
| | Bacteriostatic, bactericidal, and anti-adhesion effects against Gram-positive and Gram-negative bacteria | Zhang et al. (2014), Jahani and Jahani (2015), Rybarczyk et al. (2017), Kell et al. (2020), Berding et al. (2016), Hu et al. (2012) |
| | Promotion of wound healing and inhibition of infection | Takayama and Aoki (2012) |
| Osteopontin | Immunological signaling | Christensen et al. (2020), Ashkar et al. (2000) |
| | Intestinal cell proliferation | Buccigrossi et al. (2007) |
| | Intestinal mucosa protection | Woo et al. (2019) |
| | Cognitive development | Jiang et al. (2019) |
| Enzymes | Antibacterial, antiviral, and applications in delivering antibacterial treatments | Wang et al. (2021), Xue et al. (2021), Reiter and Harnulv (1984) |
| Folate-binding protein | Sequester folate from blood plasma | Rosenberg and Selhub (2006), Ford et al. (1975), Tani and Iwai (1984), Salter and Blakeborough (1988) |
| PP3 | Immunostimulatory and proteolysis results in antibacterial peptides | Sugahara et al. (2005), Campagna et al. (2004), Pedersen et al. (2012) |
| | Prebiotic | Etienne et al. (1994) |

may also aid in the regulation of sleep-wake cycles in infants through its high tryptophan content (Steinberg et al., 1992). Studies evaluating the impact of α-LA on gut development in infants have mainly focused on changes in the gut microbiome and on antibacterial activity (Lönnerdal and Lien, 2003, Layman et al., 2018).

α-LA and some of its fragments have bactericidal and antiviral activities. Numerous peptides are likely released during digestion of α-LA, which are also likely to possess these activities (Brück et al., 2006). A tripeptide (Gly-Leu-Phe) with immunostimulatory activity was isolated by *in vitro* digestion of milk proteins using trypsin and was found to be present in α-LA. The peptide stimulates phagocytic activity of macrophages *in vitro* and confers protection against *Klebsiella pneumoniae* infection in mice (Berthou et al., 1987). It also stimulates phagocytosis by human macrophages (Jaziri et al., 1992). Other α-LA peptides have been shown to increase the growth of bifidobacteria (Kee et al., 1998). Three polypeptide fragments from α-LA were found to have antimicrobial activity against a range of pathogens (Pellegrini et al., 1999). *In vitro* studies have also shown that α-LA-derived peptides had antiviral properties against herpes simplex virus type 1 and human immunodeficiency virus type 1 (Berkhout et al., 1997, Oevermann et al., 2003).

In adults, α-LA may have potential as a supplement to enhance neurological function and sleep, given its uniquely high tryptophan content (Markus et al., 2005). Other AAs of α-LA that may be significant in nutritional supplements include leucine, a BCAA, which stimulates protein accretion in skeletal muscle (Breen and Phillips, 2011). α-LA may also have potential in cancer treatments. Complexes of partially unfolded human α-LA with oleic acid known as HAMLET (Human Alpha-Lactalbumin Made Lethal to Tumor) have been shown to be cytotoxic to various tumor and bacterial cells (Ho et al., 2017). Similarly, BAMLET (Bovine Alpha-Lactalbumin Made Lethal to Tumor), a complex of bovine α-LA and oleic acid, was also found to kill tumor cells, and both HAMLET and BAMLET seem to penetrate the cell membrane and induce a pathway for lysosomal cell death (Rammer et al., 2010). In terms of non-nutritional applications, α-LA preparations have physical properties similar to WPIs, including clean flavor profile, heat stability, high protein quality, and high solubility in water across a pH range of 2.0–9.0, making them suitable for formulating beverages. α-LA preparations are also suited to the development of foams, emulsions, and gels and provide flexibility in product formulation (Layman et al., 2018).

### 3.2 β-Lactoglobulin

β-Lactoglobulin (β-Lg) is the major whey protein in bovine milk, making up more than 50% of total whey protein (Marshall, 2004). It predominantly occurs as a dimer with a molecular weight of 36 kDa, at the pH typical of bovine milk (Kontopidis et al., 2004), and dissociates into a monomer below a pH of 3. The complete AA sequence (162 AAs) of β-Lg is known, and 12 genetic variants of bovine β-Lg have been identified, the most common being A and B (Creamer et al., 1983, Rachagani et al., 2006). β-Lg is not detected in human milk, and the function of bovine β-Lg is not fully defined. β-Lg is believed to bind ligands and transport small hydrophobic ligands based on its structure (Pérez and Calvo, 1995). It is capable of binding to vitamins A and D, palmitic acid, and other hydrophobic compounds because of its multiple ligand-binding sites. Its calyx structure forms a central hydrophobic cavity that binds to these molecules (reviewed by Le Maux et al., 2014). Putative biological roles include increasing fatty acid absorption (Perez et al., 1992), protection of sensitive ligands against oxidation (Futterman and Heller, 1972) or other stresses, modification of the kinetics of the enzymatic hydrolysis of the protein (Puyol et al., 1993, Mandalari et al., 2009), and modification of the bioaccessibility of the ligands (Puyol et al., 1995, Riihimäki-Lampén, 2009). The binding properties of β-Lg–ligand complexes in food products may be affected by the structure of β-Lg and/or the presence of other proteins, which may compete with β-Lg for ligand binding (Le Maux et al., 2014).

Interestingly, β-Lg shares a high (83%) AA homology with human glycodelin A, which is involved in maintaining the feto-maternal immune system (van Cong et al., 1991). Glycodelin A suppresses all main immune cells, including both Th1 and Th2 responses (Scholz et al., 2008, Oggé et al., 2011). In addition, monoclonal antibodies raised against β-Lg cross-react with glycodelin A (Dutta et al., 1998). A recent review by Sawyer (2021) provides a general overview of the two proteins in the context of them belonging to the lipocalin family. Also of note, β-Lg is considered a major milk allergen, to which the majority of individuals with milk allergy are sensitized (Lam et al., 2008). Importantly, despite comparable milk consumption, only certain individuals will become allergic to milk, strongly suggesting that cofactors such as their ligands might be key in the sensitization process. Recently, Roth-Walter et al. (2021) demonstrated that β-Lg provides immune tolerance when loaded with iron-quercetin complex. The study suggested that proper loading of β-Lg prevents the antigenicity of the cow's milk allergen itself, by both masking important immune epitopes and shuttling ligands to immune cells that downregulate their antigen presentation skills. The authors suggest that spiking β-Lg with natural ligands such as quercetin or retinoic acid may pave the way to a new generation of prophylactic or therapeutic approaches against the allergy (Roth-Walter et al., 2021).

## 3.3 Immunoglobulins

Immunoglobulins (Igs) account for 10% of whey proteins in bovine milk (Feeney et al., 2019b). They are some of the most abundant bioactive proteins ranging from 61.4 g/L in colostrum to 0.8 g/L in mature bovine milk (Wheeler et al., 2007). The major Ig classes are IgA, IgG, and IgM (Korhonen and Marnila, 2009) with IgG being predominant in bovine milk, unlike human milk, where IgA predominates (Butler et al., 1986). The monomeric basic structure of all Igs is similar and is composed of four subunits of polypeptides in a Y-shaped formation. Igs consist of two identical Fab 'arms' and an Fc stem, with the Fab regions containing the antigen-binding sites. Elimination of the antigen is mediated by the Fc region through the activation of the host immune response (Recio et al., 2009). The main function of Igs is to neutralize toxins, agglutinate bacteria, inactivate viruses, and stimulate the growth of health-promoting bacteria (Bojsen et al., 2007, Kvistgaard et al., 2004). Igs in milk therefore give immunological protection against some infections. Many studies (reviewed by

Jasion and Burnett, 2015) have demonstrated that Ig preparations derived from human sera, bovine colostrum and sera survive digestion and are present in fecal material. Cross-species reactivity between human and bovine immune-related milk proteins has been shown (den Hartog et al., 2014, van Neerven et al., 2012), and bovine milk-derived IgG binds to GI pathogens that also infect humans (Korhonen et al., 2000, den Hartog et al., 2014). Recently, Arenas et al. (2021) hypothesized the use of IgG in controlling COVID-19 since COVID-19 and bovine coronavirus are very close phylogenetically. The authors suggest the use of bovine coronavirus immune milk in controlling COVID-19. The immune milk would be obtained by cow hyperimmunization using a commercial coronavirus vaccine. For a more extensive review on the effects of bovine Igs on immune function, allergy, and infection in humans, we refer the reader to Ulfman et al. (2018) where clinical studies focused on various human cohorts are summarized.

Strategies aimed at increasing Ig concentrations in milk have been recently reviewed by Hine et al. (2019). For example, studies in dairy cows have identified cows, which are high antibody-mediated immune responders, based on antigen-specific serum antibody responses to antigens. These cows have higher concentrations of total IgG in their colostrum relative to their average and low immune responder counterparts (Fleming et al., 2016). In turn, isolation of Igs from whey has been accomplished using membrane filtration and chromatography alone or in combination. Mehra and Kelly (2004) used a 0.1 μM microfiltration membrane and selective membrane separation using pH variation to produce an Ig-rich fraction from cheese whey. Korhonen et al. (1998) employed a multistep process with a series of membrane steps (UF, microfiltration, reverse osmosis) followed by a cation exchange step for concentration of Igs from colostral whey. This process resulted in final freeze-dried concentrates with levels of Ig ranging from 45% to 75%. Indeed, several chromatographic techniques can be used for purification of Igs from bovine milk and dairy streams, including size exclusion, ion exchange, expanded bed adsorption, immobilized metal affinity chromatography, metal chelate ion, hydroxyapatite, hydrophobic charge induction chromatography, and most commonly affinity chromatography (Feeney et al., 2019b). Heidebrecht et al. (2018) isolated bovine IgG from milk- and colostral whey using mixed-mode chromatography at lab and pilot scale. Serial connection of two mixed-mode resins resulted in an ingredient with an average purity of 96.1% IgG. Given the progress made in the separation and isolation of Igs from bovine colostrum and cheese whey, IgG products are manufactured at a large scale. The use of immune milks and Ig concentrates as nutritional supplements most likely leaves the normal gut microflora unchanged and provides a strategy to prevent infections, which is of major importance considering the emerging antibiotic resistance crisis.

### 3.4 Caseinomacropeptide

CMP is a highly polar casein-derived peptide which as mentioned is found in sweet whey. The addition of chymosin to milk during cheese making enzymatically hydrolyzes the milk protein *k*-casein into two peptides, an insoluble peptide (*para*-kappa casein) and a soluble hydrophilic peptide (CMP) which can be glycosylated (GMP) as mentioned. *para*-kappa-casein contains the AA residues 1–105 and is coagulated and precipitates into the cheese curd. CMP, the smaller of the two peptides, contains the AA residues 106–169 and is released into the whey (Walstra et al., 2005). CMP is the third most abundant whey protein, after β-Lg and α-LA, constituting approximately 12% of total whey protein (Thomä-Worringer et al., 2006, Layman et al., 2018). CMP has unique properties given that phenylalanine, tryptophan, tyrosine (Tyr), cysteine, histidine, and arginine residues are absent (Neelima et al., 2013). The peptide is rich, however, in BCAAs, such as isoleucine and valine (Marshall, 2004, Krissansen, 2007). The role of CMP in maintaining and improving human health was recently summarized by Feeney et al. (2019a) and Córdova-Dávalos et al. (2019). The effects of CMP include antibacterial, prebiotic, remineralizing, modulating of gastric hormone secretion and metabolism, inhibitory of gastric secretion, anti-tumoral, and immunomodulatory activities. One of the most documented properties associated with CMP is in the management of phenylketonuria (PKU) which is an autosomal recessive disorder caused by mutations in the phenylalanine hydroxylase gene. This gene encodes an enzyme that catalyzes the conversion of phenylalanine (Phe) to Tyr (Blau et al., 2010). This enzyme deficiency in untreated PKU leads to an accumulation of Phe, resulting in toxic concentrations in the brain (Vockley et al., 2014). Without treatment, individuals develop irreversible intellectual disability The main treatment for PKU is a lifelong low-Phe diet with strict control over Phe intake from natural foods. The main source of protein in the diet of PKU patients is usually supplied through special medical formulas that supply vitamins, minerals, and a mixture of EAAs that lack Phe (MacLeod et al., 2009, Singh et al., 2014). A variety of palatable medical CMP-fortified foods that are high in protein content and low in Phe are now available (Etzel, 2004, Lim et al., 2007). CMP is susceptible to enzymatic hydrolysis during digestion, although it can also be absorbed intact and distributed by blood vessels, or it can reach the large intestine to be fermented by microbiota (Córdova-Dávalos et al., 2019).

### 3.5 Serum albumin

Serum albumin is a large protein present in human and bovine milk and constitutes approximately 5–6% of total whey protein (Layman et al., 2018). It is a source of EAAs, but there is very little information available regarding its biological function (Gupta et al., 2012, Kassem, 2015). Considering its properties in milk are similar to those in blood, it is thought that it may not be synthesized by the mammary gland (Lönnerdal, 2003) and instead is believed to be transferred from the maternal circulation. Bovine serum albumin is known to bind fatty acids as well as other small molecules (Fletcher et al., 1971, van der Vusse, 2009).

## 4 Minor whey proteins and peptides

### *4.1 Lactoferrin*

Lactoferrin (Lf) belongs to the family of transferrin proteins and is a glycosylated 80-kDa iron-binding protein (reviewed by Superti, 2020). Lf is the main iron-binding protein in human milk and the second most abundant protein in human milk, with concentrations ranging from 2 to 6 g/L depending on the lactation stage (Rai et al., 2014). In bovine milk, concentrations are much lower and range from 0.8 g/L in colostrum to 0.1 g/L in mature milk (Sánchez et al., 1988). Human and bovine Lf share a high homology (77%) which suggests that supplementation of bovine Lf to IF may confer similar beneficial effects to that of Lf in human milk (Nguyen et al., 2016). Bovine Lf is known to be involved in a variety of biological functions, which include antioxidant, anti-inflammatory, anti-tumor, and antimicrobial activities (Mayeur et al., 2016, Zhang et al., 2014, Kruzel et al., 2017, Kell et al., 2020). Large-scale preparation of bovine Lf from cheese whey or skim milk has ensured this protein has been commercialized as an ingredient incorporated into health products for humans and animals (Tomita et al., 2002, Plate et al., 2006, Kete et al., 2018). Apart from IF, Lf has also been used for the supplementation of probiotic foods, skin care and oral care products, and as a nutraceutical, aimed at improving the immune system and inhibiting inflammatory responses (reviewed by Superti, 2020). Lf is also commercially used for food conservation and safety as it delays lipid oxidation and inhibits microbial growth (Medina et al., 2002, Duran and Kahve, 2017). The impact of Lf on intestinal maturation and the microbiota has been addressed in clinical trials of populations at risk of life-threatening sepsis, such as pre-term and full-term newborn infants (Manzoni, 2016). Studies conducted in piglets as an animal model of neonatal diseases including sepsis have reported the *in vivo* effects of Lf on the growth and diversity of the piglet intestinal microbiota. The benefits of Lf on the intestinal microbiota (reviewed by Vega-Bautista et al., 2019) may occur through several mechanisms, including (1) favoring the growth of beneficial organisms and inhibiting the colonization and proliferation of enteric pathogens (Berding et al., 2016, Hu et al., 2012), (2) gut-brain-microbiota axis signaling for the maturation of nerve fibers (Berding et al., 2016, Yang et al.,

2014), (3) intestinal cell maturation and differentiation, and (4) modulating immune and inflammatory responses. It should be noted, however, that Lf undergoes significant physicochemical changes when thermally processed (Goulding et al., 2021a). *In vitro* experiments have shown that these changes in Lf negatively impact some of its biological activities, for example, simulated infant gastric digestion, bactericidal, and anti-inflammatory activity (Goulding et al., 2021b). This is an important consideration for Lf applications in IF where high temperature-short time thermal processing conditions are commonly used.

### 4.2 Osteopontin

Osteopontin (OPN) consists of 300 AAs and is a phosphorylated glycoprotein with a high proportion of acidic AAs. It is a multifunctional protein involved in a wide range of bioactivities, including immunomodulation, cell proliferation and differentiation, biomineralization, as well as myelination (reviewed by Jiang and Lönnerdal, 2020). In a recent large multicenter study across a number of countries (China, Japan, the Republic of Korea, and Denmark), the human milk OPN concentration was found to vary from 99.7 to 266.2 mg/mL (Bruun et al., 2018). The median OPN level in bovine milk was recently determined as 21.9 mg/L (range of 0.4–67.8 mg/L) in a study of 661 Danish Holstein cows (Christensen et al., 2021). Standard IF contains on average 9 mg/L OPN (Schack et al., 2009). Human and bovine milk OPN are structurally homologous with regard to AA sequence and phosphorylation pattern, and both proteins are O-glycosylated in the same region of the protein. Methods for isolating OPN from bovine milk for use in IF have been developed, and bovine milk OPN is commercially available. Numerous studies have examined the effects of OPN when ingested (reviewed by Christensen and Sørensen, 2016). These studies suggest that milk OPN may play important roles in intestinal, brain, and immune system development in infants (Christensen and Sørensen, 2016, Jiang and Lönnerdal, 2016, Jiang et al., 2019).

### 4.3 Lactoperoxidase

Whey contains several enzymes such as oxidoreductases, transferases, lyases, hydrolases, isomerases, and ligases (O'Mahony et al., 2013). One of the most abundant enzymes in whey is lactoperoxidase (LPO) that accounts for approximately 1% of total whey protein and is a member of the peroxidase-cyclooxygenase superfamily. Bovine milk LPO is a calcium- and iron-containing glycoprotein, consisting of a single polypeptide chain of 612 AA residues with a molecular weight of 78 kDa (Davies et al., 2008). This enzyme reacts with thiocyanate ($SCN^-$) in the presence of hydrogen peroxide and forms antimicrobial compounds; this is termed the LPO system (Kennedy et al., 2000). The biological significance of the system is to protect the lactating mammary gland and the intestinal tract of newborn infants, thereby providing a host defense system against potential pathogens (Reiter and Harnulv, 1984). Novel methods for extraction of and purification of LPO from whey have been developed (Borzouee et al., 2016, Priyanka and Rastogi, 2018).

### 4.4 Proteose peptone component 3

Proteose peptone component-3 (PP3) or lactophorin is a hydrophilic phosphoglycoprotein not found in human milk but found at concentrations of 300 mg/L in bovine milk (Sørensen and Petersen, 1993). PP3 is the bovine homolog of murine glycosylation-dependent cell adhesion molecule-1 (Girardet et al., 1995). Sørensen et al. (1997) not only confirmed the presence of PP3 in the milk fat globule membrane (MFGM) but also detected PP3 in the whey fraction suggesting PP3 is weakly associated with the MFGM and is readily lost into the whey fraction. The peptide backbone of PP3 has been shown *in vitro* to have antibacterial activity (Campagna et al., 2004) and prebiotic properties (Etienne et al., 1994). PP3 has also been shown to have mitogenic activity in DNA synthesis in MARK3 hybridomas, a property which is lost following neuraminidase treatment, suggesting that the glycans on PP3 are involved in influencing hybridoma mitosis (Mati et al., 1993).

### 4.5 Folate binding protein

Milk contains 160–210 nmol/L of folate-binding proteins (FBPs; Forssén et al., 2000) which can bind folates at a ratio of 1:1 (Yaman et al., 2019). Bovine FBP has 222 AAs and contains eight disulfide bridges. The molecular weight is approximately 30–35 kDa (Svendsen et al., 1979). Studies have suggested that FBP prevents the uptake of folate by intestinal bacteria, protects folate from degradation at body temperature, and assists in the transport of folate to the intestinal mucosa (Ford et al., 1975, Tani and Iwai, 1984, Salter and Blakeborough, 1988). Additionally, FBP plays an important role in increasing the bioavailability of folates (Swiatlo et al., 1990). Many different methods (reviewed by Nygren-Babol and Jägerstad, 2012) with varying success exist for measuring FBP in milk and dairy products.

## 5 Case study: application of whey enriched in immunoglobulin G

Bifidobacteria and lactobacilli are the most widely used probiotics in humans, and several articles have reviewed their mechanisms of action, which include modulation of the immune system, induction of anti-inflammatory and antioxidant responses, production of antimicrobial substances, as well as competitive pathogen exclusion (Yousefi et al., 2019, Plaza-Diaz et al., 2019). These bacteria not only should survive GI transit, but they also need to colonize the GI tract. An effective probiotic must reside at the preferred target sites in the GI tract long enough, and at adequate concentrations, to confer probiotic effects. To the best of our knowledge, there are currently no commercial products available that aid or increase the attachment of health-promoting bacteria to the gut mucosal surface despite the growing market for probiotics.

The intake of colostrum and milk may be expected to have the greatest impact on microbial colonization in early life. Indeed, recent *in vitro* studies suggest that milk components may modulate the ability of bacteria to adhere to intestinal cells and also modulate the surface properties of the intestinal epithelium, priming the cell surface for colonization (Morrin et al., 2019a,b, Morrin and Hickey, 2020a, Morrin et al., 2020b). It was recently shown that bovine colostrum influences the intestinal cell surface and results in the attachment of bifidobacteria (Morrin et al., 2019a). In the study, human HT-29 intestinal cells were exposed to a bovine colostrum fraction (BCF), which resulted in increased (up to 52-fold) cellular attachment of all health-promoting bacteria examined. Importantly, the changes to the HT-29 cell surface did not result in an increase in the attachment of the undesirable enteric pathogens tested, namely, *Escherichia coli*, *Salmonella enterica*, and *Listeria monocytogenes*. Transcriptomic analysis of the intestinal cells revealed a number of genes associated with glycan synthesis were differentially expressed (Morrin et al., 2019a). Proteomic and glycomic analysis of the intestinal cells following treatment with BCF confirmed that the cell surface was altered due to changes in the levels of various glycoproteins present and through enzymatic addition of sugars to glycoconjugates. Subtle changes occurred in relation to the terminal carbohydrate moieties at the cell surface which were characterized mainly by increases in N-acetylglucosamine (Morrin et al., 2019b).

A subsequent study examined the main components of the BCF to investigate the active component/s responsible for influencing the changes in the intestinal cells (Morrin et al., 2020b). After screening the components in the BCF, IgG was identified as the main driver of the changes to the intestinal cells. The effect of IgG was concentration dependent with 16 mg/mL characterized as the optimal concentration to use for bioactivity. The glycans of IgG were important in the novel activity as their inactivation through periodate treatment of IgG resulted in the loss of activity represented by a reduction in the adhesion of bifidobacteria to the intestinal cells. IgG isolated from bovine colostrum and mature milk displayed an effect on bifidobacterial adhesion although colostral IgG was over 10 times more effective compared to IgG from mature milk at the same concentration (Morrin et al., 2020b). A patent (P11847PC00) protecting this novel application for IgG was filed, and recently, whey streams enriched in IgG were also found to increase the adhesion of bifidobacteria to intestinal cells (data unpublished). These studies (Morrin et al., 2019a,b, 2020b) build a strong case for the inclusion of IgG ingredients sourced from whey in functional foods aimed at increasing the numbers of health-promoting bacteria in the human gut.

## 6 Future trends in research

Whey protein is a safe, high-quality protein source for food and medical applications worldwide. With the growth of the market for functional proteins, there is a need to continuously develop and invest in simple cost-effective processes for the isolation, purification, production, and scalability of whey proteins and peptides in adequate amounts to meet this global demand. Whey processors must continue to demonstrate the efficacy of whey protein and its individual constituents as functional ingredients. Analyzing the latest trends in whey processing, it seems that one of a pivotal and perspective directions of whey management is beverage production (Skryplonek and Jasińska, 2017). Production of whey-based beverages appears to be the most economical and simplest solution for whey utilization in human nutrition. The industrial production of whey-based beverages dates back to 1970s, and different products (e.g., unfermented and fermented beverages, diet beverages, alcoholic beverages, high-protein sports drinks) have been developed and are currently available on the market (Zotta et al., 2020). The use of whey protein however is not limited to nutritional beverages. Whey protein is incorporated into meat and meat products, ice creams and yogurts, cheeses, bakery products, confectionery and pastry products, many of which are reduced-fat products. Traditionally, whey protein is used as a functional additive for its solubility, gelling, emulsifying, foaming, and water-binding properties. However, the biological properties of whey proteins are now increasingly exploited for food and medical applications across various industries.

Indeed, never has the need for novel food-derived bioactives been so great with the emergence of the novel human coronavirus SARS-CoV-2 which has caused a worldwide pandemic of respiratory disease (COVID-19). Recently, a whey protein from human breastmilk was found to effectively inhibit both SARS-CoV-2 and a related pangolin coronavirus by blocking viral attachment and viral replication at entry and even post-entry (Fan et al., 2020). Moreover, the human whey protein inhibited infectious virus production. Notably, the bovine counterpart from whey protein also showed anti-coronavirus properties, as did commercial bovine formula milk. Looking to whey for a solution to COVID-19 may prove to be a promising approach.

## 7 Where to look for further information

- Deeth, H. C. and Bansal, N. (2019). *Whey Proteins: From Milk to Medicine*. London, UK: Academic Press.
- Pires, A. F., Marnotes, N. G., Rubio, O. D., Garcia, A. C. and Pereira, C. D. (2021). Dairy by-products: A review on the valorization of whey and second cheese whey. *Foods* 10, 1067. https://doi.org/10.3390/foods10051067.
- Costa, C., Azoia, N. G., Coelho, L., Freixo, R., Batista, P. and Pintado, M. (2021). Proteins derived from the dairy losses and by-products as raw materials for non-food applications. *Foods* 10, 135. https://doi.org/10.3390/foods10010135.
- Minj, S. and Anand, S. (2020). Whey proteins and its derivatives: Bioactivity, functionality, and current applications. *Dairy* 1, 233–258. https://doi.org/10.3390/dairy1030016.
- Zarić, D., Bulatovic, M. and Rakin, M. (2021). Effect of whey protein on satiety in overweight and obese individuals. In: *The Diversified Benefits of Whey Protein*, 1st Edition, Ed. Zachary Hicks, Nova Science Publishers, New York, pp. 51–74.
- Fox, P. and McSweeney, P. (2009). *Advanced Dairy Chemistry*. 10.1007/978-0-387-84865-5.
- Kassem, J. M. (2015). Future challenges of whey proteins. *International Journal of Dairy Science* 10, 139–159.

## 8 Acknowledgements

Geoffrey McCarthy was in receipt of a Teagasc Walsh Scholarship. This research was funded by Teagasc (Grant No. 2017049).

## 9 References

Adar, T., Ben Ya'acov, A., Lalazar, G., Lichtenstein, Y., Nahman, D., Mizrahi, M., Wong, V., Muller, B., Rawlin, G. and Ilan, Y. 2012. Oral administration of immunoglobulin G-enhanced colostrum alleviates insulin resistance and liver injury and is associated with alterations in natural killer T cells. *Clinical and Experimental Immunology* 167(2), 252–260.

Alexander, D. D., Schmitt, D. F., Tran, N. L., Barraj, L. M. and Cushing, C. A. 2010. Partially hydrolyzed 100% whey protein infant formula and atopic dermatitis risk reduction: A systematic review of the literature. *Nutrition Reviews* 68(4), 232–245.

Ali, A., Lee, S.-J. and Rutherfurd-Markwick, K. J. 2019. Chapter 16: Sports and exercise supplements. In: Deeth, H. C. and Bansal, N. (Eds), *Whey Proteins*. Cambridge, MA: Academic Press.

Almeida, C. C., Mendonça Pereira, B. F., Leandro, K. C., Costa, M. P., Spisso, B. F. and Conte-Junior, C. A. 2021. Bioactive compounds in infant formula and their effects on infant nutrition and health: A systematic literature review. *International Journal of Food Science* 2021, 8850080.

Amirani, E., Milajerdi, A., Reiner, Ž., Mirzaei, H., Mansournia, M. A. and Asemi, Z. 2020. Effects of whey protein on glycemic control and serum lipoproteins in patients with metabolic syndrome and related conditions: A systematic review and meta-analysis of randomized controlled clinical trials. *Lipids in Health and Disease* 19(1), 209–209.

Aoe, S., Koyama, T., Toba, Y., Itabashi, A. and Takada, Y. 2005. A controlled trial of the effect of milk basic protein (MBP) supplementation on bone metabolism in healthy menopausal women. *Osteoporosis International* 16(12), 2123–2128.

Arenas, A., Borge, C., Carbonero, A., Garcia-Bocanegra, I., Cano-Terriza, D., Caballero, J. and Arenas-Montes, A. 2021. Bovine coronavirus immune milk Against COVID-19. *Frontiers in Immunology* 12, 637152.

Artym, J., Zimecki, M. and Kruzel, M. L. 2021. Lactoferrin for prevention and treatment of anemia and inflammation in pregnant women: A comprehensive review. *Biomedicines* 9(8), 898.

Ashkar, S., Weber, G. F., Panoutsakopoulou, V., Sanchirico, M. E., Jansson, M., Zawaideh, S., Rittling, S. R., Denhardt, D. T., Glimcher, M. J. and Cantor, H. 2000. Eta-1 (osteopontin): An early component of type-1 (cell-mediated) immunity. *Science* 287(5454), 860–864.

Athira, S., Mann, B., Sharma, R. and Kumar, R. 2013. Ameliorative potential of whey protein hydrolysate against paracetamol-induced oxidative stress. *Journal of Dairy Science* 96(3), 1431–1437.

Attaallah, W., Yilmaz, A. M., Erdoğan, N., Yalçin, A. S. and Aktan, A. O. 2012. Whey protein versus whey protein hydrolyzate for the protection of azoxymethane and dextran sodium sulfate induced colonic tumors in rats. *Pathology Oncology Research* 18(4), 817–822.

Azuma, N., Yamauchi, K. and Mitsuoka, T. 1984. Bifidus growth-promoting activity of a glycomacropeptide derived from human K-casein. *Agricultural and Biological Chemistry* 48(8), 2159–2162.

Badr, G., Badr, B. M., Mahmoud, M. H., Mohany, M., Rabah, D. M. and Garraud, O. 2012a. Treatment of diabetic mice with undenatured whey protein accelerates the wound healing process by enhancing the expression of MIP-1α, MIP-2, KC, CX3CL1 and TGF-β in wounded tissue. *BMC Immunology* 13, 32.

Badr, G., Ebaid, H., Mohany, M. and Abuelsaad, A. S. 2012b. Modulation of immune cell proliferation and chemotaxis towards CC chemokine ligand (CCL)-21 and CXC chemokine ligand (CXCL)-12 in undenatured whey protein-treated mice. *Journal of Nutritional Biochemistry* 23(12), 1640–1646.

Badr, G., Mohany, M. and Metwalli, A. 2011. Effects of undenatured whey protein supplementation on CXCL12- and CCL21-mediated B and T cell chemotaxis in diabetic mice. *Lipids in Health and Disease* 10, 203–203.

Balmer, S. E., Scott, P. H. and Wharton, B. A. 1989. Diet and faecal flora in the newborn: Casein and whey proteins. *Archives of Disease in Childhood* 64(12), 1678 –1684.

Bansal, N. and Bhandari, B. 2016. Functional milk proteins: production and utilization–whey-based ingredients. In: Mcsweeney, P. L. H. and O'Mahony, J. A. (Eds), *Advanced Dairy Chemistry: Volume 1B: Proteins: Applied Aspects*. New York: Springer.

Barukčić, I., Lisak Jakopović, K. and Božanić, R. 2019. Valorisation of whey and buttermilk for production of functional beverages–An overview of current possibilities. *Food Technology and Biotechnology* 57(4), 448–460.
Berding, K., Wang, M., Monaco, M. H., Alexander, L. S., Mudd, A. T., Chichlowski, M., Waworuntu, R. V., Berg, B. M., Miller, M. J., Dilger, R. N. and Donovan, S. M. 2016. Prebiotics and bioactive milk fractions affect gut development, microbiota, and neurotransmitter expression in piglets. *Journal of Pediatric Gastroenterology and Nutrition* 63(6), 688–697.
Berkhout, B., Derksen, G. C., Back, N. K., Klaver, B., De Kruif, C. G. and Visser, S. 1997. Structural and functional analysis of negatively charged milk proteins with anti-HIV activity. *AIDS Research and Human Retroviruses* 13(13), 1101–1107.
Berthou, J., Migliore-Samour, D., Lifchitz, A., Delettré, J., floc'h, F. and Jollès, P. 1987. Immunostimulating properties and three-dimensional structure of two tripeptides from human and cow caseins. *FEBS Letters* 218(1), 55–58.
Billeaud, C., Guillet, J. and Sandler, B. 1990. Gastric emptying in infants with or without gastro-oesophageal reflux according to the type of milk. *European Journal of Clinical Nutrition* 44(8), 577–583.
Blau, N., van Spronsen, F. J. and Levy, H. L. 2010. Phenylketonuria. *Lancet* 376(9750), 1417–1427.
Bojsen, A., Buesa, J., Montava, R., Kvistgaard, A. S., Kongsbak, M. B., Petersen, T. E., Heegaard, C. W. and Rasmussen, J. T. 2007. Inhibitory activities of bovine macromolecular whey proteins on rotavirus infections *in vitro* and *in vivo*. *Journal of Dairy Science* 90(1), 66–74.
Boland, M., Golding, M. and Singh, H. 2014. *Food Structures, Digestion and Health*. Amsterdam, The Netherlands: Academic Press.
Borzouee, F., Mofid, M. R., Varshosaz, J. and Samsam Shariat, S. Z. A. 2016. Purification of lactoperoxidase from bovine whey and investigation of kinetic parameters. *Advanced Biomedical Research* 5, 189–189.
Bounous, G. and Molson, J. H. 2003. The antioxidant system. *Anticancer Research* 23(2B), 1411–1415.
Brandelli, A., Daroit, D. J. and Corrêa, A. P. F. 2015. Whey as a source of peptides with remarkable biological activities. *Food Research International* 73, 149–161.
Breen, L. and Phillips, S. M. 2011. Skeletal muscle protein metabolism in the elderly: interventions to counteract the 'anabolic resistance' of ageing. *Nutrition and Metabolism* 8, 68.
Brück, W. M., Kelleher, S. L., Gibson, G. R., Graverholt, G. and Lönnerdal, B. L. 2006. The effects of alpha-lactalbumin and glycomacropeptide on the association of CaCo-2 cells by enteropathogenic *Escherichia coli*, *Salmonella typhimurium* and *Shigella flexneri*. *FEMS Microbiology Letters* 259(1), 158–162.
Bruun, S., Jacobsen, L. N., Ze, X., Husby, S., Ueno, H. M., Nojiri, K., Kobayashi, S., Kwon, J., Liu, X., Yan, S., Yang, J., Zachariassen, G., Chen, L., Zhou, W., Christensen, B. and Sørensen, E. S. 2018. Osteopontin levels in human milk vary across countries and within lactation period: Data from a multicenter study. *Journal of Pediatric Gastroenterology and Nutrition* 67(2), 250–256.
Buccigrossi, V., De Marco, G., Bruzzese, E., Ombrato, L., Bracale, I., Polito, G. and Guarino, A. 2007. Lactoferrin induces concentration-dependent functional modulation of intestinal proliferation and differentiation. *Pediatric Research* 61(4), 410–414.

Bumrungpert, A. and Pavadhgul, P., Nunthanawanich, P., Sirikanchanarod, A. and Adulbhan, A. 2018. Whey protein supplementation improves nutritional status, glutathione levels, and immune function in cancer patients: A randomized, double-blind controlled trial. *Journal of Medicinal Food* 21, 612–616.

Butler, J. E., Seawright, G. L., Mcgivern, P. L. and Gilsdorf, M. 1986. Preliminary evidence for a diagnostic immunoglobulin G1 antibody response among culture-positive cows vaccinated with *Brucella abortus* strain 19 and challenge exposed with strain 2308. *American Journal of Veterinary Research* 47(6), 1258–1264.

Campagna, S., Mathot, A. G., Fleury, Y., Girardet, J. M. and Gaillard, J. L. 2004. Antibacterial activity of lactophoricin, a synthetic 23-residues peptide derived from the sequence of bovine milk component-3 of proteose peptone. *Journal of Dairy Science* 87(6), 1621–1626.

Cermak, N. M., Res, P. T., De Groot, L. C., Saris, W. H. M. and van Loon, L. J. C. 2012. Protein supplementation augments the adaptive response of skeletal muscle to resistance-type exercise training: A meta-analysis. *American Journal of Clinical Nutrition* 96(6), 1454–1464.

Cheng, S. H., Tseng, Y. M., Wu, S. H., Tsai, S. M. and Tsai, L. Y. 2017. Selective effects of whey protein concentrate on glutathione levels and apoptosis in rats with mammary tumors. *Food and Chemical Toxicology* 107(A), 440–448.

Christensen, B., Karlsen, N. J., Jørgensen, S. D. S., Jacobsen, L. N., Ostenfeld, M. S., Petersen, S. V., Müllertz, A. and Sørensen, E. S. 2020. Milk osteopontin retains integrin-binding activity after *in vitro* gastrointestinal transit. *Journal of Dairy Science* 103(1), 42–51.

Christensen, B. and Sørensen, E. S. 2016. Structure, function and nutritional potential of milk osteopontin. *International Dairy Journal* 57, 1–6.

Christensen, B., Zachariae, E. D., Poulsen, N. A., Buitenhuis, A. J., Larsen, L. B. and Sørensen, E. S. 2021. Factors influencing milk osteopontin concentration based on measurements from Danish Holstein cows. *Journal of Dairy Research* 88(1), 89–94.

Chungchunlam, S. M., Moughan, P. J., Henare, S. J. and Ganesh, S. 2012 Effect of time of consumption of preloads on measures of satiety in healthy normal weight women. *Appetite* 59(2), 281–288.

Córdova-Dávalos, L. E., Jiménez, M. and Salinas, E. 2019. Glycomacropeptide bioactivity and health: a review highlighting action mechanisms and signaling pathways. *Nutrients* 11(3).

Creamer, L. K., Parry, D. A. and Malcolm, G. N. 1983. Secondary structure of β-lactoglobulin. *Archives of Biochemistry and Biophysics* 227(1), 98–105.

Davies, M. J., Hawkins, C. L., Pattison, D. I. and Rees, M. D. 2008. Mammalian heme peroxidases: From molecular mechanisms to health implications. *Antioxidants and Redox Signaling* 10(7), 1199–1234.

den Hartog, G., Jacobino, S., Bont, L., Cox, L., Ulfman, L. H., Leusen, J. H. and van Neerven, R. J. 2014. Specificity and effector functions of human RSV-specific IgG from bovine milk. *PLoS ONE* 9(11), e112047.

Detzel, C. J., Horgan, A., Henderson, A. L., Petschow, B. W., Warner, C. D., Maas, K. J. and Weaver, E. M. 2015. Bovine immunoglobulin/protein isolate binds pro-inflammatory bacterial compounds and prevents immune activation in an intestinal co-culture model. *PLoS ONE* 10(4), e0120278.

Duran, A. and Kahve, H. I. 2017. The use of lactoferrin in food industry. *Academic Journal of Science* 7, 89–94.

Dutta, B., Mukhopadhyay, D., Roy, N., Das, G. and Karande, A. A. 1998. Cloning, expression, purification, and immunocharacterization of placental protein-14. *Protein Expression and Purification* 14(3), 327–334.

Etienne, L., Girardet, J. M. and Linden, G. 1994. Growth promotion of *Bifidobacterium animalis* by bovine milk proteose-peptone. *Lait* 74(5), 313–323.

Etzel, M. R. 2004. Manufacture and use of dairy protein fractions. *Journal of Nutrition* 134(4), 996S–1002S.

Fan, H., Hong, B., Luo, Y., Peng, Q., Wang, L., Jin, X., Chen, Y., Hu, Y., Shi, Y., Li, T., Zhuang, H., Zhou, Y. H., Tong, Y. and Xiang, K. 2020. The effect of whey protein on viral infection and replication of SARS-CoV-2 and pangolin coronavirus *in vitro*. *Signal Transduction and Targeted Therapy* 5(1), 275.

Fardellone, P., Séjourné, A., Blain, H., Cortet, B., Thomas, T. and GRIO Scientific Committee. 2017. Osteoporosis: Is milk a kindness or a curse? *Joint Bone Spine* 84(3), 275–281.

Feeney, S., Ryan, J. T., Kilcoyne, M., Joshi, L. and Hickey, R. M. 2017. Glycomacropeptide reduces intestinal epithelial cell barrier dysfunction and adhesion of entero-hemorrhagic and entero-pathogenic *Escherichia coli in vitro*. *Foods* 6(11), 93.

Feeney, S., Joshi, L. and Hickey, R. M. 2019a. Biological roles and production technologies associated with bovine glycomacropeptide. In: Hayes, M. (Ed.), *Novel Proteins for Food, Pharmaceuticals and Agriculture*, 1–50. New Jersey: Wiley Blackwell.

Feeney, S., Morrin, S. T., Joshi, L. and Hickey, R. M. 2019b. The role of immunoglobulins from bovine colostrum and milk in human Health Promotion. In: Hayes, M. (Ed.). *Novel Proteins for Food, Pharmaceuticals and Agriculture*. 291–314. , New Jersey: Wiley Blackwell.

Fenelon, M. A., Hickey, R. M., Buggy, A., Mccarthy, N. and Murphy, E. G. 2019. Whey proteins in infant formula. In: Deeth, H. C. and Bansal, N. (Eds), *Whey Proteins: From Milk to Medicine*, 439–494, London, UK; Massachusetts: Academic Press.

Fleming, K., Thompson-Crispi, K. A., Hodgins, D. C., Miglior, F., Corredig, M. and Mallard, B. A. 2016. Short communication: variation of total immunoglobulin G and β-lactoglobulin concentrations in colostrum and milk from Canadian Holsteins classified as high, average, or low immune responders. *Journal of Dairy Science* 99(3), 2358–2363.

Fletcher, J. E., Spector, A. A. and Ashbrook, J. D. 1971. Analysis of long-chain free fatty acid binding to bovine serum albumin by determination of stepwise equilibrium constants. *Biochemistry* 10(17), 3229–3232.

Ford, J. E., Scott, K. J., Sansom, B. F. and Taylor, P. J. 1975. Some observations on the possible nutritional significance of vitamin B12-and folate-binding proteins in milk. Absorption of [58Co]cyanocobalamin by suckling piglets. *British Journal of Nutrition* 34(3), 469–492.

Forssén, K. M., Jägerstad, M. I., Wigertz, K. and Witthöft, C. M. 2000. Folates and dairy products: A critical update. *Journal of the American College of Nutrition* 19(2), 100S–110S.

Frankowski, K. M., Miracle, R. E. and Drake, M. A. 2014. The role of sodium in the salty taste of permeate. *Journal of Dairy Science* 97(9), 5356–5370.

Fransson, G. B., Keen, C. L. and Lönnerdal, B. 1983. Supplementation of milk with iron bound to lactoferrin using weanling mice: L. Effects on hematology and tissue iron. *Journal of Pediatric Gastroenterology and Nutrition* 2(4), 693–700.

Freedman, D. J., Tacket, C. O., Delehanty, A., Maneval, D. R., Nataro, J. and Crabb, J. H. 1998. Milk immunoglobulin with specific activity against purified colonization factor antigens can protect against oral challenge with enterotoxigenic *Escherichia coli*. *Journal of Infectious Diseases* 177(3), 662–667.

Futterman, S. and Heller, J. 1972. The enhancement of fluorescence and the decreased susceptibility to enzymatic oxidation of retinol complexed with bovine serum albumin, beta-lactoglobulin, and the retinol-binding protein of human plasma. *Journal of Biological Chemistry* 247(16), 5168–5172.

Gilbert, J. A., Bendsen, N. T., Tremblay, A. and Astrup, A. 2011. Effect of proteins from different sources on body composition. *Nutrition, Metabolism, and Cardiovascular Diseases* 21, B16–31.

Girardet, J. M., Coddeville, B., Plancke, Y., Strecker, G., Campagna, S., Spik, G. and Linden, G. 1995. Structure of glycopeptides isolated from bovine milk component PP3. *European Journal of Biochemistry* 234(3), 939–946.

Gong, J., Chen, Q., Yan, Y. and Pang, G. 2014. Effect of casein glycomacropeptide on subunit p65 of nuclear transcription factor-κB in lipopolysaccharide-stimulated human colorectal tumor HT-29 cells. *Food Science and Human Wellness* 3(2), 51–55.

Goulding, D. A., O'Regan, J., Bovetto, L., O'Brien, N. M. and O'Mahony, J. A. 2021a. Influence of thermal processing on the physicochemical properties of bovine lactoferrin. *International Dairy Journal* 119, 105001.

Goulding, D. A., Vidal, K., Bovetto, L., O'Regan, J., O'Brien, N. M. and O'Mahony, J. A. 2021b. The impact of thermal processing on the simulated infant gastrointestinal digestion, bactericidal and anti-inflammatory activity of bovine lactoferrin - an *in vitro* study. *Food Chemistry* 362, 130142.

Gupta, C., Prof, D., Garg, A. and Gupta, S. 2012. Whey proteins: A novel source of bioceuticals. *Middle East Journal of Scientific Research* 12, 365–375.

Ha, D. J., Kim, J., Kim, S., Go, G. W. and Whang, K. Y. 2021. Dietary whey protein supplementation increases immunoglobulin G production by affecting helper T cell populations after antigen exposure. *Foods* 10(1), 194.

Hascoët, J. M., Hubert, C., Rochat, F., Legagneur, H., Gaga, S., Emady-Azar, S. and Steenhout, P. G. 2011. Effect of formula composition on the development of infant gut microbiota. *Journal of Pediatric Gastroenterology and Nutrition* 52(6), 756–762.

Hegar, B., Munasir, Z., Suryawan, A., Sidhiarta, IgL., Wati, K. D. K., Prafiantini, E., Irene, I. and Vandenplas, Y. 2021. Partially hydrolyzed whey protein: a review of current evidence, implementation, and further directions. *World Nutrition Journal* 5(1), 53.

Heidebrecht, H. J., Kainz, B., Schopf, R., Godl, K., Karcier, Z., Kulozik, U. and Förster, B. 2018. Isolation of biofunctional bovine immunoglobulin G from milk- and colostral whey with mixed-mode chromatography at lab and pilot scale. *Journal of Chromatography A* 1562, 59–68.

Hine, B. C., Hunt, P. W. and Colditz, I. G. 2019. Production and active transport of immunoglobulins within the ruminant mammary gland. *Veterinary Immunology and Immunopathology* 211, 75–84.

Ho, J. C. S., Rydström, A., Trulsson, M., Bålfors, J., Storm, P., Puthia, M., Nadeem, A. and Svanborg, C. 2012. HAMLET: Functional properties and therapeutic potential. *Future Oncology* 8(10), 1301–1313.

Ho, J. C. S., Nadeem, A. and Svanborg, C. 2017. Hamlet–A protein-lipid complex with broad tumoricidal activity. *Biochemical and Biophysical Research Communications* 482(3), 454–458.

Hoffman, J. R. and Falvo, M. J. 2004. Protein–Which is best? *Journal of Sports Science and Medicine* 3(3), 118–130.

Hoppe, C., Andersen, G. S., Jacobsen, S., Mølgaard, C., Friis, H., Sangild, P. T. and Michaelsen, K. F. 2008. The use of whey or skimmed milk powder in fortified blended foods for vulnerable groups. *Journal of Nutrition* 138(1), 145S–161S.

Hu, W., Zhao, J., Wang, J., Yu, T., Wang, J. and Li, N. 2012. Transgenic milk containing recombinant human lactoferrin modulates the intestinal flora in piglets. *Biochemistry and Cell Biology* 90(3), 485–496.

Jahani, S. S. A. and Jahani, L. 2015. The antimicrobial effect of lactoferrin on Gram-negative and Gram-positive bacteria. *International Journal of Infectious Diseases* 2(3), e27954.

Jasion, V. S. and Burnett, B. P. 2015. Survival and digestibility of orally-administered immunoglobulin preparations containing IgG through the gastrointestinal tract in humans. *Nutrition Journal* 14, 22–22.

Jaziri, M., Migliore-Samour, D., Casabianca-Pignède, M. R., Keddad, K., Morgat, J. L. and Jollès, P. 1992. Specific binding sites on human phagocytic blood cells for Gly-Leu-Phe and Val-Glu-Pro-Ile-Pro-Tyr, immunostimulating peptides from human milk proteins. *Biochimica et Biophysica Acta* 1160(3), 251–261.

Jiang, R. and Lönnerdal, B. 2016. Biological roles of milk osteopontin. *Current Opinion in Clinical Nutrition and Metabolic Care* 19(3), 214–219.

Jiang, R. and Lönnerdal, B. 2020. Effects of milk osteopontin on intestine, neurodevelopment, and immunity. *Nestle Nutrition Institute Workshop Series* 94, 152–157.

Jiang, R., Prell, C. and Lönnerdal, B. 2019. Milk osteopontin promotes brain development by up-regulating osteopontin in the brain in early life. *FASEB Journal* 33(2), 1681–1694.

Kamau, S. M., Cheison, S. C., Chen, W., Liu, X. and Lu, R. 2010. Alpha-lactalbumin: Its production technologies and bioactive peptides. *Comprehensive Reviews in Food Science and Food Safety* 9(2), 197–212.

Kareb, O. and Aïder, M. 2019. Whey and its derivatives for probiotics, prebiotics, Synbiotics, and functional foods: A critical review. *Probiotics and Antimicrobial Proteins* 11(2), 348–369.

Kassem, J. M. 2015. Future challenges of whey proteins. *International Journal of Dairy Science* 10(4), 139–159.

Kee, H., Hong, Y. H. and Kim, E. R. 1998. Effect of enzymatically hydrolyzed alpha-LA fractions with pepsin on growth-promoting of *Bifidobacterium longum* ATCC 15707. *Korean Journal of Dairy Science* 20, 61–68.

Kell, D. B., Heyden, E. L. and Pretorius, E. 2020. The biology of lactoferrin, an iron-binding protein that can help defend against viruses and bacteria. *Frontiers in Immunology* 11, 1221–1221.

Kennedy, M., O'Rourke, A. L., Mclay, J. and Simmonds, R. 2000. Use of a ground beef model to assess the effect of the lactoperoxidase system on the growth of *Escherichia coli* O157:H7, *Listeria monocytogenes* and *Staphylococcus aureus* in red meat. *International Journal of Food Microbiology* 57(3), 147–158.

Kennedy, R. S., Konok, G. P., Bounous, G., Baruchel, S. and Lee, T. D. 1995. The use of a whey protein concentrate in the treatment of patients with metastatic carcinoma: A phase I-II clinical study. *Anticancer Research* 15(6B), 2643–2649.

Kete, M. Š. A., Lokar, B. and Zupancic Justin, M. 2018. Method for manufacturing highly purified lactoferrin and lactoperoxidase from milk, colostrum and acid or sweet whey. Patent application no: 18204674.8.

Kontopidis, G., Holt, C. and Sawyer, L. 2004. Invited review: beta lactoglobulin: binding properties, structure, and function. *Journal of Dairy Science* 87(4), 785-796.

Korhonen, H., Marnila, P. and Gill, H. S. 2000. Bovine milk antibodies for health. *The British Journal of Nutrition* 84 (Suppl. 1), S135-S146.

Korhonen, H., Syvaeoja, E., Vasara, E., Kosunen, T. and Marnila, P. 1998. Pharmaceutical composition, comprising complement proteins, for the treatment of *Helicobacter* infections and a method for the preparation of the composition. W098/00150.

Korhonen, H. J. and Marnila, P. 2009. Chapter 10: Bovine milk immunoglobulins against microbial human diseases. In: Corredig, M. (Ed.), *Dairy-Derived Ingredients*. Cambridge, UK: Woodhead Publishing.

Krissansen, G. W. 2007. Emerging health properties of whey proteins and their clinical implications. *Journal of the American College of Nutrition* 26(6), 713S-723S.

Kruzel, M. L., Zimecki, M. and Actor, J. K. 2017. Lactoferrin in a context of inflammation-induced pathology. *Frontiers in Immunology* 8, 1438.

Kvistgaard, A. S., Pallesen, L. T., Arias, C. F., López, S., Petersen, T. E., Heegaard, C. W. and Rasmussen, J. T. 2004. Inhibitory effects of human and bovine milk constituents on rotavirus infections. *Journal of Dairy Science* 87(12), 4088-4096.

Lajnaf, R., Gharsallah, H., Jridi, M., Attia, H. and Ayadi, M. A. 2020. Antioxidant and antibacterial activities, interfacial and emulsifying properties of the apo and holo forms of purified camel and bovine α-lactalbumin. *International Journal of Biological Macromolecules* 165(A), 205-213.

Lam, H. Y., van Hoffen, E., Michelsen, A., Guikers, K., an Der Tas, C. H., Bruijnzeel-Koomen, C. A. and Knulst, A. C. 2008. Cow's milk allergy in adults is rare but severe: Both casein and whey proteins are involved. *Clinical and Experimental Allergy* 38(6), 995-1002.

Layman, D. K., Lönnerdal, B. and Fernstrom, J. D. 2018. Applications for α-lactalbumin in human nutrition. *Nutrition Reviews* 76(6), 444-460.

Le Maux, S., Bouhallab, S., Giblin, L., Brodkorb, A. and Croguennec, T. 2014. Bovine β-lactoglobulin/fatty acid complexes: Binding, structural, and biological properties. *Dairy Science and Technology* 94, 409-426.

Leischner, C., Egert, S., Burkard, M. and Venturelli, S. 2021. Potential protective protein components of Cow's milk against certain tumor entities. *Nutrients* 13(6), 1974.

Lievore, P., Simões, D. R. S., Silva, K. M., Drunkler, N. L., Barana, A. C., Nogueira, A. and Demiate, I. M. 2015. Chemical characterisation and application of acid whey in fermented milk. *Journal of Food Science and Technology* 52(4), 2083-2092.

Lim, K., van Calcar, S. C., Nelson, K. L., Gleason, S. T. and Ney, D. M. 2007. Acceptable low-phenylalanine foods and beverages can be made with glycomacropeptide from cheese whey for individuals with PKU. *Molecular Genetics and Metabolism* 92(1-2), 176-178.

Lönnerdal, B. 2003. Nutritional and physiologic significance of human milk proteins. *American Journal of Clinical Nutrition* 77(6), 1537S-1543S.

Lönnerdal, B. and Lien, E. L. 2003. Nutritional and physiologic significance of α-lactalbumin in infants. *Nutrition Reviews* 61(9), 295-305.

Luhovyy, B. L., Akhavan, T. and Anderson, G. H. 2007. Whey proteins in the regulation of food intake and satiety. *Journal of the American College of Nutrition* 26(6), 704S-712S.

Ma, Z., Zhang, F., Ma, H., Chen, X., Yang, J., Yang, Y., Yang, X., Tian, X., Yu, Q., Ma, Z. and Zhou, X. 2021. Effects of different types and doses of whey protein on the physiological and intestinal flora in D-galactose induced aging mice. *PLoS ONE* 16(4), e0248329.

MacLeod, E. L., Gleason, S. T., van Calcar, S. C. and Ney, D. M. 2009. Reassessment of phenylalanine tolerance in adults with phenylketonuria is needed as body mass changes. *Molecular Genetics and Metabolism* 98(4), 331–337.

Mandalari, G., Mackie, A. M., Rigby, N. M., Wickham, M. S. and Mills, E. N. 2009. Physiological phosphatidylcholine protects bovine beta-lactoglobulin from simulated gastrointestinal proteolysis. *Molecular Nutrition and Food Research* 53 (Suppl. 1), S131–S139.

Manzoni, P. 2016. Clinical benefits of lactoferrin for infants and children. *Journal of Pediatrics* 173 (Suppl.), S43–52.

Markus, C. R., Jonkman, L. M., Lammers, J. H., Deutz, N. E., Messer, M. H. and Rigtering, N. 2005. Evening intake of alpha-lactalbumin increases plasma tryptophan availability and improves morning alertness and brain measures of attention. *American Journal of Clinical Nutrition* 81(5), 1026–1033.

Marshall, K. 2004. Therapeutic applications of whey protein. *Alternative Medicine Review* 9(2), 136–156.

Mati, A., Moulti-Mati, F., Girardet, J.-M., Fokou, E., Belleville-Nabet, F., Nabet, P. and Linden, G. 1993. Mitogenic activity of hydrophobic fractions of proteose peptone from cows', ewes' and goats' milk measured with Mark 3 hybridoma culture. *Journal of Dairy Research* 60(3), 443–448.

Mayeur, S., Spahis, S., Pouliot, Y. and Levy, E. 2016. Lactoferrin, a pleiotropic protein in health and disease. *Antioxidants and Redox Signaling* 24(14), 813–836.

Mazzuca, F., Roberto, M., Arrivi, G., Sarfati, E., Schipilliti, F. M., Crimini, E., Botticelli, A., Di Girolamo, M., Muscaritoli, M. and Marchetti, P. 2019. Clinical impact of highly purified, whey proteins in patients affected With colorectal cancer undergoing chemotherapy: preliminary results of a placebo-controlled study. *Integrative Cancer Therapies* 18, 1534735419866920.

Medina, I., Tombo, I., Satué-Gracia, M. T., German, J. B. and Frankel, E. N. 2002. Effects of natural phenolic compounds on the antioxidant activity of lactoferrin in liposomes and oil-in-water emulsions. *Journal of Agricultural and Food Chemistry* 50(8), 2392–2399.

Mehra, R. and Kelly, P. 2004. Whey protein fractionation using cascade membrane filtration. *Bulletin International Dairy Federation*, 40–44.

Miller, P. E., Alexander, D. D. and Perez, V. 2014. Effects of whey protein and resistance exercise on body composition: A meta-analysis of randomized controlled trials. *Journal of the American College of Nutrition* 33(2), 163–175.

Mølgaard, C., Larnkjær, A., Arnberg, K. and Michaelsen, K. F. 2011. Milk and growth in children: Effects of whey and casein. *Nestle Nutrition Workshop Series. Paediatric Programme* 67, 67–78.

Morel, J., MD, Zain, S. N. and Archer, R. 2022. Comparison of drying techniques for bovine lactoferrin: iron binding and antimicrobial properties of dried lactoferrin. *International Dairy Journal*, 124, 105142.

Morita, Y., Matsuyama, H., Serizawa, A., Takeya, T. and Kawakami, H. 2008. Identification of angiogenin as the osteoclastic bone resorption-inhibitory factor in bovine milk. *Bone* 42(2), 380–387.

Morrin, S. T., Lane, J. A., Marotta, M., Bode, L., Carrington, S. D., Irwin, J. A. and Hickey, R. M. 2019a. Bovine colostrum-driven modulation of intestinal epithelial cells for increased commensal colonisation. *Applied Microbiology and Biotechnology* 103(6), 2745–2758.

Morrin, S. T., Owens, R. A., Le Berre, M., Gerlach, J. Q., Joshi, L., Bode, L., Irwin, J. A. and Hickey, R. M. 2019b. Interrogation of milk-driven changes to the proteome of intestinal epithelial cells by integrated proteomics and glycomics. *Journal of Agricultural and Food Chemistry* 67(7), 1902–1917.

Morrin, S. T. and Hickey, R. M. 2020a. New insights on the colonization of the human gut by health-promoting bacteria. *Applied Microbiology and Biotechnology* 104(4), 1511–1515.

Morrin, S. T., Mccarthy, G., Kennedy, D., Marotta, M., Irwin, J. A. and Hickey, R. M. 2020b. Immunoglobulin G from bovine milk primes intestinal epithelial cells for increased colonization of bifidobacteria. *AMB Express* 10(1), 114.

Mullally, M. M., Meisel, H. and Fitzgerald, R. J. 1997. Angiotensin-I-converting enzyme inhibitory activities of gastric and pancreatic proteinase digests of whey proteins. *International Dairy Journal* 7(5), 299–303.

Naclerio, F. and Larumbe-Zabala, E. 2016. Effects of whey protein alone or as part of a multi-ingredient formulation on strength, fat-free mass, or lean body mass in resistance-trained individuals: A meta-analysis. *Sports Medicine* 46(1), 125–137.

Nakajima, K., Tamura, N., Kobayashi-Hattori, K., Yoshida, T., Hara-Kudo, Y., Ikedo, M., Sugita-Konishi, Y. and Hattori, M. 2005. Prevention of intestinal infection by glycomacropeptide. *Bioscience, Biotechnology, and Biochemistry* 69(12), 2294–2301.

Neelima, Sharma, R., Rajput, Y. S. and Mann, B. 2013. Chemical and functional properties of glycomacropeptide (GMP) and its role in the detection of cheese whey adulteration in milk: A review. *Dairy Science and Technology* 93(1), 21–43.

Ney, D. M., Stroup, B. M., Clayton, M. K., Murali, S. G., Rice, G. M., Rohr, F. and Levy, H. L. 2016. Glycomacropeptide for nutritional management of phenylketonuria: A randomized, controlled, crossover trial. *American Journal of Clinical Nutrition* 104(2), 334–345.

Nguyen, D. N., Jiang, P., Stensballe, A., Bendixen, E., Sangild, P. T. and Chatterton, D. E. W. 2016. Bovine lactoferrin regulates cell survival, apoptosis and inflammation in intestinal epithelial cells and preterm pig intestine. *Journal of Proteomics* 139, 95–102.

Nygren-Babol, L. and Jägerstad, M. 2012. Folate-binding protein in milk: a review of biochemistry, physiology, and analytical methods. *Critical Reviews in Food Science and Nutrition* 52(5), 410–425.

O'Mahony, J. A., Fox, P. F. and Kelly, A. L. 2013. Indigenous enzymes of milk. In: Mcsweeney, P. and Fox, P. (Eds), *Advanced Dairy Chemistry*. Boston, MA: Springer.

O'Riordan, N., O'Callaghan, J., Butto, L. F., Kilcoyne, M., Joshi, L. and Hickey, R. M. 2018. Bovine glycomacropeptide promotes the growth of *Bifidobacterium longum* ssp. *infantis* and modulates its gene expression. *Journal of Dairy Science* 101(8), 6730–6741.

Oevermann, A., Engels, M., Thomas, U. and Pellegrini, A. 2003. The antiviral activity of naturally occurring proteins and their peptide fragments after chemical modification. *Antiviral Research* 59(1), 23–33.

Oggé, G., Romero, R., Lee, D. C., Gotsch, F., Than, N. G., Lee, J., Chaiworapongsa, T., Dong, Z., Mittal, P., Hassan, S. S. and Kim, C. J. 2011. Chronic chorioamnionitis displays distinct alterations of the amniotic fluid proteome. *Journal of Pathology* 223(4), 553–565.

Otani, H., Horimoto, Y. and Monnai, M. 1996. Suppression of interleukin-2 receptor expression on mouse CD4+ T cells by bovine κ-Caseinoglycopeptide. *Bioscience, Biotechnology, and Biochemistry* 60(6), 1017–1019.

Pal, S., Ellis, V. and Dhaliwal, S. 2010. Effects of whey protein isolate on body composition, lipids, insulin and glucose in overweight and obese individuals. *British Journal of Nutrition* 104(5), 716–723.

Pedersen, L. R. L., Hansted, J. G., Nielsen, S. B., Petersen, T. E., Sørensen, U. S., Otzen, D. and Sørensen, E. S. 2012. Proteolytic activation of proteose peptone component 3 by release of a C-terminal peptide with antibacterial properties. *Journal of Dairy Science* 95(6), 2819–2829.

Pellegrini, A., Thomas, U., Bramaz, N., Hunziker, P. and Von Fellenberg, R. 1999. Isolation and identification of three bactericidal domains in the bovine α-lactalbumin molecule. *Biochimica et Biophysica Acta* 1426(3), 439–448.

Perez, M. D., Sanchez, L., Aranda, P., Ena, J. M., Oria, R. and Calvo, M. 1992. Effect of beta-lactoglobulin on the activity of pregastric lipase. A possible role for this protein in ruminant milk. *Biochimica et Biophysica Acta* 1123(2), 151–155.

Pérez, M. D. and Calvo, M. 1995. Interaction of beta-lactoglobulin with retinol and fatty acids and its role as a possible biological function for this protein: A review. *Journal of Dairy Science* 78(5), 978–988.

Permyakov, E. A. 2020. α-lactalbumin, Amazing calcium-Binding Protein. *Biomolecules* 10(9), 1210.

Plate, K., Beutel, S., Buchholz, H., Demmer, W., Fischer-Frühholz, S., Reif, O., Ulber, R. and Scheper, T. 2006. Isolation of bovine lactoferrin, lactoperoxidase and enzymatically prepared lactoferricin from proteolytic digestion of bovine lactoferrin using adsorptive membrane chromatography. *Journal of Chromatography A* 1117(1), 81–86.

Plaza-Diaz, J., Ruiz-Ojeda, F. J., Gil-Campos, M. and Gil, A. 2019. Mechanisms of action of probiotics. *Advances in Nutrition* 10(suppl_1), S49–66.

Priyanka, B. S. and Rastogi, N. K. 2018. Downstream processing of lactoperoxidase from milk whey by involving liquid emulsion membrane. *Preparative Biochemistry and Biotechnology* 48(3), 270–278.

Prussick, R., Prussick, L. and Gutman, J. 2013. Psoriasis improvement in patients using glutathione-enhancing, nondenatured whey protein isolate: A pilot study. *The Journal of Clinical and Aesthetic Dermatology* 6(10), 23–26.

Puyol, P., Dolores Perez, M. D., Sanchez, L., Ena, J. M. and Calvo, M. 1995. Uptake and passage of beta-lactoglobulin, palmitic acid and retinol across the Caco-2 monolayer. *Biochimica et Biophysica Acta* 1236(1), 149–154.

Puyol, P., Perez, M. D., Mata, L., Ena, J. and Calvo, M. 1993. Effect of retinol and fatty acid binding by bovine beta-lactoglobulin on its resistance to trypsin digestion. *International Dairy Journal* 3(7), 589–597.

Rachagani, S., Dayal Gupta, I. D., Gupta, N. and Gupta, S. C. 2006. Genotyping of β-lactoglobulin gene by PCR-RFLP in Sahiwal and Tharparkar cattle breeds. *BMC Genetics* 7, 31.

Rai, D., Adelman, A. S., Zhuang, W., Rai, G. P., Boettcher, J. and Lönnerdal, B. 2014. Longitudinal changes in lactoferrin concentrations in human milk: A global systematic review. *Critical Reviews in Food Science and Nutrition* 54(12), 1539–1547.

Rammer, P., Groth-Pedersen, L., Kirkegaard, T., Daugaard, M., Rytter, A., Szyniarowski, P., Høyer-Hansen, M., Povlsen, L. K., Nylandsted, J., Larsen, J. E. and Jäättelä, M. 2010. BAMLET activates a lysosomal cell death program in cancer cells. *Molecular Cancer Therapeutics* 9(1), 24–32.

Recio, I., Moreno, F. J. and López-Fandiño, R. 2009. Chapter 7: Glycosylated dairy components: their roles in nature and ways to make use of their biofunctionality in dairy products. In: Corredig, M. (Ed.) *Dairy-Derived Ingredients*. Cambridge, UK: Woodhead Publishing.

Reimer, R. A., Willis, H. J., Tunnicliffe, J. M., Park, H., Madsen, K. L. and Soto-Vaca, A. 2017. Inulin-type fructans and whey protein both modulate appetite but only fructans alter gut microbiota in adults with overweight/obesity: A randomized controlled trial. *Molecular Nutrition and Food Research* 61(11).

Reiter, B. and Harnulv, G. 1984. Lactoperoxidase antibacterial system: natural occurrence, biological functions and practical applications. *Journal of Food Protection* 47(9), 724–732.

Riihimäki-Lampén, L. 2009. Interactions of natural products with beta-lactoglobulins, members of the lipocalin family. Available at: http://urn.fi/URN:ISBN:978-952-10-5882-0.

Rocha-Mendoza, D., Kosmerl, E., Krentz, A., Zhang, L., Badiger, S., Miyagusuku-Cruzado, G., Mayta-Apaza, A., Giusti, M. M., Jimenez-Flores, R. and Garcia-Cano, I. 2021. Invited review: acid whey trends and health benefits. *Journal of Dairy Science* 104(2), 1262–1275.

Rosenberg, I. H. and Selhub, J. 2006. Folate and vitamin B12 transport systems in the developing infant. *Journal of Pediatrics* 149(5), S62–S63.

Roth-Walter, F., Afify, S. M., Pacios, L. F., Blokhuis, B. R., Redegeld, F., Regner, A., Petje, L. M., Fiocchi, A., Untersmayr, E., Dvorak, Z., Hufnagl, K., Pali-Schöll, I. and Jensen-Jarolim, E. 2021. Cow's milk protein β-lactoglobulin confers resilience against allergy by targeting complexed iron into immune cells. *Journal of Allergy and Clinical Immunology* 147(1), 321-334.e4.

Ryan, M. P. and Walsh, G. 2016. The biotechnological potential of whey. *Reviews in Environmental Science and Bio/Technology* 15(3), 479–498.

Rybarczyk, J., Kieckens, E., Vanrompay, D. and Cox, E. 2017. *In vitro* and *in vivo* studies on the antimicrobial effect of lactoferrin against *Escherichia coli* O157:H7. *Veterinary Microbiology* 202, 23–28.

Salehi, A., Gunnerud, U., Muhammed, S. J., Östman, E., Holst, J. J., Björck, I. and Rorsman, P. 2012. The insulinogenic effect of whey protein is partially mediated by a direct effect of amino acids and GIP on β-cells. *Nutrition and Metabolism* 9(1), 48.

Salter, D. N. and Blakeborough, P. 1988. Influence of goat's-milk folate-binding protein on transport of 5-methyltetrahydrofolate in neonatal-goat small intestinal brush-border-membrane vesicles. *British Journal of Nutrition* 59(3), 497–507.

Sánchez, L., Aranda, P., Pérez, M. D. and Calvo, M. 1988. Concentration of lactoferrin and transferrin throughout lactation in cow's colostrum and milk. *Biological Chemistry Hoppe-Seyler* 369(9), 1005–1008.

Sandström, O., Lonnerdal, B., Graverholt, G. and Hernell, O. 2008. Effects of a-lactalbumin-enriched formula containing different concentrations of glycomacropeptide on infant nutrition. *American Journal of Clinical Nutrition* 87(4), 921–928.

Savino, F., Cresi, F., MacCario, S., Cavallo, F., Dalmasso, P., Fanaro, S., Oggero, R., Vigi, V. and Silvestro, L. 2003. "Minor" feeding problems during the first months of life: effect of a partially hydrolysed milk formula containing fructo- and galacto-oligosaccharides. *Acta Paediatrica* 91(441), 86–90.

Sawin, E. A., De Wolfe, T. J., Aktas, B., Stroup, B. M., Murali, S. G., Steele, J. L. and Ney, D. M. 2015. Glycomacropeptide is a prebiotic that reduces Desulfovibrio bacteria, increases cecal short-chain fatty acids, and is anti-inflammatory in mice. *American Journal of Physiology. Gastrointestinal and Liver Physiology* 309(7), G590-601.

Sawyer, L. 2021. β-Lactoglobulin and Glycodelin: two sides of the same coin? *Frontiers in Physiology* 12, 678080.

Schaafsma, A., Mallee, L., van den Belt, M., Floris, E., Kortman, G., Veldman, J., van den Ende, D. and Kardinaal, A. 2021. The effect of A whey-protein and galacto-oligosaccharides based product on parameters of sleep quality, stress, and gut microbiota in apparently healthy adults with moderate sleep disturbances: A randomized controlled cross-Over study. *Nutrients* 13(7).

Schack, L., Lange, A., Kelsen, J., Agnholt, J., Christensen, B., Petersen, T. E. and Sørensen, E. S. 2009. Considerable variation in the concentration of osteopontin in human milk, bovine milk, and infant formulas. *Journal of Dairy Science* 92(11), 5378-5385.

Schmelzle, H., Wirth, S., Skopnik, H., Radke, M., Knol, J., Böckler, H. M., Brönstrup, A., Wells, J. and Fusch, C. 2003. Randomized double-blind study of the nutritional efficacy and bifidogenicity of a new infant formula containing partially hydrolyzed protein, a high beta-palmitic acid level, and nondigestible oligosaccharides. *Journal of Pediatric Gastroenterology and Nutrition* 36(3), 343-351.

Scholz, C., Toth, B., Brunnhuber, R., Rampf, E., Weissenbacher, T., Santoso, L., Friese, K. and Jeschke, U. 2008. Glycodelin A induces a tolerogenic phenotype in monocyte-derived dendritic cells *in vitro*. *American Journal of Reproductive Immunology* 60(6), 501-512.

Sharma, R. 2019. Chapter 17. Whey proteins in functional foods. In: Deeth, H. C. and Bansal, N. (Eds), *Whey Proteins: From Milk to Medicine*. Massachusetts: Academic Press.

Singh, R. H., Rohr, F., Frazier, D., Cunningham, A., Mofidi, S., Ogata, B., Splett, P. L., Moseley, K., Huntington, K., Acosta, P. B., Vockley, J. and van Calcar, S. C. 2014. Recommendations for the nutrition management of phenylalanine hydroxylase deficiency. *American College of Medical Genetics* 16(2), 121-131.

Singh, R. K., Chang, H. W., Yan, D., Lee, K. M., Ucmak, D., Wong, K., Abrouk, M., Farahnik, B., Nakamura, M., Zhu, T. H., Bhutani, T. and Liao, W. 2017. Influence of diet on the gut microbiome and implications for human health. *Journal of Translational Medicine* 15(1), 73.

Skryplonek, K. and Jasińska, M. 2017. Whey-based beverages. *Electronic Journal of Polish Agricultural Universities* 20(4) 1-8.

Sørensen, E. S. and Petersen, T. E. 1993. Phosphorylation, glycosylation and amino acid sequence of component PP3 from the proteose peptone fraction of bovine milk. *Journal of Dairy Research* 60(4), 535-542.

Sørensen, E. S., Rasmussen, L. K., Møller, L. and Petersen, T. E. 1997. The localization and multimeric nature of component PP3 in bovine milk: purification and characterization of PP3 from caprine and ovine milks. *Journal of Dairy Science* 80(12), 3176 -3181.

Steinberg, L. A., O'connell, N. C., Hatch, T. F., Picciano, M. F. and Birch, L. L. 1992. Tryptophan intake influences infants' sleep latency. *Journal of Nutrition* 122(9), 1781-1791.

Sugahara, T., Onda, H., Shinohara, Y., Horii, M., Akiyama, K., Nakamoto, K. and Hara, K. 2005. Immunostimulation effects of proteose-peptone component 3 fragment on human hybridomas and peripheral blood lymphocytes. *Biochimica et Biophysica Acta* 1725(2), 233-240.

Superti, F. 2020. Lactoferrin from bovine milk: A protective companion for life. *Nutrients* 12(9), 2562.

Svendsen, I. B., Martin, B., Pedersen, T. G., Hansen, S. I., Holm, J. and Lyngbye, J. 1979. Isolation and characterization of the folate-binding protein from cow's milk. *Carlsberg Research Communications* 44(2), 89–99.

Swiatlo, N., O'Connor, D. L., Andrews, J. and Picciano, M. F. 1990. Relative folate bioavailability from diets containing human, bovine and goat milk. *Journal of Nutrition* 120(2), 172–177.

Szajewska, H., Mrukowicz, J. Z., Stoińska, B. and Prochowska, A. 2004. Extensively and partially hydrolysed preterm formulas in the prevention of allergic diseases in preterm infants: A randomized, double-blind trial. *Acta Paediatrica* 93(9), 1159–1165.

Takada, Y., Kobayashi, N., Matsuyama, H., Kato, K., Yamamura, J., Yahiro, M., Kumegawa, M. and Aoe, S. 1997. Whey protein suppresses the osteoclast-mediated bone resorption and osteoclast cell formation. *International Dairy Journal* 7(12), 821–825.

Takayama, Y. and Aoki, R. 2012. Roles of lactoferrin on skin wound healing. *Biochemistry and Cell Biology (Biochimie et Biologie Cellulaire)* 90(3), 497–503.

Tani, M. and Iwai, K. 1984. Some nutritional effects of folate-binding protein in bovine milk on the bioavailability of folate to rats. *Journal of Nutrition* 114(4), 778–785.

Teixeira, F. J., Santos, H. O., Howell, S. L. and Pimentel, G. D. 2019. Whey protein in cancer therapy: A narrative review. *Pharmacological Research* 144, 245–256.

Thomä-Worringer, C., Sørensen, J. and López-Fandiño, R. 2006. Health effects and technological features of caseinomacropeptide. *International Dairy Journal* 16(11), 1324–1333.

Thomas, S. D. C., Morris, H. A. and Nordin, B. E. C. 2015. Acute effect of a supplemented milk drink on bone metabolism in healthy postmenopausal women is influenced by the metabolic syndrome. *Nutrition Journal* 14, 99–99.

Thorkelsson, T., Mimouni, F., Namgung, R., Fernández-Ulloa, M., Krug-Wispé, S. and Tsang, R. C. 1994. Similar gastric emptying rates for casein- and whey-predominant formulas in preterm infants. *Pediatric Research* 36(3), 329–333.

Toba, Y., Takada, Y., Matsuoka, Y., Morita, Y., Motouri, M., Hirai, T., Suguri, T., Aoe, S., Kawakami, H., Kumegawa, M., Takeuchi, A. and Itabashi, A. 2001. Milk basic protein promotes bone formation and suppresses bone resorption in healthy adult men. *Bioscience, Biotechnology, and Biochemistry* 65(6), 1353–1357.

Tomita, M., Wakabayashi, H., Yamauchi, K., Teraguchi, S. and Hayasawa, H. 2002. Bovine lactoferrin and lactoferricin derived from milk: Production and applications. *Biochemistry and Cell Biology* 80(1), 109–112.

Uenishi, K., Ishida, H., Toba, Y., Aoe, S., Itabashi, A. and Takada, Y. 2007. Milk basic protein increases bone mineral density and improves bone metabolism in healthy young women. *Osteoporosis International* 18(3), 385–390.

Ulfman, L. H., Leusen, J. H. W., Savelkoul, H. F. J., Warner, J. O. and van Neerven, R. J. J. 2018. Effects of bovine immunoglobulins on immune function, allergy, and infection. *Frontiers in Nutrition* 5, 52.

van Cong, N., Vaisse, C., Gross, M. S., Slim, R., Milgrom, E. and Bernheim, A. 1991. The human placental protein 14 (PP14) gene is localized on chromosome 9q34. *Human Genetics* 86(5), 515–518.

van der Vusse, G. J. 2009. Albumin as fatty acid transporter. *Drug Metabolism and Pharmacokinetics* 24(4), 300–307.

van Neerven, R. J., Knol, E. F., Heck, J. M. and Savelkoul, H. F. 2012. Which factors in raw cow's milk contribute to protection against allergies? *Journal of Allergy and Clinical Immunology* 130(4), 853–858.

Vandenplas, Y., Hauser, B. and Salvatore, S. 2019. Functional gastrointestinal disorders in infancy: impact on the health of the infant and family. *Pediatric Gastroenterology, Hepatology and Nutrition* 22(3), 207–216.

Vandenplas, Y., Leluyer, B., Cazaubiel, M., Housez, B. and Bocquet, A. 2013. Double-blind comparative trial with 2 antiregurgitation formulae. *Journal of Pediatric Gastroenterology and Nutrition* 57(3), 389–393.

Vandenplas, Y., Cruchet, S., Faure, C., Lee, H., Di Lorenzo, C., Staiano, A., Chundi, X., Aw, M., Gutiérrez-Castrellón, P., Asery, A., Spolidoro, J., Heine, R., Miqdady, M., Arancibia, M. and Alarcón, P. 2014. When should we use partially hydrolysed formulae for frequent gastrointestinal symptoms and allergy prevention? *Acta Paediatrica* 103(7), 689–695.

Vega-Bautista, A., De La Garza, M., Carrero, J. C., Campos-Rodríguez, R., Godínez-Victoria, M. and Drago-Serrano, M. E. 2019. The impact of lactoferrin on the growth of intestinal inhabitant bacteria. *International Journal of Molecular Sciences* 20(19), 4707.

Villarreal, M. 2017. Value Added Products Utilizing Acid Whey: Development of a Fruit Yogurt Beverage and a Sports Drink. Cornell Theses and Dissertations. New York: Cornell University.

Vockley, J., Andersson, H. C., Antshel, K. M., Braverman, N. E., Burton, B. K., Frazier, D. M., Mitchell, J., Smith, W. E., Thompson, B. H.and Berry, S. A. 2014. Phenylalanine hydroxylase deficiency: Diagnosis and management guideline. *Genetics in Medicine* 16(2), 188–200.

Von Berg, A., Filipiak-Pittroff, B., Schulz, H., Hoffmann, U., Link, E. Sußmann, M., Schnappinger, M., Brüske, I., Standl, M., Krämer, U., Hoffmann, B., Heinrich, J., Bauer, C. P., Koletzko, S. and Berdel, D. 2016. Allergic manifestation 15 years after early intervention with hydrolyzed formulas – the GINI study. *Allergy* 71, 210–219.

Walstra, P., Woulters, J. T. M., and Geurts, T. J. 2005. *Dairy Science and Technology*. Florida: CRC Press.

Walzem, R. L., Dillard, C. J. and German, J. B. 2002. Whey components: Millennia of evolution create functionalities for mammalian nutrition: What we know and what we may be overlooking. *Critical Reviews in Food Science and Nutrition* 42(4), 353–375.

Wang, Y., Xue, Y., Bi, Q., Qin, D., Du, Q. and Jin, P. 2021. Enhanced antibacterial activity of eugenol-entrapped casein nanoparticles amended with lysozyme against gram-positive pathogens. *Food Chemistry* 360, 130036.

Westermann, C. and Gleinser, M., Corr, S. C. and Riedel, C. U. 2016. A critical evaluation of bifidobacterial adhesion to the host tissue. *Frontiers in Microbiology* 7, 1220.

Wheeler, T. T., Hodgkinson, A. J., Prosser, C. G. and Davis, S. R. 2007. Immune components of colostrum and milk--a historical perspective. *Journal of Mammary Gland Biology and Neoplasia* 12(4), 237–247.

WHO (World Health Organization) 2003. *United Nations Children's Fund*, 30. Geneva, Switzerland: Global Strategy for Infant and Young Child Feeding; World Health Organization.

WHO (World Health Organization) 2004. WHO scientific group on the assessment of osteoporosis at primary health care level. *Summary Meeting Report*. Brussels, Belgium: WHO.

Wilborn, C. D., Taylor, L. W., Outlaw, J., Williams, L., Campbell, B., Foster, C. A., Smith-Ryan, A., Urbina, S. and Hayward, S. 2013. The Effects of Pre- and Post-Exercise Whey vs.

casein Protein Consumption on Body Composition and Performance Measures in Collegiate Female Athletes. *Journal of Sports Science and Medicine* 12(1), 74–79.

Wirunsawanya, K., Upala, S., Jaruvongvanich, V. and Sanguankeo, A. 2018. Whey protein supplementation improves body composition and cardiovascular risk factors in overweight and obese patients: A systematic review and meta-analysis. *Journal of the American College of Nutrition* 37(1), 60–70.

Woo, S. H., Lee, S. H., Park, J. W., Go, D. M. and Kim, D. Y. 2019. Osteopontin protects colonic mucosa from dextran sodium sulfate-induced acute colitis in mice by regulating functional distribution of occludin. *Digestive Diseases and Sciences* 64(2), 421–431.

Xinias, I., Analitis, A., Mavroudi, A., Roilides, I., Lykogeorgou, M., Delivoria, V., Milingos, V., Mylonopoulou, M. and Vandenplas, Y. 2017. Innovative dietary intervention answers to baby colic. *Pediatric Gastroenterology, Hepatology and Nutrition* 20(2), 100–106.

Xue, Y., Chen, J., Zhang, L. and Han, Y. 2021. BSA-lysozyme coated NaCa2HSi3O9 nanorods on titanium for cytocompatibility and antibacterial activity. *Journal of Materials Science and Technology* 88, 240–249.

Yaman, M., Mızrak, ÖF., Çatak, J. and Sargın, H. S. 2019. *In vitro* bioaccessibility of added folic acid in commercially available baby foods formulated with milk and milk products. *Food Science and Biotechnology* 28(6), 1837–1844.

Yang, C., Zhu, X., Liu, N., Chen, Y., Gan, H., Troy, F. A., 2ND and Wang, B. 2014. Lactoferrin up-regulates intestinal gene expression of brain-derived neurotrophic factors BDNF, UCHL1 and alkaline phosphatase activity to alleviate early weaning diarrhea in postnatal piglets. *Journal of Nutritional Biochemistry* 25(8), 834–842.

Yoneme, H., Hatakeyama, J., Danjo, A., Oida, H., Yoshinari, M., Aijima, R., Murata, N., Watanabe, T., Oki, Y. and Kido, M. A. 2015. Milk basic protein supplementation enhances fracture healing in mice. *Nutrition* 31(2), 399–405.

Yousefi, B., Eslami, M., Ghasemian, A., Kokhaei, P., Salek Farrokhi, A. and Darabi, N. 2019. Probiotics importance and their immunomodulatory properties. *Journal of Cellular Physiology* 234(6), 8008–8018.

Zhang, Y., Lima, C. F. and Rodrigues, L. R. 2014. Anticancer effects of lactoferrin: Underlying mechanisms and future trends in cancer therapy. *Nutrition Reviews* 72(12), 763–773.

Zotta, T., Solieri, L., Iacumin, L., Picozzi, C. and Gullo, M. 2020. Valorization of cheese whey using microbial fermentations. *Applied Microbiology and Biotechnology* 104(7), 2749–2764.

# Chapter 7

## Functional ingredients based on bioactive peptides from milk proteins

*Egon Bech Hansen, Technical University of Denmark, Denmark*



### 1 Introduction

Milk is a complete and sufficient source of nutrition for the newborn of any mammal. Milk is a medium that not only allows basic nutrients to be passed from mother to offspring, but also facilitates numerous maturation processes. Maturation of the digestive system includes reprogramming from digesting milk towards digesting the typical food of the particular species. These transitions will be different for herbivores, carnivores, or omnivores. Typically, complex microbiota in an anaerobic environment are established early in life.

The composition of the microbiota depends on the species and will adapt to the food available after weaning. The immune system of the newborn also needs to be mature and be programmed to distinguish between molecules originating from self, friends, or pathogens. Milk plays an important role in most of those maturation processes. Even the medical history of the mother is passed on to the newborn through antibodies in the milk. Numerous molecules with a diverse range of biological activities have been identified in milk. Proteins and carbohydrates are well represented among those molecules. The almost infinite space for possible sequence variations allow proteins, peptides, and oligosaccharides to endow milk with a multitude of bioactivities.

http://dx.doi.org/10.19103/AS.2022.0099.13

The full-length proteins of milk typically carry a primary biological activity in addition to its role as a source of amino acids after digestion. Caseins are important in keeping calcium phosphate in suspension within a casein micelle. The whey proteins are diverse and include, among others, immunoglobulins and enzymes. Eventually, all proteins are digested into amino acids and reused for protein synthesis or catabolized to generate energy. En route to complete digestion, milk proteins are initially hydrolyzed into peptides of varying lengths. Research has shown that these components of milk proteins are encrypted with important biological activities that influence the health and well-being of the offspring. The so-called bioactive milk peptides have been intensively researched for about half a century. Collectively, this body of research has given a deep insight into the health benefits of milk. Considering the importance of the dairy industry, the majority of research on milk is focused on bovine milk. Other dairy animals such as goat, sheep, camel, horse and donkey, which represent much smaller volumes in the industry, are less prominent in dairy science. Considering health aspects, human milk is also obviously interesting and intensively studied.

This chapter will focus on the biological activities of peptides derived from bovine milk, and the main emphasis will be on peptides for industrial applications or with a potential for such applications.

Before diving into the aspects of science, it is worthwhile to consider the definitions of 'peptides' and 'bioactivity'. Peptides are polymers of amino acids. Dipeptides are the shortest peptides, composed of two amino acids. There is no clear upper limit for size to distinguish between peptides and proteins, but it is usually considered to be around 50 amino acids.

'Bioactivity' is also a broad term. It typically designates a beneficial effect on human physiology, going beyond ordinary nutrition. Obviously, such effects are not necessarily limited to humans; the distinction between beneficial and harmful can be difficult; and nutrition is not always ordinary. In this chapter, bioactivity will be discussed generally but with a focus on properties with potential application.

## 2 Bioactives identified in milk-derived peptides

In 1979, Brantl et al. described the isolation of a peptide with morphine-like activity from a casein peptone (Brantl et al., 1979; Henschen et al., 1979). This pioneering discovery opened the field of research into bioactive peptides from milk.

The enzymes of the digestive tract generate peptide fragments from ingested proteins. Such peptides released from milk proteins during digestion represent an important track of research on bioactive peptides. This avenue of research has revealed insights into the fundamental aspects of nutrition and health. Two additional avenues related to the needs of the dairy industry have also been followed. One of these has been to study peptides released through fermentation. This research has identified the basis of some of the beneficial effects of fermented or probiotic dairy products. The other avenue of application has been to investigate whether peptides derived from whey proteins could have potential applications in human or animal health. The purpose has been to upgrade the value of the by-product, whey.

Table 1 lists the major biological activities for milk-derived peptides together with the protein sources. The table also lists the health conditions which could potentially be ameliorated. Listing the individual peptides or reviewing the data behind each peptide will be beyond the scope of this chapter. However, the literature references given will allow such documentation to be traced.

**Table 1** Biological activities identified for milk-derived peptides

| Bioactivity | Protein source of peptides | Possible benefit for human health | Reference or recent review |
|---|---|---|---|
| Opioid receptor-binding | β-casein, $\alpha_{S1}$-casein, κ-casein, β-lactoglobulin | Pain reduction, sleep, well-being, mood | (Brantl et al., 1979; Liu and Udenigwe, 2019; Summer et al., 2020; Nielsen et al., 2017) |
| ACE inhibition | β-casein, $\alpha_{S1}$-casein, $\alpha_{S2}$-casein, κ-casein, α-lactalbumin, β-lactoglobulin, lactoferrin | Maintenance of normal blood pressure, reduction of hypertension | (Jäkälä and Vapaatalo, 2010; Siltari et al., 2019; Nielsen et al., 2017) |
| Mineral binding | β-casein, $\alpha_{S1}$-casein, $\alpha_{S2}$-casein | Bone and dental health, reducing the risk of osteoporosis | (Savastano et al., 2019; Scholz-Ahrens and Schrezenmeir, 2000) |
| Immunomodulation | β-casein, α-lactalbumin, β-lactoglobulin, lactoferrin | Reduction of inflammation, reduction of risk of infections | (Nielsen et al., 2017) |
| DPP-IV inhibition | Lactoferrin, β-casein, β-lactoglobulin, κ-casein, α-lactalbumin, $\alpha_{S2}$-casein, $\alpha_{S1}$-casein | Maintenance of normal blood glucose levels, antidiabetic effects on type 2 diabetes | (Patil et al., 2015) |
| Antioxidant | β-casein, $\alpha_{S1}$-casein | Radical scavenging, reduction of oxidative stress | (Nielsen et al., 2017) (Pihlanto, 2006) |
| Antimicrobial | Lactoferrin, $\alpha_{S1}$-casein, $\alpha_{S2}$-casein, κ-casein | Reduce infections, prebiotic effects on microbiotas | (Nielsen et al., 2017) (Malkoski et al., 2001) |

Databases on bioactive peptides have been established and they allow easy retrieval of bioactivity data for individual peptides. Table 2 lists the currently most relevant databases with regard to bioactive peptides. A longer list of bioactive peptide databases with various scopes is given by Nongonierma and FitzGerald (2016).

## 2.1 Opioid receptor binding

Enzymatic hydrolysis of caseins releases peptides that are able to bind to opioid receptors. Most of these peptides stimulate the receptor, whereas a few peptides have a weak inhibitory effect. Opioid receptors are involved in the neurohormonal system, and they are stimulated by endogenous peptides like encephalin, dynorphin, and β-endorphin (Janecka et al., 2004). In the first study on β-casomorphin carried out in 1979, biological activity was determined to be 250 times weaker than normorphine. In the Discussion section of their paper, Brantl et al. wrote: 'Thus, it may still be regarded as an open question if these opioids could play a physiological role, e.g. in the nutrition of newborns, or if they even could have toxicological significance' (Brantl et al., 1979). Four decades of active research into this question has not changed much: A working group under the European Food Safety Authority (EFSA) and a recent review by Summer et al. (2020) both concluded that the risk and benefits seem to be negligible (EFSA, 2009; Summer et al., 2020). It is still an open question whether casein-derived peptides with opioid receptor binding play any significant physiological role. The review by Janecka et al. (2004) gives an excellent overview of opioid receptors and endogenous bioactive opioid peptides, and the authors also cover atypical opioid peptides including β-casomorphins (Janecka et al., 2004). The review by Liu and Udenigwe (2019) gives an update on the current status of research on opioid peptides derived from food proteins (Liu and Udenigwe, 2019).

## 2.2 Mineral binding

One of the numerous remarkable properties of milk is the high mineral content, allowing the newborn animal to grow rapidly and develop strong and healthy bones. The calcium phosphate content in bovine milk exceeds the solubility in water by a factor of ten. The calcium phosphate is kept in suspension in the form of casein micelles containing nanoclusters of calcium phosphate (Dalgleish, 2011; Holt, 2016). The alpha- and beta-caseins carry tracts of phosphorylated serines, and these heavily phosphorylated regions are able to bind to the divalent metal ions in the calcium phosphate nanoclusters. The kappa-casein endows the casein micelle with a hydrophilic surface, allowing the micelle with a high load of minerals to be kept in suspension. The calcium-binding

**Table 2** Databases for bioactive peptides

| Name of database | Purpose | Link | Reference |
|---|---|---|---|
| BIOPEP-UWM (formerly BIOPEP) | Bioactive peptides, allergenic epitopes, sensory peptides | http://www.uwm.edu.pl/biochemia/index.php/en/biopep | (Minkiewicz et al., 2019) |
| MBPDB | Milk bioactive peptides | https://mbpdb.nws.oregonstate.edu/ | (Nielsen et al., 2017) |
| AHTPDB | Antihypertensive peptides | http://crdd.osdd.net/raghava/ahtpdb/index.php | (Kumar et al., 2015) |

property of casein-derived peptides can be used to improve the solubility and bioavailability of Ca.

Calcium salts are generally poorly soluble, and this property limits their bioavailability. The solubility can be increased by complexing $Ca^{2+}$ ions with ligands such as citric acid, amino acids, or peptides. Complexing metal ions with short peptides is supposed to facilitate their absorption in the gastrointestinal tract (Sun et al., 2016). Peptides released through casein hydrolysis will contain peptides with tracts of phosphoserines, and such casein phosphopeptides (CPP) retain the ability to complex divalent metal ions and increase the solubility of calcium. Early research on the effect of phosphopeptides on mineral metabolism was conducted by Mellander in the 1940s and 1950s (Mellander, 1945; Mellander and Isaksson, 1950). Animal and short-term studies have demonstrated the stimulation of mineral uptake by supplements of CPP (Scholz-Ahrens and Schrezenmeir, 2000). Sun et al. (2016) have recently reviewed calcium-binding peptides derived from food proteins.

## 2.3 Angiotensin-converting enzyme (ACE) inhibition

The maintenance of blood pressure is a complicated process. The renin-angiotensin system (RAS) is central to the physiology of blood pressure control. The protease renin converts angiotensinogen to angiotensin-I. Angiotensin-I is subsequently converted to angiotensin-II by the enzyme ACE. Angiotensin-II exerts its effect through various receptors, leading to an increase in blood pressure (Iwaniak et al., 2014). Maruyama and Suzuki (1982) realized that ACE-inhibiting peptides often contain proline. They analyzed a tryptic casein hydrolysate for ACE inhibition and found one peptide able to inhibit ACE (Maruyama and Suzuki, 1982). Yamamoto et al. (1994) demonstrated that fermentation could release several potent ACE-inhibiting peptides from milk. ACE-inhibiting peptides from milk proteins have become a very productive field of research. ACE is a dipeptidase enzyme, releasing dipeptides from the

C-terminal end of the peptide substrates. It is not surprising that peptides able to inhibit this enzyme can be found among non-substrate peptides. The degree of inhibition of the enzyme can be determined by relatively simple enzyme assays and has often been expressed as the IC50 value of the inhibitor. IC50 is the concentration of the inhibitor causing 50% inhibition of the enzyme. IC50 will also depend on the concentration of the enzyme, and comparison across different assay conditions can be difficult. The AHTPDB database lists 435 different peptide sequences of milk origin showing ACE inhibition. However, the IC50 values of these peptides differ by more than 1000-fold, and one could question whether they should all be classified as ACE inhibitors. Of the 435 peptides, 151 show IC50 values in the low μM range (IC50 < 10 μM). The average peptide length is 5 amino acids; about half the peptides (69) are tripeptides, and the longest of these potent inhibitors is 19 amino acids long. The large number and short length of the most potent ACE-inhibiting peptides identified in milk protein hydrolysates indicate that ACE inhibition is not a physiological function preprogrammed into milk proteins, but an incidental consequence of the amino acid composition of milk proteins. Even if this property might be irrelevant for the young calf and other neonates, it might still be a beneficial and useful property for a middle-aged consumer. Short peptides are absorbed well into the bloodstream and short ACE inhibitors like tripeptides rich in proline have therefore attracted particular attention. Recent reviews have been published by several researchers (Saito, 2008; Jäkälä and Vapaatalo, 2010; Iwaniak et al., 2014). An enzyme analog of ACE, ACE2, was discovered by Donoghue et al. (2000). ACE2 converts angiotensin-I and -II to shorter angiotensin peptides which, via the receptor MasR, lead to a reduction in blood pressure. The ACE2 route opens new possibilities for the maintenance of normal blood pressure via peptide food supplements (Liao and Wu, 2020).

## 2.4 DPP-IV inhibition

Dipeptidyl peptidase IV (DPP-IV) is involved in the turnover and inactivation of numerous peptide hormones and cytokines (Mentlein, 1999). DPP-IV is among other substrates acting on the incretin hormones GLP-1 and GIP. The inhibition of DPP-IV will indirectly increase the glucose-dependent secretion of insulin (Nongonierma and FitzGerald, 2013; Kwok et al., 2014). DPP-IV is similar to ACE – a peptidase – and is therefore likely to be inhibited by non-substrate peptides. However, the two enzymes differ; DPP-IV is an aminopeptidase, whereas ACE is a carboxypeptidase. DPP-IV is a peptidase releasing dipeptides from the N-terminal end of the substrates. The substrates are characterized by the presence of the dipeptide X-proline or X-alanine at the N-terminus. The most potent inhibitor of DPP-IV is the tripeptide Ile-Pro-Ile, which acts as a competitive inhibitor with an IC50 of 3.5 μM (Nongonierma and FitzGerald, 2013). Larger peptides with a non-competitive mode of inhibition have also been reported for DPP-IV (Nongonierma and FitzGerald, 2013; Nongonierma et al., 2018; Jia et al., 2020), and the database MBPDB lists some of those among the most potent inhibitors with low IC50 values. However these low values do not agree with the original publication, and the general picture seems to be that potent DPP-IV inhibitors are tripeptides with proline at the second position, analogous to the IPI inhibitor peptide (Nongonierma et al., 2018). The occurrence of DPP-IV-inhibiting bioactive peptides in milk protein hydrolysates is therefore likely to be a consequence of the frequent occurrence of proline in milk proteins. The property might nevertheless be beneficial for human consumers, and this mechanism could be an explanation for the beneficial effect of milk on diabetes (Patil et al., 2015). The effect of milk on diabetes has been reported to have both advantages and disadvantages, but the disadvantage is probably modest; however, most studies show a beneficial effect of consumption of dairy products on type 2 diabetes (Alvarez-Bueno et al., 2019).

## 2.5 Antioxidant activity

Peptides with antioxidant activities derived from caseins and whey proteins have been characterized (Pihlanto, 2006). Antioxidants have several modes of action, including oxygen capture, reduction of reactive oxygen, radical scavenging, metal chelating, and inhibition of oxidative enzymes, and several different assays are usually used to characterize antioxidant activity (Sah et al., 2018). Antioxidant activity is associated with intact casein and whey proteins, hydrolysates, amino acids, and individual peptides derived from caseins and whey proteins (Power et al., 2013). Radical-scavenging peptides often contain methionine, tyrosine, arginine, proline, or hydrophobic amino acids; the casein phosphopeptides (CPP) due to their metal-chelating activity are also antioxidants; and some peptides additionally have antioxidant activity due to the sequence and structure that is not fully understood (Power et al., 2013). The antioxidant property of milk protein hydrolysates can be utilized as a preservative to protect against aerobic spoilage of lipid-containing food products, or the health benefits could be the desirable property of the food ingredient. The antioxidant peptides and amino acids present in yoghurt supposedly contribute to preserving yoghurt from oxidative spoilage (Sabeena Farvin et al., 2010) and might have beneficial effects on human health (Pihlanto, 2006).

## 2.6 Antimicrobial activity

Several antimicrobial mechanisms are present in milk. The major antimicrobial activities come from the lactoperoxidase system (Seifu et al., 2005) and the antimicrobial protein lactoferrin (Superti, 2020). In addition to the antimicrobial activities of intact lactoferrin, this protein also harbors a potent antimicrobial peptide, lactoferricin B, which can be released by the gastric enzyme pepsin (Bellamy et al., 1992). Lactoferricin B is a 25-amino-acid-long peptide with a potent and broad spectrum of antibacterial and antifungal activities (Bellamy et al., 1994; Jones et al., 1994). Lactoferricin B is the most potent of the long range of antimicrobial peptides which can be released from caseins and whey proteins (Mohanty et al., 2016).

## 3 Ingredient products derived from milk proteins

Four decades of intensive research on bioactive dairy-derived peptides have resulted in surprisingly few new products on the market. The research typically substantiates the notion that milk and fermented dairy products are healthy and have numerous health benefits. Thereby, research generally supports the market position of dairy products as promoting good health. Similarly, the research on the benefits of peptides generated during milk fermentation have generally supported the position of existing fermented dairy products.

Dairy-derived functional protein ingredients are produced by most dairy companies in order to utilize whey and milk in excess of what is used in fresh dairy products (cf. Fig. 1). A few companies have taken dairy-derived proteins to a level going far beyond valorization of by-products. Peptides and proteins for infant formula and special nutrition can be more profitable than the primary dairy products. A survey of dairy-derived peptide products from major suppliers of functional protein ingredients is presented in Table 3. The survey was conducted by contacting executives responsible for R&D in the listed companies and by extracting product information from the web portals of those companies.

Casein and whey protein products are offered by all suppliers. A number of dairy companies are also able to provide dairy protein products to be used as protein supplements in foods and beverages and to increase protein quality and content.

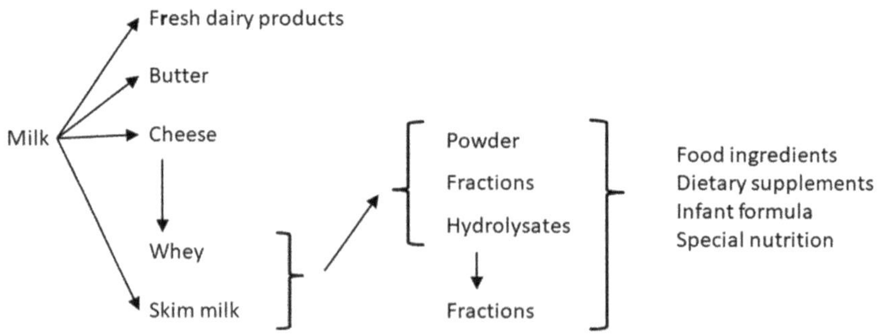


**Figure 1** Routes from milk to ingredients containing bioactive peptides.

**Table 3** Peptide ingredients and isolated proteins produced from milk

| Supplier | Country of origin | Protein and peptide ingredients | Web reference |
|---|---|---|---|
| Arla Food Ingredients | Denmark | Alpha-lactalbumin, Lacprodan®, Casein glycomacropeptide, Casein phosphopeptide, Osteopontin, Milk fat globule membrane, Whey proteins and hydrolysates | https://www.arlafoodsingredients.com |
| Asahi Group Foods, Ltd. | Japan | Lactotripeptides/AmealPeptide®/CALPIS® | https://www.amealpeptide.com/ |
| Fonterra | New Zealand | Lactoferrin, Lactoperoxidase, Caseins, Whey proteins and hydrolysates | https://www.nzmp.com |
| Friesland Campina | The Netherlands | NutriWhey™, Hyvital®, Vivinal®, Lactoferrin | https://www.frieslandcampinaingredients.com |
| Ingredia | France | Lactium®, PROFERRIN®, Pep2Dia®, OSTEUM | https://www.ingredia-nutritional.com |
| Lactalis | France | Caseins, Whey proteins, Prolacta®, Pronativ® | https://www.lactalisingredients.com |
| Morinaga Milk Industry | Japan | Lactoferrin, Peptide "MKP®", whey proteins | https://mnf-ap.com |

Purified single milk proteins, like alpha-lactalbumin, beta-lactoglobulin, lactoferrin and osteopontin, are offered by some suppliers. These full-length proteins have numerous interesting properties, including health benefits such as stimulation of mineral absorption, antimicrobial activity, and immune stimulation. The health benefits of these proteins have recently been reviewed by Permyakov (2020) and Demmelmair et al. (2017).

Mixed peptide products in the form of hydrolysates of caseins and whey proteins are produced by several suppliers. Hydrolysates can be applied in numerous food applications to provide taste and nutrition. In sports nutrition, hydrolysates are used to provide amino acids in an easily absorbable format. Extensively hydrolyzed products are essential for the production of infant formulas, specifically targeting children who are allergic to cow's milk. Such products provide an important health benefit for children suffering from allergy. The health benefit of such products is produced by removing bioactive peptides carrying intact epitopes (Von Berg et al., 2003). Peptides playing a specific role in allergy prevention and induction of tolerance also seem to be important for some of these products (Larsen et al., 2020).

Specific peptides are used only for a small range of products. However, since these types of products are the primary focus of this chapter, it is worthwhile to take a closer look at them: casein glycomacropeptide, alpha-casozepine (Lactium), CPP, lactotripeptides, peptide MKP (Met-Lys-Pro), and dipeptide AP (Arg-Pro).

### 3.1 Casein glycomacropeptide

Casein glycomacropeptide is a large peptide arising from the specific hydrolysis of kappa-casein by coagulating enzymes used to produce cheese from bovine milk. Kappa-casein is present in the milk of all mammals, whereas orthologs for any of the other caseins are missing in some species (Holt, 2016). Kappa-casein thus plays a central role in the biology of mammals. Its primary function is to stabilize the casein micelle and keep the micelle in suspension during lactation. The C-terminal half of kappa-casein is strongly hydrophilic, whereas the N-terminal half is hydrophobic and adheres to the other caseins in the casein micelle. The specific protease, chymosin, is able to cleave kappa-casein centrally, to release the hydrophilic C-terminal part from the casein micelle. The cleavage causes the micelles to coagulate. This reaction is utilized in the production of cheese to coagulate milk into cheese curd and whey. The C-terminal half of kappa-casein is hydrophilic in most mammals due to negatively charged glutamic acid residues, one or more phosphorylated serines, and several O-glycosylation sites at threonines or serines (Manguy and Shields, 2019). This part of kappa-casein is generally termed glycomacropeptide due to the heavy glycosylation. In bovine milk, glycosylation at all glycosylation sites is not complete, and approximately 40% of kappa-casein can be unglycosylated. The unglycosylated form of the macropeptide is termed caseinomacropeptide (O'Riordan et al., 2014). Casein glycomacropeptide and caseinomacropeptide can be isolated from whey. The process used depends on the desired purity of the peptide and will typically include membrane filtration and specific adsorptions. The macropeptide(s) from bovine milk has a life-saving 'bioactivity' which probably makes this peptide the most valuable bioactive peptide from bovine milk. Ironically, this activity is not one of the activities programmed by nature to benefit the newborn offspring. The amino acid phenylalanine is totally absent from the bovine macropeptides, whereas several other species have phenylalanine in this part of kappa-casein. The availability of a peptide devoid of phenylalanine serves as a good source of protein for the special nutrition needed by patients suffering from the congenital metabolic error phenylketonuria (Pena et al., 2018; Rocha and MacDonald, 2018; Manta-Vogli et al., 2020). Life and the normal brain development of children born with phenylketonuria can be maintained with a proper diet. Casein glycomacropeptide is also used in infant formula due to its high content of sialic acid, which supports the development of neurons (Wang et al., 2007), and casein glycomacropeptide is not only used in infant formula but also in food supplements due to its immune modulatory and antimicrobial properties. These activities will probably depend on the presence of glycosylation (Córdova-Dávalos et al., 2019).

### 3.2 Casein phosphopeptides

Casein phosphopeptides are casein hydrolysates enriched with peptides carrying the tracts of phosphorylated serines involved in carrying a high load of calcium phosphate in the casein micelle (Horne, 1998; Holt, 2016). Casein phosphopeptides carrying clusters of phosphoserine are able to bind divalent metal ions like $Ca^{2+}$, $Zn^{2+}$, $Mg^{2+}$, and $Mn^{2+}$ , and thereby facilitate the uptake of minerals (Scholz-Ahrens and Schrezenmeir, 2000; Mills et al., 2011).

CPP for commercial use is produced by the enzymatic hydrolysis of casein followed by precipitation, diafiltration, and chromatography (Ellegård et al., 1999). The products are used in early life nutrition and functional foods and drinks to stimulate mineral absorption in order to build strong bones and teeth or to slow down the loss of bone strength.

### 3.3 Lactotripeptides

Several peptides with proline at the C-terminal end have been identified as ACE inhibitors with the potential to reduce hypertension. The shorter peptides are absorbed more easily, and the tripeptides IPP and VPP are particularly considered to have this health benefit. Tripeptides are considered to be the active ingredient in the product Calpis from the Asahi group in Japan and the product Evolus from Valio in Finland. Morinaga also used the tripeptide MKP as part of their range of nutritional products. MKP is positioned as a product with beneficial effects on vascular health (Tada et al., 2020).

### 3.4 Dipeptide 'AP'

Ingredia has released a product, Pep2dia, enriched with the dipeptide AP, with the ability to inhibit alpha-glucosidase. The benefit of this product will be to assist the management of the glycemic response and regulate diabetic and prediabetic conditions (Sartorius et al., 2019).

### 3.5 Alpha-casozepine (Lactium)

Alpha-casozepine is a decapeptide released from alphaS1-casein by trypsin digestion. The peptide has affinity for the $GABA_A$ receptor (Miclo et al., 2001). Alpha-casozepine acts on the same receptor as benzodiazepines and is claimed to reduce anxiety without the side effects of benzodiazepines (Benoit, 2017). The EFSA evaluated the health claim of 'alleviation of psychological stress' and found that the effect relationship had not been established (EFSA, 2011). The company Ingredia produce a product, Lactium, that contains alpha-casozepine as the active ingredient. Lactium is positioned as a functional ingredient with a beneficial effect on mood and anxiety.

## 4 Challenges for peptide products derived from milk

### 4.1 Economy and yield

Cow's milk is produced in most regions of the world, and with an annual global production of 700 million tons, milk would thus seem to be readily available as a raw material for the extraction of valuable bioactive components. However, if the value of the milk has to be allocated entirely to the product produced, extraction of peptides from milk might not be competitive. The production of 1 g of peptide would typically require 5–50 kg of milk, representing a cost of US$2–20. This magnitude of raw material cost will probably make extraction from milk uncompetitive towards production by recombinant expression in bacteria or filamentous fungi.

The human version of any peptide with a potent and specific health effect would most likely be preferred in pharmaceutical applications, and recombinant production would be preferable for the same reason.

For food and food supplements, the value of peptide products might not be sufficient to allow recombinant production, and the type of products would then be limited to the ones preserving the general value of the milk required for the production. This will rarely be the case if the milk is depleted for a component associated with a recognized health-promoting property. It is therefore not surprising that several of the products on the market are products produced from whey or a by-product from the production of caseins and milk powders.

### 4.2 Availability of specific proteases

Bioactive peptides embedded within larger proteins will require specific hydrolysis at both ends of the peptide sequence, without destroying the biological activity by excessive internal hydrolysis. Such desired specificity might not be represented among commercially available proteases. If common digestive enzymes like trypsin or pepsin are able to release the bioactive peptide, one would assume this to happen during the normal digestion of dairy products. In this case, the potential health benefit of a peptide product might not be sufficiently increased to justify the increased processing cost. Commercially available industrial enzymes are either based on proteases extracted from animal tissue or highly active and relatively non-specific microbial proteases. The hydrolysis obtained with such enzymes would typically either resemble the normal digestion process or result in over-digestion and yield short peptides resistant to further hydrolysis.

Cell envelope proteinases (CEP) from lactic acid bacteria (LAB) are extracellular proteinases attached to the cell surface and often covalently attached to the peptidoglycan layer. The CEP enzymes have been found particularly useful for the release of encrypted bioactive peptides due to the diversity of specificities (Raveschot et al., 2018; Brown et al., 2017). The CEP enzymes are large multi-domain enzymes with a peptidase domain showing homology to the serine protease subtilisin (Siezen, 1999). It has been hypothesized that this class of enzymes is relatively unspecific for the bond cleaved, but selective regarding the substrate and the size of substrate in order to produce peptide products in the size range optimal for the oligopeptide transporter (Hansen and Marcatili, 2020). *Lactobacillus helveticus* strains have been particularly popular in the investigation of bioactive peptides released during milk fermentation (Raveschot et al., 2018). The study by Yamamoto et al. (1994) used a strain of *L. helveticus*. The two antihypertensive fermented milk products, Calpis® and Evolus®, are both produced with strains of *L. helveticus* (Yamamoto et al., 1994). A strain developed by Chr. Hansen A/S to produce yoghurt with high levels of ACE-inhibiting peptides, branded as Cardi04®, is also an *L. helveticus* strain (Flambard and Johansen, 2007).

The choice between enzymatic hydrolysis and fermentation for the generation of bioactive products from milk proteins or any other protein source will depend on the availability of enzymes with the desired specificity, the availability of strains with a suitable proteolytic activity, and the commercial value of the final product. Undoubtedly, further research into LAB proteases will generate a wider source of specific starter cultures for the production of 'bioactive fermented products'. This category of products seems to fit well with the current consumer trend showing a large interest in all kinds of traditional fermented products.

Although the CEP enzymes belong to the subtilisin class of enzymes (MEROPS S8), they have not been developed into industrial enzymes of commercial importance. The large-sized multi-domain structure and the apparent low proteolytic activity have made them unattractive as general-purpose proteases. The firm attachment to the cell wall, requirements for chaperones, and incompatibility with *E. coli* expression have also been factors contributing to the delay of enzyme products based on CEP enzymes. These obstacles could be turned into advantages when the function of the individual domains of the CEP enzymes are elucidated (Hansen and Marcatili, 2020). The CEP enzymes are good candidates for the development of proteases able to degrade proteins to oligopeptides rather than single amino acids. Such proteases might also be suitable for the production of hypoallergenic products and protein hydrolysates with functional or biological activity. Peptides or protein hydrolysates with preservation or texturizing activities have the potential to become the preferred future food ingredients. One example is the emulsifying peptides derived from potato protein through bioinformatics and hydrolysis (García-Moreno et al., 2020).

## 4.3 Regulatory framework

The research on milk protein-derived bioactive peptides has increased our fundamental understanding of human health and the general physiology of mammals. It has undoubtedly also contributed significantly to our understanding of infant nutrition, and it has allowed companies to constantly improve the composition of infant formula for general use or special needs.

When it comes to appraisals of the specific benefits of individual products, most countries require that the information be trustworthy. False, ambiguous, or misleading information is not allowed. This principle is quite universal, whereas the specific regulatory framework differs among regions and countries. The European countries have established common rules for regulating the area of food and feed and have established a common body, the EFSA, to oversee food safety in Europe. EFSA also regulates health claims in Europe. In the US, this role resides with the Food and Drug Administration (FDA). In the European Union, the EFSA determines whether a claim has been sufficiently substantiated to be allowed. The barrier is high, and will typically make it difficult for new products with novel benefits to obtain approval. This will protect the consumer from misleading information, but will also typically be a barrier to overcome while developing innovative products. In the USA, the FDA operates with two types of health claims, 'Authorized Health Claims' and 'Qualified Health Claims'. Authorized claims have a high barrier, similar to the European standards, whereas qualified claims are reviewed but not approved by the FDA. Qualified claims must be accompanied by a disclaimer to inform the consumer about the level of uncertainty.

In the European system, data is easily accessible and the evaluations and conclusions from the EFSA scientific panels are published in the *EFSA Journal* (https://www.efsa.europa.eu/en/publications).

So far, all the evaluations conducted by EFSA on health claims on milk-derived peptides or probiotic cultures have resulted in the claim not being authorized. Valio applied for approval of a claim of 'maintenance of normal blood pressure pursuant to Article 13 (5) of Regulation (EC) No 1924/2006' for the tripeptides IPP and VPP, the response to which was that 'the panel concludes that a cause and effect relationship has not been established' (EFSA, 2012). This was also the case for the evaluation of claims related to Valio's Evolus (EFSA, 2008), Ingredia's Lactium (EFSA, 2011), and for the peptide C12-(FFVAPFPEVFGK) (EFSA, 2010).

The EFSA's very restrictive approach to the approval of health claims will probably reduce the interest of companies in seeking EFSA approval. This might reduce the eagerness of European companies to invest in bioactive peptide product development.

## 5 New product development

Life science is evolving rapidly, and the various 'omics' technologies are advancing the capabilities available for research. Milk is common for all mammals, but also in a way is special for each species. Milk research will continue to be a source of amazing discoveries on nutrition and health, and this will certainly continue to include the field of bioactive peptides.

Predicting the future on a larger timescale will probably be futile, but it is worthwhile to take a look at the products to be expected during the next few years. This can be done by analyzing the current patent activity within the field of peptides from milk proteins. Few companies would dare to invest in expensive product development without trying to secure the intellectual property rights. Patent applications are made public after 18 months from filing, and we can therefore get a glimpse of the commercially oriented product developments via patent applications. The World Intellectual Property Organization (WIPO) maintains and offers easy access to a database, PATENTSCOPE, covering international PCT patent applications. A search was done in this database for applications on milk peptides, published from 2011 to 2020. The actual query is given in the legend to Fig. 2. The search revealed 269 PCT, EP, and US applications. A total of 12 applicants were responsible for 120 applications, and each had filed more than four applications. The distribution of the 120 patent applications between the 12 applicants is shown in the pie diagram of Fig. 2.

The companies most actively engaged in milk peptide product development are typically manufacturers of products providing nutrition for special needs, particularly infant formulas. Several of these companies are rooted in the dairy industry. The American infant nutrition producer, Mead Johnson Nutrition, is at the top of the list. Japan is represented by three companies: Megmilk Snow Brand, Calpis, and Morinaga Milk Industry. France, Israel, Switzerland, Sweden, Germany, Ireland, and Denmark are represented with one company each. Eight of the applicants are companies focusing on nutrition and three applicants (Mileutis, Hamlet Pharma, and Evonik) are pharmaceutical companies. The University of Wisconsin, Madison, is the single academic institution on the top 12 list of applicants for milk peptide patents.

The type of peptide products cover different mixed hydrolysates as well as specific defined peptides with particular properties. The specific well-defined peptides mentioned in the applications of the top 12 applicants are listed in Table 4, together with the target health property, applicant, and representative patent applications.

Allergy treatment and prevention is a target for several companies. Nestlé, N.V. Nutricia, and Evonik have patent applications defining specific peptides, and Mead Johnson has patent applications targeting allergy management with

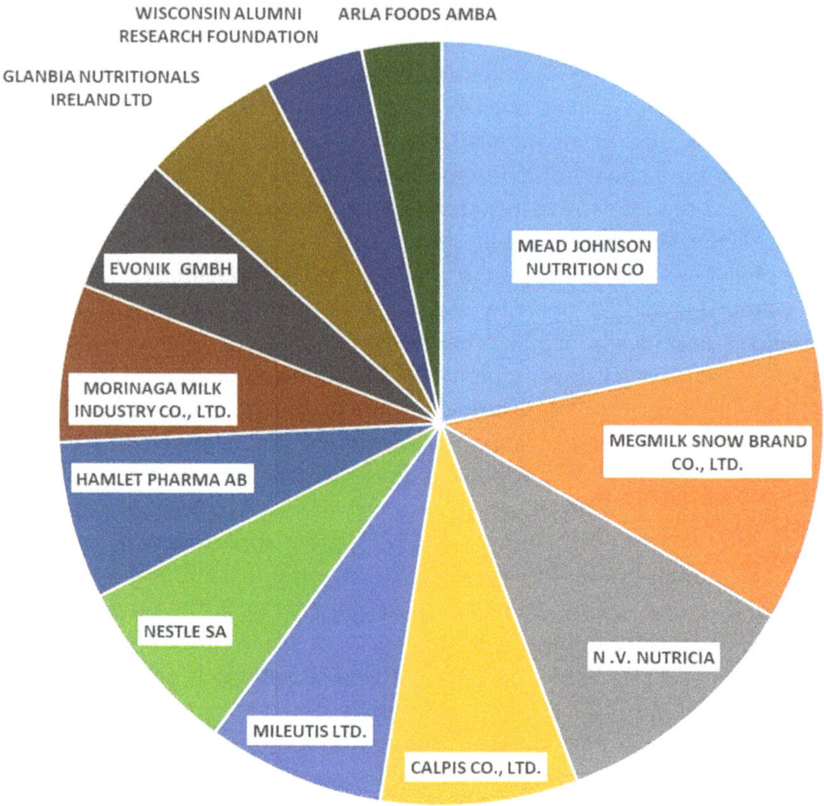


**Figure 2** The top 12 IP applicants in the field of milk-derived peptides. Relative patent application activity of the 12 most active applicants of PCT, EP, and US patents in the period from 2011 to 2020. Patents with abstracts mentioning the word 'peptide*' together with 'milk' or 'lactoferrin*' or 'casein' or 'whey*' or 'lactoglo*' or 'lactalbu*' were searched among PCT, EP, and US patent applications in the database of WIPO (https://patentscope.wipo.int). The exact query used was: 'AB:(peptide* and (milk or lactoferrin* or casein or whey* or lactoglo* or lactalbu*)) and CTR:(WO or EP or US) and DP:([from 01.01.2011 to 01.01.2021])'. The pie diagram show the 12 most active applicants.

hydrolysates. There seem to be competing interests and general agreement on the use of peptides from beta-lactoglobulin for the induction of tolerance toward bovine milk allergy. The desensitizing capacity of peptides derived from beta-lactoglobulin has also been demonstrated by researchers from the Technical University of Denmark (Bøgh et al., 2013).

Blood pressure management using casein-derived tripeptides seems to continue to be a field of active process and product development. Calpis and Nestlé have filed applications on the use of *L. helveticus* strains for the production of products rich in IPP and VPP tripeptides. The Japanese company Morinaga has filed patent applications on the tripeptide MKP, Met-Lys-Pro, a

**Table 4** Milk-derived peptides designated in patent applications by the top-12 IP applicants presented in Figure2

| Peptides | Protein source | Health target | Applicant | Patent application nr. |
|---|---|---|---|---|
| XIVTQTMKGLDIQKVAGTWYSLAMAAS<br>XIVTQTMKGLDIQKVAGTWYSLAMAASDISLL<br>TMKGLDIQKVAGTWYSLAMAASDISLL<br>TMKGLDIQKVAGTWYSLAMAASDISLLDAQ<br>DIQKVAGTWYSLAMAASDISLLDAQSAPLRVY | β-lactoglobulin | Bovine milk allergy | N.V. Nutricia | WO2013083140<br>WO2016148562<br>WO2019212332 |
| IVTQTM<br>KGLDIQK<br>DAQSAPLR<br>VEELKPTPE | β-lactoglobulin | Allergy, induction of tolerance | Nestlé | WO2015090347<br>WO2016198630 |
| IPP-VPP-LPP | Caseins | Hypertension | Nestlé | EP20599687 |
| XP, XPP, IPP, VPP | Caseins | Hypertension | Calpis | WO2012063826 |
| MKP | αS2-casein | Stroke prevention, hypertension, energy consumption | Morinaga | EP313652069<br>EP300436504 |
| GPVRGPFPIIV<br>HIQKEDVPSE<br>YPVEPFTESQ<br>QGLPQEVLNE<br>LEDSPEVIESPPEINT<br>SRYPSYGLN<br>YQKFPQYL | β-casein/ αS1-casein/ αS2-casein/ κ-casein | promoting GLP-1 secretion. | Morinaga | EP280243534 |
| NIPPLTQTPVVVPPFLQPE<br>SWMHQPHQPLPPTV-MFPPQSVL | β-casein | Brain function improvement | Calpis | WO2012176658<br>WO2012176659 |

*(Continued)*

**Table 4** (*Continued*)

| Peptides | Protein source | Health target | Applicant | Patent application nr. |
|---|---|---|---|---|
| RELEELNVPGEIVES(p)LS(p)S(p)S(p)EESITR<br>QMEAESIS(p)S(p)S(p)EEIVPDSVEQK<br>KNTMEHVS(p)S(p)S(p)EESIISNETYK<br>KVNELSKNIGS(p)ES(p)TEDQ<br>PTLNREQLS(p)TS(p)EENSKKTVD | β-casein<br>αS1-casein<br>αS2-casein | Uterine infections | Mileutis | WO2011132191 |
| KQFTKXELSQLLKDIDGYGGIALPELIXTMFHTSGYDTQA<br>LDDDITDDIMXAKKILDIKGIDYWLAHKALXTEKLEQWLXEKL<br>MAGWDIFGWFRDVLASLGLWNKH<br>MAGWDIFGWFRDVLA<br>KQFTKAELSQLLKDI<br>KQFTKAELSQLLKDIDGYGGIALPELIATM FHTSGYDTQ | α-Lactalbumin (human) | Cancer therapy | Hamlet Pharma AB | WO2014023976<br>WO2019243547 |
| KCFQWQRNMRKVRGPPVSCIKR | Lactoferrin (human) | Antigen masking agent | Evonik | EP91484196 |

tripeptide found in alphaS2-casein. This peptide is claimed to have the ability to reduce blood pressure, prevent stroke, and stimulate energy consumption (Tada et al., 2020). It is currently not clear how MKP products will be positioned in the market, and how they will be priced.

Morinaga also has patent applications on a set of casein-derived peptides able to stimulate GLP-1 secretion. Products based on such peptides could be useful for weight management and for the prevention of diabetes. The research leading to this patent application has also been published by Komatsu et al. (2019).

Calpis has filed interesting patent applications on brain function improvements using peptides derived from beta-casein. The current research in this field continues to identify novel brain-stimulating peptides (Ano et al., 2019). The potential application of such products could help in the prevention of amnesia and the general improvement of cognitive functions.

The pharmaceutical company Mileutis has filed patent applications on the antimicrobial activity of phosphopeptides of specific sequences derived from alpha and beta caseins. Hamlet Pharma has developed products for the treatment of cancer based on peptides from human alpha lactalbumin and has termed those products HAMLET (Ho et al., 2017; Nadeem et al., 2019). A peptide from human lactoferrin by the company Evonik has been used as an antigen-masking agent.

## 6 Conclusion

The research on milk protein-derived bioactive peptides has been a fertile field of fundamental research for several decades. The research has expanded our knowledge on the roles played by milk in the nutrition of newborns, and has also contributed considerably to our understanding of the health effects of milk and dairy products for the general consumer and consumers with special needs. Successful products developed based on bioactive peptides from milk have used the characteristic properties of milk proteins, which are:

- content of highly phosphorylated amino acids,
- content of proline,
- absence of phenylalanine in casein glycomacropeptide, and
- balanced amino acid composition.

These properties have allowed a range of products, typically in the form of protein hydrolysates with particular functional properties. For applications in infant nutrition, allergy prevention and management is a major field of research and development. Extensive hydrolysis can be used to remove the allergenic epitopes. Fortunately, it also seems that peptides from beta-lactoglobulin can facilitate the induction of tolerance.

Peptides derived from milk proteins have established applications in functional ingredients for managing hypertension, diabetes, mineral absorption, and antimicrobial activity. Research in product development in these fields seems to continue to produce innovations.

The release of peptides beyond the ones that will also be produced during normal digestion of dairy products will require the development of industrial enzymes with novel specificities - probably developed based on cell envelope proteases from LAB. Recombinant production of peptides with specific desirable biological functions is also likely to be implemented.

The brain continues to be the target of research and product development. Peptides interacting with the opioid receptors were among the first bioactive casein peptides to be researched, and current research reveals that peptides have the ability to improve brain function in animal models. Current products targeting the brain include a product for the reduction of anxiety and products for infant nutrition stimulating neural development.

## 7 Where to look for further information

The review paper by Nongonierma and FitzGerald (2016) would be a good place to start for acquiring a deeper understanding of biological activities of milk-derived peptides and the science behind it.

An overview of peptides of dairy origin used as functional ingredients is not readily available. The web pages of the companies listed in tables 3 and 4 are useful sources of information on current products and product innovations. The International Dairy Federation (IDF) organizes the global dairy industry and dairy research institutions. IDF establish working groups, defines industry standards, and organize scientific conferences. IDF is a good source of information on all aspects of dairy products and dairy production processes. Publications and events organized by IDF are announced on the IDF webpage (https://fil-idf.org/).

The European Food Safety Authority (EFSA) conducts evaluations of safety and health claims for food products on the European market. EFSA (https://www.efsa.europa.eu/) is a valuable source for information on safety and efficacy of all types of food and feed ingredients including dairy peptides.

## 8 Acknowledgements

I thank Dr Anne Staudt Kvistgaard from Arla Food Ingredients and Dr Ruud J. W. Schoemaker from Friesland Campina for valuable input and suggested references.

## 9 References

Alvarez-Bueno, C., Cavero-Redondo, I., Martinez-Vizcaino, V., Sotos-Prieto, M., Ruiz, J. R. and Gil, A. (2019). Effects of milk and dairy product consumption on Type 2 diabetes: overview of systematic reviews and meta-analyses. *Adv. Nutr.* 10(suppl_2), S154–S163. doi:10.1093/advances/nmy107.

Ano, Y., Kutsukake, T., Sasaki, T., Uchida, S., Yamada, K. and Kondo, K. (2019). Identification of a novel peptide from β-casein that enhances spatial and object recognition memory in mice. *J. Agric. Food Chem.* 67(29), 8160–8167. doi:10.1021/acs.jafc.9b02495.

Bellamy, W., Takase, M., Yamauchi, K., Wakabayashi, H., Kawase, K. and Tomita, M. (1992). Identification of the bactericidal domain of lactoferrin. *Biochim. Biophys. Acta* 1121(1–2), 130–136. doi:10.1016/0167-4838(92)90346-F.

Bellamy, W., Yamauchi, K., Wakabayashi, H., Takase, M., Takakura, N., Shimamura, S. and Tomita, M. (1994). Antifungal properties of lactoferricin B, a peptide derived from the N-terminal region of bovine lactoferrin. *Lett. Appl. Microbiol.* 18(4), 230–233. doi:10.1111/j.1472-765X.1994.tb00854.x.

Benoit, S. (2017). Impact of the administration of α-Casozepine, a benzodiazepine-like peptide from bovine α s1-casein, and of a proteolysis fragment on neural activity in mice. PhD thesis, Université de Lorraine, Nancy, France.

Bøgh, K. L., Barkholt, V. and Madsen, C. B. (2013). The sensitising capacity of intact β-Lactoglobulin is reduced by co-administration with digested β-Lactoglobulin. *Int. Arch. Allergy Immunol.* 161(1), 21–36. doi:10.1159/000343042.

Brantl, V., Teschemacher, H., Henschen, A. and Lottspeich, F. (1979). Novel opioid peptides derived from casein (beta-casomorphins). I. Isolation from bovine casein peptone. *Hoppe Seylers Z. Physiol. Chem.* 360(9), 1211–1216. doi:10.1515/bchm2.1979.360.2.1211.

Brown, L., Pingitore, E. V., Mozzi, F., Saavedra, L., Villegas, J. M. and Hebert, E. M. (2017). Lactic acid bacteria as cell factories for the generation of bioactive peptides. *Protein Pept. Lett.* 24(2), 146–155. doi:10.2174/0929866524666161123111333.

Córdova-Dávalos, L. E., Jiménez, M. and Salinas, E. (2019). Glycomacropeptide bioactivity and health: a review highlighting action mechanisms and signaling pathways. *Nutrients* 11(3). doi:10.3390/nu11030598.

Dalgleish, D. G. (2011). On the structural models of bovine casein micelles–review and possible improvements. *Soft Matter* 7(6), 2265–2272. doi:10.1039/C0SM00806K.

Demmelmair, H., Prell, C., Timby, N. and Lönnerdal, B. (2017). Benefits of lactoferrin, osteopontin and milk fat globule membranes for infants. *Nutrients* 9(8), 1–22. doi:10.3390/nu9080817.

Donoghue, M., Hsieh, F., Baronas, E., Godbout, K., Gosselin, M., Stagliano, N., Donovan, M., Woolf, B., Robison, K., Jeyaseelan, R., Breitbart, R. E. and Acton, S. (2000). A novel angiotensin-converting enzyme-related carboxypeptidase (ACE2) converts angiotensin I to angiotensin 1-9. *Circ. Res.* 87(5), E1–E9. doi:10.1161/01.RES.87.5.e1.

EFSA (2008). Evolus® and reduce arterial stiffness–scientific substantiation of a health claim related to Lactobacillus helveticus fermented Evolus® low-fat milk products and reduction of arterial stiffness pursuant to Article 14 of the Regulation (EC) No 1924/2006–scientific opinion of the panel on dietetic products, nutrition and allergies. *EFSA J.* 6(10), 1924, Article 14. doi:10.2903/j.efsa.2008.824.

EFSA (2009). Review of the potential health impact of β-casomorphins and related peptides. *EFSA J.* 7(2), 231r. doi:10.2903/j.efsa.2009.231r.

EFSA (2010). Scientific opinion on the substantiation of health claims related to a C12 peptide (Phe-Phe-Val-Ala-Pro-Phe-Pro-Glu-Val-Phe-Gly-Lys) and maintenance of normal blood pressure (ID 1483, 3130) pursuant to Article 13(1) of Regulation (EC) No 1924/2006. *EFSA J.* 8(2), 1–12, Article 13. doi:10.2903/j.efsa.2010.1478.

EFSA (2011). Scientific opinion on the substantiation of health claims related to "αS1-casein tryptic hydrolysate" and alleviation of psychological stress (ID 656, 1819, 4253) pursuant to. *EFSA J.* 9(6), 1–14, Article 13. doi:10.2903/j.efsa.2011.2273.

EFSA (2012). Scientific opinion on the substantiation of health claims related to isoleucine-proline-proline (IPP) and valine-proline-proline (VPP) and maintenance of normal blood pressure (ID 661, 1831, 1832, 2891, further assessment) pursuant to Article 13(1) of Regulation (EC) No 1924/2006. *EFSA J.* 10(6), Article 13. doi:10.2903/j.efsa.2012.2715.

Ellegård, K. H., Gammelgård-Larsen, C., Sørensen, E. S. and Fedosov, S. (1999). Process scale chromatographic isolation, characterization and identification of tryptic bioactive casein phosphopeptides. *Int. Dairy J.* 9(9), 639–652. doi:10.1016/S0958-6946(99)00135-1.

Flambard, B. and Johansen, E. (2007). Developing a functional dairy product: from research on Lactobacillus helveticus to industrial application of Cardi-04 ™ in novel antihypertensive drinking yoghurts. In: *Functional Dairy Products*. Elsevier, 506–520. doi:10.1533/9781845693107.3.506.

García-Moreno, P. J., Jacobsen, C., Marcatili, P., Gregersen, S., Overgaard, M. T., Andersen, M. L., Sørensen, A. M. and Hansen, E. B. (2020). Emulsifying peptides from potato protein predicted by bioinformatics: stabilization of fish oil-in-water emulsions. *Food Hydrocoll.* 101. doi:10.1016/j.foodhyd.2019.105529.

Hansen, E. B. and Marcatili, P. (2020). Modeled structure of the cell envelope proteinase of Lactococcus lactis. *Front. Bioeng. Biotechnol.* 8, 613986. doi:10.3389/fbioe.2020.613986.

Henschen, A., Lottspeich, F., Brantl, V. and Teschemacher, H. (1979). Novel opioid peptides derived from casein (beta-casomorphins). II. Structure of active components from bovine casein peptone. *Hoppe Seylers Z. Physiol. Chem.* 360(9), 1217–1224.

Ho, J. C. S., Nadeem, A. and Svanborg, C. (2017). Hamlet–a protein-lipid complex with broad tumoricidal activity. *Biochem. Biophys. Res. Commun.* 482(3), 454–458. doi:10.1016/j.bbrc.2016.10.092.

Holt, C. (2016). Casein and casein micelle structures, functions and diversity in 20 species. *Int. Dairy J.* 60, 2–13. doi:10.1016/j.idairyj.2016.01.004.

Horne, D. S. (1998). Casein interactions: casting light on the black boxes, the structure in dairy products. *Int. Dairy J.* 8(3), 171–177. doi:10.1016/S0958-6946(98)00040-5.

Iwaniak, A., Minkiewicz, P. and Darewicz, M. (2014). Food-originating ACE inhibitors, including antihypertensive peptides, as preventive food components in blood pressure reduction. *Compr. Rev. Food Sci. Food Saf.* 13(2), 114–134. doi:10.1111/1541-4337.12051.

Jäkälä, P. and Vapaatalo, H. (2010). Antihypertensive peptides from milk proteins. *Pharmaceuticals (Basel)* 3(1), 251–272. doi:10.3390/ph3010251.

Janecka, A., Fichna, J. and Janecki, T. (2004). Opioid receptors and their ligands. *Curr. Top. Med. Chem.* 4(1), 1–17.

Jia, C. L., Hussain, N., Joy Ujiroghene, O., Pang, X. Y., Zhang, S. W., Lu, J., Liu, L. and Lv, J. P. (2020). Generation and characterization of dipeptidyl peptidase-IV inhibitory peptides from trypsin-hydrolyzed α-lactalbumin-rich whey proteins. *Food Chem.* 318, 126333. doi:10.1016/j.foodchem.2020.126333.

Jones, E. M., Smart, A., Bloomberg, G., Burgess, L. and Millar, M. R. (1994). Lactoferricin, a new antimicrobial peptide. *J. Appl. Bacteriol.* 77(2), 208–214. doi:10.1111/j.1365-2672.1994.tb03065.x.

Komatsu, Y., Wada, Y., Izumi, H., Shimizu, T., Takeda, Y., Hira, T. and Hara, H. (2019). Casein materials show different digestion patterns using an in vitro gastrointestinal model and different release of glucagon-like peptide-1 by enteroendocrine GLUTag cells. *Food Chem.* 277, 423–431. doi:10.1016/j.foodchem.2018.10.123.

Kumar, R., Chaudhary, K., Sharma, M., Nagpal, G., Chauhan, J. S., Singh, S., Gautam, A. and Raghava, G. P. (2015). AHTPDB: a comprehensive platform for analysis and presentation of antihypertensive peptides. *Nucleic Acids Res.* 43(Database issue), D956–D962. doi:10.1093/nar/gku1141.

Kwok, A. J., Mashar, M., Khavandi, K. and Sabir, I. (2014). DPP-IV inhibitors: beyond glycaemic control? *Trends Cardiovasc. Med.* 24(4), 157–164. doi:10.1016/j.tcm.2013.10.002.

Larsen, J. M., Bang-Berthelsen, C. H., Qvortrup, K., Sancho, A. I., Hansen, A. H., Andersen, K. I. H., Thacker, S. S. N., Eiwegger, T., Upton, J. and Bøgh, K. L. (2020). Production of allergen-specific immunotherapeutic agents for the treatment of food allergy. *Crit. Rev. Biotechnol.* 40(6), 881–894. doi:10.1080/07388551.2020.1772194.

Liao, W. and Wu, J. (2020). The ACE2/Ang (1-7)/MasR axis as an emerging target for antihypertensive peptides. *Crit. Rev. Food Sci. Nutr.* 0, 1–15. doi:10.1080/10408398.2020.1781049.

Liu, Z. and Udenigwe, C. C. (2019). Role of food-derived opioid peptides in the central nervous and gastrointestinal systems. *J. Food Biochem.* 43(1), e12629. doi:10.1111/jfbc.12629.

Malkoski, M., Dashper, S. G., O'Brien-Simpson, N. M., Talbo, G. H., Macris, M., Cross, K. J. and Reynolds, E. C. (2001). Kappacin, a novel antibacterial peptide from bovine milk. *Antimicrob. Agents Chemother.* 45(8), 2309–2315. doi:10.1128/AAC.45.8.2309-2315.2001.

Manguy, J. and Shields, D. C. (2019). Implications of kappa-casein evolutionary diversity for the self-assembly and aggregation of casein micelles. *R. Soc. Open Sci.* 6(10), 190939. doi:10.1098/rsos.190939.

Manta-Vogli, P. D., Dotsikas, Y., Loukas, Y. L. and Schulpis, K. H. (2020). The phenylketonuria patient: a recent dietetic therapeutic approach. *Nutr. Neurosci.* 23(8), 628–639. doi:10.1080/1028415X.2018.1538196.

Maruyama, S. and Suzuki, H. (1982). A peptide inhibitor of angiotensin I converting enzyme in the tryptic hydrolysate of casein. Agr. Biol. Chem. 46(5), 1393–1394. doi:10.1271/bbb1961.46.1393.

Mellander, O. (1945). On casein from human and Cow's milk and their behaviour on hydrolysis with different proteolytic enzymes; preliminary report. (preliminary report). *Acta Paediatr.* 32(3–4), 668–683. doi:10.1111/j.1651-2227.1945.tb16854.x.

Mellander, O. and Isaksson, B. (1950). The physiological importance of the casein phosphopeptide calcium salts. I. Intravenous and peroral calcium cosage in animal experiments. *Acta Soc. Med. Ups.* 55(5–6), 239–246.

Mentlein, R. (1999). Dipeptidyl-peptidase IV (CD26)-role in the inactivation of regulatory peptides. *Regul. Pept.* 85(1), 9–24. doi:10.1016/S0167-0115(99)00089-0.

Miclo, L., Perrin, E., Driou, A., Papadopoulos, V., Boujrad, N., Vanderesse, R., Boudier, J. F., Desor, D., Linden, G. and Gaillard, J. L. (2001). Characterization of α-casozepine, a tryptic peptide from bovine α s1 -casein with benzodiazepine-like activity. *FASEB J.* 15(10), 1780–1782. doi:10.1096/fj.00-0685fje.

Mills, S., Ross, R. P., Hill, C., Fitzgerald, G. F. and Stanton, C. (2011). Milk intelligence: mining milk for bioactive substances associated with human health. *Int. Dairy J.* 21(6), 377–401. doi:10.1016/j.idairyj.2010.12.011.

Minkiewicz, P., Iwaniak, A. and Darewicz, M. (2019). BIOPEP-UWM database of bioactive peptides: current opportunities. *Int. J. Mol. Sci.* 20(23). doi:10.3390/ijms20235978.

Mohanty, D., Jena, R., Choudhury, P. K., Pattnaik, R., Mohapatra, S. and Saini, M. R. (2016). Milk derived antimicrobial bioactive peptides: a review. *Int. J. Food Prop.* 19(4), 837–846. doi:10.1080/10942912.2015.1048356.

Nadeem, A., Ho, J. C. S., Tran, T. H., Paul, S., Granqvist, V., Despretz, N. and Svanborg, C. (2019). Beta-sheet-specific interactions with heat shock proteins define a mechanism of delayed tumor cell death in response to HAMLET. *J. Mol. Biol.* 431(14), 2612–2627. doi:10.1016/j.jmb.2019.05.007.

Nielsen, S. D., Beverly, R. L., Qu, Y. and Dallas, D. C. (2017). Milk bioactive peptide database: a comprehensive database of milk protein-derived bioactive peptides and novel visualization. *Food Chem.* 232, 673–682. doi:10.1016/j.foodchem.2017.04.056.

Nongonierma, A. B., Dellafiora, L., Paolella, S., Galaverna, G., Cozzini, P. and FitzGerald, R. J. (2018). In silico approaches applied to the study of peptide analogs of Ile-Pro-Ile in relation to their dipeptidyl peptidase IV inhibitory properties. *Front. Endocrinol. (Lausanne)* 9, 329. doi:10.3389/fendo.2018.00329.

Nongonierma, A. B. and FitzGerald, R. J. (2013). Inhibition of dipeptidyl peptidase IV (DPP-IV) by proline containing casein-derived peptides. *J. Funct. Foods* 5(4), 1909–1917. doi:10.1016/j.jff.2013.09.012.

Nongonierma, A. B. and FitzGerald, R. J. (2016). Strategies for the discovery, identification and validation of milk protein-derived bioactive peptides. *Trends Food Sci. Technol.* 50, 26–43. doi:10.1016/j.tifs.2016.01.022.

O'Riordan, N., Kane, M., Joshi, L. and Hickey, R. M. (2014). Structural and functional characteristics of bovine milk protein glycosylation. *Glycobiology* 24(3), 220–236. doi:10.1093/glycob/cwt162.

Patil, P., Mandal, S., Tomar, S. K. and Anand, S. (2015). Food protein-derived bioactive peptides in management of type 2 diabetes. *Eur. J. Nutr.* 54(6), 863–880. doi:10.1007/s00394-015-0974-2.

Pena, M. J., Pinto, A., Daly, A., Macdonald, A., Azevedo, L., Rocha, J. C. and Borges, N. (2018). The use of glycomacropeptide in patients with phenylketonuria: a systematic review and meta-analysis. *Nutrients* 10(11). doi:10.3390/nu10111794.

Permyakov, E. A. (2020). α-lactalbumin, amazing calcium-binding protein. *Biomolecules* 10(9), 1–50. doi:10.3390/biom10091210.

Pihlanto, A. (2006). Antioxidative peptides derived from milk proteins. *Int. Dairy J.* 16(11), 1306–1314. doi:10.1016/j.idairyj.2006.06.005.

Power, O., Jakeman, P. and Fitzgerald, R. J. (2013). Antioxidative peptides: enzymatic production, in vitro and in vivo antioxidant activity and potential applications of milk-derived antioxidative peptides. *Amino Acids* 44(3), 797–820. doi:10.1007/s00726-012-1393-9.

Raveschot, C., Cudennec, B., Coutte, F., Flahaut, C., Fremont, M., Drider, D. and Dhulster, P. (2018). Production of bioactive peptides by lactobacillus species: from gene to application. *Front. Microbiol.* 9, 2354. doi:10.3389/fmicb.2018.02354.

Rocha, J. C. and MacDonald, A. (2018). Treatment options and dietary supplements for patients with phenylketonuria. *Expert Opin. Orphan Drugs* 6(11), 667-681. doi:10.1080/21678707.2018.1536541.

Sabeena Farvin, K. H., Baron, C. P., Nielsen, N. S., Otte, J. and Jacobsen, C. (2010). Antioxidant activity of yoghurt peptides: part 2–characterisation of peptide fractions. *Food Chem.* 123(4), 1090–1097. doi:10.1016/j.foodchem.2010.05.029.

Sah, B. N. P., Vasiljevic, T., McKechnie, S. and Donkor, O. N. (2018). Antioxidative and antibacterial peptides derived from bovine milk proteins. *Crit. Rev. Food Sci. Nutr.* 58(5), 726–740. doi:10.1080/10408398.2016.1217825.

Saito, T. (2008). Antihypertensive peptides derived from bovine casein and whey proteins. *Adv. Exp. Med. Biol.* 606, 295–317. doi:10.1007/978-0-387-74087-4_12.

Sartorius, T., Weidner, A., Dharsono, T., Boulier, A., Wilhelm, M. and Schön, C. (2019). Postprandial effects of a proprietary milk protein hydrolysate containing bioactive peptides in prediabetic subjects. *Nutrients* 11(7), 1–16. doi:10.3390/nu11071700.

Savastano, M. L., Liu, Y., Mels, J., Dittrich, D., Haus, S., Gensberger-Reigl, S. and Pischetsrieder, M. (2019). Profiling of multiphosphorylated peptides in kefir and their release during simulated gastrointestinal digestion. *ACS Omega* 4(5), 7963-7970. doi:10.1021/acsomega.8b03105.

Scholz-Ahrens, K. E. and Schrezenmeir, J. (2000). Effects of bioactive substances in milk on mineral and trace element metabolism with special reference to casein phosphopeptides. *Br. J. Nutr.* 84 (Suppl. 1), S147–S153. doi:10.1017/s0007114500002373.

Seifu, E., Buys, E. M. and Donkin, E. F. (2005). Significance of the lactoperoxidase system in the dairy industry and its potential applications: a review. *Trends Food Sci. Technol.* 16(4), 137–154. doi:10.1016/j.tifs.2004.11.002.

Siezen, R. J. (1999). Multi-domain, cell-envelope proteinases of lactic acid bacteria. *Antonie Leeuwenhoek Int. J. Gen. Mol. Microbiol.* 76(1–4), 139–155. doi:10.1023/A:1002036906922.

Siltari, A., Vapaatalo, H. and Korpela, R. (2019). Milk and milk-derived peptides combat against hypertension and vascular dysfunction: a review. *Int. J. Food Sci. Technol.* 54(6), 1920–1929. doi:10.1111/ijfs.14056.

Summer, A., Di Frangia, F., Ajmone Marsan, P., De Noni, I. and Malacarne, M. (2020). Occurrence, biological properties and potential effects on human health of β-casomorphin 7: current knowledge and concerns. *Crit. Rev. Food Sci. Nutr.* 60(21), 3705-3723. doi:10.1080/10408398.2019.1707157.

Sun, N., Wu, H., Du, M., Tang, Y., Liu, H., Fu, Y. and Zhu, B. (2016). Food protein-derived calcium chelating peptides: a review. *Trends Food Sci. Technol.* 58, 140–148. doi:10.1016/j.tifs.2016.10.004.

Superti, F. (2020). Lactoferrin from bovine milk: a protective companion for life. *Nutrients* 12(9), 1–26. doi:10.3390/nu12092562.

Tada, A. M., Hamezah, H. S., Yanagisawa, D., Morikawa, S. and Tooyama, I. (2020). Neuroprotective effects of casein-derived peptide Met-Lys-Pro (MKP) in a hypertensive model. *Front. Neurosci.* 14, 845. doi:10.3389/fnins.2020.00845.

Von Berg, A., Koletzko, S., Grübl, A., Filipiak-Pittroff, B., Wichmann, H. E., Bauer, C. P., Reinhardt, D., Berdel, D. and German Infant Nutritional Intervention Study Group

(2003). The effect of hydrolyzed cow's milk formula for allergy prevention in the first year of life: the German infant nutritional intervention study, a randomized double-blind trial. *J. Allergy Clin. Immunol.* 111(3), 533–540. doi:10.1067/mai.2003.101.

Wang, B., Yu, B., Karim, M., Hu, H., Sun, Y., McGreevy, P., Petocz, P., Held, S. and Brand-Miller, J. (2007). Dietary sialic acid supplementation improves learning and memory in piglets. *Am. J. Clin. Nutr.* 85(2), 561–569. doi:10.1093/ajcn/85.2.561.

Yamamoto, N., Akino, A. and Takano, T. (1994). Antihypertensive effect of the peptides derived from casein by an extracellular proteinase from Lactobacillus helveticus CP790. *J. Dairy Sci.* 77(4), 917–922. doi:10.3168/jds.S0022-0302(94)77026-0.

# Chapter 8

## Advances in dairy lipid science: physicochemical aspects

*Daylan A. Tzompa-Sosa, Ghent University, Belgium; and Naomi Arita-Merino, Wageningen University, The Netherlands*



## 1 Introduction

Ruminant milk fat is one of the most complex natural fats, because milk is composed of different types of lipids (glycerolipids, glycerophospholipids, sphingolipids, ceramides, sterols, liposoluble vitamins etc.) (Liu et al., 2020b). Glycerolipids are the most abundant lipid class in milk fat, accounting for 90–99% of the total lipids, of which triacylglycerols (TAG) are the most abundant type (79–95%) (Liu et al., 2018; Jensen, 2002). The other milk lipids with nutritional and industrial relevance are phospholipids (1.1 %) and cholesterol (0.25–0.46%) (Tzompa-Sosa et al., 2016b; Jensen, 2002). Lipid classes like ceramides and sphingolipids are present in small concentrations.

The complexity of milk fat is also attributed to the variety of fatty acids (FA) present, which can be produced *de novo* in the mammary gland or can be blood-derived (Bionaz and Loor, 2008). From the 400 different FA claimed to be present in milk fat (Jensen, 2002), only about 220 FA are separated by 100 m fused silica GC columns. Of these FA, only about 50 FA are typically identified and quantified. These 50 or more FA account for about 97% of the total FA (Table 1).

http://dx.doi.org/10.19103/AS.2022.0099.08

**Table 1** Identified fatty acids (g/100 g of FA) in milk fat from a Dutch Holstein Friesian cow population (N = 865) on a winter diet. The fatty acids were analyzed as fatty acid methyl esters (FAME) in a 100 m fused silica GC column

| Saturated FA | Mean | Min | Max | Unsaturated FA | Mean | Min | Max |
|---|---|---|---|---|---|---|---|
| C4:0 | 3.52 | 2.44 | 4.57 | C10:1 | 0.36 | 0.02 | 0.56 |
| C5:0 | 0.03 | 0.00 | 0.12 | C12:1 | 0.11 | 0.00 | 0.22 |
| C6:0 | 2.21 | 1.07 | 2.71 | C14:1cis9 | 1.33 | 0.24 | 2.40 |
| C7:0 | 0.03 | 0.00 | 0.14 | C16:1trans9 | 0.05 | 0.00 | 0.16 |
| C8:0 | 1.35 | 0.47 | 1.85 | C16:1cis9 | 1.41 | 0.57 | 3.34 |
| C9:0 | 0.04 | 0.00 | 0.20 | C17:1cis9 | 0.18 | 0.11 | 0.48 |
| C10:0 | 2.99 | 0.95 | 4.62 | C18:1trans6 | 0.21 | 0.11 | 0.90 |
| C11:0 | 0.07 | 0.00 | 0.36 | C18:1trans9 | 0.15 | 0.09 | 0.63 |
| C12:0 | 4.09 | 1.43 | 6.85 | C18:1trans10 | 0.24 | 0.00 | 5.23 |
| C13:0 | 0.11 | 0.02 | 0.41 | C18:1trans11 | 0.80 | 0.30 | 2.51 |
| C14:0 iso | 0.09 | 0.04 | 0.17 | C18:1trans12 | 0.13 | 0.00 | 0.61 |
| C14:0 | 11.66 | 6.17 | 14.94 | C18:1cis9 | 18.36 | 12.34 | 33.41 |
| C15:0 iso | 0.02 | 0.00 | 0.37 | C18:1trans15 | 0.22 | 0.00 | 0.51 |
| C15:0 anteiso | 0.49 | 0.25 | 0.79 | C18:1cis11 | 0.41 | 0.19 | 1.02 |
| C15:0 | 1.17 | 0.69 | 2.66 | C18:1cis12 | 0.20 | 0.07 | 0.54 |
| C16:0 iso | 0.21 | 0.02 | 0.52 | C18:1cis13 | 0.09 | 0.05 | 0.23 |
| C16:0 | 32.22 | 22.86 | 42.95 | C18:2cis9,12 | 1.21 | 0.56 | 3.42 |
| C17:0 iso | 0.30 | 0.17 | 0.46 | C18:3cis9,12,15 | 0.43 | 0.18 | 1.03 |
| C17:0 anteiso | 0.48 | 0.00 | 0.87 | C18:2cis9,trans11 | 0.41 | 0.02 | 1.25 |

| | | | | | | | |
|---|---|---|---|---|---|---|---|
| C17:0 | 0.46 | 0.30 | 0.83 | C18:2trans10,cis12 | 0.00 | 0.00 | 0.11 |
| C18:0 | 8.76 | 3.19 | 17.12 | C20:1cis11 | 0.05 | 0.00 | 0.13 |
| C19:0 | 0.06 | 0.00 | 0.19 | C20:2cis11,14 | 0.00 | 0.00 | 0.07 |
| C20:0 | 0.12 | 0.04 | 0.28 | C20:3cis8,11,14 | 0.07 | 0.00 | 0.17 |
| C21:0 | 0.03 | 0.00 | 0.08 | C22:1cis13 | 0.00 | 0.00 | 0.13 |
| C22:0 | 0.13 | 0.00 | 0.22 | C20:5cis5,8,11,14,17 | 0.03 | 0.00 | 0.13 |
| C24:0 | 0.02 | 0.00 | 0.20 | C22:5cis7,10,13,16,19 | 0.05 | 0.00 | 0.19 |
| | | | | C22:6cis4,7,10,13,16,19 | 0.02 | 0.00 | 0.10 |
| SFA | 70.65 | 49.51 | 78.44 | UFA | 26.51 | 19.36 | 45.40 |
| Unknown FA | 2.84 | 1.5 | 4.05 | | | | |

The broad range of FA identified has direct implications on the complexity of milk fat, since FA are esterified into almost every lipid class. This esterification is steered by enzymes in the milk fat synthesis pathway that respond to FA availability and to the enzymatic preferences (Bionaz and Loor, 2008; Tzompa-Sosa et al., 2014). Therefore, FA esterification is non-random (Blasi et al., 2008). Esterification of a diverse range of FA within the glycerol backbone of glycerolipids, or to other lipid classes, produces various arrangements. These individual arrangements within a lipid class are called molecular lipid species. The latest survey on lipid species in bovine milk fat accounted for 3454 molecular species of TAG and 524 molecular species of phospholipids, sphingolipids, glycolipids and ceramides (Liu et al., 2020a,b). Hardly any other natural oil or fat shows such a large number of lipid species.

The variability of milk fat FA is partly explained by the animal's diet and partly by its genetics. Heritability is an estimate that helps to understand whether an FA will be affected by the cow's diet or by the cow's genetics. The highest heritability corresponds to short- and medium-chain FA (C4:0–C16:0), whereas the lowest heritability corresponds to trans FA (Hein et al., 2018; Schennink et al., 2007). This is in agreement with the origin of the FA in milk fat. While short-chain FA are produced *de novo* in the mammary gland, trans FA and unsaturated C18 FA are blood-derived. Trans FA are produced in the rumen by biohydrogenation from dietary unsaturated C18 (Bionaz and Loor, 2008).

Regarding FA variability in milk fat in the Dutch Holstein Friesian population, it was reported that the highest yearly variation, as estimated by the coefficient of variance (CV > 10), was seen in trans FA, including C16:1 and C18:1 trans FA (Heck et al., 2009). In absolute concentration, the FA with the largest variation between feeding regimes (summer vs winter diet) and across breeds are C14:0, C16:0, C18:0 and C18:1 cis 9. These FA are the most abundant in milk fat, C16:0 being the most abundant (yearly average = 28.1–31.9 g/100 g of FA), followed by C18:1cis9 (yearly average = 17.6–23.6 g/100 g of FA), C14:0 (yearly average = 9.5–12 g/100 g of FA) and C18:0 (yearly average = 10–11.7 g/100 g of FA) (Gottardo et al., 2017; Heck et al., 2009; O'Donnell-Megaro et al., 2011).

Changes in milk fat FA composition directly impact its TAG profile, since FA are the building blocks of TAG. We recommend the study of Pacheco-Pappenheim et al. (2019) for a recent overview on the effect of feeding and genetics on the TAG profile. Although FA are relevant from a nutritional perspective, TAG are more useful to understand milk fat biosynthesis and milk fat thermal behavior. After all, TAG are the final product of milk fat synthesis and are the building blocks of fat crystals. Therefore, current studies on milk fat incorporate TAG analysis into their investigations to unravel milk fat crystallization (Staniewski et al., 2021; Tzompa-Sosa et al., 2016a).

The crystallization and melting behaviors of milk fat are important because they affect the functional and sensory attributes of products that contain it (i.e. the firmness and spreadability of butter, the structure of whipped cream, the glossiness and mouth-feel of chocolate and the smoothness of ice cream) (Campos et al., 2002; Narine and Marangoni, 1999; Marangoni et al., 2011; Wiking et al., 2009). Fat crystallization starts with nucleation, which is the formation of tiny embryonic crystals called nuclei. The nuclei grow and form crystals. TAG are the fundamental units of crystals and can pack in different crystal arrangements (polymorphs) that differ in stability (Sato, 2001). Later in this chapter, we further discuss the different structures formed during milk fat crystallization and how to study them. Milk fat crystallization is affected by external factors (e.g. cooling rate and crystallization temperature) and by the composition.

The wide variety of TAG in milk fat results in complex crystallization and melting behaviors. As the melting points and properties of the thousands of different TAG species differ, the liquid and different crystalline phases form in milk fat over a wide temperature range. In Fig. 1, we show the characteristic thermograms of anhydrous milk fat recorded during cooling and heating; in the typical crystallization and melting curves, we observe two exothermic peaks linked to the crystallization of two crystalline phases, and three melting regions. These melting regions are associated to low-, middle- and high-melting fractions. Depending on the temperature, different TAG groups of these fractions reach a supercooled state in which they can crystallize; thus, the extent of the crystallization is also determined by the composition. Moreover, the crystalline phases formed, their characteristics and their composition are affected by the formation of mixed crystals (between

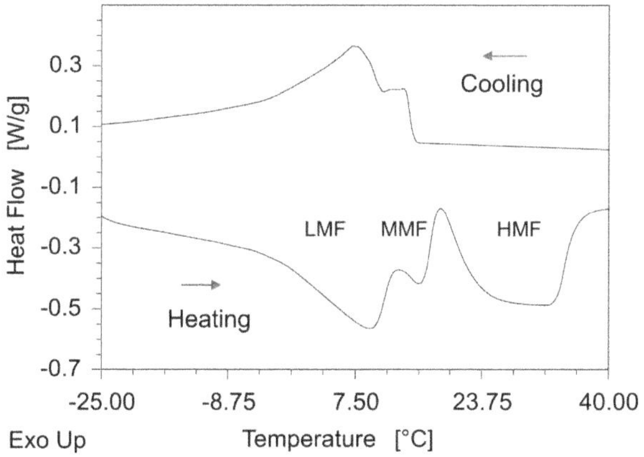


**Figure 1** Crystallization and melting DSC thermograms of anhydrous milk fat obtained during cooling at 5°C/min and heating at 10°C/ min heating rate. Low-, middle- and high-melting fractions are shown (LMF, MMF and HMF, respectively).

different TAG species), the intermiscibility of TAG and their ability to form different crystalline structures (Timms, 1984). Overall, the already complex study of fat crystallization is further complicated by the wide diversity in milk fat composition.

Although dairy lipids have been studied for decades, their industrial and nutritional relevance as well as their physicochemical complexity render their study an active research field. This chapter focuses on the important advances that have been made in this field toward three objectives: high-throughput methods for TAG profiling, the structural and physical characterization of milk fat crystal networks and the understanding of factors affecting the physicochemical properties of dairy lipids. Moreover, milk fat polymorphism and crystal networks are explained based on the milk fat TAG content. For further reading on this matter, excellent up-to-date references are included at the end of this chapter.

## 2 Chemical analysis techniques in dairy lipids

### *2.1 Established chromatographic techniques*

Glycolipids and glycerophospholipids account for most of the lipid content in milk fat (>99%) (MacGibbon, 2020). Both lipid types have FA as building blocks. The latter, coupled with their nutritional importance, have made FA profiling the most recurrent type of lipid analysis. TAG profiling based on its carbon number has also found its way into routine analysis to identify adulteration. Routine profiling of FA and TAG is typically performed by gas chromatography with flame-ionization detection (GC-FID). A practical guide for lipid analysis by gas chromatography can be found in the book by Christie (1994). Lately, infrared-based devices have been introduced as a quick method of compositional analysis of milk, including FA profiling and quantification (Rodriguez-Otero et al., 1997; Rutten et al., 2009). ISO standards have laid down the methods for milk fat extraction, the preparation of fatty acid methyl esters from milk fat and its profiling. The methods outlined in the official guidelines below are useful references for these routine analyses.

- ISO 2446, Milk - Determination of fat content (Routine method);
- ISO 14156|IDF 172, Milk and milk products - Extraction methods for lipids and liposoluble compounds;
- ISO 15884|IDF 182 Milk fat - Preparation of fatty acid methyl esters;
- ISO 15885|IDF 184 Milk fat - Determination of the fatty acid composition by gas-liquid chromatography;
- ISO 23065:2009 Milk fat from enriched dairy products - Determination of omega-3 and omega-6 fatty acid content by gas-liquid chromatography;

- ISO Standard 17678/IDF 202: 2010 Milk and milk products - Determination of milk fat purity by gas chromatographic analysis of triglycerides; and
- Commission Regulation (EC) No. 273/2008 of 5 March 2008 laying down detailed rules for the application of Council Regulation (EC) No 1255/1999 as regards methods for the analysis and quality evaluation of milk and milk products.

## 2.2 High-throughput techniques for TAG profiling

Advanced lipid analyses include identification, quantification and structural elucidation of lipid species. The aims of such detailed profiling are to increase our understanding of bovine lipid metabolism, to link lipid composition to physical behavior of milk fat and milk-fat-rich products and to associate lipid species to human health (e.g. digestion and absorption, relation with human diseases, etc.) (Tzompa-Sosa et al., 2016a, 2018; Liu et al., 2018; González-Domínguez et al., 2014; Lopez-Lopez et al., 2001; Bar-Yoseph et al., 2013; Hageman et al., 2019). Most of the attention on detailed profiling and structural analysis of milk fat has been given to FA and TAG, since these two lipid types are related to the nutritional value, the physical behavior of fats, and milk fat absorption and synthesis.

The molecular profiling of lipid species and their structural elucidation have been recently improved with the increased capabilities of mass spectrometry (MS) and with the use of novel techniques such as shotgun and time-of-flight (TOF). Moreover, open databases, such as the LIPID MAPS Consortium, have improved and quickened lipid identification (Fahy et al., 2007). With these new approaches, all the lipids from biological samples are analyzed simultaneously, limiting sample handling and experimental error. These new techniques have increased sensitivity and improved the quality of the data. However, they require handling of big data sets, where programming knowledge comes in handy. Moreover, knowledge on multivariate statistical methods is required for data analysis.

In the present section, we will focus our attention on two high-throughput techniques for TAG profiling, which have been proven useful in milk fat analysis. The techniques to be discussed are MALDI-TOF-MS and shotgun lipidomics. In addition, recent advances on the methods to study TAG structure will be presented.

### 2.2.1 MALDI-TOF-MS

Matrix-assisted laser desorption ionization (MALDI) is a technique where non-volatile compounds, such as TAG, are co-crystallized with a matrix and a cationization agent. For a brief introduction to the characterization of lipids by MALDI time-of-flight mass spectrometry (TOF-MS), we can recommend the summary in the AOCS lipid library (Lay Jr, 2020). In lipid applications, the ratio of matrix/milk fat/cationization agent used is 1:1:0.5. Appropriate MALDI matrices need to have the ability to form small crystals in the plate and to co-crystallize well with the cationization agent (0.1M NaCl solution) and the fat. In our experience, 2,5-dihydroxybenzoic acid (DHB) and 2′, 4′ 6, tri-hydroxy-aceto-phenone are MALDI matrices that co-crystallize properly with milk fat. After soft-ionization of TAG, the sodiated TAG (TAG + Na) are desorbed from the matrix. Then, TAG are separated in the TOF and identified by MS. MALDI is a soft-ionization technique, therefore the analytes are profiled as molecules in their entire form without ion fragmentation. Hence, the list of masses obtained are the molecular weight of the sodiated TAG (TAG + Na). Usually, identification of the TAG is performed using a lipidomic database, such as the one from LipidMaps Consortium or from other available databases.

MALDI-TOF-MS analysis allows profiling of TAG species based on the number of carbon atoms and the degree of unsaturation (Fig. 2) (Picariello et al., 2007). Using this technique, more than 60 TAG species with up to six unsaturation points have been identified in milk fat (Picariello et al., 2007; Tzompa-Sosa et al., 2018). TAG profiling is possible from CN26:0. Smaller TAG species cannot be properly identified with this technique because they get fragmented at a high proportion. This technique has the advantage of being fast (2–3 seconds per sample). If the equipment is used in the automatic mode, a large number of samples can be analyzed in the same run in a short time. Typically, MALDI plates have the ability to hold between 96 and 384 samples. The automatic mode of the equipment and the large number of samples held in a MALDI plate make this technique suitable for experiments with large sample sets.

MALDI-TOF-MS with post-source decay (PSD) and MS detection is used for elucidation of the FA composition of each TAG species. With this technique (MALDI-TOF-MS-MS), individual TAG species lose one FA, and in some cases, the sodiated residual (Fig. 3). Therefore, the final profile consists of a mixture of diacylglycerols (DAG). The mass of the lost FA is identified by subtraction of the DAG mass from the fragmented TAG mass (also called parent mass). The mechanism of diacyl production by PSD has not been studied yet, therefore it is assumed that the TAG lose FA in a random fashion. Therefore, it is not possible to elucidate the positioning of the lost FA, but identification of the FA within a TAG species is feasible. One limitation of PSD when analyzing milk fat is that if a neighboring TAG is equally abundant, then separation of the targeted TAG species is difficult. When this occurs, two parent masses appear in the spectrum indicating that two TAG are being fragmented. In this case, prior fragmentation of the milk fat is necessary (Yener and van Valenberg, 2019; Tzompa-Sosa et al., 2018).

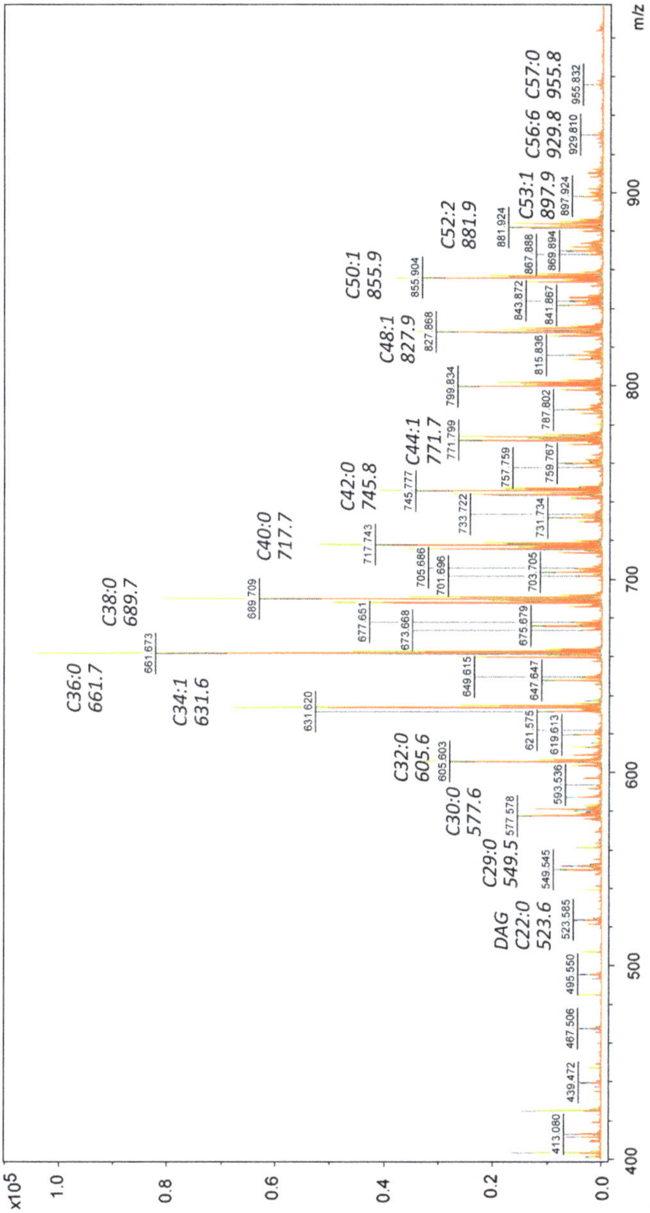


**Figure 2** Triacylglycerol (TAG +Na) profile of milk fat obtained by MALDI-TOF-MS. Target range from 400 m/z to 1000 m/z.

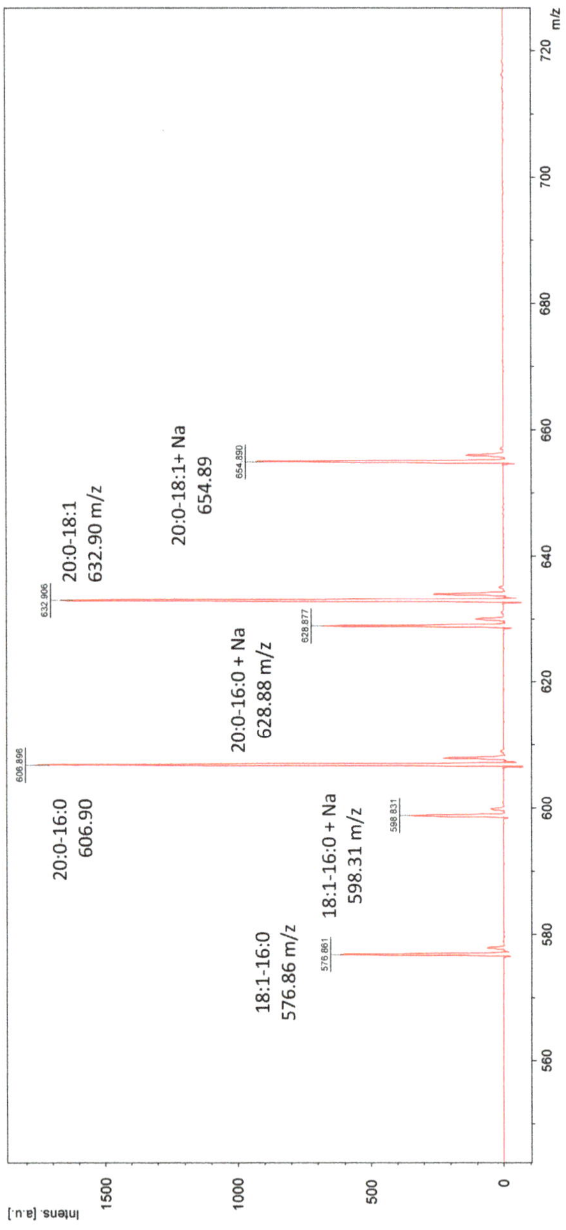


**Figure 3** Fragmentation profile of TAG standard C18:1–C20:0–C16:0 (10 mg/ml) with a parent mass of 911.2 (TAG+NA) using MALDI TOF-MS with post-source decay.

Quantification of TAG by MALDI-TOF-MS has not been reported yet in the scientific literature. Critics of this technique argue that the intensity of the signal obtained cannot be used as means to quantify these molecules, since different lipid species react in different ways to this type of ionization. However, milk fat is composed almost exclusively of TAG molecules. Therefore, in principle, every TAG will be ionized in a similar manner. Moreover, studies comparing TAG profiles obtained by MALDI-TOF-MS and by GC-FID showed that both profiles are in line (Yener and van Valenberg, 2019; Tzompa-Sosa et al., 2016a). Hence, quantification of TAG by MALDI-TOF-MS could be feasible. However, efforts on method development are still necessary.

### 2.2.2 Shotgun lipidomics

Shotgun lipidomics, also called direct-infusion lipidomics, employs electrospray ionization (ESI) techniques to separate molecular species of lipids. ESI generates ammonium adducts of TAG ($TAG+NH_4$), which are later detected by MS, allowing determination of TAG molecular weights. Quantification is possible with the use of internal standards from the same lipid class as the one intended for quantification and absent in the original sample. Typically, one or two internal standards are needed for quantification (Han and Gross, 2005). However, in milk fat, three internal standard species are required for accurate quantification of the entire TAG range (Bukowski and Picklo, 2020). Shotgun lipidomics allows identification and quantification of > 100 TAG species (Bukowski and Picklo, 2020). One drawback of the use of shotgun lipidomics for milk fat TAG profiling, compared to MALDI-TOF-MS, is that the milk sample requires isolation of TAG by solid-phase extraction prior to TAG profiling.

Similar to PSD in MALDI-TOF-MS-MS, the FA composition of individual TAG species can be elucidated by collision-induced decomposition (CID) followed by MS detection. CID results in the loss of FA-yielding DAG as fragment ions (McAnoy et al., 2005). Although CID has been proven useful in natural fats, its use in milk fat has not yet been proven.

For milk fat studies with shotgun lipidomics (ESI-MS) we recommend the work of Sokol et al. (2015). Bukowski and Picklo (2020) recently made available a method for quantification of milk fat TAG species. An overview on the workflow, methodology and fundamentals of shotgun lipidomics can be found elsewhere (Hsu, 2018; Han, 2020; McAnoy et al., 2005; Han and Gross, 2005).

TAG profiling by MALDI-TOF-MS and shotgun lipidomics have a huge potential due to its capacity to separate and identify TAG species which are unavailable by typical GC-FID or HPLC-ELSD TAG profiling. Moreover, the speed of analysis is within seconds, making these techniques suitable for large experimental set-ups. Both techniques – MALDI-TOF-MS and shotgun lipidomics

- have found acceptance in lipid analysis of various natural fats. Nonetheless, their use in milk fat studies is limited to a few papers in the scientific literature (Liu et al., 2018).

### 2.2.3 Structural analysis of TAG

The interest in elucidating the positional distribution of FA within the TAG is spurred by its relation with human digestion and absorption (Bar-Yoseph et al., 2013). Moreover, insight on the positioning of FA within the TAG will help to increase our knowledge on milk fat synthesis (Tzompa-Sosa et al., 2014). The available methods for regiospecific and stereospecific analysis of TAG (determination of sn-2/sn1-3 or sn-1, sn-2 and sn-3, respectively) are lengthy and require several reaction steps, which add large experimental error. Methods combining enzymatic or chemical degradation of TAG with high-throughput techniques will increase the accuracy and reduce the time needed for this analysis. However, these methods have not yet been developed.

Two main methods have been used until now: (1) generating a random mixture of MAG and DAG by a Grignard reagent; and (2) producing sn-2 MAG using pancreatic lipase degradation (Brockerhoff, 1965; Christie and Moore, 1969; Parodi, 1979). After TAG degradation, sn-2 MAG or sn1,2(2,3) DAG are isolated by TLC (Christie and Moore, 1969). Separation of sn1,2(2,3) DAG can also be performed by means of chiral chromatography (Itabashi et al., 1990, 1993). Partial glycerides with FA at sn-2 can change the positioning of FA to find a stable form in the outer positions (sn-1(3)) (Christie and Han, 2010; Turon et al., 2003). Therefore, after isolation, the partial glycerides formed are transformed into stable molecules such as lysophosphatidylcholine by the action of phospholipase A or as 3,5-dinitrophenylurethane derivatives. Finally, the FA of the partial glycerides are further determined by GC-FID after methylation of the FA.

A recent method for regiospecific analysis of TAG proposed the use of immobilized *Candida Antarctica* lipase B (CALB) as a means of acyl hydrolysis (Watanabe et al., 2014). This hydrolysis yields sn-2 MAG that are further isolated by TLC. Finally, the FA composition of sn-2 is determined by GC-FID and the FA composition of sn1,3 is calculated from the original lipid. This method takes into account acyl migration of sn-2 to sn1(3), but assumes that it occurs in a random way. Thus, its effect is neglected. The preference of CALB over pancreatic lipase relies on its specificity toward sn-1(3) and on its non-selectivity toward specific FA. This method for TAG regiospecific analysis is applicable for fats with short- and medium-chain FA, such as milk fat, and has been validated by several laboratories (Yoshinaga et al., 2016). The method is currently approved as a standard method for regiospecific analysis of oils and fats by the Japan Oil Chemists' Society and by the American Oil Chemists' Society (JOCS/AOCS, 2019).

# 3 Milk fat crystallization throughout the different length scales

For decades, great efforts have been made to understand the crystallization behavior of milk fat, both out of scientific interest and because it greatly influences functional and sensorial attributes of products that contain it. In this section, we discuss our current understanding of milk fat crystallization and the methods used to study it.

The crystallization behavior of milk fat is governed by the ~98% of TAG that comprise it. At temperatures below their melting points, the self-assembly of TAG in a hierarchical arrangement results in a three-dimensional crystal network that entraps other liquid TAG (Marangoni et al., 2020). When a fat is crystallized, TAG laterally pack, forming long layers called lamellae; the lamellae epitaxially stack producing nanoplatelets (Marangoni et al., 2020). Those nanoplatelets are the primary fat crystals that aggregate forming polycrystalline (composed by many crystals) particles that, in turn, cluster into flocs that shape a fractal crystal network (Marangoni et al., 2020). Thus, the hierarchical arrangement of a fat crystal network is comprised by structural elements in several length scales: microscale clusters of polycrystalline particles, crystalline nanoplatelets and lamellae, in which TAG display a longitudinal organization of a few nm and a lateral organization of a few Å. Different analytical techniques are required to characterize fat crystal networks covering all these length scales.

## *3.1 Milk fat microstructure*

The microstructure of milk fat crystal networks has been characterized using brightfield microscopy, polarized light microscopy and confocal laser scanning microscopy (CLSM). In Fig. 4, a simple brightfield microscopy image shows the microstructure of milk fat crystallized at 20°C after applying a 5°C/min cooling rate. This microstructure is typically observed in slowly cooled milk fat (at 5°C per min or lower). Several authors have reported the formation of branched spherulites of typically 10–70 μm (Batte and Marangoni, 2005; Kaufmann et al., 2012; Wiking et al., 2009) for milk fat. When milk fat is submitted to high degrees of supercooling or faster cooling rates, more nuclei form, resulting in numerous smaller polycrystalline particles forming a denser microstructure; in such networks, the individual particles are difficult or impossible to discern (Kaufmann et al., 2012; Wiking et al., 2009).

Although it has been possible to describe the microstructure of fat crystal networks in general terms, these techniques are limited by a relatively low resolution; furthermore, light microscopy allows only two-dimensional imaging of the three-dimensional structures. Three-dimensional images can be obtained with CLSM. However, CLSM requires the use of fluorescent dyes that could

**Figure 4** Brightfield microscopy image showing the microstructure of anhydrous milk fat after being cooled at 5°C/min and being kept at 20°C for 1 h.

have an effect on the crystallization process. Recently, Raman confocal imaging was used to characterize the microstructure of fat crystal networks; this is a non-invasive technique that, in combination with multivariate curve resolution, allows the acquisition of three-dimensional maps in which the distribution of the solid fat content (SFC) can be quantified (Martens et al., 2018). This recently developed application of Raman confocal microscopy has yet to be applied to study the microstructure of milk fat crystal networks.

### 3.2 Milk fat nanostructure

For decades, we have been able to characterize fat crystal networks, both at the microscale and at the molecular level (from a few angstroms to a few nanometers); however, it was not until the last decade that we could fully study the nanostructure. At the beginning of the last decade, it was shown how it was possible to characterize the nanostructure of fat crystal networks using ultra-small-angle X-ray scattering (USAXS) (Peyronel et al., 2013) and to directly visualize it using cryogenic transmission electronic microscopy (Cryo-TEM) (Acevedo and Marangoni, 2010).

Using Cryo-TEM, the visualization of milk fat lamellar structures in nano-emulsions was achieved by Truong et al. (2014) and of nanoplatelets in bulk milk fat by Ramel et al. (2016). In their study, Ramel et al. (2016) confirmed the presence of platelet-like structures and calculated an average crystalline nanoplatelet size (~1.14 µm) through the interpretation of USAXS patterns. If instead of using USAXS to study the nanostructures, X-ray techniques in wider angles are used, it is possible to study the TAG arrangements that shape those structures.

Crystallized milk fat displays two structural features related to the ability of TAG to form different crystal structures (polymorphism). The first one, more precisely referred to as polytypism, is related to the ways in which TAG can stack forming lamellae of differing thickness; this lamellar thickness can be measured using small-angle X-ray diffraction (SAXD). As an example, the time-resolved SAXD patterns of milk fat crystallizing at 17°C are included in Fig. 5a, in which the initial formation of 4.3 nm lamellae and their transition into 3.9 nm lamellae can be seen.

Typically, milk fat forms lamellae with a 3.8–4.8 nm thickness, corresponding to the length of two acyl chains (2L); at low temperatures (below ~14°C), lamellae with 5.4–7.2 nm thickness corresponding to the length of three acyl chains (3L) can also form (Lopez et al., 2001a, 2005). Lamellar thickness is also affected by the other structural feature related to the way in which the acyl chains laterally pack. In the fat research field, this second type of variation in the crystal structure of TAG is what is commonly referred to as polymorphism.

### 3.2.1 Polymorphism in milk fat

The acyl chains of TAG arrange in different subcell-packing modes referred to as polymorphs. To identify these polymorphs, we can use wide-angle X-ray diffraction (WAXD), as each polymorph is characterized by specific diffraction peaks. As an example, Fig. 5b shows the time-resolved WAXD patterns corresponding to the same crystallization process as the one in Fig. 5a. In this example, the initial formation of an α polymorph phase and its transition into

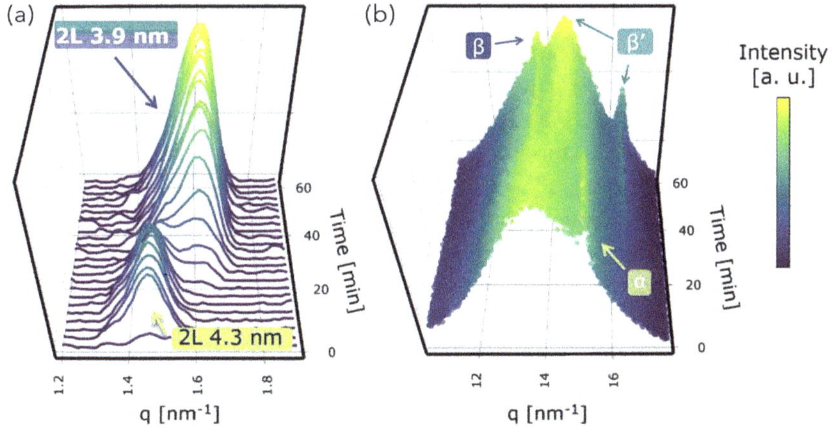


**Figure 5** Time-resolved X-ray diffraction patterns of anhydrous milk fat crystallizing at 17°C upon cooling at 5°C/min. (a) Small-angle diffraction patterns displaying the first-order diffraction peaks of lamellae with thicknesses of 4.3 nm and 3.9 nm. (b) Wide-angle diffraction (WAXD) patterns displaying the formation of α, β′ and β polymorphs. The WAXD plot was adapted from (Arita-Merino et al., 2020).

β′ and β phases can be seen (Fig. 5a). According to the Ostwald rule of stages, unstable polymorphs will form before metastable (α) and stable crystals (β′ and β) (ten Grotenhuis et al., 1999).

Milk fat predominantly crystallizes and is stable as its β′ polymorph; however, high degrees of supercooling (temperatures below ~20°C) yield the initial formation of the metastable α polymorph and, with the combination of fast cooling rates (quenching) and lower temperatures, even an unstable sub-α or γ polymorph can form (ten Grotenhuis et al., 1999; Lopez et al., 2005). Over time, or upon heating, these metastable structures (α and sub-α) transform into the more stable β′ polymorph (ten Grotenhuis et al., 1999; Lopez et al., 2005). Additionally, the formation of β polymorph has been intermittently reported in the literature, and the correlations between composition and β formation have been shown (Tzompa-Sosa et al., 2016a). As can be clearly seen in Fig. 5, the two structural features (longitudinal and lateral packing of TAG) are linked; in milk fat, the occurrence of α 2L, α 3L, β′ 2L and β′ 3L crystalline phases is often reported (Lopez, 2018). Although the polymorphism of milk fat has been extensively studied with a combination of X-ray techniques, the crystallization behavior of milk fat has been shown to be one of the most complex among naturally occurring fats.

### 3.2.2 Quantitative phase analysis

The complex crystallization behavior of milk fat and the effect that composition and external factors (e.g. shearing, temperature and cooling rate) have on it, are typically studied in a qualitative way. Quantification of coexisting polymorphic phases in complex fats is not common, and until recently, had only been done using synchrotron radiation small-angle X-ray scattering (Dewettinck et al., 2004; Lopez et al., 2001c; Mazzanti et al., 2005). In a recently published work (Arita-Merino et al., 2020), a quantitative phase analysis (QPA) method based on a laboratory-scale WAXD was applied to the study of milk fat crystallization. This QPA method is based on the decomposition of WAXD patterns into the area of each composing phase, as illustrated in Fig. 6a-b. To convert the areas into their corresponding mass fractions, phase-specific response factors are derived from time-resolved SFC values measured by nuclear magnetic resonance (NMR). As illustrated in Fig. 6c, this QPA method allows the quantification of coexisting liquid, α, β′ and β phases in milk fat, which can be a useful tool in the future study of milk fat crystallization and the factors that affect it.

## 4 Crystallization as affected by lipid composition

Processing conditions, such as thermal history (i.e. cooling rate, final crystallization temperature and storage time) and agitation greatly affect fat crystallization. Several studies have published the effects of these factors on

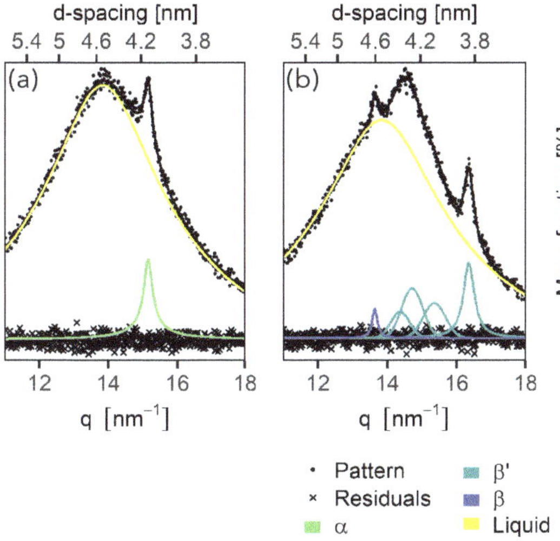


**Figure 6** Quantification of the crystalline phases formed in anhydrous milk fat during crystallization at 17°C upon cooling at 5°C/min. (a) and (b) Examples of decomposed WAXD patterns into the area contribution of each phase to the total area of the pattern. The patterns are plotted in arbitrary units. (c) Mass fractions of the crystalline phases. The error bars correspond to the confidence interval based on the uncertainty of the bootstrap estimates of the response factors (95%). Adapted from (Arita-Merino et al., 2020).

milk fat crystallization. Besides these external factors, milk fat composition directly affects its crystallization behavior. However, the wide diversity of FA and hence TAG species, as well as the many factors that affect their concentrations, complicate the study of the composition–crystallization relation in milk fat.

In early studies, differential scanning calorimetry (DSC) and SFC by NMR were used to study correlations between typical compositional variations and milk fat crystallization. In an extensive study with milk fat from 440 butter samples collected throughout the year (Precht and Frede, 1994), correlations were found between TAG groups and SFC; in particular, SFC at temperatures below 30 °C were positively correlated with the levels of TAG C36, C44 and C46 and negatively correlated with TAG C40, C52 and C54 levels. In a study of anhydrous milk fat from different regions and seasons (Shi et al., 2001), correlations were found between high nucleation and crystallization rates and the ratio of C46/C52 TAG and TAG with carbon number equal or lower than 40. In a study looking at the variation in milk fat from individual cows (Buldo et al., 2013), the melting point of the medium-melting fractions was positively correlated with C16:0 and negatively correlated to certain long-chain unsaturated FA. More recently (Larsen et al., 2014), correlations were found between low-melting fractions with higher melting points, higher levels of myristic acid (C14:0) and palmitic acid (C16:0) and lower levels of stearic acid (C18:0) and oleic acid (C18:1 *cis*-9).

While these studies reported the effect of specific FA and TAG groups on crystallization kinetics (Shi et al., 2001) or on the fractions that crystallize or melt at different temperatures (Buldo et al., 2013; Larsen et al., 2014; Precht and Frede, 1994), they do not provide information about their effect on the structures formed during crystallization. Our knowledge has advanced in this regard thanks to studies that combine X-ray diffraction with DSC. Such studies have looked at the polymorphic crystalline structures formed in milk fats with compositional variations due to specific feeding regimes (Bugeat et al., 2015; Singh et al., 2004; Smet et al., 2010). The supplementation of cows' diet with algae altered milk fat composition, the main effects were higher and lower levels of C18:1 *trans* and *cis* isomer, respectively; these compositional changes were linked to the direct formation of β' polymorph at 5°C, in contrast with the control milk fat that first crystallized into α polymorph (Singh et al., 2004). In other studies, cows were fed linseed-supplemented diets (Bugeat et al., 2015; Smet et al., 2010); the resulting higher levels of unsaturated fatty acids (UFA) in milk fat affected the crystallization kinetics, the abundance of the different melting fractions as well as the melting and formation temperatures of those fractions; however, the same polymorphic structures were formed in these UFA-enriched milk fats. Nevertheless, the use of SAXD in one of these studies (Bugeat et al., 2015), revealed that milk fat enriched in UFA via linseed supplementation resulted in thicker lamellae.

A different approach to study the compositional effect on milk fat crystalline polymorphism is to look at samples with more pronounced variations. Tzompa-Sosa et al. (2016a) studied the polymorphism of milk fat from individual cows, selecting samples with specific compositional variations. At 20°C, the formation of α and β polymorphs were linked to higher contents of saturated C34–38 TAG and unsaturated C52–54 TAG, respectively (Tzompa-Sosa et al., 2016a).

Besides the compositional changes that occur due to indirect factors (e.g., farming practices) as discussed above, milk fat is also submitted to direct modifications. Different technological processes (e.g. blending, fractionation, interesterification, supercritical fluid extraction and hydrogenation) have been applied to milk fat to tailor its crystallization by altering its composition. As these processes transform milk fat into a different system (e.g. hydrogenated milk fat and milk fat olein), the effect of milk fat transformations on its crystallization behavior will not be discussed here. Information on this topic can be found in the material suggested at the end of this chapter.

## 5 Case studies: Practical consequences of milk fat composition variability

In some countries, according to their farming practices, seasonal variation in milk fat is common (Larsen et al., 2014; Liu et al., 2017; Poulsen et al., 2020).

FA composition related to seasonal variation in milk fat composition affects their thermal behavior; consequently, milk fat structured products, as well as their processing conditions, are also submitted to changes throughout the year. In this section, we present two cases studies where the variability of FA composition comes into play. We look at this industrially relevant variation, as a case study to illustrate the combined use of some of the analytical techniques discussed in this chapter. In the first case study, a predictive model for the final melting point of milk fat is presented. In the second case study, we look at the effect that seasonal changes in composition have on milk fat polymorphism and rheology.

## 5.1 Case study 1: Predictive model for final melting temperature

The melting point of milk fat is dependent on its FA and TAG composition and on the positioning of the FA within the TAG. If we take into account that more than 400 FA have been identified in milk fat and, 3454 molecular TAG species exist (Liu et al., 2020a), then we may expect that milk fat has an extremely wide melting range. In reality, the melting temperature of milk fat ranges from –40 to 40°C. TAG with a melting point higher than 40°C do exist in milk fat, but they normally dissolve in the lower-melting TAG fraction (Walstra et al., 1992), mainly because these high-melting TAG are present in low amounts.

The variability in the chemical composition of milk fat affects its thermal behavior. Several studies have reported the thermal behavior of milk fat in bulk samples, laying down the fundamental knowledge to understand milk fat crystallization kinetics (ten Grotenhuis et al., 1999; van Aken et al., 1999; van Aken and Visser, 2000; Lopez et al., 2001a,b, 2005). An interesting feature noticed by Lopez et al. (2005) was that the initial temperature of crystallization and the shape of the crystallization and melting curves depend on the cooling rate. They showed that the initial temperature of crystallization of milk fat decreases with increasing cooling rate, which is explained by the influence of the nucleation process. A similar feature was seen for the initial temperature of melting for the peaks related to low-melting and middle-melting milk fat fractions. This feature is also related to the cooling rate and the formation of milk fat polymorphs. However, this study also pointed out that the final melting temperature of milk fat was not significantly affected by the cooling rate. A similar feature was observed by van Aken et al. (1999) in milk fat allowed to isothermally crystallize for various periods.

Although the final melting temperature of milk fat is not affected by cooling rates or by the period of isothermal crystallization, it is affected by its FA composition. Therefore, a predictive model for the final melting temperature based on its FA composition can help to design fat-rich products. The aim of this study was to develop models for predicting the final melting temperature of milk fat according to its FA composition. This model was developed and validated using milk fat from 54 cows belonging to different commercial herds (46 herds) in the Netherlands. This sample selection provided large variability in FA profiles. Moreover, it provided the best possible representation of management and feeding regimes within the population of Dutch Holstein Friesian Cows. The milk fat samples and FA compositions used in this study were a subset of the original sample set of the Dutch Milk Genomics Initiative. The FA composition of this population has been previously reported and discussed elsewhere (Heck et al., 2012; Schennink et al., 2007; Stoop et al., 2008)

Thermal analysis on each of the 54 samples was performed using a TA Q1000 DSC (New Castle, DE) calibrated with indium. Milk fat (15 ± 1 mg) was loaded in aluminum pans of 50 μL and hermetically sealed with a cap of the same material. An empty, hermetically sealed pan was used as a reference. The thermograms were analyzed using Pyris data Analysis (Perkin Elmer). The protocol applied to study the thermal behavior of milk fat was as follows: milk fat was melted to 70°C for 10 min to remove all crystals. Then, the fat was cooled at 5°C/min from 70°C to -60°C and maintained at this temperature for 10 min. Next, the melting behavior was determined by heating at 5°C/min from the -60°C to 70°C. The last offset of the melting curve was defined as the final melting temperature (offset 3) (Fig. 7).

### 5.1.1 Building a predictive model

Modeling was based on linear and quadratic correlations between the FA composition and the final melting temperature. The final models were developed by statistical regression using SPSS. First, we diagnosed the type of correlation (linear and quadratic). FA having a significant correlation ( $P \leq .05$) and a value of $R^2 > 0.1$ were retained and used to build a regression model. Out of the initial 44 different FA in the profile, 12 FA showed a linear relation (C8:0, C10:0, C10:1, C12:0, C12:1, C14:0, C16:0, C17:0, C18:1 t9, C18:1 t10, C18:1c12, C18:2c9t11) and three showed a quadratic relation (C6:0, C18:1 t15, C20:0). These FA were retained for building the predictive model.

Regression modeling was used to build the predictive equation. Both backward and forward methods were tested. Two types of models were produced, one was based only on linear FA, therefore referred to as the 'linear model'. The other was based on both linear and quadratic FA, referred to as the 'quadratic model'. Assumptions were checked for every produced model (multicollinearity of variables, homogeneity in the variation of residuals and normal distribution of the standardized residual).

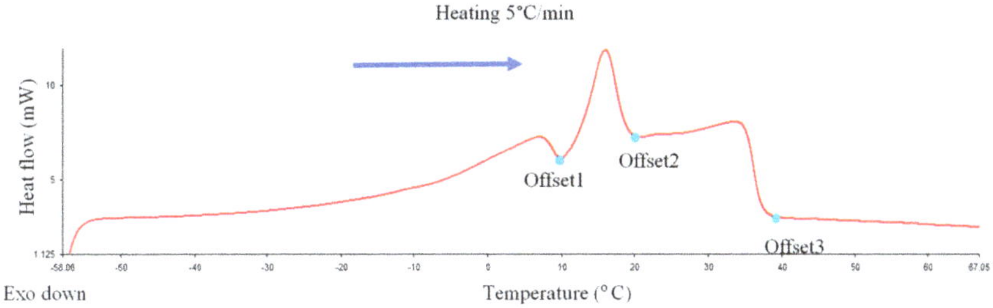


**Figure 7** Melting profile of milk fat heated at a rate of 5°C/min.

### 5.1.2 Validation of the predictive model for final melting point

The prediction model for the final melting point was validated against a new sample set (33 milk fat samples) which was analyzed under different conditions. Thermal analysis was performed using the same thermal protocol but using a different equipment and sample pan material. The equipment used was a Diamond-DSC-Autosampler (PerkinElmer) calibrated with indium. The samples were weighed (15 ± 1 mg) using stainless steel pans of 50 μL, which were hermetically sealed with a cover of the same material. The reference was an empty, hermetically sealed stainless steel pan.

After validation, the predictive model was corrected using the methodology explained in Section 5.1.1.

### 5.1.3 Predictive model, model validation and correction of the model for final melting point

The data set used to build the predictive model and to validate the model had indeed a broad range of FA, as confirmed by the wide range in the final melting temperatures recorded (Table 2). The average final melting point of both data sets was similar, and differed by only 0.47°C. The broad ranges in melting temperatures helped to build a robust predictive model.

The linear and quadratic models proposed are shown in Table 3 (A and B). Both predictive models account for 83% of variation in the final melting temperature, as expressed by its $R^2$ (0.83). Moreover, the bias and mean square error (MSE) of both predictive models were low (Table 3). Bias and MSE are estimators of the quality of a model. The bias of an estimator is the difference between the estimator's expected value and the true value. A bias as close to zero as possible is desirable since this indicates an unbiased model. The MSE measures the average of the squares of the errors, which are the differences between the predicted values and the observed values. The differences occur either due to the randomness in the measurements or due to the model lacking the information that could help produce a more accurate prediction. Therefore,

**Table 2** Variation in the final melting temperature of milk fat after heating at 5°C/min. Data obtained using the data set to build the predictive model and for its validation.

| | Predictive model dataset | Validation dataset |
|---|---|---|
| Minimum | 35.22 | 34.63 |
| Maximum | 39.26 | 38.45 |
| D (Max, min) | 4.04 | 3.82 |
| Mean | 37.11 | 36.64 |
| S.D. | 1.06 | 1.07 |

a predictive model with a small MSE will be more robust than the one with a higher MSE. Since both models showed similar estimation parameters and have the same FA as predictive parameters, it is up to the user to decide on one of them.

The quadratic and the linear model claim to account for 83% of variation in the final melting temperature ($R^2$). However, when the predictive model was challenged with a new data set, the variation explained dropped to 69% (Table 3). Moreover, the bias and MSE increased, indicating that the estimation power of the predictive model decreased when using a different data set, different equipment and a different DSC pan material. However, the estimation power remains relatively high and will predict the final melting point of milk fat with a 69% accuracy. Figure 8 (a and b) shows the observed vs the predicted values (linear model) of the final melting point in the samples used to build the model and to validate it.

In the third and last phase of this study, the predictive models for final melting point were corrected. These predictive models include the sample sets of the predictive and the validation models. Increasing the number of observations helps to increase the accuracy of the predictive models. These second predictive models had lower predictive power ($R^2$) compared to the initial model, but remain a powerful model having an $R^2 = 0.7$ (Table 3, model C and D).

### 5.1.4 Final considerations

In this work, TAG profiles were also considered as predictive variables for the final melting temperature. However, the correlations among the different TAG are very high (>0.7). This feature is not desirable for modeling based on regression analysis because it results in extreme multicollinearity between the variables and reliable models cannot be obtained (Field, 2009). Therefore, in this study, the FA were considered the best variables to build the predictive models.

The utilization of different DSC machine and pan materials for analyzing the validation samples resulted in a very different nucleation process upon

**Table 3** Predictive models for final melting point of milk fat using fatty acids (g/100 g of total FA) as predictors

| | | Predictive model | | | | Validation data set | | |
|---|---|---|---|---|---|---|---|---|
| Models | Final melting temperature | $R^2$ | Sig. | Bias | MSE | $R^2$ | Bias | MSE |
| A. Linear model | y = 33.162 + 0.258 C16:0 +21.168 C20:0 - 1.836 C6:0 - 0.268 C14:0 | 0.83 | $P < 0.001$ | 0.01 | 0.19 | 0.69 | 0.28 | 0.42 |
| B. Quadratic model | y = 31.123 + 0.256 C16:0 + 21.158 C20:0 - 0.406 $(C6:0)^2$ - 0.267 C14:0 | 0.83 | $P < 0.001$ | -0.02 | 0.19 | 0.69 | 0.26 | 0.41 |
| C. Corrected linear model | Y = 32.08 + 0.266 C16:0 + 17.214 C20:0 - 1.393 C6:0 - 0.252 C14:0 | 0.78 | $P < 0.001$ | -0.01 | 0.25 | | | |
| D. Corrected quadratic model | Y = 30.545 + 0.266 C16:0 + 17.199 C20:0 - 0.309 $(C6:0)^2$ - 0.252 C14:0 | 0.77 | $P < 0.001$ | 0.01 | 0.25 | | | |

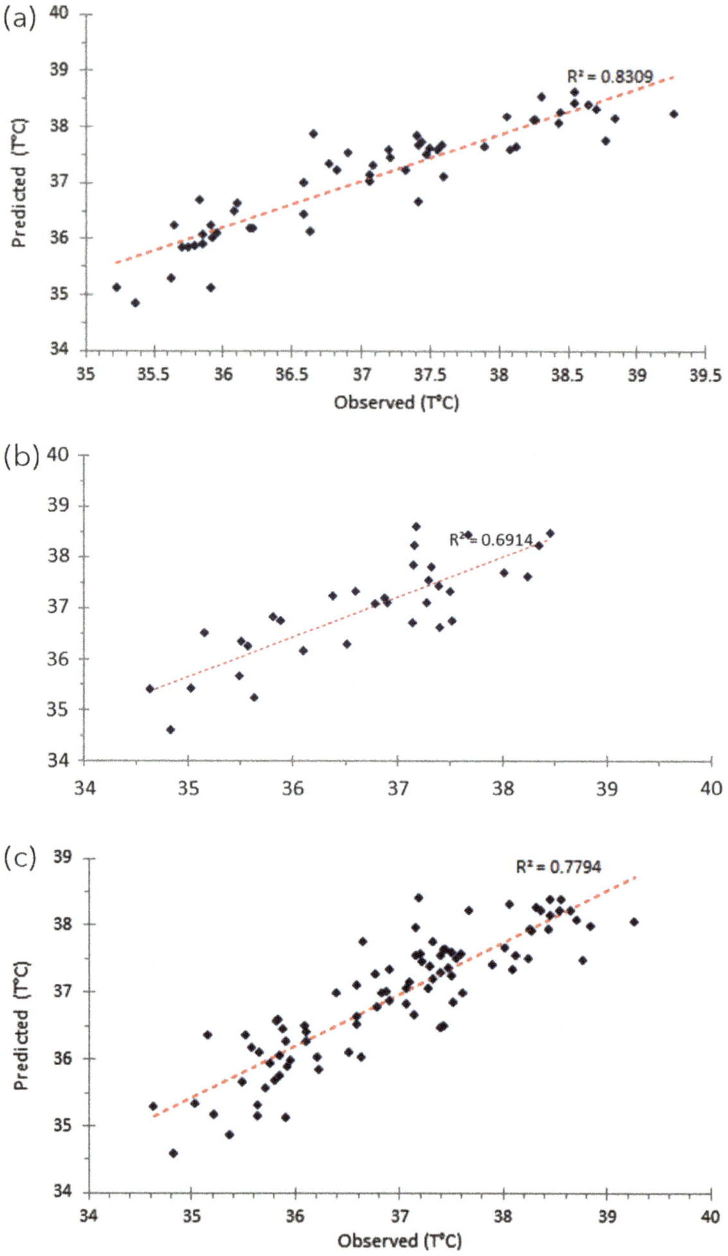


**Figure 8** Observed vs predicted values of the final melting point of milk fat using a linear predictive model (FA as g/100g of FA). (a) data set used to build the model (N = 54), (b) validation data set using the initial predictive model (N = 33), (c) corrected data set (N = 87).

crystallization. While the shapes of the thermal curves were conserved, the crystallization and melting temperatures of all samples were changed considerably, except for the final melting point. This finding proved that the final melting point was not affected by nucleation, which is in accordance with previous studies performed at low and medium cooling rates (Lopez et al., 2005; ten Grotenhuis et al., 1999). The high prediction power of the model proposed for the final melting temperature, together with its successful validation, provided a prediction model of the final melting temperature of milk fat.

This model might not be useful for extremely unsaturated milk fat samples (saturation index <2.7) and crystallized at 20°C. This is because these fat profiles tend to form β polymorphs upon crystallization at 20°C. These polymorphs are the most stable types and have a higher melting point compared to metastable polymorphs. β polymorphs might not be formed if lower crystallization temperatures are used (Tzompa-Sosa et al., 2016a).

## 5.2 Case study 2: Effect of chemical composition of milk fat on its crystallization behavior

In this case study, we compared the composition and the crystallization of characteristic winter and summer milk fats. The samples were kindly provided by FrieslandCampina (the Netherlands). Detailed FA and TAG compositions (by GC and MALDI-TOF-MS, respectively) have been published (Tzompa-Sosa et al., 2018). Quantitative TAG profiles were obtained by GC according to the ISO Standard 17678/IDF 202 (ISO-IDF, 2010).

Winter milk fat was characterized by a higher content of saturated FA (72.67% vs 69.75% in summer milk fat); the most evident differences for the main fatty acids (Fig. 9a) are the higher levels of myristic (C14:0) and palmitic (C16:0) acids and lower levels of stearic (C18:0) and oleic (C18:1) acids in winter milk fat (Tzompa-Sosa et al., 2018). Linked to these FA differences, the TAG compositions also displayed clear changes (Fig. 9b): winter milk fat has noticeably higher levels of TAG with total carbon numbers from 34 to 48 and lower levels of the higher molecular weight TAG (C50–C56). These more general TAG compositions obtained by GC, can be complemented with the detailed TAG profiles obtained by MALDI-TOF-MS.

Using a heat map, it is possible to visually compare the relative intensities of the MALDI-TOF-MS spectra for winter and summer milk fat (Fig. 10). These relative intensities are related to the relative abundances of the different TAG specifies. Thus, winter milk fat displayed a higher relative abundance for the majority of the saturated TAG species, including the most abundant TAG (i.e., C34:0, C36:0 and C38:0); summer milk fat displayed higher relative abundances for all the polyunsaturated TAG, and most of the monounsaturated species. By comparing the FA and TAG compositions by GC (Fig. 9) with the MALDI-TOF-MS

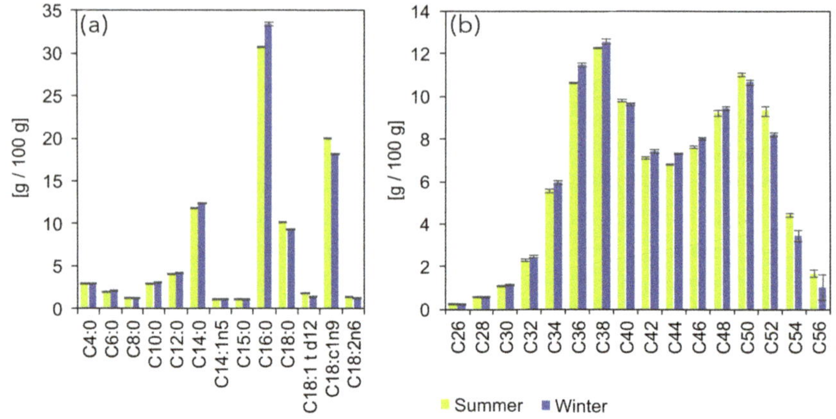


**Figure 9** (a) Concentration of the main fatty acids and (b) triacylglycerol composition of winter and summer anhydrous milk fat as obtained by GC-FID. The error bars correspond to the standard deviation. The fatty acid values were obtained from (Tzompa-Sosa et al., 2018).

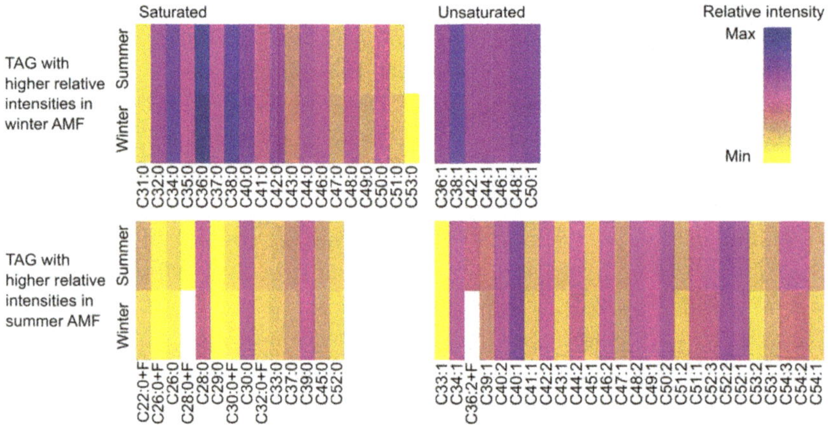


**Figure 10** TAG profiles of winter and summer anhydrous milk fat obtained by MALDI-TOF-MS. The color scale corresponds to the relative intensities in the spectra, attributable to corresponding relative abundances of the different TAG species. +F indicates a TAG fragment. Adapted from (Tzompa-Sosa et al., 2018).

TAG profiles (Fig. 10), we can conclude that the higher concentrations of high-molecular weight TAG in summer milk fat are mainly attributed to polyunsaturated species, while the higher levels of C32–C48 in winter milk fat correspond to saturated TAG. Combining quantitative GC analysis with MALDI-TOF profiling allowed us to study in detail the compositional differences between summer and winter milk fat. The next step is to evaluate how these compositional differences affect the crystallization.

### 5.2.1 *Effect of TAG and FA composition on milk fat rheological behavior*

We evaluated the effect of the seasonal compositional variation on the crystallization behavior of milk fat; we used small-deformation oscillatory rheology to characterize the crystal network and WAXD to study the crystal polymorphism. For this example, samples of summer and winter milk fat were cooled at 5°C/min to different temperatures and were kept for 3 h at those crystallization temperatures. To characterize the crystal networks formed, amplitude sweeps (0.001% –1% strain) were performed on an MCR 502 controlled stress rheometer (Anton Paar, Austria) using a concentric cylinder geometry (CC 10) at a constant frequency of 1 Hz. The liquid-like and the solid-like portions of viscoelastic response (loss modulus and storage modulus, respectively) of the crystal networks can be observed in Fig. 11.

For both summer and winter milk fat, and at all the studied temperatures, the storage modulus (G′) was 1–2 orders of magnitude greater than the loss modulus (G″) (Fig. 11), indicating that the crystal networks produced solid-like structures. The strain range of the G′ plateau corresponds to the linear viscoelastic (LVE) region in which the sample is deformed without being destroyed. The G′ in the LVE region can be interpreted as a measure of the structural strength of the network; winter milk fat produced stronger networks than summer milk fat. Sharper G′ declines at the lower temperatures, particularly in winter milk fat, indicate a more brittle fracture upon leaving the LVE region; this could be associated to certain product properties. For example, butter made with summer milk fat is likely to show better spreadability at lower temperatures as its network strength decreases slowly after the LVE region. The rheological properties of fat crystal networks are the result of the underlying structures that compose it, crystal polymorphism being a determining factor.

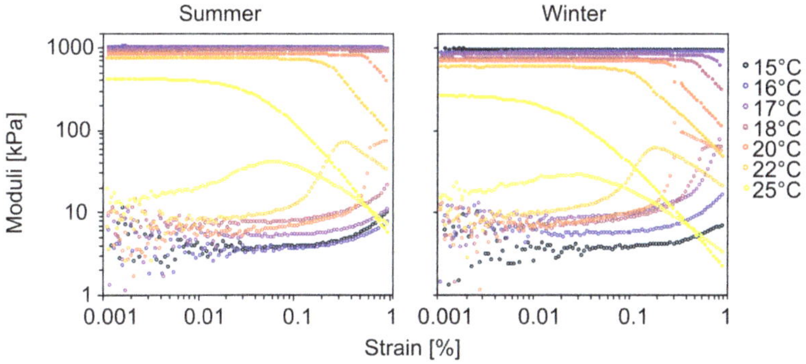


**Figure 11** Amplitude sweep of summer and winter anhydrous milk fat after isothermal crystallization for 3 h at different temperatures. Open symbols correspond to the loss modulus and closed symbols to the storage modulus.

### 5.2.2 Effect of TAG and FA composition on crystal polymorphs

We studied the evolution of the crystalline polymorphic phases formed in summer and winter milk fat during crystallization. Time-resolved WAXD patterns were recorded for 2 h at 20°C after cooling the sample at 5°C/min (Fig. 12 top panels), and the results were used to quantify the different polymorphic phases (Fig. 12 bottom panels). The response factors used for the quantification were

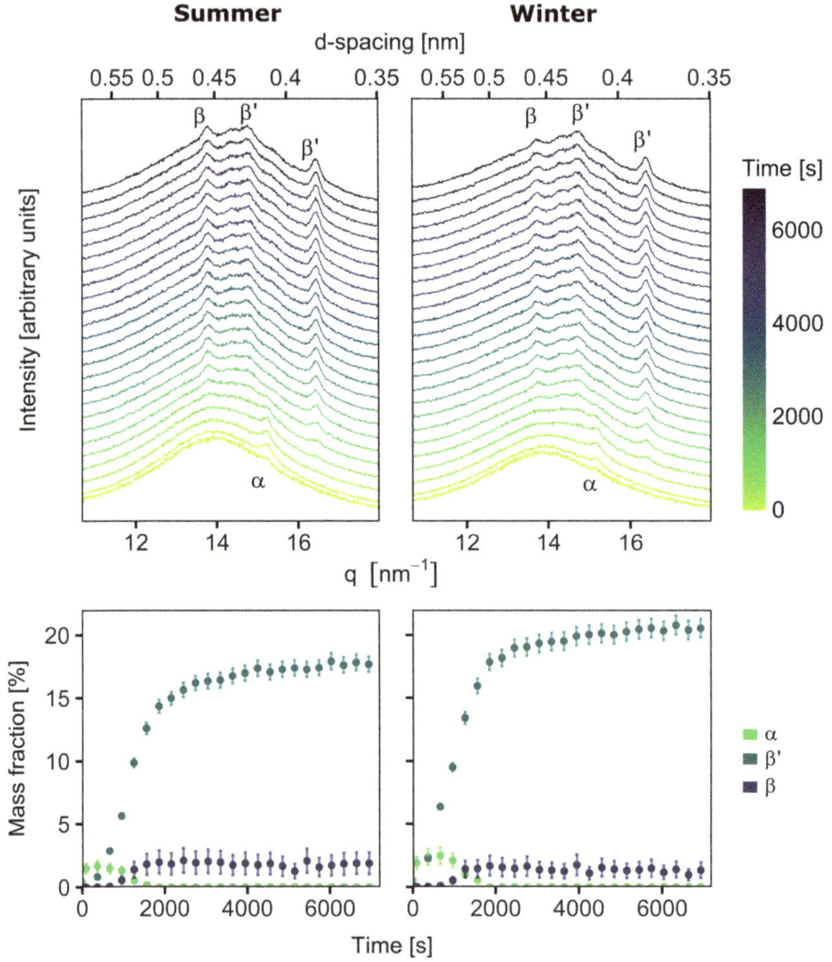


**Figure 12** Comparison of the polymorphic phases formed in summer and winter anhydrous milk fat during crystallization at 20°C. The top and bottom panels correspond to time-resolved wide-angle X-ray diffraction patterns and phase transition diagrams, respectively. The error bars correspond to the confidence interval based on the uncertainty of the bootstrap estimates of the response factors (95%).

obtained for a reference milk fat sample (not included here). The details of the data collection setup as well as the phase quantification method can be found elsewhere (Arita-Merino et al., 2020). From a qualitative perspective, the evolution of the crystalline phases in summer and winter milk fat was comparable; initially, α polymorph formed from the melt, after ~400 s, β′ and β phases appeared and grew, while the α phase decreased until disappearing after ~1800 s. With the QPA method, we could find quantitative differences between summer and winter milk fat; the amounts of α and β′ formed in winter milk fat were higher than in summer milk fat (2.5% and 20.4% vs 1.7% and 17.8%, respectively), but the amount of β was slightly lower (1.4% vs 1.8%).

In this case study, we showed how, by combining information of the chemical profile and physical characteristics of milk fat, we can increase our understanding of the effect of seasonal changes in milk fat composition and their effect on crystallization. The stronger crystal networks in winter milk fat were associated to a higher formation of β′ crystals, as well as to a higher formation of α. This crystal formation is explained by its higher content of C34–C48 TAG, identified as saturated by MALDI-TOF-MS. The moderately higher formation of β in summer milk fat seems to be related to the higher content of high-molecular weight unsaturated TAG. These findings are in agreement with a previous study that linked the formation of α and β′ to higher contents of C34–C38 TAG, and the formation of β with C52–C54 TAG (Tzompa-Sosa et al., 2016a). A more rigorous study would be required to evaluate the significance of these findings. Nevertheless, this example illustrates the potential of tools that have been used in milk fat research only recently, such as TAG profiling by MALDI-TOF-MS and QPA by WAXD.

## 6 Conclusion

Milk fat is one of the most complex lipids in nature. It contains several lipid types, which in turn are composed of several individual species. Studies on milk fat should go further than FA profiling, since only then can a full overview of the phenomenon under study be fully achieved. For instance, if the aim is to study the effect of an enzyme involved in TAG synthesis, then, efforts should be directed toward profiling TAG species and structures. The broad range of lipid types and FA in milk fat require the use of several complementary techniques. However, directing the efforts toward the lipid components will explain the phenomenon under study better. In our experience, it is always a good practice to couple the results of high-throughput lipid analysis with established chromatographic methods.

The advances in high-throughput analytical techniques to profile milk fat, discussed in this chapter, go along with developments in the field of fat crystallization. Here, we have highlighted three techniques recently applied to study crystallized fat: Cryo-TEM and USAXS to characterize the nanostructure and Raman confocal imaging to produce three-dimensional SFC maps. Additionally, we discuss a recently published method that allows the quantification of coexisting polymorphic phases in milk fat using laboratory-scale WAXD. As the reach and availability of tools to study the chemical and physical properties of milk fat grow, our understanding of the link between composition and crystallization will also move forward. Finally, the case studies illustrate the relation between the physical characteristics of milk fat and its chemical composition, which are directly related to animal husbandry.

## 7 Future trends in research

The current lengthy methods for TAG structure analysis hinder the research in this field. Elucidation of positioning of FA within the TAG using a combination of TAG degradation and high-throughput techniques could help to increase the speed of analysis, which currently prolongs to days. This development could trigger the interest of researchers toward linking milk fat synthesis with fat biosynthesis, lipid absorption, etc.

In the past decade, USAXS has been shown to be a useful tool to study the mesoscale structures in fats. A few research groups have successfully used this technique to study bulk milk fat. Studies linking the mesoscale structure of milk fat with its polymorphism and with industrial processes will help to delve the effect of nano- and mesoscale structures in milk fat on its macrostructure. These studies will benefit from the use of the laboratory-scale SAXS/USAXS equipment which are already commercially available.

With growing interest regarding the sustainability and the nutritional aspects of dairy products, the interest in developing alternative feeding sources for cows is also growing. Hence, the effect that new diets can have on the chemical and physical properties of milk lipids will be highly relevant.

## 8 Where to look for further information

The following online resources provide a good overview on lipid chemistry and have useful tools for lipid analysis

- Lipid library. Information on lipid analysis, oil processing, lipid biochemistry, human nutrition and much more essential knowledge on oils and fats (https://lipidlibrary.aocs.org/).
- Triglyceride property calculator. A melting property calculator for pure TAG which includes physical properties (https://lipidlibrary.shinyapps.io/Triglyceride_Property_Calculator/).
- Lipid Web. Introduction to the chemistry and biochemistry of individual lipid classes, compendium of information on mass spectrometry of fatty

acids, plus a lively blog on Lipid Matters written by William Christie (https://www.lipidmaps.org/resources/lipidweb/index.php?page=).

- Lipid maps. Open source database of MS precursor ions, product ions peak list, structural database, a software for statistical analysis, tutorials and webinars on Lipid Science and much more (https://lipidmaps.org/).

The following material collects up-to-date information for further reading:

- Lopez, C. Crystallization properties of milk fats, in *Crystallization of Lipids*, 2018. p. 283–321.
- Truong, T., Bansal, N. and Bhandari, B. Effect of emulsion droplet size on foaming properties of milk fat emulsions. *Food and Bioprocess Technology*, 2014. 7(12): p. 3416–3428.
- Mohan, M. S., et al. Milk fat: opportunities, challenges and innovation. *Critical Reviews in Food Science and Nutrition*, 2020.
- Liu, Z. Q., Rochfort, S. and Cocks, B. Milk lipidomics: what we know and what we don't. *Progress in Lipid Research*, 2018. 71: p. 70–85.
- McSweeney, P. L. H., Fox, P. F. and O'Mahony, J. A. *Advanced Dairy Chemistry. Volume 2 Lipids*, Fourth ed. (2020). Springer International Publisher.

Follow these profiles on twitter for the latest news about lipid science

- Journal of Lipid Research: @jlipidres.
- The Lipid Chronicles: @lipidchronicles.
- Lipid Maps: @lipidmaps.

## 9 References

Acevedo, N. C. and Marangoni, A. G. (2010). Characterization of the nanoscale in triacylglycerol crystal networks. *Crystal Growth and Design* 10(8): 3327-3333.

Arita-Merino, N., van Valenberg, H., Gilbert, E. P. and Scholten, E. (2020). Quantitative phase analysis of complex fats during crystallization. *Crystal Growth and Design* 20(8): 5193-5202.

Bar-Yoseph, F., Lifshitz, Y. and Cohen, T. (2013). Review of sn-2 palmitate oil implications for infant health. *Prostaglandins, Leukotrienes, and Essential Fatty Acids* 89(4): 139-143.

Batte, H. D. and Marangoni, A. G. (2005). Fractal growth of milk fat crystals is unaffected by microstructural confinement. *Crystal Growth and Design* 5(5): 1703-1705.

Bionaz, M. and Loor, J. J. (2008). Gene networks driving bovine milk fat synthesis during the lactation cycle. *BMC Genomics* 9: 366.

Blasi, F., Montesano, D., De Angelis, M., Maurizi, A., Ventura, F., Cossignani, L., Simonetti, M. S. and Damiani, P. (2008). Results of stereospecific analysis of triacylglycerol fraction from donkey, cow, ewe, goat and buffalo milk. *Journal of Food Composition and Analysis* 21(1): 1-7.

Brockerhoff, H. (1965). A stereospecific analysis of triglycerides. *Journal of Lipid Research* 6(1): 10–15.

Bugeat, S., Perez, J., Briard-Bion, V., Pradel, P., Ferlay, A., Bourgaux, C. and Lopez, C. (2015). Unsaturated fatty acid enriched vs. control milk triacylglycerols: solid and liquid TAG phases examined by synchrotron radiation X-ray diffraction coupled with DSC. *Food Research International* 67: 91–101.

Bukowski, M. R. and Picklo, M. J. (2020). Simple, rapid lipidomic analysis of triacylglycerols in bovine milk by infusion-electrospray mass spectrometry. *Lipids* 56(2): 243–255.

Buldo, P., Larsen, M. K. and Wiking, L. (2013). Multivariate data analysis for finding the relevant fatty acids contributing to the melting fractions of cream. *Journal of the Science of Food and Agriculture* 93(7): 1620–1625.

Campos, R., Narine, S. S. and Marangoni, A. G. (2002). Effect of cooling rate on the structure and mechanical properties of milk fat and lard. *Food Research International* 35(10): 971–981.

Christie, W. W. (1994). *Gas chromatography and Lipids : A Practical Guide*. Dundee, Escocia: The Oily Press.

Christie, W. W. and Han, X. (2010). *Positional Distribution of Fatty Acids in Glycerolipids. Lipid Analysis: Isolation, Separation, Identification and Lipidomic Analysis*. Bridgwater: Oily Press.

Christie, W. W. and Moore, J. H. (1969). A semimicro method for stereospecific analysis of triglycerides. *Biochimica et Biophysica Acta* 176(3): 445–452.

Dewettinck, K., Foubert, I., Basiura, M. and Goderis, B. (2004). Phase behavior of cocoa butter in a two-step isothermal crystallization. *Crystal Growth and Design* 4(6): 1295–1302.

Fahy, E., Sud, M., Cotter, D. and Subramaniam, S. (2007). LIPID MAPS online tools for lipid research. *Nucleic Acids Research* 35: W606–W612.

Field, A. (2009). *Discovering Statistics Using SPSS* (3rd edition). London: SAGE Publications Ltd.

González-Domínguez, R., García-Barrera, T. and Gómez-Ariza, J. L. (2014). Combination of metabolomic and phospholipid-profiling approaches for the study of Alzheimer's disease. *Journal of Proteomics* 104: 37–47.

Gottardo, P., Penasa, M., Righi, F., Lopez-Villalobos, N., Cassandro, M. and De Marchi, M. (2017). Fatty acid composition of milk from Holstein-Friesian, Brown Swiss, Simmental and Alpine Grey cows predicted by mid-infrared spectroscopy. *Italian Journal of Animal Science* 16(3): 380–389.

Hageman, J. H. J., Danielsen, M., Nieuwenhuizen, A. G., Feitsma, A. L. and Dalsgaard, T. K. (2019). Comparison of bovine milk fat and vegetable fat for infant formula: implications for infant health. *International Dairy Journal* 92: 37–49.

Han, X. (2020). Shotgun lipidomics. Available at: https://lipidlibrary.aocs.org/lipid-analysis/lipidomics/shotgun-lipidomics (accessed 14 January 2021).

Han, X. and Gross, R. W. (2005). Shotgun lipidomics: electrospray ionization mass spectrometric analysis and quantitation of cellular lipidomes directly from crude extracts of biological samples. *Mass Spectrometry Reviews* 24(3): 367–412.

Heck, J. M. L., van Valenberg, H. J. F., Bovenhuis, H., Dijkstra, J. and van Hooijdonk, T. C. (2012). Characterization of milk fatty acids based on genetic and herd parameters. *Journal of Dairy Research* 79(1): 39–46.

Heck, J. M. L., van Valenberg, H. J. F., Dijkstra, J. and van Hooijdonk, A. C. (2009). Seasonal variation in the Dutch bovine raw milk composition. *Journal of Dairy Science* 92(10): 4745–4755.

Hein, L., Sørensen, L. P., Kargo, M. and Buitenhuis, A. J. (2018). Genetic analysis of predicted fatty acid profiles of milk from Danish Holstein and Danish Jersey cattle populations. *Journal of Dairy Science* 101(3): 2148–2157.

Hsu, F. F. (2018). Mass spectrometry-based shotgun lipidomics - a critical review from the technical point of view. *Analytical and Bioanalytical Chemistry* 410(25): 6387 -6409.

ISO-IDF (2010). Milk and Milk products- Determination of milk fat purity by gas chromatographic analysis of triglycerides (Reference method). In: ISO 17678:2019. Available at: www.iso.org/standard/72284.html.

Itabashi, Y., Kuksis, A. and Myher, J. J. (1990). Determination of molecular species of enantiomeric diacylglycerols by chiral phase high performance liquid chromatography and polar capillary gas-liquid chromatography. *Journal of Lipid Research* 31(11): 2119–2126.

Itabashi, Y., Myher, J. J. and Kuksis, A. (1993). Determination of positional distribution of short-chain fatty acids in bovine milk fat on chiral columns. *Journal of the American Oil Chemists' Society* 70(12): 1177–1181.

Japan Oil Chemists' Society and American Oil Chemists' Society (2019). Ch 3a.19 Determination of the composition of fatty acids at the 2-position of oils and fats-enzymatic transesterification method using *Candida antarctica* Lipase. (JAOCS/AOCS Ch 3a-19).

Jensen, R. G. (2002). The composition of bovine milk lipids: January 1995 to December 2000. *Journal of Dairy Science* 85(2): 295–350.

Kaufmann, N., Andersen, U. and Wiking, L. (2012). The effect of cooling rate and rapeseed oil addition on the melting behaviour, texture and microstructure of anhydrous milk fat. *International Dairy Journal* 25(2): 73–79.

Larsen, M. K., Andersen, K. K., Kaufmann, N. and Wiking, L. (2014). Seasonal variation in the composition and melting behavior of milk fat. *Journal of Dairy Science* 97(8): 4703–4712.

Lay Jr., J. O. (2020). Characterization of lipids by MALDI mass spectrometry. Available at: https://lipidlibrary.aocs.org/lipid-analysis/lipidomics/characterization-of-lipids-by-maldi-mass-spectrometry (accessed 14 January 2021).

Liu, Z., Li, C., Pryce, J. and Rochfort, S. (2020a). Comprehensive characterization of bovine milk lipids: triglycerides. *ACS Omega* 5(21): 12573–12582.

Liu, Z., Li, C., Pryce, J. and Rochfort, S. (2020b). Comprehensive characterization of bovine milk lipids: phospholipids, sphingolipids, glycolipids, and ceramides. *Journal of Agricultural and Food Chemistry* 68(24): 6726–6738.

Liu, Z., Wang, J., Cocks, B. G. and Rochfort, S. (2017). Seasonal variation of triacylglycerol profile of bovine milk. *Metabolites* 7(2), 24. https://doi.org/10.3390/metabo7020024.

Liu, Z. Q., Rochfort, S. and Cocks, B. (2018). Milk lipidomics: what we know and what we don't. *Progress in Lipid Research* 71: 70–85.

Lopez, C. (2018). Crystallization properties of milk fats. In: *Crystallization of Lipids: Fundamentals and Applications in Food, Cosmetics and Pharmaceuticals* Sato, K. (Ed.) (pp. 283–321). John Wiley & Sons. ISBN 9781118593929.

Lopez, C., Lavigne, F., Lesieur, P., Bourgaux, C. and Ollivon, M. (2001a). Thermal and structural behavior of milk fat. 1. Unstable species of anhydrous milk fat. *Journal of Dairy Science* 84(4): 756–766.

Lopez, C., Lavigne, F., Lesieur, P., Keller, G. and Ollivon, M. (2001b). Thermal and structural behavior of anhydrous milk fat. 2. Crystalline forms obtained by slow cooling. *Journal of Dairy Science* 84(11): 2402–2412.

Lopez, C., Lesieur, P., Bourgaux, C., Keller, G. and Ollivon, M. (2001c). Thermal and structural behavior of milk fat: 2. *Journal of Colloid and Interface Science* 240(1): 150–161.

Lopez-Lopez, A., Castellote-Bargallo, A. I., Campoy-Folgoso, C., Rivero-Urgël, M., Tormo-Carnicé, R., Infante-Pina, D. and López-Sabater, M. C. (2001). The influence of dietary palmitic acid triacylglyceride position on the fatty acid, calcium and magnesium contents of at term newborn faeces. *Early Human Development* 65 (Suppl.): S83–S94.

Lopez, C., Lesieur, P., Bourgaux, C. and Ollivon, M. (2005). Thermal and structural behavior of anhydrous milk fat. 3. Influence of cooling rate. *Journal of Dairy Science* 88(2): 511–526.

Lu, J., Pickova, J., Vázquez-Gutiérrez, J. L. and Langton, M. (2018). Influence of seasonal variation and ultra high temperature processing on lipid profile and fat globule structure of Swedish cow milk. *Food Chemistry* 239: 848–857.

MacGibbon, A. K. H. (2020). Composition and structure of bovine milk lipids. In: *Advanced Dairy Chemistry Volume 2: Lipids* McSweeney, P. L. H., Fox, P. F. and O'Mahony, J. A. (Eds) (pp. 1-32). Cham: Springer International Publishing.

Marangoni, A. G., Acevedo, N., Maleky, F., Co, E., Peyronel, F., Mazzanti, G., Quinn, B. and Pink, D. (2011). Structure and functionality of edible fats. *Soft Matter* 8(5): 1275–1300.

Marangoni, A. G., van Duynhoven, J. P. M., Acevedo, N. C., Nicholson, R. A. and Patel, A. R. (2020). Advances in our understanding of the structure and functionality of edible fats and fat mimetics. *Soft Matter* 16(2): 289–306.

Martens, K. J. A., van Dalen, G., Heussen, P. C. M., Voda, M. A., Nikolaeva, T. and van Duynhoven, J. P. M. (2018). Quantitative structural analysis of fat crystal networks by means of Raman confocal imaging. *Journal of the American Oil Chemists' Society* 95(3): 259–265.

Mazzanti, G., Marangoni, A. G. and Idziak, S. H. J. (2005). Modeling phase transitions during the crystallization of a multicomponent fat under shear. *Physical Review. E, Statistical, Nonlinear, and Soft Matter Physics* 71(4 Pt 1): 041607.

McAnoy, A. M., Wu, C. C. and Murphy, R. C. (2005). Direct qualitative analysis of triacylglycerols by electrospray mass spectrometry using a linear ion trap. *Journal of the American Society for Mass Spectrometry* 16(9): 1498–1509.

Narine, S. S. and Marangoni, A. G. (1999). Relating structure of fat crystal networks to mechanical properties. *Food Research International* 32(4): 227–248.

O'Donnell-Megaro, A. M., Barbano, D. M. and Bauman, D. E. (2011). Survey of the fatty acid composition of retail milk in the United States including regional and seasonal variations. *Journal of Dairy Science* 94(1): 59–65.

Pacheco-Pappenheim, S., Yener, S., van Valenberg, H. J. F., Tzompa-Sosa, D. A. and Bovenhuis, H. (2019). The DGAT1 K232A polymorphism and feeding modify milk fat triacylglycerol composition. *Journal of Dairy Science* 102(8): 6842–6852.

Parodi, P. W. (1979). Stereospecific distribution of fatty-acids in bovine milk-fat triglycerides. *Journal of Dairy Research* 46(1): 75–81.

Peyronel, F., Ilavsky, J., Mazzanti, G., Marangoni, A. G. and Pink, D. A. (2013). Edible oil structures at low and intermediate concentrations. II. Ultra-small angle X-ray scattering of in situ tristearin solids in triolein. *Journal of Applied Physics* 114(23): 234902.

Picariello, G., Sacchi, R. and Addeo, F. (2007). One-step characterization of triacylglycerols from animal fat by MALDI-TOF MS. *European Journal of Lipid Science and Technology* 109(5): 511-524.

Poulsen, N. A., Hein, L., Kargo, M. and Buitenhuis, A. J. (2020). Realization of breeding values for milk fatty acids in relation to seasonal variation in organic milk. *Journal of Dairy Science* 103(3): 2434-2441.

Precht, D. and Frede, E. (1994). Determination of the solid fat-content in milk-fat by gas-chromatographic triglyceride analysis. *Fett Wissenschaft Technologie/Fat Science Technology* 96(9): 324-330.

Ramel, P. R. R., Peyronel, F. and Marangoni, A. G. (2016). Characterization of the nanoscale structure of milk fat. *Food Chemistry* 203: 224-230.

Rodriguez-Otero, J. L., Hermida, M. and Centeno, J. (1997). Analysis of dairy products by near-infrared spectroscopy: a review. *Journal of Agricultural and Food Chemistry* 45(8): 2815-2819.

Rutten, M. J. M., Bovenhuis, H., Hettinga, K. A., van Valenberg, H. J. and van Arendonk, J. A. (2009). Predicting bovine milk fat composition using infrared spectroscopy based on milk samples collected in winter and summer. *Journal of Dairy Science* 92(12): 6202-6209.

Sato, K. (2001). Crystallization behaviour of fats and lipids - a review. *Chemical Engineering Science* 56(7): 2255-2265.

Schennink, A., Stoop, W. M., Visker, M. H. P. W., Heck, J. M., Bovenhuis, H., van der Poel, J. J., van Valenberg, H. J. and van Arendonk, J. A. (2007). DGAT1 underlies large genetic variation in milk-fat composition of dairy cows. *Animal Genetics* 38(5): 467-473.

Shi, Y., Smith, C. M. and Hartel, R. W. (2001). Compositional effects on milk fat crystallization. *Journal of Dairy Science* 84(11): 2392-2401.

Singh, A. P., Avramis, C. A., Kramer, J. K. G. and Marangoni, A. G. (2004). Algal meal supplementation of the cows' diet alters the physical properties of milk fat. *Journal of Dairy Research* 71(1): 66-73.

Smet, K., Coudijzer, K., Fredrick, E., De Campeneere, S., De Block, J., Wouters, J., Raes, K. and Dewettinck, K. (2010). Crystallization behavior of milk fat obtained from linseed-fed cows. *Journal of Dairy Science* 93(2): 495-505.

Sokol, E., Ulven, T., Faergeman, N. J. and Ejsing, C. S. (2015). Comprehensive and quantitative profiling of lipid species in human milk, cow milk and a phospholipid-enriched milk formula by GC and MS/MSALL. *European Journal of Lipid Science and Technology* 117(6): 751-759.

Staniewski, B., Ogrodowska, D., Staniewska, K. and Kowalik, J. (2021). The effect of triacylglycerol and fatty acid composition on the rheological properties of butter. *International Dairy Journal* 114: 104913.

Stoop, W. M., van Arendonk, J. A. M., Heck, J. M. L., van Valenberg, H. J. and Bovenhuis, H. (2008). Genetic parameters for major milk fatty acids and milk production traits of Dutch Holstein-Friesians. *Journal of Dairy Science* 91(1): 385-394.

ten Grotenhuis, E., van Aken, G. A., van Malssen, K. F. and Schenk, H. (1999). Polymorphism of milk fat studied by differential scanning calorimetry and real-time X-ray powder diffraction. *Journal of the American Oil Chemists' Society* 76(9): 1031-1039.

Timms, R. E. (1984). Phase-behavior of fats and their mixtures. *Progress in Lipid Research* 23(1): 1-38.

Truong, T., Bansal, N. and Bhandari, B. (2014). Effect of emulsion droplet size on foaming properties of milk Fat Emulsions. *Food and Bioprocess Technology* 7(12): 3416-3428.

Turon, F., Bonnot, F., Caro, Y., Pina, M. and Graille, J. (2003). Acyl migration incidence on accuracy of a triacylglycerol regioanalysis - a theoretical evaluation. *Chemistry and Physics of Lipids* 125(1): 41–48.

Tzompa-Sosa, D. A., Meurs, P. P. and van Valenberg, H. J. F. (2018). Triacylglycerol profile of summer and winter bovine milk fat and the feasibility of triacylglycerol fragmentation. *European Journal of Lipid Science and Technology* 120(3): 1700291.

Tzompa-Sosa, D. A., Ramel, P. R., van Valenberg, H. J. F. and van Aken, G. A. (2016a). Formation of beta Polymorphs in Milk Fats with Large Differences in Triacyiglycerol Profiles. *Journal of Agricultural and Food Chemistry* 64(20): 4152–4157.

Tzompa-Sosa, D. A., van Valenberg, H. J. F., van Aken, G. A. and Bovenhuis, H. (2016b). Milk fat triacylglycerols and their relations with milk fatty acid composition, DGAT1 K232A polymorphism and milk production traits. *Journal of Dairy Science* 99(5): 3624–3631.

Tzompa-Sosa, D. A., Van Aken, G. A., Van Hooijdonk, A. C. M. and van Valenberg, H. J. (2014). Influence of C16:0 and long-chain saturated fatty acids on normal variation of bovine milk fat triacylglycerol structure. *Journal of Dairy Science* 97(7): 4542–4551.

van Aken, G. A., ten Grotenhuis, E., van Langevelde, A. J. and Schenk, H. (1999). Composition and crystallization of milk fat fractions. *Journal of the American Oil Chemists' Society* 76(11): 1323–1331.

van Aken, G. A. and Visser, K. A. (2000). Firmness and crystallization of milk fat in relation to processing conditions. *Journal of Dairy Science* 83(9): 1919–1932.

Walstra, P., van Vliet, T. and Kloek, W. (1992). Crystallization and rheological properties of milk fat. In: *Advanced Dairy Chemistry* Fox, P. F. (Ed). Elsevier, London, pp. 179–211.

Watanabe, Y., Sato, S., Sera, S., Sato, C., Yoshinaga, K., Nagai, T., Sato, R., Iwasaka, H. and Aki, T. (2014). Enzymatic analysis of positional distribution of fatty acids in solid fat by 1,3-selective transesterification with Candida antarctica lipase B. *Journal of the American Oil Chemists' Society* 91(8): 1323–1330.

Wiking, L., De Graef, V., Rasmussen, M. and Dewettinck, K. (2009). Relations between crystallisation mechanisms and microstructure of milk fat. *International Dairy Journal* 19(8): 424–430.

Yener, S. and van Valenberg, H. J. F. (2019). Characterisation of triacylglycerols from bovine milk fat fractions with MALDI-TOF-MS fragmentation. *Talanta* 204: 533–541.

Yoshinaga, K., Sato, S., Sasaki, R., Asada, M., Hori, R., Imagi, J., Miyazaki, Y., Nagai, T., Saito, K., Sano, T., Sasaki, A., Sato, C., Tsukahara, Y., Yamashita, A., Watanabe, S. and Watanabe, Y. (2016). The collaborative study on the enzymatic analysis of positional distribution of short- and medium-chain fatty acids in milk fat using immobilized Candida antarctica Lipase B. *Journal of Oleo Science* 65(4): 291–302.

# Chapter 9

## Advances in understanding the biosynthetic pathways of milk lipids, their health benefits and bioactive properties

*Nurit Argov-Argaman, The Hebrew University of Jerusalem, Israel*



### 1 Introduction

Milk is the only whole food evolved solely to provide complete nutrition to mammals. Throughout the evolution of mammals, lactation has been under constant Darwinian natural selection to meet suckling neonate's needs for nourishment, to support growth and to provide protection. Milk demonstrates adaptive plasticity by modulating its basic composition, structural complexity and nutritional value to optimize the survival chances of the mammalian infant. All aspects of the structure and bioactivity of milk should be considered in developing health-promoting products.

Among all the components in milk, the milk fat globule remains the least understood. Darwinian selective pressure has resulted in the emergence of a remarkable lipid delivery system in which particles and their surface properties are unique compared to any other biological lipid export system. For decades, research has overlooked the complexity of self-assembled milk fat globules

http://dx.doi.org/10.19103/AS.2022.0099.09

and interpreted milk fat simply as triglyceride-rich particles indistinguishable in complexity from plant-derived oils. However, milk fat globules are a remarkable dispersion of highly bioactive particles. These particles have a size distribution across the entire nanometer length scale, from globules less than 200 nm to globules greater than 15 μm. The composition of milk fat globule proteins, fatty acids, glycerophospholipids and sphingolipids differs in milk fat globules of different sizes. The mean diameter of milk fat globules is a dynamic property. It demonstrates the plasticity of the mammary gland in delivering different compositions of milk lipids. Milk fat globule size also changes over time from parturition, changing as a consequence of factors such as genetic background, maternal diet, maternal health, environmental conditions like seasons and aspects of maternal metabolism like energy balance.

Recent nutrition research is identifying how the milk fat globule and its membrane affect lipid composition and bioactivity. Given that lipid composition is closely associated with the size of the fat globule, any change in size will change the composition and bioactivity of lipids. These changes can also affect the technological and rheological properties of milk. This highlights the need to understand how milk fat globule size is regulated and the consequences of changing its size in terms of nutritional and functional properties.

The study of the effect of milk fat globule size on the functionality of milk lipids is challenging. Using common separation protocols based on centrifugation or membrane filtration to separate milk fat globules according to size can cause significant alterations in their structure and composition. This is due to their fragile nature, which stems from the multiple layers of polar lipids and protein enveloping the triglyceride core, with varying melting and freezing temperatures. It is important to acknowledge the differences in the preparation protocol for size fractions of milk fat globules, since this may change the composition of milk fat and the proteins embedded in the membrane. This challenge also applies when investigating milk fat globule membrane (MFGM) using different commercially available isolates with varying lipid and protein compositions.

Some excellent research reviews have been published recently on dairy lipids in areas such as beneficial effects in human nutrition (Thorning et al., 2017, Willett and Ludwig, 2020), the anti-enteropathogenic properties of milk lipids (Douellou et al., 2017) and the importance of milk lipids in infant development (Lee et al., 2018). The effect of polar lipid production processes on milk composition has recently been reviewed by Raza et al. (2021). This chapter focuses on the structure of the milk fat globule, as secreted by the mammary gland cells, with special attention to the regulation of lipid composition in the mammary gland and at the whole animal level.

## 2 Assessing milk lipid content and its health effects: from single components to matrices

The lipid composition of milk and dairy products has been systematically studied, with a special focus on fatty acid composition and total fat content. In terms of total milk fat content, fat content widely varies among species, partially due to species lactation strategy, and even within the same species, due to genetic, nutritional, physiological and environmental factors. For example, the fat content of donkey milk is 0.3–1.8%, in cow milk, it is 3.3–6.4%, in buffalo milk, it is 5.3–15%, in goat milk, it is 3.0–7.2% and, finally, in human milk, it is 2.1–4.0% (Gantner et al., 2015). Other species have much higher fat content, for example, 17–29% in murine milk (Gors et al., 2009), with the highest in the seal where milk can reach 61% fat (Oftedal et al., 1993). Variation in fat content within species can be found in different breeds of dairy cows (i.e., Jersey and Holstein; Palladino et al., 2010) and dairy goats (Hadaya et al., 2017). In addition, environmental factors like nutrition, season and their interaction may affect fat content as demonstrated in dairy cows (Argov-Argaman et al., 2014, Larsen et al., 2014). The composition of milk fatty acid is more dynamically regulated and depends on local and systemic endocrine and metabolic cues. In dairy animals like cows and goats, there are major metabolic differences around parturition, a time frame that is commonly referred to as the 'transition period'. A robust adaptation to the metabolic load required to support embryonic development in the last trimester of pregnancy, and the demands of milk production following parturition, changes the fatty acid composition of milk. Fatty acid composition is therefore known to be modulated during different stages of lactation (Mesilati-Stahy and Argov-Argaman, 2014, Hadaya et al., 2017). This is mostly attributed to the dependency of the mammary gland on free fatty acids released to plasma upon adipose lipolysis (Argov-Argaman et al., 2012).

The energy provided by concentrate feed or forage, and especially the inclusion of protected fatty acids that can pass the rumen without being utilized or hydrogenated by rumen bacteria, also affects the fatty acid composition of milk (Argov-Argaman et al., 2014, Hadaya et al., 2017, Lopez et al., 2008, Moallem, 2018). Moreover, including plant secondary metabolites, like polyphenols, or allowing dairy animals to graze on polyphenol-rich vegetation, increase the content of polyunsaturated fatty acids in milk (Argov-Argaman et al., 2021, 2020a, 2016).

In terms of health effects of milk lipids, the saturated fatty acid content is particularly important due to its possible role in the pathogenesis of metabolic syndrome (Johnson et al., 2008, Cnop, 2008), dyslipidemia and hypercholesterolemia (Vessby et al., 2001). However, not all saturated fatty acids have the same influence on human health (German and Dillard, 2004).

Short- and medium-chain saturated fatty acids exert beneficial effects on lipid metabolism, blood pressure and immunomodulation. In this regard, milk from goats and sheep is considered to be a rich source for medium-chain fatty acids which contribute to the unique rheological properties of dairy products from small ruminants (Chilliard and Ferlay, 2004). In order to improve the health traits of milk and dairy products, significant effort has been made over the years to increase the content of unsaturated fatty acids. One common approach is to use rumen-protected unsaturated fatty acids or oil for the mammary gland to absorb and incorporate into milk fat. However, the extent of alteration of milk composition by these products is thought to be low (Lanier and Corl, 2015). Alternatively, pasture has been found to increase the content of omega 3 fatty acids in goat milk (Hadaya et al., 2017, Hadaya et al., 2020, Argov-Argaman et al., 2021) and cow milk (Benbrook et al., 2018), and therefore, milk from animals that graze on pasture is often, but not always (Capuano et al., 2015), considered healthier.

For decades, research on the metabolic and health effects of milk fat was based on a reductionist approach, that is, trying to predict health effects based on a component such as a specific fatty acid (Thorning et al., 2017). However, fatty acids are almost never secreted into milk in their free form as free fatty acids. Instead, they are secreted in complex lipid molecules, assembled into a complex and unique macrostructure called the milk fat globule. Milk fat globule structure may account for some of the discrepancies between expected and actual health outcomes of consuming dairy fat. For example, according to the reductionist approach, one could expect that consumption of saturated fat products will exert detrimental health effects when compared with unsaturated fat products like vegetable oil. However, outcomes are often counterintuitive and consumption of dairy fat seems to exert health benefits in terms, for example, of plasma lipid composition (Vanderhout et al., 2020) and insulin sensitivity (Nestel et al., 2014, Nestel, 2012). The surprising positive effect of saturated fat content in milk could be attributed to the milk matrix and especially to the complex structure of the milk fat globule, which may impact digestion and absorption kinetics, availability of macro- and micronutrients as well as postprandial metabolism.

The milk fat globule consists of a triglyceride core which is the main source of energy in milk. This hydrophobic core is stabilized by the MFGM, which is an amphipathic envelope acquired by the milk fat globule during its secretion process. The MFGM is composed of glycerophospholipids, sphingolipids, cholesterol, proteins, enzymes and glycoconjugates, derived from the membranes of mammary gland epithelial cells (Gallier et al., 2010). While the MFGM was considered for years merely as an emulsifier of milk fat, data from the last decade provide strong evidence for the health properties associated with this unique structure, as discussed in the following sections.

## 3 Milk lipids: composition, synthesis and secretion

The milk lipid fraction is dominated by its triglyceride content, which accounts for over 95% of milk fat in all mammals. Other milk lipids in addition to triglycerides are glycerophospholipids and sphingolipids (together termed phospholipids), diacylglycerol, monoacylglycerol, cholesterol and free fatty acids at marginal concentrations. The main groups of polar lipids in milk are glycerophospholipids and sphingolipids. Milk phospholipids appear to be synthesized de novo in the mammary gland (Annison et al., 1967). Mammary fatty acids are absorbed during circulation in addition to being synthesized de novo (DeKay et al., 1976). Fatty acids with a carbon chain length of less than 16 atoms are synthesized de novo in mammary epithelial cells, whereas those with a carbon chain length of more than 16 atoms are absorbed during circulation, either from dietary sources or from adipose lipolysis. Fatty acids with a carbon chain length of 16 atoms originate from both sources (Bauman and Davis, 1974).

At the transcriptional level, a marked upregulation of various genes associated with lipid synthesis has been demonstrated during lactation. For example, genes associated with mammary fatty acid uptake from blood, intracellular fatty acid trafficking, long- and short-chain intracellular fatty acid activation, de novo fatty acid synthesis, desaturation, triacylglycerol synthesis and lipid droplet formation are all upregulated during the peak of lactation compared to the onset of lactation and the stage of established lactation (Bionaz and Loor, 2008).

Synthesis and secretion of milk fat start with a cytoplasmic lipid droplet, a triglyceride droplet formed between the endoplasmic reticulum membranes and released into the cytoplasm covered with one layer of phospholipid and specific surface-associated proteins (Mather and Keenan, 1998, McManaman et al., 2007). From their site of synthesis, cytoplasmic lipid droplets migrate toward the apical pole of the cell, where they are apocrine-secreted into the alveolar lumen, enveloped by the apical cell membrane bilayer. It has been suggested that fusion between lipid droplets during the migration process can increase their size prior to secretion (Mather and Keenan, 1998). Fusion has indeed been demonstrated in vivo in murine mammary gland epithelial cells (Masedunskas et al., 2017).

The final contact between the plasma membrane bilayer and the droplet surface is mediated through the integral protein butyrophilin (Smoczynski, 2017), which forms a secretion complex with xanthine oxidoreductase, adipophilin and other proteins on the droplet surface (McManaman, 2012, Masedunskas et al., 2017, Monks et al., 2016). This complex is considered to initiate the process of droplet coating with a membrane and its secretion outside the cell (Smoczynski, 2017).

Eventually, the secreted milk fat globule consists of a triglyceride core covered with three layers of phospholipids and proteins, derived from cellular membranes, which is known as the MFGM (Heid and Keenan, 2005, Mather and Keenan, 1998). The main proteins found in the fat fraction of raw, unprocessed milk facilitate the apocrine secretion of the milk fat globule, including xanthine dehydrogenase/oxidase, butirophilin and adipophilin, as well as some membrane proteins like mucin (MUC1) and CD36 (Dewettinck et al., 2008, Mather, 2000).

The accepted docking model involves three proteins (Monks et al., 2016):

- perilipin2/adipophilin coating the lipid droplet;
- xanthine oxidoreductase recruited from the cytoplasm; and
- the integral membrane protein butyrophilin anchored within the apical membrane.

This three-protein complex forms a bridge between the droplet and the apical membrane and executes the budding process. Oxytocin has also been suggested as playing a role in the milk fat globule secretion process (Masedunskas et al., 2017), beyond its accepted role in releasing milk from the alveolar lumen to the milk-collecting ducts.

The milk fat globule secretion process is distinct from the classical secretory pathway used for lipid secretion by hepatocytes and enterocytes (Hussain, 2014). A key additional difference between milk fat globule and other secreted lipid-protein assemblies is the milk fat globule size range. Milk fat globules are found in milk in various sizes of over three orders of magnitude (Michalski et al., 2005), whereas lipid droplets in other tissues have a very narrow size range, and deviation from the normal range is usually associated with tissue dysfunction and pathology (Liang and Nishino, 2011).

## 4 Milk fat globule structure features

Milk fat globules are secreted in a wide range of sizes, from the nanometer scale to over 15 μm in both humans (Michalski et al., 2005) and bovines (Ma and Barbano, 2000), whereas smaller milk fat globules, from 0.53 μm to 8.58 μm in diameter, are found in caprine milk (Attaie and Richter, 2000). The diameter of milk fat globules is closely associated with its lipidome (Mesilati-Stahy et al., 2011) and proteome (Lu et al., 2016), and especially with the phospholipid-to-triglyceride mass ratio, which is higher in small versus larger milk fat globules (Argov-Argaman, 2019).

The content of MFGM is roughly 0.5–2% of total milk fat, making it a micro-constituent of milk. Numerous studies have demonstrated the beneficial effects of MFGM on infant gut and cognitive development, improved microbiome and increased resilience to parasitic and pathogenic bacterial infections, resistance to cancer and improved inflammatory status in obese adults (Hernell et al., 2016). Strategies to increase its concentration in milk or dairy products have been pursued by both lactation physiologists and dairy technologists.

The content of MFGM in milk is affected by mammalian species. Argov-Argaman et al. (2021) showed that goat and sheep milks contain higher levels of phospholipids than cow milk. MFGM content in milk is regulated by genetic background. In dairy goats, for example, Mediterranean breeds were found to have higher MFGM content in their milk than their Alpine crossbred counterparts (Argov-Argaman et al., 2016). In addition, Holstein cows with a less effective diacylglycerol acyltransferase genotype were found to produce milk with higher MFGM content than the more efficient genotype (Argov-Argaman et al., 2013).

## 5 Metabolic regulation of milk fat globule structure and composition

Milk fat globule diameter is determined by the size of its precursor: intracellular lipid droplets (Heid and Keenan, 2005). These can increase in size upon direct addition of triglyceride molecules onto the droplet surface or prior to secretion during cytoplasmic fusion events between lipid droplets (Argov-Argaman, 2019). The fusion of lipid droplets is determined by the composition of the polar lipid in the cellular membranes, which is regulated by the metabolic status of mammary gland epithelial cells and their mitochondria activity and number (Argov-Argaman, 2019).

It is widely accepted that lipid droplet size correlates with the amount of triglycerides stored in the cell, as has been demonstrated in the liver, muscle cells and adipocytes (Paar et al., 2012, Lara-Castro and Garvey, 2008). Similarly, in milk, the general assumption is that milk fat globule size is positively correlated with milk fat content. However, several studies found no significant correlation between fat content and milk fat globule size. This suggests that the extent of fusion may be regulated by the polar lipid composition of cellular membranes, partially through mitochondria enzymes (Cohen et al., 2015), which contribute to the change in the lipid droplet size. Previously, a negative association was found between the concentration of non-esterified fatty acids (NEFA) in the plasma of lactating cows and the ratio between triglyceride and phospholipid content in milk (Mesilati-Stahy et al., 2012). This association was found to be affected by the estrous cycle stage, suggesting a link between the progesterone level and the triglyceride to phospholipid ratio, a surrogate used to predict milk fat globule size (Argov-Argaman et al., 2020b). This may be attributed to the effect of progesterone on lipoprotein lipase (LPL) expression or activity (Flint et al., 1984), which can modulate the availability of long-chain fatty acids to the mammary gland, affecting the mitochondria content or the balance between utilizing the long-chain fatty acids as a precursor for phospholipids or triglyceride synthesis (Argov-Argaman et al., 2013). The fact that long-chain fatty acid availability in the mammary gland can change milk fat globule size and therefore the content of MFGM was shown in a study using hyperinsulinogenic-euglycemic clamp in dairy goats (Argov-Argaman et al., 2012). One of the major influences of insulin is the inhibition of adipose lipolysis and therefore its concentration is directly associated with NEFA concentration in the blood. Hence, the results of the hyperinsulinogemic clamp study support the assumption that the availability of long-chain fatty acids from the blood can modulate the size of the secreted milk fat globule.

As these studies suggest, the changing membrane stability of the mammary epithelial cells could result in enhanced fusion of lipid droplets prior to their secretion and subsequently increased diameter of milk fat globules. The two major membrane components that change membrane stability are the mass ratios of phosphatidylcholine and phosphatidylethanolamine, The milk fat globule size of dairy cows fed a low concentrate diet was increased when compared to low concentrate fed counterparts due to the increase in the relative content of phosphtidylethanolamine as a membrane-destabilizing agent (Mesilati-Stahy et al., 2015), most probably due to altered membrane stability and enhanced lipid droplet fusion rates (Cohen et al., 2015).

A final factor that can change the milk fat globule size is the modulation of the secretion dynamics of the mammary cells. This might be achieved by remodeling the epithelial cell membrane, which is the source of the MFGM (Mather and Keenan, 1998). In *Escherichia coli*, phosphatidylethanolamine has been shown to have an essential role in the exocytotic process due to its membrane-destabilizing properties (Mikhaleva et al., 2001).

## 6 Milk fat globule size, organoleptic properties of milk and the effects of dairy processing

Milk fat globule size has significance in processing since it affects the physicochemical and sensory characteristics of dairy products (O'Mahony et al., 2005, Avramis et al., 2003). Such characteristics include curd firmness, stability and even color of dairy products. As an example, smaller milk fat globules produced in cows fed fish meal increased the churning time for butter production (Avramis et al., 2003). The butter had a softer texture and better cold-spreading ability. Camembert cheeses produced with milk with small milk fat globules had a higher melting and elastic texture, had a higher flow and was less yellow. These results were explained by the greater surface area of small MFGMs compared with large milk fat globules, at a given fat content, and by the thinner casein strands in cheeses made from milk with smaller milk fat globules, due to the smaller interglobular distance. The use of native milk fat globules with different sizes can thus lead to new products with different technological and sensory properties.

Milk processing can modify the lipidome of milk and dairy products in terms of fatty acid and phospholipid content and composition. For example, ultra-high temperature treatment, used to extend the shelf life of bovine milk, was shown to alter fatty acid and triglyceride composition (Ajmal et al., 2018). Homogenization and spray drying of bovine milk changed the MFGM and reduced the concentration of phosphatidylethanolamine (Gallier et al., 2010). Moreover, pasteurization (76°C for 15 s) and homogenization increase the concentration of sphingomyelin and reduce that of phosphatidylethanolamine, compared to raw bovine milk.

It seems that not all polar lipids in milk are susceptible to thermal or physical treatment to the same degree. In addition to the compositional changes induced by the processing of milk, a change in the bioavailability of the MFGM also occurs, mostly due to its fragile nature and the shear and physical forces employed on the milk fat structure during pasteurization or homogenization. In addition, the highly structured glycocalyx network of the glycoconjugates of the MFGM is disrupted by homogenization and pasteurization which means that the functionality of the MFGM is impaired, as shown by the reduced binding capacity to bacteria and viruses (Jensen et al., 1992). The effects of heat treatments on milk functionality are not only restricted to the protein fraction but also impair milk fat globule bioactive properties, including inhibition of enteropathogenic bacteria adhesion and propagation in the gut. Understanding how the MFGM lipids are changed during milk processing is vital to improve the nutritional and health properties of dairy products.

## 7 Nutritional and health properties of milk fat affected by fat globule size

The most remarkable change in milk fat globule size is demonstrated in breast milk during the first 72-h postpartum (Michalski et al., 2005). Within the first 12-h postpartum, milk consists of two distinct size populations of milk fat globule: either extremely large (~9 μm) or very small milk fat globules (~0.1 μm) that constitute over 50% of all milk fat globules (Michalski et al., 2005). The extremely small globules overlap with milk membrane vesicles (Argov et al., 2008, Zonneveld et al., 2021). The later size distribution and mean diameter of milk fat globules changes to that found in milk during established lactation, with a mean diameter of approximately 4 μm. This major shift in milk fat globule size is paralleled by major functional and structural differences in the newborn gut, including gut closure (Vukavic, 1984). Since the newborn gut-blood barrier is not fully matured, populating the infant gut with small milk fat globules from breast milk suggests a protective role of these globules against opportunistic pathogens that can invade the gut. Small milk fat globules may also assist gut colonization with probiotic bacteria. Some of these potential protective properties are attributed to the greater membrane content in small compared with large globules.

Supporting the protective role of small, MFGM-rich milk fat globules are the positive health effects associated with the MFGM. For example, MFGM supplementation improved neonate intestinal development and was found to protect against *Clostridium difficile*-induced colitis in baby rats (Bhinder et al., 2017). MFGM supplementation also improved cognitive performance and decreased infectious morbidity in human infants during their first year of life (Timby et al., 2015, 2014). In addition, milk replacement enriched with MFGM reduced the number of days with fever and the number of short febrile episodes in healthy preschool children aged 2.5–6 years (Veereman-Wauters et al., 2012). The MFGM has also been shown to benefit adult health. For example, in overweight and obese adults, the addition of MFGM to a high-saturated fat meal reduced postprandial insulinemic and inflammatory responses (Beals et al., 2019). Mastitis has been found to be associated with milk fat globule size, suggesting this may be a therapeutic target to ameliorate the prevalence of mastitis (Mizuno et al., 2012).

Fat globule size also affects probiotic bacterial growth and survival. Small milk fat globules have a higher content of polar lipids compared with larger globules since more membrane material (MFGM) is required to envelop small globules. This structural difference determines the composition of fatty acids, cholesterol, triglyceride, glycerophospholipids, sphingolipids and glycoconjugates. Fatty acid composition affects bacterial survival, proliferation and even biofilm formation. For example, a bactericidal effect against Gram-negative bacteria such as *E. coli* has been demonstrated for medium-chain fatty acids (Sprong et al., 2001). Fatty acids can also modulate the interaction of milk fat with bacteria. For example, short- and medium-chain fatty acids have been shown to affect growth and biofilm formation by Gram-positive bacilli (Cukkemane et al., 2015). Butyric acid was found to be a potent trigger for biofilm formation in *Bacillus* spp including *Bacillus subtilis* (Pasvolsky et al., 2014). This suggests that some lipid molecules present in milk provide benefits for specific bacterial strains, while other lipids may inhibit proliferation and even induce a stress response in bacteria.

Sphingosine has been shown to disrupt biofilm in the Gram-positive bacteria, *Staphylococcus epidermidis* (Beck et al., 2020). It was suggested that sphingomyelin and their glycoconjugates may act as a decoy for bacterial adhesion (Fantini et al., 1997), preventing their adhesion to the glycocalyx of the intestine and hence their propagation and invasion into host enterocytes or macrophages. This sphingomyelin-dependent invasion pathway is also used by enteric viruses (Nieva et al., 1994). Sphingomyelin is one of the major components of the MFGM, and the protective effect of the MFGM against pathogens is probably associated with the presence of sphingomyelin in MFGM. For example, supplementation of MFGM in adults alleviated infection after oral challenge with diarrheagenic *E. coli* (Bruggencate et al., 2016) and reduced gut colonization of *Listeria* spp. in adult rats (Sprong et al., 2012).

Ganglioside is another MFGM constituent that has been shown to actively bind bacteria and therefore reduce the chance of pathogenic bacteria adhering to epithelial cells of the intestine and propagating in the gastrointestinal tract (Rueda, 2007). These are acid glycosphingolipids embedded in the plasma membrane of mammary epithelial cells and therefore secreted as part of the MFGM. They are probably linked to some of the benefits of MFGM supplementation to infants and children (Raza et al., 2021). Sialic acid, which is part of the ganglioside structure, is another lipid component of the MFGM that may serve as a decoy by preventing adherence and propagation of pathogens in the intestine (Sprong et al., 2001). The MFGM has been shown to have general anti-adhesive effects against enteropathogens (Douellou et al., 2017), partially attributed to the carbohydrate moieties in glycoprotein and glycolipids embedded in the MFGM and targeted by enteropathogenic carbohydrate recognition domains of surface proteins (Douellou et al., 2017). It is still not clear if the benefits are attributed to the prebiotic effect of the MFGM or its bactericidal effects.

As noted earlier, lipids are not secreted in their free isolated form. A reductionist approach of studying the effect of isolated molecules on bacteria disregards the complexity of milk as a food matrix in general and especially milk lipids. It is therefore hard to predict the response of different strains of bacteria to milk fat globules as a substrate, based on studies using an isolated lipid component.

To exert physiological effects on bacterial growth or biofilm formation, lipids first need to be released from their macrostructure. Fatty acids need to be released from milk fat globules, and sphingosine has to be released from its sphingomyelin structure. These processes require a digestion step, typically hydrolysis of ester bonds, executed by specific lipases. Different lipases are required to penetrate the polar lipid envelope of the globule membrane (MFGM) and to hydrolyze its triglyceride core. Given the differences in the ratio between the MFGM and the triglyceride core, the availability of milk lipids may differ as a factor of milk fat globule size. Lipase from bacterial sources can digest the ester bonds and release free fatty acids into milk. The digestion kinetics of lipid droplets is modulated by their size (Armand et al., 1999, Ringo and Olsen, 1999). While these studies did not use milk lipids in their natural form, or lipases originating from bacteria, they do give an indication of the digestion efficiency and availability of fatty acids when administrated in different sizes of lipid particles.

## 8 Conclusion and future trends

One of the unique properties of milk is the structure of the milk fat globule. Starting from its secretion pathway, which is distinguished from that of all other lipid-protein assemblies (including its triple-layered envelope), through its variety of sizes, the milk fat globule delivers fat from the lactating mother to her suckling neonate in a unique way. This unique structure has not been well studied. The question of why and how the mammary gland has developed such a well-defined and efficient delivery system of lipid-protein assemblies (i.e., plasma lipoproteins), and developed its own delivery system, is yet to be determined. Research on lipid-protein assemblies and macromolecular structures is challenging, partially due to the basic decomposition protocols used to investigate the role played by specific nutrients and bioactive compounds. The importance of the MFGM has recently been revealed through a series of *in vivo* studies in animal models as well as in human infants and adults. However, the role played by particle size in this unique delivery system is still overlooked.

A comprehensive approach is needed to fully understand the nutritional or nutraceutical role played by the structure of the milk fat globule. There is a need for a thorough understanding of the circumstances that change milk fat globule size, for example, whether size is a function of maternal condition or is modulated by the crosstalk between the lactating mother and her suckling neonate. There is also a need for a deeper understanding of the consequences of changing the milk fat globule size on lipid availability and functionality. This will lead to an understanding of how to control the size of the milk fat globule and what implications there are in making this change. By changing milk fat globule size, it seems the bioactive content of milk can be significantly modulated, and its inflammatory, developmental and metabolic signals can be controlled and optimized.

## 9 References

Ajmal, M., Nadeem, M., Imran, M. and Junaid, M. 2018. Lipid compositional changes and oxidation status of ultra-high temperature treated milk. *Lipids Health Dis.* 17(1), 227.

Annison, E. F., Linzell, J. L., Fazakerley, S. and Nichols, B. W. 1967. Oxidation and utilization of palmitate stearate oleate and acetate by mammary gland of Fed Goat in relation to their overall metabolism and role of plasma phospholipids and neutral lipids in milk-fat synthesis. *Biochem. J.* 102(3), 637-647+.

Argov, N., Lemay, D. G. and German, J. B. 2008. Milk fat globule structure & function; nanoscience comes to milk production. *Trends Food Sci. Technol.* 19(12), 617-623.

Argov-Argaman, N. 2019. Symposium review: Milk fat globule size: Practical implications and metabolic regulation. *J. Dairy Sci.* 102(3), 2783-2795.

Argov-Argaman, N., Cohen-Zinder, M., Leibovich, H., Yishay, M., Eitam, H., Agmon, R., Hadaya, O., Mesilati-Stahy, R., Miron, J. and Shabtay, A. 2020a. Dietary pomegranate peel improves milk quality of lactating ewes: Emphasis on milk fat globule membrane properties and antioxidative traits. *Food Chem.* 313, 125822.

Argov-Argaman, N., Raz, C. and Roth, Z. 2020b. Progesterone regulation of milk fat globule size is VLDL dependent. *Front. Endocrinol. (Lausanne)* 11, 596.

Argov-Argaman, N., Glasser, T., Muklada, H., Hadaya, O., Mesilati-Stahy, R., Raz, C. and Landau, S. Y. 2021. Lipidome changes, with a focus on phospholipids, due to feeding systems and processing in goat milk. *Food Chem.* 340, 127938.

Argov-Argaman, N., Hadaya, O., Glasser, T., Muklada, H., Dvash, L., Mesilati-Stahy, R. and Landau, S. Y. 2016. Milk fat globule size, phospholipid contents and composition of milk from purebred and alpine-crossbred mid-eastern goats under confinement or grazing condition. *Int. Dairy J.* 58, 2–8.

Argov-Argaman, N., Mbogori, T., Sabastian, C., Shamay, A. and Mabjeesh, S. J. 2012. Hyperinsulinemic clamp modulates milk fat globule lipid composition in goats. *J. Dairy Sci.* 95(10), 5776–5787.

Argov-Argaman, N., Mesilati-Stahy, R., Magen, Y. and Moallem, U. 2014. Elevated concentrate-to-forage ratio in dairy cow rations is associated with a shift in the diameter of milk fat globules and remodeling of their membranes. *J. Dairy Sci.* 97(10), 6286–6295.

Argov-Argaman, N., Mida, K., Cohen, B. C., Visker, M. and Hettinga, K. 2013. Milk fat content and DGAT1 genotype determine lipid composition of the milk fat globule membrane. *PLoS ONE* 8(7), e68707.

Armand, M., Pasquier, B., Andre, M., Borel, P., Senft, M., Peyrot, J., Salducci, J., Portugal, H., Jaussan, V. and Lairon, D. 1999. Digestion and absorption of 2 fat emulsions with different droplet sizes in the human digestive tract. *Am. J. Clin. Nutr.* 70(6), 1096–1106.

Attaie, R. and Richter, R. L. 2000. Size distribution of fat globules in goat milk. *J. Dairy Sci.* 83(5), 940–944.

Avramis, C. A., Wang, H., McBride, B. W., Wright, T. C. and Hill, A. R. 2003. Physical and processing properties of milk, butter, and cheddar cheese from cows fed supplemental fish meal. *J. Dairy Sci.* 86(8), 2568–2576.

Bauman, D. E. and Davis, C. L. B. 1974. Biosynthesis of milk fat. In: *Lactation: A Comprehensive Treatise*. New York: Academic Press.

Beals, E., Kamita, S. G., Sacchi, R., Demmer, E., Rivera, N., Rogers-Soeder, T. S., Gertz, E. R., Van Loan, M. D., German, J. B., Hammock, B. D., Smilowitz, J. T. and Zivkovic, A. M. 2019. Addition of milk fat globule membrane-enriched supplement to a high-fat meal attenuates insulin secretion and induction of soluble epoxide hydrolase gene expression in the postprandial state in overweight and obese subjects. *J. Nutr. Sci.* 8, e16.

Beck, S., Sehl, C., Voortmann, S., Verhasselt, H. L., Edwards, M. J., Buer, J., Hasenberg, M., Gulbins, E. and Becker, K. A. 2020. Sphingosine is able to prevent and eliminate Staphylococcus epidermidis biofilm formation on different orthopedic implant materials in vitro. *J. Mol. Med. (Berl.)* 98(2), 209–219.

Benbrook, C. M., Davis, D. R., Heins, B. J., Latif, M. A., Leifert, C., Peterman, L., Butler, G., Faergeman, O., Abel-Caines, S. and Baranski, M. 2018. Enhancing the fatty acid profile of milk through forage-based rations, with nutrition modeling of diet outcomes. *Food Sci. Nutr.* 6(3), 681–700.

Bhinder, G., Allaire, J. M., Garcia, C., Lau, J. T., Chan, J. M., Ryz, N. R., Bosman, E. S., Graef, F. A., Crowley, S. M., Celiberto, L. S., Berkmann, J. C., Dyer, R. A., Jacobson, K., Surette, M. G., Innis, S. M. and Vallance, B. A. 2017. Milk fat globule membrane supplementation in formula modulates the neonatal gut microbiome and normalizes intestinal development. *Sci. Rep.* 7, 45274.

Bionaz, M. and Loor, J. J. 2008. Gene networks driving bovine milk fat synthesis during the lactation cycle. *BMC Genomics* 9, 366.

Bruggencate, S. J., Frederiksen, P. D., Pedersen, S. M., Floris-Vollenbroek, E. G., Lucas-Van De Bos, E., Van Hoffen, E. and Wejse, P. L. 2016. Dietary milk-fat-globule membrane affects resistance to diarrheagenic Escherichia coli in healthy adults in a randomized, placebo-controlled, double-blind study. *J. Nutr.* 146(2), 249–255.

Capuano, E., Gravink, R., Boerrigter-Eenling, R. and Van Ruth, S. M. 2015. Fatty acid and triglycerides profiling of retail organic, conventional and pasture milk: Implications for health and authenticity. *Int. Dairy J.* 42, 58–63.

Chilliard, Y. and Ferlay, A. 2004. Dietary lipids and forages interactions on cow and goat milk fatty acid composition and sensory properties. *Reprod. Nutr. Dev.* 44(5), 467–492.

Cnop, M. 2008. Fatty acids and glucolipotoxicity in the pathogenesis of Type 2 diabetes. *Biochem. Soc. Trans.* 36, 348–352.

Cohen, B. C., Shamay, A. and Argov-Argaman, N. 2015. Regulation of lipid droplet size in mammary epithelial cells by remodeling of membrane lipid composition: A potential mechanism. *PLoS ONE* 10(3), e0121645.

Cukkemane, N., Bikker, F. J., Nazmi, K., Brand, H. S., Sotres, J., Lindh, L., Arnebrant, T. and Veerman, E. C. 2015. Anti-adherence and bactericidal activity of sphingolipids against *Streptococcus mutans*. *Eur. J. Oral Sci.* 123(4), 221–227.

DeKay, D. E., Bauman, D. E. and Davis, C. L. 1976. Characterization of fatty acid synthesis by cow mammary subcellular fractions. *J. Dairy Sci.* 59(8), 1513–1517.

Dewettinck, K., Rombaut, R., Thienpont, N., Le, T. T., Messens, K. and Van Camp, J. 2008. Nutritional and technological aspects of milk fat globule membrane material. *Int. Dairy J.* 18(5), 436–457.

Douellou, T., Montel, M. C. and Thevenot Sergentet, D. 2017. Invited review: Anti-adhesive properties of bovine oligosaccharides and bovine milk fat globule membrane-associated glycoconjugates against bacterial food enteropathogens. *J. Dairy Sci.* 100(5), 3348–3359.

Fantini, J., Hammache, D., Delezay, O., Yahi, N., Andre-Barres, C., Rico-Lattes, I. and Lattes, A. 1997. Synthetic soluble analogs of galactosylceramide (GalCer) bind to the V3 domain of HIV-1 gp120 and inhibit HIV-1-induced fusion and entry. *J. Biol. Chem.* 272(11), 7245–7252.

Flint, D. J., Clegg, R. A. and Knight, C. H. 1984. Effects of prolactin, progesterone and ovariectomy on metabolic activities and insulin receptors in the mammary gland and adipose tissue during extended lactation in the rat. *J. Endocrinol.* 102(2), 231–236.

Gallier, S., Gragson, D., Cabral, C., Jimenez-Flores, R. and Everett, D. W. 2010. Composition and fatty acid distribution of bovine milk phospholipids from processed milk products. *J. Agric. Food Chem.* 58(19), 10503–10511.

Gantner, V., Mijic, P., Baban, M., Skrtic, Z. and Turalija, A. 2015. The overall and fat composition of milk of various species. *Mljekarstvo* 65, 223–231.

German, J. B. and Dillard, C. J. 2004. Saturated fats: What dietary intake? *Am. J. Clin. Nutr.* 80(3), 550–559.

Gors, S., Kucia, M., Langhammer, M., Junghans, P. and Metges, C. C. 2009. Milk composition in mice: Methodological aspects and effects of mouse strain and lactation day. *J. Dairy Sci.* 92(2), 632–637.

Hadaya, O., Bransi-Nicola, R., Shalev, Y., Azaizeh, H., Roth, Z., Muklada, H., Deutch, T., Landau, S. Y. and Argov-Argaman, N. 2020. Pistacia lentiscus extract enhances mammary epithelial cells' productivity by modulating their oxidative status. *Sci. Rep.* 10(1), 20985.

Hadaya, O., Landau, S. Y., Glasser, T., Muklada, H., Dvash, L., Mesilati-Stahy, R. and Argov-Argaman, N. 2017. Milk composition in Damascus, Mamber and F1 Alpine crossbred goats under grazing or confinement management. *Small Rumin. Res.* 153, 31–40.

Heid, H. W. and Keenan, T. W. 2005. Intracellular origin and secretion of milk fat globules. *Eur. J. Cell Biol.* 84(2-3), 245–258.

Hernell, O., Timby, N., Domellof, M. and Lonnerdal, B. 2016. Clinical benefits of milk fat globule membranes for infants and children. *J. Pediatr.* 173(Suppl), S60–S65.

Hussain, M. M. 2014. Intestinal lipid absorption and lipoprotein formation. *Curr. Opin. Lipidol.* 25(3), 200–206.

Jensen, R. G., Ferris, A. M. and Lammi-Keefe, C. J. 1992. Lipids in human milk and infant formulas. *Annu. Rev. Nutr.* 12, 417–441.

Johnson, L., Mander, A. P., Jones, L. R., Emmett, P. M. and Jebb, S. A. 2008. A prospective analysis of dietary energy density at age 5 and 7 years and fatness at 9 years among UK children. *Int. J. Obes. (Lond.)* 32(4), 586–593.

Lanier, J. S. and Corl, B. A. 2015. Challenges in enriching milk fat with polyunsaturated fatty acids. *J. Anim. Sci. Biotechnol.* 6(1), 26.

Lara-Castro, C. and Garvey, W. T. 2008. Intracellular lipid accumulation in liver and muscle and the insulin resistance syndrome. *Endocrinol. Metab. Clin. North Am.* 37(4), 841–856+.

Larsen, M. K., Andersen, K. K., Kaufmann, N. and Wiking, L. 2014. Seasonal variation in the composition and melting behavior of milk fat. *J. Dairy Sci.* 97(8), 4703–4712.

Lee, H., Padhi, E., Hasegawa, Y., Larke, J., Parenti, M., Wang, A., Hernell, O., Lonnerdal, B. and Slupsky, C. 2018. Compositional dynamics of the milk fat globule and its role in infant development. *Front. Pediatr.* 6, 313.

Liang, W. C. and Nishino, I. 2011. Lipid storage myopathy. *Curr. Neurol. Neurosci. Rep.* 11(1), 97–103.

Lopez, C., Briard-Bion, V., Menard, O., Rousseau, F., Pradel, P. and Besle, J. M. 2008. Phospholipid, sphingolipid, and fatty acid compositions of the milk fat globule membrane are modified by diet. *J. Agric. Food Chem.* 56(13), 5226-5236.

Lu, J., Argov-Argaman, N., Anggrek, J., Boeren, S., Van Hooijdonk, T., Vervoort, J. and Hettinga, K. A. 2016. The protein and lipid composition of the membrane of milk fat globules depends on their size. *J. Dairy Sci.* 99(6), 4726–4738.

Ma, Y. and Barbano, D. M. 2000. Gravity separation of raw bovine milk: Fat globule size distribution and fat content of milk fractions. *J. Dairy Sci.* 83(8), 1719–1727.

Masedunskas, A., Chen, Y., Stussman, R., Weigert, R. and Mather, I. H. 2017. Kinetics of milk lipid droplet transport, growth, and secretion revealed by intravital imaging: Lipid droplet release is intermittently stimulated by oxytocin. *Mol. Biol. Cell* 28(7), 935–946.

Mather, I. H. 2000. A review and proposed nomenclature for major proteins of the milk-fat globule membrane. *J. Dairy Sci.* 83(2), 203–247.

Mather, I. H. and Keenan, T. W. 1998. Origin and secretion of milk lipids. *J. Mammary Gland Biol. Neoplasia* 3(3), 259–273.

McManaman, J. L., Russell, T. D., Schaack, J., Orlicky, D. J. and Robenek, H. 2007. Molecular determinants of milk lipid secretion. *J. Mammary Gland Biol. Neoplasia* 12(4), 259–268.

Mesilati-Stahy, R. and Argov-Argaman, N. 2014. The relationship between size and lipid composition of the bovine milk fat globule is modulated by lactation stage. *Food Chem.* 145, 562-570.

Mesilati-Stahy, R., Malka, H. and Argov-Argaman, N. 2012. Association of plasma insulin concentration to fatty acid distribution between milk fat and membrane synthesis. *J. Dairy Sci.* 95(4), 1767-1775.

Mesilati-Stahy, R., Mida, K. and Argov-Argaman, N. 2011. Size-dependent lipid content of bovine milk fat globule and membrane phospholipids. *J. Agric. Food Chem.* 59(13), 7427-7435.

Mesilati-Stahy, R., Moallem, U., Magen, Y. and Argov-Argaman, N. 2015. Altered concentrate to forage ratio in cows ration enhanced bioproduction of specific size subpopulation of milk fat globules. *Food Chem.* 179, 199-205.

Michalski, M. C., Briard, V., Michel, F., Tasson, F. and Poulain, P. 2005. Size distribution of fat globules in human colostrum, breast milk, and infant formula. *J. Dairy Sci.* 88(6), 1927-1940.

Mizuno, K., Hatsuno, M., Aikawa, K., Takeichi, H., Himi, T., Kaneko, A., Kodaira, K., Takahashi, H. and Itabashi, K. 2012. Mastitis is associated with IL-6 levels and milk fat globule size in breast milk. *J. Hum. Lact.* 28(4), 529-534.

Moallem, U. 2018. Invited review: Roles of dietary n-3 fatty acids in performance, milk fat composition, and reproductive and immune systems in dairy cattle. *J. Dairy Sci.* 101(10), 8641-8661.

Monks, J., Dzieciatkowska, M., Bales, E. S., Orlicky, D. J., Wright, R. M. and McManaman, J. L. 2016. Xanthine oxidoreductase mediates membrane docking of milk-fat droplets but is not essential for apocrine lipid secretion. *J. Physiol.* 594(20), 5899-5921.

Nestel, P. 2012. Nutrition and metabolism: The changing face of the dairy-cardiovascular risk paradox. *Curr. Opin. Lipidol.* 23(1), 1-3.

Nestel, P. J., Straznicky, N., Mellett, N. A., Wong, G., De Souza, D. P., Tull, D. L., Barlow, C. K., Grima, M. T. and Meikle, P. J. 2014. Specific plasma lipid classes and phospholipid fatty acids indicative of dairy food consumption associate with insulin sensitivity. *Am. J. Clin. Nutr.* 99(1), 46-53.

Nieva, J. L., Bron, R., Corver, J. and Wilschut, J. 1994. Membrane fusion of Semliki Forest virus requires sphingolipids in the target membrane. *EMBO J.* 13(12), 2797-2804.

Oftedal, O. T., Bowen, W. D. and Boness, D. J. 1993. Energy-transfer by lactating hooded seals and nutrient deposition in their pups during the 4 days from birth to weaning. *Physiol. Zool.* 66(3), 412-436.

O'Mahony, J. A., Auty, M. A. and Mcsweeney, P. L. 2005. The manufacture of miniature Cheddar-type cheeses from milks with different fat globule size distributions. *J. Dairy Res.* 72(3), 338-348.

Paar, M., Jungst, C., Steiner, N. A., Magnes, C., Sinner, F., Kolb, D., Lass, A., Zimmermann, R., Zumbusch, A., Kohlwein, S. D. and Wolinski, H. 2012. Remodeling of lipid droplets during lipolysis and growth in adipocytes. *J. Biol. Chem.* 287(14), 11164-11173.

Palladino, R. A., Buckley, F., Prendiville, R., Murphy, J. J., Callan, J. and Kenny, D. A. 2010. A comparison between Holstein-Friesian and Jersey dairy cows and their F-1 hybrid on milk fatty acid composition under grazing conditions. *J. Dairy Sci.* 93(5), 2176-2184.

Pasvolsky, R., Zakin, V., Ostrova, I. and Shemesh, M. 2014. Butyric acid released during milk lipolysis triggers biofilm formation of Bacillus species. *Int. J. Food Microbiol.* 181, 19-27.

Raza, G. S., Herzig, K. H. and Leppaluoto, J. 2021. Invited review: Milk fat globule membrane: A possible panacea for neurodevelopment, infections, cardiometabolic diseases, and frailty. *J. Dairy Sci.* 104(7), 7345–7363.

Ringo, E. and Olsen, R. E. 1999. The effect of diet on aerobic bacterial flora associated with intestine of Arctic charr (*Salvelinus alpinus* L.). *J. Appl. Microbiol.* 86(1), 22–28.

Rueda, R. 2007. The role of dietary gangliosides on immunity and the prevention of infection. *Br. J. Nutr.* 98(Suppl 1), S68–S73.

Smoczynski, M. 2017. Role of phospholipid flux during milk secretion in the mammary gland. *J. Mammary Gland Biol. Neoplasia* 22(2), 117–129.

Sprong, R. C., Hulstein, M. F., Lambers, T. T. and Van Der Meer, R. 2012. Sweet buttermilk intake reduces colonisation and translocation of *Listeria monocytogenes* in rats by inhibiting mucosal pathogen adherence. *Br. J. Nutr.* 108(11), 2026–2033.

Sprong, R. C., Hulstein, M. F. and Van Der Meer, R. 2001. Bactericidal activities of milk lipids. *Antimicrob. Agents Chemother.* 45(4), 1298–1301.

Thorning, T. K., Bertram, H. C., Bonjour, J. P., De Groot, L., Dupont, D., Feeney, E., Ipsen, R., Lecerf, J. M., Mackie, A., Mckinley, M. C., Michalski, M. C., Remond, D., Riserus, U., Soedamah-Muthu, S. S., Tholstrup, T., Weaver, C., Astrup, A. and Givens, I. 2017. Whole dairy matrix or single nutrients in assessment of health effects: Current evidence and knowledge gaps. *Am. J. Clin. Nutr.* 105(5), 1033–1045.

Timby, N., Domellof, E., Hernell, O., Lonnerdal, B. and Domellof, M. 2014. Neurodevelopment, nutrition, and growth until 12 mo of age in infants fed a low-energy, low-protein formula supplemented with bovine milk fat globule membranes: A randomized controlled trial. *Am. J. Clin. Nutr.* 99(4), 860–868.

Timby, N., Hernell, O., Vaarala, O., Melin, M., Lonnerdal, B. and Domellof, M. 2015. Infections in infants fed formula supplemented with bovine milk fat globule membranes. *J. Pediatr. Gastroenterol. Nutr.* 60(3), 384–389.

Vanderhout, S. M., Aglipay, M., Torabi, N., Juni, P., Da Costa, B. R., Birken, C. S., O'connor, D. L., Thorpe, K. E. and Maguire, J. L. 2020. Whole milk compared with reduced-fat milk and childhood overweight: A systematic review and meta-analysis. *Am. J. Clin. Nutr.* 111(2), 266–279.

Veereman-Wauters, G., Staelens, S., Rombaut, R., Dewettinck, K., Deboutte, D., Brummer, R. J., Boone, M. and Le Ruyet, P. 2012. Milk fat globule membrane (INPULSE) enriched formula milk decreases febrile episodes and may improve behavioral regulation in young children. *Nutrition* 28(7–8), 749–752.

Vessby, B., Uusitupa, M., Hermansen, K., Riccardi, G., Rivellese, A. A., Tapsell, L. C., Nalsen, C., Berglund, L., Louheranta, A., Rasmussen, B. M., Calvert, G. D., Maffetone, A., Pedersen, E., Gustafsson, I. B., Storlien, L. H. and KANWU Study. 2001. Substituting dietary saturated for monounsaturated fat impairs insulin sensitivity in healthy men and women: The KANWU study. *Diabetologia* 44(3), 312–319.

Willett, W. C. and Ludwig, D. S. 2020. Milk and health. *N. Engl. J. Med.* 382(7), 644–654.

Zonneveld, M. I., Van Herwijnen, M. J. C., Fernandez-Gutierrez, M. M., Giovanazzi, A., De Groot, A. M., Kleinjan, M., Van Capel, T. M. M., Sijts, A. J. A. M., Taams, L. S., Garssen, J., De Jong, E. C., Kleerebezem, M., Nolte-'T Hoen, E. N. M., Redegeld, F. A. and Wauben, M. H. M. 2021. Human milk extracellular vesicles target nodes in interconnected signalling pathways that enhance oral epithelial barrier function and dampen immune responses. *J. Extracell. Vesicles* 10(5), e12071.

# Part 3

## Carbohydrates and other components

# Chapter 10

## Lactose in milk: properties, nutritional characteristics and role in dairy products

*Thom Huppertz, FrieslandCampina and Wageningen University & Research, The Netherlands*



## 1 Introduction

Lactose is the main carbohydrate in the milk of all mammalian species studied to date. It consists of galactose and glucose subunits linked by a β-1-4 glycosidic bond (Fox, 2009). The concentrations of lactose vary between species, with some examples shown in Table 1. For bovine milk, values in the range 4–5% (w/w) are typically reported, but for human milk, values of >7% are often reported. The interspecies comparison highlights a negative correlation between lactose content and salt osmolarity (Jenness and Sloan, 1970) and between lactose content and casein content (Holt and Jenness, 1984). The former can be related to the fact that lactose and dissolved salts are together the main determinants of the osmotic pressure of milk, which should be equal to the osmotic pressure of blood.

Compared to other dietary carbohydrates, lactose has some distinct features, namely the β-glycosidic bond, rather than the more common α-glycosidic bond between the monosaccharides, and the presence of galactose as one of the monosaccharides (Romero-Velerde et al., 2019). These features must have significance and cannot be considered to be random, particularly considering that lactose is, in the first stages of life, the only digestible carbohydrate source for humans and most other mammalian species. In addition to an important role in nutrition, development and maintenance of health, lactose has important

http://dx.doi.org/10.19103/AS.2022.0099.10

**Table 1** Concentration of lactose (in g /100 g) in milk from various species (data from Jenness, 1986)

| Species | Lactose | Species | Lactose |
|---|---|---|---|
| Cow | 4.6 | Yak | 4.6 |
| Human | 7.1 | Zebu | 4.7 |
| Goat | 4.3 | Reindeer | 2.8 |
| Sheep | 4.8 | Rat | 3.7 |
| Camel | 5.0 | Pig | 5.5 |
| Horse | 6.2 | Dog | 3.3 |
| Buffalo | 4.8 | Fur seal | 0.1 |

technical features. This chapter aims to provide an overview of the important features of lactose from both the nutritional and the technological aspects. Lactose synthesis and secretion are first discussed in Section 2, after which the physiochemical properties of lactose are discussed in Section 3. Section 4 discusses the nutritional and health aspects of lactose, whereas Section 5 discusses the role of lactose in various milk and dairy products.

## 2 Lactose synthesis and secretion

Lactose synthesis occurs in the mammary gland and involves a number of enzymatic steps, outlined in Table 2. The topic has been reviewed frequently, for example, recently by Mardones and Villagrán (2020), Stelwagen (2020) and Sadovnikova et al. (2021). Glucose is first taken up from the blood and a part of the glucose is converted to UDP-glucose (Table 2, Steps 1–3). Subsequently, UDP-glucose is converted to UDP-galactose via epimerization (Table 2, Step 4). From glucose and UDP-galactose, lactose can be synthesized by lactose synthetase (Table 2, Step 5), which is a complex between galactosyltransferase and α-lactalbumin. Steps 1–4 (outlined in Table 2) occur within the cytosol, whereas step 5, that is, the synthesis of lactose from UPD-galactose and glucose, occurs in the Golgi. The galactosyltransferase involved in this reaction only

**Table 2** Enzymatic steps involved in lactose production

| Step | Reaction | | | Enzyme |
|---|---|---|---|---|
| (1) | Glucose + ATP | → | Glucose-6P + ADP | Hexokinase |
| (2) | Glucose-6P | → | Glucose-1P | Phosphoglucomutase |
| (3) | Glucose-1P + UTP | → | UDP-Glucose + $PP_i$ | UDP-glucose pyrophosphorylase |
| (4) | UDP-Glucose | → | UDP-Galactose | UPD-galactose-4-epimerase |
| (5) | UDP-Glucose + Galactose | → | Lactose + UDP | Lactose synthetase |

attains affinity for glucose after complexation with α-lactalbumin (Stelwagen, 2020). The enzyme complex also requires divalent cations, that is, Zn, Mg or Ca, for activation. Subsequently, the lactose is included in the secretory vesicles formed on the Golgi, which are then transported through the cell toward the apical membrane where the contents are released into the milk pool inside the alveolar lumen (Stelwagen, 2020).

To attain osmotic balance between the secretory vesicles and the cytosol, water is drawn from the cytosol into the secretory vesicles. This explains why lactose content only varies a little within species but also indicates that lactose production to a large extent determines milk volume (Stelwagen, 2020). In mice, it has been shown that eliminating lactose synthesis results in a much lower milk yield but also that the milk is much more concentrated (Stacey et al. 1995).

## 3 Physicochemical properties of lactose

### *3.1 Solubility*

Compared to other carbohydrates such as sucrose or maltose, the solubility of lactose is comparatively low. Solubility is temperature-dependent and increases with increasing temperature, as shown in Fig. 1. Despite the comparatively low solubility, however, the lactose content of milk does not exceed the solubility limit, nor does it in concentrated milk products such as evaporated milk. Only in milk concentrated to high dry matter content (e.g. 50–55%, w/w) before spray-drying is the solubility of lactose exceeded. However, even here, this does not automatically lead to crystallization, because

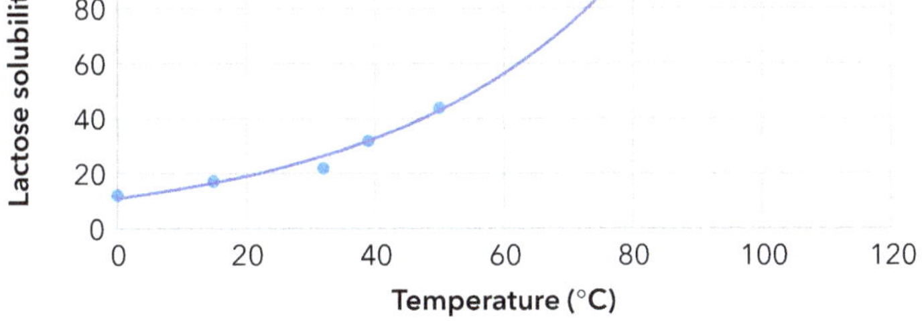


**Figure 1** Lactose solubility as a function of temperature; data from Holsinger (1997).

lactose can exist in a considerably supersaturated form before crystallization occurs. This so-called supersolubility of lactose has been estimated to be roughly equal to the solubility at a temperature of 30°C or higher (Holsinger, 1997).

It is important to keep in mind that lactose solubility is typically expressed per 100 g of water and not per 100 g of product. Although this works well for simple aqueous solutions, for more complex systems, an expression based on the available rather than the total water would appear to be more appropriate because other molecules, such as proteins, can bind certain amounts of water - which subsequently becomes unavailable as a solvent (Huppertz and Gazi, 2016).

### 3.2 Mutarotation

Two anomers of lactose are observed - α-lactose and β-lactose - with specific optical rotations $[\alpha]^{20}_{D}$ of +89.4 and +35°, respectively (Walstra and Jenness, 1984; Holsinger, 1997). The conversion between α-lactose and β-lactose occurs via the open-chain form of the glucose moiety of lactose, with an equilibrium at $[\alpha]^{20}_{D} = +55.3°$, representing ~37% α-lactose and 63% β-lactose (Walstra and Jenness, 1984; Holsinger, 1997). This equilibrium depends on various factors, including lactose concentration and temperature; increases in these parameters shift the equilibrium to a higher percentage of β-lactose. The pH does not affect the equilibrium value but strongly affects the rate at which the equilibrium establishes; the minimum in mutarotation rate is found at pH 5. In particular, pH $< 2$ and pH $> 7$ result in very rapid mutarotation (Walstra and Jenness, 1984; Holsinger, 1997). Mutarotation also occurs more rapidly at higher temperatures, with an ~2.8-fold increase in mutarotation rate for every 10°C temperature increase (Walstra and Jenness, 1984). The mutarotation rate of lactose is strongly enhanced by the presence of milk salts, and is ~2-fold higher in a simulated milk ultrafiltrate compared to that in water (Haase and Nickerson, 1966).

### 3.3 Lactose crystallization

When the solubility of lactose is exceeded in solutions, crystallization can occur. For crystallization from the liquid in dairy products under normal industrial conditions, the main crystalline form in which lactose is found is α-lactose monohydrate, which is also the most stable crystalline form. β-Lactose crystals are formed if crystallization occurs at temperatures >93.5°C, but crystalline β-lactose is unstable; over time, it will often transform into the stable α-lactose monohydrate form (Walstra and Jenness, 1984; Walstra et al., 2005; Huppertz and Gazi, 2016).

Lactose crystallization is a two-step process, consisting of (i) nucleation followed by (ii) crystal growth (Paterson, 2009). Spontaneous nucleation can occur when the supersolubility limit is exceeded, whereas at lower levels of saturation, forced nucleation is needed to initiate crystallization, either by flash cooling or by the addition of seeding crystals (Paterson, 2009). The further growth of the crystals depends on both the degree of saturation and the temperature of the solution. Temperature affects lactose solubility and hence the degree of supersaturation, which, at a given concentration would thus increase at a lower temperature (Huppertz and Gazi, 2016); however, at a given absolute degree of supersaturation, temperature does not affect the nucleation rate (McLeod et al., 2011), whereas crystallization speed increases with the absolute degree of supersaturation and temperature (Jelen and Coulter, 1973b). Furthermore, temperature also influences crystallization through mutarotation; due to the crystallization of lactose as α-lactose monohydrate, the degree of supersaturation of α-lactose in solution decreases and the ratio of α-lactose/β-lactose decreases. Conversion of β-lactose to α-lactose will occur until equilibrium is reached, increasing the supersaturation of α-lactose again. The increased rate of mutarotation at higher temperature thus facilitates the crystallization process. Likewise, pH can also affect the mutarotation rate, as can the presence of milk salts. The importance of the mutarotation step was also highlighted in the kinetic model for lactose crystallization proposed by Mimouni et al. (2009). Lactose crystallization is affected by various milk constituents and other components, an overview of which is provided in Table 3. Such effects clearly indicate the impact of the product matrix on lactose crystallization.

## 4 Nutritional and health aspects of lactose

### *4.1 Lactose digestion and absorption*

Following ingestion, lactose is not affected by the oral and gastric phase of digestion and is only digested during the intestinal phase. The fact that oral digestion does not affect lactose has benefits for dental health, as discussed further in Section 4.4. Intestinal digestion of lactose occurs via the enzyme lactase-phlorizin hydrolase (LPH; hereafter referred to as lactase) which is a glycoside hydrolase located at the brush border of the upper small intestine, duodenum and jejunum. The enzyme belongs to the class of β-glycosidases and catalyzes the hydrolysis of the β-1-4-bond between galactose and glucose in lactose. It can also catalyze the hydrolysis of other substrates, for example, cellobiose, flavonoids and isoflavone glycosides.

Following lactose digestion, D-glucose and D-galactose are absorbed from the duodenum and jejunum into the bloodstream, mainly via the transporters SGLT1 (both glucose and galactose) and SGLT2 (only glucose), which carry the

**Table 3** Effects of various components on lactose crystallization

| Component | Effect on lactose crystallization | Reference |
|---|---|---|
| Whey protein | Promotes nucleation<br>Slows down crystal growth | Mimouni et al. (2005), Gernigon et al. (2013) |
| Sodium phosphate | Promotes crystallization | Jelen and Coulter (1973a), Smart (1988), Smart and Smith (1992) |
| Potassium phosphate | Retards crystallization | Bhargava and Jelene (1996) |
| Lactic acid | Reduced growth of lactose crystals | Jelen and Coulter (1973a) |
| Lactic acid | No effect on lactose crystallization | Smart (1988), Smart and Smith (1992) and Gernigon et al. (2013), Wijayasinghe et al. (2020) |
| Sodium lactate | Promotes lactose crystallization | Smart (1988), Smart and Smith (1992) |
| Calcium lactate | Promotes lactose crystallization | Bhargava and Jelene (1996) |
| Phosphoric acid | Retards crystallization | Wijayasinghe et al. (2020) |
| Citric acid | Retards crystallization | Wijayasinghe et al. (2020) |
| Galacto-oligosaccharides | Reduce lactose crystallization | Smart and Smith (1992); Ihli and Paterson (2015) |
| Lactose phosphate | Inhibits crystal growth | Visser (1984, 1988), Lifran et al. (2007) |

molecule across the brush border membrane from the intestinal lumen into the enterocyte (Szilagyi, 2019). Subsequently, the transporter GLUT2 facilitates the transport from the enterocyte across the basolateral membrane into the circulation, allowing entry into the hepatic portal vein (Röder et al., 2014). First-pass extraction of some glucose occurs in the liver and glucose is directed toward liver glycogen synthesis or is oxidized (Szilagyi, 2019). However, much of the glucose will escape first-pass extraction and enter the systemic circulation. This results in divergence from blood glucose homeostasis and consequently insulin release. The entry of glucose into the systemic circulation results in the uptake of glucose by the tissues, for example, the skeletal muscle, where glucose is converted to glycogen (Odell and Wallis, 2021).

The metabolism of galactose is different from that of glucose. First-pass extraction of galactose also occurs, after which galactose is converted to glucose-1-phosphate via the Leloir pathway (Holden et al., 2003; Odell and Wallis, 2021). This glucose-1-phosphate is subsequently available for glucose or glycogen production. However, not all galactose is converted to glucose using this pathway, as the rate of glucogenesis is limited; some galactose also ends up in the systemic circulation. The circulating galactose will ultimately be taken up by the liver. A maximal splanchnic uptake rate of ~10 g $h^{-1}$ has been estimated for galactose (Sunehag and Haymond, 2002).

Interestingly, it has been observed that co-ingestion of glucose and galactose more than doubles the first-pass clearance of galactose (Sunehag and Haymond, 2002). This, in turn, can eliminate, or at least drastically reduce, the rise in plasma galactose after ingestion. Similar effects to those from free glucose and galactose, which would represent, for example, a lactose-hydrolyzed milk, are observed for lactose (Williams et al., 1983). This effect of co-ingestion of glucose and galactose is likely important in the glycemic response of lactose, as discussed in Section 4.3.

### *4.2 Lactase non-persistence and lactose maldigestion*

While all infants can digest lactose, approximately two-thirds of adults fail to hydrolyze lactose into its constituent monosaccharides, a process referred to as lactose maldigestion (LM). The main cause of LM is the low levels of lactase expression, or even the absence thereof, in adults. This is related to down-regulation of lactase expression past infancy. This trait, known as lactase non-persistence (LNP), occurs in approximately two-thirds of the population, whereas the other one-third is classified as having lactase persistence (LP) (Szilagyi et al., 2019). For LNP individuals, 90–95% of lactase activity is on average lost by the age of six. In some cases, LM can lead to adverse symptoms such as cramps, bloating, flatulence or affected bowel movement, which is then referred to as lactose intolerance (LI). The classification of lactose intolerance should only be done after confirmed medical diagnosis, but there are many cases of self-reported LI (Di Costanzo and Canani, 2018; Fassio et al., 2018; Vázquez et al., 2020). It is important to realize that not all cases of LM lead to LI, and that in many cases, people with LM can still consume notable amounts of lactose (e.g. 10–15 g) without experiencing adverse symptoms.

The level of LNP differs strongly with geography. Low levels of LNP are found in Western Europe and New Zealand, whereas the highest levels, in some cases exceeding 95%, are observed in Sub-Saharan Africa and South and East Asia (Storhaug et al., 2017). LP is a genetic trait, and is in fact the best-known; in many populations, it is the strongest example of recent natural selection, convergent evolution and gene-culture coevolution in humans. Within the past ten millennia, culture has adapted to the availability of milk from domestic animals, and biology has also adapted to these changes.

As outlined above, LNP does not mean that lactose, or dairy products in general, cannot be consumed; many LNP individuals can still consume some amounts of lactose. Furthermore, lactose-hydrolyzed products are also available (see Section 5.1) whereas yogurt and other fermented dairy products with live cultures can also assist in lactose digestion in LNP individuals (see Section 5.2).

### 4.3 Glycemic index and glycemic response

The glycemic index (GI) is a parameter that indicates the effect of a carbohydrate on the blood glucose level of an individual. A value of 100 is the standard, representing pure glucose, whereas GI ratings are divided as low (<55), medium (55–69) or high (>70). Lactose has a GI of 46 and is thus listed as a low-GI carbohydrate. The low GI of lactose is believed to be primarily caused by the non-insulinogenic response to galactose and is often considered to be a metabolic advantage, compared to, for example, glucose polymers, which cause a larger increase in plasma glucose and plasma insulin.

A comparison of milk or a milk protein-based test meal (both containing 25 g lactose) with a white bread meal (containing 25 g of glucose in the form of starch) confirmed a notably higher glycemic response of the latter (Gunnerud et al., 2012). Low-GI diets may also result in weight loss, reduced blood sugar levels and lowered risks of heart disease and type 2 diabetes. The potential mechanisms for reduction of type 2 diabetes include evidence that low-GI diets improve insulin sensitivity and β-cell function in people with type 2 diabetes and those at risk for type 2 diabetes (Augustin et al., 2015). As a result, lactose is considered to be more acceptable as a carbohydrate in individuals with diabetes. In a prospective cohort study to investigate the relations between dietary GI, glycemic load and other carbohydrate measures (added sugars, total sugars, glucose, sucrose, lactose, fructose, starch, carbohydrate) and depression in postmenopausal women, higher consumption of lactose was found to be significantly associated with lower odds of incident depression (Gangwisch et al., 2015).

### 4.4 Lactose and dental health

It is well established that carbohydrates are a major contributor to dental caries, which occurs when the bacteria in the mouth metabolize the carbohydrates, leading to the production of organic acids and a concomitant decrease in pH, resulting in demineralization of teeth (Touger-Decker and Van Loveren, 2003). Being a fermentable sugar, lactose is listed as having a cariogenic potential. However, this is the lowest of all fermentable sugars (Moynihan, 2002) and is not considered a risk for dental health. The low cariogenic potential of lactose is likely related to the β-1-4-glycosidic bond between galactose and lactose, which, unlike the α-glycosidic bond in many other carbohydrates, is not readily hydrolyzed by salivary amylase or the oral micro-organisms. The anticariogenic effects of milk and dairy products were already established in the 1940s (Schweigert et al., 1946) and are generally attributed to both the proteins and the salts (e.g. calcium and phosphate) in milk and dairy products. Some dairy foods, such as cheese and yogurt without added sugar, are even protective against damage to teeth (Kashket and DePaola, 2002).

# 5 Lactose in dairy products

## *5.1 Lactose in milk*

In liquid milk products, lactose is present at ~4.5%, which is well below the solubility limit. The main effect of this is that it contributes to some sweetness in the milk. Lactose is not affected by pasteurization, but if heat treatments are more extensive, as in the in-container sterilization of UHT treatment, some heat-induced changes in lactose can occur, such as (Berg and Van Boekel, 1994):

- Isomerization of some lactose to lactulose;
- Participation of lactose in the Maillard reaction; and
- Heat-induced degradation of lactose.

The Maillard reaction can affect both the color and the flavor of milk, with some browning occurring, as well as a so-called sterilized flavor which has been described as 'caramelized' and/or 'burnt' and originates mostly from the products of the Maillard reaction (Walstra et al., 2005). In addition, the interaction of lactose with the ε-amino group of lysine that forms the first step of the Maillard reaction can also have nutritional impact, as it reduces the amount of metabolically available lysine (Van Lieshout et al., 2020). Heat-induced degradation of lactose can also lead to the formation of organic acids, for example, formic and acetic acids, which can lead to irreversible heat-induced reductions in pH (Berg and Van Boekel, 1994). Such changes, however, are only observed after in-container sterilization and not after UHT treatment or other heat treatment (Huppertz, 2016).

In addition to the conventional milk products, reduced-lactose or lactose-free milk products are also available in many areas of the world, catering specifically to lactase-non-persistent consumers or others preferring not to consume lactose. Although lactose can be removed by filtration and chromatographic techniques (Harju et al., 2012; Singh et al., 2020), such processes are not feasible industrially to achieve sufficiently low residual lactose levels. The preparation of lactose-free milk products typically involves an enzymatic step, in which lactose is hydrolyzed to glucose and galactose by β-galactosidase (also referred to as lactase) (Harju et al., 2012; Singh et al., 2020).

There are two options for adding the enzyme – either before or after heat treatment. Addition of the enzyme before heat treatment involves pre-incubation of the milk with the enzyme, after which the heat treatment (e.g. pasteurization or UHT treatment) inactivates the enzyme (Harju et al., 2012). For pasteurized products, this is typically the process of choice. For UHT-treated products, however, hydrolysis prior to heat treatment leads to the risk of additional browning, due to the fact that the monosaccharides glucose and galactose are much more reactive than lactose. Adding the enzyme after heat treatment can also be done, but requires aseptic addition of the enzyme post UHT treatment (Harju et al., 2012; Jensen et al., 2015). In this case, the lactose hydrolysis occurs during the shelf-life of the product. One advantage of this processing route is that the time available for incubation is much longer, therefore lower enzyme levels can suffice. Particular care should also be taken in selecting β-galactosidase preparations with very high purity, because residual proteolytic and lipolytic enzymes can cause instability during shelf-life (Troise et al., 2016; Nielsen et al., 2018).

For lactose-hydrolyzed milk, the hydrolysis of lactose to glucose and galactose increases sweetness of the milk, because both glucose and galactose are notably sweeter than lactose. This increased sweetness is not always preferred; in such cases, partial removal of lactose via membrane filtration prior to enzyme addition can be implemented. For this method, the milk can be subjected to ultrafiltration (UF), after which the UF permeate is subjected to nanofiltration (NF). The NF permeate, which is depleted in lactose and mainly contains salts, can be added back to the UF retentate to create a lactose-reduced base product for treatment with β-galactosidase. Typically, removal of ~1/3 of lactose prior to enzyme addition gives a sweetness similar to that of the original milk (Singh et al., 2020).

## 5.2 Lactose in yogurt

Lactose plays an important role in yogurt products as it is the primary source of carbohydrate for fermentation and also for other properties including texture and sensory properties. The main source of lactose in yogurt is typically the milk, but further lactose can be included via inclusion of dry dairy ingredients such as milk powder, milk protein concentrate (MPC) or whey protein concentrate (WPC) (Huppertz, 2017). Yogurt is traditionally prepared with a mixture of the thermophilic homofermentative *Streptococcus thermophilus* and *Lactobacillus delbrueckii* ssp. *bulgaricus*, which can both metabolize lactose as well as glucose hydrolyzed from lactose; however, they do not metabolize galactose (Walstra et al., 2005). During fermentation in typical yogurt production, 20–40% of lactose is metabolized by the lactic acid bacteria, leading to products with pH in the desired range of 4.0–4.5 (Tamime and Robinson, 2007). It is important to note as well that a large amount of lactose thus remains in yogurt.

In addition to its crucial role in fermentation, lactose also has an effect on yogurt texture. Studies by Meletharayil et al. (2016) and Niki and Motoshima (2006) highlight that at higher lactose content, acid-induced coagulation commences at a higher pH and leads to gels with a higher storage modulus (G′). Furthermore, increased water-holding capacity and reduced porosity of acid-induced milk gels with increasing lactose content have been observed (Meletharayil et al., 2016). Further study appears warranted to identify the underlying mechanisms for the role of lactose on yogurt structure and texture. However, compared to the effects achievable by other means, for example, changes in protein content and composition or processing, the effects of lactose are also comparatively small.

Compared to milk, yogurt can still be consumed by most people who suffer from lactose maldigestion due to LNP (Solomons, 2002; Savaiano, 2014). This would at first appear counterintuitive, considering the considerable amounts of residual lactose in yogurt. However, it has been shown that the presence of live cultures in yogurt aids lactose digestion in the intestine through their β-galactosidase activity. For this, it is important that sufficient live cells reach the intestine, requiring survival of the high acidity in the stomach, which can reduce cell viability. However, the high buffering capacity of yogurt and the comparatively rapid gastric emptying can enhance survival of cells (Martini et al., 1987). In the intestine, the pH increases to ~6.5–7.5, which is close to optimum for the enzyme and lactose hydrolysis can occur. For heat-treated yogurt products, in which all bacterial cells are dead and the enzymes inactivated, gastrointestinal symptoms associated with lactose maldigestion are still observed (Rizkalla et al., 2000; Pelletier et al., 2001).

## 5.3 Lactose in cheese

Unlike in milk and yogurt, little or no residual lactose is found in ripened cheese products, whereas some residual lactose may occur in fresh cheese products. During cheese manufacture, most lactose (>95%) is removed with whey (Walstra et al., 2005). However, the absence of residual lactose in ripened cheese products does not mean that lactose does not have an important role in cheese. As in yogurt, the conversion of lactose to lactic acid is crucial for attaining the right pH of the cheese (Fox et al., 1990). Furthermore, the lactic acid is metabolized to propionic acid, acetic acid and $CO_2$ in Swiss-type cheeses, and is therefore crucial for both flavor development and eye formation (McSweeney and Fox, 2004). For example, in Dutch-type cheeses, the presence of some residual lactose also enables citrate fermentation, which leads to the formation of diacetyl, an important contributor to flavor (Düsterhöft et al., 2017). The citrate fermentation pathway also leads to $CO_2$ production and is responsible for the (limited) eye formation in Dutch-type cheeses (Düsterhöft et al., 2017).

## 5.4 Lactose in dairy ingredients

### 5.4.1 Milk powder and whey powder

Lactose is the main constituent of milk powder, representing more than 50% of the dry matter in skim milk powder and ~40% of the dry matter in whole milk powder. Lactose in milk powder occurs as an amorphous matrix, due to the fact that water removal during the spray-drying process is too rapid for the lactose to crystallize (Huppertz and Gazi, 2016). Amorphous lactose is hydroscopic and care thus needs to be taken during storage of the powder, because uptake of moisture by the powder matrix can lead to several undesired effects (Masum et al., 2020). One is the caking of powders, which can occur when the storage temperature of the powder exceeds the glass transition temperature ($T_g$) of the lactose matrix (Huppertz and Gazi, 2016). At temperatures lower than $T_g$, lactose behaves like a solid, but at temperatures higher than $T_g$, it starts behaving like rubber and can develop stickiness (Aguilera et al., 1995). In this rubbery state, bridges can form between powder particles, resulting in the caking of powders (Huppertz and Gazi, 2016). As $T_g$ is mainly determined by moisture content, ensuring sufficiently low moisture content is crucial for storage stability of milk powders.

In addition to caking, lactose also plays a major role in the Maillard reaction, which can occur during the storage of powders and lead to browning and loss of available lysine (Saxena et al., 2021). Unlike for liquid products, where Maillard reactions are only observed during sterilization treatments, Maillard reactions in powders are already relevant at much lower temperatures, for example, at 20–40°C (Higgs and Boland, 2014). This is due to the long storage times of milk powders (months to years) and the fact that non-enzymatic browning reactions have an optimum reaction rate at a water activity ($a_w$) ~0.6, and the reaction rate does not, like many other reactions, increase with increasing $a_w$. Maillard browning in milk powders increases with increasing $a_w$, storage time and storage temperature (Higgs and Boland, 2014).

Like for milk powder, lactose is also the main constituent of whey powder; however, for whey powder, due to the absence of caseins, lactose represents even >70% of the total dry matter in whey powder (Jelen, 2009). Therefore, the aforementioned issues with hygroscopicity, sticking and caking, both during processing and storage, are even more extreme (Hynd, 1980). This necessitates pre-crystallization of lactose in the whey powder prior to drying, to limit the amount of amorphous lactose in the product. Pre-crystallization is achieved by concentrating the whey by evaporation to >60% dry matter, followed by flash cooling, for example, from 60°C to 30°C, to initiate nucleation (Písecký, 2012). Subsequently, controlled further cooling at a rate of 2–3°C/min will ensure that the concentrate remains supersaturated with respect to lactose, and crystallization continues. Crystallization processes are typically slow and take at least several hours to achieve crystallinity values up to 80%. Following the pre-crystallization process, the concentrate is spray-dried. Further crystallization to yield crystallinity values >90°C can be achieved in the fluidized bed of the dryer. As a result, non-hygroscopic whey powders can be produced which have little or no susceptibility to stickiness and caking (Pisecky, 2012). The pre-crystallization of lactose also reduces the susceptibility of the product to Maillard reactions, because lactose in the crystal lattice does not participate in Maillard reactions.

The ease with which whey powder can be produced depends to a large extent on the type of whey used. The process outlined above can readily be applied to whey from cheese manufacture. However, for acid whey, for example, from Greek-style yogurt or quark production, lactose crystallization is hampered by the high levels of lactic acid and the high levels of salts present. Particularly, the calcium lactate crystals that form here can lead to large increases in viscosity and hinder crystallization (Mimouni et al., 2007; Chandrapala et al., 2017).

### 5.4.2 Lactose and lactose derivatives

In addition to having an important role in various dairy products, as outlined in the previous sections, lactose is also produced as an ingredient for addition to different food and non-food products. Examples are infant formula, confectionary and bakery products. In addition, lactose is also used widely as an excipient in the pharmaceutical industry. Lactose ingredients always contain at least 95% lactose on a dry-matter basis, and grades can be distinguished, such as (Huppertz and Gazi, 2017):

- Crude lactose (95–98% purity);
- Edible lactose (95–98% purity);
- Refined edible lactose (99–99.5% purity); and
- Pharmaceutical lactose (99–99.5% purity; strict requirements on particle size).

Most lactose is prepared from whey permeate and α-lactose monohydrate is the most common form of lactose produced (Hourigan et al., 2013). For lactose production, the permeate is first concentrated to at least 60% of total solids (Paterson, 2009). Subsequently, the lactose needs to be crystallized, which can be initiated by flash cooling and/or by adding seeding crystals. Flash cooling is typically achieved by lowering the temperatures to the range of 30–40°C, after which further cooling to 10–15°C at a rate of 2–3°C per minute is carried out, followed by holding for some further time to complete the crystallization process (Paterson, 2009; Wong and Hartel, 2014). Up to 80% of lactose can be crystallized in these processes.

Following crystallization, the crystals are removed from the surrounding liquid, called mother liquor, by centrifuges or decanters. This is usually followed by a washing step to remove further impurities (Paterson, 2009; Wong and Hartel, 2014). The crystals in the recovered wash are normally dried on flash driers, operating at temperatures between 120°C and 180°C (Paterson, 2009).

This is followed by a fluid-bed drying, which also cools the product. Failure of sufficient drying and/or cooling prior to packaging leads to high risks of caking (Paterson, 2017). If particular particle sizes or shapes are required, as is the case in pharmaceutical applications, sieving and milling operations can be applied.

In addition to being used as an ingredient, lactose can also be the base material for production of lactose derivatives which can be used as ingredients in food and non-food products. A complete coverage of all lactose derivatives is out of the scope of this chapter and is covered by Playne and Crittenden (2009), but some interesting examples are:

- Galacto-oligosaccharides (GOS), which are the β-linked galactose moieties with a galactose or glucose at the reducing end, typically with a degree of polymerization of 3–8. GOS are prepared using β-galactosidases at high lactose concentration, whereby, due to the limitation in water content, polymerization reactions rather than hydrolysis reactions are favored. Galacto-oligosaccharides are mainly used in infant formula for the purpose of stimulating the development of healthy gut microbiota (Crittenden and Playne, 1996; Playne and Crittenden, 2009; Fischer and Kleinschmidt, 2018).
- Lactobionic acid (β-4′galactosylglucuronic acid) can be produced from lactose by chemical oxidation at high pH over a noble catalyst. Because of its salt-binding capacity, it is used in the Wisconsin transplant solution to reduce oxidative damage by metal ions. Its other applications are also mainly outside the food sector (Playne and Crittenden, 2009; Minal et al., 2017; Cardoso et al., 2019).
- Lactulose (galactosyl-β-1→4-fructose) is a lactose isomer that is formed in small quantities upon heating of milk but can also be produced by isomerization reactions at alkaline pH. Lactulose is mainly used in the treatment of constipation and hepatic encephalopathy, but has also been approved for use as a prebiotic food additive in several countries (Playne and Crittenden, 2009; Nooshkam et al., 2018).
- Lactosucrose can be produced enzymatically from lactose by transferring the fructosyl moiety of sucrose to a lactose molecule. Lactosucrose is mainly used as a prebiotic and sweetener and is used, for example, in beverages, bakery products and confectionary (Playne and Crittenden, 2009).

## 6 Conclusion and future trends

This chapter has outlined the various roles of lactose in milk and dairy products and the physicochemical and biochemical basis thereof. The importance of lactose, unfortunately, is often underestimated. To the wider public, the main connotation with lactose is in relation to maldigestion and intolerance. However, nutritionally, lactose plays an important role, being a low-GI carbohydrate with low cariogenic potential and being the only main dietary source galactose. Technologically, lactose is crucial not only in the development of fermented dairy products but also as an ingredient in many different applications. Overall, it is clear that lactose is a unique disaccharide with important properties. Future research should be dedicated to exploring and substantiating these key functionalities and support communication on the true potential of lactose.

## 7 Where to look for further information

As outlined in this chapter, lactose is a key constituent in dairy products for various reasons, both nutritional and functional. The topic has been addressed in many publications, with excellent summaries of work provided in the Advanced Dairy Chemistry series of books as well as the recent book *Lactose: Evolutionary Role, Health Effects, and Applications* published in 2019. The topic of lactose is also an important work item in organizations like International Dairy Federation, which has published fact sheets and bulletins on the topic.

## 8 References

Aguilera, J., del Valle, J. and Karel, M. (1995). Caking phenomena in amorphous food powders. *Trends in Food Science and Technology* 6(5), 149–155.

Augustin, L. S., Kendall, C. W., Jenkins, D. J., Willett, W. C., Astrup, A., Barclay, A. W., Björck, I., Brand-Miller, J. C., Brighenti, F., Buyken, A. E., Ceriello, A., La Vecchia, C., Livesey, G., Liu, S., Riccardi, G., Rizkalla, S. W., Sievenpiper, J. L., Trichopoulou, A., Wolever, T. M., Baer-Sinnott, S. and Poli, A. (2015). Glycemic index, glycemic load and glycemic response: an International Scientific Consensus Summit from the International Carbohydrate Quality Consortium (ICQC). *Nutrition, Metabolism, and Cardiovascular Diseases: NMCD* 25(9), 795–815.

Berg, H. E. and Van Boekel, M. S. (1994). Degradation of lactose during heating of milk. I: reaction pathways. *Netherlands Milk and Dairy Journal* 48(3), 157–175.

Bhargava, A. and Jelene, P. (1996). Lactose solubility and crystal growth as affected by mineral impurities. *Journal of Food Science* 61(1), 180–184.

Cardoso, T., Marques, C., Dagostin, J. L. A. and Masson, M. L. (2019). Lactobionic acid as a potential food ingredient: recent studies and applications. *Journal of Food Science* 84(7), 1672–1681.

Chandrapala, J., Duke, M. C., Gray, S. R., Weeks, M., Palmer, M. and Vasiljevic, T. (2017). Strategies for maximizing removal of lactic acid from acid whey-addressing the un-processability issue. *Separation and Purification Technology* 172, 489–497.

Crittenden, R. G. and Playne, M. J. (1996). Production, properties and applications of food-grade oligosaccharides. *Trends in Food Science and Technology* 7(11), 353–361.

Di Costanzo, M. and Canani, R. B. (2018). Lactose intolerance: common misunderstandings. *Annals of Nutrition and Metabolism* 73(Suppl. 4), 30–37.

Düsterhöft, E. M., Engels, W. and Huppertz, T. (2017). Gouda and related cheeses. In: McSweeney, P. L. H., Fox, P. F., Cotter, P. D. and Everett, D. W. (Eds) *Cheese* (pp. 865-888). Academic Press, London.

Fassio, F., Facioni, M. S. and Guagnini, F. (2018). Lactose maldigestion, malabsorption, and intolerance: a comprehensive review with a focus on current management and future perspectives. *Nutrients* 10(11), 1599.

Fischer, C. and Kleinschmidt, T. (2018). Synthesis of galactooligosaccharides in milk and whey: a review. *Comprehensive Reviews in Food Science and Food Safety* 17(3), 678-697.

Fox, P. F. (Ed.) (2009). Lactose: chemistry and properties. In: *Advanced Dairy Chemistry* (pp. 1-15). Springer, New York.

Fox, P. F., Lucey, J. A. and Cogan, T. M. (1990). Glycolysis and related reactions during cheese manufacture and ripening. *Critical Reviews in Food Science and Nutrition* 29(4), 237-253.

Gangwisch, J. E., Hale, L., Garcia, L., Malaspina, D., Opler, M. G., Payne, M. E., Rossom, R. C. and Lane, D. (2015). High glycemic index diet as a risk factor for depression: analyses from the Women's Health Initiative. *The American Journal of Clinical Nutrition* 102(2), 454-463.

Gernigon, G., Baillon, F., Espitalier, F., Le Floch-Fouéré, C., Schuck, P. and Jeantet, R. (2013). Effects of the addition of various minerals, proteins and salts of organic acids on the principal steps of α-lactose monohydrate crystallisation. *International Dairy Journal* 30(2), 88-95.

Gunnerud, U., Holst, J. J., Östman, E. and Björck, I. (2012). The glycemic, insulinemic and plasma amino acid responses to equi-carbohydrate milk meals, a pilot-study of bovine and human milk. *Nutrition Journal* 11(1), 83.

Haase, G. and Nickerson, T. A. (1966). Kinetic reactions of alpha and beta lactose. I. Mutarotation. *Journal of Dairy Science* 49(2), 127-132.

Harju, M., Kallioinen, H. and Tossavainen, O. (2012). Lactose hydrolysis and other conversions in dairy products: technological aspects. *International Dairy Journal* 22(2), 104-109.

Higgs, K. and Boland, M. J. (2014). Changes in milk proteins during storage of dry powders. *Milk Proteins*, 343-357.

Holden, H. M., Rayment, I. and Thoden, J. B. (2003). Structure and function of enzymes of the Leloir pathway for galactose metabolism. *Journal of Biological Chemistry* 278(45), 43885-43888.

Holsinger, V. H. (1997). Physical and chemical properties of lactose. In: Fox, P. F. (Ed.) *Advanced Dairy Chemistry* (vol. 3) (pp. 1-38). Springer, Boston, MA.

Holt, C. and Jenness, R. (1984). Interrelationships of constituents and partition of salts in milk samples from eight species. *Comparative Biochemistry and Physiology. A, Comparative Physiology* 77(2), 275-282.

Hourigan, J. A., Lifran, E. V., Vu, L. T., Listiohadi, Y. and Sleigh, R. W. (2013). Lactose: chemistry, processing, and utilization. In: Smithers, G. W. and Augustin, M. A. (Eds) *Advances in Dairy Ingredients* (pp. 21-41). Wiley-Blackwell, Ames, IA.

Huppertz, T. (2016). Heat stability of milk. In: McSweeney, P. L. H. and O'Mahony, S. A. (Eds) *Advanced Dairy Chemistry* (pp. 179-196). Springer, New York.

Huppertz, T. (2017). Lactose in yogurt. In: *Yogurt in Health and Disease Prevention* (pp. 387-394). Academic Press.

Huppertz, T. and Gazi, I. (2016). Lactose in dairy ingredients: effect on processing and storage stability. *Journal of Dairy Science* 99(8), 6842-6851.

Huppertz, T. and Gazi, I. (2017). Ingredients from milk for use in food and non-food products: from commodity to value-added ingredients. In: *Burleigh Dodds Series in Agricultural Science* (pp. 121–144). Burleigh Dodds Science Publishing.

Hynd, J. (1980). Drying of whey. *International Journal of Dairy Technology* 33(2), 52–54.

Ihli, J. and Paterson, A. H. J. (2015). Effect of galacto-oligosaccharide concentration on the kinetics of lactose crystallisation. *International Dairy Journal* 41, 26-31.

Jelen, P. (2009). Dried whey, whey proteins, lactose and lactose derivative products. In: Tamime, A. Y. (Ed.) *Dairy Powders and Concentrated Products* (pp. 255-267). Blackwell Publishing Ltd, Chichester, UK.

Jelen, P. and Coulter, S. T. (1973a). Effects of certain salts and other whey substances on the growth of lactose crystals. *Journal of Food Science* 38(7), 1186–1189.

Jelen, P. and Coulter, S. T. (1973b). Effects of supersaturation and temperature on the growth of lactose crystals. *Journal of Food Science* 38(7), 1182–1185.

Jenness, R. (1986). Lactational performance of various mammalian species. *Journal of Dairy Science* 69(3), 869–885.

Jenness, R. and Sloan, R. E. (1970). The composition of milk of various species: a review. *Dairy Science Abstracts* 32, 599–612.

Jensen, S., Jansson, T., Eggers, N., Clausen, M. R., Larsen, L. B., Jensen, H. B., Ray, C., Sundgren, A., Andersen, H. J. and Bertram, H. C. (2015). Storage-induced changes in the sensory characteristics and volatiles of conventional and lactose-hydrolyzed UHT processed milk. *European Food Research and Technology* 240(6), 1247–1257.

Kashket, S. and DePaola, D. P. (2002). Cheese consumption and the development and progression of dental caries. *Nutrition Reviews* 60(4), 97–103.

Lifran, E. V., Vu, T. T. L., Durham, R. J., Hourigan, J. A. and Sleigh, R. W. (2007). Crystallisation kinetics of lactose in the presence of lactose phosphate. *Powder Technology* 179(1-2), 43–54.

Mardones, L. and Villagrán, M. (2020). Lactose synthesis. In: Guiterrez-Mendez, N. (Ed.) *Lactose and Lactose Derivatives*. IntechOpen, London.

Martini, M. C., Bollweg, G. L., Levitt, M. D. and Savaiano, D. A. (1987). Lactose digestion by yogurt beta-galactosidase: influence of pH and microbial cell integrity. *The American Journal of Clinical Nutrition* 45(2), 432–436.

Masum, A. K. M., Chandrapala, J., Huppertz, T., Adhikari, B. and Zisu, B. (2020). Production and characterization of infant milk formula powders: a review. *Drying Technology*, 1–21.

McLeod, J., Paterson, A. H. J., Jones, J. R. and Bronlund, J. E. (2011). Primary nucleation of alpha-lactose monohydrate: the effect of supersaturation and temperature. *International Dairy Journal* 21(7), 455–461.

McSweeney, P. L. H. and Fox, P. F. (2004). Metabolism of residual lactose and of lactate and citrate. *Cheese: Chemistry, Physics and Microbiology* 1, 361–371.

Meletharayil, G. H., Patel, H. A., Metzger, L. E. and Huppertz, T. (2016). Acid gelation of reconstituted milk protein concentrate suspensions: influence of lactose addition. *International Dairy Journal* 61, 107–113.

Mimouni, A., Bouhallab, S., Famelart, M. H., Naegele, D. and Schuck, P. (2007). The formation of calcium lactate crystals is responsible for concentrated acid whey thickening. *Journal of Dairy Science* 90(1), 57–65.

Mimouni, A., Schuck, P. and Bouhallab, S. (2005). Kinetics of lactose crystallization and crystal size as monitored by refractometry and laser light scattering: effect of proteins. *Le Lait* 85(4–5), 253–260.

Mimouni, A., Schuck, P. and Bouhallab, S. (2009). Isothermal batch crystallization of alpha-lactose: a kinetic model combining mutarotation, nucleation and growth steps. *International Dairy Journal* 19(3), 129–136.
Minal, N., Bharwade, Balakrishnan, S., Chaudhary, N. N. and Jain, A. K. (2017). Lactobionic acid: significance and application in food and pharmaceutical. *International Journal of Fermented Foods* 6(1), 25–33.
Moynihan, P. J. (2002). Dietary advice in dental practice. *British Dental Journal* 193(10), 563–568.
Nielsen, S. D., Zhao, D., Le, T. T., Rauh, V., Sørensen, J., Andersen, H. J. and Larsen, L. B. (2018). Proteolytic side-activity of lactase preparations. *International Dairy Journal* 78, 159–168.
Niki, R. and Motoshima, H. (2006). Influence of lactose on viscoelastic properties of acid-induced casein micelle gels: the secondary role of lactose in yogurt production. *Milchwissenschaft* 61(4), 411–415.
Nooshkam, M., Babazadeh, A. and Jooyandeh, H. (2018). Lactulose: properties, techno-functional food applications, and food grade delivery system. *Trends in Food Science and Technology* 80, 23–34.
Odell, O. J. and Wallis, G. A. (2021). The application of lactose in sports nutrition. *International Dairy Journal* 116, 104970.
Paterson, A. H. J. (2009). Production and uses of lactose. In: McSweeney, P. L. H. and Fox, P. F. (Eds) *Advanced Dairy Chemistry* (pp. 105–120). Springer, New York.
Paterson, A. H. J. (2017). Lactose processing: from fundamental understanding to industrial application. *International Dairy Journal* 67, 80–90.
Pelletier, X., Laure-Boussuge, S. and Donazzolo, Y. (2001). Hydrogen excretion upon ingestion of dairy products in lactose-intolerant male subjects: importance of the live flora. *European Journal of Clinical Nutrition* 55(6), 509–512.
Písecký, J. (2012). *Handbook of Milk Powder Manufacture*. GEA Process Engineering A/S, Soeborg.
Playne, M. J. and Crittenden, R. G. (2009). Galacto-oligosaccharides and other products derived from lactose. In: McSweeneey, P. L. H. and Fox, P. F. (Eds) *Advanced Dairy Chemistry* (pp. 121–201). Springer, New York.
Rizkalla, S. W., Luo, J., Kabir, M., Chevalier, A., Pacher, N. and Slama, G. (2000). Chronic consumption of fresh but not heated yogurt improves breath-hydrogen status and short-chain fatty acid profiles: a controlled study in healthy men with or without lactose maldigestion. *The American Journal of Clinical Nutrition* 72(6), 1474 –1479.
Röder, P. V., Geillinger, K. E., Zietek, T. S., Thorens, B., Koepsell, H. and Daniel, H. (2014). The role of SGLT1 and GLUT2 in intestinal glucose transport and sensing. *PLoS ONE* 9(2), e89977.
Romero-Velarde, E., Delgado-Franco, D., García-Gutiérrez, M., Gurrola-Díaz, C., Larrosa-Haro, A., Montijo-Barrios, E., Muskiet, F. A. J., Vargas-Guerrero, B. and Geurts, J. (2019). The importance of lactose in the human diet: outcomes of a Mexican consensus meeting. *Nutrients* 11(11), 2737.
Sadovnikova, A., Garcia, S. C. and Hovey, R. C. (2021). A comparative review of the cell biology, biochemistry, and genetics of lactose synthesis. *Journal of Mammary Gland Biology and Neoplasia* 26(2), 181–196.
Savaiano, D. A. (2014). Lactose digestion from yogurt: mechanism and relevance. *The American Journal of Clinical Nutrition* 99(Suppl.), 1251S–1255S.

Saxena, J., Adhikari, B., Brkljaca, R., Huppertz, T., Zisu, B. and Chandrapala, J. (2021). Influence of lactose on the physicochemical properties and stability of infant formula powders: a review. *Food Reviews International*, 1–15.

Schweigert, B. S., Shaw, J. H., Zepplin, M. and Elvehjem, C. A. (1946). Dental caries in the cotton rat: VI. The effect of the amount of protein, fat and carbohydrate in the diet on the incidence and extent of carious lesions. *The Journal of Nutrition* 31(4), 439–447.

Singh, P., Rao, P. S., Sharma, V. and Arora, S. (2020). Physico-chemical aspects of lactose hydrolysed milk system along with detection and mitigation of Maillard reaction products. *Trends in Food Science and Technology* 107, 57–67.

Smart, J. B. (1988). Effect of whey components on the rate of crystallization and solubility of α-lactose monohydrate. *New Zealand Journal of Dairy Science and Technology* 23(4), 275–289.

Smart, J. B. and Smith, J. M. (1992). Effect of selected compounds on the rate of α-lactose monohydrate crystallization, crystal yield and quality. *International Dairy Journal* 2(1), 41–53.

Solomons, N. W. (2002). Fermentation, fermented foods and lactose intolerance. *European Journal of Clinical Nutrition* 56(Suppl. 4), S50–S55.

Stacey, A., Schnieke, A., Kerr, M., Scott, A., McKee, C., Cottingham, I., Binas, B., Wilde, C. and Colman, A. (1995). Lactation is disrupted by alpha-lactalbumin deficiency and can be restored by human alpha-lactalbumin gene replacement in mice. *Proceedings of the National Academy of Sciences of the United States of America* 92(7), 2835–2839.

Stelwagen, K. (2020). Mammary gland, milk biosynthesis and secretion: lactose. In: *Reference Module in Food Science*. Elsevier. https://doi.org/10.1016/B978-0-08-100596-5.00885-4.

Storhaug, C. L., Fosse, S. K. and Fadnes, L. T. (2017). Country, regional, and global estimates for lactose malabsorption in adults: a systematic review and meta-analysis. *The Lancet. Gastroenterology and Hepatology* 2(10), 738–746.

Sunehag, A. L. and Haymond, M. W. (2002). Splanchnic galactose extraction is regulated by coingestion of glucose in humans. *Metabolism: Clinical and Experimental* 51(7), 827–832.

Szilagyi, A. (2019). Digestion, absorption, metabolism, and physiological effects of lactose. In: Paques, M. and Lindner, C. (Eds) *Lactose* (pp. 49–111). Academic Press, London.

Szilagyi, A., Walker, C. and Thomas, M. G. (2019). Lactose intolerance and other related food sensitivities. In: Paques, M. and Lindner, C. (Eds) *Lactose* (pp. 113–153). Academic Press, London.

Tamime, A. Y. and Robinson, R. K. (2007). *Yoghurt: Science and Technology*. Woodhead Publishing, Cambridge, UK.

Touger-Decker, R. and Van Loveren, C. (2003). Sugars and dental caries. *The American Journal of Clinical Nutrition* 78(4), 881S–892S.

Troise, A. D., Bandini, E., De Donno, R., Meijer, G., Trezzi, M. and Fogliano, V. (2016). The quality of low lactose milk is affected by the side proteolytic activity of the lactase used in the production process. *Food Research International*, 89, 514–525.

Van Lieshout, G. A. A., Lambers, T. T., Bragt, M. C. E. and Hettinga, K. A. (2020). How processing may affect milk protein digestion and overall physiological outcomes: a systematic review. *Critical Reviews in Food Science and Nutrition* 60(14), 2422–2445.

Vázquez, S. M., de Rojas, J. N., Troche, J. R., Adame, E. C., Ruíz, R. R. and Domínguez, L. U. (2020). The importance of lactose intolerance in individuals with gastrointestinal symptoms. Revista de Gastroenterología de México (English Edition).
Visser, R. A. (1984). Experiments for tracing growth retarders in lactose. *Netherlands Milk and Dairy Journal* 38, 107–133.
Visser, R. A. (1988). Crystal growth retarding of alpha-lactose hydrate by sugar phosphates - continued study. *Netherlands Milk and Dairy Journal* 42, 449–468.
Walstra, P. and Jenness, R. (1984). *Dairy Chemistry & Physics*. John Wiley & Sons, New York.
Walstra, P., Walstra, P., Wouters, J. T. and Geurts, T. J. (2005). *Dairy Science and Technology*. CRC Press, Boca Raton, FL.
Wijayasinghe, R., Bogahawaththa, D., Chandrapala, J. and Vasiljevic, T. (2020). Crystallization behavior and crystal properties of lactose as affected by lactic, citric, or phosphoric acid. *Journal of Dairy Science* 103(12), 11050–11061.
Williams, C. A., Phillips, T. and Macdonald, I. (1983). The influence of glucose on serum galactose levels in man. *Metabolism: Clinical and Experimental* 32(3), 250–256.
Wong, S. Y. and Hartel, R. W. (2014). Crystallization in lactose refining–a review. *Journal of Food Science* 79(3), R257–R272.

# Chapter 11

## Nutritional properties and biological activity of lactose and other dairy carbohydrates

*Michael Gänzle, University of Alberta, Canada*



## 1 Introduction

The consumption of milk and (fermented) dairy products dates back more than 7000 years (McClure et al., 2018) and has sustained human populations, particularly in areas that are unsuitable for plant agriculture, e.g. the Eurasian steppe (Wilkin et al., 2020), arid grasslands in Eastern Africa (Bleasdale et al., 2021) or alpine meadows, and areas with cold climates in Northern Europe. Lactose, by far the major carbohydrate in milk (Table 1), accounts for more than one-third of the dry matter content of bovine milk and for more than one-fourth of the dietary energy. The nutritional properties of lactose, however, are dependent on the amount of lactose consumed as well as the age and genetic background of human individuals (Fig. 1). The ability to digest lactose throughout adulthood has evolved relatively late in human evolution, manifests only in a minority of humans and is predated by consumption of milk or dairy products (Bleasdale et al., 2021; McClure et al., 2018). This apparent mismatch of dairy consumption and lactose indigestibility was likely compensated by consumption of fermented dairy products that not only contain lactose but also viable fermentation organisms with active β-galactosidase that aids in lactose

http://dx.doi.org/10.19103/AS.2022.0099.11

**Table 1** Carbohydrates and major glycoproteins in bovine milk[a]

| Concentration (g/L) | Colostrum | Mature milk |
|---|---|---|
| **Carbohydrates** | | |
| Lactose | 23–26 | 40–50 |
| 3′-Sialyllactose | 0.8–1.3 | 0.05–0.06 |
| 6′-Sialyllactose | 0.08–0.12 | 0.01–0.02 |
| Sialyllactosamine | 0.1–0.26 | 0.001 or less |
| Disialyllactose | 0.1–0.4 | <0.001 |
| Other acidic oligosaccharides | <0.05 | Traces |
| Neutral oligosaccharides | <0.05 | Traces |
| **Glycoproteins** | | |
| κ-Casein | 16 | 3.5-4.5 |
| Lactoferrin | 2 - 5 | 0.05 - 0.4 |

[a] Based on data from Fischer-Tlustos et al. (2020), Fong et al. (2011), McGrath et al. (2016), Nakamura et al. (2003), O'Riordan et al. (2014).

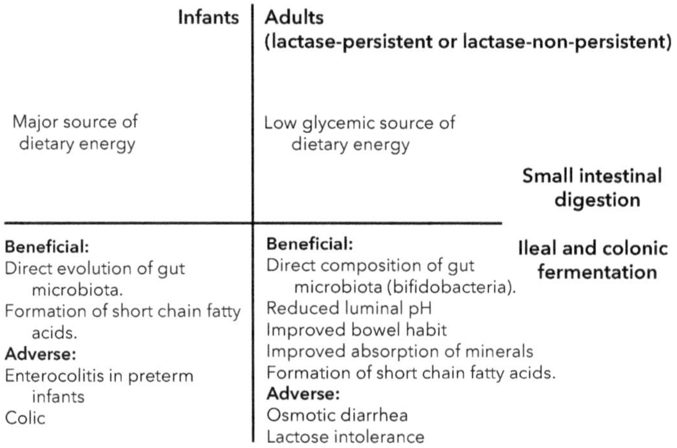


**Figure 1** Overview of the nutritional functionality of lactose in infants and lactase-persistent or lactase-non-persistent individuals.

digestion (Gänzle, 2019a). Depending on the dietary context and the genetic background of the consumer, lactose is a digestible carbohydrate, a substrate for beneficial colonic fermentation to short-chain fatty acids (SCFA) or a cause of substantial intestinal discomfort (Gänzle, 2019b). In addition to lactose, bovine milk contains glycoproteins and small amounts of oligosaccharides (Table 1). These dairy carbohydrates do not add substantially to the caloric content of milk but have other biological activities that are relevant for infants and other suckling mammals and can be exploited in food, feed or pharmaceutical applications. This chapter aims to provide an overview of the nutritional properties and biological activity of dairy carbohydrates and major glycoproteins in milk. Although human milk or the milk of dairy animals contains only trace amounts of β-galacto-oligosaccharides (GOS) other than lactose, GOS are produced with lactose as substrate (Gänzle et al., 2008), and their nutritional functionality in humans conceptually overlaps with the nutritional functionality of lactose in lactase-non-persistent adults; therefore, this chapter also provides a brief overview of GOS.

## 2 Lactose

### 2.1 Lactose digestion in infants

Lactose (β-D-galactopyranosyl-(1→4)-D-glucose) is the sole digestible carbohydrate source for breast-fed infants (Fig. 1). Lactose has a low sweetness relative to sucrose, and milk consumption thus does not condition for consumption of sweet foods after weaning. Brush border lactase-phlorizin hydrolase, a β-galactosidase that is subsequently referred to as lactase, hydrolyses lactose to glucose and galactose, which are then absorbed by the small intestinal mucosa (Hooton et al., 2015). Glucose and galactose are transported by the same monosaccharide transporters, the sodium glucose transport protein 1 and the glucose transporter 2; galactose is converted to glucose-6-phosphate in the Leloir pathway, predominantly in the liver (Suzuki et al., 2001). Brush border lactase activity develops relatively late during gestation; therefore, lactose is digested more slowly in preterm infants, and an average of 35% of lactose, with a range of 2% to more than 90%, is not hydrolysed in the small intestine but fermented in the large intestine (Kien et al., 1996). Lactose is also partially fermented in the large intestine of term infants (Lifschitz et al., 1983). Owing to the metabolic focus of bifidobacteria on lactose, undigested lactose reaching the intestine was reported to increase the proportion of *Bifidobacterium* species in infant intestinal microbiota (Francavilla et al., 2012). Colonic lactose fermentation was also linked, however, to infantile colic (Moore et al., 1988). Moreover, *Enterobacteriaceae* causing necrotizing enterocolitis also ferment lactose, and excess colonic lactose has been linked to necrotizing enterocolitis in preterm infants (Lin, 2004).

Galactosemia is a rare and inherited metabolic disorder that develops if any one of the enzymes of the Leloir pathway is dysfunctional. In European populations, the frequency of galactosemia is about 1 in 50 000 individuals (Suzuki et al., 2001). In infants, galactosemia presents as poor growth, liver dysfunction and lethargy and may favour infections by *Escherichia coli*. Unless treated by removing dietary galactose and lactose, long-term symptoms include neurological disorders including impaired cognitive function and movement disorders and reduced female fertility (Demirbas et al., 2018).

## 2.2 Lactose digestion in lactase-persistent adults

In most humans, expression of small intestinal lactase activity ceases after weaning, but mutations that maintained lactase expression throughout adulthood independently arose about 6000–8000 years ago in pastoral societies in North Africa, the Middle East and Europe (Bleasdale et al., 2021; Ségurel and Bon, 2017; Storhaug et al., 2017). In Europe, lactase persistence is attributed to the dominant allele -13910*T; the greatest diversity of alleles leading to lactase persistence occurs in Africa (Bleasdale et al., 2021; Ségurel and Bon, 2017). In lactase-persistent individuals, lactose remains a digestible disaccharide which is hydrolysed and absorbed in the small intestine throughout adulthood (Fig. 1). After weaning, however, lactose is generally no longer the major dietary carbohydrate, and the role of lactose as substrate for colonic fermentation is replaced by starch (Wang et al., 2019; Ze et al., 2012). Lactose has a lower glycemic index when compared to glucose or starch (Augustin et al., 2015). Although lactose is fermented by oral streptococci, lactose does not support the formation of the extracellular biofilm matrix by *Streptococcus mutans* (Bowen and Koo, 2011) and thus has a lower cariogenicity when compared to sucrose (Aimutis, 2012).

## 2.3 Lactose digestion in lactase-non-persistent adults: adverse physiological effects

Approximately 70% of humans are lactase-non-persistent and do not express intestinal lactase activity after weaning. In these individuals, lactose is utilized after microbial conversion to SCFA in the terminal ileum and the colon (Fig. 1) (Forsgård, 2019; Zoetendal et al., 2012). Whether lactose consumption has beneficial or adverse effects on health depends on the amount of lactose consumed, the kinetics of digesta transit in the small intestine and the composition of intestinal microbiota (Deng et al., 2015). Adverse intestinal symptoms in response to lactose consumption, termed lactose intolerance, include abdominal pain, distension of the intestine and flatus and osmotic diarrhoea (Deng et al., 2015). Mechanisms and symptoms of lactose intolerance overlap with the symptoms that present after digestion of other dietary carbohydrates that are indigestible but rapidly fermented in the terminal ileum or the proximal colon, collectively termed as 'fermentable oligosaccharides, disaccharides, monosaccharides and polyols', e.g. raffinose-family oligosaccharides or fructo-oligosaccharides (Priyanka et al., 2018; Yan et al., 2018). Lactose is not the only sugar that is conditionally digested in humans. Fructose malabsorption, which affects approximately 15% of humans, results in comparable symptoms (DiNicolantonio and Lucan, 2015; Fedewa and Rao, 2014).

The presence of alleles resulting in lactase-non-persistence, the objective measurement of lactose intolerance by a breath hydrogen test and subjective lactose intolerance as measured by indicators of intestinal discomfort only partially overlap (Deng et al., 2015). First, negative effects are dose dependent, and 10–15 g of lactose, corresponding to about 250 mL of fluid bovine milk, are typically tolerated without adverse symptoms. Second, dietary β-galactosidase, which is provided by food-fermenting lactic acid bacteria, viable probiotic bacteria or enzyme preparations that are consumed with dairy foods, sufficiently hydrolyses lactose during small intestinal transit to avoid adverse symptoms (Kies, 2014; Pelletier et al., 2001). Third, the kinetics of intestinal transit is influenced by other dietary components, e.g. dietary fibre, bitter or sweet tastants, and can differ substantially between different individuals (Depoortere, 2014). Lastly, the capacity of colonic microbiota to ferment lactose without excess gas formation also differs between individuals and may differ in one individual over time (Forsgård, 2019).

### 2.4 Lactose digestion in lactase-non-persistent adults: beneficial physiological effects

If lactose consumption by lactase-non-persistent individuals does not result in intestinal discomfort, it likely provides the same health benefits that are amply documented for other oligosaccharides, particularly GOS other than lactose (Bindels et al., 2015; Krumbeck et al., 2018; Macfarlane et al., 2008; Paganini et al., 2017). The link between lactose activity and composition of intestinal microbiota and human health is not widely documented in clinical trials but is supported by consistent clinical evidence and mechanistic considerations. The long-term persistence of a probiotic strain of *Bifidobacterium longum* in some but not all individuals was related to the lack of organisms with β-galactose activity that ferment lactose and other GOS (Maldonado-Gómez et al., 2016). In human intestinal microbiota, bifidobacteria are among the major degraders of lactose and GOS (Reichardt et al., 2018). Accordingly, a genome-wide association study aiming to identify host genes that relate to the composition of intestinal microbiota identified a strong correlation between lactase-non-persistent individuals who consume dairy products and the relative abundance of bifidobacteria (Bonder et al., 2016). Only a few studies included lactose as a control group when assessing the beneficial impact of GOS on human health. One of these screened study participants for lactose intolerance but not for lactase-non-persistence and thus included lactase-non-persistent but lactose-tolerant individuals in the lactose control group. Results suggested that the beneficial effect of lactose in these individuals was qualitatively comparable to the effect of GOS in lactase-persistent individuals (Krumbeck et al., 2018). In short, lactose can be considered a 'conditional prebiotic' that may have comparable beneficial physiological effects in lactase-non-persistent individuals as GOS.

## 3 Galacto-oligosaccharides

GOS are composed of one to nine β-linked galactose moieties with glucose or galactose at the reducing end (Gänzle et al., 2008). GOS other than lactose occur only in trace amounts in the milk of dairy animals (Table 1) but are a major product of the value-added conversion of lactose from milk (Gänzle et al., 2008; Tao et al., 2008). GOS are also one of the most widely used prebiotic oligosaccharide preparations and among the prebiotics for which health benefits are most widely documented (Bindels et al., 2015; Krumbeck et al., 2018; Macfarlane et al., 2008; Paganini et al., 2017). Most commercial GOS include 20–40% monosaccharides, 30–45% lactose and non-lactose disaccharides, 20–35% trisaccharides, less than 10% tetrasaccharides and less than 3% pentasaccharides; however, few of the commercial products are processed by simultaneous moving bed chromatography to remove mono- and disaccharides (Azcarate-Peril et al., 2017; Chen and Gänzle, 2017; Guo et al., 2018; Van Leeuwen et al., 2016). The catalytic efficiency of brush border lactase is lower for β-(1→6)- and β-(1→3)-linked galactosides when compared to β-(1→4) linkages and slower for trisaccharides when compared to disaccharides (Ferreira-Lazarte et al., 2017); commercial GOS are thus largely indigestible even in lactase-persistent individuals despite the high content of disaccharides.

The concept of prebiotics was initially based on the assumption that health benefits relate to the increased abundance of specific health-beneficial bacterial taxa after dietary intervention with non-digestible carbohydrates (Gibson and Roberfroid, 1995). In addition, the term prebiotic was predominantly used to imply or to explicitly state that selective stimulation by prebiotics relates to an increased relative abundance of bifidobacteria and lactobacilli in the human intestinal microbiome (Gibson and Roberfroid, 1995). After 1995, however, two decades of research on the composition of human intestinal microbiota and its role in health and disease not only questioned the concept of 'specificity' but also provided insight into mechanisms that account for the health benefits of prebiotic carbohydrates and dietary fibre. Owing to the individuality of human intestinal microbiota, the response to intervention with non-digestible carbohydrates also differs between individuals (Davis et al., 2011), a factor that also accounts for the individual response to lactose in lactase-non-persistent individuals discussed above. Second, 'specificity' of prebiotics is qualitatively and quantitatively undefined in terms of how many bacterial taxa should increase in abundance, how large the increase in abundance should be or at which taxonomic level (species, genus or family) these taxa should be defined (Gibson et al., 2017). Non-digestible carbohydrates including lactose and GOS generally increase the abundance of several bacterial species or genera (Davis et al., 2011; Reichardt et al., 2018).

Irrespective of the individual response of intestinal microbiota to dietary intervention with non-digestible carbohydrates, the increase in colonic formation of SCFA is relatively consistent (Reichardt et al., 2018) and accounts for most of the known health benefits of prebiotic carbohydrates. SCFA formation decreases the luminal pH, which reduces protein fermentation by *Bacteroidetes* and facilitates the uptake of minerals (Holtug et al., 2009; Paganini et al., 2017). Local effects of SCFA include an improved energy supply of the colonic mucosa, improved motility and a reduced stool transit time (Cherbut et al., 1998). The interaction of SCFA with G-protein-coupled receptors also elicits beneficial systemic effects, particularly improved satiety, glucose homeostasis and insulin sensitivity and reduced inflammation (Bindels et al., 2015; Yan et al., 2018). Recognizing the questionable usefulness of 'specificity' in defining prebiotics, and the knowledge on mechanisms that underlie documented health benefits of prebiotics, an updated definition of prebiotics was proposed as 'a non-digestible compound that, through its metabolization by microorganisms in the GI tract, modulates composition and/or activity of the GI microbiota, thus conferring a beneficial physiological effect on the host' (Bindels et al., 2015). The definition of prebiotics, however, remains subject to scientific controversy, and regulatory approval for the concept has been obtained only in a few jurisdictions.

## 4 Glycosylated proteins: lactoferrin, κ-casein and the caseinmacropeptide

κ-Casein is the major glycoprotein in bovine milk that accounts for about 15% of the casein in mature bovine milk, corresponding to a concentration in milk of 3.5–4.5 g/L (Dalgleish and Corredig, 2012; Muller-Renaud et al., 2010; O'Riordan et al., 2014). The C-terminus of the protein is glycosylated, and two to four threonine residues carry O-linked glycans (Fig. 2) (Holland et al., 2006; Hua et al., 2011). Glycosylation of κ-casein occurs preferably at the $T^{154}$ (41%) and $T^{163}$ (29%) positions, followed by $T^{152}$ (14%), $T^{142}$ (7%) and $T^{157}$ (0.1%). $S^{148}$ and/or $S^{170}$ are phosphorylated. Substitution at $T^{166}$ is only a minor isoform of κ-casein; this residue is alternatively phosphorylated or glycosylated (Holland et al., 2006; Hua et al., 2011). Glycosylation occurs with di-, tri- or tetrasaccharides that are composed of N-acetylgalactosamine (GalNAc), galactose and sialic acid Neu5Ac (Fig. 2) (Holland et al., 2006). A greater diversity of oligosaccharide structures which additionally include N-acetylglucosamine has been identified in κ-casein from colostrum (O'Riordan et al., 2014; Saito et al., 1982).

Lactoferrin is the second major glycoprotein in bovine milk. The concentration of lactoferrin in bovine milk is 10-fold lower when compared to

(a)

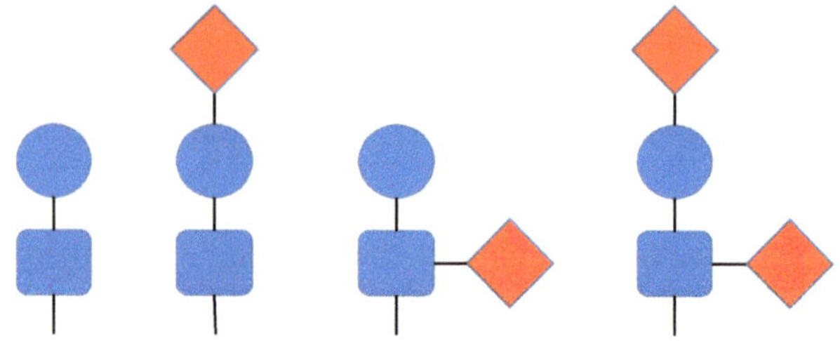


(b) S**FM**AIP$^{130}$ PKKNQDKTEI$^{140}$ P**T**INTIASGE$^{150}$ P**T**S**T**PT**T**EAV$^{160}$
I

ES**T**VA**T**LEDS$^{170}$ PEVTESPPEI$^{180}$ NTVQVTSTAV$^{190}$
A

**Figure 2** Glycan structures (Panel a) and amino acid sequence (Panel b) of the caseinmacropeptide. In panel A, monosaccharides and linkage types are designated as follows: Blue squares, N-acetylgalactosamine; blue circle, galactose; red diamonds, N-acetylneuraminic acid (sialic acid); black solid lines, β-(1→3) glycosidic linkage; black dotted lines; α-(2→3) or α-(2→6) glycosidic linkages; gray lines, link to threonine residues on the peptide backbone. The peptide sequence of the CMP is drawn based on the GenBank accession numbers NP_776719.1 (variant A) and AAA30433.1 (variant B); of variant B, only the two differing amino acid residues are shown. The chymosin cleavage site is underlined and bold; glycosylated threonine residues are printed in bold and phosphorylated serine residues are underlined. The numbering of residues is based on the sequence of the κ-casein precursor; the sequence shown corresponds to position number 104–169 of the mature κ-casein. Drawn with information from Dalgleish and Corredig (2012), Holland et al. (2006), Hua et al. (2011).

κ-casein; moreover, the lactoferrin content in mature milk varies from 0.03 to 0.49 g/L with a median concentration of about 0.1 g/L (Cheng et al., 2008). Lactoferrin is N-glycosylated at one or more of five asparagine residues with heterogeneous oligosaccharides composed of mannose, N-GlcNAc, Gal and GalNAc with fucose and sialic acids as minor components (Hua et al., 2011; O'Riordan et al., 2014). Lactoferrin is used as a functional ingredient in infant formulae, but it is unclear whether protein glycosylation contributes to its biological activity (Kaur et al., 2019; O'Riordan et al., 2014).

κ-Casein is located at the surface of the casein micelles with the glycosylated C-terminal part oriented to the outside and the N-terminal sequence oriented to the inside of the micelle. The hydrophilic and 'hairy' structure maintains the casein micelle in colloidal suspension and prevents aggregation of micelles (Dalgleish and Corredig, 2012). During digestion in the calf stomach, and during cheese-making, chymosin cleaves specifically κ-casein between positions 105 and 106 (126 and 127 in the precursor peptide, Fig. 2), thus releasing the caseinmacropeptide (CMP), also termed glycomacropeptide (GMP). Removal of the CMP from κ-casein eliminates the hydrophilic, hairy layer of the casein micelles and results in aggregation and precipitation of casein. In cheese whey from bovine milk, the CMP accounts for 20% of the protein, corresponding to a concentration of about 2 g/L (Dalgleish and Corredig, 2012). The GMP is purified experimentally or commercially from sweet or acid whey with a combination of heating, membrane filtration and ion exchange chromatography steps (Martín-Diana et al., 2002; Nakano et al., 2018; O'Riordan et al., 2014). The nutritional relevance of the CMP is threefold: First, it provides a source of sialic acid, which is an essential nutrient in infant nutrition; second, the glycan portion of the CMP provides a substrate for growth of intestinal bacteria, particularly bifidobacteria and third, the CMP prevents the adhesion of intestinal pathogens to mucosal surfaces.

Sialic acid is an essential nutrient for brain development in infants, particularly in preterm infants. Human milk contains a relatively high level of sialylated oligosaccharides; the concentration of sialylated free oligosaccharides in bovine milk or the milk of other dairy animals is less than 1% of the concentration in human milk (Tao et al., 2008; Wang, 2009; Zivkovic et al., 2011). The major sialic acid in CMP is N-acetylneuraminic acid, which differs from most other animal sources of sialic acid where N-glycolylneuraminic acid, which was suggested to have pro-inflammatory properties, is the predominant form (Samraj et al., 2015; Wang, 2009).

Multiple members of intestinal microbiota including *Bifidobacterium*, *Akkermansia* and *Bacteroides* species use O-glycans as carbon source for conversion to SCFA (González-Morelo et al., 2020). In particular, bifidobacteria including *Bifidobacterium longum* subsp. *infantis* maintain an extensive toolset for metabolism of sialylated oligosaccharides and glycoproteins; *B. longum* also grows with CMP as the sole carbon source (O'Riordan et al., 2018; Sela et al., 2008). Owing to the low yield and high cost of purification of CMP from whey, it is an unlikely competitor for prebiotic GOS, which are also produced from whey for use as a bulk prebiotic food ingredient. The commercial production of sialylated and fucosylated human milk oligosaccharides, however, is currently achieved only in low-yield/high-cost microbial cell factories (Zhou et al., 2021). In those areas of medical or infant nutrition where sialylated oligosaccharides are necessary, CMP thus represents a promising functional food ingredient (O'Riordan et al., 2014).

The virulence of most intestinal bacterial pathogens is dependent on the interaction of glycan-binding domains of bacterial surface proteins or toxins with oligosaccharide receptors on the surface of intestinal cells (Kulkarni et al., 2010; Shoaf-Sweeney and Hutkins, 2008). These glycan–protein interactions are also required for the intestinal food- or water-borne pathogens, *Vibrio cholera*, enterotoxigenic and enterohaemorrhagic *Escherichia coli* and *Salmonella enterica* (Kulkarni et al., 2010; Shoaf-Sweeney and Hutkins, 2008). Soluble oligosaccharides or glycopeptides bind to the lectin domains of bacterial surface proteins or toxins and thus prevent infection or toxin activity (Chen and Gänzle, 2017; Shoaf-Sweeney and Hutkins, 2008). CMP binds to the lectin domains of multiple pathogens or toxins including fimbriae of enterohaemorrhagic *E. coli*, the animal-specific K88 fimbriae of enterotoxigenic *E. coli,* the cholera toxin and the highly related heat-labile toxin of enterotoxigenic *E. coli* and fimbriae of *Salmonella enterica in vitro* (Feeney et al., 2017; Kawasaki et al., 1992; Nakajima et al., 2005). *In vitro* inhibition of pathogen or toxin binding is observed in the range of 1–10 g CMP/L, i.e. at concentrations that are physiologically relevant in suckling mammals and are readily achieved by the addition of CMP to food or feed (Feeney et al., 2017; Kawasaki et al., 1992; Nakajima et al., 2005). A protective effect of CMP at an inclusion level of approximately 15 g CMP/kg feed dry matter protected piglets in an enterotoxigenic *E. coli* K88 infection model (Hermes et al., 2013). Enterotoxigenic *E. coli* is a major cause of childhood mortality in developing countries (Qadri et al., 2005); in addition, animal-specific enterotoxigenic *Escherichia coli* causes significant mortality and economic losses in suckling and weanling piglets (Nagy and Fekete, 2005). The use of CMP in infant and piglet nutrition thus holds promise to reduce diarrhoeal disease without the use of antibiotics.

## 5 Bovine milk oligosaccharides

In contrast to human milk, which contains more than 10 g/L acidic and neutral oligosaccharides, the concentration of free oligosaccharides in mature bovine milk is low, typically below 0.1 g/L, and acidic, sialylated oligosaccharides are the major compounds while fucosylated oligosaccharides occur only in trace amounts (Mehra and Kelly, 2006; ten Bruggencate et al., 2014). Mature ovine and caprine milk differ from bovine milk in their oligosaccharide spectrum but have a low concentration of free oligosaccharides (Albrecht et al., 2014; Mehra and Kelly, 2006). In comparison to mature milk, bovine colostrum has a lower lactose content, about 23–26 g/L, but a higher content of sialylated oligosaccharides, about 1–1.5 g/L, which decreases within a few days after parturition (Fong et al., 2011; Nakamura et al., 2003; ten Bruggencate et al., 2014). Specifically, the concentration of the major oligosaccharides, 3′-sialyllactose (Neu5Ac-α(2→3)-Gal-β(1→4)Glc), 6′-sialyllactose (Neu5Ac-α(2→6)-Gal-β(1→4)Glc) and sialyllactoseamine (Neu5Ac-α(2→6)-Gal-β(1→4) GlcNAc) decreased from about 0.6, 0.1 and 0.14 g/L, respectively, at parturition to 0.05, 0.05 and less than 0.01 g/L, respectively, after 6 days (Fischer-Tlustos et al., 2020). Disialyllactose is also a major oligosaccharide in colostrum (Fischer-Tlustos et al., 2020; Fong et al., 2011). The concentration of oligosaccharides was substantially higher in milk from multiparous cows when compared to milk from primiparous cows (Fischer-Tlustos et al., 2020). More than 30 additional neutral and acidic oligosaccharides with a degree of polymerization of 3–7 were additionally identified but occur only in trace amounts (Albrecht et al., 2014; Aldredge et al., 2013).

Colostrum is critical for the development of calves, including the development of intestinal microbiota and the gut-associated immune system (Hammon et al., 2020; Malmuthuge and Guan, 2017); however, colostrum also contains multiple micronutrients and immune-reactive proteins (McGrath et al., 2016), and the specific contribution of free oligosaccharides to the development of calves remains unclear. 3′-Sialyllactose and 6′-sialyllactose are also major components of human milk oligosaccharides with a combined concentration of up to 0.8 g/L (Mehra and Kelly, 2006) and are produced commercially with microbial cell factories (Zhou et al., 2021). As outlined above in the section on the CMP, sialylated oligosaccharides are an important dietary source of sialic acid in early infant nutrition and are metabolized by colonic microbiota. In human infants, sialylated oligosaccharides are preferentially metabolized by bifidobacteria and *Bacteroides* spp. (Sela et al., 2008; ten Bruggencate et al., 2014). Sialylated oligosaccharides were also demonstrated to prevent pathogen adhesion. *In vivo* evidence demonstrates that sialyllactose reduced the toxicity of the cholera toxin, reduced the severity of infections with *Salmonella enterica* and rotavirus and improved the eradication of *Helicobacter pylori* (Chen and Gänzle, 2017; ten Bruggencate et al., 2014). Owing to the low concentration of free oligosaccharides in mature bovine milk, it is unclear whether these oligosaccharides have an important function in calf nutrition. Moreover, the low concentration of free oligosaccharides in milk makes it unlikely that their recovery and purification for use in infant nutrition is economically viable.

## 6 Conclusion

In conclusion, lactose is a dairy carbohydrate that is digested in the small intestine of all human infants and of about 30% of adult humans. In infants and lactase-non-persistent adults, lactose also functions as a substrate for colonic microbes and favours formation of SCFA. Negative consequences of lactose ingestion include infant colic and overgrowth of *Enterobacteriaceae* in preterm infants. Negative consequences of lactose ingestion in lactase-non-persistent adults include lactose intolerance. Lactose is thus a prime example for 'personalized nutrition' as the same compound can have very different effects in different individuals. It is unclear whether any of the glycoproteins or minor carbohydrates have relevant nutritional function or biological activity in human adults.

In addition to the presence of dairy carbohydrates in milk, lactose is also a major component of cheese whey as well as permeates that are obtained after concentration of milk proteins by ultrafiltration (Gänzle et al., 2008; Kaur et al., 2019). Cheese whey additionally contains the major glycoproteins that are present in milk (Kaur et al., 2019). The production of cheese and protein concentrates from milk generates substantially more lactose than is used in food or pharmaceutical applications–despite many decades of research, the valorization of whey (glycol-)proteins and carbohydrates largely remains an unsolved problem. The nutritional properties of lactose and the biological activities of glycoproteins and minor carbohydrates offer yet untapped opportunities to extract valuable components for use in novel food, feed and pharmaceutical applications.

## 7 Where to look for further information

A recent book on lactose provides a comprehensive coverage on lactase-persistence, nutritional properties of lactose, and its industrial use: Paques, M. and Lindner, C. (Eds). 2019. *Lactose: Evolutionary Role, Health Effects, and Applications*. Academic Press, London.

An overview on health benefits of prebiotic carbohydrates including GOS and mechanisms underlying these health benefits is provided by: Bindels, L. B., Delzenne, N. M., Cani, P. D. and Walter, J. 2015. Towards a more comprehensive concept for prebiotics. *Nat. Rev. Gastroenterol. Hepatol.* 12(5), 303–310. https://doi.org/10.1038/nrgastro.2015.47.

The following review provides information on chemical properties and biological activities of milk glycoproteins: O'Riordan, N., Kane, M., Joshi, L. and Hickey, R. M. 2014. Structural and functional characteristics of bovine milk protein glycosylation. *Glycobiology* 24(3), 220–236.

## 8 Acknowledgements

The Canada Research Chairs program is acknowledged for funding.

## 9 References

Aimutis, W. R. 2012. Lactose cariogenicity with an emphasis on childhood dental caries. *Int. Dairy J.* 22(2), 152–158. https://doi.org/10.1016/J.IDAIRYJ.2011.10.007.

Albrecht, S., Lane, J. A., Mariño, K., Al Busadah, K. A., Carrington, S. D., Hickey, R. M. and Rudd, P. M. 2014. A comparative study of free oligosaccharides in the milk of domestic animals. *Br. J. Nutr.* 111(7), 1313–1328. https://doi.org/10.1017/S0007114513003772.

Aldredge, D. L., Geronimo, M. R., Hua, S., Nwosu, C. C., Lebrilla, C. B. and Barile, D. 2013. Annotation and structural elucidation of bovine milk oligosaccharides and determination of novel fucosylated structures. *Glycobiology* 23(6), 664–676. https://doi.org/10.1093/GLYCOB/CWT007.

Augustin, L. S. A., Kendall, C. W. C., Jenkins, D. J. A., Willett, W. C., Astrup, A., Barclay, A. W., Björck, I., Brand-Miller, J. C., Brighenti, F., Buyken, A. E., Ceriello, A., La Vecchia,

C., Livesey, G., Liu, S., Riccardi, G., Rizkalla, S. W., Sievenpiper, J. L., Trichopoulou, A., Wolever, T. M. S., Baer-Sinnott, S. and Poli, A. 2015. Glycemic index, glycemic load and glycemic response: an International Scientific Consensus Summit from the International Carbohydrate Quality Consortium (ICQC). *Nutr. Metab. Cardiovasc. Dis.* 25(9), 795–815. https://doi.org/10.1016/J.NUMECD.2015.05.005.

Azcarate-Peril, M. A., Ritter, A. J., Savaiano, D., Monteagudo-Mera, A., Anderson, C., Magness, S. T. and Klaenhammer, T. R. 2017. Impact of short-chain galactooligosaccharides on the gut microbiome of lactose-intolerant individuals. *Proc. Natl. Acad. Sci. U. S. A.* 114(3), E367–E375. https://doi.org/10.1073/PNAS.1606722113.

Bindels, L. B., Delzenne, N. M., Cani, P. D. and Walter, J. 2015. Towards a more comprehensive concept for prebiotics. *Nat. Rev. Gastroenterol. Hepatol.* 12(5), 303–310. https://doi.org/10.1038/nrgastro.2015.47.

Bleasdale, M., Richter, K. K., Janzen, A., Brown, S., Scott, A., Zech, J., Wilkin, S., Wang, K., Schiffels, S., Desideri, J., Besse, M., Reinold, J., Saad, M., Babiker, H., Power, R. C., Ndiema, E., Ogola, C., Manthi, F. K., Zahir, M., Petraglia, M., Trachsel, C., Nanni, P., Grossmann, J., Hendy, J., Crowther, A., Roberts, P., Goldstein, S. T. and Boivin, N. 2021. Ancient proteins provide evidence of dairy consumption in eastern Africa. *Nat. Commun.* 12(1), 632. https://doi.org/10.1038/s41467-020-20682-3.

Bonder, M. J., Kurilshikov, A., Tigchelaar, E. F., Mujagic, Z., Imhann, F., Vila, A. V., Deelen, P., Vatanen, T., Schirmer, M., Smeekens, S. P., Zhernakova, D. V., Jankipersadsing, S. A., Jaeger, M., Oosting, M., Cenit, M. C., Masclee, A. A. M., Swertz, M. A., Li, Y., Kumar, V., Joosten, L., Harmsen, H., Weersma, R. K., Franke, L., Hofker, M. H., Xavier, R. J., Jonkers, D., Netea, M. G., Wijmenga, C., Fu, J. and Zhernakova, A. 2016. The effect of host genetics on the gut microbiome. *Nat. Genet.* 48(11), 1407–1412. https://doi.org/10.1038/ng.3663.

Bowen, W. H. and Koo, H. 2011. Biology of *Streptococcus mutans*-derived glucosyltransferases: role in extracellular matrix formation of cariogenic biofilms. *Caries Res.* 45(1), 69–86. https://doi.org/10.1159/000324598.

Chen, X. Y. and Gänzle, M. G. 2017. Lactose and lactose-derived oligosaccharides: more than prebiotics? *Int. Dairy J.* 67, 61–72. https://doi.org/10.1016/j.idairyj.2016.10.001.

Cheng, J. B., Wang, J. Q., Bu, D. P., Liu, G. L., Zhang, C. G., Wei, H. Y., Zhou, L. Y. and Wang, J. Z. 2008. Factors affecting the lactoferrin concentration in bovine milk. *J. Dairy Sci.* 91(3), 970–976. https://doi.org/10.3168/JDS.2007-0689.

Cherbut, C., Ferrier, L., Rozé, C., Anini, Y., Blottière, H., Lecannu, G. and Galmiche, J. P. 1998. Short-chain fatty acids modify colonic motility through nerves and polypeptide YY release in the rat. *Am. J. Physiol.* 275(6), G1415–G1422. https://doi.org/10.1152/ajpgi.1998.275.6.G1415.

Dalgleish, D. G. and Corredig, M. 2012. The structure of the casein micelle of milk and its changes during processing. *Annu. Rev. Food Sci. Technol.* 3, 449–467. https://doi.org/10.1146/ANNUREV-FOOD-022811-101214.

Davis, L. M. G., Martínez, I., Walter, J., Goin, C. and Hutkins, R. W. 2011. Barcoded pyrosequencing reveals that consumption of galactooligosaccharides results in a highly specific bifidogenic response in humans. *PLoS ONE* 6(9), e25200. https://doi.org/10.1371/JOURNAL.PONE.0025200.

Demirbas, D., Coelho, A. I., Rubio-Gozalbo, M. E. and Berry, G. T. 2018. Hereditary galactosemia. *Metabolism* 83, 188–196. https://doi.org/10.1016/J.METABOL.2018.01.025.

Deng, Y., Misselwitz, B., Dai, N. and Fox, M. 2015. Lactose intolerance in adults: biological mechanism and dietary management. *Nutrients* 7(9), 8020–8035. https://doi.org/10.3390/NU7095380.

Depoortere, I. 2014. Taste receptors of the gut: emerging roles in health and disease. *Gut* 63(1), 179–190. https://doi.org/10.1136/gutjnl-2013-305112.

DiNicolantonio, J. J. and Lucan, S. C. 2015. Is fructose malabsorption a cause of irritable bowel syndrome? *Med. Hypo.* 85(3), 295–297. https://doi.org/10.1016/j.mehy.2015.05.019.

Fedewa, A. and Rao, S. S. C. 2014. Dietary fructose intolerance, fructan intolerance and FODMAPs. *Curr. Gastroenterol. Rep.* 16(1), 370. https://doi.org/10.1007/s11894-013-0370-0.

Feeney, S., Ryan, J. T., Kilcoyne, M., Joshi, L. and Hickey, R. 2017. Glycomacropeptide reduces intestinal epithelial cell barrier dysfunction and adhesion of entero-hemorrhagic and entero-pathogenic *Escherichia coli* in vitro. *Foods* 6(11), 93. https://doi.org/10.3390/FOODS6110093.

Ferreira-Lazarte, A., Olano, A., Villamiel, M. and Moreno, F. J. 2017. Assessment of *in vitro* digestibility of dietary carbohydrates using rat small intestinal extract. *J. Agric. Food Chem.* 65(36), 8046–8053. https://doi.org/10.1021/acs.jafc.7b01809.

Fischer-Tlustos, A. J., Hertogs, K., van Niekerk, J. K., Nagorske, M., Haines, D. M. and Steele, M. A. 2020. Oligosaccharide concentrations in colostrum, transition milk, and mature milk of primi- and multiparous Holstein cows during the first week of lactation. *J. Dairy Sci.* 103(4), 3683–3695. https://doi.org/10.3168/JDS.2019-17357.

Fong, B., Ma, K. and McJarrow, P. 2011. Quantification of bovine milk oligosaccharides using liquid chromatography-selected reaction monitoring-mass spectrometry. *J. Agric. Food Chem.* 59(18), 9788–9795. https://doi.org/10.1021/JF202035M.

Forsgård, R. A. 2019. Lactose digestion in humans: intestinal lactase appears to be constitutive whereas the colonic microbiome is adaptable. *Am. J. Clin. Nutr.* 110(2), 273–279. https://doi.org/10.1093/ajcn/nqz104.

Francavilla, R., Calasso, M., Calace, L., Siragusa, S., Ndagijimana, M., Vernocchi, P., Brunetti, L., Mancino, G., Tedeschi, G., Guerzoni, E., Indrio, F., Laghi, L., Miniello, V. L., Gobbetti, M. and De Angelis, M. 2012. Effect of lactose on gut microbiota and metabolome of infants with cow's milk allergy. *Pediatr. Allergy Immunol.* 23(5), 420–427. https://doi.org/10.1111/j.1399-3038.2012.01286.x.

Gänzle, M. 2019a. Fermented foods. In: Doyle, M. P., Diez-Gonzalez, F. and Hill, C. (Eds), *Food Microbiology: Fundamentals and Frontiers. Wiley*, pp. 855–900. https://doi.org/10.1128/9781555819972.ch33.

Gänzle, M. 2019b. Lactose–a conditional prebiotic? In: Paques, M. and Lindner, C. (Eds), *Lactose*. Elsevier, pp. 155–173. https://doi.org/10.1016/b978-0-12-811720-0.00004-0.

Gänzle, M. G., Haase, G. and Jelen, P. 2008. Lactose: crystallization, hydrolysis and value-added derivatives. *Int. Dairy J.* 18(7), 685–694. https://doi.org/10.1016/j.idairyj.2008.03.003.

Gibson, G. R., Hutkins, R., Sanders, M. E., Prescott, S. L., Reimer, R. A., Salminen, S. J., Scott, K., Stanton, C., Swanson, K. S., Cani, P. D., Verbeke, K. and Reid, G. 2017. Expert consensus document: the International Scientific Association for Probiotics and Prebiotics (ISAPP) consensus statement on the definition and scope of prebiotics. *Nat. Rev. Gastroenterol. Hepatol.* 14(8), 491–502. https://doi.org/10.1038/nrgastro.2017.75.

Gibson, G. R. and Roberfroid, M. B. 1995. Dietary modulation of the human colonic microbiota: introducing the concept of prebiotics. *J. Nutr.* 125(6), 1401–1412. https://doi.org/10.1093/JN/125.6.1401.

González-Morelo, K. J., Vega-Sagardía, M. and Garrido, D. 2020. Molecular insights into O-linked glycan utilization by gut microbes. *Front. Microbiol.* 11, 591568. https://doi.org/10.3389/FMICB.2020.591568.

Guo, Q., Goff, H. D. and Cui, S. W. 2018. Structural characterisation of galacto-oligosaccharides (VITAGOS™) synthesized by transgalactosylation of lactose. *Bioact. Carbohydr. Diet. Fibre* 14, 33–38. https://doi.org/10.1016/J.BCDF.2017.07.007.

Hammon, H. M., Liermann, W., Frieten, D. and Koch, C. 2020. Review: importance of colostrum supply and milk feeding intensity on gastrointestinal and systemic development in calves. *Animal* 14(S1), s133–s143. https://doi.org/10.1017/S1751731119003148.

Hermes, R. G., Molist, F., Francisco Pérez, J., De Segura, A. G., Ywazaki, M., Davin, R., Nofrarías, M., Korhonen, T.K., Virkola, R. and Martín-Orúe, S. M. 2013. Casein glycomacropeptide in the diet may reduce Escherichia coli attachment to the intestinal mucosa and increase the intestinal lactobacilli of early weaned piglets after an enterotoxigenic *E. coli* K88 challenge. *Br. J. Nutr.* 109(6), 1001–1012. https://doi.org/10.1017/S0007114512002978.

Holland, J. W., Deeth, H. C. and Alewood, P. F. 2006. Resolution and characterisation of multiple isoforms of bovine κ-casein by 2-DE following a reversible cysteine-tagging enrichment strategy. *Proteomics* 6(10), 3087–3095. https://doi.org/10.1002/PMIC.200500780.

Holtug, K., Clausen, M. R., Hove, H., Christiansen, J. and Mortensen, P. B. 2009. The colon in carbohydrate malabsorption: short-chain fatty acids, pH, and osmotic diarrhoea. *Scand. J. Gastroenterol.* 27(7), 545-552. https://doi.org/10.3109/00365529209000118.

Hooton, D., Lentle, R., Monro, J., Wickham, M. and Simpson, R. 2015. The secretion and action of brush border enzymes in the mammalian small intestine. In: *Rev. Physiol. Biochem. Pharmacol.* 168, 59–118. https://doi.org/10.1007/112_2015_24.

Hua, S., Nwosu, C. C., Strum, J. S., Seipert, R. R., An, H. J., Zivkovic, A. M., German, J. B. and Lebrilla, C. B. 2011. Site-specific protein glycosylation analysis with glycan isomer differentiation. *Anal. Bioanal. Chem.* 403(403), 1291–1302. https://doi.org/10.1007/S00216-011-5109-X.

Kaur, N., Sharma, P., Jaimni, S., Kehinde, B. A. and Kaur, S. 2019. Recent developments in purification techniques and industrial applications for whey valorization: a review. *Chem. Eng. Commun.* 207(1), 123–138. https://doi.org/10.1080/00986445.2019.1573169.

Kawasaki, Y., Isoda, H., Tanimoto, M., Dosako, S., Idota, T. and Ahiko, K. 1992. Inhibition by lactoferrin and κ-casein glycomacropeptide of binding of cholera toxin to its receptor. *Biosci. Biotechnol. Biochem.* 56(2), 195–198. https://doi.org/10.1271/BBB.56.195.

Kien, C. L., McClead, R. E. and Cordero, L. 1996. In vivo lactose digestion in preterm infants. *Am. J. Clin. Nutr.* 64(5), 700–705. https://doi.org/10.1093/AJCN/64.5.700.

Kies, A. K. 2014. Authorised EU health claims related to the management of lactose intolerance: reduced lactose content, dietary lactase supplements and live yoghurt cultures. In: Sadler, M. J. (Ed.), *Foods, Nutrients and Food Ingredients with Authorised*

*EU Health Claims*. Elsevier Ltd, Amsterdam, pp. 177–211. https://doi.org/10.1533/9780857098481.3.177.

Krumbeck, J. A., Rasmussen, H. E., Hutkins, R. W., Clarke, J., Shawron, K., Keshavarzian, A. and Walter, J. 2018. Probiotic Bifidobacterium strains and galactooligosaccharides improve intestinal barrier function in obese adults but show no synergism when used together as Synbiotics. *Microbiome* 6(1), 121. https://doi.org/10.1186/s40168-018-0494-4.

Kulkarni, A. A., Weiss, A. A. and Iyer, S. S. 2010. Glycan-based high-affinity ligands for toxins and pathogen receptors. *Med. Res. Rev.* 30(2), 327–393. https://doi.org/10.1002/MED.20196.

Lifschitz, C. H., Smith, E. O. and Garza, C. 1983. Delayed complete functional lactase sufficiency in breast-fed infants. *J. Pediatr. Gastroenterol. Nutr.* 2(3), 478–482. https://doi.org/10.1097/00005176-198302030-00014.

Lin, J. 2004. Too much short chain fatty acids cause neonatal necrotizing enterocolitis. *Med. Hypo.* 62(2), 291–293. https://doi.org/10.1016/S0306-9877(03)00333-5.

Macfarlane, G. T., Steed, H. and Macfarlane, S. 2008. Bacterial metabolism and health-related effects of galacto-oligosaccharides and other prebiotics. *J. Appl. Microbiol.* 104(2), 305–344 https://doi.org/10.1111/j.1365-2672.2007.03520.x.

Maldonado-Gómez, M. X., Martínez, I., Bottacini, F., O'Callaghan, A., Ventura, M., van Sinderen, D., Hillmann, B., Vangay, P., Knights, D., Hutkins, R. W. and Walter, J. 2016. Stable engraftment of *Bifidobacterium longum* AH1206 in the human gut depends on individualized features of the resident microbiome. *Cell Host Microbe* 20(4), 515–526.

Malmuthuge, N. and Guan, L. L. 2017. Understanding the gut microbiome of dairy calves: opportunities to improve early-life gut health. *J. Dairy Sci.* 100(7), 5996–6005. https://doi.org/10.3168/JDS.2016-12239.

Martín-Diana, A., Fraga, M. and Fontecha, J. 2002. Isolation and characterisation of caseinmacropeptide from bovine, ovine, and caprine cheese whey. *Eur. Food Res. Technol.* 214(4), 282–286. https://doi.org/10.1007/S00217-001-0452-3.

McClure, S. B., Magill, C., Podrug, E., Moore, A. M. T., Harper, T. K., Culleton, B. J., Kennett, D. J. and Freeman, K. H. 2018. Fatty acid specific $^{\delta 13C}$; values reveal earliest Mediterranean cheese production 7,200 years ago. *PLoS ONE* 13(9), e0202807. https://doi.org/10.1371/journal.pone.0202807.

McGrath, B. A., Fox, P. F., McSweeney, P. L. H. and Kelly, A. L. 2016. Composition and properties of bovine colostrum: a review. *Dairy Sci. Technol.* 96(2), 133–158. https://doi.org/10.1007/s13594-015-0258-x.

Mehra, R. and Kelly, P. 2006. Milk oligosaccharides: structural and technological aspects. *Int. Dairy J.* 16(11), 1334–1340. https://doi.org/10.1016/J.IDAIRYJ.2006.06.008.

Moore, D. J., Robb, T. A. and Davidson, G. P. 1988. Breath hydrogen response to milk containing lactose in colicky and noncolicky infants. *J. Pediatr.* 113(6), 979–984. https://doi.org/10.1016/S0022-3476(88)80567-5.

Muller-Renaud, S., Dupont, D. and Dulieu, P. 2010. Quantification of κ-casein in milk by an optical immunosensor. *Food and Agricultural Immunology* 15(3–4), 265–277. https://doi.org/10.1080/09540100400013401.

Nagy, B. and Fekete, P. Z. 2005. Enterotoxigenic *Escherichia coli* in veterinary medicine. *Int. J. Med. Microbiol.* 295(6–7), 443–454. https://doi.org/10.1016/J.IJMM.2005.07.003.

Nakajima, K., Tamura, N., Kobayashi-Hattori, K., Yoshida, T., Hara-Kudo, Y., Ikedo, M., Sugita-Konishi, Y. and Hattori, M. 2005. Prevention of intestinal infection by glycomacropeptide. *Biosci. Biotechnol. Biochem.* 69(12), 2294–2301. https://doi.org/10.1271/BBB.69.2294.

Nakamura, T., Kawase, H., Kimura, K., Watanabe, Y., Ohtani, M., Arai, I. and Urashima, T. 2003. Concentrations of sialyloligosaccharides in bovine colostrum and milk during the prepartum and early lactation. *J. Dairy Sci.* 86(4), 1315–1320. https://doi.org/10.3168/jds.S0022-0302(03)73715-1.

Nakano, T., Ozimek, L. and Betti, M. 2018. Separation of bovine κ-casein glycomacropeptide from sweet whey protein products with undetectable level of phenylalanine by protein precipitation followed by anion exchange chromatography. *J. Dairy Res.* 85(1), 110–113. https://doi.org/10.1017/S0022029917000838.

O'Riordan, N., Kane, M., Joshi, L. and Hickey, R. M. 2014. Structural and functional characteristics of bovine milk protein glycosylation. *Glycobiology* 24(3), 220–236. https://doi.org/10.1093/GLYCOB/CWT162.

O'Riordan, N., O'Callaghan, J., Buttò, L. F., Kilcoyne, M., Joshi, L. and Hickey, R. M. 2018. Bovine glycomacropeptide promotes the growth of *Bifidobacterium longum* ssp. *infantis* and modulates its gene expression. *J. Dairy Sci.* 101(8), 6730–6741. https://doi.org/10.3168/JDS.2018-14499.

Paganini, D., Uyoga, M. A., Cercamondi, C. I., Moretti, D., Mwasi, E., Schwab, C., Bechtler, S., Mutuku, F. M., Galetti, V., Lacroix, C., Karanja, S. and Zimmermann, M. B. 2017. Consumption of galacto-oligosaccharides increases iron absorption from a micronutrient powder containing ferrous fumarate and sodium iron EDTA: a stable-isotope study in Kenyan infants. *Am. J. Clin. Nutr.* 106(4), 1020–1031. https://doi.org/10.3945/ajcn.116.145060.

Pelletier, X., Laure-Boussuge, S. and Donazzolo, Y. 2001. Hydrogen excretion upon ingestion of dairy products in lactose-intolerant male subjects: importance of the live flora. *Eur. J. Clin. Nutr.* 55(6), 509–512. https://doi.org/10.1038/sj.ejcn.1601169.

Priyanka, P., Gayam, S. and Kupec, J. T. 2018. The role of a low fermentable oligosaccharides, disaccharides, monosaccharides, and polyol diet in nonceliac gluten sensitivity. *Gastroenterol. Res. Pract.* 2018 113.

Qadri, F., Svennerholm, A. M., Faruque, A. S. G. and Sack, R. B. 2005. Enterotoxigenic *Escherichia coli* in developing countries: epidemiology, microbiology, clinical features, treatment, and prevention. *Clin. Microbiol. Rev.* 18(3), 465–483. https://doi.org/10.1128/CMR.18.3.465-483.2005.

Reichardt, N., Vollmer, M., Holtrop, G., Farquharson, F. M., Wefers, D., Bunzel, M., Duncan, S. H., Drew, J. E., Williams, L. M., Milligan, G., Preston, T., Morrison, D., Flint, H. J. and Louis, P. 2018. Specific substrate-driven changes in human faecal microbiota composition contrast with functional redundancy in short-chain fatty acid production. *ISME J.* 12(2), 610–622. https://doi.org/10.1038/ismej.2017.196.

Saito, T., Itoh, T., Adachi, S., Suzuki, T. and Usui, T. 1982. A new hexasaccharide chain isolated from bovine colostrum κ-casein taken at the time of parturition. *Biochim. Biophys. Acta (B.B.A) Gen. Subj.* 719(2), 309–317. https://doi.org/10.1016/0304-4165(82)90104-0.

Samraj, A. N., Pearce, O. M. T., Läubli, H., Crittenden, A. N., Bergfeld, A. K., Banda, K., Gregg, C. J., Bingman, A. E., Secrest, P., Diaz, S. L., Varki, N. M. and Varki, A. 2015. A red meat-derived glycan promotes inflammation and cancer progression. *Proc. Natl. Acad. Sci. U. S. A.* 112(2), 542–547. https://doi.org/10.1073/PNAS.1417508112.

Ségurel, L. and Bon, C. 2017. On the evolution of lactase persistence in humans. *Annu. Rev. Genomics Hum. Genet.* 18, 297–319. https://doi.org/10.1146/annurev-genom-091416-035340.

Sela, D. A., Chapman, J., Adeuya, A., Kim, J. H., Chen, F., Whitehead, T. R., Lapidus, A., Rokhsar, D. S., Lebrilla, C. B., German, J. B., Price, N. P., Richardson, P. M. and Mills, D. A. 2008. The genome sequence of *Bifidobacterium longum* subsp. *infantis* reveals adaptations for milk utilization within the infant microbiome. *Proc. Natl. Acad. Sci. U. S. A.* 105(48), 18964–18969. https://doi.org/10.1073/PNAS.0809584105.

Shoaf-Sweeney, K. D. and Hutkins, R. W. 2008. Chapter 2. Adherence, anti-adherence, and oligosaccharides: preventing pathogens from sticking to the host. In: *Adv. Food Nutr. Res.* Academic Press 55, 101–161. https://doi.org/10.1016/S1043-4526(08)00402-6.

Storhaug, C. L., Fosse, S. K. and Fadnes, L. T. 2017. Country, regional, and global estimates for lactose malabsorption in adults: a systematic review and meta-analysis. *Lancet Gastroenterol. Hepatol.* 2(10), 738–746. https://doi.org/10.1016/S2468-1253(17)30154-1.

Suzuki, M., West, C. and Beutler, E. 2001. Large-scale molecular screening for galactosemia alleles in a pan-ethnic population. *Hum. Genet.* 109(2), 210–215. https://doi.org/10.1007/S004390100552.

Tao, N., DePeters, E. J., Freeman, S., German, J. B., Grimm, R. and Lebrilla, C. B. 2008. Bovine milk glycome. *J. Dairy Sci.* 91(10), 3768–3778. https://doi.org/10.3168/JDS.2008-1305.

ten Bruggencate, S. J., Bovee-Oudenhoven, I. M., Feitsma, A. L., van Hoffen, E. and Schoterman, M. H. 2014. Functional role and mechanisms of sialyllactose and other sialylated milk oligosaccharides. *Nutr. Rev.* 72(6), 377–389. https://doi.org/10.1111/NURE.12106.

Van Leeuwen, S. S., Kuipers, B. J. H., Dijkhuizen, L. and Kamerling, J. P. 2016. Comparative structural characterization of 7 commercial galacto-oligosaccharide (GOS) products. *Carbohydr. Res.* 425, 48–58. https://doi.org/10.1016/j.carres.2016.03.006.

Wang, B. 2009. Sialic acid is an essential nutrient for brain development and cognition. *Annu. Rev. Nutr.* 29, 177–222. https://doi.org/10.1146/ANNUREV.NUTR.28.061807.155515.

Wang, W., Hu, H., Zijlstra, R. T., Zheng, J. and Gänzle, M. G. 2019. Metagenomic reconstructions of gut microbial metabolism in weanling pigs. *Microbiome* 7(1), 48. https://doi.org/10.1186/s40168-019-0662-1.

Wilkin, S., Ventresca Miller, A., Taylor, W. T. T., Miller, B. K., Hagan, R. W., Bleasdale, M., Scott, A., Gankhuyg, S., Ramsøe, A., Uliziibayar, S., Trachsel, C., Nanni, P., Grossmann, J., Orlando, L., Horton, M., Stockhammer, P. W., Myagmar, E., Boivin, N., Warinner, C. and Hendy, J. 2020. Dairy pastoralism sustained eastern Eurasian steppe populations for 5,000 years. *Nat. Ecol. Evol.* 4(3), 346–355. https://doi.org/10.1038/s41559-020-1120-y.

Yan, Y. L., Hu, Y. and Gänzle, M. G. 2018. Prebiotics, FODMAPs and dietary fiber – conflicting concepts in development of functional food products? *Curr. Opin. Food Sci.* 20, 30–37. https://doi.org/10.1016/j.cofs.2018.02.009.

Ze, X., Duncan, S. H., Louis, P. and Flint, H. J. 2012. Ruminococcus bromii is a keystone species for the degradation of resistant starch in the human colon. *ISME J.* 6(8), 1535–1543. https://doi.org/10.1038/ismej.2012.4.

Zhou, W., Jiang, H., Wang, L., Liang, X. and Mao, X. 2021. Biotechnological production of 2′-fucosyllactose: a prevalent fucosylated human milk oligosaccharide. *ACS Synth. Biol.* 10(3), 447–458. https://doi.org/10.1021/acssynbio.0c00645.

Zivkovic, A. M., German, J. B., Lebrilla, C. B. and Mills, D. A. 2011. Human milk glycobiome and its impact on the infant gastrointestinal microbiota. *Proc. Natl. Acad. Sci. U. S. A.* 108(Suppl. 1), 4653–4658. https://doi.org/10.1073/pnas.1000083107.

Zoetendal, E. G., Raes, J., van den Bogert, B., Arumugam, M., Booijink, C. C. G. M., Troost, F. J., Bork, P., Wels, M., de Vos, W. M. and Kleerebezem, M. 2012. The human small intestinal microbiota is driven by rapid uptake and conversion of simple carbohydrates. *ISME J.* 6(7), 1415–1426. https://doi.org/10.1038/ismej.2011.212.

# Chapter 12

## Advances in understanding of indigenous milk enzymes

*Lotte Bach Larsen, Søren Drud-Heydary Nielsen and Nina Aagaard Poulsen, Aarhus University, Denmark; and Alan L. Kelly, University College Cork, Ireland*



### 1 Introduction

The indigenous enzymes in milk are of great importance for the properties of milk, in terms of both its processing qualities and its final product quality, as well as its shelf life and perception by the consumer. Furthermore, knowing their potential biological effects is important for understanding both infant nutrition and their significance for the calf, but in many cases the knowledge is very limited. The indigenous enzymes form a complex and varying system, in terms of origin, level and significance (i.e. their biological role in and significance for dairy processing). In this chapter, we will present briefly the most important indigenous enzymes present in milk and their significance for dairy products. The focus will be on bovine milk enzymes. However, in some instances, enzymes in milk of other origins, like human milk, will also be covered.

The enzymes of significance in milk and dairy products can originate from a variety of sources, including the cow, bacteria that contaminate and grow in the milk (e.g. proteases and lipases produced by psychrotrophic bacteria such as *Pseudomonas fluorescens*), and by deliberate addition of enzymes for product manufacture (such as the addition of rennet in cheesemaking).

http://dx.doi.org/10.19103/AS.2022.0099.12

Different terminologies have been proposed over the years to categorise these three types of enzymes, and in this chapter, we will use the term 'indigenous' for those originating from the mammal producing the milk (e.g. for bovine or human enzymes), 'endogenous' for those arising from bacterial contamination and 'exogenous' for those deliberately added, and we will focus here on the indigenous enzymes.

The indigenous enzymes in milk cover different classes of enzymes (Webb, 1992), including oxidoreductases (class 1), transferases (class 2), hydrolases (class 3) and isomerases (class 5). Enzymes representing the remaining enzyme classes, the lyases (class 4) and the ligases (class 6) have, to our knowledge, not been identified in milk. The hydrolases (e.g. proteases, phosphatases and lipases) are probably of the greatest importance for milk and dairy quality, and are, together with the oxidoreductases - the major classes being covered here. The major emphasis is on proteases, which are typically used in our examples, as these are of current ongoing research and industry relevance, and are also one of the major areas of focus in the authors' research. The hydrolases include proteases/proteinases like plasmin and other indigenous proteases, but also lipase, lysozyme, phosphatases and glycosidases (e.g. N-acetyl-galactosamine, NAGase), thus representing enzymes that are able to cleave and convert many of the major constituents in milk and dairy products, that is, the proteins, lipids and carbohydrates. The oxidoreductases present include lactoperoxidase, sulphydryl oxidase, catalase, polyamine oxidase (in human milk) and some enzymes from the milk fat globule membrane (MFGM) (like xanthine oxidase), whereas an example of a milk transferase is γ-glutamyl transferase. Examples of type, origin and overall significance of the indigenous enzymes present in milk are further covered in Section 2.

## 2 Main types of indigenous enzymes in milk

Bovine milk contains a very diverse and complex complement of enzymes, and almost 100 enzymes have been reported since the first discovery of an enzyme in milk, which was lactoperoxidase in 1881. By the turn of the twentieth century, it was recognized that a wide range of enzyme activities, such as peroxidase, amylase, protease, lipase and arylesterase are present in milk. The historical development of knowledge on the enzymology of milk was reviewed in detail by Fox and Kelly (2006a,b).

The indigenous enzymes in milk originate from four main sources, largely related to the physiology of secretion of milk, as follows:

1 Blood plasma, through leaky junctions between mammary cells;
2 White blood (somatic) cells, which are present at a low level in milk from healthy cows but are found in greatly increased numbers in case

of mastitis, as they are attracted to the mammary gland in response to invading microorganisms;
3 Secretory epithelial cell cytoplasm; and
4 The MFGM, which is partly derived from the apical membrane of epithelial mammary cells.

The sources of some of the principal indigenous enzymes in milk are summarised in Table 1. The presence of many enzymes may be an indirect consequence of milk secretion (instances 3 and 4 above). The exact functions of many of the enzymes in milk are unclear, as in many cases substrates suitable for them are absent from that environment, or the conditions are unfavourable for their action.

The indigenous enzymes in milk have been of interest as topics of study for decades, due to the different ways by which they impact the quality of milk and dairy products, including the following:

1 Negative consequences for product quality - such as lipase activity, which, if not fully inactivated, leads to rancidity in high fat products such as cream and butter, or the protease plasmin contributing to the gelation of UHT milk (Anema, 2019);
2 Positive consequences for dairy product quality, such as the role of plasmin in the early stages of proteolysis during ripening of many cheese varieities;
3 Use as indicators of processing, such as testing for residual activity of alkaline phosphatase activity as an indicator of the efficacy of high-temperature short-time pasteurisation of milk, or testing for lactoperoxidase activity to evaluate more severe heat treatment, such as that applied to cream; and
4 Use in preservation of milk in certain circumstances - such as when the antimicrobial effect of lactoperoxidase is enhanced by addition of hydrogen peroxide to raw milk for short-term preservation, when a cold chain for milk storage cannot be guaranteed.

**Table 1** Origins of principal indigenous milk enzymes

| Enzyme origin | Examples |
|---|---|
| Blood | Plasmin, lipoprotein lipase |
| Lysosomes (mammary epithelial cells or somatic in milk) | Cathepsins D, B, G, elastase, catalase, plasminogen activators, glycosidases* |
| Mammary gland milk fat globule membrane and extracellular vesicles | Xanthine oxidoreductase, alkaline phosphatase, sulphydryl oxidase, lactoperoxidase, γ-glutamyl transferase |

* may also be derived from microorganisms.

The properties of the main families of milk enzymes will be summarised in turn here.

## 2.1 Proteases

The key protease present in milk is plasmin (EC 3.4.21.7), an alkaline protease with a trypsin-like character (cleaves after lysine and arginine residues, except when the next residue is proline) that originates from blood, where it plays a key physiological role in the dissolution of blood clots (fibrinolysis). Due to the key role of plasmin in a major biological function, it is not surprising that it forms part of a complex regulatory system that ensures that this activity can be up- or down-regulated as required in the body. The blood plasmin system, actually called the plasminogen system, includes the active enzyme itself, its inactive zymogen plasminogen (which is typically present in milk 10 times more than its active form), plasminogen activators (PA) that convert plasminogen to plasmin (of which there are two families, tissue- and urokinase-type, i.e. tPA and uPA), and inhibitors of both plasmin itself and PA (Heegaard et al., 1994). The concentration of plasminogen in bovine milk has been estimated to be on average 1.5 mg/L (Benfeldt et al., 1995).

The entire blood plasmin system is found in milk, which it likely enters through paracellular leakage. In milk, plasmin acts on a range of caseins (the whey proteins being relatively resistant) and in particular hydrolyses β-casein at three different sites, giving rise to products called γ-caseins and proteose peptones (Bastian and Brown, 1996). Raw milk always contains these products, showing that plasmin is ubiquitous in milk and has already acted by the time milk is secreted from the udder.

Plasmin is optimally active at pH 7.5 and at 37°C, and is relatively heat-stable, with a certain proportion of activity being retained even after UHT treatment. In addition, PA are heat-stable, but the inhibitors of both plasmin and PA are heat-labile; therefore lower-temperature treatments such as HTST pasteurisation can actually result in increased plasmin activity in milk, due to increased activation of plasminogen and less inhibition of the active enzyme. For a recent review on the plasmin system in milk, see that of France et al. (2021).

Milk was recognised in the 1970s to contain another protease, active at lower pH values, which was subsequently identified as the lysosomal aspartyl protease cathepsin D (for a review, see Hurley et al., 2000a). The enzyme action on caseins is somewhat similar to that of chymosin, and can coagulate milk, although not as easily and specifically as chymosin (McSweeney et al., 1995; Larsen et al., 1996). The presence of lysosomal cathepsin D means that other lysosomal proteases should also be present in milk; indeed, evidence for the presence of cathepsin B, elastase and other non-plasmin proteases has been presented (see review by Larsen et al., 2021). Immunological evidence for the presence of cathepsin B in milk was reported by Magboul et al. (2001).

The lysosomal origins of these enzymes mean that their activity should be increased when somatic cell count (SCC) increases, perhaps such that the relative contribution of plasmin to proteolysis in milk decreases, with non-plasmin proteases dominating the pattern of proteolysis; indeed, this has been seen experimentally (Larsen et al., 2006).

Proteomic and peptidomic studies (Larsen et al., 2010; Hinz et al., 2012) have also suggested – through identification of peptides and consideration of the cleavage sites involved in their production – the presence of amino- and carboxy-peptidase activities, although these have only been limitedly studied or identified (Reimerdes et al., 1979; D'Alessandro et al., 2010).

### 2.2 Lipase

Bovine milk contains lipoprotein lipase (Deeth, 2006), and it has been estimated that milk should be naturally rancid due to the combination of enzyme and suitable substrate present, potentially leading to the production of short-chain fatty acids with strong flavours. However, thankfully, this is not the case, due to the role of the MFGM in keeping enzyme and substrate apart. Before any step is taken that damages the MFGM and risks inducing rancidity by allowing the two to mix, sufficient heat treatment (usually a little more severe than that applied for the pasteurisation of milk) must be applied to ensure enzyme inactivation; this is principally relevant in the production of butter. The properties and significance of bovine lipoprotein lipase were recently reviewed by Deeth (2021).

Human milk also contains a bile salt-simulated lipase (BSSL), which is considered to be very important for the digestion of lipids by babies, who secrete low levels of bile salts and pancreatic lipase.

### 2.3 Lysozyme

This enzyme was originally identified from egg white but has been detected in the milk from several species; there is more lysozyme in human milk than in bovine milk, with only trace activity being found in bovine milk (Benkerroum, 2008). The level in human milk was estimated as 39 mg/100 mL using 117 samples, which is about 3000 times higher than that in bovine milk (Chandan et al., 1964).

Bovine milk lysozyme has a molecular weight of ~18 kDa compared to ~15 kDa for human milk and egg white lysozyme; and its amino acid composition and immunological properties are considerably different from those of the latter two lysozymes (Eitenmiller et al., 1971). According to the complete amino acid sequence study by Jollès and Jollès (1972), human milk lysozyme contains 130 amino acid residues.

## 2.4 Phosphatases

Alkaline phosphatase (EC 3.1.3.1) is perhaps the most widely studied and measured indigenous milk enzyme. However, this is not for reasons of its significance for dairy products (in which it has relatively little impact, perhaps apart from dephosphorylation of peptides during the ripening of some cheese), but rather due to its thermal inactivation kinetics, which happen to approximate those of pathogenic bacteria such as *Mycobacterium tuberculosis* reasonably closely - meaning that it can be used as an indicator of the efficacy of pasteurisation. Therefore, dairy plants operating HTST pasteurization routinely assay milk using rapid fluorometric or colorimetric tests to validate whether it has been heated sufficiently to inactivate alkaline phosphatase, and hence, by proxy, render the milk safe for consumption (for a review, see Farkye, 2021).

Milk also contains an acid phosphatase (EC 3.1.3.2), which is present at lower levels than alkaline phosphatase but might contribute to cheese-ripening reactions such as dephosphorylation (particularly for cheese made from pasteurised milk) to a greater extent, as it is more heat-stable and also has a lower pH optimum (4.0) (Farkye, 2021).

## 2.5 Lactoperoxidase

The enzyme lactoperoxidase (EC 1.11.17) catalyses the reaction of oxidisable substrates with hydrogen peroxide. It can exhibit potent bactericidal activity even in the presence of low levels of hydrogen peroxide (reviewed by Bansal et al., 2021). In some countries, it is permissible to add hydrogen peroxide for this reason, where cold-chain maintenance of milk quality is not reliably possible.

Lactoperoxidase is also of interest as an indicator of heat treatments more severe than HTST pasteurisation. As it is typically inactivated only at temperatures above 76°C, it could be used to determine over-pasteurisation of milk or to validate the efficiency of heating of the products that require such heat treatments (e.g. high-fat products such as cream) (Bansal et al., 2021).

## 2.6 Milk fat globule membrane enzymes

The enzymes associated with the MFGM and with extracellular vesicles have recently been reviewed (Hansen and Rasmussen, 2021). The enzyme present in milk at the highest levels by weight is xanthine oxidoreductase (XOR, EC 1.13.22), which is found as part of the MFGM. Xanthine oxidoreductase can exist in two forms: xanthine oxidase (XO, EC 1.1.3.22) and xanthine dehydrogenase (XDH, EC 1.1.1.204). These were originally thought to be separate enzymes, but it is now known that they can inter-convert in the presence of sulphydryl reagents, while XDH can be converted irreversibly to XO by proteolysis.

XOR can exhibit bactericidal activity, due to production of hydrogen peroxide or peroxynitrite. In cheese, it may function to reduce nitrate (added in some cases to suppress the growth of *Clostridium tyrobutyricum*, which can cause gas blowing) and is known to have a significant physiological role in the secretion of fat globules from mammary cells. XOR activity is very sensitive to processing, and it can be released into skimmed milk following homogenisation, while it is extensively inactivated only by heating to temperatures in the range of 80–90°C. Apart from XOR, a large range of other enzymes have been ascribed to the MFGM, including lactoperoxidase, γ-glutamyl transferase and sulphydryl oxidase. Catalase (EC1.11.1.6) catalyses the cleavage of hydrogen peroxide to generate water and oxygen, and is also present in milk. It was earlier ascribed to the MFGM, but recent proteomic studies suggest that it is not an MFGM-associated enzyme, but a milk enzyme, increasing during mastitis (Hansen and Rasmussen, 2021).

The enzyme GGT (GGT, sometimes called γ-glutamyl transpeptidase, EC 2.3.2.3) catalyses the transfer of γ-glutamyl residues from glutathione or from γ-glutamyl-containing peptides in milk to other amino acids, other peptides or even water, forming glutamate. In milk, GGT may be associated with the MFGM or membrane material in skimmed milk, and may have a role in the biosynthesis of milk proteins; GGT activity also increases with advancing lactation (Calamari et al., 2015). GGT has its optimal activity at pH 8.5–9.0 and at 45°C. It may be of interest as an indicator of heat treatments in the range of 70–80°C and has been shown to influence the production of kokumi-enhancing γ-glutamyl dipeptides in Parmesan cheese, such activity being destroyed by pasteurisation (Hillmann et al., 2016).

Sulphydryl oxidase (SOX, EC 1.8.3) catalyses the oxidation of the SH-groups of cysteine residues to form disulphide bridges, and has been isolated from milk (Hansen and Rasmussen, 2021). SOX may be involved in the non-random formation of disulphide bonds in proteins during biosynthesis or in heated milk; the disulphide bonds in the latter may be reduced to sulphydryl groups in hard cheeses during ripening, which may play a role in cheese flavour.

## 2.7 NAGases and other milk glycosidases

The activities of nine different glycosidases in milk have been reported, including β-N-acetylglucosaminidase (NAGase), α-fucosidase, α-galactosidase, β-galactosidase (lactase), α-glucosidase, β-glucosidase, β-glucuronidase, α-mannosidase and N-acetyl-α-neuraminidase (sialidase). The glycosidases have been recently reviewed (Sunds et al., 2021a). These generally show higher activities in colostrum compared to mature milk (O'Riordan et al., 2014). The activity of NAGase is a well-documented indicator of subclinical and clinical mastitis (Hovinen et al., 2016), but other milk glycosidases also vary significantly with udder inflammation (Sunds et al., 2021b). The origin of milk glycosidases has still not been fully elucidated. Besides lysosomal origin, milk glycosidases may also partly be introduced by microorganisms. As the milk glycosidases hydrolyse glycosidic bonds, increasing activities can affect milk glycans, including the free oligosaccharides and glyco-conjugates such as O- and N-linked glycoproteins. The milk glycosidases are generally present at low levels and their activities are commonly measured by activity assays, for example, by using fluorescent 4-methylumbelliferyl substrates (O'Riordan et al., 2014).

### 2.8 Other enzymes

α-Amylase (also called glycogenase, EC 3.2.1.1) has been identified in human milk and primarily hydrolyses the α-1,4-linked glucose from starch. It has not been shown to have an effect in milk, as activity towards the α-1,4-linked glucose on the free/bound oligosaccharides in milk is lacking (Chan et al., 2021). Other enzymes mentioned as potentially present in milk but which lack definite characterisation include the protein disulphide isomerase (PDI, EC 5.3.4.1) and the neutrophil-associated enzyme myeloperoxidase (EC 1.11.2.2).

## 3 Advances in measurement and detection of indigenous milk enzymes

As mentioned, more than 100 indigenous enzymes or enzymatic activities have so far been detected in bovine milk, but the number is increasing with refined methods, permitting improved limits of detection. The new methods include proteomic and mass spectrometry-based methodologies. However, the more traditional methods for detection of enzyme activities using specific substrates or immunological methods are also versatile tools for measuring enzyme activities in milk. An overview of the various methods used for detection of indigenous milk enzymes can be found in Le et al. (2017).

When detecting or measuring enzymes, it is critical to discriminate between the measurement of enzyme *activities*, the detection of enzyme *presence* on some molecular basis of identification, or the measurement of the *result* of enzymatic activity in a matrix where enzyme and substrate are present together, like in a milk or dairy system. The different approaches give complementary but significantly different types of information on the presence, activity or significance of enzymes in milk.

Measurement of enzyme activity indicates the presence of enzymes able to produce measurable products based, for example, on a substrate releasing coloured products. However, the enzyme activity detected is not definite proof of a specific enzyme being present, as the same activities may derive from different enzymes. Enzymatic methods measuring enzyme activities will always reflect the use of specific substrates and the specific conditions of the assay, permitting the expression of the enzyme in activity units. For example, this can be in raw milk or in relation to heat treatment or other processing. This contrasts with the detection of enzymes by their molecular identification by immunoblotting, also called western blotting, where the molecular forms of the enzyme are separated by gel electrophoresis and subsequently electroblotted onto more stable membranes and detected using specific antibodies (e.g. Larsen et al., 2006). The strength of this method is that it permits the assessment of apparent molecular mass, in contrast to other immunologically based methods such as ELISA and radio-immunoassays. Those methods, on the other hand, present other advantages in terms of being plate-based and permitting handling of many samples at the same time, or even automation.

### *3.1 Detection of indigenous milk enzymes through assessing activities by specific substrates*

As mentioned in Section 2.4, the most versatile indicator of heat treatment and correct pasteurisation of milk is the indigenous alkaline phosphatase (phosphomonoesterase). The enzyme is easily assessed due to its cleavage of monoesters of phosphoric acid from substrates at alkaline pH, called a phosphatase test (Schlimme et al., 1998). Another well-known enzyme in milk is the oxidative lactoperoxidase, which is often assessed based on its ability to oxidise $H_2O_2$ and produce thiocyanate (Turner and Thompson, 2007). Lysozyme, being a widespread enzyme in biological secretions, is normally assessed by its enzymatic activity against peptidoglycans of Gram-positive bacteria (Carlsson and Björck, 1987).

In terms of measurement of plasmin activity, a number of different fluorometric (e.g. that of Richardson and Pearce, 1981) and colorimetric methods (e.g. that of Politis et al., 1992) have been used, which can render comparison of the results of different studies difficult. In addition, a number of different methods exist for the preparation of milk samples for assay. The methods for analysing plasmin activity in milk and dairy products, and their relative advantages and disadvantages, were compared by Chove et al. (2011).

Previous methods for measurement of plasmin and plasminogen-derived activities were widely based on measurements in skimmed milk, and therefore included a centrifugation step to remove the cream or did not specify the fat content of the samples analysed. Furthermore, in the case of plasmin activity determination, a number of assay preparation methods have been developed, including cases where the enzyme is essentially removed from inhibitory factors in the milk before being measured (Politis et al., 1993). Such an approach will give an indication of the 'true' plasmin activity, as affected by factors such as mammary physiology. Arguably, this does not reflect 'effective' activity, because although two samples of milk could have identical amounts of plasmin, but higher levels of inhibitors in one significantly diminish the consequences of the enzyme being present. Finally, a milk sample could have a certain effective level of plasmin activity, which, if sampled after different durations of storage (during which the cumulative breakdown of protein will have increased), could exhibit very different consequences for the quality of dairy products. Thus, from this example, it can be seen that multiple approaches can exist which give useful and complementary, but not identical information, and the reason why an enzyme is of interest should be the major driving factor influencing the selection of an appropriate methodology. Relatively recently, improved spectrophotometric methods for the determination of plasmin and plasminogen activities have been reported in turbid samples of milk or dairy product extracts like cheese (Saint-Dennis et al., 2001; Rauh et al., 2014a).

In addition, modifications to the structure of the substrate as well as to the activity of the enzyme have to be considered when evaluating enzyme activity in dairy products. For example, heat treatment can induce lactosylation of milk proteins that inhibits their proteolysis by plasmin, even if the enzyme is present following the heating (Dalsgaard et al., 2007).

Apart from the plasmin/plasminogen system in bovine milk, an increasing range of other proteolytic enzymes have been identified in milk by enzymatic assays. This includes the cathepsin D/procathepsin D system (Larsen et al., 1993). Both synthetic substrates (Larsen et al., 2006) and protein substrates have been used, with haemoglobin (Larsen et al., 1993) or casein, including azocasein, being useful substrates for cathepsin D. For the protein substrates, digestion is typically set up at an appropriate pH and the enzymatic activity is determined by measuring the polypeptide-containing supernatant at A280 or visible light after precipitation with trichloroacetic acid. Often, the enzymatic activity is then calculated based on subtraction of a blank (eventually including inhibitor, like pepstatin) and the difference is expressed as delta absorbance (units). Again, it is emphasized that such activity measurements will give an indication of activity belonging to a specific class; for example, in the case of cathepsin D, activity measured as described will give an indication of aspartic protease activity, which then requires more molecular methods for unambiguous identification of the actual enzyme present.

### 3.2 Application and advances in mass spectrometry-based methods

Recent advances in mass spectrometry-based methods, databases and bioinformatics have increasingly permitted the increased detection of a larger number of enzymes present in milk and dairy products, based on their molecular identification or even transcriptional profiling. Some examples of this include the detection and identification of the proteolytic enzymes in milk, which are present at low levels (from less than one to a few mg/L). By transcriptomics, the expression of a whole range of bovine proteolytic enzymes (e.g. cathepsins B, C, D, G, H, L, K, S and Z), as well as polymorphonuclear (PMN) elastase, have been identified (Wickramasinghe et al., 2012). Several of these (cathepsins B, D and Z) were furthermore identified in a proteomic study of bovine milk (D'Alessandro et al., 2011). Nissen et al. (2017) identified a number of enzymes and inhibitors, including complement factors and serpins, in a proteomic study of bovine colostrum.

In a study of human milk, a wide range of milk enzymes was summarised and included in an enzyme network analysis (D'Alessandro et al., 2010). The enzymes identified by various studies, including proteomics, included oxido-reductases like xanthine oxidase, protein disulphide isomerase (PDI), myeloperoxidase and lactoperoxidase. The list further included a wide range of hydrolases like lipases (lipoprotein lipase, carboxyl ester lipase and bile salt-stimulated lipase), endo- and exo-proteases (such as thrombin, cathepsins D, H and S, dipeptidyl peptidase 1, metallopeptidase, aminopeptidase N, complement components, endoplasmic reticulum aminopeptidase 1), in addition to lysozyme, γ-glutamyl transferase, α-amylase and ribonuclease.

Consequently, and especially in relation to proteolytic enzymes, the use of MS-based methods has found even more versatility in the assessment and detection of the *result* of enzyme activities in milk and dairy. This includes different types of peptidomic studies on milk or dairy products (Dallas and Nielsen, 2018). Examples include the detection of peptides in relation to mastitis or SCC (Guerrero et al., 2015), lactation stage (Dallas et al., 2014), the pre-digestion of milk proteins by indigenous proteases in human milk (Nielsen et al., 2017a; Gan et al., 2019) or studies of cheese characteristics and ripening (Robinson et al., 2021). Common to these studies is the profiling of peptides resulting from proteolytic enzymes, which, depending on source and set-up, can represent indigenous, endogenous and exogenous proteases. By using knowledge of specificities and cleavage sites (e.g. via MEROPS database www.https://www.ebi.ac.uk/merops/), potentially responsible proteases can be deciphered.

## 4 Advances in understanding the biological significance of indigenous enzymes in milk

There is a growing list of indigenous enzymes identified or measured in milk, but no biological role has been ascribed to them yet. They have, in some instances, been suggested to be some kind of 'spill-over' from the blood, mammary gland or immune cells. In several cases, the substrates of the enzymes are not present in milk, or the conditions for activity are not met, like pH or presence of co-factors. In more recent years, however, the potential biological roles of milk indigenous enzymes have been extended, as a broader perspective has been applied to the whole biological system that milk is a part of, including udder health, milk secretion, tissue remodelling and other roles of importance for the lactating animal. In addition, roles relating to the use of the secreted milk as nutrition for the newborn, where an interplay between enzymes indigenous to the milk and enzymes in the recipient's digestive tract has been suggested (Demers-Mathieu et al., 2018; Nielsen et al., 2018a), are probably more generally accepted.

### *4.1 Proteases*

Plasmin has been suggested to play a part in the remodelling of the udder after lactation ends and the cow enters the dry period, and later in connection with a new pregnancy; Shamay et al. (2003) reported that infusion of casein hydrolysates into the bovine mammary gland can disrupt tight-junction integrity and induce involution. Plasmin is known to be able to degrade, for example, the extracellular matrix in the human system, and extensive tissue remodelling is expected to be required in connection with these biological processes. Initially, analogous to its role in blood, where plasmin has a pivotal role in relation to the dissolution of blood clots, it was suggested that plasmin could play a role in avoiding generation of milk clumps in the mammary ducts (Horie et al., 1987). This was due to the fact that casein easily precipitates or forms coagula when different conditions are met, and therefore may also, in some instances, lead to clots in the alveolar ducts, which are suspected to possibly lead to mastitis later. On the other hand, it was suggested that the presence of plasminogen activator inhibitors on the casein micelle itself also could prevent more extensive casein micelle degradation in the mammary gland (Dallas et al., 2015a).

More recently, focus has been drawn towards a potential role of native milk proteases in pre-digestion of milk proteins (Nielsen et al., 2018a; Demers-Mathieu et al., 2018). Pre-digestion of the milk proteins could be an important biological function to assist the infant's digestion of the complex macrostructure of milk proteins. Such mechanisms may be especially important for preterm infants compared to term infants, as their underdeveloped gastrointestinal system may impede their ability to sufficiently break down milk proteins. Hence, studies have indicated that milk peptide release is greater in milk from preterm-delivering mothers compared to term-delivering mothers (Dallas et al., 2015b; Nielsen et al., 2020). Furthermore, the indigenous milk proteases could play a role in relation to the generation of bioactive peptides which may have functional properties in the mammary glands and/or in newborns (Nielsen et al., 2018b). Recently, a major compilation of the reported bioactivities of peptides from the milk proteins released by different proteases (including indigenous, endogenous and exogenous) has been created (Nielsen et al., 2017b).

As the major milk proteins are present in various genetic variants (Farrell et al., 2004), the genetic and structural changes behind these isoforms (amino acid exchanges, deletions etc.) may lead to changes in the patterns of cleavage by the indigenous milk proteases. As an example of the effect of this, an *in silico* investigation was conducted using plasmin as a blueprint of this potential implication, as shown in Table 2. Here, the known genetic variants of the major bovine milk proteins were selected to represent variation, where changes in the genetic code would result in an implication for cleavage by plasmin. With the specificity of plasmin in mind (it only cleaves after lysine or arginine residues), different genetic variants of the caseins where a lysine or an arginine residue has been introduced or deleted will affect the cleavage pattern and thereby the peptide pattern obtained. In the table, instances where these peptides are involved in potential bioactivities and in their possible bitter taste are indicated (Table 2). This example for plasmin illustrates the complexity of the enzyme-substrate interactions that may occur in raw milk or during processing and storage.

The other suggested roles of, for example, the lysosomal proteases identified in milk, should be evaluated in the light of their presence as active and/or inactive pro-forms, such as in the case of cathepsin/procathepsin D. In many instances, it is actually not known whether the active or inactive precursor, or both forms, are present, which in practice may relate to their origin of secretion/leakage and to the different cell types being part of the somatic cell pool (Larsen et al., 2006). In the case of cathepsin D, given that small amounts of the active enzyme seem to be present in milk, the optimal pH for activity (probably pH 5 or below) is not met by milk (pH 6.7), but it may still exert low activities at native conditions. Furthermore, the conditions required for activation should be considered, which, in the case of cathepsin D, is low pH - at which it autoactivates into a proteolytically active intermediate form.

With regard to cathepsin B, also shown to be present in milk, the situation is different, being a cysteine protease. Firstly, it is not completely clear which forms (active processed forms or pro-forms) are prevalent. Furthermore, cysteine proteases need a reducing milieu in order to maintain activity, which is not really the case with milk. Therefore, even though cathepsin B is able to cleave the

**Table 2** Potential implications of bovine milk protein genetic variations on plasmin-derived peptides from the major bovine milk proteins and their impact on bitter peptide production and formation of bioactive peptides

| Protein | Peptides in reference variant[a] | Predicted bitterness[b] | Potential bioactivity[c] | Genetic variant[d] | Position and change introduced[d] | Change to | Change in peptides induced by genetic variant | Predicted bitterness[b] | Significance |
|---|---|---|---|---|---|---|---|---|---|
| $\alpha_{S1}$-casein | HPIKHQGLPQEVLNENLLR | Non-Bitter | | A | 14-26 (deleted) | → | HPIKHQGLPQPFPEVFGK | Bitter | Potential plasmin cleavage site removed in A variant |
| | FFVAPFPEVFGK | Bitter | ACE-inhibitory | | | | | | |
| $\alpha_{S1}$-casein | DIGSESTEDQAMEDIK | Non-Bitter | | H | 51-58 (deleted) | 8 | DIGSESTEQMEAESISSSE EIVPNSVEQK | Non-Bitter | Potential plasmin cleavage site removed in H variant |
| | QMEAESISSSEEIVPNSVEQK | Non-Bitter | Immuno-modulatory | | | | | | |
| $\alpha_{S1}$-casein | QMEAESISSSEEIVPNSVEQK | Non-Bitter | Immuno-modulatory | E | 59 (Q → K) | → | K | - | Potential plasmin cleavage site added in E variant<br>A lysine is present before the peptide N-terminal, which could impact cleavage |
| | | | | | | | MEAESISSSEEIVPNSVEQK | Bitter | |
| β-casein | RELEELNVPGEIVESLSSS EESITR | Bitter | Immuno-modulatory | D | 18 (S → K) | → | RELEELNVPGEIVESLSK | Bitter | Potential plasmin cleavage site added in D variant |
| | | | | | | | SEESITR | Non-Bitter | |

| | | | | | | | | |
|---|---|---|---|---|---|---|---|---|
| β-casein | RELEELNVPGEIVESLSSS EESITR<br>INK | Bitter<br>Bitter | H1 | 25 (R → C) | → | RELEELNVPGEIVESLSSS EESITCINK | Bitter | Potential plasmin cleavage site removed in H1 variant |
| β-casein | FQSEEQQQTEDELQDK | Non-Bitter | E | 36 (E → K) | → | FQSK<br>EQQQTEDELQDK | Non-Bitter<br>Non-Bitter | Potential plasmin cleavage site added in E variant |
| β-casein | FQSEEQQQTEDELQDK | Non-Bitter | C | 37 (E → K) | → | FQSEK<br>QQQTEDELQDK | Non-Bitter<br>Non-Bitter | Potential plasmin cleavage site added in C variant |
| β-casein | YPVEPFTESQSLTLTDVENL HLPLPLLQSWMHQPHQPLPP TVMFPPQSVLSLSQ-SK | Non-Bitter | B | 122 (S → R) | → | YPVEPFTER<br>QSLTLTDVENLHLPLPLLQS WMHQPHQPLPP-TVM FPPQSVLSLSQSK | Bitter<br>Non-Bitter | Potential plasmin cleavage site added in B variant |
| κ-casein | QEQNQEQPIR<br>CEK | Bitter<br>Non-Bitter | F2 | 10 (R → H) | → | QEQNQEQPIHCEK | Non-Bitter | Potential plasmin cleavage site removed in F2 variant |
| κ-casein | SCQAQPTTMAR<br>HPHPHLSFMAIPPK | Non-Bitter<br>Non-Bitter | C | 97 (R → H) | → | SCQAQPTTMAHHPHP HLSFMAIPPK | Non-Bitter | Potential plasmin cleavage site removed in C variant |

(*Continued*)

**Table 2** *(Continued)*

| Protein | Peptides in reference variant[a] | Predicted bitterness[b] | Potential bioactivity[c] | Genetic variant[d] | Position and change introduced[d] | Change to | Change in peptides induced by genetic variant | Predicted bitterness[b] | Significance |
|---|---|---|---|---|---|---|---|---|---|
| κ-casein | SCQAQPTTMAR<br>HPHPHLSFMAIPPK | Non-Bitter<br>Non-Bitter | | G1 | 97 (R → C) | → | SCQAQPTTMACHPHP HLSFMAIPPK | Non-Bitter | Potential plasmin cleavage site removed in G1 variant |
| B-lactoglobulin | K<br>IIAEK | -<br>Bitter | | H | 70 (K → N) | → | NIIAEK | Bitter | Potential plasmin cleavage site removed in H variant A lysine is present before the peptide N-terminal, which could impact cleavage |
| α-lactalbumin | CEVFR<br>ELK | Non-Bitter<br>Non-Bitter | | A | 10 (R → Q) | → | CEVFQELK | Non-Bitter | Potential plasmin cleavage site removed in A variant |

[a] Reference variants are: $\alpha_{s1}$-casein variant B, β-casein variant A2, κ-casein variant A, β-lactoglobulin variant B and α-lactalbumin variant B.
[b] Bitternes predicted using iBitter-SCM (http://camt.pythonanywhere.com/iBitter-SCM) (Charoenkwan et al., 2020). Peptides with a core greater than 331 were assigned as bitter, or otherwise as non-bitter.
[c] Identified by searching the milk bioactive peptide database (https://mbpdb.nws.oregonstate.edu/).
[d] According to Farrell et al., 2004.

caseins (Magboul et al., 2001) in the presence of added reducing agents (as in model studies), it is not clear if, or to what extent, this happens in milk.

### 4.2 Lipases

The main role of lipases in milk is in aiding digestion, particularly for preterm infants whose digestive system is immature. In general, lipases are probably less intensively studied than proteases in milk. However, in recent years, much of the focus on lipases has been around measuring their activity in milk and the factors affecting this. For example, the effects of breed (Bar et al., 2020) and seasonal variation (Glantz et al., 2020) on lipase activity in milk have been recently reported.

A continuing topic of interest is the fact that some milk from some cows appears more susceptible to 'spontaneous' lipolysis (and subsequent development of off-flavours) than that of others, being more likely to show significant increases in lipolysis and free fatty acid production during storage. The likelihood of spontaneous lipolysis occurring seems to depend on the balance between factors that activate or inhibit the reaction in milk (Deeth, 2006).

Factors that affect the incidence of spontaneous lipolysis seem to include the use of robotic milking systems; diet, with pasture feeding resulting in lower levels of lipolysis in milk (Vanbergue et al., 2017, 2018, 2020), as well as season (Tousova et al., 2013). Vanbergue et al. (2016) reported an association between polymorphism of the DGAT1 K232A gene (which encodes for diacylglycerol acyltransferase 1, an enzyme involved in triglyceride synthesis) and the incidence of spontaneous lipolysis, but with interactive effects of factors such as milking frequency. Cold storage of milk also affects the association of lipoprotein lipase (LPL) with the MFGM, leading to concentration of LPL in the cream phase of raw milk, while storage at warmer temperatures can reduce the activities of LPL in both cream and skimmed-milk (Dickow et al., 2011).

The impact of lipolysis due to LPL activity remains a key parameter to control as part of dairy processing. While some lipolysis is certainly beneficial in some cheese varieties, inadequate thermal inactivation of LPL before butter-making can result in rancidity in the product. Low-pressure homogenization to break the MFMG and allow increased access of LPL has been shown to increase levels of free fatty acids in Cheddar cheese, both initially and during ripening (Deegan and McSweeney, 2013).

### 4.3 Antimicrobial enzymes

A particular function of indigenous enzymes which has received significant attention in recent years has been in activities which represent antimicrobial defences in milk (Bansal et al., 2021). It is known that bacterial growth in pasteurised milk is significantly accelerated compared to that in raw milk, which is due to the inactivation of such enzymes and other defensive mechanisms in milk. In addition, in some countries, the antimicrobial enzyme systems can be actively enhanced with a view to short-term preservation of raw milk, especially where cold chain cannot be maintained; such an approach involves the addition of hydrogen peroxide as a substrate for lactoperoxidase activity.

Indigenous enzymes can show antimicrobial functions in two broad ways. Firstly, hydrolases may target components of the cell walls, that is, the carbohydrate/peptidoglycan building blocks of microorganisms, by degradation, which inactivates the cells. For example, lysozyme may target cell wall peptidoglycans of Gram-positive bacteria. Secondly, oxidoreductases exert their effect through the *in situ* generation of reactive molecules that can destroy vital proteins in bacterial cells; examples of enzymes that work in this way are xanthine oxidase (XO) and lactoperoxidase.

### 4.4 Glycosidases

Variation in the gene expression of bovine milk glycosidases has been documented to vary across lactation, which may play a role in milk oligosaccharide metabolism (Wickramasinghe et al., 2011). The importance of milk glycosidases on free or bound glycan substrates in milk (Sunds et al., 2021a) is, however, scarcely explored. Milk glycosidases may be of importance for the neonate by affecting biological activities of milk carbohydrates, but may also affect glycans in milk during processing by hydrolysing these, depending on their degree of abundance, activity and stability towards heat treatment.

## 5 Advances in understanding the significance of indigenous milk enzymes on dairy quality

### 5.1 Impact of processing on milk enzymes

One area of intense study for many years in terms of the significance of indigenous milk enzymes is their role in the quality of dairy products. In terms of the quality of raw or fluid milk, the main focus has been on lipolysis or potential antimicrobial (preservative) functions, while for long-shelf-life products (like UHT milk), the focus has been on the potential role of plasmin in influencing product stability and phenomena such as age-gelation (Rauh et al., 2014b; Anema, 2019; Zhang et al., 2020). There are different mechanisms by which plasmin and bacterial proteases - both of which are sufficiently heat-stable to retain at least some residual activity after UHT treatment - have been elucidated in recent years; plasmin, through the hydrolysis of β-casein is more likely to lead to a diffuse particulate gel (Rauh et al., 2014c), whereas the chymosin-like activity of bacterial proteases on κ-casein is more likely to lead to a system-spanning rennet-like gel filling some or all of the container (see Datta and Kelly, 2021, for a recent review).

The role of such enzymes in cheese ripening has also been extensively studied, with enzymes playing a range of roles depending on variety and processes applied.

While enzymes such as alkaline phosphatase have been used for many years as indicators of the efficacy of thermal processes such as HTST pasteurisation applied to milk (Staniuck et al., 2011), the growing interest in alternative and frequently non-thermal treatments applied to milk has led to interest in alternative enzymatic indices of the extent of processing. For example, the application of high-pressure treatment to raw milk is of interest in a number of countries with a view to producing a more fresh-like product, and in such cases, the very different sensitivities of enzymes to heat and pressure mean that alternative indices of whether sufficient process severity has been achieved to reflect the target levels of safety or stability are required.

In addition, novel ways of thermal processing of milk are being evaluated, and their impact on the indigenous enzymes requires examination. Hammershoj et al. (2010) reported the effect of instant infusion pasteurization (at 72–110°C) but for very short holding times (from 100 ms to 700 ms) on a range of milk enzymes (alkaline phosphatase, lactoperoxidase, xanthine oxidase and γ-glutamyl transpeptidase) and found different inactivation kinetics, which might represent new ways of determining indicators for novel thermal processes for application to milk preservation.

Ahmad et al. (2019) comprehensively reviewed the sensitivity of milk enzymes to a range of technologies, including ultrasound, pulsed electric fields, high-pressure processing and ultraviolet irradiation. Compared to other enzymes, alkaline phosphatase was actually very resistant to many non-thermal processes, emphasising the importance of the careful selection of suitable indicator enzymes for these emerging processes. Munir et al. (2019) also reviewed the inactivation of enzymes by high-pressure treatment, microwave heating and ultrasound processing, and reported that some processes, under low-intensity treatment, can actually enhance enzyme activity (in terms of enhancing the activity of proteases against unfolded substrates).

A number of other innovative processes have also been examined in recent studies in terms of their impact on indigenous enzyme activities. For example, Liao et al. (2019) studied the impact of high-pressure carbon dioxide on the activity of alkaline phosphatase, which showed complete inactivation at 50°C when combined with high-pressure carbon dioxide. Martinez-Monteagudo et al. (2014) showed that combined high pressure and temperature (up to 600 MPa at 60–90°C) resulted in inactivation of the enzyme, with pressure increasing the relative rates of inactivation at any temperature.

High-pressure homogenization (HPH) of raw milk has been shown to result in inactivation of alkaline phosphatase (Picart et al., 2006), while the relative levels of inactivation of several enzymes (alkaline phosphatase, lactoperoxidase, plasmin) by HPH have been shown to be greater than the level of inactivation expected due to the heating component of that process (Datta et al., 2005).

## 5.2 Impact of plasmin and other proteases on dairy product quality

To date, much of the significance of plasmin in dairy products has been focused on products such as cheese and UHT milk, where it arguably exerts positive and negative effects, respectively: its proteolytic action in cheese leads to early-stage protein breakdown during ripening, but in UHT milk it may destabilize casein micelles in such a way that gelation ensues.

However, in recent years, there has been increased interest in the activity of plasmin in other dairy product systems. For example, many dairy processors have developed an increasing range of fractionated or purified protein products from milk, such as micellar casein concentrates (MCC) and milk protein concentrates (MPC). A key technology for the manufacture of such products is membrane filtration, which may separate enzymes and their substrate (i.e. plasmin associated with casein micelles) from serum-phase inhibitors which are lost in permeates, leading to up-regulated plasmin activity in the concentrated protein products and potentially leading to issues with off-flavours and physicochemical instability in the final products which include these fractions.

Gazi et al. (2014) compared the plasmin activities in a range of milk protein ingredients and found that milk protein concentrates and isolates had higher activity than sodium or calcium caseinates, and incubation of reconstituted protein samples resulted in extensive proteolysis, as well as plasminogen activation and the autolysis of plasmin. Li et al. (2020) showed that Cheddar cheese made from micellar casein concentrate showed higher plasmin activity and faster proteolysis during storage than that made from milk, presumably due to the up-regulation of that activity during production of the protein ingredient. Residual protease activity has also been recently linked to aggregation of casein micelles and thickening in concentrated milk protein suspensions during storage (Nair and Corredig, 2021).

Lysosomal proteinases have been less extensively studied, even though they actively hydrolyse milk caseins and consequently can affect the milk processing properties and sensorial quality of dairy products. Several of the lysosomal proteases have acidic pH optima, and therefore may not be active in milk with a pH of 6.7, but may in turn have proportionately higher activity in more acidic dairy products like cheese and yoghurt. Cathepsin D is quite heat-sensitive and is even thought to be largely inactivated by processes such as the cooking of Swiss cheese (Hayes et al., 2001); procathepsin D has also been detected in heated milk and rennet-free UF-feta cheese (Larsen et al., 2000), and the activity of this acid protease has also been seen in quarg cheese (Hurley et al., 2000b). The specificity of this enzyme against the caseins was described in a model system by Considine et al. (2004), but whether the enzyme is active in milk is still unclear due to the requirement for a reducing milieu.

### 5.3 Relationship between plasmin and bacterial protease activities in milk

In recent years, there has been extensive study of the diversity of proteolytic and lipolytic enzymes produced by psychrotrophic bacterial species growing in raw milk, with a particular focus on their consequences for the physical stability of UHT milk, related to their extreme heat-stability – which means that residual activity can persist even after treatment at 135–145°C. In the case of bacterial proteases, this can lead to a slow but eventually irreversible gelation of stored UHT milk through the hydrolysis of κ-casein.

The relationship between such enzymes and the plasmin system has been of interest in two contexts, relating to their relative thermal stability and the potential interactions between the enzyme types in raw milk. Zhang et al. (2018) compared the relative types of destabilisation of UHT milk by plasmin and the protease AprX from *Pseudomonas fluorescens*. In terms of comparative thermal inactivation kinetics, plasmin is somewhat more susceptible to heat, in particular at lower temperatures (80–90°C) where so-called 'low-temperature inactivation' can take place. In terms of controlling protease activity in UHT milk, it seems that in the case of plasmin, the most suitable point of control is in the pre-heating stage around the optimal temperature, often applied in such processes for the inactivation mentioned above, whereas in terms of bacterial proteases, the difficulty in inactivating these means that prevention is better than cure, putting the focus on good hygiene of raw milk, very low temperature storage and short holding times which do not allow the bacteria to grow to numbers where enzyme production is significant. The relative heat stabilities of plasmin and *Pseudomonas* proteases, and the strategies for their control on this basis by thermal or other approaches were compared by Stoeckel et al. (2016).

In terms of the interaction between the enzyme systems in raw milk, it has been shown that psychrotrophic bacterial proteases can result in the release of plasmin from the casein micelle, and thereby influence the partitioning of the enzyme between streams such as cheese and whey (Frohbieter et al., 2005); some bacterial enzymes, such as *Pseudomonas fluorescens* M3/6 protease (Frohbieter et al., 2005) and *Bacillus polymyxa* protease (Larson et al., 2006) have been shown to enhance plasmin activity by stimulating plasminogen activation, thereby enabling a potential synergistic effect between the enzyme systems.

## 6 Future trends in research

Kelly and Fox (2006) and Kelly et al. (2021) previously considered future challenges and opportunities relating to indigenous enzymes in milk. In the former article, the key points for future consideration included improving our understanding of the enzyme complement of milk, the factors affecting the variation in enzyme activity, the impact of processing, and the practical significance of enzymes for dairy product quality, food safety or consumer health. Topics revisited in the current chapter were also raised, including inconsistencies between studies in the methodologies used. It was concluded that better co-operation between international research groups would be beneficial in terms of better standardisation of research methods, for example. Interestingly, it was also suggested that the study of milk enzymes could benefit from better collaboration between research teams studying the same enzymes but in different biological milieus, such as blood or tissue.

In the later (2021) review, the authors revisited the challenges raised in the article of Kelly and Fox (2006) and systematically evaluated progress against the points raised then, which were based on discussions at an International Dairy Federation conference specifically dedicated to the topic, and included the views expressed by an industry panel. It was broadly concluded that much progress had been made, but the issues of standardisation of methods across studies remains acute, while the literature on indigenous milk enzymes arguably grows at a slower rate than that relating to the enzymes of dairy-related bacteria - including the psychrotrophic bacteria and the lactic acid bacteria used in production of fermented dairy products.

In addition, changing animal husbandry practices (e.g. genetically-informed breeding strategies, new approaches to feeding) and changing external environmental factors (e.g. driven by climate change) may change bovine physiology with unquestionable implications for the enzymology of milk. In addition, breeding or selection strategies for specific traits (e.g. the recent interest in milk secreted by cows selected for having the ancestral A2 gene of β-casein) may change the relationship between milk enzymes and their natural substrates, as illustrated in this chapter for the example of bioactive peptides.

As also discussed earlier, developments in the processing strategies for milk and the potential implementation of novel processing technologies require a continuing consideration of the implications on milk enzymes, whether that involves new non-thermal processing techniques that might activate or inactivate enzymes or new fractionation techniques that separate or combine enzymes, substrates, inhibitors and activators, reshuffling the complex

**Table 3** A brief summary of the main enzymes in milk and other products, their significance and some current research challenges and knowledge gaps

| Enzyme | Type of reactions catalysed | Principal significance | Research challenges and knowledge gaps |
|---|---|---|---|
| Acid phosphatase | Phosphomonoesterase<br>Dephosphorylation of phosphoproteins | Cheese ripening | Impact on PTMs in milk proteins at low pH. |
| Alkaline phosphatase | Phosphomonoesterase<br>Dephosphorylation of phosphoproteins | Indicator of heat treatment | Activity on PTMs on proteins, especially in relation to eventual reactivation. |
| α-Amylase | Hydrolyses α-1,4 linked glucose in starch but apparently not in milk oligosaccharides | Starch hydrolysis | Role on milk oligosaccharides. Role in new blends with plant material. |
| Catalase | Catalyses breakdown of $H_2O_2$ to $H_2O$ and $O_2$ | Pro-oxidant, index of mastitis | Role in preventing accumulation of oxidative products in milk and udder. Use of peroxidase to convert nitrite to nitrate. |
| Cathepsin D | Aspartic protease activity Specificity in line with many other aspartic proteases, cleaving peptide bonds especially involving hydrophobic residues at the N- or C-terminal of the scissile bond. Broad specificity | Proteolysis in milk, cheese ripening | Effects in whey-based products. Effect for bioactive peptide formation and other biological significances. Activation aspects. |
| γ-Glutamyl transferase | Catalyses the transfer of γ-glutamyl functional groups from, for example, glutathione to a free amino acid or a residue in a protein or peptide or to water (forming glutamate) | Transfer of γ-glutamyl residues | Potential role in formation of glutamate (umami taste). |
| Glycosidases | Hydrolysis of glycosidic bonds in disaccharides, oligosaccharides and glycoconjugates (with proteins or fatty acids) | NAGase as an indicative trait of mastitis | Contribution from microbial vs lysosomal origin in milk. Significance of activities on free vs bound oligosaccharides. Heat stabilities and potential effects in dairy ingredients. |
| Lactoperoxidase | Formation of hypothiocyanite and water from hydrogen peroxide and thiocyanate | Antimicrobial activity | Contribution to non-immune biological defence. |

*(Continued)*

**Table 3** (*Continued*)

| Enzyme | Type of reactions catalysed | Principal significance | Research challenges and knowledge gaps |
|---|---|---|---|
| Lipase | Hydrolysis of fatty acids from triglycerides | Release of free fatty acids. Flavour and off-flavour development | Biological significances in relation to pre-digestion. Partitioning and activity in relation to association with dairy fractions. |
| Lysozyme | Hydrolyses β-1,4 bonds between the C-1 of N-acetylmuramic acid and the C-4 of N-acetylglucosamine of peptidoglycan polymers in Gram-positive bacterial cell walls | Antimicrobial activity<br>Also called N-acetylmuramidase | Clarification of actual significance in milk and dairy. |
| Plasmin | Serine protease activity<br>Cleavage of peptide bonds involving lysine or arginine residues at P1 | Proteolysis in milk, impact on product yields, rennet coagulation, cheese ripening, UHT milk stability | Principal form for active plasmin in bovine milk in comparison with human milk.<br>Role in mammary tissue remodelling and in (anti-microbial) bioactive peptide formation. Control of partitioning and binding to casein. |
| Sulphydryl oxidase | Catalyses the oxidation of protein thiols (SH-groups of cysteine residues) to form disulphide bonds with the simultaneous reduction of $O_2$ to $H_2O_2$ | Reduction of cooked flavour | Activity in milk in relation to formation of disulphide bonds. |
| Xanthine oxidoreductase | Oxidase form: oxidation of xanthine, formation of $NAD^+$, oxidation of $O_2$ to $O_2^-$ and $H_2O_2$<br>Reductase form: reduction of xanthine, formation of NADH, reduction of $H_2O_2$ and $O_2^-$ to $O_2$ | Antimicrobial activity | Contribution to antimicrobial effects or potentially to off-flavours in dairy products. |

P1: Indicates the residue a proteolytic enzyme cleaves after, that is, the residue on the left side of the scissile bond.

deck of cards that is the multi-component milk enzyme system. There is also the possibility of a growing interest in enzymes driven by the parallel influences of a greater understanding of the enzymology of human milk and how that differs from that of bovine milk, and the drive to develop infant formulae that more closely mimic the biological impact of milk, which may in the future include enzyme activities.

The rise of interest in non-bovine milk, including cellular agriculture, precision fermentation and in blended plant-dairy products, is also of interest, given differences in enzymology between species and the potential plant-milk enzyme-based interactions (e.g. the amylase in milk finally meeting a suitable substrate, or plant-derived enzyme inhibitors changing the enzyme profile of milk). A number of companies are also investigating the production of such animal-free milk (driven by the rise in vegan diets), whereby milk constituents are produced in cell culture, leading to products which may have enzyme profiles that are very different from the mammalian milk archetype.

In conclusion, research on indigenous milk enzymes has been a rich field of direct relevance to dairy science and dairy processing for many years. Yet, many questions remain, which will challenge researchers in the future. Some of the key points covered in this chapter, and some of these future challenges, are summarised in Table 3.

## 7 Where to look for further information

The following book and collection of articles provide a good overview of the subject:

- The collection of articles in a Special Issue of *International Dairy Journal* (Volume 16, Issue 6, 2006) arising from the First IDF Symposium on Indigenous Enzymes in Milk, Cork, Ireland, 20-22 April 2005.
- Kelly, A. L. and Larsen, L. B. (eds) (2021). Agents of change: enzymes in milk and dairy products. *Food Engineering Series*, ed G. V. Barboso-Cánuvas. Springer. Jan 2021.

There are no dedicated conferences relating to the topic of indigenous milk enzymes (the one mentioned above, from Ireland in 2005, was the first and the only major conference dedicated to the topic), but all major dairy conferences (e.g. International Dairy Federation or the American Dairy Science Association) typically feature a number of talks on the subject.

The major international centres of active research on the topic include those of the authors (Aarhus University, Denmark and University College Cork, Ireland) as well essentially all active dairy research organisations throughout the world due to the cross-cutting nature of research into this topic.

Useful websites include:

- Expasy enzyme database (https://enzyme.expasy.org/);
- MEROPS peptidase database (https://www.ebi.ac.uk/merops/); and
- Milk bioactive peptide database (https://mbpdb.nws.oregonstate.edu/).

## 8 References

Ahmad, T., Butt, M. Z., Aaadil, R. M., Inam-ur-Raheem, M., Abdullah, Bekhit, A. E., Guimraes, J. T., Balthazaar, C. F., Rochar, R. S., Esmerino, E. A., Freitas, M. Q., Silva, M. C., Sameen, A. and Cruz, A. G. (2019). Impact of nonthermal processing on different milk enzymes, *Int. J. Dairy Technol.* 72(4), 481–495.

Anema, S. G. (2019). Age gelation, sedimentation, and creaming in UHT milk: a review, *Compr. Rev. Food Sci. Food Saf.* 18(1), 140–166.

Bansal, N., Zhang, J. and Zou, Z. (2021). Antimicrobial enzymes in milk, and their role in human milk. In: *Agents of Change: Enzymes in Milk and Dairy Products* Kelly, A. L. and Larsen, L. B. (Eds), pp 101–126, New York; Springer Nature.

Bar, C., Sutter, M., Kopp, C., Neuhaus, P., Oortmann, R., Egger, L., Reidy, B. and Bisig, W. (2020). Impact of herbage proportion, animal breed, lactation stage and season on the fatty acid and protein composition of milk, *Int. Dairy J.* 109, 104785.

Bastian, E. D. and Brown, R. J. (1996). Plasmin in milk and dairy products: an update, *Int. Dairy J.* 6(5), 435–457.

Benfeldt, C., Larsen, L. B., Rasmussen, J. T., Andreasen, P. A. and Petersen, T. E. (1995). Isolation and characterization of plasminogen and plasmin from bovine milk, Int. Dairy J. 5(6), 577–592.

Benkerroum, N. (2008). Antimicrobial activity of lysozyme with special rerefence to milk, Afr. J. Biotechnol. 7(22), 4856–4867.

Calamari, L., Gobbi, L., Russo, F. and Cappelli, F. P. (2015). Pattern of gamma-glutamyl transferase activity in cow milk throughout lactation and relationships with metabolic conditions and milk composition, *J. Anim. Sci.* 93(8), 3891–3900.

Carlsson, A. and Björck, L. (1987). The effect of indigenous antibacterial factors in milk on the growth of *Bacillus strearothermophilis* var. calidolactis, *Milchwissenschaft* 42, 282–285.

Chan, L. E., Beverly, R. L. and Dallas, D. C. (2021). The enzymology of human milk. In: *Agents of Change: Enzymes in Milk and Dairy Products* Kelly, A. L. and Larsen, L. B. (Eds), pp 209–243, New York; Springer Nature.

Chandan, R. C., Shahani, K. M. and Holly, R. G. (1964). Lysozyme content of human milk, Nature 204(495), 76.

Charoenkwan, P., Yana, J., Schaduangrat, N., Nantasenamat, C., Hasan, M. M. and Shoombuatong, W. (2020). iBitter-SCM: identification and characterization of bitter peptides using a scoring card method with propensity scores of dipeptides, *Genomics* 112(4), 2813–2822.

Chove, L. M., Grandison, A. S. and Lewis, M. J. (2011). Comparison of methods for analysis of proteolysis by plasmin in milk, *J. Dairy Res.* 78(2), 184–190.

Considine, T., Healy, Á, Kelly, A. L. and McSweeney, P. L. H. (2004). Hydrolysis of bovine caseins by cathepsin B, a cysteine proteinase indigenous to milk, *Int. Dairy J.* 14(2), 117–124.

D'Alessandro, A., Scaloni, A. and Zolla, L. (2010). Human milk proteins: an interactomics and updated functional overview, *J. Proteome Res.* 9(7), 3339–3373.

D'Alessandro, A., Zolla, L. and Scaloni, A. (2011). The bovine milk proteome: cherishing, nourishing and fostering molecular complexity: an interactomics and functional overview, *Mol. Biosyst.* 7(3), 579–597.

Dallas, D. C., Guerrero, A., Parker, E. A., Garay, L. A., Bhandari, A., Lebrilla, C. B., Barile, D. and German, J. B. (2014). Peptidomic profile of milk of Holstein cows at peak lactation, *J. Agric. Food Chem.* 62(1), 58–65.

Dallas, D. C., Murray, N. M. and Gan, J. (2015a). Proteolytic systems in milk: perspectives on the evolutionary function within the mammary gland and the infant, *J. Mammary Gland Biol. Neoplasia* 20(3–4), 133–147.

Dallas, D. C. and Nielsen, S. D. (2018). Milk peptidomics to identify functional peptides and for quality control of dairy products, *Meth. Mol. Biol.* 1719, 223–240.

Dallas, D. C., Smink, C. J., Robinson, R. C., Tian, T., Guerrero, A., Parker, E. A., Smilowitz, J. T., Hettinga, K. A., Underwood, M. A., Lebrilla, C. B., German, J. B. and Barile, D. (2015b). Endogenous human milk peptide release is greater after preterm birth than term birth, *J. Nutr.* 145(3), 425–433.

Dalsgaard, T. K., Nielsen, J. H. and Larsen, L. B. (2007). Proteolysis of milk proteins lactosylated in model systems, *Mol. Nutr. Food Res.* 51(4), 404–414.

Datta, N., Hayes, M. G., Deeth, H. C. and Kelly, A. L. (2005). Significance of frictional heating for effects of high-pressure homogenization on milk, *J. Dairy Res.* 72(4), 393 –399.

Datta, N. and Kelly, A. L. (2021). The role of proteases in the stability of UHT-treated milk. In: *Agents of Change: Enzymes in Milk and Dairy Products* Kelly, A. L. and Larsen, L. B. (Eds), pp 311–348, New York; Springer Nature.

Deegan, K. C. and McSweeney, P. L. H. (2013). Effects of low-pressure homogenisation pre-treatment of cheesemaking on the ripening of Cheddar cheese, *Dairy Sci. Technol.* 93(6), 641–655.

Deeth, H. C. (2006). Lipoprotein lipase and lipolysis in milk, *Int. Dairy J.* 16(6), 555–562.

Deeth, H. C. (2021). Lipases from milk and other sources. In: *Agents of Change: Enzymes in Milk and Dairy Products* Kelly, A. L. and Larsen, L. B. (Eds), pp 245–286, New York; Springer Nature.

Demers-Mathieu, V., Nielsen, S. D., Underwood, M. A., Borghese, R. and Dallas, D. C. (2018). Changes in proteases, antiproteases, and bioactive proteins From mother's breast milk to the premature infant stomach, *J. Pediatr. Gastroenterol. Nutr.* 66(2), 318–324.

Dickow, J. A., Larsen, L. B., Hammershoj, M. and Wiking, L. (2011). Cooling causes changes in the distribution of lipoprotein lipase and milk fat globule membrane proteins between the skim milk and cream phase, *J. Dairy Sci.* 94(2), 646–656.

Eitenmiller, R. R., Barnhart, H. M. and Shahani, K. M. (1971). Effects of gamma-irradiation on bovine and human milk lysozymes, J. Food Sci. 36(7), 1127.

Farkye, N. F. (2021). Phosphatases in milk. In: *Agents of Change: Enzymes in Milk and Dairy Products* Kelly, A. L. and Larsen, L. B. (Eds), pp 85–100, New York; Springer Nature.

Farrell, H. M., Jr., Jimenez-Flores, R., Bleck, G. T., Brown, E. M., Butler, J. E., Creamer, L. K., Hicks, C. L., Hollar, C. M., Ng-Kwai-Hang, K. F. and Swaisgood, H. E. (2004). 'Nomenclature of the proteins of cows' milk--sixth revision, *J. Dairy Sci.* 87(6), 1641–1674.

Fox, P. F. and Kelly, A. L. (2006a). Indigenous enzymes in milk: overview and historical aspects - part 1, *Int. Dairy J.* 16(6), 500–516.

Fox, P. F. and Kelly, A. L. (2006b). Indigenous enzymes in milk: overview and historical aspects - part 2, *Int. Dairy J.* 16(6), 517–532.

France, T. C., O'Mahony, J. A. and Kelly, A. L. (2021). The plasmin system in milk and dairy products. In: *Agents of Change: Enzymes in Milk and Dairy Products* Kelly, A. L. and Larsen, L. B. (Eds), pp 11–56, New York; Springer Nature.

Frohbieter, K. A., Ismail, B., Nielsen, S. S. and Hayes, K. D. (2005). Effects of *Pseudomonas fluorescens* M3/6 bacterial protease on plasmin system and plasminogen activation, *J. Dairy Sci.* 88(10), 3392–3401.

Gan, J., Randall, R. C., Wang, J., Krishnakumar, N., Manning, C. J., Lor, Y., Breck, M., Barile, D. and German, J. B. (2019). Peptidomic profiling of human milk with LC-MS/MS reveals pH-specific proteolysis of milk proteins, *Food Chem.* 274, 766–774.

Gazi, I., Vitalava, I. C. and Huppertz, T. (2014). Plasmin activity and proteolysis in milk protein ingredients, *Int. Dairy J.* 38(2), 208–212.

Glantz, M., Rosenlöw, M., Lindmark-Månsson, H., Buhelt Johansen, L., Hartmann, J., Höjer, A., Waak, E., Löfgren, R., Hallin Saedén, K., Svensson, C., Svensson, B., Lindau, J., Rauh, V. and Paulsson, M. (2020). Impact of protease and lipase activities on quality of Swedish raw milk, *Int. Dairy J.* 107, 104724.

Guerrero, A., Dallas, D. C., Contreras, S., Bhandari, A., Canovas, A., Islas-Trejo, A., Medrano, J. F., Parker, E. A., Wang, M., Hettinga, K., Chee, S., German, J. B., Barile, D. and Lebrilla, C. B. (2015). Peptidomic analysis of healthy and subclinically mastitic bovine milk, *Int. Dairy J.* 46, 46–52.

Hammershoj, M., Hougaard, A. B., Vestergaard, J. S., Poulson, O. and Ipsen, R. H. (2010). Instant infusion pasteurisation of bovine milk. II. Effects on indigenous milk enzymes activity and whey protein denaturation, *Int. J. Dairy Technol.* 63(2), 197–208.

Hansen, M. S. and Rasmussen, J. T. (2021). Enzymes associated with milk phospholipid membrane structures: milk fat globule membranes and extracellular vesicles. In: *Agents of Change: Enzymes in Milk and Dairy Products* Kelly, A. L. and Larsen, L. B. (Eds), pp 127–162, New York; Springer Nature.

Hayes, M. G., Hurley, M. J., Larsen, L. B., Heegaard, C. W., Magboul, A. A. A., Oliveira, J. C., McSweeney, P. L. H. and Kelly, A. L. (2001). Thermal inactivation kinetics of bovine cathepsin D, *J. Dairy Res.* 68(2), 267–276.

Heegaard, C. W., Rasmussen, L. K. and Andreasen, P. A. (1994). The plasminogen activation system in bovine milk: differential localization of tissue-type plasminogen activator and urokinase in milk fractions is caused by binding to casein and urokinase receptor, *Biochim. Biophys. Acta* 1222(1), 45–55.

Hillmann, H., Behr, J., Ehrmann, M. A., Vogel, R. F. and Hoffman, T. (2016). Formation of kokumi-enhancing gamma-glutamyl dipeptides in Parmesan cheese by means of gamma-glutamyltranspeptidase activity and stable isotope double-labelling studies, *J. Agric. Food Chem.* 64(8), 1784–1793.

Hinz, K., Larsen, L. B., Wellnitz, O., Bruckmaier, R. M. and Kelly, A. L. (2012). Proteolytic and proteomic changes in milk at quarter level following infusion with *Escherichia coli* lipopolysaccharide, *J. Dairy Sci.* 95(4), 1655–1666.

Horie, N., Okamoto, U. and Wijngaards, G. (1987). Demonstration of two types of plasminogen activators in human milk by SDS-PAGE co-polymerizing fibrinogen and plasminogen: different variations of the activators in pueriperium, *Thromb. Res.* 45(5), 703–707.

Hovinen, M., Simojoki, H., Poso, R., Suolaniemi, J., Kalmus, P., Suojala, L. and Pyorala, S. (2016). N-acetyl-beta-D-glucosaminidase activity in cow milk as an indicator of mastitis, *J. Dairy Res.* 83(2), 219–227.

Hurley, M. J., Larsen, L. B., Kelly, A. L. and McSweeney, P. L. H. (2000a). Cathepsin D activity in quarg, *Int. Dairy J.* 10(7), 453–458.

Hurley, M. J., Larsen, L. B., Kelly, A. L. and McSweeney, P. L. H. (2000b). The milk acid proteinase cathepsin D: a review, *Int. Dairy J.* 10(10), 673–681.

Jollés, J. and Jollés, P. (1972). Comparison between human and bird lysozyme - note concerning previously observed deletion, FEBS Lett. 22(1), 31.

Kelly, A. L., Crowley, S. V., Crotty, A., Meng, F. and Larsen, L. B. (2021). Future opportunities and challenges in dairy enzymology. In: *Agents of Change: Enzymes in Milk and Dairy Products* Kelly, A. L. and Larsen, L. B. (Eds), pp 525–542, New York; Springer Nature.

Kelly, A. L. and Fox, P. F. (2006). Indigenous enzymes in milk: a synopsis of future research requirements, *Int. Dairy J.* 16(6), 707–715.

Larsen, L. B., Benfeldt, C., Rasmussen, L. K. and Petersen, T. E. (1996). Bovine milk procathepsin D and cathepsin D: coagulation and milk protein degradation, *J. Dairy Res.* 63(1), 119–130.

Larsen, L. B., Boisen, A. and Petersen, T. E. (1993). Procathepsin D cannot autoactivate to cathepsin D at acid pH, *FEBS Lett.* 319(1–2), 54–58.

Larsen, L. B., Drud-Heydary Nielsen, S., Paludetti, L. and Kelly, A. L. (2021). Lysosomal and other indigenous non-plasmin proteases in bovine milk. In: *Agents of Change: Enzymes in Milk and Dairy Products* Kelly, A. L. and Larsen, L. B. (Eds), pp 57–84, New York; Springer Nature.

Larsen, L. B., Hinz, K., Jorgensen, A. L. W., Moller, H. S., Wellnitz, O., Bruckmaier, R. M. and Kelly, A. L. (2010). Proteomic and peptidomic study of proteolysis in quarter milk after infusion with lipoteichoic acid from *Staphylococcus aureus*, *J. Dairy Sci.* 93(12), 5613–5626.

Larsen, L. B., McSweeney, P. L. H., Hayes, M. G., Andersen, J. B., Ingvartsen, K. L. and Kelly, A. L. (2006) Variation in activity and heterogeneity of bovine milk proteases with stage of lactation and somatic cell count, *Int. Dairy J.* 16(1), 1–8.

Larsen, L. B., Wium, H., Benfeldt, C., Heegaard, C. W., Ardo, Y., Qvist,K. B. and Petersen, T. E. (2000). Bovine milk procathepsin D: presence and activity in heated milk and in extracts of rennet-free UF-Feta cheese, *Int. Dairy J.* 10(1–2), 67–73.

Larson, N. K., Ismail, B., Nielsen, S. S. and Hayes, K. D. (2006). Activity of *Bacillus polymyxa* protease on components of the plasmin system in milk, *Int. Dairy J.* 16(6), 586–592.

Le, T. T., Zhao, D. and Larsen, L. B. (2017). Chapter 2: analytical methods for measuring or detecting whey proteins. In: From Milk to Medicine Deeth, H. and Bansal, N. (Eds), pp 155–184, London: Elsevier.

Li, B., Waldron, D. S., Tobin, J. T., Subhir, S., Kelly, A. L. and McSweeney, P. L. H. (2020). Evaluation of production of Cheddar cheese from micellar casein concentrate, *Int. Dairy J.* 107, 104711.

Liao, H. M., Zhong, K., Hu, X. and Liao, X. J. (2019). Effect of high pressure carbon dioxide on alkaline phosphatase activity and quality characteristics of raw bovine milk, *Innov. Food Sci. Emerg. Technol.* 52, 457–462.

Magboul, A. A. A., Larsen, L. B., McSweeney, P. L. H. and Kelly, A. L. (2001). Cysteine protease activity in bovine milk, *Int. Dairy J.* 11, 865–872.

Martinez-Monteagudo, S. I., Ganzle, M. G. and Saldana, M. D. A. (2014). High-pressure and temperature effects on the inactivation of *Bacillus amyloliquefaciens*, alkaline phosphatase and storage stability of conjugated linoleic acid, *Innov. Food Sci. Emerg. Technol.* 26, 59–66.

McSweeney, P. L. H., Fox, P. F. and Olson, N. F. (1995). Proteolysis of bovine caseins by cathepsin-D- preliminary-observations and comparison with chymosin, *Int. Dairy J.* 5(4), 321–336.

Munir, M., Nadeem, M., Quresi, T. M., Leong, T. S. H., Gamlath, C. J., Martin, G. J. O. and Ashokumar, M. (2019). Effects of high pressure, microwave and ultrasound processing on proteins and enzyme activity in dairy systems – a review, *Innov. Food Sci. Emerg. Technol.* 57, 102192.

Nair, P. K. and Corredig, M. (2021). Time dependent aggregation of casein micelle concentrates, J. Dairy Sci. 104(1), 92–101.

Nielsen, S. D., Beverly, R. L. and Dallas, D. C. (2017a). Peptides released from foremilk and hindmilk proteins by breast milk proteases are highly similar, *Front. Nutr.* 4, 4.

Nielsen, S. D., Beverly, R. L., Qu, Y. and Dallas, D. C. (2017b). Milk bioactive peptide database: a comprehensive database of milk protein-derived bioactive peptides and novel visualization, *Food Chem.* 232, 673–682.

Nielsen, S. D., Beverly, R. L. and Dallas, D. C. (2018a). Milk proteins are predigested Within the human mammary gland, *J. Mammary Gland Biol. Neoplasia* 22(4), 251–261.

Nielsen, S. D., Beverly, R. L., Underwood, M. A. and Dallas, D. C. (2018b). Release of functional peptides from mother's milk and fortifier proteins in the premature infant stomach, *PLoS ONE* 13(11), e0208204.

Nielsen, S. D., Beverly, R. L., Underwood, M. A. and Dallas, D. C. (2020). Differences and similarities in the peptide profile of preterm and term mother's milk, and preterm and term infant gastric samples, *Nutrients* 12(9), 2825.

Nissen, A., Andersen, P. H., Bendixen, E., Ingvartsen, K. L. and Røntved, C. M. (2017). Colostrum and milk protein rankings and ratios of importance to neonatal calf health using a proteomics approach, *J. Dairy Sci.* 100(4), 2711–2728.

O'Riordan, N., Kane, M., Joshi, L. and Hickey, R. M. (2014). Glycosidase activities in bovine milk over lactation, *Int. Dairy J.* 35(2), 116–121.

Picart, L., Thiebaud, M., Rene, M., Guiraid, J. P., Cheftel, J. C. and Dumay, E. (2006). Effects of high-pressure homogenisation of raw bovine milk on alkaline phosphatase and microbial inactivation. A comparison with continuous short-time thermal treatments, J. Dairy Res. 73, 454–463.

Politis, I., Barbano, D. M. and Gorewit, R. C. (1992). Distribution of plasminogen and plasmin in fractions of bovine-milk, *J. Dairy Sci.* 75(6), 1402–1410.

Politis, I., Zavizion, B., Barbano, D. M. and Gorewit, R. C. (1993). Enzymatic assay for the combined determination of plasmin plus plasminogen in milk- revisited, *J. Dairy Sci.* 76(5), 1260–1267.

Rauh, V. M., Bakman, M., Ipsen, R., Paulsson, M., Kelly, A. L., Larsen, L. B. and Hammershøj, M. (2014a). The determination of plasmin and plasminogen-derived activity in turbid samples from various dairy products using an optimized spectrophotometric method, *Int. Dairy J.* 38(1), 74–80.

Rauh, V. M., Johansen, L. B., Ipsen, R., Paulsson, M., Larsen, L. B. and Hammershoj, M. (2014b). Plasmin activity in UHT milk: relationship between proteolysis, age gelation, and bitterness, *J. Agric. Food Chem.* 62(28), 6852–6860.

Rauh, V. M., Sundgren, A., Bakman, M., Ipsen, R., Paulsson, M., Larsen, L. B. and Hammershoj, M. (2014c). Plasmin activity as a possible cause for age gelation in UHT milk produced by direct steam infusion, *Int. Dairy J.* 38(2), 199–207.

Reimerdes, E. H., Petersen, F. and Kielwein, G. (1979). Milchproteinasen 9. Proteinasespektren von Caseinmicellen, Milchserum, Rinderblutserum und *Pseudomonas fluorescens*, *Milchwissenschaft* 34, 548-551.

Richardson, B. C. and Pearce, K. N. (1981). The determination of plasmin in dairy products, N.Z. J. Dairy Sci. Technol. 16, 209-220.

Robinson, R. D., Nielsen, S. D., Dallas, D. C. and Barile, D. (2021). Can cheese mites, maggots and molds enhance bioactivity? Peptidomic investigation of functional peptides in four traditional cheeses, *Food Funct.* 12, 533.

Saint-Dennis, B. T., Humbert, G. and Gaillard, J. L. (2001). Enzymatic assays for native plasmin, plasminogen and plasminogen activators in bovine milk, *J. Dairy Res.* 68(3), 437-449.

Schlimme, E., Kiesner, C., Lorenzen, P. C. and Martin, D. (1998). Influence of heat treatment of milk on the activities of the indigenous milk enzymes alkaline phosphatase and adenosine deamindase', IDF, *Bull* 332, 25-31.

Shamay, A., Shapiro, F., Leitner,G. and Silanikove, N. (2003). Infusions of casein hydrolysates in to the mammary gland disrupt tight junction integrity and induce involution in cows, *J. Dairy Sci.* 86(4), 1250-1258.

Staniuck, N., Ardelean, A., Diaconu, V., Rapeanu, G., Stanciu, S. and Nicolau, A. (2011). Kinetic and thermodynamic parameters of alkaline phosphatase and gamma-glutamyl transferase inactivation in bovine milk, *Dairy Sci. Technol.* 91(6), 701-717.

Stoeckel, M., Lidolt, M., Stressler, T., Fischer, L., Wenning, M. and Hinrichs, J. (2016). Heat stability of indigenous milk plasmin and proteases from *Pseudomonas*: a challenge in the production of ultra-high-temperature milk products, *Int. Dairy J.* 61, 250-261.

Sunds, A. V., Nielsen, S. D.-H., Larsen, L. B. and Poulsen, N. A. (2021a). Milk and other glycosidases. In: *Agents of Change: Enzymes in Milk and Dairy Products* Kelly, A. L. and Larsen, L. B. (Eds), pp 163-180, New York; Springer Nature.

Sunds, A. V., Sundekilde, U. K., Poulsen, N. A. and Larsen, L. B. (2021b). Association between udder inflammation and glycosidase activities and free sugar levels in bovine milk, *Int. Dairy J.* 120, 105093.

Tousova, R., Stadnik, L. and Duchacek, J. (2013). Effects of season and time of milking on spontaneous and induced lipolysis in bovine milk fat, *Czech J. Food Sci.* 31(1), 20-26.

Turner, S. A. and Thompson, N. A. (2007). Variation of lactoferrin and lactoperoxidase in bovine milk and the impact of level of pasture intake, *N. Z. J. Agric. Res.* 50(1), 33-40.

Vanbergue, E., Delaby, L., Peyraud, J. L., Colette, S., Gallard, Y. and Hurtaud, C. (2017). Effects of breed, feeding system, and lactation stage on milk fat characteristics and spontaneous lipolysis in dairy cows, *J. Dairy Sci.* 100(6), 4623-4636.

Vanbergue, E., Peyraud, J. L., Ferlay, A., Miranda, G., Martin, P. and Hurtaud, C. (2018). Effects of feeding level, type of forage and milking time on milk lipolytic system in dairy cows, *Livest. Sci.* 217, 116-126.

Vanbergue, E., Peyradud, J. L., Guinard-Flament, J., Charton, C., Barbey, S., Lefebvre, R., Gallard, Y. and Hurtaud, C. (2016). Effects of DGAT1 K232A polymorphism and milking frequency on milk composition and spontaneous lipolysis in dairy cows, *J. Dairy Sci.* 99(7), 5739-5749.

Vanbergue, E., Poulet, J. L., Peyraud, J. L. and Hurtaud, C. (2020). Cow milk lipolysis updates: determining features and biochemical mechanisms, *INRA Product. Anim.* 33, 41-51.

Webb, E. C. (1992). *Enzyme nomenclature 1992: Recommendations of the Nomenclature Committee of the International Union of Biochemistry and Molecular Biology on the Nomenclature and Classification of Enzymes*, New York: Academic Press. ISBN 978-0-12-227164-9.

Wickramasinghe, S., Hua, S., Rincon, G., Islas-Trejo, A., German, J. B., Lebrilla, C. B. and Medrano, J. F. (2011). Transcriptome profiling of bovine milk oligosaccharide metabolism genes using RNA-sequencing, *PLoS ONE* 6(4), e18895.

Wickramasinghe, S., Rincon, G., Islas-Trejo, A. and Medrano, J. F. (2012). Transcriptional profiling of bovine milk using RNA sequencing, *BMC Genomics* 13, 45.

Zhang, C. Y., Bijl, E. and Hettinga, K. (2018). Destabilization of UHT milk by protease AprX from *Pseudomonas fluorescens* and plasmin, *Food Chem.* 263, 127–134.

Zhang, C. Y., Bijl, E., Muis, K. E. and Hettinga, K. (2020). Stability of fat globules in UHT milk during proteolysis by the *AprX* protease from *Pseudomonas fluorescens* and by plasmin, *J. Dairy Sci.* 103(1), 179–190.

# Chapter 13

## Advances in understanding milk salts

*Xiao-Chen Liu and Leif H. Skibsted, University of Copenhagen, Denmark*



### 1 Introduction

Calcium in the form of phosphates constitutes the basis for skeletons of mammals, while calcium carbonates are a major structure builder for many organisms ranking lower in the evolution hierarchy (Holt and Carver, 2012). The structural disorder and randomness which are typical of proteins involved in biomineralization seem important for the transport of minerals with low aqueous solubility to newborn animals, as well as for controlled crystallization and for ordered crystal growth (Kalmar et al., 2012). Caseins, the proteins of the milk micelles, and other disordered proteins also enable soft tissue and hard tissue to coexist in the same organism without ectopic calcification of soft tissue and without demineralization of tissue in bones and teeth (Holt, 2013). With their hydrophilic aspartate and glutamate side chains, caseins (especially the phosphorylated serines) enable sequestration and transport of calcium and magnesium phosphates without precipitation (Grewal et al., 2021). Amorphous calcium phosphates (ACP), whether pure or as calcium combined with magnesium, are important for many physiological functions (Gelli et al., 2019). During mineralization to create or repair hard tissue, ACP sequestered by structurally disordered proteins transports calcium and magnesium prior to bone structure repair or bone growth.

http://dx.doi.org/10.19103/AS.2022.0099.14

Caseins form the basis for transport of calcium and magnesium phosphates in milk. Calcium and magnesium are sequestered by 'competent' phosphopolypeptides which solubilize minerals and prevent precipitation in the udder (Holt et al., 2014). Calcium binds more strongly and selectively to caseins than magnesium, despite the smaller size and higher charge density of magnesium. This is due to highly charged and more flexible binding pockets involving combinations of several phosphorylated serines in conjugation with neighbouring carboxylate side chains of glutamate and aspartate (Jing et al., 2018). During digestion, calcium and magnesium may also self-assemble into amorphous magnesium-calcium phosphate particles with unknown functions in the intestines which may be important for immune defence or for mineral adsorption (Gelli et al., 2019)

Calcium and magnesium are essential nutrients. Magnesium deficiency is rare as leafy green vegetables and grains are rich in magnesium. Calcium is the fifth most abundant element in the Earth's crust, yet calcium deficiency is common mainly due to low bioavailability of calcium from many foods and beverages (Skibsted, 2016). Milk products are an important or even the main source of calcium for many populations. One litre of bovine milk contains around the recommended daily intake of about 1 g of calcium depending on age and gender. The high frequency of osteoporosis, the disease in which bones of the skeleton become fragile and fracture more easily, is now epidemic in the ageing population of Northwestern Europe, despite a high intake of dairy products. Osteoporosis now affects around 30 million people in the EU and more than 75 million people in Europe, the US and Japan taken together (Hernlund et al., 2013, EFFO, 1997).

A better understanding of mechanisms of calcium absorption from foods and beverages could lead to changed dietary recommendations to prevent conditions like osteoporosis. Calcium bioavailability from milk and dairy products is known to depend on the distribution of calcium and phosphates between the milk phases and on calcium speciation, especially in the homogeneous aqueous phase of milk (Gaucheron, 2005, Vavrusova and Skibsted, 2014). Increasing the calcium content of milk by addition of calcium salts, in order to improve calcium uptake to prevent osteoporosis and other conditions like hypertension, clearly needs increased knowledge of calcium balances and dynamics in milk in order not to induce instability and phase separation and precipitation in the enriched milk (Wang et al., 2020b). More than magnesium, addition of calcium reduces the electrostatic repulsion in different caseins and their mixtures, in effect increasing hydrophobic interaction which initiates phase transitions and precipitation (Grewal et al., 2021).

Global trade of milk products involves expensive transportation, the costs of which may be reduced by evaporation or by membrane filtration of milk prior to transportation. Milk minerals may, however, alter during the upconcentration of milk by evaporation and by membrane filtration techniques, with negative effects on product quality (Christiansen et al., 2021). Scaling during evaporation is often caused by calcium salts, reducing heating efficiency and fouling membranes to reduce filtration efficiency. Solubility, rate of precipitation and robustness of supersaturation, all affected by temperature, become important factors in process design and optimization.

## 2 Mineral distribution in milk

Distribution of minerals between caseins and the homogeneous serum phase depends on temperature and pH, with low temperature and low pH favouring dissociation from micelles (Holt, 2004). For the typical salt concentrations of bovine milk shown in Table 1, only about 30% of calcium is present in the aqueous phase of fresh milk and is diffusible (Gaucheron, 2005, Holt, 2004). At a pH below 3.5, all ionic calcium is dissociated from the micelles due to protonization of phosphate and phosphorylated casein (Fig. 1). The non-diffusible salts of milk are in the form of micellar calcium phosphates (MCP) as part of calcium phosphate nanoclusters (CPN). The diffusible calcium phosphates are supersaturated in the serum phase. The constancy of the ionic product (Q) based on ion activities at varying pH has been used to estimate the

**Table 1** Typical salt concentration in bovine milk and their diffusible fraction

| Compound | Total concentration (mM) | Diffusible fraction[a] (%) |
|---|---|---|
| Calcium | 26–32 | 30 |
| Magnesium | 4–6 | 65 |
| Phosphate | 30–32 | 50 |
| Citrate | 7–11 | 90 |
| Sodium | 17–28 | 95 |
| Potassium | 31–43 | 95 |
| Chloride | 22–34 | 100 |

[a] Corresponds to concentration in the aqueous phase of milk having an ionic strength of 0.080 M (Gaucheron, 2005, Holt, 2004).

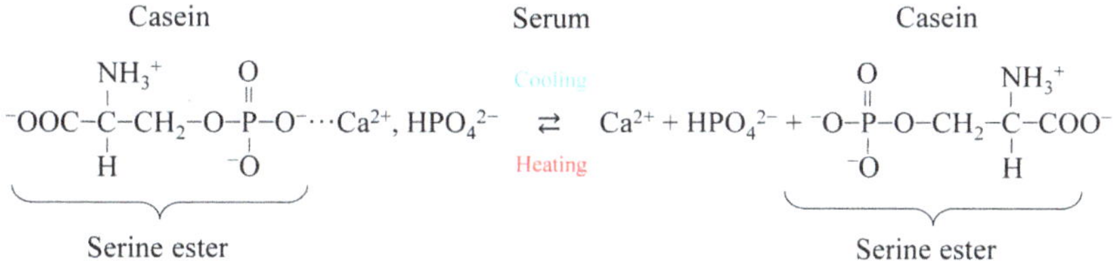


**Figure 1** Overall equilibrium between calcium and phosphate ions in the serum phase and the caseins (Jiang et al., 2021b).

**Table 2** Solubility products ($K_{sp}$) of stochiometric well-defined calcium phosphates in water at 25°C based on ion activities ordered according to increasing saturation index (*S*) for typical milk ultrafiltrates including values for non-stochiometric calcium phosphates also important in milk. $S = 1$ indicates saturated solution and $S > 1$ supersaturation (Chow, 2001, Gaucheron, 2005, Holt, 2004)

| Compound | Formula | Abbreviation | Ca/P ratio | −log ($K_{sp}$) | *S* |
|---|---|---|---|---|---|
| Calcium dihydrogen phosphate | $Ca(H_2PO_4)_2$ | | 0.5 | | <<0 |
| Calcium dihydrogen phosphate monohydrate | $Ca(H_2PO_4)_2 \cdot H_2O$ | | 0.5 | | <<0 |
| Amorphous calcium phosphate | $Ca_3(HPO_4)_{0.2}(PO_4)_{1.87} \cdot xH_2O$ | ACP | 1.45 | (17.6) | 0.2 |
| Micellar calcium phosphate hydrate | $Ca(HPO_4)_{0.7}(PO_4)_{0.2} \cdot xH_2O$ | MCP | 1.1 | (6.80) | 1 |
| Dicalcium phosphate | $CaHPO_4$ | DCP | 1.0 | 6.90 | 3 |
| Dicalcium phosphate dihydrate | $CaHPO_4 \cdot 2H_2O$ | DCPD | 1.0 | 6.59 | 6 |
| Octacalcium phosphate | $Ca_8H_2(PO_4)_6 \cdot 5H_2O$ | OCP | 1.33 | 96.6 | 60 |
| β-Tricalcium phosphate | $Ca_3(PO)_2$ | β-TCP | 1.5 | 28.9 | 200 |
| Hydroxyapatite | $Ca_5(OH)(PO_4)_3$ | HAP | 1.67 | 58.4 | $8 \times 10^8$ |

composition of the MCP, as seen in Table 2 with a saturation index of one for MCP (Holt, 2004).

Not all equilibria in milk adjust rapidly. Precipitation of calcium phosphates involves the formation of ACP with saturation indices less than one, as shown in Fig. 2 and Table 2. Increasing pH decreases the stability of milk systems and increases the risk of precipitation of calcium phosphates depending on the ratio between calcium and the number of o-phosphoserine binding sites available (Bijl et al., 2019). Fortification of milk by adding calcium salts may also reduce the stability of milk.

Calcium phosphates vary widely in living organisms. The Ca/P ratio varies from 1:1 as for dicalcium phosphate (monetite without water, or brushite with two crystal waters and monohydrate (Lu et al., 2020)), close to the ratio in MCP, up to 1.67 in hydroxyapatite (HAP). Precipitation seems to depend on nucleation rate. The nucleation rate seems higher for more simple and soluble calcium phosphates like dicalcium phosphate dihydrate (DCPD) as compared to less soluble phosphates like HAP (Lu and Leng, 2005). Equilibration is further complicated in the presence of casein because inorganic phosphate and calcium interact with phosphorylated peptides through o-phosphoserine side chains, as shown in Fig. 1. Citrate also decreases the rate of formation both of ACP as a precursor for octacalcium phosphate (OCP) and of thermodynamically stable HAP due to complexing of calcium and absorption on crystal surfaces (Spanos et al., 2007).

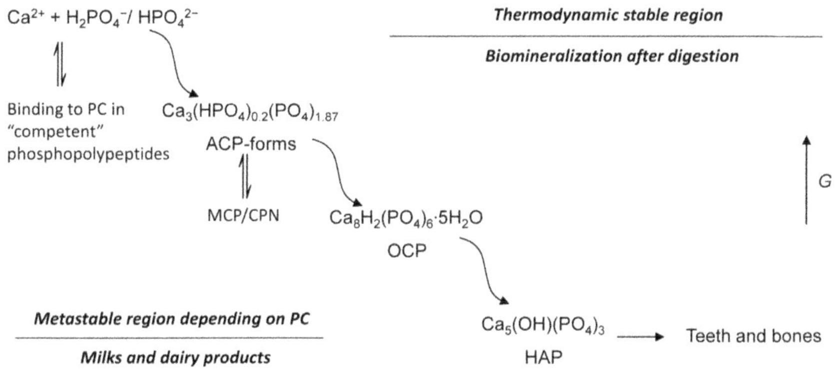


**Figure 2** Precipitation of calcium phosphates depending on the presence of phosphate centres (PC) in peptides forming metastable amorphous calcium phosphates (ACP, in material science, often known as 'calcium phosphate cement') of varying acidity with different $H_2PO_4^-$/ $HPO_4^{2-}$ ratios and water content in agreement with Ostwalds rule of stages (Van Kemenade and De Bruyn, 1987). Micellar calcium phosphate (MCP) as a non-diffusible salt is part of calcium phosphate nanoparticles (CPN). Octacalcium phosphate (OCP) and hydroxyapatite (MAP) are formed following digestion of dairy products in the biomineralization process as part of the metabolism (Holt, 2013, Mekmene et al., 2012).

## 3 Solubility and supersaturation

Calcium is present in three main forms in milk:

- free calcium ions;
- calcium complexes of mainly phosphate and citrate; and
- micellar calcium in or associated with casein micelles.

The serum phase of milk is supersaturated in calcium phosphate (Table 2). The caseins act as calcium buffers, releasing or absorbing calcium in response to environmental changes such as temperature, ionic strength and pH during digestion or milk processing. The complex binding of calcium to phosphate, citrate and o-phosphoserine prevents precipitation of calcium as phosphates and citrates; it increases with increasing pH as seen in Tables 3 and 4. The serine derivative o-phosphoserine is included as a monomeric model compound for the phosphate centres (PC) of the caseins and is important for calcium binding (Jiang et al., 2021b). Increasing pH increases calcium and magnesium binding. Calcium binding increases the acidity of the citrates and, to a lesser degree, the acidity of the phosphates with smaller effects on the o-phosphoserines (Fig. 3).

From the correlation between binding of protons and calcium ions to each of the three ligands (citrate, phosphate and o-phosphoserine), it can be seen that calcium binding to citrate is more dependent on pH than calcium binding to phosphate and especially o-phosphoserine. The stronger calcium binding to citrate with increasing pH, which is important in preventing precipitation of calcium salts in the intestines during milk digestion, is evident from the linear relationship between the $pK_a$ value for the three forms $H_3Citr$, $H_2Citr^-$, $HCitr^{2-}$ and the association constants of $Ca^{2+}$ binding to $H_2Citr^-$, $HCitr^{2-}$ and $Citr^{3-}$ with a slope close to one:

$$\log K_{ass} = 1.16 \times pK_a - 2.5 \quad (1)$$

**Table 3** Association constants $K_{ass}$ and thermodynamic parameters for calcium binding to phosphates and basic forms of o-phosphoserine, $H_3OPS$, in 1:1 complexes at 25°C in water for the phosphates based on activities and in 0.20 M NaCl for o-phosphoserine based on concentration (Chughtai et al., 1968, Jiang et al., 2021b)

| Reaction | $K_{ass}$ | $\Delta H^0$ (kJ·mol$^{-1}$) | $\Delta S^0$ (J·mol$^{-1}$·K$^{-1}$) |
|---|---|---|---|
| $Ca^{2+} + H_2PO_4^-$ | 26 | 14 | 75 |
| $Ca^{2+} + HPO_4^{2-}$ | 681 | 14 | 95 |
| $Ca^{2+} + PO_4^{3-}$ | $2.9 \times 10^6$ | 13 | 168 |
| $Ca^{2+} + H_2OPS^-$ | 29 M$^{-1}$ | 16 | 79 |
| $Ca^{2+} + HOPS^{2-}$ | 30 M$^{-1}$ | −7 | 4 |
| $Ca^{2+} + OPS^{3-}$ | 118 M$^{-1}$ | −11 | 4 |

**Table 4** Thermodynamics of calcium binding to hydroxycarboxylate ions in water at 25°C in 1:1 complexes. $K_{ass}$ is association constant based on activity. Molecular structures were shown in Scheme 1

| Compound | $K_{ass}$ | $\Delta H^0$ (kJ·mol$^{-1}$) | $\Delta S^0$ (J·mol$^{-1}$·K$^{-1}$) | Reference |
|---|---|---|---|---|
| Calcium citrate | $3.6\times10^4$ | −5 [b] | −70 | (Vavrusova and Skibsted, 2016, Liu and Skibsted, 2022) |
| Calcium isocitrate | $6.8\times10^3$ | −28 | −7 | (Cheng et al., 2018) |
| Calcium hydroxycitrate | $8.5\times10^3$ M$^{-1}$ [a] | | | (Ahmed et al., 2021) |
| Calcium saccharate | $1.0\times10^3$ | −34 | −55 | (Garcia et al., 2016) |
| Calcium gluconate | 88 | 34 | 150 | (Vavrusova et al., 2014) |
| Calcium lactate | 49 | 31 | 135 | (Vavrusova et al., 2014) |
| Calcium lactobionate | 140 | 29 | 137 | (Vavrusova et al., 2014) |

[a]Based on concentration for aqueous 0.16 M NaCl.
[b]Strongly dependent on temperature (Liu and Skibsted, 2022).

Binding of $H^+$ and $Ca^{2+}$ is apparently controlled by the same probably electrostatic factors, and the three carboxylates seem to contribute independently to calcium binding (Davies and Hoyle, 1953). It is also important to note the acidity of the calcium complexes of the citrates:

$$CaH_2Citr^+ \rightleftharpoons CaHCitr + H^+ \qquad pK_a = 2.96 \tag{2}$$

$$CaHCitr \rightleftharpoons CaCitr^- + H^+ \qquad pK_a = 4.55 \tag{3}$$

These are also affected by binding of calcium when compared to the acids $H_2Citr^-$ and $HCitr^{2-}$. Acidity corresponds to a decrease in the $pK_a$-value around 1.9 for $H_2Citr^-$ with $pKa_2 = 4.85$ and with 1.8 for $HCitr^{2-}$ with $pKa_3 = 6.39$, further confirming the electrostatic nature of citrate binding and independent effect of the three carboxylates in calcium binding. The linear free energy relationship seen in Fig. 3 for binding of protons and calcium to citrates is in agreement with the shift between calcium and acid-induced curd formation in cheese production with decreasing pH (Koutina and Skibsted, 2015). Coupled diffusion between the calcium citrate complexes, citric acid and the deprotonated forms of citric acid has, however, been demonstrated to be non-linear, which is of importance for membrane filtration processes and electrodialysis technology in the dairy industry (Rodrigo et al., 2019).

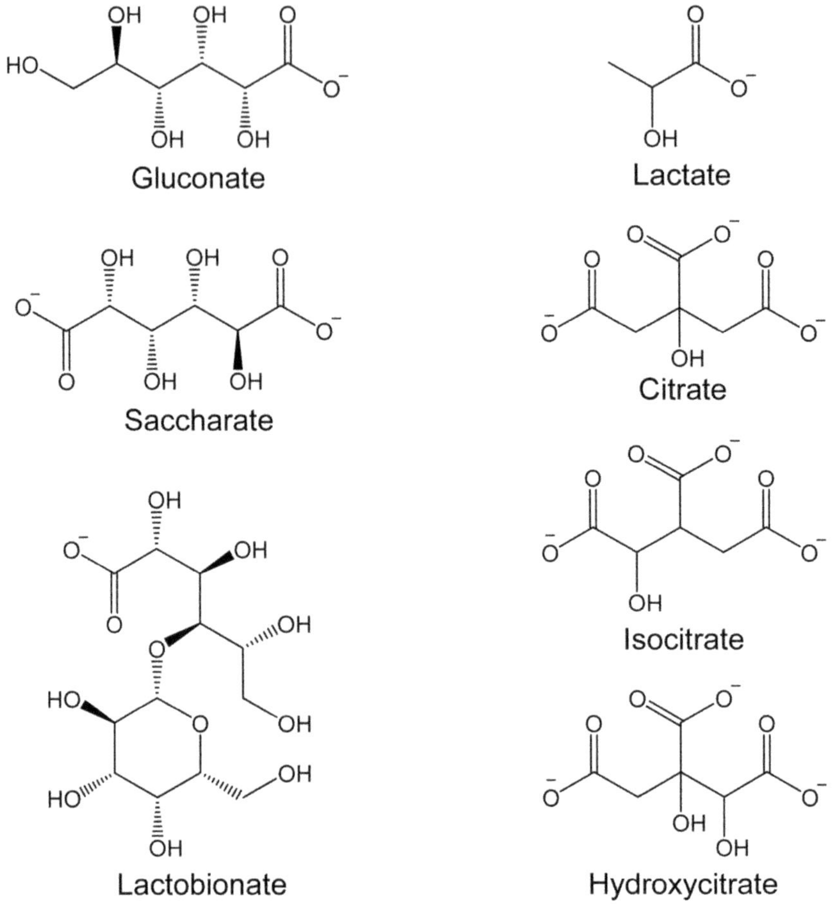


**Scheme 1** Molecule structures of indicated chemicals.

For phosphoric acid the corresponding relationship for $pK_a$-value of $H_3PO_4$, $H_2PO_4^-$ and $HPO_4^{2-}$ and the association constant of $Ca^{2+}$ binding to $H_2PO_4^-$, $HPO_4^{2-}$ and $PO_4^{3-}$ (Chughtai et al., 1968) is:

$$\log K_{ass} = 0.50 \times pK_a - 0.050 \quad (4)$$

For the $pK_a$-values of the o-phosphoserines $H_3OPS$, $H_2OPS^-$ and $HOPS^{2-}$ and the association constant of $Ca^{2+}$ binding to $H_2OPS^-$, $HOPS^{2-}$ and $OPS^{3-}$, the relationship is (Jiang et al., 2021b):

$$\log K_{ass} = 0.10 \times pK_a + 1.07 \quad (5)$$

Together these relationships indicate that charge neutralization through protonization is important in controlling calcium binding along the series

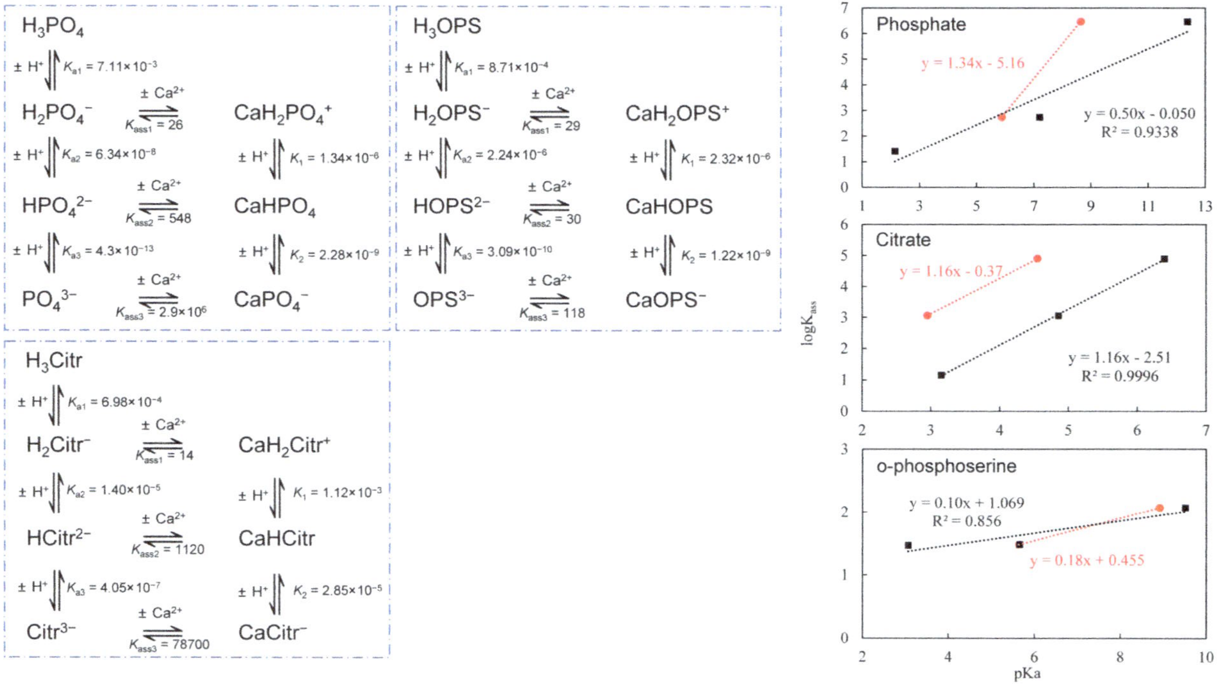


**Figure 3** Calcium binding to phosphates, citrates and anions of o-phosphoserine in aqueous solution at 25°C (Chughtai et al., 1968, Davies and Hoyle, 1953, Jiang et al., 2021b). $pK_a$-values and association constants ($K_{ass}$) for calcium binding are based on activity except for o-phosphoserine, where the constants are based on concentration, and have the unit $M^{-1}$. For citrate, a linear free energy relationship is seen between citrate affinity for protons and calcium (black regression line) and for acidity of calcium complexes and their stability (parallel red line). The decreasing slope for the dependence of proton and calcium affinity along the series citrate > phosphate > o-phosphoserine indicates decreasing selectivity.

of phosphoric acid species, especially for the o-phosphoserine species and probably also for the phosphorylated caseins. Calcium binding notably increases the acidity of the phosphoric acid anions less systematically than for citric acid:

$$CaH_2PO_4^+ \rightleftharpoons CaHPO_4 + H^+ \quad pK_a = 5.87 \tag{6}$$

$$CaHPO_4 \rightleftharpoons CaPO_4^- + H^+ \quad pK_a = 8.64 \tag{7}$$

This can be compared with $pK_{a2}$ = 7.20 and $pK_{a3}$ = 12.37 for $H_2PO_4^-$ and $HPO_4^{2-}$, respectively. For o-phosphoserine the acidity of the calcium complexes increases even less:

$$CaH_2OPS^+ \rightleftharpoons CaHOPS + H^+ \quad pK_a = 5.63 \tag{8}$$

$$CaHOPS \rightleftharpoons CaOPS^- + H^+ \quad pK_a = 8.91 \tag{9}$$

This can be seen from the $pK_a$-values for the calcium complexes compared with $pK_{a2} = 5.65$ for $H_2OPS^-$ and $pK_{a3}$ = 9.51 for $HOPS^{2-}$, respectively.

As compared to citrate, other factors like charge neutralization are important in controlling calcium binding in phosphate and especially o-phosphoserine. This is shown by the decreasing slope in the relationship between calcium binding and acidity for citrate, phosphate and o-phosphoserine shown in Fig. 3. Calcium has little if any effect on the acidity of $H_2OPS^-$, as the o-phosphoserine form dominating at the pH of milk. Calcium binding to o-phosphorylated caseins may follow the same pattern seen in monomeric o-phosphoserine, indicating a pH interval of biological relevance with constant calcium-binding affinity (Jiang et al., 2021b).

The saturation index (*S*) shown in Table 2 relates to the solubility product of the actual compound, for example in the case of dicalcium phosphate, often abbreviated to DCP as shown in Table 2 (Holt, 2004):

$$CaHPO_4\left(s\right) \rightleftarrows Ca^{2+} + HPO_4^{2-}, \quad K_{sp} = \left[Ca^{2+}\right]\cdot\left[HPO_4^{2-}\right] \tag{10}$$

The saturation index also relates to the product of concentrations of ions in milk at 25°C:

$$Q = c\left(Ca^{2+}\right)\cdot c\left(HPO_4^{2-}\right) \tag{11}$$

The saturation index is defined as:

$$S = Q / K_{sp} \tag{12}$$

*Q* is also a parameter for the driving force for precipitation of any salts and may be calculated for DCP by Eq. (11).

For calcium citrate:

$$Ca_3Citr_2\left(s\right) \rightleftarrows 3Ca^{2+} + 2Citr^{3-},\quad K_{sp} = \left[Ca^{2+}\right]^3 \cdot \left[Citr^{3-}\right]^2$$

the driving force may be calculated according to:

$$Q = c\left(Ca^{2+}\right)^3 \cdot c\left(Citr^{3-}\right)^2 \tag{13}$$

For $S$ higher than one the solution is supersaturated in the salt. Supersaturation is accordingly a thermodynamic concept, while the robustness of the supersaturated solutions is a kinetic concept often expressed as the length of the lag phase for initiation of precipitation. For precipitation from solutions like milk supersaturated in calcium phosphates, one or more precursors for the final, thermodynamically stable product HAP seem always to be formed in agreement with Ostwald's rule of stages (Van Kemenade and De Bruyn, 1987). These initial precipitates have been shown to be hydrated forms of ACP with a Ca/P ratio close to 1.5 after drying as in $Ca_3(PO_4)_2$ (Mekmene et al., 2012).

Citrate seems important for prevention of precipitation of calcium phosphates. Citrate can even spontaneously induce supersaturation, as shown for a number of calcium salts, including the slightly soluble phosphates. One example is shown in Fig. 4, where the equilibria in a saturated solution of the moderately soluble calcium lactate pentahydrate are disturbed by the addition of solid sodium citrate dihydrate. More calcium lactate dissolves rapidly and the solution becomes clear, depending on the amount of calcium lactate present, the amount of sodium citrate added and on temperature (Vavrusova et al., 2017a, 2018). The spontaneous supersaturation phenomena described in Fig. 4 have now been confirmed for several salts including dicalcium phosphate and aged calcium phosphates like HAP in dried whey mineral products in combination with gluconolactone or hydrogen citrate (Cheng and Skibsted, 2018, De Zawadzki and Skibsted, 2019, De Zawadzki et al., 2020). The spontaneous supersaturation results in solutions with $S$ larger than 1 for calcium citrate and with robustness against precipitation of calcium citrate, depending on the degree of supersaturation and with increasing time for undistributed solutions, especially of citrate in combination with the isomeric isocitrate (Cheng and Skibsted, 2018). Critical for this spontaneous supersaturation is the citrate-assisted dissolution of the calcium salt of low solubility and especially the slow precipitation of calcium citrate hydrates. Similar dynamics seem important for the prevention of precipitation of calcium phosphates in biological fluids although slow ligand exchange reactions for calcium complexes demonstrated by mathematically modelling of precipitation kinetics still need to be identified and further confirmed experimentally (Garcia et al., 2020).

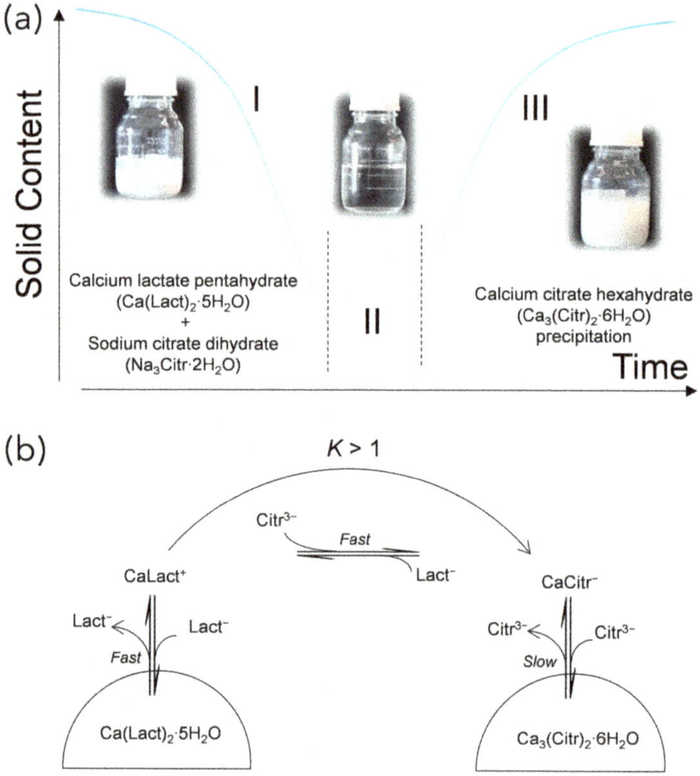


**Figure 4** An example of spontaneous supersaturation facilitated by citrate: shown by pictures of solutions (a) and explained by chemical reactions (b). Calcium lactate dissolves in sodium citrate through three phases: (i) citrate assisted dissolution of calcium lactate, (ii) a metastable homogeneous supersaturation phase and (iii) crystallization of calcium citrate hexahydrate (Vavrusova et al., 2017a, 2018). Citrate also dissolves calcium hydrogenphosphate and HAP in dried whey minerals residues spontaneously creating supersaturated solutions.

Calcium binding to phosphates is moderately endothermic as seen in Table 3 with $\Delta H^0$ around 14 kJ·mol$^{-1}$. The increasing binding strength along the series $H_2PO_4^- < HPO_4^{2-} < PO_4^{3-}$ is accordingly an entropy effect (Chughtai et al., 1968). For pH of relevance for milk, calcium binding to o-phosphoserine shifts from endothermic to exothermic as pH increases (Jiang et al., 2021b). Cooling of milk is known to increase the concentration of free calcium in the serum phase as the result of two effects:

- cooling increases dissociation of calcium from micelles (as shown in Fig. 1); and
- cooling decreases calcium binding to dissolved phosphates and hydroxycarboxylates in the serum phase (as seen in Fig. 5 for skim milk of varying pH).

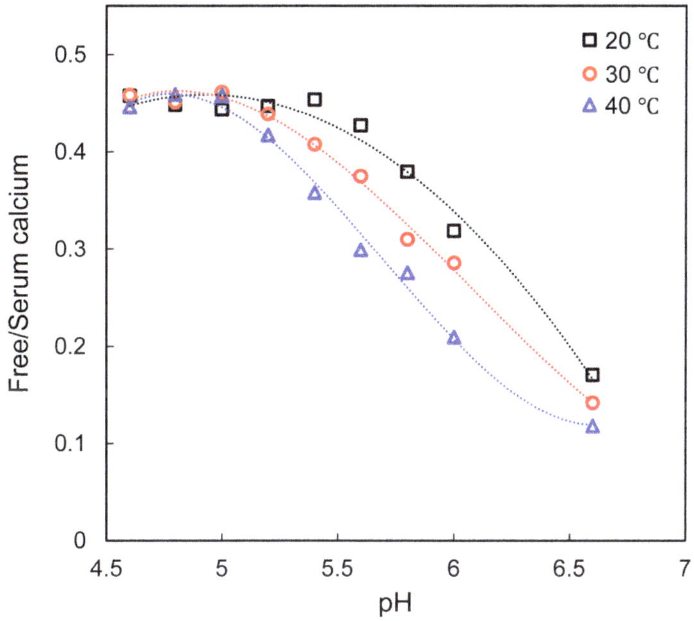


**Figure 5** For intermediate pH during acidification of skim milk cooling increases calcium dissociation in the serum phase in agreement with the endothermic binding of calcium to phosphates, the mono-negative o-phosphoserine and hydroxycarboxylates like lactate, based on Koutina and Skibsted (2015).

## 4 Roles of hydroxycarboxylates

Citrate has long been known to increase bioavailability of calcium in diets and dietary supplements (Pak et al., 1987). Calcium citrate is only slightly soluble in water but excess citrate (as sodium salt) increases absorption of calcium, as clearly shown in human intervention studies. The term 'solubilized' calcium citrate was coined for the mixture of calcium citrate and sodium citrate (Pak et al., 1987). What looked like a solubility paradox, from the perspective of a common ion effect, was later explained by spontaneous supersaturation of calcium citrate by excess citrate during early stages of isothermal dissolution, followed by slow reprecipitation from the supersaturated solution to approach equilibrium solubility (Vavrusova and Skibsted, 2016). As well as citrate, other hydroxycarboxylates as their sodium salts are now known to dissolve slightly soluble calcium salts like the phosphates. These form solutions initially strongly supersaturated in calcium hydroxycarboxylate, followed by a lag phase prior to precipitation initiated to reach the equilibrium solubility, as depicted in Fig. 4. Increased complexation by excess citrate as a thermodynamic effect does not explain the initial increase above equilibrium solubility, while a kinetic description of supersaturation does explain the 'overshooting' of solubility

(Garcia et al., 2020, Vavrusova et al., 2018). Citrate has other applications in dairy production. Acidification with citric acid or with lemon juice in the traditional production of Mozzarella cheese lowers calcium activity in the cheese more than acidification with lactic acid, due to the stronger binding of calcium by citrate lowering calcium ion activity and improving the meltability of cheese (Thybo et al., 2020).

Many hydroxycarboxylates bind calcium in an endothermic reaction, as seen for lactobionate, gluconate and lactate in Table 4. Lactobionate is the oxidation product of lactose and may form in dairy products during heating and processing. Lactate is formed during fermentation of milk and is present in many cheeses such as cheddar, while the delta lactone of gluconic acid, known as GDL in the dairy industry, is used for acidification in producing fresh cheeses like cottage cheese. Gluconate remains in the acid whey from fresh cheeses and binds calcium in whey more strongly than lactate. It may interfere with calcium recovery by electrodialysis, a technology of increasing importance in a circular economy (Nielsen et al., 2021, Bazinet, 2005). Calcium binding to citrate is almost athermal but depends on temperature due to a large difference in heat capacity between citrate and the calcium citrate complex (Liu et al., 2021b).

Isocitrate binds calcium with less affinity than citrate (Cheng et al., 2018), although the combination of citrate and isocitrate increases the inertness of supersaturated solutions dramatically. No explanation seems at present available for this observation nor for the positive effect of sodium gluconate addition in preventing precipitation of calcium lactate and other calcium salts in cheddar cheese, despite the lower solubility of calcium gluconate compared to calcium lactate, although formation of mixed lactate/gluconate complexes has been suggested (Phadungath and Metzger, 2011, Vavrusova et al., 2013).

Calcium saccharate has an even more dramatic effect on inertness of supersaturated calcium hydroxycarboxylate solutions (Garcia et al., 2016). Binding of calcium to saccharate is exothermic and stronger than binding to lactate, gluconate and lactobionate which all demonstrate endothermic and entropy-controlled binding of calcium. For supersaturated solutions of calcium gluconate, addition of saccharate followed by heating increases the supersaturation inertness and prevents crystallization and precipitation during cooling, apparently for years. The contrasting enthalpy in calcium binding in calcium exchange reactions in hydroxycarboxylate complexes, such as the exchange between the stronger binding saccharate and the weaker binding gluconate, can be shown as:

$$\mathrm{CaSac} + \mathrm{Gl}^- \underset{\text{cooling}}{\overset{\text{heating}}{\rightleftarrows}} \mathrm{CaGl}^+ + \mathrm{Sac}^{2-} \quad (14)$$

This may create reaction barriers upon cooling (see Scheme 1 for chemical structures).

## 5 Bioaccessibility of milk minerals

Citrate increases the bioavailability of calcium as shown in intervention studies and in product formulation, including both *in vivo* and *in vitro* digestion experiments (Pak et al., 1987, Wang et al., 2020a). The solubility of calcium citrate is accordingly of special interest for formulation of dairy products with enhanced calcium bioavailability (Liu et al., 2021c).

In addition to dehydrated calcium citrate, three hydrates of calcium citrate are of importance for food products in general and for calcium bioavailability in dairy and other food products (Kaduk, 2018, Liu et al., 2021b). Calcium citrate hexahydrate is the stable form at room temperature and has a positive enthalpy of dissolution in contrast to tetrahydrate, dihydrate and anhydrate, which all show negative enthalpy of dissolution and thus decreasing solubility with increasing temperature (Table 5). The dissolution process for this series of calcium citrates shows a clear enthalpy-entropy compensation with a parallel decrease in enthalpy of dissolution and entropy of dissolution corresponding to:

$$\Delta H^0{}_{\text{dissol}} = T_{\text{isoeq}} \Delta S^0{}_{\text{dissol}} + \text{constant} \tag{15}$$

The isoequilibrium temperature $T_{\text{isoeq}}$ = 49°C is very close to the temperature where the hydrates of calcium citrates have the same solubility. The dehydration of solid-state calcium citrate hydrates has a similar enthalpy-entropy relationship (Table 5) confirming homologue transfer along the series of the solid calcium citrates (Liu et al., 2021b). The enthalpy of dehydration for the first two water molecules in the hexahydrate is very similar to the heat of dehydration for the next two water molecules, while the sum of the first four water molecules is similar to the heat of dehydration of the last two water molecules, which are strongly bound.

The interconversion between solid hexahydrate and solid tetrahydrate is very slow in an aqueous solution in equilibrium with each of the hydrates (Fig. 6). The sluggishness of this conversion seems important for bioavailability of calcium from calcium citrate and may be a consequence of Ostwalds rule of stages. The solid-state or 'precipitate' of a saturated solution of calcium citrate tetrahydrate does not transfer into the more stable and less soluble hexahydrate, even after 4 months in contact with water at room temperature (Liu et al., 2021b).

Mineral fractions from sweet whey are often collected mainly as calcium phosphates including HAP, after correcting whey pH to 7.3 prior to evaporation. In contrast, calcium citrate insolubilization results in fouling whilst processing

**Table 5** Thermodynamics of dissolution and dehydration of calcium hydroxycarboxylates, formed in or added to dairy products. $K_{sp}$ is a solubility product in water at 25°C based on activity

| | | Dissolution | | Stepwise dehydration | | |
|---|---|---|---|---|---|---|
| Compound | $-\log(K_{sp})$ | $\Delta H^0$ (kJ·mol$^{-1}$) | $\Delta S^0$ (J·mol$^{-1}$·K$^{-1}$) | $\Delta H^0$ (kJ·mol$^{-1}$) | $\Delta S^0$ (J·mol$^{-1}$·K$^{-1}$) | Reference |
| Calcium citrate hexahydrate | 17.60 | 21.3 | −266 | 43.6 | 139 | (Liu et al., 2021b) |
| Calcium citrate tetrahydrate | 17.13 | −20.4 | −396 | 43.8 | 131 | (Liu et al., 2021b) |
| Calcium citrate dihydrate | 17.06 | −21.7 | −399 | 88.1 | 230 | (Liu et al., 2021b) |
| Calcium citrate anhydrate | 16.64 | −40.2 | −453 | | | (Liu et al., 2021b) |
| Calcium isocitrate tetrahydrate | 16.06 | 30.1 | −206 | | | (Cheng et al., 2018) |
| Calcium saccharate tetrahydrate | 6.21 | 48 | 42 | | | (Garcia et al., 2016) |
| Calcium gluconate | 4.26 | 16 | −26 | | | (Vavrusova et al., 2014) |
| Calcium lactate | 3.05 | ~ 0 | | | | (Vavrusova et al., 2014) |
| Calcium lactobionate | 2.78 | ~ 0 | | | | (Vavrusova et al., 2014) |

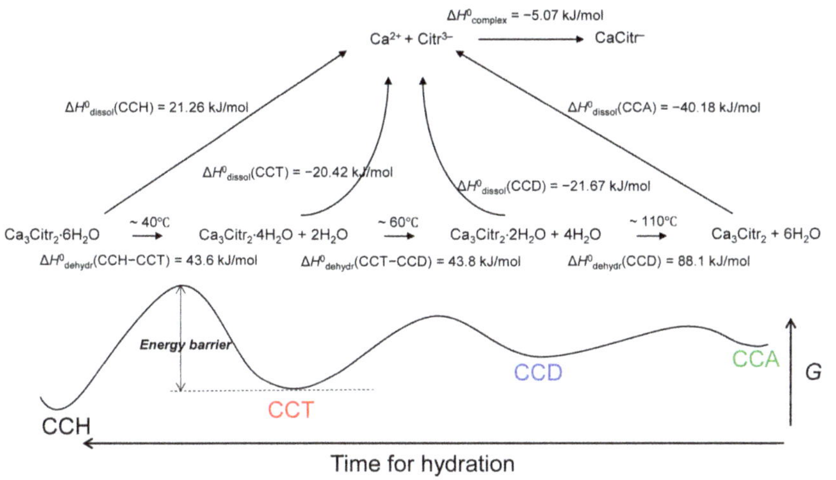


**Figure 6** Hydrates of calcium citrates and their conversion by dehydration in the solid state or through dissolution in aqueous solution followed by precipitation. For stepwise hydration in equilibrated aqueous solutions with solid calcium citrate, the tetrahydrate is kinetically preferred rather than the thermodynamic stable hexahydrate, based on Liu et al. (2021b).

acid whey in falling film evaporators at lower pH (Tanguy et al., 2019). Calcium citrate tetrahydrate (rather than hexahydrate) was also collected after descaling whey evaporators with aqueous citric acid, confirming a kinetic preference for precipitation of the less stable tetrahydrate during crystallization in the cooled flushing liquid (Vavrusova et al., 2017b).

Ion equilibria of minerals in milk have been modelled for simulated milk ultrafiltrate (SMUF), in order to clarify the influence of precipitation of calcium phosphate on bioavailability of calcium, magnesium and phosphate (Jenness and Koops, 1962, Gao et al., 2009, 2010). SMUF was prepared from purified chemicals and water. Calcium phosphate precipitates spontaneously in fresh SMUF, while calcium citrate stays dissolved. Citrate is accordingly a calcium vehicle and promotes calcium bioavailability due to formation of calcium complexes and persistent supersaturation.

The milk ultrafiltrate and SMUF have typically been used to study ion species distribution in milk serum. Absence of micelles reduces any effect of slow reactions of insoluble micellar calcium and most dissolved calcium species can be considered in fast equilibria (Dumpler et al., 2017, Mekmene et al., 2012, Wang et al., 2020b). Ionic strength needs to be considered during the calculations as ionic strength affects the activity coefficient. Ion concentration $c(X^{z\pm})$ cannot be measured directly by electrochemical methods, but depends on an activity coefficient to modify measured activity to obtain the concentration of ion $X^{z\pm}$ with the charge $z$:

$$c\left(X^{z\pm}\right) = a\left(X^{z\pm}\right) / \gamma^{z\pm} \tag{16}$$

where $\gamma^{z\pm}$ is the activity coefficient often calculated from an extended Debye-Hückel equation known as the Davies equation (Davies, 1962):

$$\log\gamma^{z\pm} = -A_{DH}z^2\left(\sqrt{I} / \left(1+\sqrt{I}\right) - 0.3I\right) \tag{17}$$

where $A_{DH}$ is 0.510 for 25°C (Davies, 1962, Manov et al., 1943). $I$ is ionic strength calculated by

$$I = 0.5\Sigma\left(z^2\left[X^{z\pm}\right]\right) \tag{18}$$

Ion speciation for milk serum and SMUF may be calculated based on the relevant complex equilibria, and citric and phosphoric acids dissociation equilibria (Gao et al., 2010, Wang et al., 2020b).

Here, $K_{ass,a}$ is the association constant based on activity for complex formation like:

$$Ca^{2+} + Citr^{3-} \rightleftarrows CaCitr^{-},\ K_{ass,a} = a\left(CaCitr^{-}\right) / \left(a\left(Ca^{2+}\right) \cdot a\left(Citr^{3-}\right)\right) \tag{19}$$

$$Ca^{2+} + HPO_4^{2-} \rightleftarrows CaHPO_4,\quad K_{ass,a} = a\left(CaHPO_4\right) / \left(a\left(Ca^{2+}\right) \cdot a\left(HPO_4^{2-}\right)\right) \tag{20}$$

$$Mg^{2+} + Citr^{3-} \rightleftarrows MgCitr^{-},\ K_{ass,a} = a\left(MgCitr^{-}\right) / \left(a\left(Mg^{2+}\right) \cdot a\left(Citr^{3-}\right)\right) \tag{21}$$

$$Mg^{2+} + HPO_4^{2-} \rightleftarrows MgHPO_4,\quad K_{ass,a} = a\left(MgHPO_4\right) / \left(a\left(Mg^{2+}\right) \cdot a\left(HPO_4^{2-}\right)\right) \tag{22}$$

and $K_{a1,a}$, $K_{a2,a}$, $K_{a3,a}$ are acid dissociation constants based on activity for citric and phosphoric acids:

$$H_3PO_4 \rightleftarrows H_2PO_4^{-} + H^{+},\ K_{a1,a} = a\left(H^{+}\right) \cdot a\left(H_2PO_4^{-}\right) / a\left(H_3PO_4\right) \tag{23}$$

$$H_2PO_4^{-} \rightleftarrows HPO_4^{2-} + H^{+},\ K_{a2,a} = a\left(H^{+}\right) \cdot a\left(HPO_4^{2-}\right) / a\left(H_2PO_4^{-}\right) \tag{24}$$

$$HPO_4^{2-} \rightleftarrows PO_4^{3-} + H^{+},\ K_{a3,a} = a\left(H^{+}\right) \cdot a\left(PO_4^{3-}\right) / a\left(HPO_4^{2-}\right) \tag{25}$$

$$H_3Citr \rightleftarrows H_2Citr^{-} + H^{+},\ K_{a1,a} = a\left(H^{+}\right) \cdot a\left(H_2Citr^{-}\right) / a\left(H_3Citr\right) \tag{26}$$

$$H_2Citr^- \rightleftarrows HCitr^{2-} + H^+,\ K_{a2,a} = a\left(H^+\right) \cdot a\left(HCitr^{2-}\right) / a\left(H_2Citr^-\right) \quad (27)$$

$$HCitr^{2-} \rightleftarrows Citr^{3-} + H^+,\ K_{a3,a} = a\left(H^+\right) \cdot a\left(Citr^{3-}\right) / a\left(HCitr^{2-}\right) \quad (28)$$

The calculation based on the iterative procedure shown in Fig. 7 gives the results shown in Fig. 8 and Table 6. The mass balances for total calcium and total magnesium were included in the calculations:

$$[Ca]_{total} = \left[Ca^{2+}\right] + \left[CaCitr^-\right] + \left[CaHPO_4\right] \quad (29)$$

$$[Mg]_{total} = \left[Mg^{2+}\right] + \left[MgCitr^-\right] + \left[MgHPO_4\right] \quad (30)$$

together with the principles of electroneutrality:

$$\Sigma x \cdot \left[X^{x+}\right] - \Sigma y \cdot \left[X^{y-}\right] = 0 \quad (31)$$

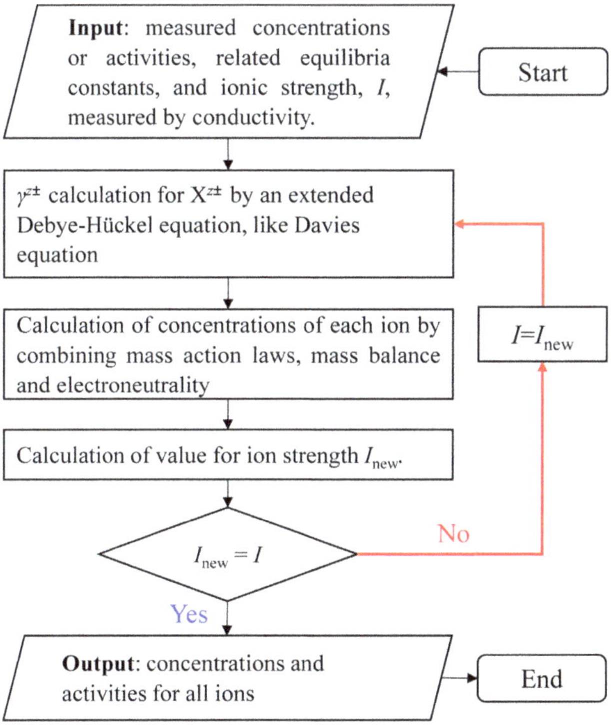


**Figure 7** Iterative calculation procedure for speciation in milk and SMUF at constant temperature (Gao et al., 2010, Holt et al., 1981, Lyster, 1981, Mekmene et al., 2009, 2010, Liu et al., 2021b).

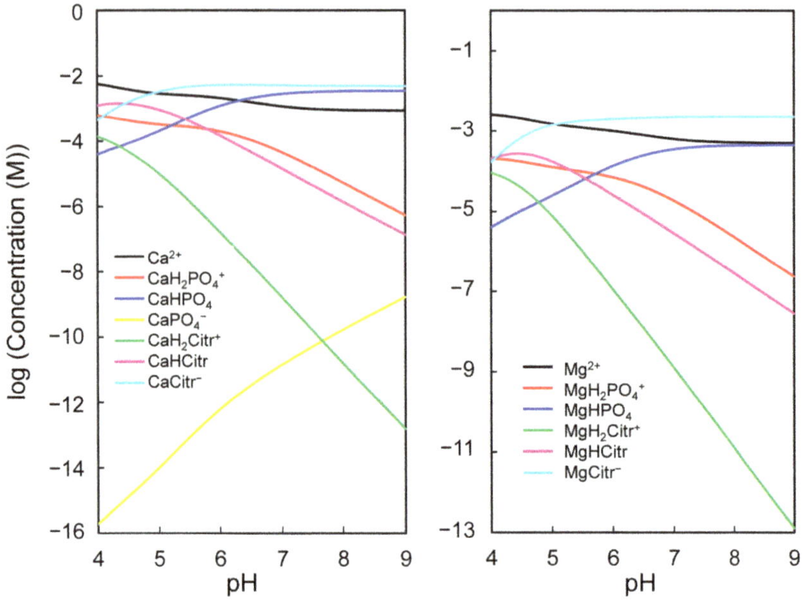


**Figure 8** Speciation of calcium and magnesium in a simple milk serum model of 9.2 mM total calcium and 3.3 mM total magnesium in water at 25°C based on iterative calculation for total citrate/citric acid of 8.2 mM and total phosphate/phosphoric acid of 11.2 mM. Concentrations for calculations adapted from Gaucheron (2005), Holt (2004).

where *x*, *y* are valency for ion $X^{x+}$, $X^{y-}$ with positive and negative charges, respectively. The numerical values for the $pK_a$-values and association constants are based on ion activity (not ion concentration) (Gao et al., 2010, Holt et al., 1981, McDowell et al., 1971, Vavrusova and Skibsted, 2016).

For calculation for dairy products with added salts with higher ionic strength, the extended Debye–Hückel equations may not work satisfactorily and need to be replaced by individual relationships obtained by parameter fitting to experimental data (Jiang et al., 2021b). Calculations for dairy products should be kept below an ionic strength of 0.5 M when the effects on milk of additives (such as calcium chloride, sodium citrate and pyrophosphates) and pH are investigated (Mekmene et al., 2009, Davies, 1962, Gao et al., 2010). Ionic strength may also sometimes be calculated directly for milk-based liquids from measurement of electric conductivity.

The speciation of calcium and magnesium shows that $CaCitr^-$, $CaPO_4^-$ and $MgHPO_4$ dominates at high pH while, at milk pH, several calcium compounds are important including $CaHPO_4$, CaHCitr and $Ca^{2+}$. Speciation is different for calcium and magnesium. Calcium phosphate has a higher tendency for precipitation than calcium citrate at higher pH while calcium citrate will precipitate at lower pH as in acid whey.

**Table 6** The driving force ($Q$) supersaturation degree, $S$, for $Ca_3Citr_2$ (s) and $CaHPO_4$ (s) precipitation based on ion speciation for milk diffusate and SMUF of their natural pH (~7) and 25°C (Gao et al., 2010, Holt et al., 1981)

| | $c(Ca^{2+})$ mM | $c(Citr^{3-})$ mM | $c(HPO_4^{2-})$ mM | Ionic Strength M | $Q_c(Ca_3Citr_2)$ $M^5$ | $Q_c(CaHPO_4)$ $M^2$ | $S = Q_c/K_c$ $(Ca_3Citr_2)$ | $S = Q_c/K_c$ $(CaHPO_4)$ |
|---|---|---|---|---|---|---|---|---|
| Milk diffusate | 2.00 | 0.26 | 2.65 | 0.08 | $5.41\times10^{-16}$ | $5.30\times10^{-6}$ | 0.21 | 6.6 |
| SMUF | 1.59 | 0.38 | 3.18 | 0.069 | $5.80\times10^{-16}$ | $5.06\times10^{-6}$ | 0.32 | 6.9 |

* The solubility products based on concentrations are $K_{sp,c}(Ca_3Citr_2{\cdot}6H_2O) = 2.56 \times 10^{-15}$ and $1.84 \times 10^{-15}$ $M^5$, $K_{sp,c}(CaHPO_4) = 7.98 \times 10^{-7}$ and $7.32 \times 10^{-7}$ $M^2$ for ionic strength of 0.080 and 0.069 M, respectively, which were calculated from the solubility products based on activities, $K_{sp,a}(Ca_3Citr_2{\cdot}6H_2O) = 2.50 \times 10^{-18}$, $K_{sp,a}(CaHPO_4) = 1.26 \times 10^{-7}$, with equations of $K_{sp,a}(Ca_3Citr_2) = K_{sp,c}(Ca_3Citr_2{\cdot}6H_2O) \times (\gamma^{2+})^3 \times (\gamma^{3-})^2$ and $K_{sp,a}(CaHPO_4) = K_{sp,c}(CaHPO_4) \times \gamma^{2+} \times \gamma^{2-}$. Coefficents ($\gamma$) are calculated using Davies equation (Mcdowell et al., 1971, Gao et al., 2010, Liu et al., 2021b).

The importance of supersaturation during transfer between precipitates of calcium phosphates and calcium citrate in equilibrium with a common saturated aqueous phase like in milk is easily seen when pH is adjusted for a saturated calcium phosphate solution also containing citrate from conditions of high pH, corresponding to evaporation of sweet whey to a lower pH characteristic of acid whey (Fig. 9). The long lag phase seen for precipitation of $Ca_3Citr_2{\cdot}xH_2O$ may relate to formation of citrate complexes of calcium and sodium (Garcia et al., 2018, Kutus et al., 2020). Citrate is a structure making or kosmotropic anion, which increases the structure of water. Citrate is an efficient stabilizer of proteins in the context of the Hofmeister series.

Improvements of the SMUF and milk diffusate model need to take account of protein and peptide binding of milk minerals. $\alpha$-lactalbumin binds calcium, while $\beta$-lactoglobulin has little affinity for calcium or magnesium, at least when not precipitated as a result of denaturation (Johns et al., 2021, Liu et al., 2021a). Especially in fermented dairy products, peptides from hydrolysis of whey proteins and caseins contribute to calcium binding (Zhao et al., 2014, Huang et al., 2015). Several peptides have been isolated from hydrolysates of milk proteins and characterized for their capacity for calcium binding and may be compared to peptides from hydrolysis of other food proteins (Choi et al., 2012, Cui et al., 2019). The peptides considered as calcium vehicles carry a varying amount of calcium like 73 μg Ca/mg peptide for the peptide Phe-Asp (Zhao et al., 2014). The effect of calcium binding on calcium bioavailability may, however, be quantified by association constants (as for aspartate) as a measure of binding affinity (Vavrusova and Skibsted, 2013):

$$Ca^{2+} + Asp^{-} \rightleftarrows CaAsp^{+} \qquad K_{ass} = 7.0\,M^{-1} \tag{32}$$

Binding of calcium to amino acids, peptides and proteins depends like calcium binding to citrate, phosphate and o-phosphoserine on pH, temperature and ionic strength (Mekmene and Gaucheron, 2011). The thermodynamic characterization of calcium binding also needs to be correlated with molecular structure. Quantum mechanical calculations have been introduced using the density functional theory (DFT) and have, for more simple systems like citrate and isocitrate, single amino acids and smaller peptides, been found to give fair estimates of the relative stability of series of calcium complexes together with structure information (Tang and Skibsted, 2016, Cheng et al., 2018, Jiang et al., 2021b, Vavrusova et al., 2014). A comparison of citrate and isocitrate using DFT calculations predicts a ratio between the binding constant for calcium binding of 3.4 with citrate being the stronger binder, a ratio close to the experimental value (Fig. 10).

The amino acids with the strongest affinity for calcium are glutamate (Glu) and aspartate (Asp) with their carboxylate side chains and o-phosphoserine (Vavrusova and Skibsted, 2013). DFT-calculations confirm that Asp binds

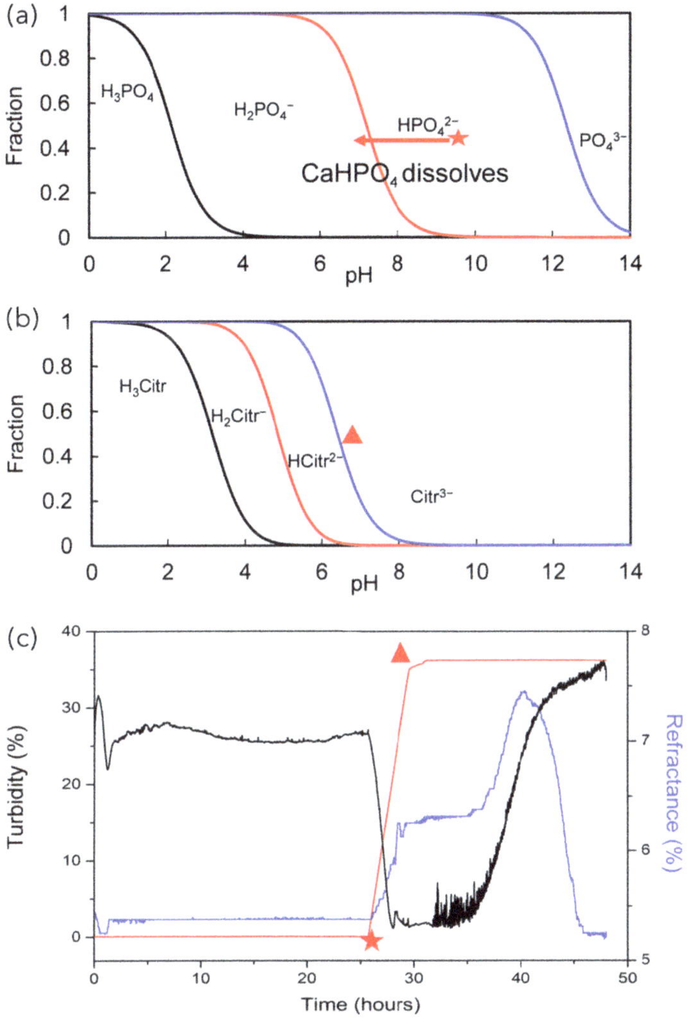


**Figure 9** Acid/base distribution for phosphoric acid (a) and citric acid (b) for varying pH. For an equimolar aqueous solution of $CaCl_2$, $Na_3Citr$, and $Na_2HPO_4$, each 10 mM, the initial precipitate formed at 25°C rapidly dissolves when pH is lowered to below 7. The red curve in (c) indicates addition of HCl to lower pH and the black curve the turbidity initially due to precipitate of $CaHPO_4$ and later of $Ca_3Citr_2{\cdot}xH_2O$ after a lag phase of 10 h. Blue curve is the refractance (Garcia et al., 2018).

calcium more strongly than Glu and the ratio between the binding constants for Asp/Glu calculated by DFT to have a value of 3.5, which is very similar to the experimental value (Fig. 11). For serine and o-phosphoserine a stronger pH-dependence shifts the preferred binding from serine at low pH to o-phosphoserine at higher pH and with a similar binding affinity for the two

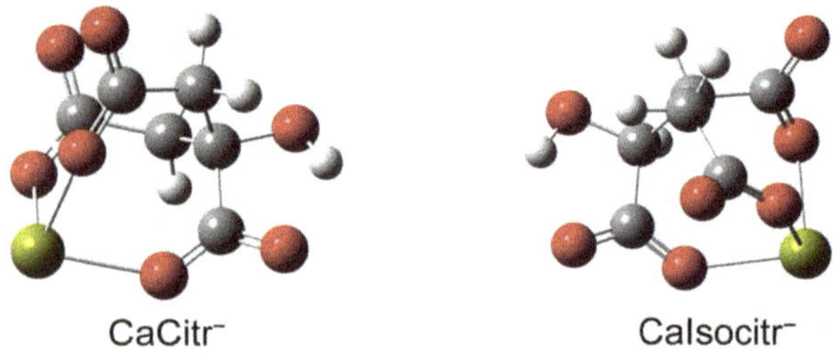


| | Calculated binding enthalpy $\Delta H$ (kJ/mol) | Number of Ca-O bonds | Calculated relative $K_{ass}$ | Experimental $K_{ass}$ |
|---|---|---|---|---|
| $CaCitr^-$ | −321 | 3 | 3.4 | 34300 |
| $CaIsocitr^-$ | −319 | 3 | 1 | 9068 |

**Figure 10** Calculated binding enthalpy for citrate and isocitrate complexes of calcium using the DFT method in comparison with experimental values of $K_{ass}$ (Cheng et al., 2018).

amino acids around milk pH close to 7 as seen in Fig. 12. For o-phosphoserine the phosphate group is involved in calcium binding at low pH, but not at higher pH, where the amino group and the carboxylate coordinate calcium together. For serine the free hydroxy group chelates calcium together with the carboxylate group throughout the pH region of relevance for milk (Jiang et al., 2021b).

Dipeptides are known to bind calcium more strongly than amino acids. Synergistic effects have been demonstrated as some dipeptides bind calcium more strongly than expected from the addition of individual binding affinity

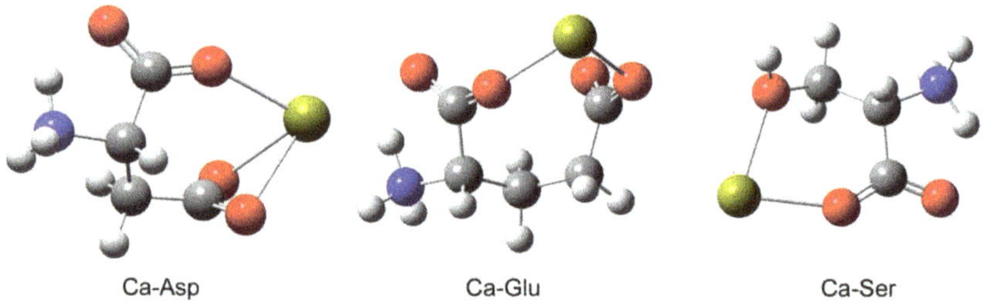


| | Calculated binding enthalpy $\Delta H$ (kJ/mol) | Number of Ca-O bonds | Calculated relative $K_{ass}$ | Experimental $K_{ass}$ ($M^{-1}$) |
|---|---|---|---|---|
| Ca-Asp | −172 | 3 | 3.5 | $7.0 \pm 0.9$ |
| Ca-Glu | −169 | 2 | 1 | $3.0 \pm 0.8$ |
| Ca-Ser | −99 | 2 | $5.7 \times 10^{-3}$ | $32 \pm 3$ |

**Figure 11** Calculated enthalpy for aspartate, glutamate and serine binding of calcium using the DFT method in comparison with experimental value (Tang and Skibsted, 2016, Vavrusova and Skibsted, 2013, Jiang et al., 2021b).

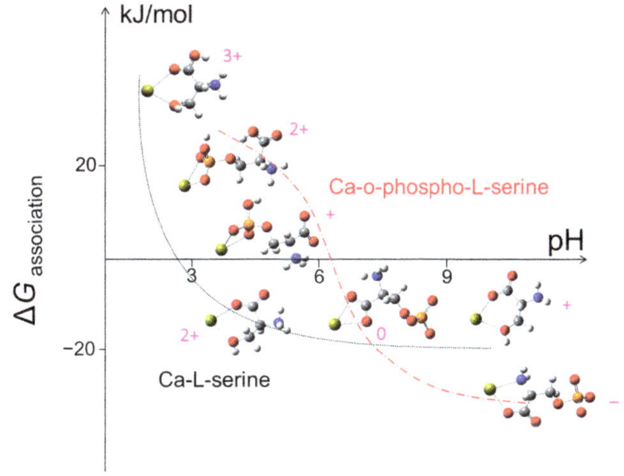


**Figure 12** Calculated binding affinity for calcium chelates by serine and o-phosphoserine using the DFT method for varying solution pH, based on Jiang et al. (2021b).

for each amino acid (Vavrusova and Skibsted, 2013). DFT calculations confirm such binding synergism for dipeptides of Asp and Glu (Fig. 13). The peptide of o-phosphoserine and glutamic acid SerP-Glu likewise shows synergism for calcium binding (Fig. 14).

Other milk components interact with the minerals. Lactose has often been suggested as an enhancer of calcium absorption and lactose has been

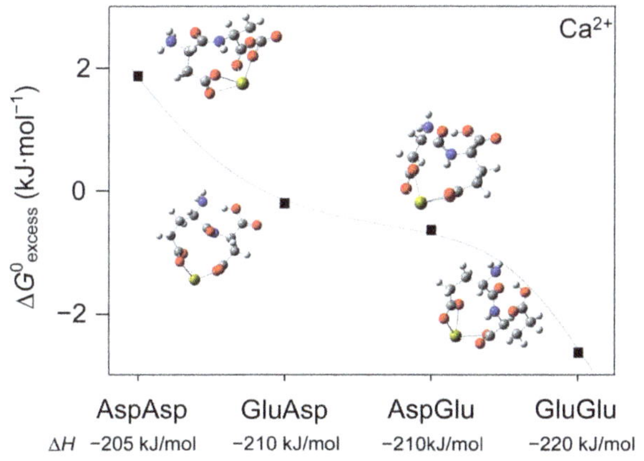


**Figure 13** Calcium chelation by the four dipeptides of glutamate (Glu) and aspartate (Asp). ΔH is calculated using the DFT method and compared to excess free energy for calcium binding to the dipeptides when compared to the sum of the free energy for binding to the two amino acids, based on Tang and Skibsted (2016), Vavrusova and Skibsted (2013).

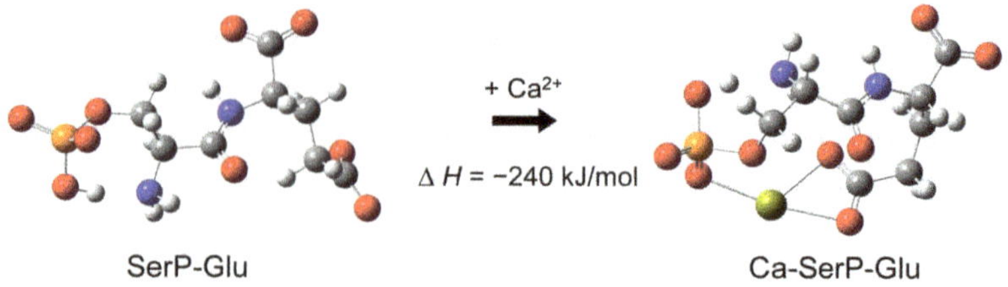


**Figure 14** Enthalpy of calcium binding to the dipeptide SerP-Glu, which is a common sequency in calcium binding proteins.

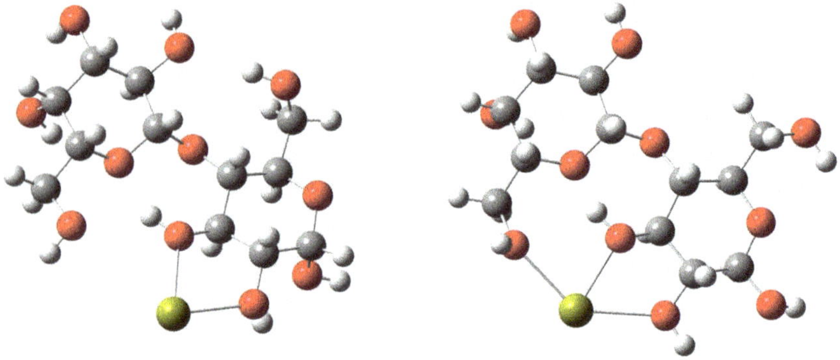


**Figure 15** Calcium binding to α-lactose involves two oxygens while binding to β-lactose involves three oxygens, a difference seen in the experimental association constants and assigned to differences in association entropy, based on Jiang et al. (2021a).

confirmed to bind calcium (Jiang et al., 2021a). DFT calculations have confirmed such binding and also explain why β-lactose binds calcium with higher affinity than α-lactose. The binding of calcium to lactose is entropy-controlled and β-lactose coordinates calcium ions through three oxygen atoms, while α-lactose only coordinates calcium through two oxygen atoms (Fig. 15). Any effect of this difference on lactose digestion or calcium bioavailability still remains to be clarified.

## 6 Designing functional foods

Milk has a positive image in many societies as a basis for developing foods and beverages with added nutritional value (Palacios et al., 2021). As for milk minerals, such functional foods have been based on addition of calcium salts to milk products in order to increase uptake of calcium and phosphate with a positive effect on bone health (Barone et al., 2020, 2021). Excessive calcium intake, however, may increase the risks of cardiovascular disease, and the bioavailability of calcium in such functional foods should be carefully controlled and adjusted (Li et al., 2018, Fardet et al., 2019).

Calcium absorption mainly occurs in the intestines by two mechanisms depending on the concentration of free calcium. Transcellular, saturable transport through cells is regulated by vitamin D, while non-saturable paracellular transport between cells depends on diffusion (Bronner, 1987). Most calcium salts dissolve and dissociate into free calcium ions at low gastric pH during digestion but calcium reprecipitate as phosphates, phytates and carbonate as pH increases when chyme is transferred to the intestines (Skibsted, 2016). Binding of calcium by peptides and hydroxycarboxylates will prevent precipitation of calcium salt. The complex binding outlined in Fig. 16 for citrate may explain the positive effect of citrate and peptides from hydrolyzed milk proteins on calcium bioavailability. For simple diffusion, the concentration of free calcium should be above the mM level.

The positive effect of excess citrate on absorption of calcium may be understood by a fast recombination of calcium and citrate dissociated at the low pH of the stomach when chyme enters the intestines (Pak et al., 1987, Sakhaee and Pak, 2013). High bone mass and bone resorption have now been found to correlate with high levels of circulating citrate in humans, supporting the role of citrate in preventing precipitation of calcium salts in the intestines. Solid calcium citrate has even been found to have positive effects on bone growth and fracture healing (Wang et al., 2012, Palermo et al., 2019). Supersaturation of calcium citrate induced by excess citrate seems to explain these observations (Skibsted, 2016). Design of functional foods for enhanced calcium uptake could benefit from addition of excess citric acid, perhaps in the form of lime or

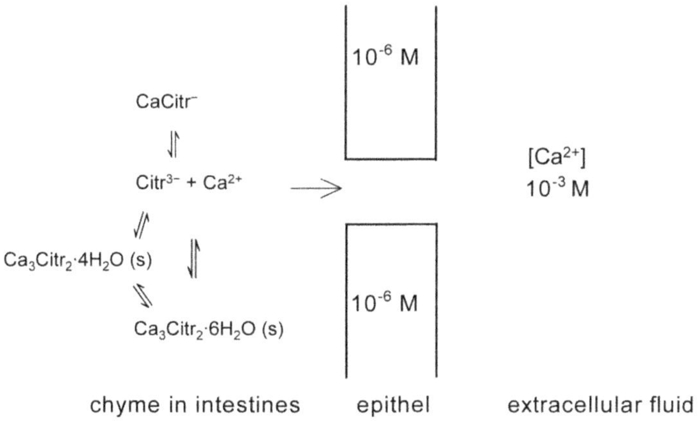


**Figure 16** Chyme with calcium citrate during intestinal digestion. Competition between complex binding of calcium to citrate and peptides prevents precipitation but also lowers calcium ion activity.

lemon juice (Wang et al., 2020a). The use of saccharate to increase the length of the lag phase for precipitation for supersaturated solutions as part of functional foods seems unexplored (Garcia et al., 2016).

Precipitation of calcium has been suggested to occur in two steps involving free calcium ions and the calcium citrate complex (Liu et al., 2021b):

$$Ca^{2+} + 2CaCitr^{-} + xH_2O \rightleftarrows Ca_3Citr_2 \cdot xH_2O \tag{33}$$

According to the speciation of calcium salt at varying pH shown in Fig. 8, free calcium ions and the $CaCitr^-$ complex dominate around pH 5. Free calcium ions and $CaCitr^-$ have similar concentration in the pH region slightly below 5, typical for fermented milk. The slow precipitation of calcium citrate hydrates from supersaturated solutions as documented in Fig. 4 may relate to the involvement of a third-order reaction as shown in Eq. (33). A dimer complex of the calcium citrate equilibrates with the monomer as shown by DFT calculations (Fig. 17), and facilitates the precipitation as a second-order process. The rate of precipitation of calcium citrates from supersaturated solution created by dissolving calcium lactate (like in Fig. 4), calcium gluconate or calcium citrate in sodium citrate was found to depend on a balance between an increased rate of dissolution and increasing supersaturation with a decreasing lag phase for precipitation as seen in Fig. 18. Quantification of the model behind this relationship between degree of supersaturation and length of lag phase for initiation of precipitation of calcium citrate is needed for a practical application in product development (Vavrusova et al., 2017a).

Hydroxycitrate (Scheme 1) also binds to calcium and protects supersaturation of slightly soluble calcium salts (Ahmed et al., 2021). Calcium binds to hydroxycitrate with less affinity than to citrate but with higher affinity than to isocitrate (Table 4). The combination of the hydroxycarboxylates lactate, gluconate and hydroxycitrate showed clear synergistic effects. The observed synergies need further investigation in comparison with the supersaturation synergies observed for combinations of citrate and isocitrate (Cheng et al., 2018).

$$2CaCitr^{-} \rightleftharpoons Ca_2Citr_2^{2-}$$

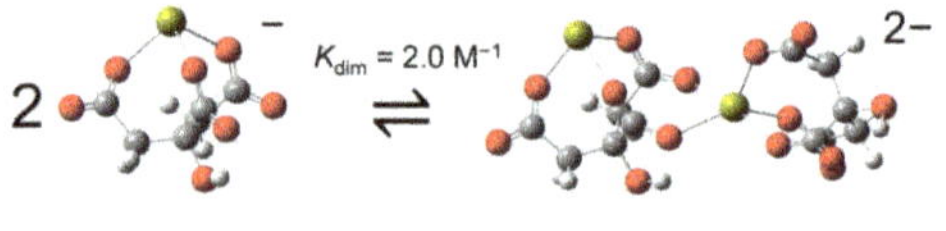


$$Ca_2Citr_2^{2-} + Ca^{2+} + xH_2O \rightleftharpoons Ca_3Citr_2 \cdot xH_2O$$

$x = 4$ for above 51 °C; $x = 6$ for below 51 °C

**Figure 17** Involvement of a dimer in precipitation of calcium citrate hydrates as been evidenced by DFT calculations, based on Liu et al. (2021c).

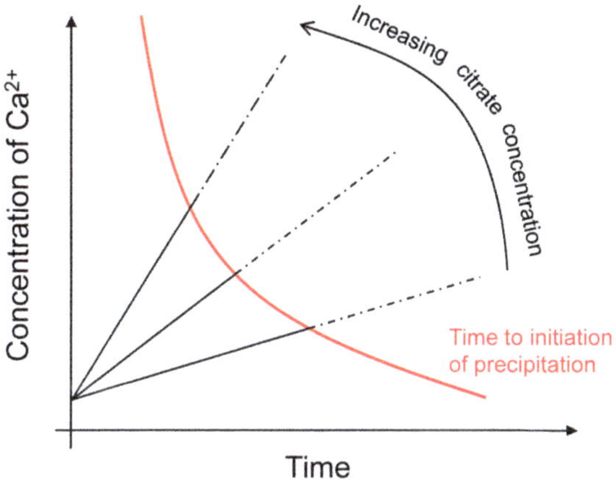


**Figure 18** Increasing calcium concentration during dissolution of slightly soluble calcium salts in aqueous sodium citrate. The red curve is time for initiation of precipitation as dependent on the calcium supersaturation with the crossing indicating the maximal supersaturation for individual citrate concentration, based on Vavrusova et al. (2017a).

## 7 Future trends

The dairy industry is currently challenged globally to achieve milk production with less environmental impact as measured by indices such as global warming potential (GWP). Energy consumption and greenhouse gas (GHG) emission should be kept low and carbon footprint (CF) registered for different cattle breeds under different production conditions and processing technologies (Della et al., 2014).

Mineral fractions are among the most important by-products for potential re-use in dairy production. Calcium-rich byproducts are whey from cheese production and the permeate from ultrafiltration and microfiltration of milk (Tanguy et al., 2016). Both acidic whey from acid-induced curd formation and sweet whey from curd formation by enzymatic hydrolysis have a rich mineral content, which can be collected by electrodialysis and used as a food ingredient after removal of whey proteins and lactose (Nielsen et al., 2021).

The evaporation of water from whey permeates leaves a residue which has an optimal calcium/phosphate ratio for human nutrition (Kato et al., 2002). Calcium is present in such powders as HAP and ACP, which both have low solubility and dissolve slowly. Gluconate combined with gluconolactone dissolves such whey mineral residues rapidly, spontaneously forming supersaturated solution with a lag phase of several weeks (De Zawadzki and Skibsted, 2019). The supersaturated solutions contain up to 7 g calcium /L corresponding to supersaturation by more than a factor of 100. Hydrogencitrate also dissolves such dry mineral residues, forming supersaturated solutions (De Zawadzki and Skibsted, 2020). Hydrogencitrate is less efficient than gluconic acid from hydrolysis of gluconolactone in creating supersaturated solutions and also has shorter lag phases. Combinations of hydroxycarboxylates increase the lag phase for precipitation. Combinations of saccharate, citrate and isocitrate, in particular, seem promising in creating supersaturated solutions with long lag phases for precipitation. These can transform a mineral residue previously considered as waste by the dairy industry into a valuable food ingredient (De Zawadzki et al., 2020). A commercial mineral residue from production of cheese has been shown to increase the nutritional value of yogurt-based dry snacks when combined with lime juice rich in citrate/citric acid in an *in vitro* protocol (Wang et al., 2020a).

The potential of citrate in increasing bioavailability of calcium from dairy products is related to strong supersaturation of calcium phosphate solutions in the presence of citrate. Milk content of citrate and calcium varies and depends on animal feed among other factors (Akkerman et al., 2019). Citrate-rich milk may form the basis for milk and dairy products with increased calcium bioavailability. It has been suggested that increasing calcium bioavailability from fermented milk products depends on the formation of specific peptides with a high affinity for calcium. Future research should focus on characterization of the individual peptides formed by the association constants for their calcium binding rather than their binding capacity (Huang et al., 2015, Zhao et al., 2014). A detailed thermodynamic characterization should include $\Delta H_{ass}$ for calcium binding and especially the pH dependence of $K_{ass}$ (Fig. 19). The pH dependence of binding will allow for identification of peptides which will prevent calcium precipitation in the intestines at increasing pH (Jiang et al., 2021b). Quantum mechanical calculations seem promising for predicting or confirming calcium binding to peptides (Tang and Skibsted, 2016). The DFT method used for smaller peptides has the potential for characterizing longer and longer peptides as computing power increases.

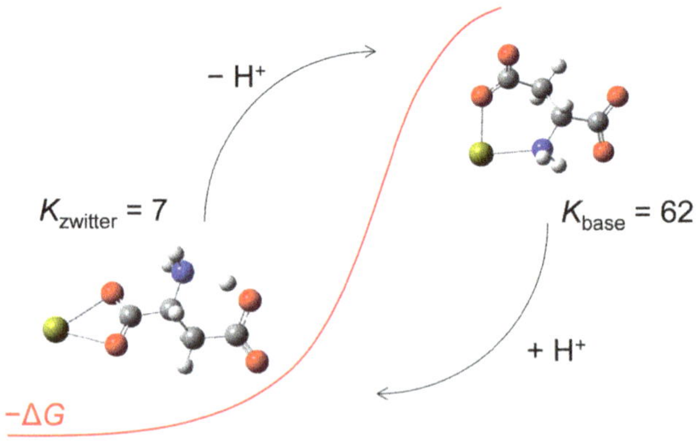


**Figure 19** Binding site shift for calcium and aspartate. Yin and Yang combination of electrochemistry and quantum mechanical calculations explains the increase in binding affinity by structural changes, based on Tang and Skibsted (2016).

## 8 Where to look for further information

Calcium nutrition should be seen both in the perspective of traditional dairy production, and in the perspective of future shift from animal-based food to plant-based food which may exacerbate the situation of insufficient availability of calcium. A current update on challenges in mineral nutrition may be found at:

- A website for dairy production international perspective: https://www.fao.org/dairy-production-products/production/en/, Gateway to dairy production and products, Food and Agriculture Organization of the United Nations.
- Micronutrient Information Center, Linus Pauling Institute, Oregon State University provides updated information on human nutrition and micronutrients including calcium at: https://lpi.oregonstate.edu/mic/minerals/calcium.

Optimising of bioavailability of calcium will depend on maximizing the concentration of specific calcium species. The calcium speciation in food is accordingly of special importance, and the mathematical background for establishing calcium speciation in milk and dairy products may be found in this recent review and using software like Sol-Eq and MINTEQA2:

- The review is: Kiss, T., Enyedy, É. A. and Jakusch, T. (2017). Development of the application of speciation in chemistry. *Coordination Chemistry Reviews, 352*, 401–423. https://doi.org/10.1016/j.ccr.2016.12.016.
- The software Sol-Eq was introduced by and can be obtained from '[Anon.] (2002). New SIT software makes speciation calculations easier. *Chemistry International – Newsmagazine for IUPAC*, 24(6), 18–19. https://doi.org/10.1515/ci.2002.24.6.18'.
- The software MINTEQA2 can be obtained from: https://www.epa.gov/ceam/minteqa2-equilibrium-speciation-model.

The DFT method introduced will help to understand the structural requirement for optimal bioaccessibility for small peptides and their calcium complexes in synergism with hydroxycarboxylates. A good introduction to such quantum mechanical calculations and a review related to the thermodynamics properties may be found in:

- Jalbout, A. F., Nazari, F. and Turker, L. (2004). Gaussian-based computations in molecular science. *Journal of Molecular Structure: THEOCHEM, 671*(1–3), 1–21. https://doi.org/10.1016/s0166-1280(03)00347-6.
- Van Speybroeck, V., Gani, R. and Meier, R. J. (2010). The calculation of thermodynamic properties of molecules. *Chemical Society Reviews, 39*(5), 1764. https://doi.org/10.1039/b809850f.

- Introduction to the use of density functional theory on metal complexes is found at: https://gaussian.com/man/.

## 9 Acknowledgement

The continuing support from the Danish Dairy Research Council is gratefully acknowledged.

## 10 References

Ahmed, A., Rodgers, A. L., Elmagbari, F. and Jackson, G. E. 2021. In support of hydroxycitrate being clinically investigated as a potential therapy of calcium nephrolithiasis: theoretical modelling and in vitro investigation of thermodynamic effects. *Journal of Crystal Growth* 558, 125956.

Akkerman, M., Larsen, L. B., Sørensen, J. and Poulsen, N. A. 2019. Natural variations of citrate and calcium in milk and their effects on milk processing properties. *Journal of Dairy Science* 102(8), 6830-6841.

Barone, G., Moloney, C., O'Regan, J., Kelly, A. L. and O'Mahony, J. A. 2020. Influence of calcium fortification on physicochemical properties of whey protein concentrate solutions enriched in α-lactalbumin. *Food Chemistry* 317, 126412.

Barone, G., O'Regan, J., Kelly, A. L. and O'Mahony, J. A. 2021. Physicochemical and bulk handling properties of micronised calcium salts and their application in calcium fortification of whey protein-based solutions. *Journal of Food Engineering* 292, 110213.

Bazinet, L. 2005. Electrodialytic phenomena and their applications in the dairy industry: a review. *Critical Reviews in Food Science and Nutrition* 45(4), 307-326.

Bijl, E., Huppertz, T., Van Valenberg, H. and Holt, C. 2019. A quantitative model of the bovine casein micelle: ion equilibria and calcium phosphate sequestration by individual caseins in bovine milk. *European Biophysics Journal* 48(1), 45-59.

Bronner, F. 1987. Intestinal calcium absorption: mechanisms and applications. *The Journal of Nutrition* 117(8), 1347-1352.

Cheng, H., Garcia, A. C., Tang, N., Danielsen, B. P. and Skibsted, L. H. 2018. Combinations of isocitrate and citrate enhance calcium salt solubility and supersaturation robustness. *International Dairy Journal* 85, 225-236.

Cheng, H. and Skibsted, L. H. 2018. Dissolution of calcium hydrogen phosphate in aqueous δ-gluconolactone; long-lasting supersaturation increasing calcium availability. *International Dairy Journal* 84, 62-71.

Choi, D.-W., Lee, J.-H., Chun, H.-H. and Song, K. B. 2012. Isolation of a calcium-binding peptide from bovine serum protein hydrolysates. *Food Science and Biotechnology* 21(6), 1663-1667.

Chow, L. C. 2001. Solubility of calcium phosphates. *Monographs in Oral Science* 18, 94-111.

Christiansen, M. V., Skibsted, L. H. and Ahrné, L. 2021. Control of viscosity by addition of calcium chloride and glucono-δ-lactone to heat treated skim milk concentrates produced by reverse osmosis filtration. *International Dairy Journal* 114, 104916.

Chughtai, A., Marshall, R. and Nancollas, G. H. 1968. Complexes in calcium phosphate solutions. *Journal of Physical Chemistry* 72(1), 208-211.

Cui, P., Lin, S., Han, W., Jiang, P., Zhu, B. and Sun, N. 2019. Calcium delivery system assembled by a nanostructured peptide derived from the sea cucumber ovum. *Journal of Agricultural and Food Chemistry* 67(44), 12283–12292.

Davies, C. W. and Hoyle, B. E. 1953. The interaction of calcium ions with some phosphate and citrate buffers. *Journal of the Chemical Society (Resumed)* 4134, 842.

Davies, W. C. 1962. *Ion Association*, Washington, Butterworths.

De Zawadzki, A., Paganelli, M. O., Garcia, A. C. and Skibsted, L. H. 2020. Hydroxycarboxylate combinations for increasing solubility and robustness of supersaturated solutions of whey mineral residues. *Food Research International* 136, 109525.

De Zawadzki, A. and Skibsted, L. H. 2019. Increasing calcium solubility from whey mineral residues by combining gluconate and delta-gluconolactone. *International Dairy Journal* 99, 104538.

De Zawadzki, A. and Skibsted, L. H. 2020. Calcium availability from whey mineral residues increased by hydrogen citrate. *Food Research International* 137, 109372.

Della, R. A., Kristensen, T., De Marchi, M., Kargo, M., Jensen, J. and Cassandro, M. 2014. Carbon footprint from dairy farming system: comparison between Holstein and Jersey cattle in Italian circumstances. *Acta Agraria Kaposváriensis* 18, 75–80.

Dumpler, J., Kieferle, I., Wohlschlager, H. and Kulozik, U. 2017. Milk ultrafiltrate analysis by ion chromatography and calcium activity for SMUF preparation for different scientific purposes and prediction of its supersaturation. *International Dairy Journal* 68, 60–69.

EFFO 1997. Who are candidates for prevention and treatment for osteoporosis? *Osteoporosis International* 7(1), 1–6.

Fardet, A., Dupont, D., Rioux, L. E. and Turgeon, S. L. 2019. Influence of food structure on dairy protein, lipid and calcium bioavailability: a narrative review of evidence. *Critical Reviews in Food Science and Nutrition* 59(13), 1987–2010.

Gao, R., Temminghoff, E. J. M., Van Leeuwen, H. P., Van Valenberg, H. J. F., Eisner, M. D. and Van Boekel, M. A. J. S. 2009. Simultaneous determination of free calcium, magnesium, sodium and potassium ion concentrations in simulated milk ultrafiltrate and reconstituted skim milk using the Donnan Membrane Technique. *International Dairy Journal* 19(8), 431–436.

Gao, R., Van Halsema, F. E. D., Temminghoff, E. J. M., Van Leeuwen, H. P., Van Valenberg, H. J. F., Eisner, M. D., Giesbers, M. and Van Boekel, M. A. J. S. 2010. Modelling ion composition in simulated milk ultrafiltrate (SMUF). I: influence of calcium phosphate precipitation. *Food Chemistry* 122(3), 700–709.

Garcia, A. C., Hansen, J. S., Bailey, N. and Skibsted, L. H. 2020. Slow lactate gluconate exchange in calcium complexes during precipitation from supersaturated aqueous solutions. *Food Research International* 137, 109539.

Garcia, A. C., Vavrusova, M. and Skibsted, L. H. 2016. Calcium d -Saccharate: aqueous solubility, complex formation, and stabilization of supersaturation. *Journal of Agricultural and Food Chemistry* 64(11), 2352–2360.

Garcia, A. C., Vavrusova, M. and Skibsted, L. H. 2018. Supersaturation of calcium citrate as a mechanism behind enhanced availability of calcium phosphates by presence of citrate. *Food Research International* 107, 195–205.

Gaucheron, F. 2005. The minerals of milk. *Reproduction, Nutrition, Development* 45(4), 473–483.

Gelli, R., Ridi, F. and Baglioni, P. 2019. The importance of being amorphous: calcium and magnesium phosphates in the human body. *Advances in Colloid and Interface Science* 269, 219–235.

Grewal, M. K., Vasiljevic, T. and Huppertz, T. 2021. Influence of calcium and magnesium on the secondary structure in solutions of individual caseins and binary casein mixtures. *International Dairy Journal* 112, 104879.

Hernlund, E., Svedbom, A., Ivergård, M., Compston, J., Cooper, C., Stenmark, J., Mccloskey, E. V., Jönsson, B. and Kanis, J. A. 2013. Osteoporosis in the European Union: medical management, epidemiology and economic burden: a report prepared in collaboration with the International Osteoporosis Foundation (IOF) and the European Federation of Pharmaceutical Industry Associations (EFPIA). *Archives of Osteoporosis* 8, 136.

Holt, C. 2004. An equilibrium thermodynamic model of the sequestration of calcium phosphate by casein micelles and its application to the calculation of the partition of salts in milk. *European Biophysics Journal* 33(5), 421-434.

Holt, C. 2013. Unfolded phosphopolypeptides enable soft and hard tissues to coexist in the same organism with relative ease. *Current Opinion in Structural Biology* 23(3), 420-425.

Holt, C. and Carver, J. A. 2012. Darwinian transformation of a 'scarcely nutritious fluid' into milk. *Journal of Evolutionary Biology* 25(7), 1253-1263.

Holt, C., Dalgleish, D. G. and Jenness, R. 1981. Calculation of the ion equilibria in milk diffusate and comparison with experiment. *Analytical Biochemistry* 113(1), 154-163.

Holt, C., Lenton, S., Nylander, T., Sorensen, E. S. and Teixeira, S. C. M. 2014. Mineralisation of soft and hard tissues and the stability of biofluids. *Journal of Structural Biology* 185(3), 383-396.

Huang, S. L., Zhao, L. N., Cai, X., Wang, S. Y., Huang, Y. F., Hong, J. and Rao, P. F. 2015. Purification and characterisation of a glutamic acid-containing peptide with calcium-binding capacity from whey protein hydrolysate. *Journal of Dairy Research* 82(1), 29-35.

Jenness, R. and Koops, J. 1962. Preparation and properties of a salt solution which simulates milk ultrafiltrate. *Nederlands Melk-en Zuiveltijdschrift* 16, 153-164.

Jiang, Y., Liu, X.-C., Ahrné, L. M. and Skibsted, L. H. 2021a. Calcium binding to lactose, inulin and their constituting monosaccharides and perspective for calcium bioaccessibility. *International Dairy Journal* 118, 105042.

Jiang, Y., Liu, X.-C., De Zawadzki, A. and Skibsted, L. H. 2021b. Binding of calcium to L-serine and o-phospho-L-serine as affected by temperature, pH and ionic strength under milk processing conditions. *International Dairy Journal* 112, 104875.

Jing, Z., Liu, C., Qi, R. and Ren, P. 2018. Many-body effect determines the selectivity for Ca2+ and Mg2+ in proteins. *Proceedings of the National Academy of Sciences of the United States of America* 115(32), E7495-E7501.

Johns, P. W., Strozier, D. C., Mckenna, R. J., Veldhuis, J., Weber, L. E. and Thompson, J. J. 2021. Evaluations of protein-metal association in nutritional products. *International Dairy Journal* 114, 104912.

Kaduk, J. A. 2018. Crystal structures of tricalcium citrates. *Powder Diffraction* 33(2), 98-107.

Kalmar, L., Homola, D., Varga, G. and Tompa, P. 2012. Structural disorder in proteins brings order to crystal growth in biomineralization. *Bone* 51(3), 528-534.

Kato, K., Toba, Y., Matsuyama, H., Aoki, T., Aoe, S. and Takada, Y. 2002. A novel type of milk calcium has higher calcium bioavailability compared with the hydroxyapatite type of milk calcium. *Journal of Nutritional Science and Vitaminology* 48(5), 390-394.

Koutina, G. and Skibsted, L. H. 2015. Calcium and phosphorus equilibria during acidification of skim milk at elevated temperature. *International Dairy Journal* 45, 1-7.

Kutus, B., Dudás, C., Friesen, S., Peintler, G., Pálinkó, I., Sipos, P. and Buchner, R. 2020. Equilibria and dynamics of sodium citrate aqueous solutions: the hydration of citrate and formation of the Na3Cit0 ion aggregate. *The Journal of Physical Chemistry B* 124(43), 9604-9614.

Li, K., Wang, X. F., Li, D. Y., Chen, Y. C., Zhao, L. J., Liu, X. G., Guo, Y. F., Shen, J., Lin, X., Deng, J., Zhou, R. and Deng, H. W. 2018. The good, the bad, and the ugly of calcium supplementation: a review of calcium intake on human health. *Clinical Interventions in Aging* 13, 2443-2452.

Liu, W., Chen, X. D., Jeantet, R., André, C., Bellayer, S. and Delaplace, G. 2021a. Effect of casein/whey ratio on the thermal denaturation of whey proteins and subsequent fouling in a plate heat exchanger. *Journal of Food Engineering* 289, 110175.

Liu, X. C., Kirkensgaard, J. J. K. and Skibsted, L. H. 2021b. Hydrates of calcium citrate and their interconversion in relation to calcium bioaccessibility. *Food Research International* 140, 109867.

Liu, X. C., Kirkensgaard, J. J. K. and Skibsted, L. H. 2021c. Temperature effects on spontaneous supersaturation of calcium citrate in presence of lactate. *International Dairy Journal* 112, 105023.

Liu, X. C. and Skibsted, L. H. 2022. Strotium increasing calcium accessibility from calcium citrate. *Food Chemistry* 367,130674.

Lu, B. Q., Willhammar, T., Sun, B. B., Hedin, N., Gale, J. D. and Gebauer, D. 2020. Introducing the crystalline phase of dicalcium phosphate monohydrate. *Nature Communications* 11(1), 1546.

Lu, X. and Leng, Y. 2005. Theoretical analysis of calcium phosphate precipitation in simulated body fluid. *Biomaterials* 26(10), 1097-1108.

Lyster, R. L. J. 1981. Calculation by computer of individual concentrations in a simulated milk salt solution .2: an extension to the previous model. *Journal of Dairy Research* 48(1), 85-89.

Manov, G. G., Bates, R. G., Hamer, W. J. and Acree, S. F. 1943. Values of the constants in the Debye–Hückel equation for activity coefficients1. *Journal of the American Chemical Society* 65(9), 1765-1767.

McDowell, H., Brown, W. E. and Sutter, J. R. 1971. Solubility study of calcium hydrogen phosphate ion-pair formation. *Inorganic Chemistry*, 10(8), 1638-1643.

Mekmene, O. and Gaucheron, F. 2011. Determination of calcium-binding constants of caseins, phosphoserine, citrate and pyrophosphate: a modelling approach using free calcium measurement. *Food Chemistry* 127(2), 676-682.

Mekmene, O., Le Graet, Y. and Gaucheron, F. 2009. A model for predicting salt equilibria in milk and mineral-enriched milks. *Food Chemistry* 116(1), 233-239.

Mekmene, O., Le Graet, Y. and Gaucheron, F. 2010. Theoretical model for calculating ionic equilibria in milk as a function of ph: comparison to experiment. *Journal of Agricultural and Food Chemistry* 58(7), 4440-4447.

Mekmene, O., Leconte, N., Rouillon, T., Quillard, S., Bouler, J. M. and Gaucheron, F. 2012. Physicochemical characterisation of calcium phosphates prepared from milk ultrafiltrates: effect of the mineral composition. *International Journal of Dairy Technology* 65(3), 334-341.

Nielsen, E. N., Merkel, A., Yazdi, S. R. and Ahrné, L. 2021. The effect of acid whey composition on the removal of calcium and lactate during electrodialysis. *International Dairy Journal* 117, 104985.

Pak, C. Y., Harvey, J. A. and Hsu, M. C. 1987. Enhanced calcium bioavailability from a solubilized form of calcium citrate. *Journal of Clinical Endocrinology and Metabolism* 65(4), 801-805.

Palacios, C., Hofmeyr, G. J., Cormick, G., Garcia-Casal, M. N., Peña-Rosas, J. P. and Betrán, A. P. 2021. Current calcium fortification experiences: a review. *Annals of the New York Academy of Sciences* 1484(1), 55-73.

Palermo, A., Naciu, A. M., Tabacco, G., Manfrini, S., Trimboli, P., Vescini, F. and Falchetti, A. 2019. Calcium citrate: from biochemistry and physiology to clinical applications. *Reviews in Endocrine and Metabolic Disorders* 20(3), 353-364.

Phadungath, C. and Metzger, L. E. 2011. Effect of sodium gluconate on the solubility of calcium lactate. *Journal of Dairy Science* 94(10), 4843-4849.

Rodrigo, M. M., Ribeiro, A. C. F., Verissimo, L. M. P., Esteso, M. A. and Leaist, D. G. 2019. Coupled diffusion in aqueous citric acid + calcium citrate solutions. *The Journal of Chemical Thermodynamics* 131, 314-321.

Sakhaee, K. and Pak, C. 2013. Superior calcium bioavailability of effervescent potassium calcium citrate over tablet formulation of calcium citrate after Roux-en-Upsilon gastric bypass. *Surgery for Obesity and Related Diseases* 9(5), 743-748.

Skibsted, L. H. 2016. Mineral nutrient interaction: improving bioavailability of calcium and iron. *Food Science and Biotechnology* 25(5), 1233-1241.

Spanos, N., Patis, A., Kanellopoulou, D., Andritsos, N. and Koutsoukos, P. G. 2007. Precipitation of calcium phosphate from simulated milk ultrafiltrate solutions. *Crystal Growth and Design* 7(1), 25-29.

Tang, N. and Skibsted, L. H. 2016. Calcium binding to amino acids and small glycine peptides in aqueous solution: toward peptide design for better calcium bioavailability. *Journal of Agricultural and Food Chemistry* 64(21), 4376-4389.

Tanguy, G., Siddique, F., Beaucher, E., Santellani, A.-C., Schuck, P. and Gaucheron, F. 2016. Calcium phosphate precipitation during concentration by vacuum evaporation of milk ultrafiltrate and microfiltrate. *LWT - Food Science and Technology* 69, 554-562.

Tanguy, G., Tuler-Perrone, I., Dolivet, A., Santellani, A. C., Leduc, A., Jeantet, R., Schuck, P. and Gaucheron, F. 2019. Calcium citrate insolubilization drives the fouling of falling film evaporators during the concentration of hydrochloric acid whey. *Food Research International* 116, 175-183.

Thybo, C. D., Lillevang, S. K., Skibsted, L. H. and Ahrné, L. 2020. Calcium balance during direct acidification of milk for Mozzarella cheese production. *LWT* 131, 109677.

Van Kemenade, M. J. J. M. and De Bruyn, P. L. 1987. A kinetic study of precipitation from supersaturated calcium phosphate solutions. *Journal of Colloid and Interface Science* 118(2), 564-585.

Vavrusova, M., Danielsen, B. P., Garcia, A. C. and Skibsted, L. H. 2018. Codissolution of calcium hydrogenphosphate and sodium hydrogencitrate in water. Spontaneous supersaturation of calcium citrate increasing calcium bioavailability. *Journal of Food and Drug Analysis* 26(1), 330-336.

Vavrusova, M., Garcia, A. C., Danielsen, B. P. and Skibsted, L. H. 2017a. Spontaneous supersaturation of calcium citrate from simultaneous isothermal dissolution of sodium citrate and sparingly soluble calcium hydroxycarboxylates in water. *RSC Advances* 7(6), 3078-3088.

Vavrusova, M., Johansen, N. P., Garcia, A. C. and Skibsted, L. H. 2017b. Aqueous citric acid as a promising cleaning agent of whey evaporators. *International Dairy Journal* 69, 45–50.
Vavrusova, M., Liang, R. and Skibsted, L. H. 2014. Thermodynamics of dissolution of calcium hydroxycarboxylates in water. *Journal of Agricultural and Food Chemistry* 62(24), 5675–5681.
Vavrusova, M., Munk, M. B. and Skibsted, L. H. 2013. Aqueous solubility of calcium L-lactate, calcium D-gluconate, and calcium D-lactobionate: importance of complex formation for solubility increase by hydroxycarboxylate mixtures. *Journal of Agricultural and Food Chemistry* 61(34), 8207–8214.
Vavrusova, M. and Skibsted, L. H. 2013. Calcium binding to dipeptides of aspartate and glutamate in comparison with Orthophosphoserine. *Journal of Agricultural and Food Chemistry* 61(22), 5380–5384.
Vavrusova, M. and Skibsted, L. H. 2014. Calcium nutrition: bioavailability and fortification. *LWT - Food Science and Technology* 59(2), 1198–1204.
Vavrusova, M. and Skibsted, L. H. 2016. Aqueous solubility of calcium citrate and interconversion between the tetrahydrate and the hexahydrate as a balance between endothermic dissolution and exothermic complex formation. *International Dairy Journal* 57, 20–28.
Wang, J., Aalaei, K., Skibsted, L. H. and Ahrné, L. M. 2020a. Lime juice enhances calcium bioaccessibility from yogurt snacks formulated with whey minerals and proteins. *Foods* 9(12).
Wang, L. M., Wang, W., Li, X. C., Peng, L., Lin, Z. Q. and Xv, H. Z. 2012. Calcium citrate: a new biomaterial that can enhance bone formation in situ. *Chinese Journal of Traumatology* 15(5), 291–296.
Wang, Q., Holt, C., Nylander, T. and Ma, Y. 2020b. Salt partition, ion equilibria, and the structure, composition, and solubility of micellar calcium phosphate in bovine milk with added calcium salts. *Journal of Dairy Science* 103(11), 9893–9905.
Zhao, L., Huang, S., Cai, X., Hong, J. and Wang, S. 2014. A specific peptide with calcium chelating capacity isolated from whey protein hydrolysate. *Journal of Functional Foods* 10, 46–53.

# Part 4

## Optimising milk composition and quality

# Chapter 14

# Advances in instrumental analysis of dairy products

*Paul Andrewes and Stephen E. Holroyd, Fonterra Research and Development Centre, New Zealand; Brendon Gill and Richard Johnson, Fonterra Co-operative Group Ltd, New Zealand; and Andrew Lewis, Andrew MacKenzie and Mikhail Vyssotski, Callaghan Innovation, New Zealand*



## 1 Introduction

Milk is a complex biologically derived matrix with many constituents. Combined, these afford milk a healthy, nutritious profile that is key to its value as a globally recognised food. The effective use of analytical science as a key component in a quality system contributes markedly. Techniques for the measurement of fat, protein, lactose, minerals and micronutrients have been developed over many years using a variety of analytical technologies. In this chapter, we specifically focus on the routinely applied technologies that are used for the assessment of both suppliers' milk and the composition of the many products that are made from it. We do not cover the wide range of analytical technology that is used to assess milk and dairy products for specific contaminants or those approaches that routinely assess the microbiological profile of milk and dairy products.

The measurement of raw milk as part of the milk collection process is well established in most countries. Traditional wet chemistry quality tests have increasingly been supplanted with both rapid and more advanced methods

http://dx.doi.org/10.19103/AS.2022.0099.22

**Table 1** Summary of performance of analytical techniques

| Technique | IR | ICP | XRF | NMR | LC–MS | GC–MS |
|---|---|---|---|---|---|---|
| Speed | High | Low | High | High | Low | Low |
| Cost | Low | High | Low | High/Low | High | Medium |
| Sensitivity | Low | High | Medium | Medium | High | High |
| Specificity | Composition | Minerals | Minerals | Composition, specific nuclei | Micronutrients, specific proteins | Specific fats, flavour compounds |

for assessing the composition and microbiological quality. These tests are also extended right through the value chain, from the raw milk after collection to the great variety of dairy-based ingredients and foods that are manufactured globally. The ability to test for milk composition at levels of detail beyond total fat, protein and lactose has also benefited from significant developments in recent years and is clearly an area of great future value.

Milk arrives at a processing facility and is typically split into its component streams and then recombined into a range of products. Extensive measurement occurs before, during and after milk processing. Effective sampling and accurate measurement, which depend on the nature of the product, are important aspects of routine testing. Effective analysis then provides results that underpin the quality specifications of each product batch. Adherence to international standards in the application of testing is also a crucial factor that ensures the maintenance of consistent quality. In this chapter, we cover established analytical technologies that are applied for milk composition across the value chain, including infrared (IR) spectroscopy, inductively coupled plasma (ICP), X-ray fluorescence (XRF) and nuclear magnetic resonance (NMR) spectrometry, along with liquid and gas chromatography with mass spectrometric detection (LC–MS and GC–MS). Each technique is widely used in the dairy industry and performs a different role, depending on its specific analytical capability. We have summarised our views with respect to their performance in Table 1, with factors such as speed and cost of analysis along with sensitivity (can the technique measure analytes at low concentration?) and specificity (what sort of analytes can be measured?).

In this chapter, we focus on areas in which there have been significant advances in the last 10–20 years, either in the analytical technology itself or, in some cases, where existing technology has found new areas of application within the dairy value chain.

## 2 Infrared spectroscopy

The principle of IR spectroscopy relies on the precise measurement of the absorption of IR radiation by specific chemical bonds in the sample. The fat, protein, lactose and total solids content in every sample of milk supplied by farmers is typically analysed prior to milk processing. In most regions, these values determine the quality of the milk, as the basis on which farmers are paid. Both mid-infrared (MIR) spectroscopy and near-infrared (NIR) spectroscopy, often using the Fourier transform infrared (FTIR) principle, and with transmsission, reflectance or transflectance format instruments, have succeeded in replacing both the more laborious, time-consuming and expensive laboratory test methods and the earlier filter-based MIR spectrometers. We note the recent publication by the International Dairy Federation of two standards covering the application of IR spectroscopy to milk and milk products and an associated bulletin that details the practical performance of NIR spectroscopy in milk processing (International Dairy Federation, 2019; International Standards Organization, 2020a,b). We note that these publications cover the application of NIR spectroscopy to a wide variety of milk and milk products, along with how best to apply IR spectroscopy in-line to control milk processing. The former area is extensive and has been reviewed (Cattaneo and Holroyd, 2013); however, our perspective is that the major advances in the application of IR spectroscopy in the dairy area in the last decade centre more upon the application of MIR spectroscopy to liquid milk than milk products.

The current routine use of IR spectroscopy for the analysis of milk components is due largely to the development of purpose-built milk analysers, enabling rapid analysis of every sample of milk, along with effective collation of the spectral data collected. Whereas the technology of gathering IR spectral data has not changed significantly since the advent of FTIR spectroscopy, much research has been directed towards examining factors that affect the accuracy of the results, and also at increasing the range of components in milk that can be analysed by these instruments. In the last two decades, there has been an increasing level of research into applications of IR spectroscopy for milk in addition to gross composition, as indicated by the number of publications, as shown in Fig. 1.

Broadly, the areas of advances in the application of MIR spectroscopy to milk fall into five categories.

- Assessment of detailed milk composition where, rather than total fat or protein being measured, some subset of these components, such as a group of fatty acids, whey proteins or caseins or a specific fatty acid or protein in some cases, is measured. Included in this area are studies seeking to understand the impact of feed types on milk composition and traits in milk that are useful for determining how it should be processed.

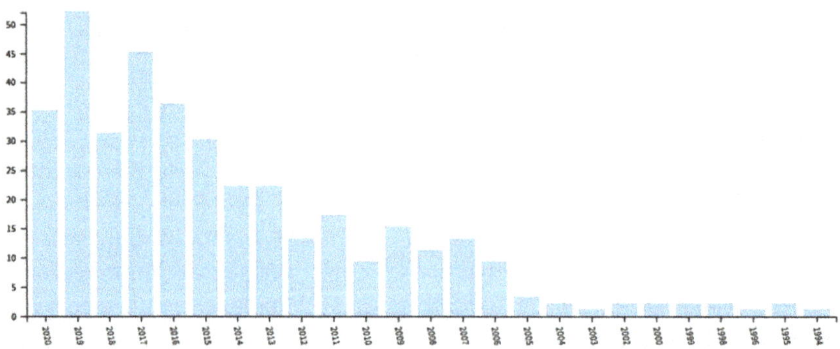


**Figure 1** Topic search Web of Science core collection 'infrared and dairy', 1994–2020.

- Investigation of the use of genetic information from herd management systems and correlating this with IR spectral data gathered from the milk of individual cows.
- Measurement of other micronutrient parameters of milk such as vitamins or mineral content.
- Indication of cow physiological status.
- Assessment of milk authenticity or geographical origin.

FTIR spectral data along with other data such as laboratory results have been used in all these cases to build models; in many cases, these models have then been applied to FTIR data collected when routine measurement is carried out upon milk collection. This can mean that the results from many millions of milk samples are rapidly analysed. Figure 2 shows the type of IR spectral absorbance data across the wavelength range of 1000 $cm^{-1}$ to 1600 $cm^{-1}$ that are collected by FTIR instruments and illustrates the variety of absorbances that are typical.

We note that, in some cases, there has been a need to move further from the traditional creation of a correlation between IR spectral data and laboratory results, and thus, data as diverse as on-farm feed quality, methane emissions and genetic parameters have all been used to create correlations that provide information of value to the dairy industry. In these cases, it is important to have some measure of assessing the utility of these calibrations. For example, can they be used to predict new results, or should they be used only for screening purposes so that high-value testing can be better targeted? In order to address this, we note many of the papers referenced

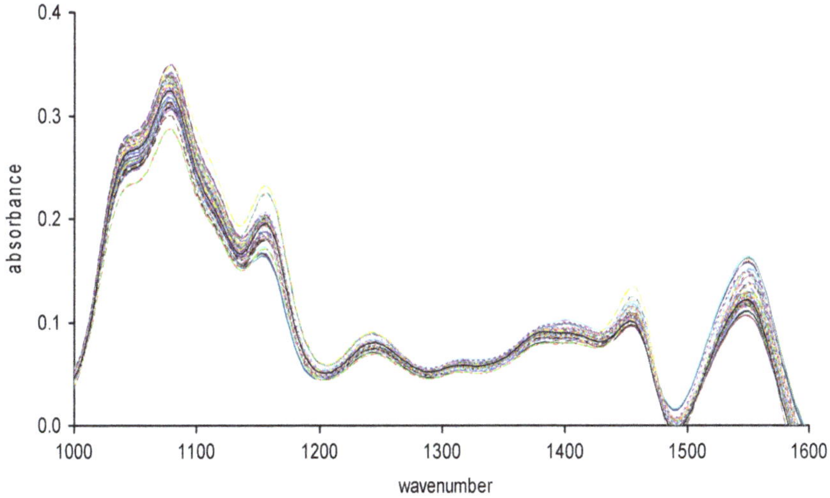


**Figure 2** FTIR spectra of 100 different milk samples.

in this area use independent calibration and validation sets. This is coupled with the use of statistical approaches such as Pearson correlations and the use of the ratio performance deviation (RPD) statistic, that is the ratio of the standard error of prediction to the standard deviation of each characteristic. This affords some measure of how such a calibration can be used and how robust the predictions derived from it can be expected to be in the context of its application.

The following discussion gives some examples of each of the five applications mentioned above; the references used are not exhaustive but collectively cover the wide range of new applications of IR spectroscopy to milk that have occurred over the last decade.

Some of the earliest work on detailed milk composition looked at using FTIR data to predict bovine milk fatty acids across multiple breeds, production systems and countries (Soyeurt et al., 2011). Numerous studies have followed this trend, with both groups of fatty acids and individual fatty acids in milk being predicted with a range of accuracies. Several of these studies have shown a clear trend; that is, prediction is more reliable when the statistics are expressed in terms of quantity of fatty acid/quantity of milk rather than in the more common laboratory-reported result of grams of fatty acid/grams of extracted fat. Our assumption is that this is due to FTIR spectroscopy obtaining spectral data from the milk samples and a laboratory result expressed in the same terms are better aligned. Detailed protein compositions have also been explored in the last decade (Bonfatti et al., 2011); however, in contrast, the chemical complexity of dairy proteins has made relative accuracy challenging (Sørensen et al., 2003; De Marchi et al., 2009b).

More recently, FTIR data have allowed prediction models to be applied to over 90 000 routinely collected milk spectra to investigate correlations of the predicted protein and free amino acid values with cow and environmental factors, including time of milking, stage of lactation and breed of cow. This work amply demonstrates that MIR spectroscopy models that could be regarded as having relatively low prediction accuracy still have considerable utility for application as a screening tool at a herd or national level (McDermott et al., 2017). Several studies have analysed fine milk composition by FTIR spectroscopy to assess the impact of feed types on milk composition (Coppa et al., 2012; Capuano et al., 2014). FTIR spectroscopy has also been used successfully to predict technological traits of potential relevance to milk processing, including milk coagulation properties, titratable acidity and casein micelle size (Dal Zotto et al., 2008; Cecchinato et al., 2009; De Marchi et al., 2009a; Visentin et al., 2015).

An area that extends the application of IR spectroscopy beyond that of correlating with laboratory data is the prediction of genetic traits, including genome-wide association traits and various heritable features such as fatty acid profiles, methane emissions, milk calcium content and resistance to infection (Bittante et al., 2014; Cecchinato et al., 2015; Vanlierde et al., 2016). Allied to this, one of the most valuable uses of IR spectra for dairy management lies in the modelling of milk components, by exploiting the inferential nature of the relationship between the trait of interest and the MIR spectrum. For example, in predicting the methane, ketosis or energy balance from the MIR spectra of milk, there is no single IR peak or group of peaks that changes; rather, subtle changes in several spectral regions when combined in a multivariate manner can signal these matrix changes. Being able to predict these traits from IR data can be used to define best practices, adjust feeding and health management, improve animal welfare, improve milk quality and mitigate environmental impact, as has been well illustrated by Gengler et al. (2016).

IR spectroscopy has shown poor prediction abilities for minerals such as sodium and magnesium, but has been effective in estimating calcium and phosphate contents (Soyeurt et al., 2009; Toffanin et al., 2015). However, it is not possible to directly observe peaks associated with the presence of minerals; most of the calibrations are based on the indirect impact of the ionic strength of the solution on spectroscopic observations. Further work by Visentin et al. (2016) has investigated predicting the macro milk minerals such as calcium, potassium, magnesium, sodium and phosphorus. The prediction of vitamins naturally present in milk presents a unique challenge. They are chemically complex and are present at concentrations well below those that could be expected to be quantified by IR spectroscopy. Although several studies have claimed some ability to assess vitamins, it remains an area where other analytical technologies would seem to be more reliable.

In addition, other minor components related to cow health and productivity, such as ketones, body energy status, milk colour, citrate and blood metabolites, have also been successfully predicted by FTIR spectroscopy (de Roos et al., 2007; van Knegsel et al., 2010; Calamari et al., 2016; Benedet et al., 2019; Smith et al., 2019).

Milk authenticity has been an area of focus since melamine was deliberately used in 2008 to adulterate milk to augment apparent protein values. Some work has focused on the questionable ability of IR spectroscopy to detect low concentrations of adulterants in either liquid milk or products such as milk powder. More promising has been the development of both targeted and non-targeted applications using FTIR spectroscopy to detect the presence of some specific adulterants, as well as models that can indicate, with a high probability of success, whether a milk sample is different from a representative range of previous milk samples, and thus may be investigated further to detect any potential adulteration (Hansen and Holroyd, 2019). We also note that Raman spectroscopy has shown some success in this area (Yazgan et al., 2020).

Raman spectroscopy has been widely applied to other dairy systems, but its main advantages would appear when allied with imaging techniques that can differentiate dairy product microstructural information. IR spectroscopy has also been used to determine the geographical origin of milk and dairy products, such as cheese from a certain region (Caredda et al., 2017; El Jabri et al., 2020).

Given the ease of use of IR spectroscopy and its widespread deployment, there has been a wealth of new applications to milk prior to processing in the last decade. From assessment of fine milk composition to correlation with genetic information from herd management systems and indications of a cow's physiological status, each application adds value and ensures the provision of safe nutrition by the global dairy industry.

## 3 Inductively coupled plasma spectrometry

ICP-atomic emission spectrometry (ICP-AES) and ICP-mass spectrometry (ICP-MS) rely on the ionisation of the sample via the plasma to create atomic and polyatomic ions that, when detected, are representative of the parent compound. They have been the techniques of choice for macro mineral and trace mineral analyses for more than a decade, largely replacing flame and graphite furnace atomic absorption spectrometry, because of their simultaneous multi-element capability.

Although there is wide application of these techniques, there have been few significant recent advances in the application of ICP spectrometry to the analysis of dairy products, beyond the publication of two standards by the International Dairy Federation. These provide for the analysis of calcium (Ca), copper (Cu), iron (Fe), magnesium (Mg), manganese (Mn), phosphorus (P), potassium (K), sodium (Na), zinc (Zn), chromium (Cr), molybdenum (Mo) and selenium (Se) by ICP-MS; and Ca, Cu, Fe, Mg, Mn, P, K, Na and Zn by ICP-AES (International Standards Organization, 2019, 2020c).

The majority of the recent published research (over 60 papers in the last 10 years) relates to the application of ICP spectrometry as a means to survey mineral elements of interest in regional milks and other dairy products rather than to develop the analytical technique, with some other examples of this work being examined statistically to infer dairy product authenticity (Bandoniene et al., 2020).

## 4 X-ray fluorescence spectrometry

XRF spectrometry depends upon the emission of characteristic secondary X-rays from a sample that is bombarded with high-energy X-rays. The application of XRF spectrometry for the mineral analysis of dairy matrices has increased markedly over the last two decades. This has coincided with the relatively rapid development of detectors utilised in energy dispersive (EDXRF) instruments, which has facilitated the analytical performance of bench-top analysers to approach and, for some elements, to surpass the performance of larger laboratory-based wavelength dispersive instruments.

Although XRF spectrometry lacks the sensitivity of ICP spectrometry for trace element determination, which is particularly relevant to the measurement of potential toxic contaminants such as lead (Pb), cadmium (Cd) and mercury (Hg), it is well suited to the measurement of nutritionally important minerals in milk products including Na, Ca, P, K, Mg, Zn and Fe. Total reflectance XRF spectroscopy provides the potential to determine trace elements at sub parts-per-million (ppm) concentrations (Araújo et al., 2019); however, spectral overlaps with macro elements mean that some important potential contaminants, such as Pb, cannot be determined at useful limits of detection.

One of the challenges of XRF measurement is the means of calibration. The technique is commonly a comparative method that requires a set of calibration samples. Options include the use of reference materials, synthetic standards, samples analysed by reference methods, standard addition and internal standard; for further information, readers are referred to the more comprehensive review of Pashkova et al. (2018).

A major advantage of XRF spectrometry is that many samples require little preparation prior to analysis. Powder samples can be pressed into pellets or packed loosely into cups with an X-ray film base, such as Mylar. Liquid samples can be analysed directly in similar cups or can be freeze-dried. These non-destructive approaches contrast with the need to mineralise samples for ICP spectrometry and atomic absorption spectrometry by either wet ashing or dry ashing.

The simplicity of sample preparation, routine instrument operation, speed of analysis and relatively low cost make EDXRF spectroscopy a particularly suitable tool for the control of minerals in milk-based powders (Perring and Andrey, 2003). Instruments can be located in processing plants (at-line), can be operated with minimal training and can provide results within minutes, facilitating an effective feedback loop for process control.

A recent novel application was for the determination of the ash content of milk-based powders (Perring and Tschopp, 2019). EDXRF spectrometry provided results in a few minutes, compared with the hours required for classical determination in a furnace. Rather than evaluation and summation of the individual elemental components representing the ash content, the calibration was developed using a region of interest that included the major elements and a second region of interest representing the matrix composition. Overall, XRF spectrometry has become an important rapid technique for the analysis of nutritionally important minerals in dairy products.

## 5 Nuclear magnetic resonance spectroscopy

Hardly any other analytical method offers such a broad range of applications as does NMR spectroscopy. Because of the amount of chemical information provided by an NMR spectrum, which informs the analyst about the number, type and chemical environment of the nuclei being investigated, and even the concentration of the compounds present, NMR spectroscopy is the primary technique that is used to determine the precise chemical structure of molecules. Consequently, NMR spectroscopy has been an essential technique in the pharmaceutical industry and most chemical research facilities for decades. In recent years, NMR spectroscopy has also begun to be applied to the analysis of foods, food ingredients and related components – such as proteins, lipids and phospholipids – in many products, including dairy products.

Although numerous chromatographic and MS-based techniques have been perfected for the qualitative and quantitative analysis of different classes of natural and artificial components found in dairy products, these techniques are targeted and therefore highly discriminatory (e.g. the setup used for the analysis of fatty acids may not be directly applied to the determination of the protein content of the same sample on the same instrument). These techniques usually involve direct contact of the samples with the equipment (so that the sample is destroyed) and often require lengthy or complicated sample preparation to make the compounds amenable to analysis (e.g. derivatisation of esterified and free fatty acids present in the sample lipids into volatile methyl, *iso*-propyl or butyl esters for gas chromatographic analysis) or to remove unwanted components, such as water, protein, ions or sugars, that might interfere with the analysis or damage the equipment. In contrast, NMR samples are contained in glass tubes (so that they are recoverable) and NMR spectrometers can perform literally tens of different experiments on each sample, each one designed to provide specific information. Because many elements in the periodic table have NMR-active nuclei including hydrogen, carbon, nitrogen, phosphorus and sodium, which are present in dairy products, a huge variety of NMR measurements is possible, especially considering that many common NMR experiments use combinations of nuclei (e.g. $^{1}$H and $^{13}$C or $^{1}$H and $^{31}$P) to look for specific compounds or chemical subunits; and some can be used to investigate the presence of a few compounds in complex mixtures such as milk. Advances in automation and software now permit the routine analysis of many types of samples using different experiments on the same NMR instrument, often without the need to manually change the setup between samples.

Although often used for qualitative analysis (such as detecting the presence or absence of compounds) or for determining chemical structures, with proper procedures, NMR spectroscopy is fundamentally a quantitative technique. The

NMR response can be calibrated using chemicals that are different from the compound being analysed (which may be difficult or impossible to obtain commercially or in pure form), and the chemical being used for the calibration may be added in the sample being analysed (internal standard), or contained in a separate sample (external standard), and is not required every time the experiments are run (which can result in significant cost and time savings). Another important attribute of NMR spectroscopy is that it is highly reproducible, so that spectra acquired on different days or are run on NMR instruments operating in different laboratories can be virtually indistinguishable, opening up the possibility to build libraries of reference spectra from, for example, milk samples, for use in chemometric modelling to predict geographical origin or animal source (Bertram, 2018). The non-discriminatory nature of NMR spectroscopy means that it detects all of the compounds containing, for example, hydrogen or phosphorus nuclei in a mixture, which is extremely helpful in the determination of food adulteration and economically motivated fraud (for a recent review, see Sobolev et al., 2019).

The progress in the advancement of NMR techniques during the last decade is astounding. Hardware such as cryogenically cooled NMR probes that allow reductions in the sample amount from tens of milligrams down to micrograms is a good example. In addition, the use of gradient coils, selective excitation, clever sampling paradigms that allow data collection simultaneously rather than sequentially and advanced data processing techniques and algorithms that permit non-uniform sampling of data or prediction to improve sensitivity and resolution have combined to dramatically reduce analysis times from hours to minutes, or even seconds. Other developments in experimental protocols and software have occurred; they now allow measurements of the total fat, the spatial distribution of water in the product, the detection of unknown impurities and the quantification of processing-related artefacts. Even minor components of interest (e.g. different isomers of reportedly biologically active, conjugated polyunsaturated fatty acids) can now be detected using modern NMR approaches (e.g. Papaemmanouil et al., 2015).

Small, relatively inexpensive bench-top NMR spectrometers have also been developed and are increasingly being designed and sold to automate the analysis of samples in industrial settings either at-line (beside processing lines) or off-line (e.g. in testing laboratories in production facilities). These generally incorporate permanent magnets operating at much lower magnetic field strengths (1–2 T versus the 7–20 T traditional superconducting NMR magnets) and can be quite compact in size and modestly priced. The spectral resolution and sensitivity of these bench-top NMR spectrometers are lower than that of high-field NMR spectrometers, but are often sufficient for answering processing-related questions or measuring quality parameters. One more technological development, so-called in-line NMR spectroscopy, looks especially appealing for industrial applications. Whereas the 'classic' NMR spectrometer requires placing each individual sample into a special glass tube followed by loading it into the instrument's magnet, manually or with the use of an auto-sampler, in-line NMR spectrometers are made to be used with a conveyor belt. For example, a side-field NMR spectrometer might be added to an existing transporter as a stand-alone device. Apart from being a non-destructive technique, in-line NMR spectroscopy has the additional advantage of being able to analyse the product without the need to open the packaging (Colnago et al., 2014).

$^{1}$H and $^{13}$C are still the most common nuclei that are studied by NMR spectroscopy, mainly because $^{1}$H is the most sensitive NMR-active nucleus, and the majority of the compounds of interest contain both hydrogen and carbon atoms. Consequently, many NMR techniques have been developed for their study. $^{1}$H NMR spectroscopy has been used for 'fingerprinting' dairy products, and as a screening tool to validate the nutrition labelling of milk (Monakhova et al., 2012). $^{1}$H NMR analysis has been used to measure the metabolites that are present (and their concentrations) in a variety of dairy products (Scano et al., 2019), and has also been used to estimate the quality, origin or physical properties of milk such as its coagulation properties (e.g. Sundekilde et al., 2014). The adulteration of butter with margarine has been investigated using $^{1}$H NMR spectroscopy and principal component analysis (Obeidat, 2020), and formaldehyde has been measured in raw milk using time-domain NMR spectroscopy (Coimbra et al., 2020).

$^{31}$P NMR spectroscopy is one of the more routine NMR techniques because $^{31}$P has an isotopic abundance of 100% and a relatively high gyromagnetic ratio, making it relatively easy to detect. $^{31}$P chemical shifts are affected by changes in the chemical structure of the compounds, often many chemical bonds-removed, which allows many different types of quite similar compounds to be distinguished in a mixture (Fig. 3). For dairy products, the analysis of major phospholipid classes by $^{31}$P NMR spectroscopy is now routinely performed by some commercial analytical laboratories. A key feature of dairy phospholipids (as opposed to soy lecithin) is the similarity of the phospholipid class profile to that of human milk (e.g. the presence of phosphatidylserine and sphingomyelin), which makes it more suitable for applications such as infant formula.

Some ingredients (e.g. buttermilk powder) are rich enough in phospholipids that measurement can be accomplished in an aqueous detergent with no extraction or enrichment steps required (MacKenzie et al., 2009), whereas finished products such as infant formula generally require at least a lipid extraction prior to analysis (Zhu et al., 2019). $^{31}$P NMR spectroscopy can also quantify phospholipid subclasses (e.g. ethanolamine plasmalogen) that are not as abundant in cow milk as in human milk (Garcia et al., 2012). Artefacts of processing

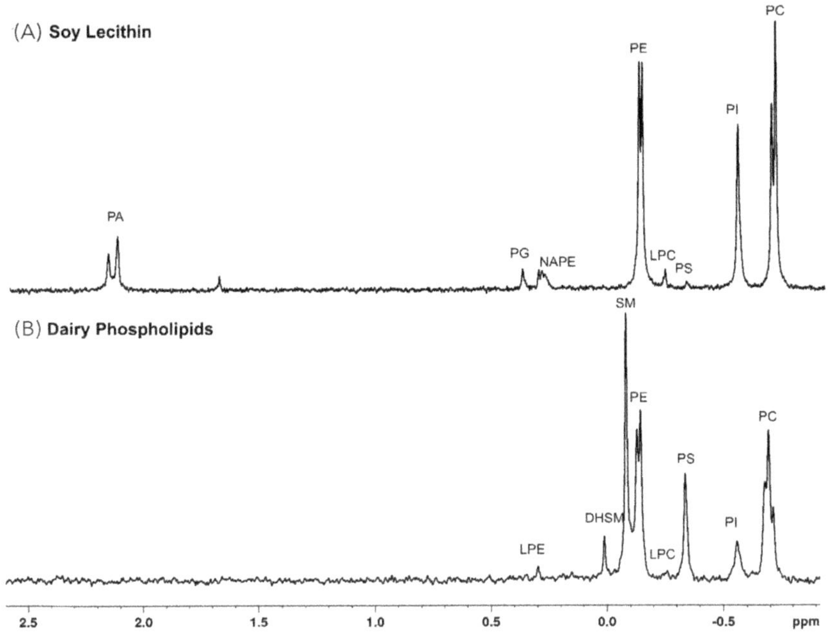


**Figure 3** $^{31}P$ NMR analysis of (A) soy lecithin and (B) dairy phospholipids in a sodium cholate detergent at pH 7.3. PA, phosphatidic acid; PG, phosphatidylglycerol; LPE, lysophosphatidylethanolamine; NAPE, N-acyl phosphatidylethanolamine; DHSM, dihydrosphingomyelin; PE, phosphatidylethanolamine; SM, sphingomyelin; PS, phosphatidylserine; LPC, lysophosphatidylcholine; PI, phosphatidylinositol; PC, phosphatidylcholine.

(e.g. the Amadori product of lactose and phosphatidylethanolamine formed during the spray-drying or storage of powders) may also be measured by this technique. Other phosphorus-containing compounds present in milk (e.g. *N*-acetyl-D-glucosamine 1-phosphate and glycerophosphocholine) have been observed using $^{31}P$ NMR spectroscopy (Belloque et al., 2000). Moreover, the measurement of phosphoserine in milk has been used to quantify its casein content (Belloque and Ramos, 2002) (Fig. 4). In addition, $^{31}P$ spectroscopy has been used to study the hydrolysis of polyphosphate additives in cheese (Barth et al., 2017).

Although mainly applied to liquid samples or solutions, solid samples such as cheese can also be analysed directly by solid-state NMR spectroscopy, using the magic angle spinning experiment. Such an approach has been used successfully to assess both the quality and the traceability of mozzarella cheese (Mazzei and Piccolo, 2012).

An interested reader is referred to the review article by Maher and Rochfort (2014), 'Applications of NMR in dairy research'. An older review, 'Application of NMR spectroscopy to milk and dairy products' by Belloque and Ramos (1999),

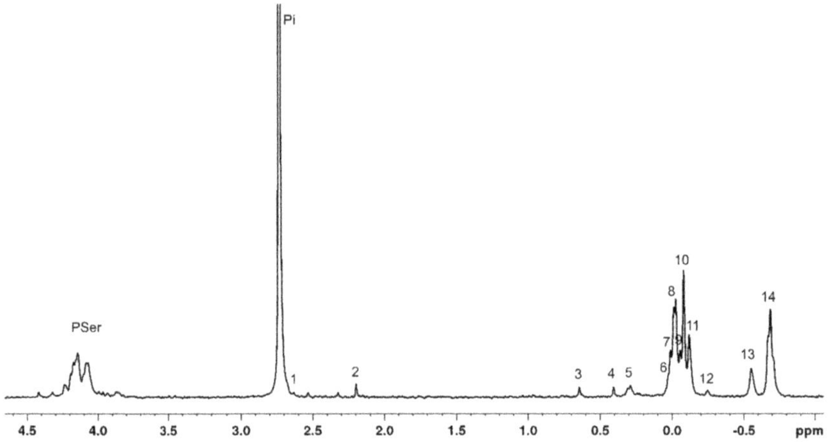


**Figure 4** $^{31}P$ NMR analysis of beta serum powder in a sodium cholate detergent at pH 9. PSer, protein-bound phosphoserine; Pi, inorganic phosphate; 1, galactose 1-phosphate; 2, N-acetyl-D-glucosamine; 3, glycerophosphorylethanolamine; 4, lysophosphatidylethanolamine; 5, LacPE (Amadori product of phosphatidylethanolamine and lactose); 6, glycerophosphorylinositol; 7, dihydrosphingomyelin; 8, phosphat-idylethanolamine; 9, glycerophosphorylcholine; 10, sphingomyelin; 11, phosphatidylserine; 12, lysophosphatidylcholine; 13, phosphatidylinositol; 14, phosphatidylcholine.

is still very useful and can be recommended to those unfamiliar with NMR because it presents helpful basics in a convenient form.

## 6 Liquid chromatography-mass spectrometry

Liquid chromatography-mass spectrometry (LC–MS) is a sophisticated analytical technique that enables the determination of the identity and concentration of analytes in complex matrices. LC–MS consists of three major components: a chromatographic system, an atmospheric pressure ion source that acts as the interface between the liquid and gas phases, and a mass analyser.

LC–MS is a key analytical technology on which the evolving '-omics' methodologies are based. MS provides detection and quantification of proteins, lipids, and biologically active metabolites down to ultra-trace levels, in either targeted mode or untargeted mode. In the past decade, ongoing developments in MS instrumentation, combined with advancements in the handling of large data sets, have afforded researchers a powerful tool to obtain a better understanding of the plethora of complex compounds in biological samples. Over this time, LC–MS has also become ubiquitous in analytical laboratories as the technique of choice for many quantitative analytical applications, providing a unique capability for rapid, cost-effective and quantitative measurements of organic compounds in a wide variety of applications (Di Girolamo et al., 2013; Sargent, 2013).

The continuing demands for increased milk production and the development of new products deriving from milk require a deep insight into the composition and technological properties of milk. The use of mass spectrometry has greatly expanded our understanding of milk, enabling identification, quantification, and characterisation of milk proteins (Siciliano et al., 2013; Bär et al., 2019), carbohydrates (Fong et al., 2011), and lipids (Liu et al., 2015). LC–MS has become a vital tool for food safety in demonstrating milk authenticity and traceability (Caira et al., 2016; Chung et al., 2019), and detecting contamination (Flores-Flores and González-Peñas, 2017) and adulteration (MacMahon et al., 2012; Abernethy and Higgs, 2013; Hrbek et al., 2014) in milk and dairy products.

MS can be divided into two categories: low-resolution MS (LRMS) and high-resolution MS (HRMS). The key parameters differentiating the two categories are mass accuracy and resolution. Mass accuracy is a comparison of a measured value for the mass of an analyte with its exact mass, and is expressed in ppm. Resolution is expressed as the value of the full width at half of the peak's maximum value (FWHM) at a specific mass-to-charge (m/z) ratio. In general, low-resolution instruments resolve peaks with resolution values of less than 10 000 FWHM, whereas high-resolution instruments offer resolution exceeding 100 000 FWHM (Righetti et al., 2016; Moldoveanu and David, 2017; Rochat, 2019).

Low-resolution quadrupole mass spectrometers have been the first choice for quantitative analysis, because of their high sensitivity and selectivity. Single quadrupole instruments are less commonly used as they lack the selectivity required for detection in complex matrices and have increasingly been superseded by tandem MS (MS/MS) utilising triple quadrupole (QqQ) mass spectrometers.

A QqQ mass spectrometer consists of three quadrupoles arranged in series, with the first (Q1) and third (Q3) quadrupoles acting as first and second mass analysers respectively, and the ion fragmentation taking place in the second quadrupole (q2). The basic principle of MS/MS consists of the filtering of a precursor ion in Q1, fragmentation of this ion – usually by collision-induced dissociation – in q2, and measurement of the product ion(s) formed in Q3. By employing selected reaction monitoring (SRM), both Q1 and Q3 are set at a specific mass, allowing only a distinct fragment ion from a certain precursor ion to be detected, resulting in a better signal-to-noise ratio and thereby enhancing sensitivity and affording lower limits of quantification. If Q1 and/or Q3 is set to more than a single mass, this configuration is called multiple reaction monitoring (MRM). As scanning over the m/z range is slow on QqQ instrumentation and suffers from low sensitivity, analyses are typically performed with quadrupoles working in filtering mode only. SRM and MRM acquisition modes require the analyst to pre-select the precursor and product ions and the collision energies, and therefore do not allow for retrospective analysis of the sample (Sargent, 2013; Moldoveanu and David, 2017; Rochat, 2019).

HRMS mass analysers have significant advantages over LRMS mass analysers, particularly in the collection of full-scan spectra. The high scan rate, high resolution and high mass accuracy of these instruments enable (1) the screening of unknown compounds with a full MS scan, (2) the construction of large databases for detecting targeted compounds, and (3) the elimination of many interferences during analysis, thereby simplifying sample preparation. HRMS is gaining prominence in dairy analysis as analytical instruments become cheaper and more available to a broader audience, particularly with the development of the Orbitrap MS platform and the improvements to time-of-flight (TOF) instruments (Bianchi et al., 2018).

A TOF mass analyser separates ions based on their velocities as they travel over a fixed length. An ion with a large mass travels more slowly than an ion with a small mass, and the time it takes for an ion to move between the source and the detector allows its m/z ratio to be calculated. TOF MS is a non-scanning technique whereby a full mass spectrum is acquired simultaneously rather than by sequentially sweeping through a series of m/z values while acquiring the data. A TOF mass analyser allows for the detection of all ions over a nearly unlimited m/z range with a high acquisition speed and high mass accuracy, allowing for the detection and identification of both targeted and non-targeted analytes (Moldoveanu and David, 2017).

The Orbitrap mass analyser is an ion trap instrument in which moving ions are trapped in an electrostatic field in an orbital motion around the central electrode. The axial component of these oscillations is independent of the initial energy, position and angle, and depends only on the ion's m/z and the instrument. Because ions of different mass and charge oscillate at different frequencies, Fourier transform is employed to obtain oscillation frequencies, resulting in an accurate measurement of their m/z. The Orbitrap mass analyser can achieve very high resolutions of up to 480 000 FWHM with sub-ppm mass accuracy and provides analysts with the ability to unambiguously discriminate analytes over a wide mass range (Michalski et al., 2012).

Hybrid MS/MS instruments use various combinations of two or more mass analysers to provide the benefits of each into one mass instrument, thereby enhancing their ability to analyse molecules (Glish and Burinsky, 2008). A quadrupole TOF (Q-TOF) mass spectrometer has a similar configuration to that of a QqQ mass spectrometer, except that Q3 is replaced by a TOF tube. Q-TOF MS successfully combines the performance characteristics of both quadrupole and TOF instrumentation, offering high sensitivity, mass resolution and mass accuracy of precursor and product ions. The development of a quadrupole-Orbitrap (Q-Orbitrap) mass spectrometer combines high-performance quadrupole precursor selection with high-resolution and accurate-mass

Orbitrap detection and provides both broad screening capabilities and targeted quantification experiments. LC–Q-Orbitrap and LC–Q-TOF platforms have become popular amongst researchers for proteomics and high-throughput screening because of their high performance in the screening of complex matrices efficiently and accurately (Glish and Burinsky, 2008; Zubarev and Makarov, 2013; Senyuva et al., 2015; Allen and McWhinney, 2019).

A recent review of LC–HRMS and QqQ platforms showed that similar quantitative performance and suggestions of better sensitivity from QqQ instrumentation were not related to the mass analyser itself; instead, the result of the latest hardware developments being first introduced to QqQ instruments (Allen and McWhinney, 2019; Rochat, 2019). Quantitative analyses in HRMS operate in a similar manner as in LC–QqQ–MS. Firstly, LC is used to separate the components in a mixture and the separated components are introduced into the mass spectrometer. A list of target compounds is created, exact mass filtering and fragmentation are performed and then retention time and MS/MS fragmentation spectra are matched to targets. The extent to which different compounds ionise in the ion source is influenced by the properties of the compound and the mobile phase. To accurately quantify unknown concentrations of analyte, the instrument response needs to be calibrated against samples containing known concentrations (Sargent, 2013). Despite the large number of compounds that can be detected and identified by HRMS, the ability to obtain accurate quantitative information is limited by the availability of analytical standards. The application of statistical techniques to the quantification of non-targeted LC–MS screening without the use of standards is one approach that is used to overcome this barrier (Liigand et al., 2020).

In qualitative LC–MS/MS, there are two acquisition modes that may be used: data-dependent acquisition (DDA) and data-independent acquisition (DIA). In DDA, a full spectrum is acquired and selected precursor ions emerging from the first mass analyser are fragmented into smaller product ions, which are subsequently scanned in the second mass analyser to create product ion spectra. DDA uses a target list or criteria set around a detected response of a precursor ion of interest (e.g. when precursor ions exceed a predefined level of intensity, isotope pattern, mass defect or the presence of a diagnostic ion) and collects MS/MS fragment data for confirmation (Fenaille et al., 2017). Hence, the second MS/MS scan is dependent on the first MS scan. This technique eliminates the necessity to identify the target precursor ions in a sample in MS mode and then to re-run the sample in MS/MS mode to acquire product ion data from each of these precursors. Although DDA can be performed on low-resolution QqQ instruments, the slow scanning speed and long cycle time severely limit the number of product ion spectra that can be collected.

In DIA, all analytes within a predefined window are fragmented and analysed in the second stage of MS/MS, not just the precursors that are high in abundance. This results in fragment ions obtained from all compounds present in the sample, making data analysis challenging. The relationship between a precursor ion and its fragment ions is lost because the fragment ions in DIA spectra may potentially result from multiple precursor ions, and large spectral libraries are typically required to interpret the multiplexed fragment spectra.

Variable DIA (vDIA) has variably sized mass windows that capture product ion spectra from a smaller range of precursor ions. Selectivity and sensitivity are improved with vDIA, compared with DDA, while also generating complete full scan and fragmentation data of the sample (Jia et al., 2020).

Sequential window acquisition of all theoretical fragment ion spectra (SWATH) allows untargeted qualitative and quantitative analyses of multiple components. In SWATH MS, all ionised components that fall within the isolated precursor mass range are fragmented and recorded in a systematic and unbiased manner. The mass spectrometer rapidly cycles through the entire mass range within a few seconds, with a series of product ion scans recorded continually over the course of chromatographic peak elution. The information in SWATH MS data needs to be extracted from highly multiplexed product ion spectra, which requires advanced processing strategies often utilising known fragment spectra from large spectral libraries. SWATH MS records all fragment ions of the detectable precursors present in a biological sample, and combines the advantages of the high throughput associated with DDA with the reproducibility and sensitivity of targeted methods (Ludwig et al., 2018).

Studies of milk utilising MS have focused on analysing the composition of different fractions such as proteins, milk fat globule membrane and lipids (Verma and Ambatipudi, 2016). Milk contains a relatively small number of major proteins; however, a complete profile is complex because of the presence of protein variants arising from genetic differences between species, polymorphism and post-translational modification (Abd El-Salam, 2014). MS and proteomic techniques have been applied extensively to the analysis of milk proteins derived from casein, whey and milk fat globule membrane (Zhang et al., 2011; Le et al., 2013; Verma and Ambatipudi, 2016; Vincent et al., 2016; Wang et al., 2017, 2021).

The analysis of lipids from milk can be challenging as they range from polar (phospholipids and other membrane lipids) to non-polar (triacylglycerols). Polar lipids are readily separated by hydrophilic interaction liquid chromatography, with subsequent detection by MS (Liu et al., 2015). Profiles of non-polar triacylglycerols and long-chain polyunsaturated fatty acids in milk have been surveyed using the Orbitrap, QqQ and Q-TOF–MS techniques (Zhou et al., 2014; Liu et al., 2017; Tu et al., 2017).

The authenticity of milk products is an issue of extreme importance to the dairy industry and MS has proven to be an invaluable technique in detecting milk adulteration. For example, an untargeted LC–Q-Orbitrap strategy applying chemometric techniques has been used to observe subtle differences in the molecular profiles of raw milk from different animal species (Jia et al., 2020), and a targeted LC-Q-Orbitrap MS method was developed to use lipid biomarkers to differentiate milk of different species as a means to detect adulteration (Li et al., 2017).

The paradigm for performing LC-MS analysis has been that QqQ MS is generally suited for routine quantitative analyses, whereas complex and expensive HRMS instruments are the reserve of qualitative and research analyses. Although QqQ MS remains the mainstay of analytical laboratories, a significant advantage of using HRMS accrues from the ability to record a full scan and detect all ions in a sample, either in real time or by retrospective analysis of archived data. This capability is significant if new knowledge on a previously unknown or suspected compound becomes a concern (Di Girolamo et al., 2013; Rochat, 2019; Aydoğan, 2020).

The ability of LC-HRMS to perform qualitative analysis is becoming increasingly important, such that LC-HRMS can undertake the core task of quantitative analysis with additional benefit of being able to monitor other compounds of interest, both at the time of sample analysis and in post-acquisition data mining, without the need for any extra work involved in reanalysing the sample (Fung et al., 2013).

The improvements in the performance of HRMS for quantitative analysis, comparable with that of QqQ MS in many instances, coupled with the additional benefits in qualitative and untargeted analyses, provide a versatility that makes HRMS an attractive option for dairy analyses. In the future, it is likely that most quantitative analyses using LC-MS will be accomplished using HRMS, with QqQ MS limited to those assays that require the sensitivity to measure at the lowest concentrations.

## 7 Gas chromatography-mass spectrometry

Gas chromatography (GC) (and, by extension, GC-MS) is considered to be a core technique in instrumental chemical analysis (Dorman et al., 2010). However, there have been few significant advances in the industrial application of this technique to dairy products.

The obvious reason for GC playing only a minor role in the dairy industry is that it is a technique that is intended for gases (including compounds that can be made into gases as vapours or derivatives). With most dairy components (Lucey et al., 2017) being non-gases (proteins, carbohydrates, lipids, nutrients, salts etc.), GC is reserved for specialised dairy analyses.

In the industrial dairy laboratory, GC (including GC-MS) is used to monitor contaminants (e.g. pesticide residues), perform specialised nutritional analyses (e.g. fatty acids as their methyl ester derivatives) and monitor flavour volatiles.

The use of GC and GC–MS to monitor contaminants in foods is not specific to the dairy industry and tends to advance slowly, because of understandably high regulation. As such, it is a subject that is best covered elsewhere (Hernández et al., 2013).

The derivatisation of fatty acids to their methyl esters and their subsequent analysis by GC, GC–MS or two-dimensional GC is a fascinating application of analytical chemistry to lipid science, including the so-called lipidomics (Jurowski et al., 2017). However, its application in the dairy industry to conduct nutritional analyses, or to monitor for milk fat adulteration, although useful, does not tend to lie at the cutting-edge of analytical science. Indeed, using a fatty acid analysis to detect adulteration is not even considered to be the best technique by some authors (Cossignani et al., 2019).

Flavour research is probably the most common application of GC and GC–MS that is described in dairy science journals. However, such research has not been translated into industrial applications. This is primarily because most problems that flavour analyses attempt to solve can also be solved by sensory evaluations (Schiano et al., 2017). In fact, many problems have demanded sensory input for their resolution, e.g. GC coupled with olfactometry.

Invariably, dairy flavour research has been about using very well-established techniques to determine relationships between flavour volatiles (or related diagnostic volatiles) and sensory attributes as a function of variables such as milk quality, ingredient composition or different processes. As such, dairy flavour researchers were early adopters of advances in instrumental analyses, but these advances were often not pivotal to the completion of the work. For instance, the analytical performance requirements of a flavour analysis are often not high, because concentration differences between samples are of no sensory significance below an order of magnitude; that is, a reproducibility RSD over 100% can be considered to be fit for purpose. Hence, relatively crude techniques, such as solid-phase microextraction, that might not be adopted elsewhere, are well established (Jalili et al., 2020).

Because most flavour research advances have been made in the wider area of food research, and considering the above comments, readers can obtain a deeper understanding of the use of GC and GC–MS in food and dairy flavour analysis from Chambers and Koppel (2013).

## 8 Conclusions

The advances reviewed above cover many of the analytical techniques that are routinely applied to the analysis of milk and dairy products. Although, in some areas, the technology has changed markedly and we have seen a significant increase in its application in dairy laboratories, in other areas, the technology

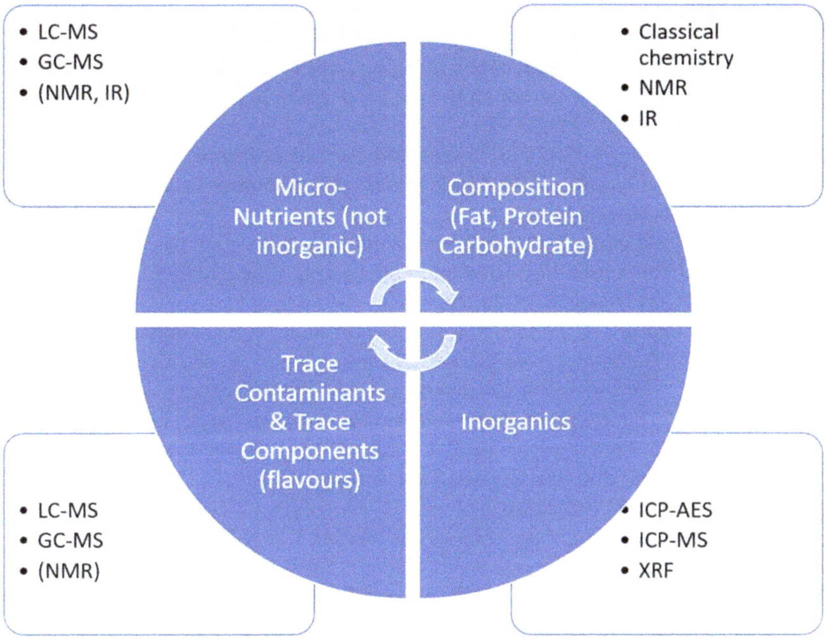


**Figure 5** Summary of techniques and areas of application of dairy products.

has remained similar, but the scope of application has been broadened. Figure 5 summarises how these techniques are applied to the range of compositional analyses routinely performed.

We also appreciate that dairy is a subset of the wider global food chain and has benefited from advances made in other areas. The only certainty that we predict for the future is that these analytical tools will continue to be widely used and further developed as part of a robust and safe dairy food chain, in conjunction with the associated international standards.

## 9 Where to look for further information

A consistent trend in recent years has been towards the miniaturization of a range of analytical technologies. This suggests that while the actual techniques may not change how and where they are applied may as the equipment becomes more mobile and easier to deploy. The dairy analytical standards development organisations quoted above, such as the International Dairy Federation, International Organisation for Standards and AOAC International, all have comprehensive work programmes devoted to new analytical standards for testing a wide range of dairy products.

## 10 References

Abd El-Salam, M. H. (2014). Application of proteomics to the areas of milk production, processing and quality control–a review, *Int. J. Dairy Technol.* 67(2), 153–166. doi: 10.1111/1471-0307.12116.

Abernethy, G. and Higgs, K. (2013). Rapid detection of economic adulterants in fresh milk by liquid chromatography-tandem mass spectrometry, *J. Chromatogr. A* 1288, 10–20. doi: 10.1016/j.chroma.2013.02.022.

Allen, D. R. and McWhinney, B. C. (2019). Quadrupole time-of-flight mass spectrometry: a paradigm shift in toxicology screening applications, *Clin. Biochem. Rev.* 40(3), 135-146. doi: 10.33176/AACB-19-00023.

Araújo, U. B., da Costa, A. C. M., de Oliveira, D. F., de Jesus, E. F. O., Anjos, M. J., Mársico, E. T., da Silva Carneiro, C., de Oliveira Resende Ribeiro, R. and Lopes, R. T. (2019). Analysis of milk trace elements with a home-made portable automated total reflection x-ray fluorescence system, *Radiat. Phys. Chem.* 156, 216–221.

Aydoğan, C. (2020). Recent advances and applications in LC-HRMS for food and plant natural products: a critical review, *Anal. Bioanal. Chem.* 412(9), 1973–1991. doi: 10.1007/s00216-019-02328-6.

Bandoniene, D., Walkner, C., Ringdorfer, F. and Meisel, T. (2020). Authentication of meat and dairy products using rare earth element labeling and detection by solution based and laser ablation ICP-MS, *Food Res. Int.* 132, 109106. doi: 10.1016/j.foodres.2020.109106.

Bär, C., Mathis, D., Neuhaus, P., Dürr, D., Bisig, W., Egger, L. and Portmann, R. (2019). Protein profile of dairy products: simultaneous quantification of twenty bovine milk proteins, *Int. Dairy J.* 97, 167–175, doi: 10.1016/j.idairyj.2019.01.001.

Barth, A. P., Tormena, C. F. and Viotto, W. H. (2017). pH influences hydrolysis of sodium polyphosphate in dairy matrices and the structure of processed cheese, *J. Dairy Sci.* 100(11), 8735–8743.

Belloque, J., de la Fuente, M. A. and Ramos, A. M. (2000). Qualitative and quantitative analysis of phosphorylated compounds in milk by means of 31P-NMR, *J. Dairy Res.* 67(4), 529–539.

Belloque, J. and Ramos, M. (1999). Application of NMR spectroscopy to milk and dairy products, *Trends Food Sci. Technol.* 10(10), 313–320.

Belloque, J. and Ramos, M. (2002). Determination of the casein content in bovine milk by 31P-NMR, *J. Dairy Res.* 69(3), 411–418.

Benedet, A., Franzoi, M., Penasa, M., Pellattiero, E. and De Marchi, M. (2019). Prediction of blood metabolites from milk mid-infrared spectra in early-lactation cows, *J. Dairy Sci.* 102(12), 11298–11307.

Bertram, H. C. (2018). NMR-based metabolomics: quality and authenticity of milk and meat. In: Webb, G. A. (Ed), *Modern Magnetic Resonance* (2nd edn.), Springer, Cham, Switzerland, pp. 1729–1741.

Bianchi, F., Riboni, N., Termopoli, V., Mendez, L., Medina, I., Ilag, L., Cappiello, A. and Careri, M. (2018). MS-based analytical techniques: advances in spray-based methods and EI-LC-MS applications, *J. Anal. Methods Chem.* 2018, 1308167. doi: 10.1155/2018/1308167.

Bittante, G., Ferragina, A., Cipolat-Gotet, C. and Cecchinato, A. (2014). Comparison between genetic parameters of cheese yield and nutrient recovery or whey loss traits measured from individual model cheese-making methods or predicted from unprocessed bovine milk samples using Fourier-transform infrared spectroscopy, *J. Dairy Sci.* 97(10), 6560–6572.

Bonfatti, V., Di Martino, G. and Carnier, P. (2011). Effectiveness of mid-infrared spectroscopy for the prediction of detailed protein composition and contents of protein genetic variants of individual milk of Simmental cows, *J. Dairy Sci.* 94(12), 5776–5785.

Caira, S., Gabriella Pinto, G., Nicolai, M. A., Chianese, L. and Addeo, F. (2016). Simultaneously tracing the geographical origin and presence of bovine milk in Italian water buffalo Mozzarella cheese using MALDI-TOF data of casein signature peptides, *Anal. Bioanal. Chem.* 408(20), 5609–5621. doi: 10.1007/s00216-016-9663-0.

Calamari, L., Ferrari, A., Minuti, A. and Trevisi, E. (2016). Assessment of the main plasma parameters included in a metabolic profile of dairy cow based on Fourier Transform mid-infrared spectroscopy: preliminary results, *BMC Vet. Res.* 12, 4. doi: 10.1186/s12917-015-0621-4.

Capuano, E., Rademaker, J., van den Bijgaart, H. and M. van Ruth, S. (2014). Verification of fresh grass feeding, pasture grazing and organic farming by FTIR spectroscopy analysis of bovine milk, *Food Res. Int.* 60, 59–65.

Caredda, M., Addis, M., Ibba, I., Leardi, R., Scintu, M. F., Piredda, G. and Sanna, G. (2017). Building of prediction models by using mid-infrared spectroscopy and fatty acid profile to discriminate the geographical origin of sheep milk, *LWT* 75, 131–136.

Cattaneo, T. M. P. and Holroyd, S. E. (2013). New applications of near infrared spectroscopy on dairy products, *J. Near Infrared Spectrosc.* 21(5), 307–310.

Cecchinato, A., Albera, A., Cipolat-Gotet, C., Ferragina, A. and Bittante, G. (2015). Genetic parameters of cheese yield and curd nutrient recovery or whey loss traits predicted using Fourier-transform infrared spectroscopy of samples collected during milk recording on Holstein, Brown Swiss, and Simmental dairy cows, *J. Dairy Sci.* 98(7), 4914–4927.

Cecchinato, A., De Marchi, M., Gallo, L., Bittante, G. and Carnier, P. (2009). Mid-infrared spectroscopy predictions as indicator traits in breeding programs for enhanced coagulation properties of milk, *J. Dairy Sci.* 92(10), 5304–5313.

Chambers, E. and Koppel, K. (2013). Associations of volatile compounds with sensory aroma and flavor: the complex nature of flavor, *Molecules* 18(5), 4887–4905.

Chung, I. M., Kim, J. K., Yarnes, C. T., An, Y. J., Kwon, C., Kim, S. Y., Yang, Y. J., Chi, H. Y. and Kim, S. H. (2019). Fatty acid- and amino acid-specific isotope analysis for accurate authentication and traceability in organic milk, *J. Agric. Food Chem.* 67(2), 711–722. doi: 10.1021/acs.jafc.8b05063.

Coimbra, P. T., Bathazar, C. F., Guimarães, J. T., Coutinho, N. M., Pimentel, T. C., Neto, R. P. C., Esmerino, E. A., Freitas, M. Q., Silva, M. C., Tavares, M. I. B. and Cruz, A. G. (2020). Detection of formaldehyde in raw milk by time domain nuclear magnetic resonance and chemometrics, *Food Control* 110. doi: 10.1016/j.foodcont.2019.107006, 107006.

Colnago, L. A., Andrade, F. D., Souza, A. A., Azeredo, R. B. V., Lima, A. A., Cerioni, L. M., Osán, T. M. and Pusiol, D. J. (2014). Why is inline NMR rarely used as industrial sensor? challenges and opportunities, *Chem. Eng. Technol.* 37(2), 191–203.

Coppa, M., Martin, B., Agabriel, C., Chassaing, C., Sibra, C., Constant, I., Graulet, B. and Andueza, D. (2012). Authentication of cow feeding and geographic origin on milk using visible and near-infrared spectroscopy, *J. Dairy Sci.* 95(10), 5544–5551.

Cossignani, L., Pollini, L. and Blasi, F. (2019). Invited review: authentication of milk by direct and indirect analysis of triacylglycerol molecular species, *J. Dairy Sci.* 102(7), 5871–5882.

Dal Zotto, R., De Marchi, M., Cecchinato, A., Penasa, M., Cassandro, M., Carnier, P., Gallo, L. and Bittante, G. (2008). Reproducibility and repeatability of measures of milk coagulation properties and predictive ability of mid-infrared reflectance spectroscopy, *J. Dairy Sci.* 91(10), 4103–4112.

De Marchi, M., Fagan, C. C., O'Donnell, C. P., Cecchinato, A., Dal Zotto, R., Cassandro, M., Penasa, M. and Bittante, G. (2009a). Prediction of coagulation properties, titratable acidity, and pH of bovine milk using mid-infrared spectroscopy, *J. Dairy Sci.* 92(1), 423–432.

De Marchi, M., Bonfatti, V., Cecchinato, A., Di Martino, G. and Carnier, P. (2009b). Prediction of protein composition of individual cow milk using mid-infrared spectroscopy, *Ital. J. Anim. Sci.* 8(Suppl. 2), 399–401.

de Roos, A. P. W., van den Bijgaart, H. J. C. M., Hørlyk, J. and de Jong, G. (2007). Screening for subclinical ketosis in dairy cattle by Fourier transform infrared spectrometry, *J. Dairy Sci.* 90(4), 1761–1766.

Di Girolamo, F., Lante, I., Muraca, M. and Putignani, L. (2013). The role of mass spectrometry in the "omics" era, *Curr. Org. Chem.* 17(23), 2891–2905.

Dorman, F. L., Whiting, J. J., Cochran, J. W. and Gardea-Torresdey, J. (2010). Gas chromatography, *Anal. Chem.* 82(12), 4775–4785.

El Jabri, M., Trossat, P., Wolf, V., Beuvier, E., Rolet-Répécaud, O., Gavoye, S., Gaüzère, Y., Belysheva, O., Gaudillière, N., Notz, E., Grosperrin, P., Laithier, C. and Delacroix-Buchet, A. (2020). Mid-infrared spectrometry prediction of the cheese-making properties of raw Montbéliarde milks from herds and cheese dairy vats used for the production of protected designation of origin and protected geographical indication cheeses in Franche-Comté, *J. Dairy Sci.* 103(7), 5992–6002.

Fenaille, F., Saint-Hilaire, P. B., Rousseau, K. and Junot, C. (2017). Data acquisition workflows in liquid chromatography coupled to high resolution mass spectrometry-based metabolomics: where do we stand?, *J. Chromatogr. A* 1526, 1–12. doi: 10.1016/j.chroma.2017.10.043.

Flores-Flores, M. E. and González-Peñas, E. (2017). An LC-MS/MS method for multi-mycotoxin quantification in cow milk, *Food Chem.* 218, 378–385. doi: 10.1016/j.foodchem.2016.09.101.

Fong, B., Ma, K. and McJarrow, P. (2011). Quantification of bovine milk oligosaccharides using liquid chromatography-selected reaction monitoring-mass spectrometry, *J. Agric. Food Chem.* 59(18), 9788–9795. doi: 10.1021/jf202035m.

Fung, E. N., Jemal, M. and Aubry, A. F. (2013). High-resolution MS in regulated bioanalysis: where are we now and where do we go from here?, *Bioanalysis* 5(10), 1277–1284. doi: 10.4155/bio.13.81.

Garcia, C., Lutz, N. W., Confort-Gouny, S., Cozzone, P. J., Armand, M. and Bernard, M. (2012). Phospholipid fingerprints of milk from different mammalians determined by 31P NMR: towards specific interest in human health, *Food Chem.* 135(3), 1777–1783.

Gengler, N., Soyeurt, H., Dehareng, F., Bastin, C., Colinet, F., Hammami, H., Vanrobays, M. L., Lainé, A., Vanderick, S., Grelet, C., Vanlierde, A., Froidmont, E. and Dardenne, P. (2016). Capitalizing on fine milk composition for breeding and management of dairy cows, *J. Dairy Sci.* 99(5), 4071–4079.

Glish, G. L. and Burinsky, D. J. (2008). Hybrid mass spectrometers for tandem mass spectrometry, *J. Am. Soc. Mass Spectrom.* 19(2), 161–172. doi: 10.1016/j.jasms.2007.11.013.

Hansen, P. W. and Holroyd, S. E. (2019). Development and application of Fourier transform infrared spectroscopy for detection of milk adulteration in practice, *Int. J. Dairy Technol.* 70(3), 321–331.

Hernández, F., Cervera, M. I., Portolés, T., Beltrán, J. and Pitarch, E. (2013). The role of GC-MS/MS with triple quadrupole in pesticide residue analysis in food and the environment, *Anal. Methods* 5(21), 5875–5894.

Hrbek, V., Vaclavik, L., Elich, O. and Hajslova, J. (2014). Authentication of milk and milk-based foods by direct analysis in real time ionization-high resolution mass spectrometry (DART-HRMS) technique: a critical assessment, *Food Control* 36(1), 138–145. doi: 10.1016/j.foodcont.2013.08.003.

International Dairy Federation (2019). *Applications of near infrared spectrometry for the analysis of milk and milk products*, Bulletin No. 497, International Dairy Federation, Brussels, Belgium.

International Standards Organization (2019). ISO 15151/IDF 229, *milk, milk products, infant formula and adult nutritionals–determination of minerals and trace elements–inductively coupled plasma atomic emission spectrometry (ICP-AES) method*. International Standards Organization, Geneva, Switzerland.

International Standards Organization (2020a). ISO 21543/IDF 201, *Milk and milk products–guidelines for the application of near infrared spectrometry*. International Standards Organization, Geneva, Switzerland.

International Standards Organization (2020b). ISO 23291/IDF 248, *Milk and milk products–guidelines for the application of in-line and on-line infrared spectrometry*. International Standards Organization, Geneva, Switzerland.

International Standards Organization (2020c). ISO 21424/IDF 243, *Milk, milk products, infant formula and adult nutritionals–determination of minerals and trace elements–inductively coupled plasma mass spectrometry (ICP-MS) method*. International Standards Organization, Geneva, Switzerland.

Jalili, V., Barkhordari, A. and Ghiasvand, A. (2020). A comprehensive look at solid-phase microextraction technique: a review of reviews, *Microchem. J.* 152. doi: 10.1016/j.microc.2019.104319.

Jia, W., Dong, X., Shi, L. and Chu, X. (2020). Discrimination of milk from different animal species by a foodomics approach based on high-resolution mass spectrometry, *J. Agric. Food Chem.* 68(24), 6638–6645. doi: 10.1021/acs.jafc.0c02222.

Jurowski, K., Kochan, K., Walczak, J., Barańska, M., Piekoszewski, W. and Buszewski, B. (2017). Analytical techniques in lipidomics: state of the art, *Crit. Rev. Anal. Chem.* 47(5), 418–437.

Le, T. T., Debyser, G., Gilbert, W., Struijs, K., Van Camp, J., Van de Wiele, T., Devreese, B. and Dewettinck, K. (2013). Distribution and isolation of milk fat globule membrane proteins during dairy processing as revealed by proteomic analysis, *Int. Dairy J.* 32(2), 110–120.

Li, Q., Zhao, Y., Zhu, D., Pang, X., Liu, Y., Frew, R. and Chen, G. (2017). Lipidomics profiling of goat milk, soymilk and bovine milk by UPLC-Q-exactive orbitrap mass spectrometry, *Food Chem.* 224, 302–309. doi: 10.1016/j.foodchem.2016.12.083.

Liigand, J., Wang, T., Kellogg, J., Smedsgaard, J., Cech, N. and Kruve, A. (2020). Quantification for non-targeted LC/MS screening without standard substances, *Sci. Rep.* 10(1), 5808. doi: 10.1038/s41598-020-62573-z.

Liu, Z., Moate, P., Ezerniks, V., Cocks, B. G. and Rochfort, S. (2015). Identification and quantification of triacylglycerols containing n-3 long-chain polyunsaturated fatty acids in bovine milk, *J. Dairy Sci.* 98(12), 8473–8485.

Liu, Z., Wang, J., Cocks, B. G. and Rochfort, S. (2017). Seasonal variation of triacylglycerol profile of bovine milk, *Metabolites* 7(2), 24. doi: 10.3390/metabo7020024.

Lucey, J. A., Otter, D. and Horne, D. S. (2017). A 100-year review: progress on the chemistry of milk and its components, *J. Dairy Sci.* 100(12), 9916–9932.

Ludwig, C., Gillet, L., Rosenberger, G., Amon, S., Collins, B. C. and Aebersold, R. (2018). Data-independent acquisition-based SWATH-MS for quantitative proteomics: a tutorial, *Mol. Syst. Biol.* 14(8), e8126. doi: 10.15252/msb.20178126.

MacKenzie, A., Vyssotski, M. and Nekrasov, E. (2009). Quantitative analysis of dairy phospholipids by $^{31}$P NMR, *J. Am. Oil Chem. Soc.* 86(8), 757–763.

MacMahon, S., Begley, T. H., Diachenko, G. W. and Stromgren, S. A. (2012). A liquid chromatography-tandem mass spectrometry method for the detection of economically motivated adulteration in protein-containing foods, *J. Chromatogr. A* 1220, 101–107. doi: 10.1016/j.chroma.2011.11.066.

Maher, A. D. and Rochfort, S. J. (2014). Applications of NMR in dairy research, *Metabolites* 4(1), 131–141.

Mazzei, P. and Piccolo, A. (2012). $^{1}$H HRMAS-NMR metabolomic to assess quality and traceability of mozzarella cheese from Campania buffalo milk, *Food Chem.* 132(3), 1620–1627.

McDermott, A., De Marchi, M., Berry, D. P., Visentin, G., Fenelon, M. A., Lopez-Villalobos, N. and McParland, S. (2017). Cow and environmental factors associated with protein fractions and free amino acids predicted using mid-infrared spectroscopy in bovine milk, *J. Dairy Sci.* 100(8), 6272–6284.

Michalski, A., Damoc, E., Lange, O., Denisov, E., Nolting, D., Müller, M., Viner, R., Schwartz, J., Remes, P., Belford, M., Dunyach, J. J., Cox, J., Horning, S., Mann, M. and Makarov, A. (2012). Ultra-high resolution linear ion trap Orbitrap mass spectrometer (Orbitrap Elite) facilitates top-down LC MS/MS and versatile peptide fragmentation modes, *Mol. Cell. Proteomics* 11(3), O111.013698. doi: 10.1074/mcp.O111.013698.

Moldoveanu, S. C. and David, V. (2017). Short overviews of analytical techniques not containing an independent separation step. In: *Selection of the HPLC Method in Chemical Analysis*, Elsevier, Boston, MA, pp. 31–53.

Monakhova, Y. B., Kuballa, T., Leitz, J., Andlauer, C. and Lachenmeier, D. W. (2012). NMR spectroscopy as a screening tool to validate nutrition labeling of milk, lactose-free milk, and milk substitutes based on soy and grains, *Dairy Science & Technol.* 92(2), 109–120.

Obeidat, S. M. (2020). Investigation of butter adulteration with margarine using $^{1}$H NMR and PCA, *Appl. Magn. Reson.* 51(8), 691–701.

Papaemmanouil, C., Tsiafoulis, C. G., Alivertis, D., Tzamaloukas, O., Miltiadou, D., Tzakos, A. G. and Gerothanassis, I. P. (2015). Selective one-dimensional total correlation spectroscopy nuclear magnetic resonance experiments for a rapid identification of minor components in the lipid fraction of milk and dairy products: toward spin chromatography?, *J. Agric. Food Chem.* 63(22), 5381–5387.

Pashkova, G. V., Smagunova, A. N. and Finkelshtein, A. L. (2018). X-ray fluorescence of milk and dairy products: a review, *Trends Anal. Chem.* 106, 183–189.

Perring, L. and Andrey, D. (2003). ED-XRF as a tool for rapid minerals control in milk-based products, *J. Agric. Food Chem.* 51(15), 4207–4212.

Perring, L. and Tschopp, A. (2019). Determination of ash content of milk-based powders by energy dispersive X-ray fluorescence, *Microchem. J.* 145, 162–167.

Righetti, L., Paglia, G., Galaverna, G. and Dall'Asta, C. (2016). Recent advances and future challenges in modified mycotoxin analysis: why HRMS has become a key instrument in food contaminant research, *Toxins* 8(12), 361. doi: 10.3390/toxins8120361.

Rochat, B. (2019). Quantitative and qualitative LC-high-resolution MS: the technological and biological reasons for a shift of paradigm. In: Ince, M. and Ince, O. K. (Eds), *Recent Advances in Analytical Chemistry*, IntechOpen, London, UK, pp. 3–20.

Sargent, M. (Ed.) (2013). *Guide to Achieving Reliable Quantitative LCMS Measurements*, RSC Analytical Methods Committee, London, UK.

Scano, P., Cusano, E., Caboni, P. and Consonni, R. (2019). NMR metabolite profiles of dairy: a review, *Int. Dairy J.* 90, 56–67.

Schiano, A. N., Harwood, W. S. and Drake, M. A. (2017). A 100-year review: sensory analysis of milk, *J. Dairy Sci.* 100(12), 9966–9986.

Senyuva, H. Z., Gökmen, V. and Sarikaya, E. A. (2015). Future perspectives in Orbitrap™-high-resolution mass spectrometry in food analysis: a review, *Food Addit. Contam. Part A Chem. Anal. Control Expo. Risk Assess.* 32(10), 1568–1606. doi: 10.1080/19440049.2015.1057240.

Siciliano, R. A., Mazzeo, M. F., Arena, S., Renzone, G. and Scaloni, A. (2013). Mass spectrometry for the analysis of protein lactosylation in milk products, *Food Res. Int.* 54(1), 988–1000. doi: 10.1016/j.foodres.2012.10.044.

Smith, S. L., Denholm, S. J., Coffey, M. P. and Wall, E. (2019). Energy profiling of dairy cows from routine milk mid-infrared analysis, *J. Dairy Sci.* 102(12), 11169–11179.

Sobolev, A. P., Thomas, F., Donarski, J., Ingallina, C., Circi, S., Cesare Marincola, F., Capitani, D. and Mannina, L. (2019). Use of NMR applications to tackle future food fraud issues, *Trends Food Sci. Technol.* 91, 347–353.

Sørensen, L. K., Lund, M. and Juul, B. (2003). Accuracy of Fourier transform infrared spectrometry in determination of casein in dairy cows' milk, *J. Dairy Res.* 70(4), 445–452.

Soyeurt, H., Bruwier, D., Romnee, J. M., Gengler, N., Bertozzi, C., Veselko, D. and Dardenne, P. (2009). Potential estimation of major mineral contents in cow milk using mid-infrared spectrometry, *J. Dairy Sci.* 92(6), 2444–2454.

Soyeurt, H., Dehareng, F., Gengler, N., McParland, S., Wall, E., Berry, D. P., Coffey, M. and Dardenne, P. (2011). Mid-infrared prediction of bovine milk fatty acids across multiple breeds, production systems, and countries, *J. Dairy Sci.* 94(4), 1657–1667.

Sundekilde, U. K., Gustavsson, F., Poulsen, N. A., Glantz, M., Paulsson, M., Larsen, L. B. and Bertram, H. C. (2014). Association between the bovine milk metabolome and rennet-induced coagulation properties of milk, *J. Dairy Sci.* 97(10), 6076–6084.

Toffanin, V., De Marchi, M. M., Lopez-Villalobos, N. and Cassandro, M. (2015). Effectiveness of mid-infrared spectroscopy for prediction of the contents of calcium and phosphorus, and titratable acidity of milk and their relationship with milk quality and coagulation properties, *Int. Dairy J.* 41, 68–73.

Tu, A., Ma, Q., Bai, H. and Du, Z. (2017). A comparative study of triacylglycerol composition in Chinese human milk within different lactation stages and imported infant formula by SFC coupled with Q-TOF-MS, *Food Chem.* 221, 555–567.

van Knegsel, A. T. M., van der Drift, S. G. A., Horneman, M., de Roos, A. P., Kemp, B. and Graat, E. A. (2010). Short communication: ketone body concentration in milk determined by Fourier transform infrared spectroscopy: value for the detection of hyperketonemia in dairy cows, *J. Dairy Sci.* 93(7), 3065–3069.

Vanlierde, A., Vanrobays, M.-L., Gengler, N., Dardenne, P., Froidmont, E., Soyeurt, H., McParland, S., Lewis, E., Deighton, M. H., Mathot, M. and Dehareng, F. (2016). Milk mid-infrared spectra enable prediction of lactation-stage-dependent methane emissions of dairy cattle within routine population-scale milk recording schemes, *Anim. Prod. Sci.* 56(3), 258–264.

Verma, A. and Ambatipudi, K. (2016). Challenges and opportunities of bovine milk analysis by mass spectrometry, *Clin. Proteomics* 13, 8. doi: 10.1186/s12014-016-9110-4.

Vincent, D., Elkins, A., Condina, M. R., Ezernieks, V. and Rochfort, S. (2016). Quantitation and identification of intact major milk proteins for high-throughput LC-ESI-Q-TOF MS analyses, *PLoS ONE* 11(10), e0163471. doi: 10.1371/journal.pone.0163471.

Visentin, G., McDermott, A., McParland, S., Berry, D. P., Kenny, O. A., Brodkorb, A., Fenelon, M. A. and De Marchi, M. (2015). Prediction of bovine milk technological traits from mid-infrared spectroscopy analysis in dairy cows, *J. Dairy Sci.* 98(9), 6620–6629.

Visentin, G., Penasa, M., Gottardo, P., Cassandro, M. and De Marchi, M. (2016). Predictive ability of mid-infrared spectroscopy for major mineral composition and coagulation traits of bovine milk by using the uninformative variable selection algorithm, *J. Dairy Sci.* 99(10), 8137–8145.

Wang, X., Yu, Z., Zhao, X., Han, R., Huang, D., Yang, Y. and Cheng, G. (2021). Comparative proteomic characterization of bovine milk containing β-casein variants A1A1 and A2A2, and their heterozygote A1A2, *J. Sci. Food Agric.* 101(2), 718–725. doi: 10.1002/jsfa.10684.

Wang, X., Zhao, X., Huang, D., Pan, X., Qi, Y., Yang, Y., Zhao, H. and Cheng, G. (2017). Proteomic analysis and cross species comparison of casein fractions from the milk of dairy animals, *Sci. Rep.* 7, 43020. doi: 10.1038/srep43020.

Yazgan, N. N., Genis, H. E., Bulat, T., Topcu, A., Durna, S., Yetisemiyen, A. and Boyaci, I. H. (2020). Discrimination of milk species using Raman spectroscopy coupled with partial least squares discriminant analysis in raw and pasteurized milk, *J. Sci. Food Agric.* 100(13), 4756–4765.

Zhang, L.-Y., Wang, J.-Q., Yang, Y.-X., Bu, D., Li, S. and Zhou, L. (2011). Comparative proteomic analysis of changes in the bovine whey proteome during the transition from colostrum to milk, *Asian-Australas. J. Anim. Sci.* 24(2), 272–278.

Zhou, Q., Gao, B., Zhang, X., Xu, Y., Shi, H. and Yu, L. L. (2014). Chemical profiling of triacylglycerols and diacylglycerols in cow milk fat by ultra-performance convergence chromatography combined with a quadrupole time-of-flight mass spectrometry, *Food Chem.* 143, 199–204.

Zhu, D., Hayman, A., Kebede, B., Stewart, I., Chen, G. and Frew, R. (2019). $^{31}$P NMR-based phospholipid fingerprinting of powdered infant formula, *J. Agric. Food Chem.* 67(36), 10265–10272.

Zubarev, R. A. and Makarov, A. (2013). Orbitrap mass spectrometry, *Anal. Chem.* 85(11), 5288–5296. doi: 10.1021/ac4001223.

# Chapter 15

## Genetic factors affecting the composition and quality of cow's milk

*Nina Aagaard Poulsen and Lotte Bach Larsen, Aarhus University, Denmark*



## 1 Introduction

Milk secures the cognitive development, health and growth of mammalian neonates by the mother, offering a tailored cocktail of fat, protein and lactose together with essential vitamins and minerals, as well as an overwhelming number of minor bioactive components (German et al., 2002). Since the domestication of cattle, dairying has played a pivotal role for humans, which the strong selective sweep for lactase persistence (lactose tolerance) also documents.

So why does one cow produce milk with, for example, a high casein number and superior coagulation properties, compared to other cows? It is well known that variances in milk composition result not only from feeding and management, but also from genetic factors, as has been established, for example, for the milk protein fraction (Fig. 1). These variances may to a large extent relate to the genetic predisposition of specific cows, given that the genome has a major impact on the specific milk quality trait of interest. Hereditary traits can be improved through selective breeding, where the goal is to improve traits like cow health, fertility, milk yield and milk quality. Bulls

http://dx.doi.org/10.19103/AS.2022.0099.15

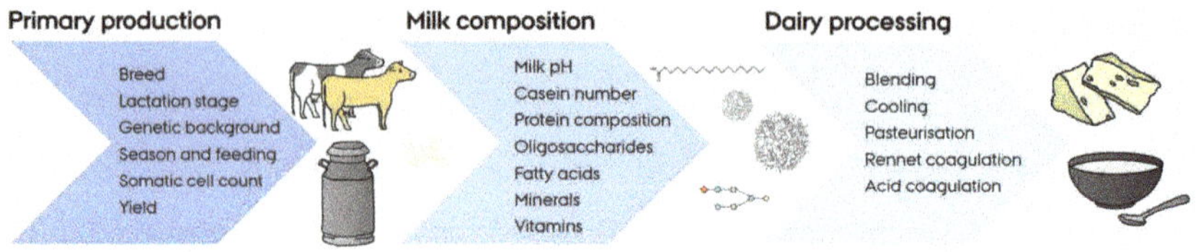


**Figure 1** Factors which may affect milk composition, and thus milk functionality, during dairy processing.

with premium genetic merit are selected to increase the genetic gain for traits of interest. For a long time, effects of genetic variants of milk proteins on milk composition and quality have been documented. However, with the current use of more sensitive methods for milk phenotyping, combined with the reduced cost and optimised bioinformatics for genotyping array, numerous genes and genomic regions affecting both lactation performance, specific components of milk and even functional properties have been identified.

Performing genome-wide association studies (GWAS) can identify single-nucleotide polymorphisms (SNPs) contributing to the phenotypic trait of interest. This leads to the identification of genomic regions that may contain candidate genes associated with the trait of interest or specific pathways, which may influence the trait. Many of these traits are polygenic, complex traits, and therefore it is often hard to identify causal genes underlying these. However, GWAS provide valuable insight into the molecular basis of the traits and enhance our understanding of underlying factors, but in terms of dairy breeding programs, the estimation of genetic merit is more important (Goddard and Hayes, 2009). Heritability provides an estimate of how much of the phenotypic variance can be ascribed to additive genetic variance, which in other words, defines the part of the variation that can be changed through selective breeding or genomic selection. Apart from heritability, the estimation of genetic correlations among traits may also be of interest in order to pinpoint potential correlated effects on other traits, especially when these are also included in the breeding goal.

The phenotypic variance is the sum of genetic variance and environmental variance. Heritability ($h^2$) is defined as the proportion of phenotypic variance, which can be explained by additive genetic variance. Heritability estimates envisage the potential for genetic progress, and thus the possibility of changing a specific trait through selective breeding. Within breed, variance components are estimated to calculate heritability. In the past, especially economically valuable traits such as protein, fat and milk yield have been the targets for selection, which over the years have resulted in large genetic gain achieved as a result of selection for these productivity traits. Both milk yield and concentration of fat, protein or even lactose exhibit significant heritabilities, whereas urea and somatic cell count (SCC), for example, which are more closely related to environmental factors, are less heritable (Stoop et al., 2008).

Ogorevc et al. (2009) developed a quantitative trait loci database based on literature, identifying cattle candidate genes and genetic markers for milk production and mastitis. Over the past 20 years, milk genomics studies have identified a number of major QTL regions, including, to a profound extent, the regions around *DGAT1* on BTA14, *SCD1* on BTA26, *FASN* on BTA19, as well genes for major milk proteins such as the casein gene cluster on BTA6, *PAEP* (encoding for β-lactoglobulin (β-LG)) on BTA 11 and *ALBA* (encoding α-lactalbumin (α-LA)) on BTA5. These specific genes or regions seem to influence numerous milk traits, including total yields of milk, protein and fat (Ogorevc et al., 2009).

The genomic organisation of the casein loci on BTA6 spans 250 kb (Threadgill and Womack, 1990). As 80% of the proteins in bovine milk are caseins, this region is expected to have a large influence on milk protein and milk protein-related traits, and is found to play a significant role for the caseins (Schopen et al., 2009; Buitenhuis et al., 2016). Likewise, this region has been identified to have a strong association with rennet- and acid-induced coagulations (Gregersen et al., 2014; Gregersen et al., 2015). The genomic region on BTA11, which harbours the *PAEP* gene encoding β-LG, has been associated with milk compositional traits in a number of studies (Gregersen et al., 2014; Poulsen et al., 2018), and affects β-LG in milk (Schopen et al., 2009).

*DGAT1* codes for acyl-CoA:diacylglycerol acyltransferase 1, an enzyme that plays an important role in triacylglycerol synthesis by catalysing the esterification of a fatty acyl-CoA to the sn-3 position of a diacylglycerol in the synthesis of triglycerides (Grisart et al., 2002). The effect of the *DGAT* K232A polymorphism has been intensively studied (Grisart et al., 2002; Bovenhuis et al., 2016), but the gene, or the chromosomal region(s) around *DGAT1,* also affects a range of economically important milk compositional traits like milk yield, concentrations of fat and protein, as well as casein number, milk metabolites, vitamins and fatty acid composition (Schennink et al., 2007; 2009; Stoop et al., 2009; Schopen et al., 2009; Buitenhuis et al., 2013; 2014; 2016; Melzer et al., 2013; Poulsen et al., 2015b). The enzyme stearoyl-coenzyme A desaturase 1 (SCD1) catalyses the conversion of the group of C10:0–C18:0 saturated fatty acids into their mono-unsaturated counterparts. In a number of studies, the *SCD1* gene is associated with milk fatty acid composition and desaturase indexes (Mele et al., 2007; Schennink et al., 2008; Stoop et al., 2009; Conte et al., 2010; Buitenhuis et al., 2014).

## 2 Breed

Today, high-yielding Holstein–Friesians are widespread, and in many areas, the breed of choice for modern dairy production. However, since domestication around 11 000 years ago, diversification of animal genetic resources has resulted in over 1000 cattle breeds worldwide, that through selection and isolation represent unique genetic make-up (FAO, 2015). Due to low census population sizes, many indigenous breeds are classified as endangered and at risk of extinction (FAO, 2015). International commitments and guided breeding programs are pivotal to conserve these genetic resources for the future, but the documentation of phenotypic and performance characteristics of the breeds is also central (FAO, 2015). Studies have documented genetic distinctiveness and phylogenetic relationship among local and modern breeds (Kantanen et al., 2000; Tapio et al., 2006; Iso-Touru et al., 2016), but in many areas, local breeds are mainly used for either low-input farming or maintenance of grasslands and open landscapes. Thus, the characterisation of milk composition from local indigenous breeds in a global perspective is lacking. However, given their unique genetic variation, these breeds may possess distinct milk composition with possibilities for exploitation into niche dairy products, in order to promote awareness, economic incitement and thus sustainable use of these genetic resources. Apart from local breeds, commercial modern dairy breeds such as Jersey, Ayrshire and Simmental also possess well-documented differences in milk composition utilised by dairies worldwide. This relates, for example, to higher protein and fat concentrations in milk from the overall lower-yielding Jersey cows (Poulsen et al., 2012), but differences in genetic casein variants that may be manifested into the milk quality are also documented for both modern and local breeds, as discussed in Section 3.

## 3 Milk proteins and genetic variants

Milk from cattle is characterised by the six major milk proteins, including the four caseins ($\alpha_{S1}$-CN, $\alpha_{S2}$-CN, β-CN, and κ-CN), and the two major whey proteins, α-LA and β-LG. At the DNA level, synonymous mutations are silent mutations, which do not change the protein sequence, whereas non-synonymous mutations lead to amino acid changes. Functional polymorphisms leading to amino acid changes in the peptide backbone, or even deletions, result in heterogeneity of the major milk proteins, summarising into $\alpha_{S1}$-CN (8), $\alpha_{S2}$-CN (4), β-CN (12), κ-CN (11), β-LG (11) and α-LA (3) variants (Farrell et al., 2004). Due to linkage disequilibrium (LD) at the casein cluster, the casein variants are often presented as haplotypes or composite genotypes instead of individual variants (Bonfatti et al., 2010; Vallas et al., 2012; Poulsen et al., 2013). Furthermore, variations in the non-coding regions, for example, in regulatory areas, may affect the expression of protein genetic variants, and thereby the level or proportion of specific proteins found in milk. Thereby, the genetic imprint of protein genetic variants into milk phenotypes includes both differences directly related to the individual variants due to amino acid changes, and varying expression of the proteins characteristic for different protein variants.

Milk protein variants have been characterised in many breeds and population variant frequencies estimated, identifying a number of common and rare variants (Caroli et al., 2009). This results in high variant richness and genetic distinctiveness, for example, as observed for 70 breeds across Europe (Beja-Pereira et al., 2003). For $\alpha_{S1}$-CN, the most common and widespread variant is variant B; for $\alpha_{S2}$-CN it is variant A; whereas for β-CN both $A^1$, $A^2$ and B are widespread and common in most breeds; while for the κ-CN, variants A and B are most common (Caroli et al., 2009). For β-LG, variants A and B are common, whereas for α-LA, variant B is predominant in *Bos taurus* (Caroli et al., 2009). As exemplified across common and indigenous Danish and Swedish breeds, large variation exists in terms of both the number of variants present as well as their frequencies (Table 1). It should be noted, however, that given the low sample set of the indigenous breeds, these frequencies may not present true population averages but give a good indication.

The protein genetic variants are strongly associated with the compositional traits of milk and the relative protein profile is particularly affected (Heck et al., 2008; Gustavsson et al., 2014a). Higher κ-CN concentration is associated with κ-CN B (Hallén et al., 2008; Bonfatti et al., 2010), whereas lower β-LG concentration is associated with β-LG B (Hallén et al., 2008). This means that a higher casein number (casein/total protein) is associated both with β-LG B and k-CN B (Hallén et al., 2008; Bonfatti et al., 2010). Devold et al. (2000) found $\alpha_{S1}$-CN BB genotypes to have significantly larger micelles than BC genotypes and k-CN AB genotypes to have significantly smaller micelles than AA and AE genotypes, whereas no significant effect was found from β-LG variants. Glantz et al. (2011) confirmed that k-CN B was associated with smaller casein micelles. Apart from the specific effects on milk proteins, the genetic variants also affect milk yield, as well as fat and protein concentration and yield (Ikonen et al., 1999). In relation to protein composition, Buitenhuis et al. (2016) identified three major QTL regions common to the Holstein and Jersey breeds. These included BTA6 (for k-CN) covering the casein gene complex, BTA14 (for protein percent and CN percent) covering the *DGAT* gene, and BTA11 (for β-LG) covering the *PAEP* gene. This is in line with the findings of Schopen et al. (2011) identifying the same major regions of importance for bovine milk protein composition. The heritabilities of major milk proteins are generally moderate to high (Schopen et al., 2009; Bonfatti et al., 2014; Buitenhuis et al., 2016). The strong negative genetic correlation between β-LG and casein number (Schopen et al., 2009) suggests that selection for increasing casein content will decrease β-LG in milk.

In addition to the allelic genetic protein variants, a number of splice variants have also been identified in bovine milk, especially for $\alpha_S$-CNs (Martin et al., 2013). Documentation includes, among others, minor non-allelic isoforms of

**Table 1** Variant frequencies of $\alpha_{S1}$-CN, β-CN, κ-CN and β-LG in 455 Danish Holstein (DH) and 433 Danish Jersey (DJ), 12 Jutland, 28 Danish Red anno 1970 (RDM70), 367 Swedish Red (SR), 23 Swedish Mountain cattle (SM) and 8 Swedish Red Polled (SRP) cows. Results are summarised from Poulsen et al. (2013), Poulsen et al. (2016b), Poulsen et al. (2017a) and unpublished data

| Protein | Variant | DH (*n* = 455) | DJ (*n* = 433) | Jutland (*n* = 12) | RDM70 (*n* = 28) | SR (*n* = 367) | SM (*n* = 23) | SRP (*n* = 8) |
|---|---|---|---|---|---|---|---|---|
| $\alpha_{S1}$-CN | B | 0.995 | 0.57 | 1.000 | 0.982 | 0.996 | 0.848 | 0.938 |
| | C | 0.005 | 0.43 | - | 0.018 | 0.004 | 0.152 | 0.062 |
| β-CN | $A^1$ | 0.254 | 0.081 | 0.500 | 0.286 | 0.474 | 0.305 | 0.250 |
| | $A^2$ | 0.621 | 0.629 | 0.292 | 0.482 | 0.526 | 0.565 | 0.750 |
| | $A^3$ | 0.004 | - | - | - | - | - | - |
| | B | 0.044 | 0.22 | - | - | - | 0.130 | - |
| | F | 0.008 | - | 0.208 | 0.232 | NA | - | - |
| | I | 0.069 | 0.07 | - | - | - | - | - |
| κ-CN | A | 0.697 | 0.203 | 0.833 | 0.643 | 0.696 | 0.304 | 0.625 |
| | B | 0.235 | 0.797 | 0.167 | 0.321 | 0.240 | 0.696 | 0.188 |
| | E | 0.068 | - | - | 0.036 | 0.064 | - | 0.188 |
| β-LG | A | 0.538 | 0.546 | 0.583 | 0.107 | NA | 0.283 | 0.188 |
| | B | 0.462 | 0.422 | 0.417 | 0.893 | NA | 0.717 | 0.813 |
| | C | - | 0.032 | - | - | NA | - | - |

- = not observed in the dataset.
NA, Not available.

$\alpha_{S1}$-CN, where specific exons or Gln have been skipped (Miranda et al., 2020). The effect of such splice variants on milk composition and functionality is, however, scarcely explored.

The different structural features embedded in the sequences of the protein genetic variants can give rise to differences in digestibility and, for example, the release of bioactive peptides. In relation to the gastro-intestinal digestion of genetic variants of β-CN, there has been a lot of controversy due to the potential health effects of the β-CN $A^1$ variant vs. $A^2$ (Truswell, 2005; He et al., 2017; Ramakrishnan et al., 2020). The difference between the $A^1$ and $A^2$ variants of β-CN is one amino acid exchange at position 67, where the $A^2$ variant holds a Proline (Pro), while the $A^1$ variant holds a Histidine (His) in the protein sequence. The hypothesis around the $A^1$/$A^2$ milk case is that the amino acid exchange from Pro to His enables enzymatic release of the peptide called BCM-7 (Y-P-F-P-G-P-I) upon digestion of $A^1$ protein but not of $A^2$. Based on the variation at position 67, the β-CN variants can be grouped into 'families', with, for example, $A^1$, B and F variants belonging to the $A^1$ family and the $A^2$, $A^3$ and I belonging to the $A^2$ family, a fact that is often overlooked in the debate and interpretation of the frequencies and the impact of β-CN variants on the population level.

The BCM-7 peptide is claimed to be linked with stomach discomfort in some persons (Jianqin et al., 2016), but it is still very much debated. The stomach discomfort discussed resembles the discomfort experienced by individuals with lactose intolerance, but recent studies may indicate that some of these individuals may benefit from drinking $A^2$ milk instead of milk containing $A^1$, even when the lactose has been hydrolysed (He et al., 2017). The release of the BCM-7 peptide by digestion of either purified β-CN or of β-CN in milk from homozygous cows has been studied in two recent studies (Asledottir et al., 2017, 2018). Using mass spectrometry for BCM-7 detection, it was found that BCM-7 was released from the $A^1$ family milk as expected, but also from the $A^2$ family milk, though at lower levels (around 4–5 times lower). It is therefore still a matter of consideration, whether this release of BCM-7 has an impact on gastro-intestinal discomfort in some persons, and whether there is a significant difference between the $A^1$ and $A^2$ milks. Future controlled studies are needed to settle this.

Petrat-Melin et al. (2015) studied the *in vitro* gastro-intestinal digestion and generated bioactivities via isolated β-CN B, $A^1$, $A^2$ and I from homozygous milk and found that, in particular, one ≈4 kDa peptide with the N-terminal sequence 106H-K-E-M-P-F-P-K- was absent from digest of β-CN variant B, but present in the digests of the other variants. This is likely a result of the 122Ser to Arg substitution in β-CN B, introducing a novel trypsin cleavage site, leading to the changed digestion pattern. Further, both antioxidant and angiotensin-converting enzyme (ACE) inhibitory capacities of the hydrolysates increased significantly for all variants, with the B variant reaching the highest ACE inhibitory activity (Petrat-Melin et al., 2015). In another study, *in vitro* digests of the same purified β-CN genetic variants were studied and BCM-7 precursor peptides identified, representing V-Y-P-F-P-G-P-I-H-N from β-CN $A^1$/B, and V-Y-P-F-P-G-P-I-P-N from β-CN $A^2$/I (Petrat-Melin et al., 2017). The bioactivities of these peptides were assessed in an *in vitro* system of intestinal CaCo-2 cells, and it was found that the BCM-7 precursor peptide from β-CN $A^1$/ B exhibited strong ACE inhibitory activity, both before and after exposure to the Caco-2 cellular layer. In contrast, the equivalent peptide from β-CN $A^2$/I had lower initial ACE inhibitory activity but doubled its activity after exposure to the CaCo-2 intestinal cells after trimming off the C-terminal N residue, possibly due to the activity of brush-border enzyme carboxypeptidases into a BCM-9 peptide.

Petrat-Melin et al. (2016) further studied the digestion and bioactivities of isolated κ-CN A, B and E from homozygous milk. It was found that the variants displayed different *in vitro* gastro-intestinal digestion patterns and the antioxidant capacity of *in vitro*-digested κ-CN E was significantly lower than for κ-CN A or B. Lisson et al. (2013) also found that the genetic polymorphisms of both β-CN ($A^1$, $A^2$, B) and κ-CN (A, B, E) affected the *in vitro* digestion pattern of the different purified genetic variants and that this was further reflected in the IgE-binding epitopes of the digests. Combined, these results document that genetic polymorphisms affect the digestion of these proteins, but also that variants of β- and κ-CN could affect the allergenicity and bioactivity of the proteins.

## 4 Milk proteins and post-translational modifications

Apart from genetic variation, the heterogeneity of the milk proteins is high due to post-translational modifications (PTM), including disulphide bonds, phosphorylations and glycosylations (Fig. 2). Phosphate groups are esterified to the casein molecules during synthesis, via hydroxyl groups of mainly serine (Ser) residues, making P-Ser. Even though the structure of casein micelles in milk is not fully solved, there is a consensus that the P-Ser molecules of the caseins are anchor points for the micellar-bound calcium (colloidal calcium phosphate). The number of phosphorylations commonly found are in the ranges of: $\alpha_{S1}$-CN (8-9P), $\alpha_{S2}$-CN (10-13P), β-CN (5P) and κ-CN (1-2P). However, by the use of more sensitive mass spectrometry instruments, additional low-abundance isoforms have been identified (e.g. Miranda et al., 2020). In addition to phosphorylations, O-glycosylation of κ-CN results in glycan attachment at threonine (Thr) residues, resulting in attachment of 1-6 glycans at one or more of the specific sites: Thr121, Thr131, Thr133, Thr136 (only in A and E variants), Thr142, Thr145 and Thr165. In addition, there are two cysteine residues (Cys11 and Cys88) in κ-CN and two cysteines in $\alpha_{S2}$-CN (Cys36 and Cys40). These are engaged in disulphide bonds in native bovine

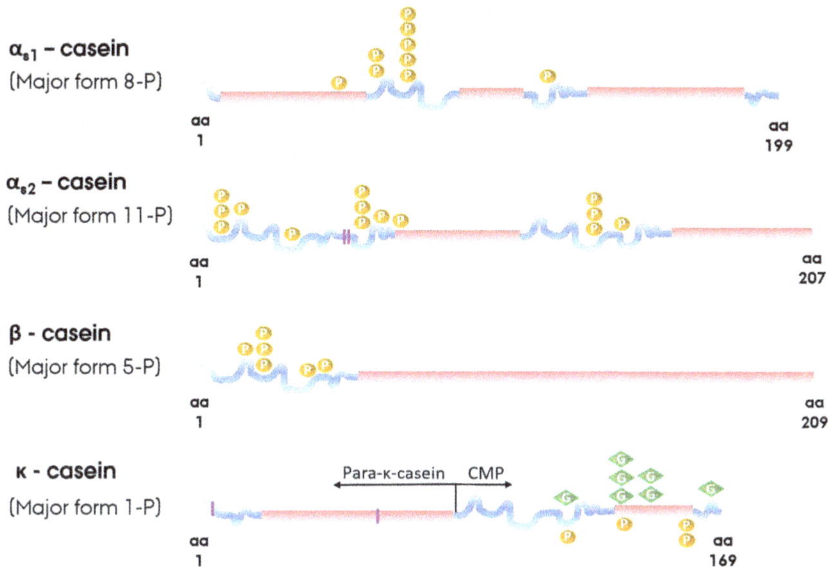


**Figure 2** Post-translational modifications of caseins based on the review by Le et al. (2020). Hydrophilic and hydrophobic regions of the mature protein are indicated. Blue stretches: hydrophilic regions, red stretches: hydrophobic regions. Purple lines indicate positions of Cys. All possible phosphorylation and glycosylation sites are highlighted. P, phosphorylation; G, glycosylation. Note that stacked phosphorylations and glycosylation are due to PTM sites in close proximity within the protein sequence, which appear to be at similar positions due to scale.

CN micelles, in the form of κ-CN homo-multimers with little or no content of free SH-groups, and in the case of $\alpha_{S2}$-CN, in the form of dimers which are connected in either parallel or antiparallel pattern of the S-S bonds (Rasmussen et al., 1992a,b).

The PTMs of the CNs affect CN micelle stability, and especially the highly glycosylated, hydrophilic part of κ-CN, the casein-macropeptide (CMP) part, ensures electrostatic and steric repulsion between micelles in bovine milk (Dziuba and Minkiewicz, 1996). Improved proteomic separation in later years, combined with quantitative determinations of the PTM isoforms, have shed light on their specific importance and the underlying genetic influence of PTMs on milk proteins. The two major phosphorylation isoforms of $\alpha_{S1}$-CN ($\alpha_{S1}$-CN 8P and 9P) were recently reported to be regulated by different sets of genes (Bijl et al., 2014b). It was shown that both isoforms were regulated by a specific region on BTA6. Furthermore, the $\alpha_{S1}$-CN 8P isoform was associated by a region harbouring *PAEP* on BTA11, while the $\alpha_{S1}$-CN 9P isoform was additionally influenced by a region harbouring *DGAT1* on BTA14 (Bijl et al., 2014b). In Buitenhuis et al. (2016), a significant region on BTA12 was identified to influence $\alpha_{S1}$-CN 8P percentage in Holstein, but regions on BTA6 and BTA11 were also identified; however, these were not significant. Interestingly, the phenotypic and genetic correlations between $\alpha_{S1}$-CN 8P and 9P isoforms were low (Bijl et al., 2014b), which supports somewhat of an independency between these isoforms. Heritability estimates for $\alpha_{S2}$-CN phosphorylation isoforms (10P-14P) were low to moderate, whereas heritabilities for $\alpha_{S1}$-CN 8P and 9P isoforms were high (Fang et al., 2017).

For glycosylation of κ-CN, BB k-CN genotypes have higher content (g/L) of both glycosylated (G) and unglycosylated (UG) κ-CN than AA genotypes (Bonfatti et al., 2014), and high heritabilities have been found for G κ-CN, for UG κ-CN and for the glycosylation degree (GD, measured as the ratio of G κ-CN to total κ-CN) (Bonfatti et al., 2014). For relative proportions of κ-CN, κ-CN BB had higher proportions of both UG κ-CN and G κ-CN relative to total protein, compared with κ-CN AA, whereas κ-CN AB showed intermediate results (Poulsen et al., 2016a). From a gel-based study, it was found that 95% of the κ-CN molecules in a pooled milk sample based on milk from individual cows were phosphorylated (1 or 2 P), while 36 % were glycosylated (identified with 1, 2, or 3 O-glycosylations) in Holstein. For Jersey, the numbers were 96% and 34 %, respectively (Jensen et al., 2012a). Using liquid chromatography electrospray ionisation mass spectrometry, quadrupole time of flight (LC-ESI/MS Q-TOF) on a small subset of samples with distinct κ-CN genotypes, the proportion of κ-CN relative to total protein was higher in milk from κ-CN BB cows, as compared with either κ-CN AA, AB, EE or AE genotypes in Holstein and with AB or AA genotypes in Jersey (Jensen et al., 2015). This further documents that a single cow produces a variety of κ-CN molecular isoforms, comprising variation in levels of both the number of phosphorylations and glycosylations, as well as the specific glycan structures of O-glycosylations, which relates to five different glycan structures (Saito and Itoh, 1992).

Several studies have documented that variation in the κ-CN content is associated with casein micelle size (Frederiksen et al., 2011; Day et al., 2015), and especially the glycosylated part seems to play a role (Bijl et al., 2014a). Thus, casein micelle size was strongly negatively correlated with G κ-CN, but not with the content of UG κ-CN or total κ-CN, in milk from Montbéliarde cows (Bijl et al., 2014a). In a study by Buitenhuis et al. (2016), there was a profound difference in the heritabilities between the Danish Holstein and the Danish Jersey (Holstein > Jersey,) for k-CN percentage-related traits. These heritability differences could potentially explain differences in the number of QTL detected and reported for G k-CN and UG k-CN between the breeds. A comparison of GWAS results for k-CN%, UG k-CN% and G k-CN% relative to total protein revealed a number of especially G k-CN percentage-specific QTL peaks, indicating that there is a potential to genetically differentiate between the proportions of G k-CN and UG k-CN relative to total protein in bovine milk. Among the genes assigned to the significant SNP markers for G k-CN, two genes were found to be related to the PTMs of caseins, including casein kinase 1 gamma 3 (*CSNK1G3*) on BTA7 and protein kinase C theta (*PRKCQ*) on BTA13 , both potentially involved in the PTMs of caseins (Buitenhuis et al., 2016).

## 5 Milk coagulation and other functional properties

Milk coagulation is a major economic trait in cheese manufacturing and important for both the quality and yield of cheese (Wedholm et al., 2006), as well as for process control at the dairies. Therefore, the possibilities for genetic improvement of this trait have received substantial attention, especially after finding a relatively high prevalence of non-coagulating milk at the individual cow level in several breeds. Rennet-induced milk coagulation, which is the first step in cheese manufacturing, is a two-phase process initiated by the enzymatic cleavage of k-CN by the addition of chymosin (rennet), which hydrolyses k-CN at Phe105-Met106 into CMP and para-k-CN resulting in the loss of the electrostatic and steric hindrance between casein micelles promoting aggregation and curd formation. The rennet coagulation time (RCT) defines the onset of milk coagulation, followed by changes in firmness, as the protein network develops and elasticity increases, defined by curd-firming rate and curd firmness at specific time points. From a dairy perspective, a relatively short clot time and a high gel firmness are desired. Bittante et al. (2012) identified a number of direct and indirect genetic effects affecting milk coagulation properties. These included breeds, genetic variants (especially k-CN) of the major proteins and other genes. In general, comparison among studies can be hard due to the use of different analytical methodologies and trait definitions. However, as summarised by Bittante et al., (2012), favourable milk coagulation properties have consistently been reported in milk from, for example, Brown Swiss, Simmental, Montbeliarde and several Alpine breeds (De Marchi et al., 2007), together with Jersey (Poulsen et al., 2013), compared to Holstein-Friesian. In contrast, Finnish Ayrshire (Tyrisevä et al., 2004), Swedish Red (Gustavsson et al., 2014b) and local Scandinavian breeds (Poulsen et al., 2016b) generally possess impaired milk coagulation properties with longer RCT and lower curd firmness, as measured at individual cow level on fresh, unpasteurised and non-standardised milk.

Given that milk protein content and composition play a significant role in milk coagulation, the influence of genetic protein variants on overall cheese-making capacity and milk coagulation properties has strong support (see reviews by Jakob and Puhan, 1992; Caroli et al., 2009). The influence of additive genetic variation is evident from β-LG, $\alpha_{S1}$-CN, k-CN, β-CN variants or by casein haplotypes or composite genotypes in various breeds (Comin et al., 2008; Bonfatti et al., 2010; Poulsen et al., 2013). Especially, the positive association of k-CN B with milk coagulation properties is well-documented (Hallén et al., 2007;

Joudu et al., 2007; Glantz et al., 2011; Ikonen et al., 1999; Vallas et al., 2012), as well as the positive association of $\alpha_{S1}$-CN C and β-CN B and the negative association of β-CN $A^2$ (Hallén et al., 2007; Poulsen et al., 2013). For β-LG, Glantz et al. (2011) documented favorable effects of β-LG A on rennet-induced gel strength, gelation time and yield stress, whereas others have documented the positive effect of β-LG B on cheese yield (Wedholm et al., 2006) and casein number (Hallén et al., 2008). In indigenous Danish Red anno 1970 and Jutland breeds, high frequencies of the otherwise rare β-CN variant F (Table 1) were associated with impaired milk coagulation properties (Poulsen et al., 2016b).

The results from Bonfatti et al. (2014) suggest that variation in RCT is mainly attributed to G k-CN and not to UG k-CN. Jensen et al. (2015) documented that chymosin-induced hydrolysis of k-CN was lower for glycosylated isoforms compared to non-glycosylated isoforms, suggesting that the carbohydrate attachments to some extent hinder the access of chymosin. Higher proportions of total k-CN to total protein, G k-CN to total k-CN and $\alpha_{S1}$-CN 8P to total $\alpha_{S1}$-CN are all related to good milk coagulation, that is, milk with low RCT and high CFR (Jensen et al., 2012b; Poulsen et al., 2016a).

Gregersen et al. (2014) used a GWAS to identify genomic regions affecting RCT and CFR in Holstein. In total, 19 genomic regions on 10 different chromosomes affected RCT or CFR. Apart from the major regions already described, the study identified *CTSD* on BTA29, coding for cathepsin D, as well as potential novel candidate genes such as *CWC15*, *MAP2K5* and *LMAN1*. In Swedish Red, a GWAS for rennet-induced coagulation also showed a major QTL around the casein gene cluster on BTA6, as well as a region encompassing *GALNT1* on BTA24, coding for UDP-N-acetyl-α-D-galactosamine:polypeptide N-acetylgalactosaminyl-transferase 1 (Gregersen et al., 2015). This enzyme plays a role in the O-glycosylation of κ-CN. Furthermore, regions containing *CTSZ* on BTA13 and *CTSC* on BTA29 were identified, which support that cathepsins may play a role in milk coagulation, probably through proteolysis (Gregersen et al., 2015). Poulsen et al. (2017b) confirmed the association between haplotypes identified from the GWAS study by Gregersen et al. (2014) and rennet-induced milk coagulation. The identified haplotypes were located within a major QTL on BTA6, where three significant SNPs located in intron regions or downstream *CSN3* (encoding k-CN) were identified, suggesting the genetic variation to be related to the regulation of k-CN.

With the identification of non-coagulating milk among samples from different breeds, including several Scandinavian breeds, and also Holstein (Ikonen et al., 2004; Tyrisevä et al., 2004; Cecchinato et al., 2011; Gustavsson et al., 2014b), genomic evaluation of non-coagulation has been a target. While no universal definition exists for non-coagulation, it is generally defined as milk with no recordable firmness after 30–60 min, but variation in temperature, enzyme concentration and analytical methodologies may affect the rheological evaluation, and thereby, definition. Furthermore, the evaluation and inclusion of non-coagulating milk samples in genetic parameter estimation and GWAS will affect the outcome, as discussed by Bittante et al. (2012), and should always be considered. Tyrisevä et al. (2008) used Finnish Ayrshire to identify marker genes for non-coagulation and identified two potential candidate genes, *LOC538897* (BTA 2), which encodes a nonspecific serine/threonine kinase, and *SIAT4B* (BTA18), which encodes a sialyltranferase involved in the last step of k-CN glycosylation, by catalysing the addition of sialic acid to galactose. This suggests that PTMs of caseins play a role in the ability of milk to coagulate. In Swedish Red cows, where the prevalence of non-coagulation was estimated to be as high as 18%, Duchemin et al. (2016) found a significant QTL region on BTA18, where *VPS35* was identified as the candidate gene of interest for non-coagulation. This gene relates to a mammary gene set and is in close proximity to the BTA18 region identified by Tyrisevä et al. (2008).

Milk coagulation properties are heritable, with moderate to high heritabilities estimated and reported (Ikonen et al., 2004; Cassandro et al., 2008; Vallas et al., 2010). In Jersey, the heritability for curd-firming rate (CFR) was found to be relatively low (0.15) compared with the high heritability in Holstein cows (0.75). In contrast, the heritability estimates were more similar for RCT, ranging from 0.28 in Holstein to 0.45 in Jersey (Poulsen et al., 2015a). The very high heritability for CFR in Holstein compared to Jersey and other studies may relate to study population, trait variability and underlying design. Generally, gel firmness and clot time are highly phenotypically and genetically correlated, and milk coagulation traits have further been documented to be genetically correlated to protein and casein contents, as well as to SCC, pH and acidity (Cassandro et al., 2008; Vallas et al., 2010; Gustavsson et al., 2014b; Poulsen et al., 2015a). In the Swedish Red, heritability for non-coagulation has been estimated to 0.28 (Duchemin et al., 2020), and is thought to be a multifactorial trait, but despite this, may be changed through selective breeding.

For acid-induced coagulation, genetic polymorphism especially of β-LG plays a role, which relates to casein number and β-LG concentration, where a higher β-LG concentration (associated with β-LG A) is preferable for fermented products (Allmere et al., 1998; Hallén et al., 2009). For acid-induced milk gels, Glantz et al. (2015) found QTLs containing *CSN3* (κ-CN) and *PAEP* (β-LG) as well as regions containing other potential candidate genes.

## 6 Fatty acids and minor milk components

The genetic influence on milk proteins is strong and the variations are less affected by, for example, feeding. For other milk traits, the influence of the genome may be smaller and less direct with many QTLs, each explaining a small proportion of the genetic variation. Important pathways/genes can, however, still be identified. For fatty acids, increasing the proportion of unsaturated fatty acids may be the target for selective breeding, even though the discussion of beneficial over unfavourable fatty acids relative to human health is complex (German and Dillard, 2006). Furthermore, when it comes to functional properties, an increasing proportion of unsaturated fatty acids may not be desired, as it relates to softer fat/higher spreadability and risk of increased lipid oxidation. Palmitic acid (C16:0), which is the main fatty acid in milk, derived partly from feed and partly from the mammary *de novo* synthesis, has been suggested to be the main objective for genetic selection due to its suggested negative effect on human health (Givens, 2010).

Heritability estimates for individual fatty acids are generally found to be low to moderate, but with the *de novo* fatty acids, which are synthesised in the mammary gland, exhibiting higher estimates than feed-derived fatty acids (Schennink et al., 2007; Krag et al., 2013). In contrast, desaturase indices, which reflect a proxy for $\Delta^9$-desaturase activity in the mammary gland, generally exhibit high heritabilities (Krag et al., 2013). The heritabilities for specific fatty acids indicate that it may be possible to alter the composition of fatty acids in bovine milk through selective breeding. Current selection for fat percent will mainly increase C16:0 (Stoop et al., 2008). Generally the genetic correlations between C4:0 to C14:0 fatty acids are positive and high (Stoop et al., 2008). Due to the negative correlation between groups of saturated fatty acids and unsaturated fatty acids (Krag et al., 2013), it may be possible to reduce the concentration of the less healthy saturated fatty acids and increase the concentration of unsaturated fatty acids. For unsaturated fatty acids, the results from Krag et al. (2013) suggest that the largest improvements can be achieved through changes in feeding. However, the heritability results also suggest that some achievement can be reached through selective breeding.

As stated earlier, major effects on milk fatty acid composition relate to genomic regions harbouring especially *FASN*, *DGAT1* and *SCD1*, though other regions also play a significant role (Schennink et al., 2009). When comparing QTLs for C6:0, C8:0, C10:0, C12:0, C14:0, C16:0, C18:0, C14:1, C16:1 and CLA, there was little overlap between the QTL found in the Dutch Holstein (Bouwman et al., 2011) and those found in Danish Holstein (Buitenhuis et al., 2014). For the Danish Holstein, significant SNP markers were detected for C14:1, C16:1 and CLA on the same chromosomes as those detected by Bouwman et al. (2011).

Undoubtedly, bovine milk is one of the best sources of several vitamins and minerals in human nutrition, including riboflavin (vitamin B2), cobalamin (vitamin B12), phosphorus and calcium. The primary origin of riboflavin and other B vitamins is through microbial biosynthesis in the rumen (Schwab et al., 2006). Poulsen, et al. (2015b) found substantial interbreed differences in milk riboflavin content. Milk from Jersey cows contained significantly higher levels of riboflavin (1.93 mg/L milk) than milk from Holstein cows (1.40 mg/L milk)

Published by Burleigh Dodds Science Publishing Limited, 2022.

and reported high heritabilities for riboflavin in both breeds, suggesting that the riboflavin content in milk is under significant genetic influence, and could be changed through selective breeding. Like riboflavin, vitamin B12 in milk is of microbial origin, and in Dutch Holstein Friesian cows vitamin B12 had an estimated heritability of 0.37, which also suggests that vitamin B12 content in milk can be improved by genetic selection (Rutten et al., 2013). A GWAS conducted on milk riboflavin content identified significant SNPs associated with *DGAT1* (Poulsen et al., 2015b). The most significant SNP for Holstein was located on BTA13 and assigned to *SLC52A3*, which is a riboflavin transporter gene. This gene could be a good candidate gene responsible for the genetic regulation of riboflavin content in milk (Poulsen et al., 2015a). For B12, none of the significant SNPs could be assigned to known candidate genes (Rutten et al., 2013).

The mineral fraction in milk constitutes approximately 7.1–7.4 g/L and comprises cations, including, among others, calcium (Ca), magnesium (Mg), sodium (Na), potassium (K), and anions, including, for example, phosphorus (P) and chloride (Cl) (Lindmark-Månsson et al., 2003; Hermansen et al., 2005). Bovine milk minerals and trace elements are essential for both human nutrition and dairy product quality. Several minerals occur in both colloidal and soluble states, which are important for heat stability of milk and milk coagulation properties, mainly related to their role in the structure and stability of casein micelles and levels of ionic $Ca^{2+}$.

Previous results documented considerable variation in milk mineral content from Swedish and Danish herds, especially due to season, but also to breed (Lindmark-Månsson et al., 2003; Hermansen et al., 2005). Results by van Hulzen et al. (2009) and Buitenhuis et al. (2015) also showed considerable variation for mineral content in milk and moderate to high heritabilities for Ca, P, Mg, K and Zn in Dutch Holstein (van Hulzen et al., 2009), for Ca, Cu, P and Zn in both Danish Holstein and Danish Jersey and for Mg, K and Mn in Danish Jersey (Buitenhuis et al., 2015). In both studies, non-existent to low heritabilities were found for Se, suggesting that Se is mainly affected by management and feeding and dietary manipulation. GWAS performed on milk minerals identified several significant SNPs for minerals, including BTA6 and BTA14 (Buitenhuis et al., 2015).

Apart from lactose, which is the main sugar component in bovine milk, milk also contains free milk oligosaccharides (OS) in small quantities. In human milk, OS are bioactive molecules that act as prebiotics and provide numerous health benefits to developing infants by stimulating the growth of beneficial intestinal bacteria, preventing the binding of pathogens and contributing to brain development (Pacheco et al., 2015; Jacobi et al., 2016). Bovine milk contains several OS structures in common with human milk (Aldredge et al., 2013; Kirmiz et al., 2018), and OS from dairy streams may thus act as value-added ingredients for infant formula or nutraceutical applications (Barile et al., 2009; Mehra et al., 2014). Heritability estimates for 15 individual OS in Danish

Holstein and Danish Jersey were generally high, documenting a solid breeding potential for bovine milk oligosaccharides, which ultimately could facilitate large-scale extraction of bovine milk OS. By GWAS, genomic regions with a major impact on milk OS were identified including specific glycosyltransferases as candidate loci for the synthesis of milk OS (Poulsen et al., 2019).

Milk metabolites reflect metabolic processes of the cow and may thus be important as indicator traits for cow health or play a role due to association with other traits important for the technological properties of milk. Genetic parameter estimation of milk metabolites has revealed moderate to high heritabilites for a number of milk metabolites including choline, citrate, glycerophosphocholine and orotate (Buitenhuis et al., 2013). Furthermore, urea, lactose and citrate display significant heritabilities (Stoop et al., 2007; Miglior et al., 2007; Mucha and Strandberg, 2011) and genetic improvement of these traits may be of interest.

## 7 Mid-infrared spectroscopy as large-scale phenotyping for genetic parameter estimation

In order to implement specific milk compositional traits in the breeding programmes, large-scale phenotyping and indirect ways to predict milk compositional traits or functional properties are needed. Thus, low-cost, accurate and reliable routine phenotyping is required, and as mid-infrared spectroscopy (MIRS) is widely used in milk recordings, there has been a great interest for population-level phenotyping of detailed milk composition by MIRS, as reviewed by De Marchi et al. (2014). Especially, MIRS-based prediction of fatty acids and groups of fatty acids have been of interest (Soyeurt et al., 2006; Rutten et al., 2009; Eskildsen et al., 2014) despite the somewhat indirect measurement (Eskildsen et al., 2014). In the Danish dairy cattle population, MIRS is used for routine screening of individual fatty acids (C14:0, C16:0, C18:0 and C18:1) and groups of fatty acids, which opens for the possibility of genetic selection as drivers for milk fatty acid changes (Hein et al., 2018). Also prediction of other important traits, including protein profile and genetic variants (Bonfatti et al., 2011; Rutten et al., 2011; Eskildsen et al., 2016), minerals (Soyeurt et al., 2009) and specific metabolites (Zaalberg et al., 2020) has been evaluated with varying degrees of predictabilty. MIRS-based prediction of milk coagulation has also received a lot of attention, both to be used in the milk recordings with the aim of breeding against non- and poor milk coagulation, and at the dairies for process control (Dal Zotto et al., 2008; De Marchi et al., 2009; Cecchinato et al., 2009), but to our knowledge, this has not been implemented. The majority of the MIR wavenumbers have moderate to high heritabilities (Wang et al., 2016), and Wang and Bovenhuis (2018) identified genomic regions associated with individual milk MIR wavenumbers. Apart from identifying known regions affecting the fat, protein and lactose percent, new regions possibly associated with milk phosphorus, citrate and orotic acid were also identified.

## 8 Possibilities for genetic improvement in relation to dairy milk

As outlined, the genetic influence on milk compositional traits is large, and for most traits, there is a good possibility for improvement through selective breeding. Different breeds differ phenotypically in their milk compositions, as well as in their genetic make-up, as these breeds have been separated for many generations (Kantanen et al., 2000), and have undergone strong selective breeding. This may also be reflected in the difference in the LD structure, where, for example, the Danish Jersey has higher bin-wise LD across the genome than the Danish Holstein (Buitenhuis et al., 2016). Across studies and across different breeds, differences in results may reflect the underlying design, the actual study population, or the analytical methods applied. However, for the more closely examined traits like milk proteins and fatty acids, strong and consistent evidence of the genetic influence is clear, both within and among breeds, and GWAS has confirmed major QTL regions and identified novel candidate genes that contribute to the variation of a high number of milk compositional traits.

From a dairy perspective, the desired improvement of different traits may relate to increased quality and output for specific valuable products and/or ingredients and the best use of natural genetic variation (Fig. 3). Such breeding programmes can go hand in hand with feeding and management to secure high-quality milk that meets future challenges in terms of sustainable production and consumer's needs.

Consumer milk holds a potential for differentiation with regard to, for example, production system or management or special local breeds. Differentiation of drinking milk based on genetic $A^1$ and $A^2$ variants of β-CN is significant in some parts of the world. Despite being still controversial, this relates to potential claimed health-related differences between $A^1$ and $A^2$ genetic variants, as have been discussed (see e.g. Truswell, 2005; Jianqin et al., 2016; He et al., 2017). For UHT milk, quality and stability are mainly related to enzymatic load and probably less to the raw milk protein composition as such. However, milk calcium, and especially ionic $Ca^{2+}$ influences stability of long-shelf-life UHT milk negatively (Lewis, 2011; Lewis et al., 2011). Overall, the total calcium level closely follows the total protein (Bijl et al., 2013), as a tight linkage between total protein to casein and further on to micellar calcium exists. The role of calcium in the casein micelle structure and in the milk coagulation process is fundamental (Holt, 1992). Impaired clotting properties have, among others, been linked to a lower content of the main cations in milk, calcium, and magnesium, along with the main anion, inorganic phosphate (van Hooydonk

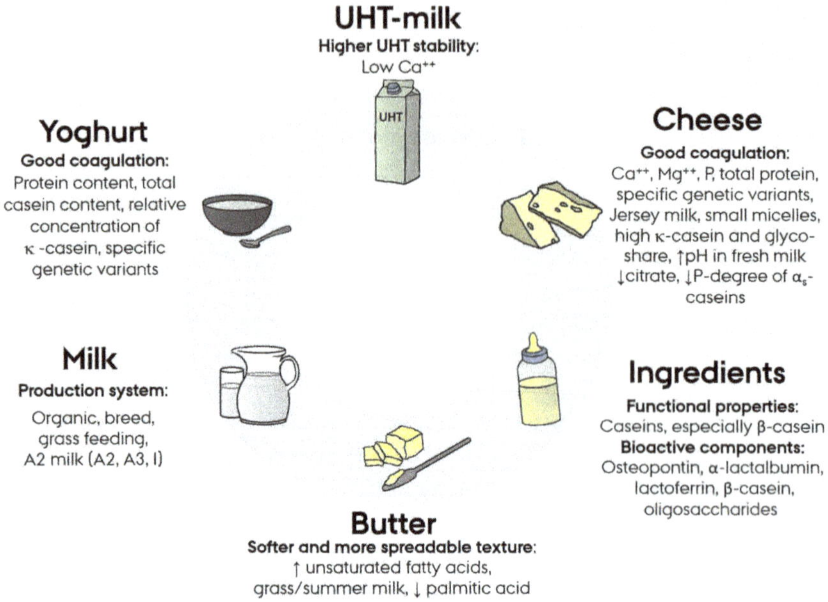


**Figure 3** Potentials of utilising variances in raw milk composition for specific dairy products, as based on the traits covered in the present chapter.

et al., 1986; Tsioulpas et al., 2007; Frederiksen et al., 2011), and a higher citrate content (Sundekilde et al., 2011). The minerals in milk exist in dynamic equilibrium between a soluble serum phase and a colloidal micellar phase. This mineral distribution is influenced by both cooling, lower milk pH and organic acids like citrate, which all lead to the dissolution of calcium phosphate from the micellar phase and may influence the integrity of the casein micelles and thereby potentially negatively influence the coagulation properties (Gaucheron, 2005). In milk, a strong positive correlation between nonsedimentable calcium and citrate has been documented (Bijl et al., 2013). However, whether natural variation in citrate, which is expected to be related to the *de novo* synthesis of fatty acids, is enough to affect milk quality during processing remains unclear. Overall, it seems that a certain level of calcium is needed to saturate and stabilise the casein micelles, and make them optimal for the cheese-making process, while on the other hand, a certain level of ionic calcium is needed in order to diminish the repulsion between the casein micelles in the second phase of the rennet induced milk coagulation. However, the mechanisms governing the calcium distribution in raw milk are less understood.

Changes in the proportions of specific proteins and of their specific PTMs may also be a strategy to improve milk technological properties, as a higher fraction of the least phosphorylated forms of both $\alpha_{S1}$- CN and of $\alpha_{S2}$- CN, and a higher fraction of glycosylated κ-CN are all positively associated with good rennet coagulation properties (Frederiksen et al., 2011; Jensen et al., 2012b). The implications of these well-documented variations in the coagulation properties of raw bovine milk point towards possibilities for improvement of raw milk composition for cheese production. However, the extent to which these variances in the coagulation properties of the raw milk are manifested onto properties of cheese milk after mixing and cooling in silo, heat-treatment and the addition of starter cultures, remains to be established. This will have to be the focus of future studies. Furthermore, non-coagulating milk is clearly important to monitor, but also a trait that is still not fully understood, despite the identification of a number of risk factors and significant trait heritability. Hallén et al. (2010) documented that the addition of $CaCl_2$ within the approved range for dairy milk could somewhat restore milk and eliminate the variation between poorly/non-coagulating and well-coagulating milk.

As with rennet-induced coagulation, several traits related to milk composition (e.g. major milk proteins, fat and lactose content) have all been linked to properties resulting from gels formed by acid-induced coagulation (Lucey and Singh, 1998; Hallén et al., 2009). Gustavsson et al. (2014b) found a positive phenotypic correlation between acid-induced coagulation and protein content (including both total CN as well as relative κ-CN concentration), whereas negative correlations were found for gel strength or yield stress with milk yield, in milk from Swedish Red cows. However, a general pattern was observed, where the observed genetic correlations were stronger than the phenotypic correlations, and findings have suggested only a weak genetic association between rennet- and acid-induced coagulation properties (Gustavsson et al., 2014b). Using this knowledge on the genetic parameters associated with poor acid-induced coagulation characteristics might ultimately benefit the production of yoghurt and fermented dairy products.

The rapidly growing milk ingredient industry exploits the vast nutritional and biological potential of milk. The main drivers are globalisation, demand for high-quality nutrients and stagnation at the European market for traditional yellow cheese. Thus, it is mandatory to explore the possibility of tailoring the content of high-value proteins in bovine milk to ensure the most optimal value proposition of milk at the global ingredient market. High-value proteins, phospholipids, and OS exert nutritional and health-promoting effects for consumers. Such value has traditionally been exploited through more traditional dairy products, but a growing market for specialised milk additives and ingredients with special functionalities (physiologically and technologically) adds great value to raw milk. An example of this is α-LA, one of the main proteins in human milk, which is added to infant formula in order to balance the composition from bovine towards that of human milk. Other functional important proteins include one of the major milk proteins, β-CN, as well as the less abundant but bioactive proteins, like osteopontin and lactoferrin. For β-CN and α-LA, large biological variations in their contents in individual cow's milk were revealed, showing potentials for exploiting this to elevate the levels of these valuable components through natural means (Le et al., 2020). Production of β-CN $A^2$ infant formulae has also gained interest, which may lead to an increase of β-CN $A^2$ cows/farms. For many breeds, β-CN $A^2$ is the predominant genetic variant. With regard to OS, milk from Danish Jersey cows contained higher relative amounts of both sialylated and the more complex neutral fucosylated OS, whereas milk from Danish Holstein had a higher abundance of smaller and simpler neutral OS (Sundekilde et al., 2012, Robinson et al., 2018a,b). The results point at the possibility of exploiting Jersey milk for the enrichment of products with fucosylated oligosaccharides, which bear more resemblance to those in human milk.

As discussed, the composition of milk fat from dairy cows is governed by both genetic and environmental factors. From a nutritional point of view, it is desirable to increase the level of unsaturated fatty acids in milk, which depends on the interplay between the feed composition, rumen metabolism and mammary synthesis. The nutritional value of milk is related to essential fatty acids (linoleic and α-linolenic acid), as well as specific fatty acids showing beneficial health effects, such as conjugated linoleic acid (CLA) and omega-3 fatty acids (German and Dillard, 2006). An increase in healthy fatty acids in human food sources and, likewise, a decrease in fatty acids with a negative impact on human health (in particular palmitic acid) are thus desirable (Mensink et al., 2003; Givens, 2010).

Generally, for butter production and quality, the more unsaturated fat that is present, the softer and more spreadable the butter appears. Apart from the fatty acid profile and the share of unsaturated to saturated fatty acids, the fat globule size and crystallisation are also important for the texture of butter products. It should also be noted that a higher level of unsaturated fatty acids generally can also lead to increased oxidation and potential development of off-flavour. As vitamin E can have antioxidant activities, variances in vitamin E can potentially counteract this effect to some degree, by combining high vitamin E levels with high-unsaturated fatty acid milk for butter production. Elevated vitamin E has been shown to be correlated with feeding of grass products (Poulsen et al., 2012).

As also outlined by Miglior et al. (2017), it is of importance that there is 1) an economic benefit for the farmer, 2) an economic interest for the dairy industry and 3) a reliable method to easily measure the milk phenotypes at a large scale in order to implement heritable milk compositional traits, either in the national breeding goal or for milk differentiation at specific farms. Furthermore, it is preferable that large-scale phenotyping can be implemented in the current infrastructure of national milk recording systems, by, for example, MIRS or similar easy and reliable fast methods. This will ensure that breeding values genetic correlations to current breeding goal traits can be established, increasing the potential economic benefit for both farmer and dairy.

## 9 Where to look for further information

Reports from the Food and Agriculture Organization of the United Nations (FAO) are important for global assessment of the current livestock genetic resources and state. A good overview of especially milk proteins and associated genetic variants and their effect on milk composition and quality can be found in McSweeney, P. L. H. and Fox, P. F. (Eds.) Advanced Dairy Chemistry: Volume 1A: Proteins: Basic Aspects, (4th ed., 2003). Springer Science and Boland, M. and Singh, H. (Eds) Milk Proteins - From Expression to Food (3rd ed., 2020). Elsevier. Important conferences include World Congress on Genetics Applied to Livestock Production (WCGALP), which is held every fourth year as well as dairy conferences, for example, from the American Dairy Science Association. Further, the International Milk Genomics Consortium (www.milkgenomics.org) and their annual symposium also cover lactation biology, comparative genetics and the effect of genetic parameters on milk quality.

## 10 References

Aldredge, D. L., Geronimo, M. R., Hua, S., Nwosu, C. C., Lebrilla, C. B. and Barile, D. 2013. Annotation and structural elucidation of bovine milk oligosaccharides and determination of novel fucosylated structures. *Glycobiology* 23(6):664–676.

Allmere, T., Andren, A., Lindersson, M. and Bjorck, L. 1998. Studies on rheological properties of stirred milk gels made from milk with defined genetic variants of kappa-casein and beta-lactoglobulin. *International Dairy Journal* 8(10–11):899–905.

Asledottir, T., Le, T. T., Petrat-Melin, B., Devold, T. G., Larsen, L. B. and Vegarud, G. E. 2017. Identification of bioactive peptides and quantification of β-casomorphin-7 from bovine β-casein A1, A2 and I after ex vivo gastrointestinal digestion. *International Dairy Journal* 71:98–106.

Asledottir, T., Le, T. T., Poulsen, N. A., Devold, T. G., Larsen, L. B. and Vegarud, G. E. 2018. Release of beta-casomorphin-7 from bovine milk of different beta-casein variants after ex vivo gastrointestinal digestion. *International Dairy Journal* 81:8–11.

Barile, D., Tao, N., Lebrilla, C. B., Coisson, J. D., Arlorio, M. and German, J. B. 2009. Permeate from cheese whey ultrafiltration is a source of milk oligosaccharides. *International Dairy Journal* 19(9):524–530.

Beja-Pereira, A., Luikart, G., England, P. R., Bradley, D. G., Jann, O. C., Bertorelle, G., Chamberlain, A. T., Nunes, T. P., Metodiev, S., Ferrand, N. and Erhardt, G. 2003. Gene-culture coevolution between cattle milk protein genes and human lactase genes. *Nature Genetics* 35(4):311–313.

Bijl, E., de Vries, R., van Valenberg, H., Huppertz, T. and van Hooijdonk, T. 2014a. Factors influencing casein micelle size in milk of individual cows: genetic variants and glycosylation of κ-casein. *International Dairy Journal* 34(1):135–141.

Bijl, E., van Valenberg, H. J. F., Huppertz, T., van Hooijdonk, A. C. M. and Bovenhuis, H. 2014b. Phosphorylation of αS1-casein is regulated by different genes. *Journal of Dairy Science* 97(11):7240–7246.

Bijl, E., van Valenberg, H. J., Huppertz, T. and van Hooijdonk, A. C. 2013. Protein, casein, and micellar salts in milk: current content and historical perspectives. *Journal of Dairy Science* 96(9):5455–5464.

Bittante, G., Penasa, M. and Cecchinato, A. 2012. Invited review: genetics and modeling of milk coagulation properties. *Journal of Dairy Science* 95(12):6843–6870.

Bonfatti, V., Chiarot, G. and Carnier, P. 2014. Glycosylation of κ-casein: genetic and nongenetic variation and effects on rennet coagulation properties of milk. *Journal of Dairy Science* 97(4):1961–1969.

Bonfatti, V., Di Martino, G. and Carnier, P. 2011. Effectiveness of mid-infrared spectroscopy for the prediction of detailed protein composition and contents of protein genetic variants of individual milk of Simmental cows. *Journal of Dairy Science* 94(12):5776–5785.

Bonfatti, V., Di Martino, G., Cecchinato, A., Vicario, D. and Carnier, P. 2010. Effects of β-κ-casein (CSN2-CSN3) haplotypes and β-lactoglobulin (BLG) genotypes on milk production traits and detailed protein composition of individual milk of Simmental cows. *Journal of Dairy Science* 93(8):3797–3808.

Bouwman, A. C., Bovenhuis, H., Visker, M. H. and van Arendonk, J. A. 2011. Genome-wide association of milk fatty acids in Dutch dairy cattle. *BMC Genetics* 12:43.

Bovenhuis, H., Visker, M. H. P. W., Poulsen, N. A., Sehested, J., van Valenberg, H. J. F., van Arendonk, J. A. M., Larsen, L. B. and Buitenhuis, A. J. 2016. Effects of the diacylglycerol o-acyltransferase 1 (DGAT1) K232A polymorphism on fatty acid, protein, and mineral composition of dairy cattle milk. *Journal of Dairy Science* 99(4):3113–3123.

Buitenhuis, A. J., Sundekilde, U. K., Poulsen, N. A., Bertram, H. C., Larsen, L. B. and Sørensen, P. 2013. Estimation of genetic parameters and detection of quantitative trait loci for metabolites in Danish Holstein milk. *Journal of Dairy Science* 96(5):3285–3295.

Buitenhuis, B., Janss, L. L. G., Poulsen, N. A., Larsen, L. B., Larsen, M. K. and Sørensen, P. 2014. Genome-wide association and biological pathway analysis for milk-fat composition in Danish Holstein and Danish Jersey cattle. *BMC Genomics* 15:1112.

Buitenhuis, B., Poulsen, N. A., Gebreyesus, G. and Larsen, L. B. 2016. Estimation of genetic parameters and detection of chromosomal regions affecting the major milk proteins and their post translational modifications in Danish Holstein and Danish Jersey cattle. *BMC Genetics* 17:114.

Buitenhuis, B., Poulsen, N. A., Larsen, L. B. and Sehested, J. 2015. Estimation of genetic parameters and detection of quantitative trait loci for minerals in Danish Holstein and Danish Jersey milk. *BMC Genetics* 16:52.

Caroli, A. M., Chessa, S. and Erhardt, G. J. 2009. Invited review: milk protein polymorphisms in cattle: effect on animal breeding and human nutrition. *Journal of Dairy Science* 92(11):5335–5352.

Cassandro, M., Comin, A., Ojala, M., Zotto, R. D., De Marchi, M., Gallo, L., Carnier, P. and Bittante, G. 2008. Genetic parameters of milk coagulation properties and their relationships with milk yield and quality traits in Italian Holstein cows. *Journal of Dairy Science* 91(1):371–376.

Cecchinato, A., De Marchi, M., Gallo, L., Bittante, G. and Carnier, P. 2009. Mid-infrared spectroscopy predictions as indicator traits in breeding programs for enhanced coagulation properties of milk. *Journal of Dairy Science* 92(10):5304–5313.

Cecchinato, A., Penasa, M., De Marchi, M., Gallo, L., Bittante, G. and Carnier, P. 2011. Genetic parameters of coagulation properties, milk yield, quality, and acidity estimated using coagulating and noncoagulating milk information in Brown Swiss and Holstein-Friesian cows. *Journal of Dairy Science* 94(8):4205–4213.

Comin, A., Cassandro, M., Chessa, S., Ojala, M., Dal Zotto, R., De Marchi, M., Carnier, P., Gallo, L., Pagnacco, G. and Bittante, G. 2008. Effects of Composite {beta}- and {kappa}-Casein genotypes on milk coagulation, quality, and yield traits in Italian holstein cows. *Journal of Dairy Science* 91(10):4022–4027.

Conte, G., Mele, M., Chessa, S., Castiglioni, B., Serra, A., Pagnacco, G. and Secchiari, P. 2010. Diacylglycerol acyltransferase 1, stearoyl-CoA desaturase 1, and sterol regulatory element binding protein 1 gene polymorphisms and milk fatty acid composition in Italian Brown cattle. *Journal of Dairy Science* 93(2): 753–763.

Dal Zotto, R., De Marchi, M., Cecchinato, A., Penasa, M., Cassandro, M., Carnier, P., Gallo, L. and Bittante, G. 2008. Reproducibility and repeatability of measures of milk coagulation properties and predictive ability of mid-infrared reflectance spectroscopy. *Journal of Dairy Science* 91(10):4103–4112.

Day, L., Williams, R. P. W., Otter, D. and Augustin, M. A. 2015. Casein polymorphism heterogeneity influences casein micelle size in milk of individual cows. *Journal of Dairy Science* 98(6):3633–3644.

De Marchi, M., Dal Zotto, R., Cassandro, M. and Bittante, G. 2007. Milk coagulation ability of five dairy cattle breeds. *Journal of Dairy Science* 90(8):3986–3992.

De Marchi, M., Fagan, C. C., O'Donnell, C. P., Cecchinato, A., Dal Zotto, R., Cassandro, M., Penasa, M. and Bittante, G. 2009. Prediction of coagulation properties, titratable acidity, and pH of bovine milk using mid-infrared spectroscopy. *Journal of Dairy Science* 92(1):423–432.

De Marchi, M., Toffanin, V., Cassandro, M. and Penasa, M. 2014. Invited review: mid-infrared spectroscopy as phenotyping tool for milk traits. *Journal of Dairy Science* 97(3):1171–1186.

Devold, T. G., Brovold, M. J., Langsrud, T. and Vegarud, G. E. 2000. Size of native and heated casein micelles, content of protein and minerals in milk from Norwegian Red Cattle–effect of milk protein polymorphism and different feeding regimes. *International Dairy Journal* 10(5–6):313–323.

Duchemin, S. I., Glantz, M., De Koning, D. J., Paulsson, M. and Fikse, W. F. 2016. Identification of QTL on chromosome 18 associated with non-coagulating milk in Swedish Red cows. *Frontiers in Genetics* 7:57.

Duchemin, S. I., Nilsson, K., Fikse, W. F., Stålhammar, H., Buhelt Johansen, L., Stenholdt Hansen, M., Lindmark-Månsson, H., de Koning, D. J., Paulsson, M. and Glantz, M. 2020. Genetic parameters for noncoagulating milk, milk coagulation properties, and detailed milk composition in Swedish Red Dairy Cattle. *Journal of Dairy Science* 103(9):8330–8342.

Dziuba, J. and Minkiewicz, P. 1996. Influence of glycosylation on micelle-stabilizing ability and biological properties of C-terminal fragments of cow's κ-casein. *International Dairy Journal* 6(11–12):1017–1044.

Eskildsen, C. E., Rasmussen, M. A., Engelsen, S. B., Larsen, L. B., Poulsen, N. A. and Skov, T. 2014. Quantification of individual fatty acids in bovine milk by infrared spectroscopy and chemometrics: understanding predictions of highly collinear reference variables. *Journal of Dairy Science* 97(12):7940–7951.

Eskildsen, C. E., Skov, T., Hansen, M. S., Larsen, L. B. and Poulsen, N. A. 2016. Quantification of bovine milk protein composition and coagulation properties using infrared spectroscopy and chemometrics: a result of collinearity among reference variables. *Journal of Dairy Science* 99(10):8178-8186.

Fang, Z. H., Bovenhuis, H., Delacroix-Buchet, A., Miranda, G., Boichard, D., Visker, M., and Martin, P. (2017). Genetic and nongenetic factors contributing to differences in αS-casein phosphorylation isoforms and other major milk proteins. *Journal of dairy science*, 100(7), 5564-5577. https://doi.org/10.3168/jds.2016-12338.

Farrell, H. M., Jr., Jimenez-Flores, R., Bleck, G. T., Brown, E. M., Butler, J. E., Creamer, L. K., Hicks, C. L., Hollar, C. M., Ng-Kwai-Hang, K. F. and Swaisgood, H. E. 2004. Nomenclature of the proteins of cow's milk–sixth revision. *Journal of Dairy Science* 87(6):1641-1674.

Food and Agriculture Organization of the United Nations (FAO) 2015. The second report on the state of the world's animal genetic resources for food and agriculture. B. D. Scherf (Ed.). FAO, Rome, Italy. Accessed 27 January 2021. Available at: The Second Report on the State of the World's Animal Genetic Resources for Food and Agriculture (fao.org).

Frederiksen, P. D., Andersen, K. K., Hammershøj, M., Poulsen, H. D., Sørensen, J., Bakman, M., Qvist, K. B. and Larsen, L. B. 2011. Composition and effect of blending of noncoagulating, poorly coagulating, and well-coagulating bovine milk from individual Danish Holstein cows. *Journal of Dairy Science* 94(10):4787-4799.

Gaucheron, F. 2005. The minerals of milk. *Reproduction, Nutrition, Development* 45(4):473-483.

German, J. B. and Dillard, C. J. 2006. Composition, structure and absorption of milk lipids: a source of energy, fat-soluble nutrients and bioactive molecules. *Critical Reviews in Food Science and Nutrition* 46(1):57-92.

German, J. B., Dillard, C. J. and Ward, R. E. 2002. Bioactive components in milk. *Current Opinion in Clinical Nutrition and Metabolic Care* 5(6):653-658.

Givens, D. I. 2010. Milk and meat in our diet: good or bad for health? *Animal: An International Journal of Animal Bioscience* 4(12):1941-1952.

Glantz, M., Gustavsson, F., Bertelsen, H. P., Stålhammar, H., Lindmark-Månsson, H., Paulsson, M., Bendixen, C. and Gregersen, V. R. 2015. Bovine chromosomal regions affecting rheological traits in acid-induced skim milk gels. *Journal of Dairy Science* 98(2):1273-1285.

Glantz, M., Lindmark Månsson, H., Stålhammar, H. and Paulsson, M. 2011. Effect of polymorphisms in the leptin, leptin receptor, and acyl-coenzyme A:diacylglycerol acyltransferase 1 (DGAT1) genes and genetic polymorphism of milk proteins on cheese characteristics. *Journal of Dairy Science* 94(7):3295-3304.

Goddard, M. E. and Hayes, B. J. 2009. Mapping genes for complex traits in domestic animals and their use in breeding programmes. *Nature Reviews. Genetics* 10(6):381-391.

Gregersen, V. R., Bertelsen, H. P., Poulsen, N. A., Larsen, L. B., Gustavsson, F., Glantz, M., Paulsson, M., Buitenhuis, A. J. and Bendixen, C. 2014. Genomic regions affecting cheese making properties identified in Danish Holsteins. Abstract from 10th World Congress on Genetics Applied to Livestock Production (WCGALP).

Gregersen, V. R., Gustavsson, F., Glantz, M., Christensen, O. F., Stålhammar, H., Andrén, A., Lindmark-Månsson, H., Poulsen, N. A., Larsen, L. B., Paulsson, M. and Bendixen,

C. 2015. Bovine chromosomal regions affecting rheological traits in rennet-induced skim milk gels. *Journal of Dairy Science* 98(2):1261-1272.

Grisart, B., Coppieters, W., Farnir, F., Karim, L., Ford, C., Berzi, P., Cambisano, N., Mni, M., Reid, S., Simon, P., Spelman, R., Georges, M. and Snell, R. 2002. Positional candidate cloning of a QTL in dairy cattle: identification of a missense mutation in the bovine DGAT1 gene with major effect on milk yield and composition. *Genome Research* 12(2):222-231.

Gustavsson, F., Buitenhuis, A. J., Johansson, M., Bertelsen, H. P., Glantz, M., Poulsen, N. A., Lindmark Månsson, H., Stålhammar, H., Larsen, L. B., Bendixen, C., Paulsson, M. and Andrén, A. 2014a. Effects of breed and casein genetic variants on protein profile in milk from Swedish Red, Danish Holstein, and Danish Jersey cows. *Journal of Dairy Science* 97(6):3866-3877.

Gustavsson, F., Glantz, M., Poulsen, N. A., Wadsö, L., Stålhammar, H., Andrén, A., Lindmark Månsson, H., Larsen, L. B., Paulsson, M. and Fikse, W. F. 2014b. Genetic parameters for rennet- and acid-induced coagulation properties in milk from Swedish red dairy cows. *Journal of Dairy Science* 97(8):5219-5229.

Hallén, E., Allmere, T., Lundén, A. and Andrén, A. 2009. Effect of genetic polymorphism of milk proteins on rheology of acid-induced milk gels. *International Dairy Journal* 19(6-7):399-404.

Hallén, E., Allmere, T., Naslund, J., Andren, A. and Lunden, A. 2007. Effect of genetic polymorphism of milk proteins on rheology of chymosin-induced milk gels. *International Dairy Journal* 17(7):791-799.

Hallén, E., Lundén, A., Tyriseva, A. M., Westerlind, M. and Andrén, A. 2010. Composition of poorly and non-coagulating bovine milk andeffect of calcium addition. *Journal of Dairy Research* 77(4):398-403.

Hallén, E., Wedholm, A., Andren, A. and Lunden, A. 2008. Effect of beta-casein, kappa-casein and beta-lactoglobulin genotypes on concentration of milk protein variants. *Journal of Animal Breeding and Genetics* 125(2):119-129.

He, M., Sun, J., Jiang, Z. Q. and Yang, Y. X. 2017. Effects of cow's milk beta-casein variants on symptoms of milk intolerance in Chinese adults: a multicentre, randomised controlled study. *Nutrition Journal* 16(1):72.

Heck, J. M. L., Schennink, A., van Valenberg, H. J. F., Bovenhuis, H., Visker, M. H. P. W., van Arendonk, J. A. M. and van Hooijdonk, A. C. 2008. Effects of milk protein variants on the protein composition of bovine milk. *Journal of Dairy Science* 92(3):1192-1202.

Hein, L., Sørensen, L. P., Kargo, M. and Buitenhuis, A. J. 2018. Genetic analysis of predicted fatty acid profiles of milk from Danish Holstein and Danish Jersey cattle populations. *Journal of Dairy Science* 101(3):2148-2157.

Hermansen, J. E., Badsberg, J. H., Kristensen, T. and Gundersen, V. 2005. Major and trace elements in organically or conventionally produced milk. *Journal of Dairy Research* 72(3):362-368.

Holt, C. 1992. Structure and stability of bovine casein micelles. *Advances in Protein Chemistry*. AP 43:63-151.

Ikonen, T., Morri, S., Tyriseva, A. M., Ruottinen, O. and Ojala, M. 2004. Genetic and phenotypic correlations between milk coagulation properties, milk production traits, somatic cell count, casein content, and pH of milk. *Journal of Dairy Science* 87(2):458-467.

Ikonen, T., Ojala, M. and Ruottinen, O. 1999. Associations between milk protein polymorphism and first lactation milk production traits in Finnish Ayrshire cows. *Journal of Dairy Science* 82(5):1026-1033.

Iso-Touru, T., Tapio, M., Vilkki, J., Kiseleva, T., Ammosov, I., Ivanova, Z., Popov, R., Ozerov, M. and Kantanen, J. 2016. Genetic diversity and genomic signatures of selection among cattle breeds from Siberia, eastern and northern Europe. *Animal Genetics* 47(6):647–657.

Jacobi, S. K., Yatsunenko, T., Li, D. P., Dasgupta, S., Yu, R. K., Berg, B. M., Chichlowski, M. and Odle, J. 2016. Dietary isomers of sialyllactose increase ganglioside sialic acid concentrations in the corpus callosum and cerebellum and modulate the colonic microbiota of formula-fed piglets. *Journal of Nutrition* 146(2):200–208.

Jakob, E. and Puhan, Z. 1992. Technological properties of milk as influenced by genetic polymorphism of milk proteins–a review. *International Dairy Journal* 2(3): 157–178.

Jensen, H. B., Holland, J. W., Poulsen, N. A. and Larsen, L. B. 2012a. Milk protein genetic variants and isoforms identified in bovine milk representing extremes in coagulation properties. *Journal of Dairy Science* 95(6):2891–2903.

Jensen, H. B., Poulsen, N. A., Andersen, K. K., Hammershøj, M., Poulsen, H. D. and Larsen, L. B. 2012b. Distinct compositon of bovine milk from Jersey and Holstein-Friesian cows with good, poor or non-coagulation properties as reflacted in protein genetic variants and isoforms. *Journal of Dairy Science* 95(12):6905–6917.

Jensen, H. B., Pedersen, K. S., Johansen, L. B., Poulsen, N. A., Bakman, M., Chatterton, D. E. W. and Larsen, L. B. 2015. Genetic variation and posttranslational modification of bovine κ-casein: effects on caseino-macropeptide release during renneting. *Journal of Dairy Science* 98(2):747–758.

Jianqin, S., Leiming, X., Lu, X., Yelland, G. W., Ni, J. and Clarke, A. J. 2016. Effects of milk containing only A2 beta casein versus milk containing both A1 and A2 beta casein proteins on gastrointestinal physiology, symptoms of discomfort, and cognitive behavior of people with self-reported intolerance to traditional cows' milk. *Nutrition Journal* 15:35.

Joudu, I., Henno, M., Varv, S., Kaart, T., Kart, O. and Kalamees, K. 2007. Milk protein genotypes and milk coagulation properties of Estonian Native cattle. *Agricultural and Food Science* 16(3):222–231.

Kantanen, J., Olsaker, I., Holm, L. E., Lien, S., Vilkki, J., Brusgaard, K., Eythorsdottir, E., Danell, B. and Adalsteinsson, S. 2000. Genetic diversity and population structure of 20 north European cattle breeds. *Journal of Heredity* 91(6):446–457.

Kirmiz, N., Robinson, R. C., Shah, I. M., Barile, D. and Mills, D. A. 2018. Milk glycans and their interaction with the infant-gut microbiota. *Annual Review of Food Science and Technology* 9(9):429–450.

Krag, K., Poulsen, N. A., Larsen, M. K., Larsen, L. B., Janss, L. L. and Buitenhuis, B. 2013. Genetic parameters for milk fatty acids in Danish Holstein cattle based on SNP markers using a Bayesian approach. *BMC Genetics* 14:79.

Le, T. T., Poulsen, N. A., Kristiansen, G. H. and Larsen, L. B. 2020. Quantitative LC-MS/MS analysis of high-value milk proteins in Danish Holstein cows. *Heliyon* 6(9):e04620.

Lewis, M., Grandison, A., Lin, M. and Tsioulpas, A. 2011. Ionic calcium and pH as predictors of stability of milk to UHT processing. *Milchwissenschaft* 66:197–200.

Lewis, M. J. 2011. The measurement and significance of ionic calcium in milk–a review. *International Journal of Dairy Technology* 64(1):1–13.

Lindmark-Månsson, H., Fonden, R. and Pettersson, H. E. 2003. Composition of Swedish dairy milk. *International Dairy Journal* 13(6):409–425.

Lisson, M., Lochnit, G. and Erhardt, G. 2013. Genetic variants of bovine beta- and kappa-casein result in different immunoglobulin E-binding epitopes after in vitro gastrointestinal digestion. *Journal of Dairy Science* 96(9):5532–5543.

Lucey, J. A. and Singh, H. 1998. Formation and physical properties of acid milk gels: a review. *Food Research International* 30(7):529–542.

Martin, P., Bianchi, L., Cebo, C. and Miranda, G. 2013. Genetic polymorphism of milk proteins. In McSweeney, P. L. H. and Fox, P. F. (Eds.) *Advanced Dairy Chemistry: Volume 1A: Proteins: Basic Aspects*, (4th ed. pp. 463–514). Springer Science.

Mehra, R., Barile, D., Marotta, M., Lebrilla, C. B., Chu, C. and German, J. B. 2014. Novel high-molecular weight fucosylated milk oligosaccharides identified in dairy streams. *PLoS ONE* 9(5):e96040.

Mele, M., Conte, G., Castiglioni, B., Chessa, S., Macciotta, N. P. P., Serra, A., Buccioni, A., Pagnacco, G. and Secchiari, P. 2007. Stearoyl-coenzyme A desaturase gene polymorphism and milk fatty acid composition in Italian holsteins. *Journal of Dairy Science* 90(9):4458–4465.

Melzer, N., Wittenburg, D., Hartwig, S., Jakubowski, S., Kesting, U., Willmitzer, L., Lisec, J., Reinsch, N. and Repsilber, D. 2013. Investigating associations between milk metabolite profiles and milk traits of Holstein cows. *Journal of Dairy Science* 96(3):1521–1534.

Mensink, R. P., Zock, P. L., Kester, A. D. and Katan, M. B. 2003. Effects of dietary fatty acids and carbohydrates on the ratio of serum total to HDL cholesterol and on serum lipids and apolipoproteins: a meta-analysis of 60 controlled trials. *American Journal of Clinical Nutrition* 77(5):1146–1155.

Miglior, F., Fleming, A., Malchiodi, F., Brito, L. F., Martin, P. and Baes, C. F. 2017. A 100-year Review: identification and genetic selection of economically important traits in dairy cattle. *Journal of Dairy Science* 100(12):10251–10271.

Miglior, F., Sewalem, A., Jamrozik, J., Bohmanova, J., Lefebvre, D. M. and Moore, R. K. 2007. Genetic analysis of milk urea nitrogen and lactose and their relationships with other production traits in Canadian Holstein cattle. *Journal of Dairy Science* 90(5):2468–2479.

Miranda, G., Bianchi, L., Krupova, Z., Trossat, P. and Martin, P. 2020. An improved LC–MS method to profile molecular diversity and quantify the six main bovine milk proteins, including genetic and splicing variants as well as post-translationally modified isoforms. *Food Chemistry: X* 5:100080.

Mucha, S. and Strandberg, E. 2011. Genetic analysis of milk urea nitrogen and relationships with yield and fertility across lactation. *Journal of Dairy Science* 94(11):5665–5672.

Ogorevc, J., Kunej, T., Razpet, A. and Dovc, P. 2009. Database of cattle candidate genes and genetic markers for milk production and mastitis. *Animal Genetics* 40(6):832–851.

Pacheco, A. R., Barile, D., Underwood, M. A. and Mills, D. A. 2015. The impact of the milk glycobiome on the neonate gut microbiota. *Annual Review of Animal Biosciences* 3:419–445.

Petrat-Melin, B., Andersen, P., Rasmussen, J. T., Poulsen, N. A., Larsen, L. B. and Young, J. F. 2015. In vitro digestion of purified β-casein variants A1, A2, B, and I: effects on antioxidant and angiotensin-converting enzyme inhibitory capacity. *Journal of Dairy Science* 98(1):15–26.

Petrat-Melin, B., Kristiansen, G. H., Le, T. T., Poulsen, N. A., Larsen, L. B. and Young, J. F. 2016. In vitro gastrointestinal digestion of purified bovine kappa-casein variants

A, B, and E: effects on antioxidant and angiotensin 1-converting enzyme inhibitory capacity. *International Dairy Journal* 57:44–51.

Petrat-Melin, B., Le, T. T., Møller, H. S., Larsen, L. B. and Young, J. F. 2017. Short communication: inhibition of angiotensin 1-converting enzyme by peptides derived from variants of bovine β-casein upon apical exposure to a Caco-2 cell monolayer. *Journal of Dairy Science* 100(2):881–885.

Poulsen, N. A., Bertelsen, H. P., Jensen, H. B., Gustavsson, F., Glantz, M., Mansson, H. L., Andren, A., Paulsson, M., Bendixen, C., Buitenhuis, A. J. and Larsen, L. B. 2013. The occurrence of noncoagulating milk and the association of bovine milk coagulation properties with genetic variants of the caseins in 3 Scandinavian dairy breeds. *Journal of Dairy Science* 96(8):4830–4842.

Poulsen, N. A., Buitenhuis, A. J. and Larsen, L. B. 2015a. Phenotypic and genetic associations of milk traits with milk coagulation properties. *Journal of Dairy Science* 98(4):2079–2087.

Poulsen, N. A., Rybicka, I., Larsen, L. B., Buitenhuis, A. J. and Larsen, M. K. 2015b. Short communication: genetic variation of riboflavin content in bovine milk. *Journal of Dairy Science* 98(5):3496–3501.

Poulsen, N. A., Glantz, M., Rosengaard, A. K., Paulsson, M. and Larsen, L. B. 2017a. Comparison of milk protein composition and rennet coagulation properties in native Swedish dairy cow breeds and high-yielding Swedish Red cows. *Journal of Dairy Science* 100(11):8722–8734.

Poulsen, N. A., Gregersen, V. R., Maciel, G. M., Madsen, L. B., Buitenhuis, B., Hansen, M. S., Bendixen, C. and Larsen, L. B. 2017b. Novel genetic variation associated to CSN3 strongly affects rennet-induced milk coagulation. *International Dairy Journal* 71:122–130.

Poulsen, N. A., Gustavsson, F., Glantz, M., Paulsson, M., Larsen, L. B. and Larsen, M. K. 2012. The influence of feed and herd on fatty acid composition in 3 dairy breeds (Danish Holstein, Danish Jersey, and Swedish Red). *Journal of Dairy Science* 95(11):6362–6371.

Poulsen, N. A., Le, T. T., Gregersen, V. R., Christensen, B., Sørensen, E. S., Larsen, L. B. and Buitenhuis, B. 2018. Genetic parameter estimation of high value milk proteins and whey to casein ratio in Danish Holstein. Proceedings of the paper 312, 11th World Congress of Genetics Applied to Livestock Production, Auckland, New Zealand.

Poulsen, N. A., Jensen, H. B. and Larsen, L. B. 2016a. Factors influencing degree of glycosylation and phosphorylation of caseins in individual cow milk samples. *Journal of Dairy Science* 99(5):3325–3333.

Poulsen, N. A., Rosengaard, A. K., Szekeres, B. D., Gregersen, V. R., Jensen, H. B. and Larsen, L. B. 2016b. Protein heterogeneity of bovine beta-casein in Danish dairy breeds and association of rare beta-casein F with milk coagulation properties. *Acta Agriculturae Scandinavica, Section A–Animal Science* 66(4):190–198.

Poulsen, N. A., Robinson, R. C., Barile, D., Larsen, L. B. and Buitenhuis, B. 2019. A genome-wide association study reveals specific transferases as candidate loci for bovine milk oligosaccharides synthesis. *BMC Genomics* 20(1):404.

Ramakrishnan, M., Eaton, T. K., Sermet, O. M. and Savaiano, D. A. 2020. Milk containing A2 β-casein only, as a single meal, causes fewer symptoms of lactose intolerance than milk containing A1 and A2 β-caseins in subjects with lactose maldigestion and intolerance: A randomized, double-blind, crossover trial. *Nutrients* 12(12):3855.

Rasmussen, L. K., Hojrup, P. and Petersen, T. E. 1992a. Localization of two interchain disulfide bridges in dimers of bovine alphas2-casein. Parallel and antiparallel alignments of the polypeptide chains. *European Journal of Biochemistry* 203(3):381–386.

Rasmussen, L. K., Hojrup, P. and Petersen, T. E. 1992b. The multimeric structure and disulfide-bonding pattern of bovine kappa-casein. *European Journal of Biochemistry* 207(1):215–222.

Robinson, R. C., Poulsen, N. A. and Barile, D. 2018a. Multiplexed bovine milk oligosaccharide analysis with aminoxy tandem mass tags. *PLoS ONE* 13(4):e0196513.

Robinson, R. C., Poulsen, N. A., Colet, E., Duchene, C., Larsen, L. B. and Barile, D. 2018b. Profiling of aminoxyTMT-labeled bovine milk oligosaccharides reveals substantial variation in oligosaccharide abundance between dairy cattle breeds. *Scientific Reports* 9(1):5465.

Rutten, M. J. M., Bouwman, A. C., Sprong, R. C., van Arendonk, J. A. M. and Visker, M. H. P. W. 2013. Genetic variation in vitamin B-12 content of bovine milk and its association with SNP along the bovine genome. *PLoS ONE* 8(4):e62382.

Rutten, M. J. M., Bovenhuis, H., Heck, J. M. L. and van Arendonk, J. A. M. 2011. Predicting bovine milk protein composition based on Fourier transform infrared spectra. *Journal of Dairy Science* 94(11):5683–5690.

Rutten, M. J. M., Bovenhuis, H., Hettinga, K. A., van Valenberg, H. J. F. and van Arendonk, J. A. M. 2009. Predicting bovine milk fat composition using infrared spectroscopy based on milk samples collected in winter and summer. *Journal of Dairy Science* 92(12):6202–6209.

Saito, T. and Itoh, T. 1992. Variations and Distributions of O-Glycosidically Linked Sugar Chains in Bovine κ-Casein. *Journal of Dairy Science* 75(7):1768–1774.

Schennink, A., Heck, J. M. L., Bovenhuis, H., Visker, M. H. P. W., van Valenberg, H. J. F. and van Arendonk, J. A. M. 2008. Milk fatty acid unsaturation: genetic parameters and effects of stearoyl-CoA desaturase (SCD1) and acyl CoA: diacylglycerol acyltransferase 1 (DGAT1). *Journal of Dairy Science* 91(5):2135–2143.

Schennink, A., Stoop, W. M., Visker, M. H. P. W., Heck, J. M. L., Bovenhuis, H., van der Poel, J. J., van Valenberg, H. J. F. and van Arendonk, J. A. M. 2007. DGAT1 underlies large genetic variation in milk-fat composition of dairy cows. *Animal Genetics* 38(5):467–473.

Schennink, A., Stoop, W. M., Visker, M. H. P. W., van der Poel, J. J., Bovenhuis, H. and van Arendonk, J. A. M. 2009. Short communication: Genome-wide scan for bovine milk-fat composition. II. Quantitative trait loci for long-chain fatty acids. *Journal of Dairy Science* 92(9):4676–4682.

Schopen, G. C., Visker, M. H., Koks, P. D., Mullaart, E., van Arendonk, J. A. and Bovenhuis, H. (2011). Whole-genome association study for milk protein composition in dairy cattle. *Journal of dairy science*, 94(6), 3148–3158. https://doi.org/10.3168/jds.2010-4030.

Schopen, G. C. B., Heck, J. M. L., Bovenhuis, H., Visker, M. H. P. W., van Valenberg, H. J. F. and van Arendonk, J. A. M. 2009. Genetic parameters for major milk proteins in Dutch Holstein-Friesians. *Journal of Dairy Science* 92(3):1182–1191.

Schwab, E. C., Schwab, C. G., Shaver, R. D., Girard, C. L., Putnam, D. E. and Whitehouse, N. L. 2006. Dietary forage and nonfiber carbohydrate contents influence B-vitamin intake, duodenal flow, and apparent ruminal synthesis in lactating dairy cows. *Journal of Dairy Science* 89(1):174–187.

Soyeurt, H., Bruwier, D., Romnee, J. M., Gengler, N., Bertozzi, C., Veselko, D. and Dardenne, P. 2009. Potential estimation of major mineral contents in cow milk using mid-infrared spectrometry. *Journal of Dairy Science* 92(6):2444–2454.

Soyeurt, H., Dardenne, P., Dehareng, F., Lognay, G., Veselko, D., Marlier, M., Bertozzi, C., Mayeres, P. and Gengler, N. 2006. Estimating fatty acid content in cow milk using mid-infrared spectrometry. *Journal of Dairy Science* 89(9):3690–3695.

Stoop, W. M., Bovenhuis, H. and van Arendonk, J. A. 2007. Genetic parameters for milk urea nitrogen in relation to milk production traits. *Journal of Dairy Science* 90(4):1981–1986.

Stoop, W. M., Schennink, A., Visker, M. H. P. W., Mullaart, E., van Arendonk, J. A. M. and Bovenhuis, H. 2009. Genome-wide scan for bovine milk-fat composition. I. Quantitative trait loci for short- and medium-chain fatty acids. *Journal of Dairy Science* 92(9):4664–4675.

Stoop, W. M., van Arendonk, J. A. M., Heck, J. M. L., van Valenberg, H. J. F. and Bovenhuis, H. 2008. Genetic parameters for major milk fatty acids and milk production traits of Dutch Holstein-Friesians. *Journal of Dairy Science* 91(1):385–394.

Sundekilde, U. K., Barile, D., Meyrand, M., Poulsen, N. A., Larsen, L. B., Lebrila, C. B., German, J. B. and Bertram, H. C. 2012. Natural variability in bovine milk oligosaccharides from Danish Jersey and Holstein-Friesian breeds. *Journal of Agricultural and Food Chemistry* 60(24):6188–6196.

Sundekilde, U. K., Frederiksen, P. D., Clausen, M. R., Larsen, L. B. and Bertram, H. C. 2011. Relationship between the metabolite profile and technological properties of bovine milk from two dairy breeds elucidated by NMR-based metabolomics. *Journal of Agricultural and Food Chemistry* 59(13):7360–7367.

Tapio, I., Värv, S., Bennewitz, J., Maleviciute, J., Fimland, E., Grislis, Z., Meuwissen, T. H. E., Miceikiene, I., Olsaker, I., Viinalass, H., Vilkki, J. and Kantanen, J. 2006. Prioritization for conservation of northern European cattle breeds based on analysis of microsatellite data. *Conservation Biology: the Journal of the Society for Conservation Biology* 20(6):1768–1779.

Threadgill, D. W. and Womack, J. E. 1990. Genomic analysis of the major bovine milk protein genes. *Nucleic Acids Research* 18(23):6935–6942.

Truswell, A. S. 2005. The A2 milk case: a critical review. *European Journal of Clinical Nutrition* 59(5):623–631.

Tsioulpas, A., Lewis, M. J. and Grandison, A. S. 2007. Effect of minerals on casein micelle stability of cows' milk. *Journal of Dairy Research* 74(2):167–173.

Tyriseva, A. M., Elo, K., Kuusipuro, A., Vilva, V., Janonen, I., Karjalainen, H., Ikonen, T. and Ojala, M. 2008. Chromosomal regions underlying noncoagulation of milk in Finnish Ayrshire cows. *Genetics* 180(2):1211–1220.

Tyriseva, A. M., Vahlsten, T., Ruottinen, O. and Ojala, M. 2004. Noncoagulation of milk in Finnish Ayrshire and Holstein-Friesian cows and effect of herds on milk coagulation ability. *Journal of Dairy Science* 87(11):3958–3966.

Vallas, M., Bovenhuis, H., Kaart, T., Pärna, K., Kiiman, H. and Pärna, E. 2010. Genetic parameters for milk coagulation properties in Estonian Holstein cows. *Journal of Dairy Science* 93(8):3789–3796.

Vallas, M., Kaart, T., Värv, S., Pärna, K., Jõudu, I., Viinalass, H. and Pärna, E. 2012. Composite β-κ-casein genotypes and their effect on composition and coagulation of milk from Estonian Holstein cows. *Journal of Dairy Science* 95(11):6760–6769.

Published by Burleigh Dodds Science Publishing Limited, 2022.

van Hulzen, K. J., Sprong, R. C., van der Meer, R. and van Arendonk, J. A. 2009. Genetic and nongenetic variation in concentration of selenium, calcium, potassium, zinc, magnesium, and phosphorus in milk of Dutch Holstein-Friesian cows. *Journal of Dairy Science* 92(11):5754–5759.

van Hooydonk, A. C. M., Hagedoorn, H. G. and Boerrigter, I. J. 1986. The effect of various cations on the renneting of milk. *Netherlands Milk and Dairy Journal* 40: 369–390.

Wang, Q. and Bovenhuis, H. 2018. Genome-wide association study for milk infrared wavenumbers. *Journal of Dairy Science* 101(3):2260–2272.

Wang, Q., Hulzebosch, A. and Bovenhuis, H. 2016. Genetic and environmental variation in bovine milk infrared spectra. *Journal of Dairy Science* 99(8):6793–6803.

Wedholm, A., Larsen, L. B., Lindmark-Mansson, H., Karlsson, A. H. and Andren, A. 2006. Effect of protein composition on the cheese-making properties of milk from individual dairy cows. *Journal of Dairy Science* 89(9):3296–3305.

Zaalberg, R. M., Buitenhuis, A. J., Sundekilde, U. K., Poulsen, N. A. and Bovenhuis, H. 2020. Genetic analysis of orotic acid predicted with Fourier transform infrared milk spectra. *Journal of Dairy Science* 103(4):3334–3348.

# Chapter 16

## Dietary factors affecting the composition of cow's milk

*Martin Auldist, Agriculture Victoria Research, Australia*



### 1 Introduction

The yield and quality of all dairy products depend to a large extent on the concentrations of the various components in the original milk. Each of the many milk components has different physicochemical properties and behaves in different ways during processing so, as a consequence, each component makes a different contribution to the yield and characteristics of the final product. Some components are clearly of greater importance for particular products than others. This raises the possibility of purposefully altering milk composition to make it more suitable not only for processing in general but also for specific products. A great deal of research effort has been devoted to this field, with milk composition having been shown to vary markedly in response to a range of factors including cow breed and genetics, stage of lactation, time of year, cow health and welfare and cow diet. Of these, it is cow diet that can be most easily manipulated by the herd manager on a day-to-day basis with a view to making rapid and meaningful changes to milk composition.

Cow diet can affect not only the overall concentrations of the main components of economic significance (i.e. fat and protein) but also can influence the different types of fats and proteins in milk and their abundance relative

http://dx.doi.org/10.19103/AS.2022.0099.16

to one another. Some components are more easily manipulated than others owing to the manner in which their precursors are derived and the way in which they are synthesised in the mammary gland. This chapter outlines some of the ways in which cow diet and nutrition can be used to alter milk composition and explores the potential impact on the manufacturing properties of milk. It is important to note that the focus is on changes to the concentrations and proportions of the various components in milk rather than their yields, because it is these changes that affect the yield and functionality of dairy products made from a given volume of milk.

## 2 Milk fat

### 2.1 Origins of milk fat

Fat is typically one of the most abundant constituents of bovine milk, if not the most abundant, but marked variations in milk fat concentrations can occur due to a range of factors including cow diet. For example, in New Zealand, Mackle et al. (1997) reported that milk fat concentrations of individual cows ranged from 2.9% to 7.7%.

Approximately 98% of milk fat exists as triglycerides, which are composed of three fatty acids and a glycerol backbone. Small amounts of diglycerides, free fatty acids, cholesterol and phospholipids are also present (Mendelson et al., 1977). The triglycerides in cow's milk consist primarily of fatty acids with a chain length between C4 and C22 (Moate et al., 2007). Of these, those with a chain length of between C4 and C14, as well as a proportion of C16, are derived from *de novo* synthesis within the mammary gland from acetate and, to a smaller extent, β-hydroxybutyrate. The remainder of the C16 fatty acids and those with longer chain lengths are derived from pre-formed fatty acids in the circulation originating either from dietary lipids or the mobilisation of body adipose reserves (Bauman and Griinari, 2001).

Milk fat occurs in the form of small globules that vary in diameter from 2.5 μm to 5.7 μm (Logan et al., 2014). These globules are formed by the aggregation of triglycerides in the rough endoplasmic reticulum of the alveolar epithelial cells, following esterification of fatty acids in the same organelle. Thereafter, the globules move to the apical membrane of the cell and are released into the alveolar lumen, enveloped in apical membrane that becomes the milk fat globule membrane. Cell membrane lost through the extrusion of the fat globules is replaced via the exocytosis of the Golgi vesicles during the secretion of protein and lactose (Murphy, 2000).

Not only is fat one of the most abundant milk components, but it is also one of the most economically important, even though many milk processing companies pay substantially less for milk fat than they do for milk protein.

Somewhat obviously, fat is the main milk constituent in butter, but it also co-precipitates with casein during cheesemaking forming the basis of the curd and subsequent cheese. In addition, milk fat is an important contributor to the yield and functionality of whole milk powder, ice cream, yoghurt and many other products.

## 2.2 Milk fat depression

It is generally acknowledged by researchers that milk fat is much more easily manipulated by cow diet than milk protein (De Peters and Cant, 1992). In particular, milk fat concentrations can decline markedly in response to certain combinations of dietary components in a phenomenon known variously as 'milk fat depression (MFD)', 'milk fat reduction' or 'low milk fat syndrome'. Rapid declines in milk fat concentrations of 50% or more are possible without any accompanying changes in protein or lactose concentrations (Bauman and Griinari, 2001; Bauman et al., 2008). Diets coincident with such a change are generally those with a high ratio of concentrate to forage, such that intake of starchy concentrates is high but intake of either neutral detergent fibre or physically effective fibre is low. This is an important point in that MFD can still occur when roughage in the diet is technically adequate, but particle size is too small to maintain chewing, saliva production and ruminal pH (Banakar et al., 2018). This has become more common in recent years on dairy farms in which mixed rations and automated feed chopping systems predominate.

Consistent with the above, Griinari et al. (1998) identified that for MFD to occur there needs to be a high dietary intake of polyunsaturated fatty acids and an alteration in rumen fermentation. Such alteration in fermentation could include, for example, reduced ruminal pH and lower pH nadirs following bouts of feeding, and/or a change in the rumen fermentation pattern towards propionate and away from acetate and butyrate. Both these conditions would commonly occur in the high concentrate diets described above. The reduction in milk fat is able to be reversed or lessened by changing the diet to provide more fibre and less concentrate, by changing to a more slowly digestible source of starch so as to lessen changes in rumen fermentation, or by removing or reducing the amount of unsaturated fatty acids in the diet (see below).

A recent example of cow diet affecting milk fat concentration is provided by data from the experiment of Auldist et al. (2013; Fig. 1), in which grazing cows offered increasing amounts of cereal grain (and therefore increasing proportions of grain relative to forage) exhibited rapid declines in milkfat concentrations from over 5.0% to around 4.0%. These declines were not observed in cows that were instead offered equivalent amounts of mixed rations that included maize grain, a more slowly digestible starch source that

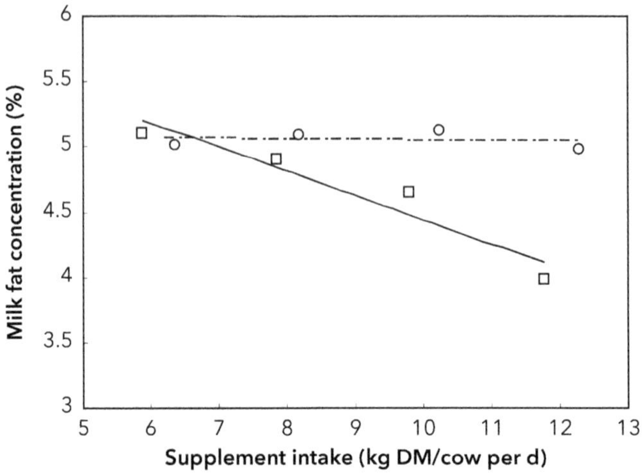


**Figure 1** Mean concentrations of milk fat for cows grazing pasture supplemented with different amounts of barley grain and pasture silage (□, solid line) or an equivalent amount of metabolisable energy fed as a mixed ration containing maize grain, barley grain, pasture silage and maize silage (○, dashed line). Cows fed the mixed ration had higher ($P < 0.01$) concentrations of fat at supplement intakes of between 9 kg and 11 kg of DM/cow per day than cows barley grain and pasture silage. Adapted from Auldist et al. (2013).

resulted in less extreme declines in ruminal pH throughout the day (Greenwood et al., 2014).

Several theories have been proposed as to the cause of the reduction in milk fat concentration and yield in such situations. The first relates simply to the fact that reduced ruminal production and absorption of acetate as a result of reduced fibre digestion would result in fewer precursors for fatty acid synthesis in the mammary gland and therefore reduced milk fat production overall. In a similar vein, it has been proposed that the increased ruminal production and absorption of propionate from increased starch digestion can in turn lead to increased glucose and then insulin production, which is known to promote body adipose reserves at the expense of milk production, including milk fat. Both these theories have been discredited at least to some extent by controlled experiments involving the infusion of acetate or propionate (or glucose) in cows exhibiting MFD, with little to no effect on milk fat concentrations reported (Bauman and Griinari, 2001).

Today the most widely accepted theory for the cause of MFD is the 'biohydrogenation theory' (Bauman and Griinari, 2001; Bauman et al., 2008). According to this theory, under certain conditions in the rumen (including reduced ruminal pH), the ruminal biohydrogenation pathways of polyunsaturated fatty acids can be disrupted and altered, leading to an accumulation of certain trans fatty acid biohydrogenation intermediates. These intermediates are absorbed in the small intestine and transported to the mammary gland where they are potent inhibitors of milk fat synthesis. The first to be proven responsible for the inhibition of milk fat synthesis in this way was trans 10, cis 12 C18:2, one of several known isomers of conjugated linoleic acid (CLA; Baumgard et al., 2000). Recently two other isomers have also been identified as possibly having a similar effect. These are trans 9, cis 11 CLA and cis 10, trans 12 CLA (Saebø et al., 2005; Perfield et al., 2007). Although the precise mechanism for the inhibition of milk fat synthesis caused by these intermediates has not been fully elucidated, it is likely to be related to a reduction in the mammary expression of genes controlling key lipogenic enzymes (Bauman et al., 2008).

Stark examples of the effectiveness of trans 10, cis 12 C18:2 in reducing milk fat synthesis can be found in experiments involving infusions of CLA (including trans 10, cis 12 C18:2 either alone or as a mixture) into the abomasums of cows consuming both total mixed ration (Baumgard et al., 2000) and pasture (Mackle et al., 2003; Kay et al., 2007; Fig. 2), which caused rapid and substantial but transitory declines in the concentrations and yields of milk fat. These experiments indicate the potential of using trans 10, cis 12 C18:2 to purposefully reduce milk fat during periods of feed shortage or nutritional stress, given that milk fat is energetically expensive to produce.

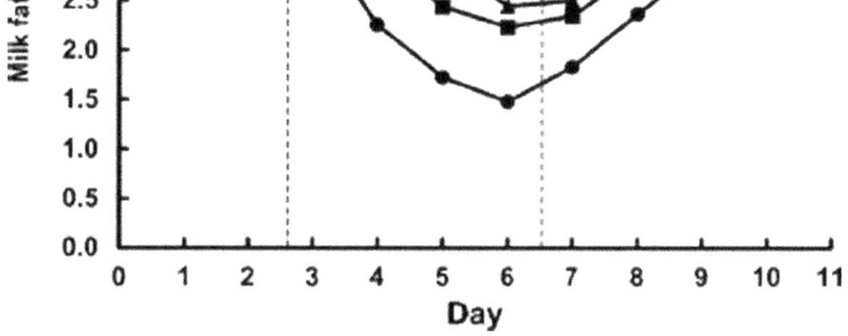


**Figure 2** Temporal pattern of milk fat concentration across an 11-day period. Abomasal infusion of a commercial CLA source was performed at either 0 (○), 20 (●), 40 (■), or 80 (▲) g/d. Infusions began following the d-2 PM milking and continued through the d-6 PM milking (shown by dashed lines). Values are means of four cows and the standard error of the differences were 0.22% and 51 g/d for milk fat concentration and yield, respectively (from Mackle et al., 2003).

When milk fat concentrations are reduced during MFD there is a concurrent alteration in the fatty acid profile of the milk fat. Specifically, *de novo* synthesised fatty acids are reduced in line with the direct inhibition of their synthesis in the mammary gland, such that the content of short- and medium-chain fatty acids are reduced, while long-chain fatty acids, which originate from either feed or adipose tissue, increase.

A meta-analysis by Iwaniuk and Erdman (2015) showed that the use of rumen buffers such as $NaHCO_3$ and $K_2CO_3$ was commonly associated with an increase in milk fat concentration. They also reported that rumen buffers have the effect of increasing dietary DCAD and increasing ruminal pH, and have been shown to reduce the duodenal outflow of biohydrogenation intermediates such as trans 18:1. Thus it is possible that the effect of rumen buffers on milk fat concentration is mediated via changes in biohydrogenation, though none of the reported studies measured the fatty acid profiles of milk to prove the effect.

## 2.3 Perceived health benefits of unsaturated vs saturated fatty acids

Feeding a diet high in polyunsaturated fatty acids increases the likelihood of MFD occurring if there is a coincident alteration to biohydrogenation pathways, because such a diet provides more potential substrate for the production of the inhibitors of milk fat synthesis described above. It is noteworthy that grazed pasture has the potential to provide sufficient polyunsaturated fatty acids for MFD to occur because most fresh forages contain between 2% and 10% lipid, most of which is linolenic acid (C18:3). However, provision of polyunsaturated fatty acids can also be done deliberately to alter the fatty acid profile of the resultant milk. This has been seen as desirable for human health because of concerns that dietary saturated fats can increase the incidence of cardiovascular disease, although there are also reports that this is not the case (Forouhi et al., 2018). Altering milk fatty acid composition can also have the effect of altering the functional properties of milk fat.

In particular, lipid supplements have commonly been offered to cattle as oil from oilseeds such as canola, linseed, sunflower, flaxseeds and soya beans, for the purposes of increasing proportions of C18:1, C18:2 or C18:3 in milk fat, depending on the fatty acid of the oilseed (Kliem and Shingfield, 2016). This effect is usually enhanced by rumen-protecting the lipid with formaldehyde or heat against biohydrogenation in the rumen, thus ensuring the ultimate absorption of more unsaturated fatty acids (Palmquist and Jenkins, 2017). Not only does increasing the proportions of these mono- and polyunsaturated fatty acids in milk fat have perceived benefits for human health, but also these long-chain unsaturated fatty acids have lower melting points, resulting in lower solid fat contents. In turn, such milk fat can be manufactured into butter that is naturally spreadable straight from the refrigerator.

## 2.4 Conjugated linoleic acid

As noted above, diets high in polyunsaturated fatty acids can be used to produce milk that also contains high levels of polyunsaturated fatty acids, thereby increasing the perceived 'healthiness' of the milk overall. However, cow diet can also be used to increase the proportions in milk fat of particular fatty acids with known health benefits. One such fatty acid is the CLA isomer cis 9, trans 11 C18:2 (note that this is a different CLA isomer to the one involved in MFD described earlier).

The cis 9, trans 11 C18:2 isomer is the major CLA in cow's milk, comprising 75% to 90% of total CLA. In in vitro cell culture systems and in in vivo animal models it has been shown to have anti-carcinogenic and anti-atherogenic properties, and has therefore generated significant research interest as a milk component with potential benefits for human health. The cis 9, trans 11 CLA isomer occurs in the rumen as an intermediate of the biohydrogenation of linoleic acid and this was originally assumed to be the source of the same isomer in milk fat. However, Griinari et al. (2000) demonstrated that endogenous synthesis of cis 9, trans 11 CLA in the mammary gland also occurred, involving the Δ9 desaturase enzyme and with the precursor being vaccenic acid (trans 11 C18:1), an intermediate in the biohydrogenation of both linoleic and linolenic acid. Subsequent experiments involving the inhibition of Δ9 desaturase activity in the mammary gland showed that endogenous synthesis was the major source of cis 9, trans 11 CLA both in cows fed TMR (Corl et al., 2002) and grazing pasture (Kay et al., 2004). As a consequence, efforts to increase cis 9, trans 11 CLA in milk have focused on enhancing this pathway.

In a review of opportunities to enhance CLA proportions in milk fat, Lock and Bauman (2004) concluded that the keys were to maximise vaccenic acid output from the rumen by increasing the amount of PUFA precursors in the diet, and then inhibit the reduction of vaccenic acid to stearic acid. The first step can be achieved by the incorporation of a range of oilseeds into the diet. Kelly et al. (1998) showed that the sunflower oil, peanut oil and linseed oil all resulted in elevated proportions of CLA (with sunflower oil being the most effective). Meanwhile, Chouinard et al. (2001) showed that feeding calcium salts of canola and soya bean oil could be similarly effective in producing milk fat with enhanced levels of cis 9, trans 11 CLA. Other researchers have shown that adding fish oil to these types of diets could increase proportions of cis 9, trans 11 CLA in milk beyond what was achievable using oilseeds alone (Abu-Ghazaleh et al., 2002), presumably because of a reduction in the final biohydrogenation step of vaccenic acid to stearic acid in the rumen. Lastly, milk

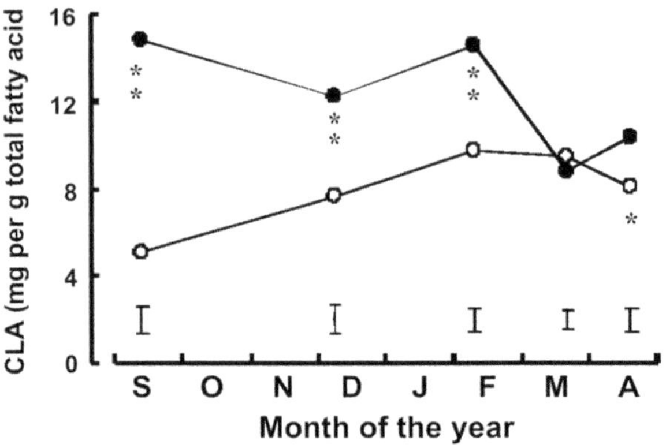


**Figure 3** Concentrations of conjugated linoleic acid (CLA) throughout lactation in milk from cows grazing pasture (●) and cows fed a typical total mixed ration (○) for an entire lactation. Vertical brackets indicate standard errors of the differences. Asterisks indicate the significance of differences at each sampling point (from Auldist et al., 2002).

from cows grazing pasture commonly has greater concentrations of cis 9, trans 11 CLA than cows consuming TMR, presumably because of their high intakes of linolenic acid from grass (Auldist et al., 2002; Fig. 3). Lynch et al. (2005) reported that enhancement of CLA concentrations in milk by dietary means such as these did not change the sensory characteristics of the milk.

In a landmark experiment, Ip et al. (1999) incorporated sunflower oil into the diets of cows consuming a TMR. Milkfat from cows that exhibited the greatest response in terms of enhanced proportions of cis 9, trans 11 CLA was then manufactured into butter, which when fed to rats was able to reduce the incidence of mammary tumours compared to butter from cows fed TMR with sunflower oil.

## 2.5 Spreadable butter

A compelling case study showing the effectiveness of managing cow diet to alter the properties of milk fat can be found in the research of Ann Fearon and her colleagues at Queens University Belfast in Northern Ireland. These researchers aimed to address the fact that at certain times of the year, especially in winter when housed animals were provided with substantial quantities of concentrates and grass silage, the milk fat produced was highly saturated with a high solid fat content and which resulted in butter with poor spreadability. Fearon and co-workers showed that adding canola oil to the diets of dairy cattle increased the proportions of unsaturated fatty acids in milk while decreasing the proportions of saturated fats. Specifically, there was an increase in oleic acid (C18:1) and a decrease in palmitic acid (C16:0). This resulted in a reduction in

**Table 1** Effect of level of dietary canola oil inclusion on solid fat content and oxidative stability of milk fat from cows in mid and late lactation (adapted from Fearon et al., 2004)

| | Level of oil inclusion (g oil/cow per d) | | | | |
|---|---|---|---|---|---|
| | 0 | 200 | 400 | 600 | SEM |
| Solid fat content at 5°C (%) | 52.0 | 47.2 | 45.2 | 40.7 | 0.44 |
| Solid fat content at 10°C (%) | 41.9 | 37.9 | 36.5 | 32.6 | 0.41 |
| Solid fat content at 20°C (%) | 16.7 | 15.2 | 15.2 | 14.5 | 0.30 |
| Total saturated fatty acids[a] | 641 | 594 | 572 | 522 | 2.9 |
| Total unsaturated fatty acids[a] | 346 | 391 | 412 | 459 | 2.9 |

[a] g per kg of total fatty acids.

the solid fat content at 5°C, but with little change at 20°C, meaning that the fat could be manufactured into butter that was spreadable at refrigeration temperatures, but which maintained its integrity at room temperatures (Fearon et al., 2004; Table 1). Moreover, the product was able to be legally labelled as butter because it had not had vegetable oils added or been otherwise modified in the factory like other spreads. The technology was employed by around 150 farmers in Northern Ireland and the spreadable butter was marketed through Marks and Spencers.

## 3 Milk protein

### 3.1 Origins of milk protein

Protein is a major component of milk but is generally less variable and difficult to change than milk fat. For example, in New Zealand, Mackle et al. (1999b) reported that milk protein concentrations of individual cows ranged from 2.9% to 4.4%.

The proteins in bovine milk can be broadly categorised into two main groups, these being the casein proteins and the whey proteins. Of these it is the caseins that are most abundant, usually comprising around 80% of total milk protein (Dalgleish, 1992). In turn, there are several different types of caseins, these being $\beta$, $\alpha_1$, $\alpha_2$, $\kappa$ and $\gamma$ (Huppertz et al., 2018) all of which result from *de novo* synthesis within the secretory cells of the mammary epithelia ($\gamma$-casein being derived from the post secretory hydrolysis of beta-casein by plasmin and other native milk proteases; Ismail and Nielsen, 2010). At least 95% of casein occurs in the form of colloidal particles or micelles (Holt et al., 1986). These micelles also contain complexes of calcium, magnesium, phosphate and citrate, and vary in diameter from 50 nm to 300 nm depending on the proportions of the different caseins contained within them. Dalgleish et al. (1989) showed that the outer surface of the micelle is enriched with $\kappa$-casein and the interior with β-casein, while α-casein is distributed within both. It is the coat of κ-casein that stabilises the micelles and governs the properties of the micelle overall (Huppertz et al., 2018).

The caseins, as with all proteins synthesised in the mammary gland, are made from amino acids absorbed directly into the gland from the blood, using glucose as an energy source. These amino acids originate from either diet or tissue (Lean, 1987). Protein synthesis then occurs in the endoplasmic reticulum of the alveolar epithelial cells via the usual pathways (Linn, 1988). The synthesised proteins pass to the Golgi apparatus and are secreted to the outside of the cell via exocytosis, encapsulated in the same vesicles as lactose. It is in the Golgi apparatus that the casein molecules aggregate into micelles (Dalgleish, 1992).

The whey protein group includes mostly proteins synthesised in the mammary gland but also some that enter from the blood. The proteins synthesised *de novo* include α-lactalbumin, β-lactoglobulin and lactoferrin. The synthesis of these proteins is similar to that for casein. The major serum proteins are bovine serum albumin, transferrin and a number of different immunoglobulins. These proteins must almost all be transported either actively or passively into the milk (or colostrum) from the blood (the exception is some of the immunoglobulins, as below).

The predominant immunoglobulin in bovine mammary secretions is IgG, which can be further categorised into two subclasses, $IgG_1$ and $IgG_2$ (Brandon et al., 1971). These are both derived from the blood, but it is primarily $IgG_1$ that is selectively transferred by the mammary epithelial cells from the interstitial fluid into the lumen of the alveoli. Under normal circumstances the tight junctions between the adjacent epithelial cells present a barrier that is impermeable to molecules even as small as lactose (Lascelles and McDowell, 1974). This means that transport of $IgG_1$ must take place through the epithelial cells rather than between them. The selective transfer of $IgG_1$ relies on the presence of receptor sites on basal membranes of the epithelial cells, to which the $IgG_1$ is bound in clusters. Pinocytotic engulfment of these clusters then occurs and the protein travels through the cell in vesicles, the contents of which are discharged into the alveolar lumen upon reaching the apical membrane on the opposite side of the cell. At the same time, proteins such as $IgG_2$ and BSA, present in the interstitial fluid, are entrapped within the same vesicles and passively transferred into the milk or colostrum. Bovine milk also contains smaller amounts of IgA and IgM, which are synthesised either locally in the mammary gland or passively transferred into the milk from the blood (Larson et al., 1980).

Of all the components in milk, protein is of greatest economic importance with many companies paying more than twice as much per kilogram for protein as for fat. This reflects the key role of protein, and the caseins in particular, in the manufacture of cheese and milk powders as well as many other more specialised products. Not surprisingly there has been a great deal of research devoted to finding strategies for increasing the concentration of protein in milk through dietary means. Unfortunately, results have been mostly unremarkable, with milk protein proving difficult to manipulate via dietary means.

### 3.2 Energy intake

The changes that it is possible to induce in milk protein concentration via alterations in cow diet are widely considered to be much smaller than for milk fat (Linn, 1988; Murphy and O'Mara, 1993). Nevertheless, one dietary strategy for increasing milk protein concentration is to ensure sufficient intake of energy. Several studies have shown that increasing the intake of metabolisable energy of dairy cattle, either by increasing their intake of pasture or supplement, can in certain circumstances increase milk protein concentration (Cragle et al., 1986; De Peters and Cant, 1992; Malossini et al., 1996; O'Brien et al., 1997; Mackle et al., 1999a; Auldist et al., 2000; Wales et al., 2001; Auldist et al., 2016; Gulati et al., 2019). For example, Mackle et al. (1999a) showed that adding a concentrate supplement to the diet of cows grazing a restricted pasture allowance increased milk protein concentration by between 0.2% and 0.3% depending on the stage of lactation of the cows (Table 2).

These increases are presumably because the higher energy intake would have increased glucose supply and in turn spared amino acids from gluconeogenesis, thereby increasing the amino acid supply for milk protein synthesis in the mammary gland. For example, cereal grain supplements would

**Table 2** Effect of diet on concentrations of milk proteins (g/kg) for cows offered a restricted pasture intake (RPI) and for cows on pasture plus a supplementary concentrate (P+C). Adapted from Mackle et al. (1999a)

| | Early lactation | | | Late lactation | | |
|---|---|---|---|---|---|---|
| | RPI | P+C | SED[a] | RPI | P+C | SED |
| Total protein | 31.5 | 34.6** | 0.40 | 34.1 | 35.7** | 0.19 |
| Casein protein | 26.3 | 28.6** | 0.39 | 28.3 | 29.6** | 0.17 |
| α-casein | 11.4 | 12.2** | 0.17 | 12.7 | 13.3** | 0.18 |
| β-casein | 11.0 | 11.7** | 0.20 | 11.5 | 11.9* | 0.18 |
| κ-casein | 2.82 | 3.54** | 0.08 | 3.10 | 3.33** | 0.06 |
| γ-casein | 0.19 | 0.36** | 0.04 | 0.25 | 0.21 | 0.02 |
| Whey protein | 5.12 | 6.05** | 0.08 | 5.74 | 6.04** | 0.06 |
| α-lactalbumin | 0.97 | 1.19** | 0.03 | 0.88 | 1.08** | 0.04 |
| β-lactoglobulin | 4.36 | 4.90** | 0.08 | 4.94 | 5.20** | 0.06 |

[a] Standard error of the difference.

likely increase the ruminal accumulation of propionate, which when absorbed into the bloodstream is a major precursor for glucose, meaning less amino acids would be required for gluconeogenesis. Despite this, feeding starch-based concentrates does not appear to increase milk protein more than digestible fibre-based supplements (Murphy and O'Mara, 1993). It is also of note that the response also depends upon energy source. For example, milk protein concentrations can be negatively affected if the additional energy is supplied via high amounts of dietary fat (De Peters and Ferguson, 1992). While this information is useful in showing the increases in milk protein that can be achieved by increasing the energy intake of underfed cows, strategies to further increase milk protein concentration by dietary means above 'normal' levels in fully fed cows remain elusive.

## 3.3 Dietary protein

In controlled experiments in which protein, usually casein, has been infused into the abomasum of dairy cattle, increases in milk protein concentrations have often been observed (Vik-Mo et al., 1974; Clark et al., 1977). Similarly, infusion of various mixtures of amino acids either abomasally or intravenously has also led to increases in milk protein concentration (MacRae et al., 2000; Aikman et al., 2002). This has included mixtures of all individual amino acids, mixtures of only those found in milk protein or those deemed to be essential including methionine and lysine. Collectively the results of these experiments raise the prospect that increasing the intestinal supply of amino acids via dietary means may have similar results (Linn, 1988).

In practice, however, increasing milk protein concentrations by manipulating dietary protein supply has proved more difficult. Increasing the amount of total crude protein in the diet generally has little or no effect on milk protein concentrations (Linn, 1988; Murphy and O'Mara, 1993), although an early report by Emery (1978) indicated that a 10 g/kg increase in dietary crude protein resulted in a 0.2 g/kg increase in milk protein concentration. It is of note that it is sometimes difficult to separate the effects of dietary protein from those of increased energy intake because some protein supplements, notable canola meal, can increase appetite in cows and therefore dry matter and energy intake (Auldist et al., 2016).

In a literature review spanning 12 years, Santos et al. (1998) reported on the effects of increasing the proportions of rumen undegradable protein in diets of dairy cattle by replacing soya bean meal with feedstuffs such as heated and chemically treated soya bean meal, corn gluten meal, distillers grains, brewers grains, blood meal, meat and bone meal and feather meal. Results showed that none of these strategies had any effect in increasing milk protein concentrations, and in fact many reduced protein concentrations, possibly by reducing microbial protein (or by causing the diversion of energy needed to then eliminate excess N as ammonia in urine). There are, however, several reports of experiments in Ireland in which feeding fish meal (which has a low rumen degradability) resulted in a moderate increase in milk protein concentrations. These experiments are summarised by Murphy and O'Mara (1993), who put the effect down to fishmeal being high in rumen undegradable protein and supplying more of the amino acids that are limiting for milk protein synthesis than other sources of undegradable protein. Also encouragingly, there are several reports that feeding rumen-protected methionine or lysine, either in transition cows or during established lactation, can induce small increases in milk protein concentration (Murphy and O'Mara, 1993; Zanton et al., 2014; Schwab and Broderick, 2017). This strategy therefore appears to offer one hope for increasing milk protein concentrations using nutritional supplements, but further research is needed.

### *3.4 Milk protein composition*

In addition to increasing the yield and concentration of protein in milk, there has also been considerable research interest in manipulating the proportions of the different milk proteins relative to one another. This is because of their different properties, the way they behave during processing, and their perceived value for the manufacture of different products (Wedholm et al., 2006). For example, the casein proteins are generally considered to be more valuable than the whey proteins owing largely to the fact that they, along with fat, form the basis of the curd during cheesemaking while the whey proteins are expelled. Casein is also a major component of various milk powders and is an important food ingredient. In addition, it has industrial uses in cosmetics, plastics, glues and other products (Huppertz et al., 2018). Increasing the proportion of total protein that is casein, or the 'casein number' as it is known, would therefore be a desirable outcome of dietary manipulation in dairy cattle.

The individual casein proteins have different properties (Huppertz et al., 2018) and dietary manipulation of their proportions could potentially be used to alter the manufacturing properties of milk. For example, changing the ratio of α-casein to β-casein could alter the rennet clotting time during cheesemaking, which in turn would have effects on syneresis, moisture expulsion and cheese moisture, an important quality parameter (Auldist et al., 1996). Similarly, some of the whey proteins, immunoglobulin and lactoferrin, for example, can be valuable as human health supplements.

Unfortunately, just as total protein concentration is difficult to consistently alter via dietary modification, so too are the proportions of different proteins. For example, energy intake is seen to affect all mammary-synthesised proteins equally, such that while casein concentration can also be increased or decreased by altering energy intake, the whey proteins will be affected to the same degree such that there will be no change to the proportion of proteins relative to one another.

While some feeding experiments in which plane of nutrition has been altered have led to small changes in either casein number or the proportions of the different caseins relative to one another (Kefford et al., 1995; Christian et al., 1999; Mackle et al., 1999a), others have shown no change (Coulon et al., 2001; Cowley et al., 2014; Gulati et al., 2019). Overall, therefore, current dietary means for manipulating 'casein number' or the proportions of the different caseins are limited (MacRae et al., 2000).

## 4 Lactose

### *4.1 Origins and uses of lactose*

Lactose is the main carbohydrate in the milk of most mammals. In dairy cattle, it is also the most abundant solid, with an average concentration in cows' milk of 4.7% (Fox et al., 2015). As such, lactose is an important solid in the manufacture of whole milk and skim milk powders. During cheesemaking, however, lactose is primarily expelled in the whey, which has traditionally been used as stockfeed, and especially for pigs. More recently, uses have been found for lactose isolated from whey for pharmaceutical purposes or as a base for confectionery, ice cream and baked goods. Some lactose separated from whey is also re-cycled to ensure that international standards for the proportions of lactose in milk powders are met (Costa et al., 2019).

### *4.2 Regulator of milk volume*

As well as being the most abundant, lactose is also one of the least variable components of milk. Mackle et al. (1997) reported that lactose concentration was less variable than either fat or protein concentration, with a reported range of between 4.2% and 5.0%. This lack of variation is presumably due to its role as the osmotic regulator of milk volume. Lactose is synthesised in the Golgi apparatus of the epithelial cells of the mammary gland from glucose absorbed directly from the blood as it passes through the gland. Lactose results from the union of one glucose molecule and one galactose molecule, though the galactose must first be manufactured from glucose itself. The membrane of the Golgi apparatus is readily permeable to monosaccharides and to water but not to lactose. Glucose therefore rapidly enters the secretory vesicles but the resulting lactose cannot leave. Water then enters the vesicles in order to maintain osmotic equilibrium and the entire contents of the vesicles are ultimately dispelled into the alveolar lumen via exocytosis at the apical membrane of the secretory cell (Mepham, 1993). Consequently, the rate of lactose synthesis determines the rate of water secretion and thus milk volume. It can be seen, therefore, that even if lactose synthesis is reduced by dietary or other factors, the actual concentration of lactose in milk will remain relatively constant since milk yield will fall accordingly.

### *4.3 Little effect of diet*

In spite of the above, lactose concentrations do vary slightly due to a range of factors including stage of lactation, genetics, cow breed and disease. This is due to changes in the contributions of some minerals (particularly Na, K and Cl) to the maintenance of osmotic pressure, and to the escape of lactose from milk through ruptured tight junctions between mammary epithelial cells such as occurs during mastitis and other diseases. As far as the use of cow diet to alter milk lactose concentrations is concerned, results from relevant experiments have been inconsistent and conflicting. For example, Ouweltjes et al. (2007) and Xue et al. (2011) both reported small increases in the lactose concentrations of milk as the energy content of the cows' diet increased. In contrast, Mackle et al. (1999a) and Auldist et al. (2000) both reported no significant difference in lactose concentrations in milk from cows offered diets with either high or low metabolisable energy contents. Overall, it is generally accepted that changes to lactose concentrations due to diet, even if possible, are too small to be meaningful either biologically or economically.

## 5 Minerals

Milk contains a range of minerals that enter milk either via passive diffusion through the mammary epithelium, or by active pumps on the epithelial and vesicular membranes, depending on the mineral of interest. The most common minerals in milk include the cations Ca, Mg, K and Na, along with the anions P, citrate and Cl (Gaucheron, 2005). These ions play important roles in the structure and stability of the casein micelles and therefore can influence the functional properties of products such as cheese and yoghurt (Dunshea et al., 2019). Milk also contains a range of up to 41 microminerals and trace elements (Zwierzchowski and Ametaj, 2018), for example Cu, Fe, Mn, Se, and Zn. Many of these more minor minerals have attracted research interest because of their influence on human health (Dunshea et al., 2019).

In terms of the manipulation of the concentrations of minerals in milk via cow diet, there has been relatively little research done. Most Ca and P, along with much Mg, in milk is incorporated into micellar casein (Dunshea et al., 2019), so any dietary strategies that increase either total milk protein or casein would be expected to increase Ca and P concurrently. As discussed above, however, such strategies are limited.

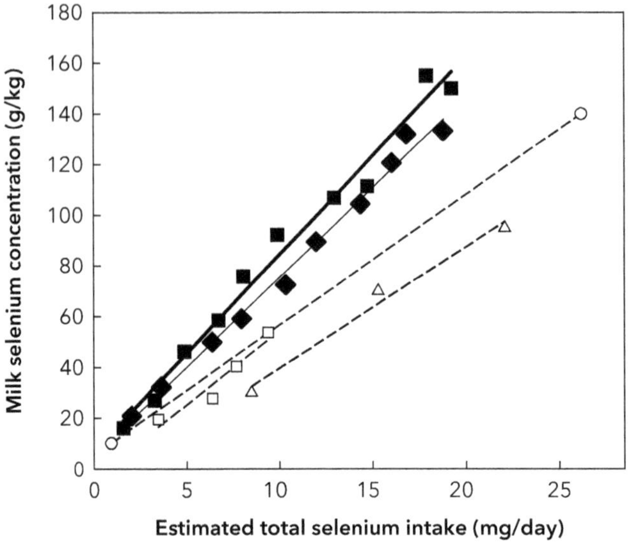


**Figure 4** Relationships between total selenium (Se) intake and the concentration of Se in milk in experiment 1 (■; average of d 15 to 42) and experiment 2 (◆; average of d 9 to 36). Results from other experiments (Δ Givens et al., 2004; ○ Heard et al., 2004; □ Juniper et al., 2006). Reproduced from Heard et al. (2007).

There are reports that concentrations of particular minerals in milk can be increased by mineral supplementation. For example, Sobhanirad et al. (2010) reported that supplementing cows with zinc sulphate or zinc methionine could increase zinc concentrations in milk. Similarly, Franke et al. (1983) reported increases in milk iodine concentrations when as little as 4 ppm of organic iodine were added to the diet. Heard et al. (2007), Doyle et al. (2011) and Stockdale et al. (2011) conducted a series of experiments in response to interest from dairy companies in producing milk with consistent and reliable concentrations of selenium. These authors reported that offering Se supplements to dairy cattle either as high Se triticale or Se yeast and found significant positive, linear relationships between Se intake and the concentrations of Se in milk (Fig. 4). Milk produced from the study of Doyle et al. (2011) was used to produce a milk protein concentrate with a Se concentration of 5.4 mg/kg.

## 6 Conclusion

The ability to change milk composition via alterations in cow diet is desirable as it can lead to increases in the yield and functional properties of dairy products, either at an industry level or for the production of specialty products for niche markets. Inducing such changes via diet on dairy farms can lead to more rapid and reversible changes to milk composition than other contributing factors such as cow genetics or stage of lactation. However, some milk components are more readily changed than others. The concentration and composition of milk fat, for example, is readily altered by changing the ratio of forage to concentrates or by feeding oilseeds containing high levels of polyunsaturated fatty acids. Such changes can result in higher yields of products such as butter, its functional properties and perceived healthiness. Similarly, the mineral composition can be altered by strategic mineral supplementation, which has possible implications for both product functionality and human health. Unfortunately for the most valuable component of milk, protein, alterations of the amount and type of milk protein are more difficult to achieve via dietary manipulation. Maintaining an adequate supply of metabolisable energy to ensure maximum protein concentrations is probably the only reliable strategy in this regard.

## 7 Future trends

Future research in this field should be divided between identifying and implementing on-farm strategies that would provide general improvements to the composition of the milk supply overall, and developing specific dietary interventions that would increase or modify the production or particular milk components for adding value to niche products. One challenge that needs to be overcome in order to unlock the full potential of altering milk composition to suit the manufacture of particular products is that many dairy industries around the world are set up for high volume manufacture of commodity products and appropriate segregation of modified milk on a large scale is problematic. The modification of milk fat so as to produce dairy products with greater functionality or benefits for human health has proved fruitful in the past and should continue to be so. Similarly, the dietary alteration of the mineral composition of milk for similar purposes is also likely to be both worthwhile and achievable in the future. Improving the yield and composition of milk protein is likely to remain difficult in future, given such scant progress has been made despite the large amount of research effort to date, though the feeding of fish meal is worthy of further attention.

## 8 Where to look for further information

General overviews of the subject can be found in the following three publications:

- Linn, J. G. (1988). 'Factors affecting the composition of milk from dairy cows'. In: *Designing Foods; Animal Product Options in the Marketplace*. National Research Council committee on technological options to improve the nutritional attributes of animal products. National Academies Press, Washington.

- Fuquay, J. W. (Ed.) (2011). *Encyclopedia of Dairy Science* (2nd Edition). Academic Press, USA, ISBN 978-0-12-374407-4.
- Agnew, R. E., Agnew, K. W. and Fearon, A. M. (Eds) (1999). *Milk Composition.*British Society of Animal Science, Occasional Publication No. 25. Proceedings of an international symposium held at Belfast, Northern Ireland.

Centres of expertise include:

- Moorepark Teagasc Food Research Centre (www.teagasc.ie).
- Agriculture Victoria Research Ellinbank Smartfarm (www.agriculture.vic.gov.au).
- DairyNZ (www.dairynz.co.nz).
- Cornell Dairy Center of Excellence (www.dairy.cornell.edu).
- Department of Animal Science, Penn State University (www.animalscience.psu.edu).

## 9 References

Abu-Ghazaleh, A. A., Schingoethe, D. J., Hippen, A. R.and Whitlock, L. A. (2002). Feeding fish meal and extruded soybeans enhances the conjugated linoleic acid (CLA) content of milk. *J. Dairy Sci.* 85(3), 624–631.

Aikman, P. C., Reynolds, C. K., Humphries, D. J., Beever, D. E. and MacRae, J. C. (2002). Milk protein response to abomasal or mesenteric vein essential amino acid infusion in lactating dairy cows. *J. Dairy Sci.* 85, 1079–1084.

Auldist, M. J., Coats, S., Sutherland, B. J., Mayes, J. J., McDowell, G. H. and Rogers, G. L. (1996). Effects of somatic cell count and stage of lactation on raw milk composition and the yield and quality of Cheddar cheese. *J. Dairy Res.* 63(2), 269–280.

Auldist, M. J., Kay, J. K., Thomson, N. A., Napper, A. R. and Kolver, E. S. (2002). Concentrations of conjugated linoleic acid in milk from cows grazing pasture or fed a total mixed ration for an entire lactation. *Proc. N.Z. Soc. Anim. Prod.* 62, 240–241.

Auldist, M. J., Marett, L. C., Greenwood, J. S., Hannah, M., Jacobs, J. L. and Wales, W. J. (2013). Effects of different strategies for feeding supplements on milk production responses in cows grazing a restricted pasture allowance. *J. Dairy Sci.* 96(2), 1218–1231.

Auldist, M. J., Marett, L. C., Greenwood, J. S., Wright, M. M., Hannah, M., Jacobs, J. L. and Wales, W. J. (2016). Milk production responses to different strategies for feeding supplements to grazing dairy cows. *J. Dairy Sci.* 99(1), 657–671.

Auldist, M. J., Thomson, N. A., Mackle, T. R., Hill, J. P. and Prosser, C. G. (2000). Effects of pasture allowance on the yield and composition of milk from cows of different β-lactoglobulin phenotypes. *J. Dairy Sci.* 83(9), 2069–2074.

Banakar, P. S., Anand Kumar, N., Shashank, C. G. and Lakhani, N. (2018). Physically effective fibre in ruminant nutrition: a review. *J. Pharm. Phytochem.* 7, 303–308.

Bauman, D. E. and Griinari, J. M. (2001). Regulation and nutritional manipulation of milk fat: low-fat milk syndrome. *Livest. Prod. Sci.* 70, 15–29.

Bauman, D. E., Perfield, J. W., Harvatine, K. J. and Baumgard, L. H. (2008). Regulation of fat Synthesis by conjugated linoleic acid: lactation and the Ruminant Model. *J. Nutr.* 138(2), 403–409.

Baumgard, L. H., Corl, B. A., Dwyer, D. A., Saebø, A. and Bauman, D. E. (2000). Identification of the conjugated linoleic acid isomer that inhibits milk fat synthesis. *Am. J. Physiol. Regul. Integr. Comp. Physiol.* 278(1), R179–R184.

Brandon, M. R., Watson, D. L. and Lascelles, A. K. (1971). The concentrations of immunoglobulins in the blood serum and whey of dairy cows. *Aust. J. Dairy Technol.* 26, 143–144.

Chouinard, P. Y., Corneau, L., Butler, W. R., Chilliard, Y., Drackley, J. K. and Bauman, D. E. (2001). Effect of dietary lipid source on conjugated linoleic acid concentrations in milk fat. *J. Dairy Sci.* 84(3), 680–690.

Christian, M. P., Grainger, C., Sutherland, B. J., Mayes, J. J., Hannah, M. C. and Kefford, B. (1999). Managing diet quality for Cheddar cheese manufacturing milk. 1. The influence of protein and energy supplements. *J. Dairy Res.* 66(3), 341–355.

Clark, J. H., Spires, H. R., Derrig, R. G. and Bennik, M. R. (1977). Milk production, nitrogen utilization and glucose synthesis in lactating cows infused post-ruminally with sodium caseinate and glucose. *J. Nutr.* 107(4), 631–644.

Corl, B. A., Baumgard, L. H., Griinari, J. M., Delmonte, P. P., Morehouse, K. M., Yurawecz, M. P. and Bauman, D. E. (2002). Trans-7, cis-9 CLA is synthesized endogenously by Δ9-desaturase in dairy cows. *Lipids* 37(7), 681–688.

Costa, A., Lopez-Villalobos, N., Sneddon, N. W., Shalloo, L., Franzoi, M., De Marchi, M. and Penasa, M. (2019). Invited review: milk lactose-Current status and future challenges in dairy cattle. *J. Dairy Sci.* 102(7), 5883–5898.

Coulon, J. B., Dupont, D., Pochet, S., Pradel, P. and Dupolyer, H. (2001). Effect of genetic potential and level of feeding on milk protein composition. *J. Dairy Res.* 68(4), 569–577.

Cowley, F. C., Barber, D. G., Anstis, A. R., Houlihan, A. and Poppi, D. P. (2014). Immediate, cumulative and residual effects of short- and long-term low plane of nutrition on milk protein and casein composition. *Anim. Prod. Sci.* 54(9), 1374–1380.

Cragle, R. G., Murphy, M. R., Williams, S. W. and Clark, J. H. (1986). Effects of altering milk production and composition by feeding on multiple component milk pricing systems. *J. Dairy Sci.* 69(1), 282–289.

Dalgleish, D. G. (1992). Bovine milk protein properties and the manufacturing quality of milk. *Livest. Prod. Sci.* 35(1–2), 75–93.

Dalgleish, D. G., Horne, D. S. and Law, A. J. R. (1989). Size-related differences in bovine casein micelles. *Biochim. Biophys. Acta* 991(3), 383–387.

De Peters, E. J. and Cant, J. P. (1992). Nutritional factors influencing the nitrogen composition of bovine milk: a review. *J. Dairy Sci.* 75(8), 2043–2070.

De Peters, E. J. and Ferguson, J. D. (1992). Nonprotein nitrogen and protein distribution in the milk of cows. *J. Dairy Sci.* 75(11), 3192–3209.

Doyle, P. T., Stockdale, C. R., Jenkin, M. L., Walker, G. P., Dunshea, F. R., Shields, P. M. and McKenna, A. (2011). Producing milk with uniform high selenium concentrations on commercial dairy farms. *Anim. Prod. Sci.* 51(2), 87–94.

Dunshea, F. R., Walker, G. P., Williams, R. and Doyle, P. T. (2019). Mineral and citrate concentrations in milk are affected by seasons, stage of lactation and management practices. *Agriculture* 9(2), 25–42.

Emery, R. S. (1978). Feeding for increased milk protein. *J. Dairy Sci.* 61(6), 825–828.
Fearon, A. M., Sinclair Mayne, C. S., Beattie, J. A. M. and Bruce, D. W. (2004). Effect of level of oil inclusion in the diet of dairy cows at pasture on animal performance and milk composition and properties. *J. Sci. Food Agric.* 84(6), 497–504.
Forouhi, N. G., Krauss, R. M., Taubes, G. and Willett, W. (2018). Dietary fat and cardiometabolic health: evidence, controversies, and consensus for guidance. *BMJ* 361, k2139.
Fox, P. F., Uniacke-Lowe, T., McSweeney, P. L. H. and O'Mahoni, J. A. (2015). *Dairy Chemistry and Biochemistry*. Springer International Publishing, Basel, Switzerland.
Franke, A. A., Bruhn, J. C. and Osland, R. B. (1983). Factors affecting iodine concentration of milk of individual cows. *J. Dairy Sci.* 66(5), 997–1002.
Gaucheron, F. (2005). The minerals of milk. *Reprod. Nutr. Dev.* 45(4), 473–483.
Givens, D. I., Allison, R., Cottrill, B. and Blake, J. S. (2004). Enhancing selenium content of bovine milk through alteration of the form and concentration of selenium in the diet of dairy cows. *J. Sci. Food Agric.* 84, 811–817.
Greenwood, J. S., Auldist, M. J., Marett, L. C., Hannah, M. C., Jacobs, J. L. and Wales, W. J. (2014). Ruminal pH and whole-tract digestibility in dairy cows consuming fresh cut herbage plus concentrates and conserved forage fed either separately or as a partial mixed ration. *Anim. Prod. Sci.* 54(8), 1056–1063.
Griinari, J. M., Corl, B. A., Lacy, S. H., Chouinard, P. Y., Nurmela, K. V. V., Bauman, D. E. (2000). Conjugated linoleic acid is synthesized endogenously in lactating dairy cows by delta 9 -desaturase. *J. Nutr.* 130, 2285–2291.
Griinari, J. M., Dwyer, D. A., McGuire, M. A., Bauman, D. E., Palmquist, D. L. and Nurmela, K. V. V. (1998). Trans-octadecenoic acids and milk fat depression in lactating dairy cows. *J. Dairy Sci.* 81(5), 1251–1261.
Gulati, A., Galvin, N., Kennedy, E., Lewis, E., McManus, J. J., Fenelon, M. A. and Guinee, T. P. (2019). Effect of reducing daily herbage allowance during early lactation on composition and processing characteristics of milk from spring-calved herds. *Int. Dairy J.* 92, 69–76.
Heard, J. W., Stockdale, C. R., Walker, G. P., Leddin, C. M., Dunshea, F. R., McIntosh, G. H., Shields, P. M., McKenna, A., Young, G. P. and Doyle, P. T. (2007). Increasing selenium concentration in milk: effects of amount of selenium from yeast and cereal grain supplements. *J. Dairy Sci.* 90(9), 4117–4127.
Heard, J. W., Walker, G. P., Royle, P. J., McIntosh, G. M., and Doyle, P. T. (2004). Effects of short-term supplementation with selenised yeast on milk production and composition of lactating cows. *Aust. J. Dairy Technol.* 59, 199–203.
Holt, C., Davies, D. T. and Law, A. J. R. (1986). Effects of colloidal calcium phosphate content and free calcium ion concentration in the milk serum on the dissociation of bovine casein micelles. *J. Dairy Res.* 53(4), 557–572.
Huppertz, T., Fox, P. F. and Kelly, A. L. (2018). The caseins: structure, stability, and functionality. In: *Proteins in Food Processing* (ed. RY Yada). Woodhead Publishing Series in Food Science, Technology and Nutrition, 2nd edn, pp. 49–92. Cambridge, UK: Woodhead Publishing.
Ip, C., Banni, S., Angioni, E., Carta, G., McGinley, J., Thompson, H. J., Barbano, D. and Baumam, D. (1999). Conjugated linoleic acid-enriched butter fat alters mammary gland morphogenesis and reduces cancer risk in rats. *J. Nutr.* 129(12), 2135–2142.
Ismail, B. and Nielsen, S. S. (2010). Invited review: plasmin protease in milk: Current knowledge and relevance t dairy industry. *J. Dairy Sci.* 93(11), 4999–5009.

Iwaniuk, M. E. and Erdman, R. A. (2015). Intake, milk production, ruminal, and feed efficiency responses to dietary cation-anion difference by lactating dairy cows. *J. Dairy Sci.* 98(12), 8973–8985.

Juniper, D. T., Phipps, R. H., Jones, A. K. and Bertin, G. (2006). Selenium supplementation of lactating cows: effect on selenium concentration in blood, milk, urine, and feces. *J. Dairy Sci.* 89, 3544–3551.

Kay, J. K., Mackle, T. R., Auldist, M. J., Thomson, N. A. and Bauman, D. E. (2004). Endogenous synthesis of cis-9, trans-11 conjugated linoleic acid in dairy cows fed fresh pasture. *J. Dairy Sci.* 87(2), 369–378.

Kay, J. K., Mackle, T. R., Bauman, D. E., Thomson, N. A. and Baumgard, L. H. (2007). Effects of a supplement containing trans-10, cis-12 conjugated linoleic acid on bioenergetic and milk production parameters in grazing dairy cows offered ad libitum or restricted pasture. *J. Dairy Sci.* 90(2), 721–730.

Kefford, B., Christian, M. P., Sutherland, B. J., Mayes, J. J. and Grainger, C. (1995). Seasonal influences on Cheddar cheese manufacture: influence of diet quality and stage of lactation. *J. Dairy Res.* 62(3), 529–537.

Kelly, M. L., Berry, J. R., Dwyer, D. A., Griinari, J. M., Chouinard, P. Y., Van Amburgh, M. E. and Bauman, D. E. (1998). Dietary fatty acid sources affect conjugated linoleic acid concentrations in milk from lactating dairy cows. *J. Nutr.* 128(5), 881–885.

Kliem, K. E. and Shingfield, K. J. (2016). Manipulation of milk fatty acid composition in lactating cows: opportunities and challenges. *Eur. J. Lipid Sci. Technol.* 118(11), 1661–1683.

Larson, B. L., Heary, H. L. and Devery, J. E. (1980). Immunoglobulin production and transport by the mammary gland. *J. Dairy Sci.* 63(4), 665–671.

Lascelles, A. K. and McDowell, G. H. (1974). Localised humoral immunity with particular reference to ruminants. *Transplant. Rev.* 19, 170–208.

Lean, I. (1987). *Nutrition of Dairy Cows*. The University of Sydney Post-graduate Foundation in Veterinary Science, Sydney, NSW, Australia.

Linn, J. G. (1988). Factors affecting the composition of milk from dairy cows. In: Designing Foods; Animal Product Options in the Marketplace, National Research Council committee on technological options to improve the nutritional attributes of animal products, National Academies Press, Washington.

Lock, A. L. and Bauman, D. E. (2004). Modifying milk fat composition of dairy cows to enhance fatty acids beneficial to human health. *Lipids* 39(12), 1197–1206.

Logan, A., Auldist, M., Greenwood, J. and Day, L. (2014). Natural variation of bovine milk fat globule size within a herd. *J. Dairy Sci.* 97(7), 4072–4082.

Lynch, J. M., Lock, A. L., Dwyer, D. A., Noorbakhsh, R., Barbano, D. M.. and Bauman, D. E. (2005). Flavor and stability of pasteurized milk with elevated levels of conjugated linoleic acid and vaccenic acid. *J. Dairy Sci.* 88(2), 489–498.

Mackle, T. R., Bryant, A. M., Petch, S. F., Hill, J. P. and Auldist, M. J. (1999a). Nutritional influences on the composition of milk from cows of different protein phenotypes in New Zealand. *J. Dairy Sci.* 82(1), 172–180.

Mackle, T. R., Bryant, A. M., Petch, S. F., Hooper, R. J. and Auldist, M. J. (1999b). Variation in the composition of milk protein from pasture-fed dairy cows in late lactation and the effect of grain and silage supplementation. *N. Z. J. Agr. Res.* 42(2), 147–154.

Mackle, T. R., Kay, J. K., Auldist, M. J., McGibbon, A. K. H., Philpott, B. A., Baumgard, L. H. and Bauman, D. E. (2003). Effects of abomasal infusion of conjugated linoleic acid on milk fat concentration and yield from pasture-fed dairy cows. *J. Dairy Sci.* 86(2), 644–652.

Mackle, T. R., Petch, S. F., Bryant, A. M., Auldist, M. J., Henderson, H. V. and MacGibbon, A. K. H. (1997). Variation in the characteristics of milkfat from pasture-fed dairy cows during late spring and the effects of grain supplementation. *N. Z. J. Agr. Res.* 40(3), 349-359.

MacRae, J. C., Bequette, B. J. and Crompton, L. A. (2000). Synthesis of milk protein and opportunities for nutritional manipulation. In: *British Society of Animal Science* (vol. 25) Agnew, R. E., Agnew, K. W. and Fearon, A. M. (Eds), Milk Composition, Occasional Publication, Belfast, Northern Ireland, 179-199.

Malossini, F., Bovolenta, S., Piras, C., Dalla Rosa, M. and Ventura, W. (1996). Effect of diet and breed on milk composition and rennet coagulation properties. *Ann. Zootech.* 45(1), 29-40.

Mendelson, C. R., Zinder, O., Blanchette-Mackie, E. J., Chernick, S. S. and Scow, R. O. (1977). Lipoprotein lipase and lipid metabolism in the mammary gland. *J. Dairy Sci.* 60(4), 666-676.

Mepham, T. B. (1993). The development of ideas on the role of glucose in regulating milk secretion. *Aust. J. Agric. Res.* 44(3), 509-522.

Moate, P. J., Chalupa, W., Boston, R. C. and Lean, I. J. (2007). Milk fatty acids. I. Variation in the concentration of individual fatty acids in bovine milk. *J. Dairy Sci.* 90(10), 4730-4739.

Murphy, J. J. (2000). Synthesis of milk fat and opportunities for nutritional manipulation. In: *British Society of Animal Science* (vol. 25) Agnew, R. E., Agnew, K. W. and Fearon, A. M. (Eds), Milk Composition, Occasional Publication, Belfast, Northern Ireland, 201-223.

Murphy, J. J. and O'Mara, F. (1993). Nutritional manipulation of milk protein concentration and its impact on the dairy industry. *Livest. Prod. Sci.* 35(1-2), 117-134.

O'Brien, B., Murphy, J. J., Connolly, J. F., Mehra, R., Guinee, T. P. and Stakelum, G. (1997). Effect of altering the daily herbage allowance in mid lactation on the composition and processing characteristics of bovine milk. *J. Dairy Res.* 64(4), 621-626.

Ouweltjes, W., Beerda, B., Windig, J. J., Calus, M. P. L. and Veerkamp, R. F. (2007). Effects of management and genetics on udder health and milk composition in dairy cows. *J. Dairy Sci.* 90(1), 229-238.

Palmquist, D. L. and Jenkins, T. C. (2017). A 100-year review: fat feeding of dairy cows. *J. Dairy Sci.* 100(12), 10061-10077.

Perfield, J. W., Lock, A. L., Griinari, J. M., Sæbø, A., Delmonte, P., Dwyer, D. A. and Bauman, D. E. (2007). Trans-9, *cis*-11 conjugated linoleic acid reduces milk fat synthesis in lactating dairy cows. *J. Dairy Sci.* 90(5), 2211-2218.

Saebø, A., Saebø, P. C., Griinari, J. M. and Shingfield, K. J. (2005). Effect of abomasal infusions of geometric isomers of 10,12 conjugated linoleic acid on milk fat synthesis in dairy cows. *Lipids* 40(8), 823-832.

Santos, F. A. P., Santos, J. E. P., Theurer, C. B. and Huber, J. T. (1998). Effects of rumen-undegradable protein on dairy cow performance: a 12-year literature review. *J. Dairy Sci.* 81(12), 3182-3213.

Schwab, C. G. and Broderick, G. A. (2017). A 100-year review: protein and amino acid nutrition in dairy cows. *J. Dairy Sci.* 100(12), 10094-10112.

Sobhanirad, S., Carlson, D. and Kashani, R. B. (2010). Effect of zinc methionine or zinc sulfate supplementation on milk production and composition of milk in lactating dairy cows. *Biol. Trace Elem. Res.* 136(1), 48-54.

Stockdale, C. R., Shields, P. M., McKenna, A., Walker, G. P., Dunshea, F. R. and Doyle, P. T. (2011). Selenium levels in cows fed pasture and concentrates or a total mixed

ration and supplemented with selenized yeast to produce milk with supra-nutritional selenium concentrations. *J. Dairy Sci.* 94(1), 262–272.
Vik-Mo, L., Emery, R. S. and Huber, J. T. (1974). Milk protein production in cows abomasally infused with casein or glucose. *J. Dairy Sci.* 57(8), 869–877.
Wales, W. J., Stockdale, C. R. and Doyle, P. T. (2001). Increasing energy intake improve milk protein concentration in cows in spring and summer. *Aust. J. Dairy Tech.* 54, 154.
Wedholm, A., Larsen, L. B., Lindmark-Mansson, H., Karlsson, A. H. and Andrén, A. (2006). Effect of protein composition on the cheese-making properties of milk from individual dairy cows. *J. Dairy Sci.* 89(9), 3296–3305.
Xue, B., Yan, T., Ferris, C. F. and Mayne, C. S. (2011). Milk production and energy efficiency of Holstein and Jersey-Holstein crossbred dairy cows offered diets containing grass silage. *J. Dairy Sci.* 94(3), 1455–1464.
Zanton, G. I., Bowman, G. R., Vázquez-Añón, M. and Rode, L. M. (2014). Meta-analysis of lactation performance in dairy cows receiving supplemental dietary methionine sources or postruminal infusion of methionine. *J. Dairy Sci.* 97(11), 7085–7101.
Zwierzchowski, G. and Ametaj, B. N. (2018). Minerals and heavy metals in the whole raw milk of dairy cows from different management systems and countries of origin: a meta-analytical study. *J. Agric. Food Chem.* 66(26), 6877–6888.

# Chapter 17

## Sensory properties of milk: understanding and analysis

*Tim Coolbear, Nathalie Janin, Rachel Traill and Rebecca Shingleton, Fonterra Research and Development Centre, New Zealand*



### 1 Introduction

For humans, milk of other species has been a key nutritional source for millennia; which species farmed being dependent on familiarity, geography, culture and lifestyle. Industrialisation, however, has scaled up cows' milk production to the point where one estimate of global production was 532 million metric tons in 2020 (Statistica, 2021). This represents nearly 81% of total world milk, with most of the remainder comprising that from buffaloes (~15%), goats and sheep (~4%) (OECD-FAO, 2021) and milk from camels, donkeys, horses, reindeer and yaks together contributing less than 1% (Faccia et al., 2020). It is inevitable that, because of commercial imperatives, the vast majority of sensory work has been done with cows' milk and derivative products, and this directly impacts the balance of the content of this chapter. Another aspect of dairy research that affects the balance of content on bovine and non-bovine milk is that much of the research has centred on increasing milk yield and on processing ability and efficiency to end products. Often, research on sensory aspects has been comparative and consequential, focusing on the end products, rather than on the raw materials. In contrast, research on human milk has focused on infant feeding and comparisons of human breast milk with infant formula (bovine

http://dx.doi.org/10.19103/AS.2022.0099.17

and non-bovine) in the context of nutrition, health and development of the child.

One of the key aspects of industrialisation of milk production has been its impact on milk flavour; what was once a short distance and time from producer animal to human consumption has, especially for cows' milk, become significantly longer from both perspectives. On-farm storage, bulk collection systems, farm-to-factory transport distances and at-factory silo storage prior to first unit operation all contribute to increased opportunities for compromising the initial flavour profile of the milk.

In this chapter, the sensory profiles of cows' milk will be covered first, with attention paid to differences imparted by different feeding regimes and the changes imparted by storage, pasteurisation and high heat treatments. One of the commercial imperatives for large-scale commercial milk production is the preservation of nutritional value and quality, and such heat treatments are currently the key preservation technologies for liquid milk. All other preservation technologies (for example, powder production, cheese, yoghurt) are out of scope for this chapter. The sensory profiles of non-bovine milk will be covered next, followed by commentary on what defects can arise and how. This leads to a discussion on the sensory analysis of milk and the determination of sensory profiles and discrimination between them.

Before continuing, some definitions and background in sensory profiles will be useful. Flavour is the perception of the combination of taste compounds (non-volatile tastants) and odour compounds (volatile), the former detected by receptors on the tongue and the latter detected by receptors on the lining of the nasal cavity at the olfactory bulb. In ortho-nasal olfaction, volatiles reach the olfactory bulb via the nostrils, while in retro-nasal olfaction, volatiles are released in the mouth from the product matrix and pass through to the olfactory bulb via the back of the throat (Meilgaard et al., 2016; Prinz and De Wijk, 2004). For humans, taste has been thought to be limited to sweet, sour, salt, bitter and umami (which are well-known), but there is evidence that further tastes such as kokumi, fat and metallic can be perceived, with milk products recently being used to show neural responses to fat taste processing (Andersen et al., 2020). In contrast, the human has about 400 nasal odour receptors that respond to approximately 230 (known) odour-active compounds (Buck and Axel, 1991; Dunkel et al., 2014), a small fraction of the approximately 10,000 volatile compounds in food.

In sensory analysis, 'attributes' are the sensory characteristics of a food detectable by the human senses, that is, aroma/odour, taste, texture (including mouthfeel), appearance or sound, and are usually describable in one, sometimes two, words (Meilgaard et al., 2016). With regard to flavour, some flavour compounds can be perceived to have different flavours at different concentrations; others can be desirable at low concentrations but undesirable at higher concentrations. At low concentrations, some flavour compounds may not be detected but enhance the perception of other flavours. Flavour compounds can also interact with each other in a manner that is not additive but delivers the perception of a completely different flavour. Further, cross-modality can come into play, where one sensory mode affects the perception of a different mode (Poinot et al., 2013).

Attributes can be deemed positive or negative in terms of overall appreciation of a flavour profile, and individual liking of a flavour profile can be modified with familiarity. To any individual, the overall flavour of a food depends on the types of flavour compounds present, their relative concentrations and the ability of the individual to perceive those flavours; liking or disliking of a food is also individualistic. The factors that determine this individuality can include genetics, familiarity, learning, culture, psychology and physiology.

A fundamental question that could be posed is: 'What is the normal flavour of milk?'. There is no single answer to this, even if applied to the concept of the flavour of the milk of a mammal direct from the teat. In 2014, the Food and Agriculture Organisation (FAO) of the United Nations, the International Dairy Federation (IDF) and the IFCN (originally the International Farm Comparison Network) Dairy Research Network published a collaborative work mapping dairy animal (cows, sheep, goats and buffalo) feeding systems (FAO, IDF and IFCN, 2014) and characterised some 47 feeding systems differing in total amount and composition in terms of levels of roughage, by-products and concentrate fed. What the animal is feeding on and the quality of that feed (both impacted by season), what supplements, if any, are provided, what length of time there has been between feeding a supplement or feed ration and milking, proximity to the sea, the physiological and health states of the animal and so on, all lead to changes in the flavour profile of the milk. Essentially, though, milk is quite mild flavoured with a sweetish taste and a smooth, rich mouthfeel, but it is this mildness that renders milk prone to acquire regional flavour profiles that consumers are used to and prefer, but also to acquire flavour defects consequent on the detail of production, processing and storage.

All of this is a commodity-based construct for milk from a human perspective. It is worthwhile, though, to briefly reflect from a different perspective on this highly nutritious food that contains a myriad of macro- and microcomponents and continues to rightly play such a valuable role in human health. Essentially, what we perceive as the macroflavour of a commoditised nutritional product has deeper constructs. In nature, mammalian neonate survival is highly dependent on maternal recognition, a learning process begun by the foetus in utero via the amniotic fluid and continued through colostrum and milk. As well as pheromone-based mechanisms (Schaal et al., 2003), olfactory cues ('signature odours') initiate recognition and suckling (Logan et al., 2012); a recent review by Loos et al. (2019) provides an overview of the behavioural effects of human milk odour in the human neonate. However, for a chapter on the flavour of commoditised milk products, these deeper aspects of milk odour compounds are out of scope, and while a later section will provide some commentary on human milk flavour, this covers only the basics for comparison with milk from other mammals.

## 2 Bovine milk flavour profiles

Overall, aroma compound concentrations in cows' milk are very low and the fundamental flavour characteristics are mild and sweet, overlaid with attributes derived from breed, feed and physiological condition. 'Seasonality' is often also called out as a factor, but, in reality, the actual climatic conditions during the seasons of any year determine both the quality and type of feed.

In the natural environment, cows are grazing animals, and so the natural flavour profiles of cows' milk come from the pastures being grazed. During the seasons, the quality of the pasture can vary widely, and during drought periods, significant amounts of supplement feed may have to be introduced to prevent the cows from drying off. Climate can vary widely between regions in some countries, and so feedstocks in one region can vary widely from those in another. In developing countries, herd sizes are generally small, and the animals are relatively low yielding; most of the feed comes from locally produced roughage. In contrast, in developed countries, in general, herd sizes are large and comprise high-yielding animals in seasonal or year-round confinement production systems in which feed is stored forage and purchased grains and concentrate. Whilst this is often termed 'conventional feeding', cows are naturally pasture animals. Key exceptions to the confinement production in developed countries are New Zealand and Ireland, where milk production with large herds is predominately pasture-based (<10% non-pasture).

Figure 1 presents a simple flavour wheel for quality milk from grass- and grain-fed cows, with the characteristic differentiating flavours together with potential off-flavours that can arise. The core creamy, sweet notes common to all cows' milk are overlaid with lactone, canned corn and feed/silage notes in milk from grain-fed cows, compared with an overlay of cowy, grassy and barny notes in milk from grass-fed cows. The potential off-flavours are common to milk from both grain- and grass-fed cows, that is, acid, unclean, musty, fruity, rancid/oxidised and medicinal notes.

In saying this, a key issue arises: several attribute terms have been used, but these are words initially agreed by panellists that have become general usage in some countries but are potentially meaningless in others. Do the terms 'creamy', 'canned corn', 'cowy', 'barny', 'musty' resonate with, for example, Chinese, Indonesian, Malay or Japanese consumers? There is no universal lexicon, and translation between lexicons is fraught. The term 'lactone' is more of a specific

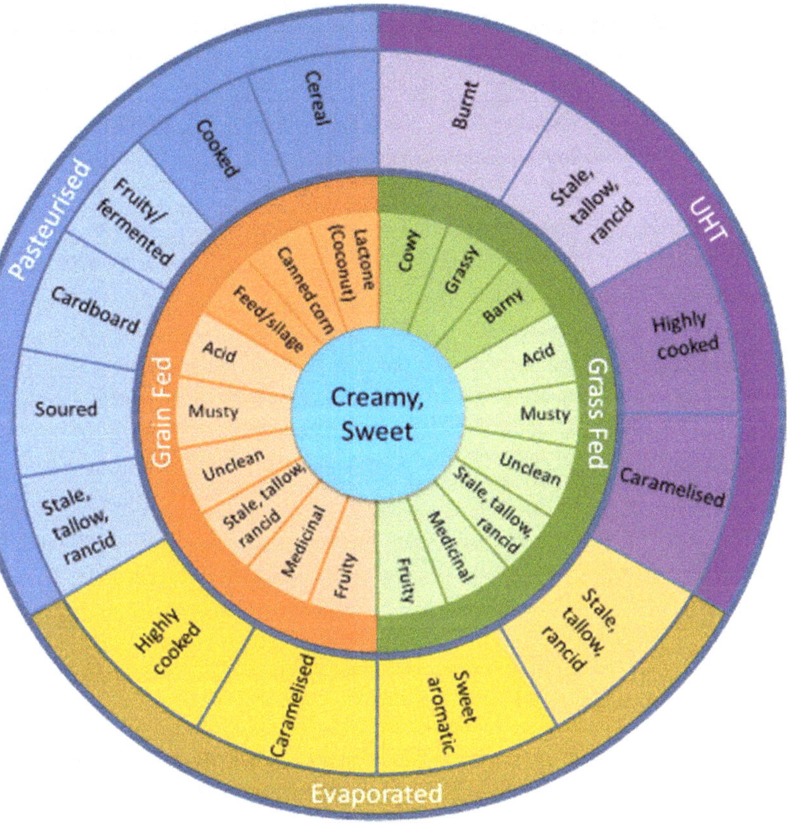


**Figure 1** Characteristic and potential off-flavours of pasteurised, UHT and evaporated cows' milk. Characteristic flavours are shaded darker, off-flavours are shaded lighter. Stale, tallow and rancid (also linseed, fishy) are often designated as 'oxidised', but this term is mechanistic and erroneously assumptive (as is 'lightstruck'). Note that there is no indication of relative intensities of flavours in this diagram. *Source:* Adapted from a diagram prepared for Fonterra by Jessica Somerton (nee Childs).

chemical name, but it is often elaborated to explain that the milk fat flavour often tastes like coconut or fruity/apricot. In some instances, specific chemicals can be used as references to help understand a flavour note or to calibrate panellists on attribute intensity, but this comes with risks; the perception of a flavour is seldom the consequence of a single chemical, and the reference chemicals used are not often intrinsic to milk (and therefore confusing, misleading or even alarming to a consumer or customer). Table 1 has been constructed to explain further some of the attribute terms used but again has the limitations noted above.

The fat component of milk has a significant contribution to the overall sensory profile of different milk types, and within bovine milk, the range is largely due to the breed, feed and lactation stage of the animals. For this

**Table 1** Chemical references for dairy flavours[a]

| Chemical | Flavour descriptor | Chemical | Flavour descriptor |
|---|---|---|---|
| Acetic acid | Sour | 2,4,6-Trichloroanisol | Musty |
| Lactic acid | Sour | 2-Ethyl fenchol | Musty, dusty |
| Sodium chloride | Salt | trans-2-Nonenal | Wet cardboard |
| Dimethyl sulphide | Creamed corn | 2-Acetyl-2-thiazole | Cereal |
| Dimethyl disulphide | Cooked cabbage | Tannic acid | Astringent |
| Dimethyl trisulphide | Onion, garlic | Alum | Astringent |
| cis-3-Hexanol | Grassy | Quinine | Bitter |
| Ferrous sulphate | Metallic | Caffeine | Bitter |
| Benzaldehyde | Almond | Butyric acid | Rancid, cheesy |
| Ethyl butyrate | Pineapple | Trans-2-4-heptadienal | Fishy |
| Ethyl hexanoate | Estery | 4-Ethyl phenol | Cow-like |
| Ethyl isovalerate | Fermented | *o*-Cresol | Cows' breath |
| Isoamyl alcohol | Over ripe fruit | *p*-Cresol | Animalic, barnyard |
| 2,3-Butanedione | Buttery | Skatole | Animalic |
| δ-Dodecalactone | Coconut, buttery | | |

[a] The taste and odour of many flavour compounds is dependent on concentration and, to a certain extent, the matrix in which they are tasted; the descriptors presented are common descriptions, many other terms can be ascribed to the compounds.

reason, there are various ranges for fat content presented in the literature. The typical fat content averaging 3.6–4.8% contributes to the mild creamy texture and mouthcoating of bovine milk and contrasts with milk such as that from buffalo with much higher fat content (6.4%) with a richer, creamier mouthfeel. Creaminess is generally the term used to describe the variation in sensory profile due to fat content. It is a complex multi-modal term that combines the appearance, aroma, taste and textural features of many products including milk (Chen and Eaton, 2012; Jervis et al., 2014) and has been related to the fat content of products (Chojnicka-Paszun et al., 2012).

Bovine milk is typically processed to provide consumers with a range of milk options containing different fat content (for example, whole, skimmed, semi-skimmed), and the removal of the fat component significantly alters the sensory experience of the milk. Several studies, including work more than 50 years ago by Rose Marie Pangborn (Pangborn and Dunkley, 1964a,b), have demonstrated the effects of fat on the sensory properties of milk (colour, texture and flavour). Phillips et al. (1995) showed that appearance (that is, whiteness) was the key differentiator of milk with different fat levels, to the extent the perception of the sensory properties of fat-free milk was improved towards those of fatty milk by making it appear whiter through the addition of titanium dioxide (Phillips and Barbano, 1997).

Visual cues are extremely important in relation to the perception of creaminess and the fat content. It has been shown in a number of studies that when evaluating samples for taste alone a high degree of change is required to perceive a difference due to fat content. For example, Pangborn et al. (1985) showed that if participants were not able to visually observe the milk they were consuming, then they could not discriminate milk samples containing between 0% and 3.5% fat, while McCarthy et al. (2017) showed that the degree of difference that was able to be perceived as fat content increased was smaller when prompted by visual cues alone versus tasting without visual cues. As the fat component is reduced, the protein and lactose components of the milk play a larger part in the overall sensory profile.

In trained sensory panel studies conducted at the Fonterra Research and Development Centre in New Zealand (Abraham, Turnbull, and Shingleton, unpublished) on milk from grass-fed cows, reduction in fat resulted in decreased creaminess, increased sweetness from the lactose and increased perception of cardboard and animalic flavours contributed from the protein; the texture became thinner and more watery, losing the creaminess and mouthcoating effects from the fat component. In contrast, however, in a study on blended fractions from milk from grain-fed cows to give 0.1% fat (equivalent to skimmed milk), 2% fat, 3.25% fat (equivalent to whole milk) and 6.6% fat products, while a trained sensory panel reported an increase in fat/lactone flavour, viscosity, opacity and yellow colour as fat content increased, sweet taste also increased (McCarthy et al., 2017). In the same study, consumers described the differences as an increase in thickness, richer and creamier with an overall stronger flavour. In a later study on milk beverages systematically blended from milk fractions to give a series of formulations of different contents of fat, true protein, and casein as a percentage of true protein, Cheng et al. (2019) showed that casein as a percentage of true protein explained most of the variation in the perception of sweetness and cardboard notes but had no effect on milk fat flavour, which was almost totally explained by fat concentration.

The flavour compounds present in the fat component of milk play a large part in the overall experience. Zhang et al. (2016) used an enzyme modification approach to intensify volatile flavours in skimmed milk and so impart whole milk flavour characteristics. How milk is stored directly after milking impacts the overall flavour profile, and the main levers are described in the defects section of this chapter. It is however important to note that a small degree of lipid oxidation is normal, even for very fresh milk, where it makes an important contribution to the overall flavour (Bendall and Olney, 2001).

Along with the flavour and texture characteristics (thickness, mouthcoating and residual mouthcoating) contributed by milk fat, the appearance is also significantly impacted. As fat content increases and the milk appears whiter, this can also lead to consumers perceiving the milk to be creamier (Phillips et al., 1995).

Preferences for bovine milk with different fat content have been studied and highlight the variation in consumer preferences. Age, familiarity with a particular product and other non-sensory aspects such as health concerns all contribute to different preferences and groups of consumers with different expectations and liking. Lara-Zamudio et al. (2013) reported that amongst children aged 6–16 years, who were habitual consumers of whole, semi-skimmed (the authors used the term partially defatted) and skimmed (defatted) milk, the overall preference was higher for skimmed milk than for whole milk. However, the study used reconstituted powder products for whole and semi-skimmed milk, whereas the skimmed milk was a liquid product. McCarthy et al. (2017) using blind taste tests found consumers preferred the sensory profile of milk of higher fat content than the milk they typically purchased.

In Fig. 1, the core flavour notes are elaborated to those derived from different heat treatments commonly used to extend the shelf life of the liquid product. Pasteurisation has the aim of controlling pathogens and spoilage organisms without affecting the nutritional and organoleptic characteristics of the milk, but the food safety aims take precedence. Typical pasteurisation conditions for cows' milk fall into two categories, low temperature long time (LTLT), that is, 63°C for 30 min or high temperature short time (HTST), that is, 72°C for 15 s, both of which provide the same microbial lethality ($F_0$). The sensory consequences for these conditions are relatively minor. Characteristic notes introduced centre on cereal and cooked flavours (reflected by low levels of Maillard compounds, for example, furosine), and off-flavours that can be introduced include fruity/fermented, cardboard, sour and oxidised, but these need not be a consequence of heat so much as of total time.

To extend the shelf life of liquid milk, extended shelf life (ESL), ultrahigh temperature (UHT) and evaporation processes have been developed, the first utilising temperatures of, for example, 125°C for 5 s (Deeth, 2017) with equal microbial lethality at 135°C for 0.5 s and 127°C for 1 s (Rysstad and Kolstad, 2006), the second utilising temperatures of 135–150°C for a few seconds to achieve commercial sterility (as defined by Codex, 1993), the third involving vacuum evaporation of pasteurised milk at 45–50°C to derive a 30–40% total solids concentrate. These processes introduce new characteristic highly cooked and caramelised flavours, but the concentrated evaporated milk also acquires sweet aromatic notes. Oxidised off-flavours can intensify and burnt notes can be introduced in the case of UHT.

Alothman et al. (2019) comprehensively reviewed how pasture feeding impacts cows' milk composition. They noted how the feed system (pasture versus total mixed ration) impacts colour (due to variation in β-carotene concentration), susceptibility to oxidative changes before and after processing (due to fatty acid composition) and the core flavour differences between the milk from grass-fed cows and that from cows fed total mixed ration but also noted that consumers may not be as discriminatory towards some of these differences as trained sensory panels or instrumental analysis, as found by Croissant et al. (2007). Clarke et al. (2020) also looked at the impact of grass, grass/clover and total mixed ration feeding systems on the volatile profiles and sensory perception of the resulting milk, confirming sensory differentiation but also implicating changes in rumen metabolism.

## 3 Non-bovine milk flavour profiles

In certain parts of the world where there is no history of bovine husbandry due to climate (for example, arid regions) or geography (for example, mountainous regions), it is the milk of non-bovine species that provides dairy nutrition and subsistence. In other parts of the world, the milk of non-bovine species is growing in importance due to perceived benefits to health and nutrition and, to a certain extent, due to its inclusion into, or presentation as, specialty foods. Much of the research on non-bovine milk centres on Protected Designation of Origin (PDO) and detection of adulteration - almost inevitably, if cows' milk is cheaper and more available than that of non-bovine species in an area, blending bovine and non-bovine milk in the production of a milk product has advantages in terms of both cost of production and potential sales volume. If done transparently and with full consumer awareness, then the issues of consumer deception do not arise; however, the economic impact on the true artisanal producer remains.

Research on adulteration tends to focus on composition and detecting marker components; any sensory data tends to relate to quality assessment rather than product differentiation. In addition, focus is on the end product, not on the component raw materials. Of the non-bovine milk types, that from sheep and goats has received the most attention and the sensory characteristics are the most defined. For other species, sensory reports on the milk itself are extremely limited; reports that do exist tend to focus on the milk products, especially since preservation of the milk is still a vital part of extending its nutritional availability for the human consumer. Another idiosyncrasy is that, just as bovine milk has an associated 'cowy' attribute, sheep and goat milk especially have been ascribed 'sheepy' and 'goaty' attributes, respectively.

### *3.1 Sheep milk*

As for bovine milk, some of the characteristic flavours of sheep milk (often described as sweeter than cow or goat milk) derive from the fat fraction and are dependent on the integrity of that fraction. Generalisations are that the total solids and mineral levels are higher in sheep milk than in cow milk, and the average fat globule size is smaller. The C10:0, C14:0, C16:0, C18:0 and C18:1 fatty acids account for more than 75% of the total fatty acids, a similar profile to that of cows' milk. Whilst some studies have suggested that caproic (C6:0), caprylic (C8:0) and capric (C10:0) acids are significantly higher in sheep milk (Park et al., 2007), a recent review has noted that the caproic and caprylic levels are similar, only the level of capric acid significantly differs and is almost double that in cow milk (Markiewicz-Kęszycka et al., 2013). In the free form, the branched-chain fatty acids 4-ethyl octanoic acid (4-ethylcaprylic acid) and 4-methyl octanoic acid also contribute to the animalic flavour that can be present (Brennand et al., 1989; Ha and Lindsay, 1991; Kaffarnik et al., 2014). The presence of the free form of a fatty acid (unless derived from amino acid metabolism) is dependent on lipolysis of the triacylglyceride, and so, the levels of lipolytic enzymes in the milk determine the concentration of the fatty acids and hence the flavour intensity of the attributes imparted.

The overall balance of flavour notes is consequent on the fatty acid composition of the milk, which is linked to breed and feed, and the intensity of the notes is dependent on the amount of lipase activity inherent in the milk, which is in turn dependent on diet, physiological status and health of the animals and on the storage time and temperature of the milk. Similarly, the types and levels of the phenolic compounds (such as cresols, ethylphenols and dimethyl phenols; Ha and Lindsay, 1991) that impart other characteristic (sheepy) flavours derive from feed and amino acid catabolism and have breed dependencies. O'Connell and Fox (2001) reviewed the significance of phenolic compounds of milk and noted the caramel-like, smoky and sweet attributes imparted and noted the changes in perception (and description) of the attributes at higher concentrations.

Studies have been undertaken on alternative feeds for sheep, both for overall nutrition and targeting modification of fatty acid composition. When cakes of *Camelina sativa* (a member of the Brassicaceae) containing unsaturated fatty acids and antioxidants was used as a supplement, the subsequent ewes' milk had distinct animal, cereal and processed aromas that were, in part, intensified by pasteurisation (Cais-Sokolińska et al., 2011).

### 3.2 Goat milk

Goats have a folklore reputation for eating almost anything, which has spawned humorous song on the one hand (Lee and Weston, 1917), but also the query as to whether the browsing habits of goats are 'a potentially overlooked "treasure trove" with respect to health promoting components' (Silanikove et al., 2010) on the other. From a sensory perspective, the flavour of goat milk has a base complexity, described as more robust, waxy and animalic than cows' milk (Le

Quere et al., 1996; Whetstine and Drake, 2006) that is built on by the diverse range of diet that goats can consume. Further, this range is seasonal; Fedele et al. (2005) reported that hay-like, green and cheesy flavours characterised winter milk, while summer milk had fruity, citrus/mint-like notes. Transhumance of goatherds from valley to mountain is common pastoralism in some regions, and this, therefore, imparts variation in the milk and derived products. For Norwegian goat milk, a rancid, tart off-flavour is common during the mountain grazing season; Eknæs and Skeie (2006) investigated the effect of roughage availability on flavour and showed that energy deficit increased the frequency of rancid/tart flavour, and Eknæs et al. (2009) showed that saturated long-chain fatty acid supplements reduced the frequency of rancid flavours. Further, Inglingstad et al. (2017) gained decreased rancid, tart off-flavour intensities by replacing palm oil with rapeseed oil in supplement use. In other regions, the different breeds are continuously grazed at different altitudes (for example, Saanen and Alpine breeds in Slovenian highlands and alpine regions, respectively) with differences presented in the fatty acid profiles of the milk (Žan et al., 2006). Summer milk of the Italian Garfagnina goat is better accepted than that of winter milk because of decreased levels of short-chain fatty acids responsible for the development of unpleasant aromas (Salari et al., 2016); and both season- and farm-specific effects have also been found for goat milk (Siefarth and Buettner, 2014).

The contribution of lipid, protein and carbohydrate to the characteristic goaty or animal flavour profile of goat milk is still unclear (Polowsky et al., 2017). The capric acid (C10:0) concentration in goat milk is triple that in cow milk (Markiewicz-Kęszycka et al., 2013), and Ha and Lindsay (1991, 1993) noted the characteristic levels of the branched-chain ethyl octanoic, methyl octanoic and methyl nonanoic acids. Again, it is the free forms of the fatty acids, hydrolysed away from the triacylglycerides, that impact the flavour profiles and the intensities of the flavours, and so the activity levels of inherent lipases are key, as well as the effects of any processing and storage of the milk.

As is the case for sheep, studies have been undertaken on alternative feeds for goats, including cacti (Catunda et al., 2016) and manicoba silage (Costa et al., 2008) without significant impact on the sensory profiles of the milk, but with the 'goaty' flavour increased to unacceptable levels with vegetable oil, especially cotton seed supplementation (Queiroga et al., 2009). Skjevdal (1979) reported low flavour intensities for goat milk at the extremes of the lactation cycle being impacted by the practice of feeding concentrates at the beginning of lactation increasing the overall flavour intensity of the milk. Cais-Sokolinska et al. (2019), paralleling their studies with ewes' milk, supplemented goat feed rations with maize dried distillers grains, giving more polyunsaturated fatty acids and conjugated linoleic acid and better atherogenic and thrombogenic indices. Sensory analysis indicated that of the descriptors used, sweetness was scored lower in the milk from the goats fed the supplement than in the milk from the control group, while grainy and processed notes were scored higher. On pasteurisation, these differences were no longer statistically significant; but cooked flavour was distinctly higher in the milk from the goats fed the supplement.

Processing can impact in ways that both increase lipolysis or prevent further lipolysis and therefore either increase or stabilise the intensity of the derived flavour compounds. The former is due to shearing, for example, during pumping, that enhances the availability of substrate due to disruption of the protective milk fat membrane, while the latter is due to denaturation of milk lipase by, for example, pasteurisation (Pereira et al., 2008). Storage of raw milk especially results in increased levels of free fatty acids due to both endogenous lipases and contaminant microbial lipases, and there is an increased risk of the development of oxidation products that give rancid notes.

Heilig et al. (2008) assessed the impact of pasteurisation, ultra-pasteurisation and UHT-heating goat milk on sensory properties and storage stability. Pasteurisation did not appear to alter sensory properties, but the higher heat treatments introduced some deficiencies, with indirect heating giving poorer results than direct heating and skimmed milk giving better sensory quality results than whole milk.

## *3.3 Buffalo milk*

Buffalo milk production comprises some 11% of the worlds' milk production. The two subspecies of the domestic water buffalo are the river buffalo, mainly of India and Pakistan, and the swamp buffalo mainly of Eastern Asia and bred primarily for their draught power (FAO, 2020). In the western world, however, buffalo milk is mostly associated with Italy and with the production of buffalo mozzarella, differentiated from bovine mozzarella by the mushroom aroma of 1-octene-3-ol, nonanal and indole (Moio et al., 1993). The fat content of buffalo milk is higher than that of cows' milk (close to double), regardless of breed, while protein, lactose and ash contents are similar, and the milk is characteristically 'white and smooth', tasting creamy and mild; nutty, cooked flavours are imparted by boiling due to the release of sulfhydryl compounds (Murtaza et al., 2017).

One study (Khan et al., 2007) has shown significant differences between sensory scores for colour, flavour and taste of Swamp buffalo milk and Water buffalo milk, with the former having higher levels of fat, total solids and protein. Off-flavours in buffalo milk, recently reviewed by Muhammad et al. (2017), can be associated with feed components that can be seasonal, for example, certain weeds.

### 3.4 Camel, yak, reindeer, horse and donkey milk

Commenting on camel, yak, reindeer, donkey and horse milk under a combined heading is, in part, aligned to the major reason for the domestication of these particular species - sustaining nomadic lifestyles. Many of these lifestyles are now threatened, to a greater or lesser extent, and the traditions enveloping the milk and milk products are also threatened. It is rather sad and ironic that many of these traditionally sustenance-based milk types are being developed in the western world at the expensive gourmet and luxury ends of the markets, and the relevant sensory features are no longer the taste of the milk and the products created to preserve the nutrition but lean more towards texture and perceptions of wealth, indulgence and cosmetic efficacy (for example, deer milk being incorporated into face creams).

#### 3.4.1 Camel milk

The one-humped Arabian camel or Dromedary generally has a much higher milk yield than the less common two-humped Bactrian camel; both can produce more milk from poor feed than any other dairy species (FAO, 2020), and the chemical composition of the camel milk and its products have been recently reviewed (Brezovecki et al., 2015), as has its therapeutic value (Sharma and Singh, 2014; Yadav et al., 2015); it can be considered as being a good alternative to cows' milk for children with cows' milk protein allergy, partly because of its acceptable taste (Navarrete-Rodriguez et al., 2018).

Whilst camels are raised for milk, meat, fibre and transport, for the nomad, milk is often the most important of these, providing the staple and often the only food (FAO, 2020). The macro-composition (water, protein, lactose, fat, minerals) of Dromedary camel milk is very similar to that of the cow, whereas Bactrian milk tends to have more fat and less water (Medhammar et al., 2012). Camel milk can be described as richer, more filling, more easily digested, and more satisfying than cow, goat or sheep milk.

Ahmed et al. (2014) noted a relationship between the frequency distribution of coliforms and sensory score; samples with the lowest total coliforms presented the best sensory scores including overall acceptability, whereas chemical and sensory parameters were completely independent. Lund et al. (2020) investigated heat treatment of camel milk with a view to extending the shelf life; while treatment for 10 min at 100°C gave the best overall keeping quality, a longer time at a lower temperature (30 min at 63°C) gained the highest sensory scores, including in comparison with untreated samples.

### 3.4.2 Yak milk

Yak milk has been a major traditional staple food for yak herders and their families in the highland plains of Tibet, especially consumed by children and the elderly and mostly immediately processed into indigenous products to preserve the nutritional quality (Dong et al., 1999, 2003). Yaks only produce as much milk as goats and production is seasonal. The milk is richer in fat (6–11%) than cows' milk and has more protein, calcium and iron. The milk, which can have a golden tinge, has been described as clean tasting and mild, sweeter and richer than cows' milk; some have described a pleasant sharpness with aromas derived from the particular feed. Ecoregional differences in the flavour of yak milk due to differences in feed (essentially plateau and valley grasslands versus alpine vegetation) can be inferred, but not identified, from the differences in flavour volatile composition found in fermented samples (Fang et al., 2020).

### 3.4.3 Reindeer milk

For the Sami, the indigenous people of northern Scandinavia and the Kola Peninsula of Russia, and the Duhka of north central Mongolia, in the taiga along the border with Russia, reindeer milk is, traditionally, central to their nutrition. Reindeer are one of their main resources and the only source of milk; no other dairy animals can survive the arctic climate. The female reindeer (vaja to the Sami) give only a very low yield of milk, about 200 grams a day (that is, about 100× less than typical bovine production), but the milk has a very high total solids content (30 g 100 $g^{-1}$). At over 20%, the fat content of reindeer milk is 6 times higher than that in cows' milk and 20 times higher than that in donkeys' milk. At 10%, the protein content is three times higher than that in cows' milk, but the lactose content, however, is half that of cows' milk. Overall, reindeer milk is thick, very creamy and sweet.

### 3.4.4 Horse milk

For the peoples of the Central Asian steppes, such as the Kazakh nomad from the Natali grasslands, located in the northern section of the Chinese province of Xinjiang that borders Kazakhstan, horse milk is traditional and cows' milk would be strange and unpalatable. While much mares' milk is fermented to kumis (derived from the simple expedient of storing the fresh milk in leather bags - saba), fresh mares' milk (saumal) is in some ways a more suitable nourishment for human infants than cows' milk, high in vitamin C and iron but low in fat, with levels of lactose and casein that are closer to human breast milk than cows' milk (Kushugulova et al., 2018). It has a sweet, slightly nutty flavour and perceived to be slightly more watery than cows' milk.

### 3.4.5 Donkey milk

If Cleopatra really did bathe daily in the milk from 700 donkeys to preserve the youth and beauty of her skin (Mansueto et al., 2012), then perhaps this was the precedent for the modern trend towards the use of artisanal milk in the most expensive luxury markets. Of more import, donkeys' milk is regarded as one of the better alternatives for infants with food allergies (Carminati and Tidona, 2017; Mansueto et al., 2012) because of its similarity to human milk in terms of lactose, protein and mineral content and its hypoallergenicity and digestibility (higher whey protein, lower casein contents compared with cows' milk). However, the low fat content requires compensating, especially with respect to essential fatty acids. The nutritional and nutraceutical qualities of donkey milk have been reviewed quite recently (Martini et al., 2018).

It is perhaps because of the long use of donkey milk for human nutrition, health and indulgence that donkey domestication bridges nomadic tradition and farming. Donkey milk essentially has the consistency of skimmed cows' milk and has been described as thin, with a pleasantly sweet taste and aroma (though not as sweet as mares' milk), slightly nutty flavour and no persistent aftertaste (Malissiova et al., 2016; Salimie and Fantuz, 2012).

## 4 Human milk flavour profiles

Mothers who are themselves used to cows' milk tend to describe breast milk as being thinner and lighter and having a milder and sweeter taste (which correlates with the compositional differences), and sometimes a soapy smell comes through that is due to lipase activity. Whilst research on commercial milk types has centred on optimising yield or on understanding flavour profiles for differentiation (for example, of PDO milk products), root cause analysis for off-flavours, or adulteration, much focus on human breast milk has been on infant feeding habits and flavour learning, acceptance and preference. Because of the range of food types that humans eat and the range of processing options of food that humans have developed, the breadth and depth of exposure to flavours is orders of magnitude greater for humans than for any other species. Exposure to different flavours begins during gestation via the amniotic fluid, and this is later reflected in the potential for the flavour profile of breast milk to change on a meal-by-meal basis. A recent study by Mastorakou et al. (2019) engaged mothers to keep a food diary with self-assessment of their fore and hind milk; they described the milk odour as neutral, creamy and sweet, the taste as sweet and bitter, and the mouthfeel as thin, watery, smooth and fatty. They also reported differences in fore and hind milk; essentially the former was more dilute and also tended to more reflect bitterness in the diet. Debong and

Loos (2020) very recently reviewed how diet influences the flavour of human milk.

In particular, the consumption of strongly flavoured foods, such as those containing garlic, spices or soy, results in the concentration of derivative flavours rising and declining in breast milk with time, which is noticed and responded to by the infant. Mennella and Beauchamp (1991) showed that garlic odour in breastmilk peaked 2 h after the mothers' ingestion of garlic capsules, and their 3- to 4-month-old babies spent more time feeding than the babies of mothers who had ingested placebo capsules. A similar effect was found with vanilla (Mennella and Beauchamp, 1996), with the infants feeding longer and consuming more milk from mothers who had consumed vanilla prior to feeding than from those who had not. Different flavours reside longer in the breast milk than others with differential transfer from the mother's diet to her milk (Hausner et al., 2008). Further, there is evidence that the early exposure to a flavour in breast milk can be manifested in enhanced acceptance of that flavour in a solid food months later (Mennella et al., 2017).

It has been hypothesised that children's preference for sweet foods over bitter foods is an evolutionary survival mechanism towards high-energy foods and away from potentially harmful foods, but there is also learning through breast milk. The fascinating area of babies learning about flavours in breast milk and recognition of these flavours on transition to solid foods is way beyond the scope of this chapter, but it is salient to the industry, from the perspective of impact for infant formula manufacturers. Infants who were breastfed are more likely than formula-fed infants to accept new foods (Forestell, 2017) and have varied diets as they get older, including earlier transition and acceptance of vegetables (de Wild et al., 2018). In assessing milk-, soy- and hydrolysate-based formula, Mennella and Beauchamp (2002) showed that, compared with children who were fed milk-based formulas, children fed hydrolysate formulas were more likely to prefer sour-flavoured juices, while those fed soy formulas preferred bitter-flavoured juice. It may be that different formulas influence the development of food preferences in different ways.

With storage of human milk in milk banks, one of the key issues is the development of a characteristic fish-like/metallic off-flavour, even at −19°C (Spitzer and Buettner, 2010, 2013), but not at −80°C (Sandgruber et al., 2012). A consistent relationship has been found between microbial load and off-flavour (Novak et al., 2008) and heating (72°C, 2 min) prior to frozen storage appears to mitigate (Spitzer et al., 2010). The defect does not have a parallel in cows' milk, which suggests that the lipolytic and peroxidative processes that appear to be involved (Spitzer et al., 2010) are the result of different and more heat-labile microbial enzymes than those that can compromise cows' milk.

## 5 Flavour defects: origins and impact

The first thing to define in a section on flavour defects in milk is, what is a defect? It is very clear that with any milk from any species, diet has a major impact on the flavour nuances of the milk. So, for cows' milk, for instance, the flavour notes present in milk from grass-fed cows would be perceived as defects by those used to milk from grain-fed cows but not by those used to the milk from grass-fed cows; the same flavour notes would be expected and appreciated. Similarly, seasonal- or transhumance-based variation in the flavour of milk due to changes in feed would be expected. Further, for those used to the flavour of milk from one species, such as horse, may well perceive the flavours of the milk from other species as unacceptable, even unpalatable – as commented above. So, for the purposes of this section, the definition of a flavour defect is a flavour that is outside of the expected range of flavours and flavour intensities for the milk of a species under usual conditions of health, nutrition and welfare. Whether or not this defect is significant depends on its acceptability to the consumer.

Consideration of cows' milk best covers the gamut of defect origins and impact because the contexts can be not only the raw milk but also the introduction of defects through processing of raw milk to other liquid milk products such as skimmed, semi-skimmed, pasteurised, homogenised, ESL, UHT, and so on, and the transportation of milk from farm to factory, storage before processing and in the supply chain to the consumer after product manufacture. Some processing of cows' milk will have relevance to the milk from other species that is similarly processed, such as sheep's milk and goats' milk and so parallels can be drawn. However, for all other species only the defects in raw cows' milk will have parallels because further processing is to gels, semi-solids, solids and powders that are out of scope for this chapter on liquid milk and liquid milk products. With regard to sources of defects in milk from non-bovine species, some commentary has been given in the specific sections above and will not be revisited in this overview section.

It is worth noting that when milk from cows fed total mixed rations (representing conventional feeding) was compared with that from cows fed grass pasture (Bendall, 2001), the majority of the odour compounds identified (66 of 71 found) were the same in the milk from both sources. This indicated that concentration differences between the compounds, rather than differences in identity of the compounds, gave the different background flavour profiles. The exception (Bendall, 2001) was γ-12:2 lactone, present in the milk of cows fed total mixed rations but absent from the milk of grass-fed cows (other lactones were present in both). The paper of Bendall (2001) contains a table that includes aroma descriptions of the compounds extracted using solvent-assisted flavour evaporation and separated using gas chromatography-olfactometry. So, with this in mind, defects can arise either from an unusual imbalance of the usual aroma compounds in milk or from the introduction of unusual aroma compounds. The recent work of Kilcawley et al. (2018) elaborates on factors influencing the flavour of the milk of cows fed on grass and non-grass-based diets.

The origins of defects in raw milk can include extraordinary climate events, physiological status of the animal, feed, water contamination, microbial loading, enzyme loading, heat, light, time and combinations of these. However, while the health and stage of lactation of the cow impacts milk composition, the effects on flavour are relatively minor. Diet has far more significant effects.

The milk samples used for analysis in the Bendall (2001) paper were from bulk milk, that is, the overall milk from a herd rather than milk from individual cows. This raises the point that for large-scale milk production, disruptions to the normal flavour profile of the milk from an individual animal due to, for instance, health (for example, facial eczema; Fraser et al., 2006) or biology (for example, a fishy off-flavour associated with unusual levels of trimethylamine and linked to a genetic mutation; Lunden et al., 2002a,b) will be diluted by the normal bulk milk from the entire herd - so the risk of defects appearing in the milk from a small herd is greater. Dietary causes of milk flavour variability (Bassette et al., 1986; Cadwallader and Singh, 2009; Kilcawley et al., 2018; Shipe et al., 1978), however, affects all animals in any herd size.

In a recent paper by Braggins et al. (2020) surveying the milk collected from New Zealand farms through the 2017/18 season, it was shown that senses testing-based outliers were rare: out of 66 250 samples from approximately 10 000 farms, only 313 outliers were found, and these presented little sensory diversity. Most contained elevated concentrations of butyric acid or ethanol and esters. At relatively low and high concentration, respectively, the butyric acid gave buttery and cheese defects, presumably, although not entirely, resulting from lipolysis. The elevated ethanol and ester concentration were characterised by a fruity defect, tentatively attributed to supplement feeding of silage or baleage. Other aroma defects, such as cowy, fishy, feedy and onion could not be categorically ascribed to particular chemical compounds but were thought to reflect weed taint, especially during periods of peak weed growth relative to grass growth. Braggins et al. (2020) noted that many of the odour outliers were also vat temperature outliers (rather than feed outliers) and presented elevated concentrations of ethanol, ethyl esters, butyric acid, *p*-cresol, diacetyl and Strecker aldehydes (such as 3-methylbutanal that is diagnostic of warm milk and microbial defects; Shipe et al., 1978). Braggins et al. (2020) also noted that their results differed from those found in a survey in Prince Edward Island, Canada, by Mounchili et al. (2004), in which 69% of all off-flavours were derived from the feed. In addition, oxidised flavours were also prevalent in the study conducted by Mounchili et al. (2004), whereas in the Braggins et al. (2020) study, oxidised flavours were absent.

Larsen et al. (2013) showed that changing the ratio of maize silage to lucerne silage fed to cows led to increased concentrations of carotenoids and reduced stale aroma and creamy flavour.

Many studies on feeding systems are geared to determine the usability of available or waste organic material and their conversion efficiency, but not all of them consider the sensory consequences on the milk or consumer acceptance of them. For instance, while a study on the use of an aquaculture pest as a protein source for dairy cows included sensory analysis to show no detectable fishy odours when up to 20% fish silage was used as the protein source (Castillo-Mercado et al., 2020), another on feeding dried olive pomace used only chemical analysis but noted sensory analysis would be needed in further work (Castellani et al., 2019). Other studies are geared towards changing the nutritive profile of cows' milk by supplement feeding. For instance, reduction in the concentration of total saturated fatty acids and atherosclerosis index with rumen-protected tuna oil supplement (Kitessa et al., 2004) and increased conjugated linoleic acid content in milk from cows fed with soybean oil and/or fish oil supplement (Lynch et al., 2005) without significant changes in flavour profiles as assessed by untrained panels have been reported. However, other studies using protected and unprotected fish oil supplements have shown concentration-dependent flavour defects with both supplement types (Lacasse et al., 2002), so the potential for flavour defects to arise needs to be carefully considered.

With respect to raw milk processing, even the most modern dairy factory equipped for aseptic processing is not the sterile environment of pharmaceutical manufacture, and the unit operations can be numerous and complex. Vulnerability to contamination is inevitable in, for example, silos, pipes, plate heat exchangers, evaporators, driers and fillers/packers, hence the need for regular, extensive cleaning-in-place, sterilisation and inspection. Once packed, there are further vulnerabilities to the quality of the liquid milk product in the supply chain to the consumer, including packaging materials, light and storage temperature.

Gram-negative and Gram-positive bacteria are killed by heat treatments at 75°C and at 85–100°C, respectively, so with this level of heat sensitivity they are killed by pasteurisation; defects due to further growth of these microorganisms is due to post heat treatment contamination. Gram-positive spores, however, are resistant to heat treatment and can survive both pasteurisation and inadequate cleaning of processing plant. Further, they can be activated in UHT processing, so initiating lipolytic and proteolytic enzyme activity. Inadequate cleaning can also result in biofilm build up that is therefore also a source of potential defect formation.

The consequences of bacterial activity are irreversible: lactose is metabolised to lactic acid, fats are hydrolysed to fatty acids, proteins are hydrolysed to peptides and amino acids. The reaction products are then the precursors for a vast range of flavour compounds. Heat treatments to kill bacterial contaminants do not solve the problem – bacterial enzymes can be released from the dead cells and if heat stable (such as those from *Pseudomonas* and *Bacillus*), the enzymes (assuming no cofactor dependency) remain active. Hence, raw milk quality needs to be controlled and monitored with timely and appropriate analysis.

Changes in the colour of a liquid milk stream during processing are affected by composition, homogenisation pressure, heat treatment and storage; non-enzymatic Maillard reactions are the key, complex set of reactions between amines and carbonyls (generally reducing sugars) that are responsible. The browning impact on the liquid milk stream is accompanied by flavour changes, whereby cereal, caramel and burnt notes are introduced, depending on the nature and degree of the reactions that have occurred.

Czerny and Schieberle (2007) showed how the overall aroma of UHT-milk packed in polyethylene can change during storage both through adsorption and permeation processes.

Light-induced changes in milk with exposure to near-UV and visible light leading to fat oxidation (Finley and Shipe, 1969) and degradation of sulphur-containing amino acid (Harper and Brown, 1964) are well known but are still being fully characterised (Zardin et al., 2016). Photooxidation off-flavours are readily identified by both trained panellists and consumers (Chapman et al., 2002). Different light sources have different effects, as shown by Brothersen et al. (2016), and different packaging materials mitigate differently (Moyssiadi et al., 2004). However, the use of heat treatments in the preservation of milk also introduces new flavour compounds, and non-thermal technologies are of interest. UV light is effective in reducing coliform growth, and although Engin and Karagül-Yüceer (2012) found new volatiles produced under UV treatment, no major differences in flavour were noted. Recently, Hassan et al. (2020) showed that sensory characteristics of buffalo milk were significantly changed by heat treatment but not by UV treatment.

For over 50 years, the periodic appearance of a flavour problem in milk termed spontaneous oxidised flavour has been reported. Recent work, however, using GC/MS-olfactometry analysis and sensory recombination studies has linked the non-oxidative microbially derived compounds endo-borneol, 2-methylisoborneol and alpha-terpineol to the green, musty and unclean off-flavours (Potts and Petersen, 2018). This illustrates that care should be taken in making assumptions about the origin of flavour defects.

The focus of this section has been human sensory determination of flavour defects. Human consumer recognition of flavour differences between different types of milk and samples of milk of different qualities has commonalities, but the thresholds of detection and acceptance/rejection are individualistic and governed by the relatively limited sense of taste and smell that humans possess. This does not mean that other odours are present that are diagnostic of defects – for instance, sniffer dogs have been trained to identify the pathogen *Staphylococcus aureus* in milk samples (Fischer-Tenhagen et al., 2018).

## 6 Sensory analysis of milk

Given the range of factors that can impact the sensory characteristics of milk, for example, animal species, breed, physiological condition, lactation state, feed and supplements, processing temperature/time combinations, the question of how to go about the sensory analysis of milk then arises. Instrumental analysis can indicate what taste and aroma compounds are present at what concentration in a milk sample, but it cannot determine or predict how the human will perceive the flavour of the milk sample; this requires human sensory analysis (Andrewes et al., 2021).

Just as there is no single milk flavour, there is no single method of sensory analysis; what method is applied wholly depends on the question that is being asked, the number and amount of samples to be analysed and the number of analysts. Just a few of the questions that may be asked are as follows:

i What are the various attributes of the milk sample and what are their intensities?
ii Are these different milk samples different in their flavour profiles, and if so, what are the differences?
iii Will consumers be able to tell the difference between these milk samples, and if so, what will they think of these differences?
iv How acceptable are these different milk samples to consumers and which do they prefer?
v Do any sensory attributes change over time, and what does this mean in terms of shelf-life acceptability?
vi Do any sensory attributes accumulate or change during consumption, and what does this mean in terms of liking and acceptability for a full serve?
vii How will consumers react emotionally to these different milk samples – will they evoke different emotions?

These questions cover qualitative, quantitative and temporal aspects of sensory and consumer analysis, and there are many different methods that can be applied to answer the questions posed; these methods are summarised in Tables 2 and 3 (which are by no means exhaustive) for sensory methods

**Table 2** Sensory methods

| Method | Comments |
|---|---|
| Discrimination tests - overall difference | |
| • Triangle<br>• Paired Comparison (Same/Different)<br>• A not A<br>• Duo Trio<br>• Difference from Control (DFC)<br>• 2-out-of-5 | Discrimination tests can be divided into two groups depending on whether the objective is to determine if an overall sensory difference exists between samples or if there is a specific attribute difference. Apart from the Ranking test that involves multiple samples, discrimination tests compare two samples. |
| Discrimination tests - attribute difference | |
| • Ranking<br>• 2-out-of-5, 2- and 3-AFC<br>• Tetrad | |
| Quality tests | |
| • Difference from Control (DFC)<br>• In/Out<br>• Quality ratings | These methods are used to monitor and maintain the production of consistent products within a defined specification and are usually conducted with specifically trained panel or personnel. |
| Descriptive analysis methods | |
| • Quantitative Descriptive Analysis (QDA)<br>• Consensus Profiling<br>• Spectrum | These methods essentially quantify the intensity of sensory characteristics, providing directional data to meet project objectives. They are often utilised in new product development, benchmarking and sensory forensics. Generally, these methods need a lot of time and resources. |
| Rapid methods | |
| • Check-All-That-Apply (CATA)<br>• Rate-All-That-Apply (RATA)<br>• Sorting<br>• Napping/Mapping (with and without CATA)<br>• Free Choice Profiling (FCP)<br>• Flash Profiling (FP)<br>• Polarised Sensory Positioning (PSP) | In general, rapid methods are used to gain fast and relatively informative general observations. Some do not provide quantitative intensity data; others are difficult to use for unstable products and can be limited to the sample set with more sophisticated statistical analysis required. Some tests, for example, napping, are more cognitive. |
| Temporal methods | |
| • Time Intensity (TI)<br>• Dual Attribute Time Intensity (DATI)<br>• Multiple Ingestion Time Intensity (MITI)<br>• Progressive Profiling (PP)<br>• Temporal Dominance of Sensations (TDS)<br>• Temporal Order of Sensations (TOS)<br>• Temporal CATA (TCATA)<br>• Sequential Profiling<br>• Single Sip versus Multiple Sips | Temporal methods capture the dynamic changes in the sensory profile of a product during (or after) consumption; the initial response to the sample changes with time. These methods can be categorised into the evaluation of the intensity of an attribute or the dominance or presence of attributes over time. |

and consumer methods, respectively. Sensory methods can be categorised in many ways and some fall into more than one classification, but in the tables, the classifications used are based on some of the more common and typical questions posed (for this reason, and to minimise repetition, Table 3 refers back to Table 2). In a new product development context, which methods are applied depends not only on factors such as time, cost, sample availability and panellist availability but also on the breadth and depth of data required by the developers to be comfortable (from a business risk perspective) in making decisions based on what becomes known and what remains unknown.

The sensory-based methodologies presented in Table 2 rely on standardised approaches and alignment of definitions and measurements to provide repeatable, clear understanding of the sensory properties of the samples being evaluated. In contrast, consumer methodologies (Table 3) are designed to understand the variation amongst consumers in their preferences, their descriptions of products and the features that are important to them. A consideration that is important for both types of approach is the preparation of the samples for evaluation. This also includes the temperature of the samples, for example, chilled, ambient, heated (in some cultures, milk is always consumed warm) and the method of delivery of the samples for tasting, for example, sample cup, glass, through a straw/bottle, which can all have significant impact on perception and acceptance. Consumer methodologies to unlock understanding of consumer behaviour are invaluable in helping to define these parameters and to ensure studies are applicable to the product and consumers of interest, and hence of genuine value.

In Table 2, the methods are arranged as discriminatory, quality control (that is, mainly quality assurance in commercial production), descriptive analyses, rapid (but not necessarily cheap) and temporal methods. Only some general commentary is provided since detailed descriptions of the methods, their application and their advantages and disadvantages are out of scope for this chapter; there are some very good books available for more information in this respect (Ares and Varela, 2018; Delarue et al., 2016; Jaeger and MacFie, 2010; Kemp et al., 2009, 2018; Lawless and Heymann, 2010; Meilgaard et al., 2016; Moskowitz et al., 2006; Munoz et al., 2013; O'Sullivan, 2017; Resurreccion, 1998; Stone et al., 2010). In addition, there are various ISO/IDF sensory methods available, such as ISO 22935-1 (ISO, 2009) and ISO 22935-2 (ISO, 2012). With all of the methods, there are considerations that have to be taken into account for that particular method, such as the type of assessor needed, the number of assessors needed to give the statistical power required, sensory fatigue/carry over depending on the type and number of samples, the risk of missing sensory nuances within a sample set that contains large sensory differences and the most appropriate method of data analysis.

**Table 3** Consumer methods

| Method | Comments |
|---|---|
| Acceptance and preference | |
| • Hedonic scaling<br>• Preference testing | Consumers are asked for their overall acceptance of a product along with their liking of specific characteristics. Often run in a central location (CLT) where the environment and product preparation is controlled or in the participant's home or other venue (HUT) where the context of the environment in which the product will be consumed and the preparation and application are also investigated. |
| Discrimination | |
| • Discrimination methods (see Table 2) | A range of discrimination methods can be conducted with consumer groups to determine if differences can be perceived between products or the degree of difference. |
| Product characterisation | |
| • Descriptive analysis methods (see Table 2)<br>• Rapid methods (see Table 2)<br>• Just About Right (JAR)<br>• Ideal scaling<br>• Preference mapping<br>• Emotional response | A range of descriptive tools can be used with consumers to describe the sensory characteristics of products; the attributes are defined by the consumer and are not calibrated between participants (in contrast to trained sensory panel approaches). JAR and Ideal scaling are designed to capture detail on the perception of key sensory attributes and direction for improvement. In preference mapping, a range of measurements can be made to describe the relationship between the product features and an attribute such as hedonic liking, thus defining ideal sensory profiles. Cluster analysis can be used to match the ideal sensory profile to the particular preferences of specific groups of consumers. Measuring emotional responses provides insights into consumer liking and food choice behaviour; there are many different approaches available. |
| Hierarchy of product features/ consumer needs | |
| • Conjoint (Consumer Needs Analysis)<br>• Kano | The hierarchy of needs within product features is determined to understand the relative importance to consumers and viable trade-offs, hence guide new product development. |
| Usage or behavioural understanding | |
| • Qualitative focus groups<br>• Ethnography studies<br>• Video diaries/online communities | Qualitative techniques are used to understand the consumer and both their relationship with products and their behaviour. These studies can help unlock how products are described, the importance of various features and how the products are used in real-life settings. |

What methods are ultimately chosen should be entirely dependent on the questions that are being asked or the problems to resolve. For instance, if the question is one of straightforward consumer acceptance, then there's little gain in a detailed quantitative analysis of the various attributes of the milk. However, such an approach is essential to obtain directional data on the origin of a flavour that the consumer perceives as 'odd' or unacceptable, compared with what they are expecting.

## 7 Conclusions

Milk, as consumed by humans, has core creamy and sweet notes and a smooth, relatively rich mouthfeel; acceptance and rejection of the milk of different species by consumers is strongly influenced by custom and familiarity with the milk. Species-specific flavour differences are overlaid within and between species by flavours that are mainly derived from the feed, which can vary through the year. Consumer familiarity and expectations determine whether these flavour differences are appreciated and desired or regarded as off-flavours or defects.

There are many methods available to determine how consumers perceive the sensory profiles of milk and to quantitatively define those profiles in terms of sensory lexicons and the intensities of the various attributes; the method of choice is determined by defining the problem statement.

The impact on the sensory profile of the milk and the consumer acceptance of that milk should be front of mind in any studies on milk production or on aspects of husbandry, methane production, and so on that may impact the flavour of the milk; it is far too important a nutritional resource to be compromised by consumer rejection.

## 8 Where to look for further information

In critiquing the information available on the sensory profiles of milk, it was clear that there are underlying factors and attitudes towards milk and milk products that are socio-economically, attitudinally and emotionally based. At one extreme, there are groups that are anti-dairy in all its forms; at the other extreme dairy is viewed as a core, essential and desired part of everyday nutrition.

Drinking milk from other animals is thought to be relatively new to humans, emerging with early pastoralisation about 10 millennia ago, with lactase persistence gradually evolving in some populations, but not in others (Ségurel and Bon, 2017). Currently, there are some impressions given that milk is declining in popularity in industrialised nations, with plant-based extracts taking over. But the scientific data do not support this; in Asian countries that do not have a history of consuming dairy products the consumption of milk and milk products is increasing, while in the USA, for instance, the decline in milk drinking is not commensurate with an increase in the consumption of plant-based extracts, but mirrored by an increase in consumption of carbonated beverages and juices (Kearney, 2010; Marshall, 2019).

Other factors relevant to milk consumption in industrial countries are attitudes and beliefs towards raw milk and milk from farms using processes that enable organic certification, both of which acquire consumer perceptions of 'healthier', 'tastier', 'more natural' and 'more nutritious'. There is no robust scientific evidence that raw milk is healthier or tastier than pasteurised milk. With respect to the former, while some studies claim associations between raw milk consumption and reduction in the risk of developing allergies, no cause-effect relationships have been identified and authorities continue to place strict controls over raw milk sales because of pathogen risks (MPI, 2013). Without heat-treatment *Listeria monocytogenes*, pathogenic toxin-producing *Escherichia coli* and *Campylobacter* spp. are potential hazards (Verraes et al., 2015). With respect to taste, studies using blind testing and implicit association testing have shown that consumers do not differentiate the flavour of milk from organically certified and non-organically certified farms; differentiation and preference of milks is significantly affected by perceptions and beliefs related to milk type (Harwood and Drake, 2020). Harwood and Drake (2020) note that their results 'demonstrate a need for greater education related to conventional milk offerings to dispel factors influencing negative implicit bias'. This is very pertinent, given that the information available on dairy ranges from misinformation on the social media to extensive and robust science-based studies. It is unfortunate that the former is more readily accessible to the general public in both its ease of availability and its communication style. In the arena of emotion and belief versus logic and science, the former almost invariably wins.

For sensory and consumer studies, it is essential that robust scientific methods continue to be applied and that these methods are based on human sensory perception, not on instrumental analyses that cannot measure human perception (Andrews et al., 2021). There are many quality books that provide sound advice on method choice, participant numbers and statistics-based analysis to gain robust data; many of these have already been referenced in this chapter (Ares and Verela, 2018; Delarue et al., 2016; Jaeger and McFie, 2010; Kemp et al., 2009, 2018; Lawless and Haymann, 2010; Meilgaard et al., 2016; Munoz et al., 2013; Moskowitz et al., 2006; O'Sullivan, 2017; Resurreccion, 1998; Stone et al., 2010).

Additional general sources include https://www.iso.org/standard/65519.html (ISO 6658:2017. Sensory analysis–Methodology–General guidance) and ASTM International sensory standards https://www.astm.org/Standards/sensory-evaluation-standards.html. For analyses specific to milk and milk products, there are also the IDF/ISO international standards, ISO 22935-1 and ISO 22935-2 referenced in this chapter (ISO, 2009 and ISO, 2012, respectively) and these are extended by https://www.iso.org/standard/41218.html (ISO 22935-3: Milk and milk products – Sensory analysis. Part 3: Guidance on a method for evaluation of compliance with product specifications for sensory properties by scoring).

However, along with the traditional approaches, the field of sensory and consumer science is continually developing and, as it does, it will build greater understanding of consumers' perceptions of, and interactions with, milk and milk products.

## 9 References

Ahmed, A. A. H., Sayed, R. G. and Sayed, M. (2014). Nutritional value and sanitary evaluation of raw Camel's milk. *Emirates J. Food Agric.* 26: 317-326.

Alothman, M., Hogan, S. A., Hennessy, D., Dillon, P., Kilcawley, K. N., O'Donovan, M., Tobin, J., Fenelon, M. A. and O'Callaghan, T. F. (2019). The "grass fed" milk story: understanding the impact of pasture feeding on the composition and quality of bovine milk. *Foods* 8(8): Article 350.

Andersen, C. A., Nielsen, L., Møller, S. and Kidmose, P. (2020). Cortical response to fat taste. *Chem. Senses* 45(4): 283-291.

Andrewes, P., Bullock, S., Turnbull, R. and Coolbear, T. (2021). Chemical instrumental analysis versus human evaluation to measure sensory properties of dairy products: what is fit for purpose? *Int. Dairy J.* 121: 105098

Ares, G. and Varela, P. (2018). *Methods in Consumer Research. New Approaches to Classic Methods*. Kidlington, UK: Woodhead Publishing.

Bassette, R., Fung, D. Y. C., Mantha, V. R. and Marth, E. H. (1986). Off-flavors in milk. *Crit. Rev. Food Sci. Nutr.* 24(1): 1-52.

Bendall, J. G. (2001). Aroma compounds of fresh milk from New Zealand cows fed different diets. *J. Agric. Food Chem.* 49(10): 4825-4832.

Bendall, J. G. and Olney, S. D. (2001). Hept-*cis*-4-enal: analysis and flavour contribution to fresh milk. *Int. Dairy J.* 11(11-12): 855-864.

Braggins, T., Jamieson, P., Luckman, M., Nickless, E., Scott, G., Yang, D. and Andrewes, P. (2020). Variability between farms of New Zealand raw bovine milk flavour: identification and characterisation of odorous outliers. *Int. Dairy J.* 111: Article 104835.

Brennand, C. P., Ha, J. K. and Lindsay, R. C. (1989). Aroma properties and thresholds of some branched-chain and other minor volatile fatty acids occurring in milk fat and meat lipids. *J. Sens. Stud.* 4(2): 105-120.

Brezovecki, A., Cagalj, M., Dermit, Z. F., Mikulec, N., Ljoljic, D. B. and Antunac, N. (2015). Camel milk and milk products. *Mljekarstvo* 65: 81-90.

Brothersen, C., McMahon, D. J., Legako, J. and Martini, S. (2016). Comparison of milk oxidation by exposure to LED and fluorescent light. *J. Dairy Sci.* 99(4): 2537-2544.

Buck, L. and Axel, R. (1991). A novel multigene family may encode odorant receptors: a molecular basis for odor recognition. *Cell* 65(1): 175-187.

Cadwallader, K. R. and Singh, T. K. (2009). Flavours and off-flavours in milk and dairy products. In: McSweeney, P. L. H. and Fox, P. F. (Eds), *Advanced Dairy Chemistry (Vol. 3). Lactose, Water, Salts and Minor Constituents*. New York, NY, USA: Springer Science+Business Media, LLC.

Cais-Sokolińska, D., Majcher, M., Pikul, J., Bielińska, S., Czauderna, M. and Wójtowski, J. (2011). The effect of *Camelina sativa* cake diet supplementation on sensory and volatile profiles of ewe's milk. *Afr. J. Biotechnol.* 10: 7245–7252.

Cais-Sokolinska, D., Wojtowski, J., Pikul, J., Dankow, R., Majcher, M., Teichert, J. and Bagnicka, E. (2019). The effect of unsaturated fatty acid concentration on the aroma profile of goat's milk. *Ann. Anim. Sci.* 19(2): 483–498.

Carminati, D. and Tidona, F. (2017). Nutritional value and potential health benefits of donkey milk. In: Watson, R. R., Collier, R. J. and Preedy, V. R. (Eds), *Nutrients in Dairy and Their Implications on Health and Disease*. San Diego, CA, USA: Academic Press, 407–414.

Castellani, F., Vitali, A., Bernardi, N., Marone, E., Grotta, L. and Martino, G. (2019). Lipolytic volatile compounds in dairy products derived from cows fed with dried olive pomace. *Eur. Food Res. Technol.* 245(1): 159–166.

Castillo-Mercado, R. A., Bucio-Galindo, A., Salinas-Hernandez, R. M., Aranda-Ibanez, E. M., Izquierdo-Reyes, F. and Ramos-Juarez, J. A. (2020). Milk quality of dual-purpose cows supplemented with biological fish silage (*Pterygoplichthys* sp.) as a protein source. *Rev. Colomb Cienc Pecu* 33(4): 252–263.

Catunda, K. L., de Aguiar, E. M., de Góes Neto, P. E., da Silva, J. G., Moreira, J. A., do Nascimento Rangel, A. H. and de Lima Júnior, D. M. (2016). Gross composition, fatty acid profile and sensory characteristics of Saanen goat milk fed with cacti varieties. *Trop. Anim. Health Prod.* 48(6): 1253–1259.

Chapman, K. W., Whited, L. J. and Boor, K. J. (2002). Sensory threshold of light-oxidized flavor defects in milk. *J. Food Sci.* 67(7): 2770–2773.

Chen, J. and Eaton, L. (2012). Multimodal mechanisms of food creaminess sensation. *Food Funct.* 3(12): 1265–1270.

Cheng, N., Barbano, D. M. and Drake, M. A. (2019). Effects of milkfat, casein and serum protein concentrations on sensory properties of milk-based beverages. *J. Dairy Sci.* 102(10): 8670–8690.

Chojnicka-Paszun, A., de Jongh, H. H. J. and de Kruif, C. G. (2012). Sensory perception and lubrication properties of milk: influence of fat content. *Int. Dairy J.* 26(1): 15–22.

Clarke, H. J., Griffin, C., Rai, D. K., O'Callaghan, T. F., O'Sullivan, M. G., Kerry, J. P. and Kilcawley, K. N. (2020). Dietary compounds influencing the sensorial, volatile and phytochemical properties of bovine milk. *Molecules* 25(1): 26.

Codex (1993). CXC 40-1993. Code of Hygienic Practice for Aseptically Processed and Packaged Low-Acid Foods. Rome, Italy: FAO/WHO Codex Alimentarius.

Costa, R. G., Mesquita, I. V. U., Queiroga, R. D. R. D., de Medeiros, A. N., de Carvalho, F. F. R. and Beltrao, E. M. (2008). Sensorial and chemical characteristics of the milk of Moxoto goats fed with manicoba silage. *Braz. J. Anim. Sci.* 37: 694–702.

Croissant, A. E., Washburn, S. P., Dean, L. L. and Drake, M. A. (2007). Chemical properties and consumer perception of fluid milk from conventional and pasture-based production systems. *J. Dairy Sci.* 90(11): 4942–4953.

Czerny, M. and Schieberle, P. (2007). Influence of the polyethylene packaging on the adsorption of odour-active compounds from UHT-milk. *Eur. Food Res. Technol.* 225(2): 215–223.

Debong, M. W. and Loos, H. M. (2020). Diet-induced flavor changes in human milk: update and perspectives. *J. Agric. Food Chem.* 68(38): 10275–10280.
Deeth, H. (2017). Optimum thermal processing for extended shelf-life (ESL) milk. *Foods* 6(11): Article 102.
Delarue, J., Lawler, J. B. and Rogeaux, M. (2016). *Rapid Sensory Profiling Techniques and Related Methods. Applications in New Product Development and Consumer Research*. Cambridge, UK: Woodhead Publishing.
de Wild, V. W. T., Jager, G., Olsen, A., Costarelli, V., Boer, E. and Zeinstra, G. G. (2018). Breast-feeding duration and child eating characteristics in relation to later vegetable intake in 2–6-year-old children in ten studies throughout Europe. *Public Health Nutr.* 21(12): 2320–2328.
Dong, S. K., Long, R. J. and Dikken, G. D. (1999). Yak farming on the Qinghai-Tibetan Plateau of China. *NCTR Livest. News* 1–2: 10–14.
Dong, S. K., Long, R. J. and Kang, M. Y. (2003). Milk and milk processing: traditional technologies in the yak farming system of the Qinghai-Tibetan Plateau, China. *Int. J. Dairy Technol.* 56: 86–93.
Dunkel, A., Steinhaus, M., Kotthoff, M., Nowak, B., Krautwurst, D., Schieberle, P. and Hofmann, T. (2014). Nature's chemical signatures in human olfaction: a foodborne perspective for future biotechnology. *Angew. Chem. Int Ed Engl* 53(28): 7124 -7143.
Eknæs, M., Havrevoll, Ø., Volden, H. and Hove, K. (2009). Fat content, fatty acid profile and off-flavours in goats milk. Effects of feed concentrates with different fat sources during the grazing season. *Anim. Feed Sci. Technol.* 152(1–2): 112–122.
Eknæs, M. and Skeie, S. (2006). Effect of different level of roughage availability and contrast levels of concentrate supplementation on flavour of goat milk. *Small Rumin. Res.* 66(1–3): 32–43.
Engin, B. and Karagül-Yüceer, Y. (2012). Effects of ultraviolet light and ultrasound on microbial quality and aroma-active components of milk. *J. Sci. Food Agric.* 92(6): 1245–1252.
Faccia, M., D'Alessandro, A. G., Summer, A. and Hailu, Y. (2020). Milk products from minor species: a review. *Animals* 10(8): Article 1260.
Fang, X., Guo, L. W., Chen, H., Ke, W. C., Guo, W., Guo, X. S. and Zhang, Y. (2020). Characteristics of volatile flavor components in traditional fermented yak milk produced in different ecoregions of the Qinghai-Tibetan Plateau. *J. Dairy Sci.* 103(1): 191–200.
FAO (2020). Gateway to dairy production and products. Available at: http://www.fao.org /dairy-production-products/production/dairy-animals/buffaloes/en/. Rome, Italy: Food and Agriculture Organisation.
FAO, IDF and IFCN (2014). *World Mapping of Animal Feeding Systems in the Dairy Sector*. [Food and Agriculture Organization of the United Nations. Food and Agriculture Organization, International Dairy Federation (FIL-IDF; Brussels, Belgium), IFCN Dairy Research Network (Kiel, Germany)]. Rome, Italy: Food and Agriculture Organization.
Fedele, V., Rubino, R., Claps, S., Sepe, L. and Morone, G. (2005). Seasonal evolution of volatile compounds content and aromatic profile in milk and cheese from grazing goat. *Small Rumin. Res.* 59(2–3): 273–279.
Finley, J. W. and Shipe, W. F. (1969). Isolation of a flavor producing fraction from light exposed milk. *J. Dairy Sci.* 54(1): 15–20.

Fischer-Tenhagen, C., Theby, V., Kromker, V. and Heuwieser, W. (2018). Detecting *Staphylococcus aureus* in milk from dairy cows using sniffer dogs. *J. Dairy Sci.* 101(5): 4317–4324.

Forestell, C. A. (2017). Flavor perception and preference development in human infants. *Ann. Nutr. Metab.* 70(Suppl. 3): 17–25.

Fraser, K., Lane, G. A., Morris, C. A. and Cullen, N. G. (2006). Effects of facial eczema on indole flavour compounds in dairy cows. *Proc. N.Z. Soc. Anim. Prod.* 66: 315 -319.

Ha, J. K. and Lindsay, R. C. (1991). Contributions of cow, sheep and goats milk to characterizing branched chain fatty acid and phenolic flavours in varietal cheeses. *J. Dairy Sci.* 74: 3267–3274.

Ha, J. K. and Lindsay, R. C. (1993). Release of volatile branched-chain and other fatty acids from ruminant milk fat by various lipase. *J. Dairy Sci.* 76(3): 677–690.

Harper, W. J. and Brown, D. H. (1964). Light induced degradation of cysteine. *J. Dairy Sci.* 47(5): 548–549.

Harwood, W. S. and Drake, M. A. (2020). The influence of automatic associations on preference for milk type. *J. Dairy Sci.* 103(12): 11218–11227.

Hassan, A., Randhawa, M. A., Abbas, S. Q. and Zia, M. A. (2020). Exploring the effect of UV treatments on the fatty acid profile of the buffalo milk. *Pak. J. Agric. Sci.* 57: 1165–1170.

Hausner, H., Bredie, W. L. P., Mølgaard, C., Petersen, M. A. and Møller, P. (2008). Differential transfer of dietary flavour compounds into human breast milk. *Physiol. Behav.* 95(1-2): 118–124.

Heilig, A., Çelik, A. and Hinrichs, J. (2008). Suitability of Dahlem Cashmere goat milk towards pasteurisation, ultrapasteurisation and UHT-heating with regard to sensory properties and storage stability. *Small Rumin. Res.* 78(1-3): 152–161.

Inglingstad, R. A., Skeie, S., Vegarud, G. E., Devold, T. G., Chilliard, Y. and Eknaes, M. (2017). Feeding a concentrate rich in rapeseed oil improves fatty acid composition and flavor in Norwegian goat milk. *J. Dairy Sci.* 100(9): 7088–7105.

ISO (2009). *Milk and Milk Products–Sensory Analysis–Part 1: General Guidance for the Recruitment, Selection, Training and Monitoring of Assessors (ISO 22935-1:2009).* Geneva, Switzerland: International Standardisation Organisation.

ISO (2012). *Milk and Milk Products–Sensory Analysis–Part 2: Recommended Methods for Sensory Evaluation (ISO 22935-2:2012).* Geneva, Switzerland: International Standardisation Organisation.

Jaeger, S. R. and MacFie, H. (2010). *Consumer Driven Innovation in Food and Personal Care Products.* Cambridge, UK: Woodhead Publishing Ltd.

Jervis, S. M., Gerard, P., Drake, S., Lopetcharat, K. and Drake, M. A. (2014). The perception of creaminess in sour cream. *J. Sens. Stud.* 29(4): 248–257.

Kaffarnik, S., Kayademir, Y., Heid, C. and Vetter, W. (2014). Concentrations of volatile 4-alkyl-branched fatty acids in sheep and goat milk and dairy products. *J. Food Sci.* 79(11): C2209-C2214.

Kearney, J. (2010). Food consumption trends and drivers. *Phil. Trans. R. Soc. B.* 365: 2793–2807.

Kemp, S. E., Hollowood, T. and Hort, J. (2009). *Sensory Evaluation. A Practical Handbook.* Chichester, UK: Wiley Blackwell.

Kemp, S. E., Hort, J. and Hollowood, T. (2018). *Descriptive Analysis in Sensory Evaluation.* Chichester, UK: Wiley Blackwell.

Khan, M. A. S., Islam, M. N. and Siddiki, M. S. R. (2007). Physical and chemical composition of swamp and water buffalo milk: a comparative study. *Ital. J. Anim. Sci.* 6(sup2): 1067–1070.

Kilcawley, K. N., Faulkner, H., Clarke, H. J., O'Sullivan, M. G. and Kerry, J. P. (2018). Factors influencing the flavour of bovine milk and cheese from grass based versus non-grass based milk production systems. *Foods* 7(3): Article 37.

Kitessa, S. M., Gulati, S. K., Simos, G. C., Ashes, J. R., Scott, T. W., Fleck, E. and Wynn, P. C. (2004). Supplementation of grazing dairy cows with rumen-protected tuna oil enriches milk fat with n-3 fatty acids without affecting milk production or sensory characteristics. *Br. J. Nutr.* 91(2): 271–278.

Kushugulova, A., Kozhakhmetov, S., Sattybayeva, R., Nurgozhina, A., Ziyat, A., Yadav, H. and Marotta, F. (2018). Mare's milk as a prospective functional product. *Funct. Foods Health Dis.* 8(11): 548–554.

Lacasse, P., Kennelly, J. J., Delbecchi, L. and Ahnadi, C. E. (2002). Addition of protected and unprotected fish oil to diets for dairy cows. I. Effects on the yield, composition and taste of milk. *J. Dairy Res.* 69(4): 511–520.

Lara-Zamudio, Y., Villalpando, S., Shamah-Levy, T., Mundo-Rosas, V., Contreras-Manzano, A. and Lamadrid-Figueroa, H. (2013). Defatted milk is preferred by Mexican school-age children over whole milk in a sensorial study. *Ann. Nutr. Metab.* 62(3): 214–222.

Larsen, M. K., Kidmose, U., Kristensen, T., Beaumont, P. and Mortensen, G. (2013). Chemical composition and sensory quality of bovine milk as affected by type of forage and proportion of concentrate in the feed ration. *J. Sci. Food Agric.* 93(1): 93–99.

Lawless, H. T. and Heymann, H. (2010). *Sensory Evaluation of Food*. New York, NY, USA: Springer Science+Business Media.

Lee, B. and Weston, R. P. (1917). *Paddy McGinty's Goat*. London, UK: Francis, Day and Hunter. Available at: https://en.wikisource.org/wiki/Paddy_McGinty%27s_Goat.

Le Quere, J. L., Septier, C., Demaizieres, D. and Salles, C. (1996). Identification and sensory evaluation of the character impact compounds of goat cheese flavor. In: Taylor, A. J. and Mottram, D. S. (Eds), *Flavor Science: Recent Developments*. Cambridge, UK: The Royal Society of Chemistry, 325–330.

Logan, D. W., Bruner, L. J., Webb, W. R., Cutforth, T., Ngai, J. and Stowers, L. (2012). Learned recognition of maternal signature odors mediates the first suckling episode in mice. *Curr. Biol.* 22(21): 1998–2007.

Loos, H. M., Reger, D. and Schaal, B. (2019). The odour of human milk: its chemical variability and detection by newborns. *Physiol. Behav.* 199: 88–99.

Lund, A. K., Shah, A. H., Jatoi, A. S., Khaskheli, G. B., Malhi, M. C. and Khaskheli, A. A. (2020). Effect of heating on shelf life and sensory characteristics of camel milk. *Pure Appl. Biol.* 9(1): 74–79.

Lunden, A., Gustafsson, V., Imhof, M., Gauch, R. and Bosset, J. O. (2002a). High trimethylamine concentration in milk from cows on standard diets is expressed as fishy off-flavour. *J. Dairy Res.* 69(3): 383–390.

Lunden, A., Marklund, S., Gustafsson, V. and Andersson, L. (2002b). A nonsense mutation in the FMO3 gene underlies fishy off-flavor in cow's milk. *Gen. Res.* 12(12): 1885–1888.

Lynch, J. M., Lock, A. L., Dwyer, D. A., Noorbakhsh, R., Barbano, D. M. and Bauman, D. E. (2005). Flavor and stability of pasteurized milk with elevated levels of conjugated linoleic acid and vaccenic acid. *J. Dairy Sci.* 88(2): 489–498.

Malissiova, E., Arsenos, G., Papademas, P., Fletouris, D., Manouras, A., Aspri, M., Nikolopoulou, A., Giannopoulou, A. and Arvanitoyannis, I. S. (2016). Assessment of donkey milk chemical, microbiological and sensory attributes in Greece and Cyprus. *Int. J. Dairy Technol.* 69(1): 143–146.

Mansueto, P., Iacono, G., Seidita, A., D'Alcamo, A., Iacono, S. and Carrocio, A. (2012). Ass's milk in allergy to cow's milk protein: a review. *J. Food All* 1: 181–191.

Markiewicz-Kęszycka, M., Czyżak-Runowska, G., Lipińska, P. and Wójtowski, J. (2013). Fatty acid profile of milk–a review. *Bull. Vet. Inst. Pulawy* 57(2): 135–139.

Marshall, M. (2019). Why humans have evolved to drink milk. *BBC Future* 2019: Article 0218 (https://www.bbc.com/future/article/20190218-when-did-humans-start-drinking-cows-milk).

Martini, M., Altomonte, I., Licitra, R. and Salari, F. (2018). Nutritional and nutraceutical quality of donkey milk. *J. Eq. Vet. Sci.* 65: 33-37.

Mastorakou, D., Ruark, A., Weenen, H., Stahl, B. and Stieger, M. (2019). Sensory characteristics of human milk: association between mothers' diet and milk for bitter taste. *J. Dairy Sci.* 102(2): 1116–1130.

McCarthy, K. S., Lopetcharat, K. and Drake, M. A. (2017). Milk fat threshold determination and the effect of milk fat content on consumer preference for fluid milk. *J. Dairy Sci.* 100(3): 1702–1711.

Medhammar, E., Wijesinha-Bettoni, R., Stadlmayr, B., Nilsson, E., Charrondiere, U. R. and Burlingame, B. (2012). Composition of milk from minor dairy animals and buffalo breeds: a biodiversity perspective. *J. Sci. Food Agric.* 92(3): 445–474.

Meilgaard, M. C., Civille, G. V. and Carr, B. T. (2016). *Sensory Evaluation Techniques* (5th edn). Boca Raton, FL, USA: CRC Press, Taylor and Francis Group.

Mennella, J. A. and Beauchamp, G. K. (1991). Maternal diet alters the sensory qualities of human milk and the nursling's behaviour. *Pediatrics* 19: 13–19.

Mennella, J. A. and Beauchamp, G. K. (1996). The human infants' response to vanilla flavors in mother's milk and formula. *Behav. Devel.* 88: 737–744.

Mennella, J. A. and Beauchamp, G. K. (2002). Flavor experiences during formula feeding are related to preferences during childhood. *Early Hum. Dev.* 68(2): 71–82.

Mennella, J. A., Daniels, L. M. and Reiter, A. R. (2017). Learning to like vegetables during breastfeeding: a randomized clinical trial of lactating mothers and infants. *Am. J. Clin. Nutr.* 106(1): 67–76.

Moio, L., Langois, D., Etievant, P. X. and Addeo, F. (1993). Powerful odorants in water buffalo and bovine mozzarella cheese by use of extract dilution sniffing analysis. *Ital. J. Food Sci.* 3: 227–237.

Moskowitz, H. R., Beckley, J. H. and Resurreccion, A. V. A. (2006). *Sensory and Consumer Research in Food Product Design and Development*. Ames, IO: Blackwell Publishing Professional.

Mounchili, A., Wichtela, J. J., Dohoo, I. R., Keefe, G. P. and Halliday, L. J. (2004). Risk factors for milk off-flavours in dairy herds from Prince Edward Island, Canada. *Prev. Vet. Med.* 64(2–4): 133–145.

Moyssiadi, T., Badeka, A., Kondyli, E., Vakirtzi, T., Savvaidis, I. and Kontominas, M. G. (2004). Effect of light transmittance and oxygen permeability of various packaging materials on keeping quality of low fat pasteurized milk: chemical and sensorial aspects. *Int. Dairy J.* 14(5): 429–436.

MPI. (2013). An assessment of the effects of pasteurisation on claimed nutrition and health benefits of raw milk MPI Technical Paper No: 2014/13. Wellington, New

Zealand: Ministry for Primary Industries. https://www.mpi.govt.nz/dmsdocument/1119-An-assessment-of-the-effects-of-pasteurisation-on-claimed-nutrition-and-health-benefits-of-raw-milk.

Muhammad, G., Rashid, I., Firyal, S. and Saqib, M. (2017). Trouble shooting off-flavour (bad odor) and bad taste in milk. *Buff Bull.* 36: 15–19.

Munoz, A. M., Civille, G. V. and Carr, T. (2013). *Sensory Evaluation in Quality Control.* New York, NY, USA: Springer Science+Business Media.

Murtaza, M. A., Pandya, A. J. and Khan, M. M. (2017). Buffalo milk utilization for dairy products. In: Park, Y. W., Haenlein, G. F. W. and Wendorff, W. L. (Eds), *Handbook of Milk of Non-Bovine Mammals* (2nd edn). Hoboken, NJ, USA: John Wiley and Sons, Ltd., 284–342.

Navarrete-Rodriguez, E. M., Rios-Villalobos, L. A., Alcocer-Arreguin, C. R., Del-Rio-Navarro, B. E., Del Rio-Chivardi, J. M., Saucedo-Ramirez, O. J., Sienra-Monge, J. J. L. and Frias, R. V. (2018). Cross-over clinical trial for evaluating the safety of camel's milk intake in patients who are allergic to cow's milk protein. *Allerg. Immunopath. (Madr)* 46(2): 149–154.

Novak, F. R., Junqueira, A. R., Dias, Mde S. and Almeida, J. A. G. (2008). Sensorial analysis of expressed human milk and its microbial load. *J. Pediatr. (Rio J)* 84(2): 181–184.

O'Connell, J. E. and Fox, P. F. (2001). Significance and applications of phenolic compounds in the production and quality of milk and dairy products: a review. *Int. Dairy J.* 11(3): 103–120.

OECD-FAO (2021). Dairy and dairy products. In Chapter 7: OECD-FAO Agricultural Outlook 2020–2029. Paris, France: Organisation for Economic Cooperation and Development-Food and Agriculture Organisation. Available at: https://www.oecd-ilibrary.org/sites/aa3fa6a0-en/index.html?itemId=/content/component/aa3fa6a0-en.

O'Sullivan, M. G. (2017). *A Handbook for Sensory and Consumer-Driven New Product Development. Innovative Technologies for the Food and Beverage Industry.* Kidlington, UK: Woodhead Publishing.

Pangborn, R. M., Bos, K. E. O. and Stern, J. S. (1985). Dietary fat intake and taste responses to fat in milk by under-, normal, and overweight women. *Appetite* 6(1): 25–40.

Pangborn, R. M. and Dunkley, W. L. (1964a). Sensory discrimination of fat and solids-not-fat in milk. *J. Dairy Sci.* 47(7): 719–726.

Pangborn, R. M. and Dunkley, W. L. (1964b). Difference-preference evaluation of milk by trained judges. *J. Dairy Sci.* 47: 414–420.

Park, Y. W., Juárez, M., Ramos, M. and Haenlein, G. F. W. (2007). Physico-chemical characteristics of goat and sheep milk. *Small Rumin. Res.* 68(1–2): 88–113.

Pereira, R. N., Martins, R. C. and Vincente, A. A. (2008). Goat milk free fatty acid characterization during conventional and ohmic heating pasteurization. *J. Dairy Sci.* 91(8): 2925–2937.

Phillips, L. G. and Barbano, D. M. (1997). The influence of fat substitutes based on protein and titanium dioxide on the sensory properties of low fat milks. *J. Dairy Sci.* 80(11): 2726–2731.

Phillips, L. G., McGiff, M. L., Barbano, D. M. and Lawless, H. T. (1995). The influence of fat on the sensory properties, viscosity and color of low-fat milks. *J. Dairy Sci.* 78(6): 1258–1266.

Poinot, P., Arvisenet, G., Ledauphin, J., Gaillard, J.-L. and Prost, C. (2013). How can aroma-related cross-modal interactions be analysed? A review of current methodologies. *Food Qual. Pref.* 28(1): 304–316.

Polowsky, P., Coudé, B., Jiménez-Maroto, L. A., Johnson, M. and Park, Y. W. (2017). Flavor and sensory characteristics of non-bovine species milk and their dairy products. In: Park, Y. W., Haenlein, G. F. W. and Wendorff, W. L. (Eds), *Handbook of Milk of Non-Bovine Mammals* (2nd edn.). Hoboken, NJ, USA: John Wiley and Sons, Ltd., 595-623.

Potts, D. M. and Petersen, D. G. (2018). Identification of objectionable flavors in purported spontaneous oxidized flavor bovine milk. *J. Dairy Sci.* 101(12): 10877-10885.

Prinz, J. F. and De Wijk, R. (2004). The role of oral processing in flavour perception. In: Taylor, A. J. and Roberts, D. D. (Eds), *Flavour Perception*. Oxford, UK: Blackwell Publishing, 39-56.

Queiroga, R. C. R. E., Fernandes, M. F., Medeiros, A. N., Costa, R. G., Oliveira, C. J. B., Bomfim, M. A. D. and Guerra, I. C. D. (2009). Physicochemical and sensory effects of cotton seed and sunflower oil supplementation on Moxoto goat milk. *Small Rumin. Res.* 82(1): 58-61.

Resurreccion, A. V. A. (1998). *Consumer Sensory Testing for Product Development*. Gaithersburg, MD, USA: Aspen Publishers, Inc.

Rysstad, G. and Kolstad, J. (2006). Extended shelf life milk–advances in technology. *Int. J. Dairy Tech.* 59(2): 85-96.

Salari, F., Altomonte, I., Ribeiro, N. L., Ribeiro, M. N., Bozzi, R. and Martini, M. (2016). Effects of season on the quality of Garfagnina goat milk. *Ital. J. Anim. Sci.* 15(4): 568-575.

Salimie, E. and Fantuz, F. (2012). Equid milk for human consumption. *Int. Dairy J.* 24(2): 130-142.

Sandgruber, S., Much, D., Amann-Gassner, U., Hauner, H. and Buettner, A. (2012). Sensory and molecular characterisation of the protective effect of storage at -80 °C on the odour profiles of human milk. *Food Chem.* 130(2): 236-242.

Schaal, B., Coureaud, G., Langlois, D., Giniès, C., Sémon, E. and Perrier, G. (2003). Chemical and behavioural characterization of the rabbit mammary pheromone. *Nature* 424(6944): 68-72.

Ségurel, L. and Bon, C. (2017). On the evolution of lactase persistence in humans. *Ann. Rev. Genom. Human Gen.* 18: 297-319.

Sharma, C. and Singh, C. (2014). Therapeutic value of camel milk–a review. *Advanced J. Pharm. Life Sci. Res.* 2: 7-13.

Shipe, W. F., Bassette, R., Deane, D. D., Dunkley, W. L., Hammond, E. G., Harper, W. J., Kleyn, D. H., Morgan, M. E., Nelson, J. H. and Scanlan, R. A. (1978). Off flavors of milk: nomenclature, standards, and bibliography. *J. Dairy Sci.* 61(7): 855-869.

Siefarth, C. and Buettner, A. (2014). The aroma of goat milk: seasonal effects and changes through heat treatment. *J. Agric. Food Chem.* 62(49): 11805-11817.

Silanikove, N., Leitner, G., Merin, U. and Prosser, C. G. (2010). Recent advances in exploiting goat's milk: quality, safety and production aspects. *Small Rumin. Res.* 89(2-3): 110-124.

Skjevdal, T. (1979). Flavour of goat's milk: a review of studies on the sources of its variations. *Livest. Prod. Sci.* 6(4): 397-405.

Spitzer, J. and Buettner, A. (2010). Characterization of aroma changes in human milk during storage at -19°C. *Food Chem.* 120(1): 240-246.

Spitzer, J. and Buettner, A. (2013). Monitoring aroma changes during human milk storage at -19°C by quantification experiments. *Food Res. Int.* 51(1): 250-256.

Spitzer, J., Doucet, S. and Buettner, A. (2010). The influence of storage conditions on flavour changes in human milk. *Food Qual. Pref.* 21(8): 998-1007.

Statistica (2021). *Cow Milk Production Worldwide from 2015 to 2019*. Available at: https://www.statista.com/statistics/263952/production-of-milk-worldwide/.

Stone, H., Bleibaum, R. N. and Thomas, H. A. (2010). *Sensory Evaluation Practices* (4th edn.). San Diego, CA, USA: Academic Press.

Verraes, C., Uyttendaele, M., Clinquart, A., Daube, G., Sindic, M., Berkvens, D., and Herman, L. (2015). Microbiological safety and quality aspects of the short supply chain: SWOT analysis of the Belgian case study. *Brit. Food J.* 117(9): 2250–2264.

Whetstine, M. E. C. and Drake, M. A. (2006). Flavor characteristics of goat milk and other minor species milk products. In: Park, Y. W. and Haenlein, G. F. W. (Eds), *Handbook of Milk of Non-Bovine Mammals*. Oxford, UK: Blackwell Publishing, 107–120.

Yadav, A. K., Kumar, R., Priyadarshini, L. and Singh, J. (2015). Composition and medicinal properties of camel milk: a review. *Asian J. Dairy Food Res.* 34(2): 83–91.

Žan, M., Stibilj, V. and Rogelj, I. (2006). Milk fatty acid composition of goats grazing on alpine pasture. *Small Rumin. Res.* 64(1–2): 45–52.

Zardin, E., Silcock, P., Siefarth, C., Bremer, P. J. and Beauchamp, J. (2016). Dynamic changes in the volatiles and sensory properties of chilled milk during exposure to light. *Int. Dairy J.* 62: 35–38.

Zhang, X. M., Ai, N. S., Wang, J., Tong, L. J., Zheng, F. P. and Sun, B. G. (2016). Lipase-catalyzed modification of the flavor profiles in recombined skim milk products by enriching the volatile components. *J. Dairy Sci.* 99(11): 8665–8679.

# Chapter 18

# Microbial quality and spoilage of raw cow's milk

*Maria Kazou, Alkmini Gavriil, Chryssi Kounenidaki and Effie Tsakalidou, Agricultural University of Athens, Greece*



## 1 Introduction

### *1.1 The microbial diversity of raw cow's milk*

Milk is among the most valuable natural foods and has been a basic component of the human diet since early history. It is highly nutritious containing proteins, carbohydrates, lipids, vitamins, minerals, and amino acids all at a near-neutral pH and high water activity, providing an excellent growth substrate for both beneficial and harmful microorganisms. According to the European Union (EU) legislation, raw milk is defined as the milk produced by the secretion of the mammary gland of farmed animals that has not been heated to more than 40°C or undergone any treatment that has an equivalent impact. Moreover, the microbial load of raw cow's milk on plate count agar (30°C) should not exceed 100 000 CFU $mL^{-1}$, whereas, for somatic cell count (SCC), the upper limit goes to 400 000 cell $mL^{-1}$ (Regulation 853/2004/EC).

Milk in a healthy mammary gland is considered to be sterile, but during lactation and milking, it is contaminated by several groups of microorganisms deriving from a variety of sources. The first source of contamination is the teat apex, followed by udder surface, farm personnel, milking equipment, and storage tanks. Additional factors that affect the raw milk microbiota are the type of animal feed, water, surrounding air, soil, and bedding material. Furthermore, studies suggest that both altitude and climate conditions may affect the raw milk microbiota (Tilocca et al., 2020).

http://dx.doi.org/10.19103/AS.2022.0099.18

Milk obtained aseptically or by complying with good hygiene practices has usually low bacterial counts, often less than 4 log CFU mL$^{-1}$ (Dahlberg et al., 2019). On the other hand, in the case of mastitis, which is an inflammatory reaction of the udder tissue due to physical trauma or microbial infections, bacterial counts reach approximately 6–8 log CFU mL$^{-1}$ (Rainard, 2017). It should be stressed, however, that the belief that the microorganisms found in raw milk derive only from external contamination has been recently challenged. Young et al. (2015) analyzed the bacterial microbiota of blood, milk, and fecal samples collected from healthy lactating cows to see whether translocation of intestinal bacteria to the mammary secretions is possible. However, although this hypothesis has been documented in women and mice (Perez et al., 2007), an entero-mammary route in cows was not confirmed based on the results of Young et al. (2015). Therefore, further research is needed to clarify the mechanisms by which microorganisms could be transferred from other body sites to the mammary gland of ruminants (Addis et al., 2016).

The indigenous milk microbiota reflects the lifestyle of the animal and the environmental conditions under which it is produced, as mentioned earlier, and can be roughly divided into three main categories (Olajide and LaPointe, 2020): (i) the so-called pro-technological microorganisms, which mostly comprise lactic acid bacteria (LAB) that are involved in the transformation of milk into dairy products, such as cheese, yogurt, and fermented milk. These products have new and improved sensorial characteristics and nutritional value compared to raw milk. Among LAB, species of the genera *Lactococcus*, *Streptococcus*, *Lactobacillus*, *Pediococcus*, and *Leuconostoc* are the protagonists in the production of dairy fermented foods. It should be noted though that some mold species present in raw milk are also used in dairy fermentations. For example, *Penicillium camemberti*, *Penicillium roqueforti*, and *Penicillium glaucum* are significant members of the pro-technological microbiota of Camembert, Roquefort and Gorgonzola cheeses, respectively, (ii) the spoilage microorganisms, which can cause milk putrefaction revealed as off-flavors and odors, gas formation, as well as changes in texture and appearance (Bautista, 2014; Samaržija et al., 2012; Postollec et al., 2012). The main milk spoilage microorganisms are psychrotrophic microbes, for example, species of the genera *Pseudomonas*, *Acinetobacter*, *Aeromonas*, *Serratia*, and *Hafnia*, with most of them being eliminated during pasteurization; however, some species, such as *Pseudomonas fluorescens* and *Pseudomonas fragi*, produce proteolytic and lipolytic extracellular enzymes, which are heat-stable and can cause post-pasteurization spoilage. Moreover, some species or strains of the genera *Bacillus* and *Clostridium* can survive pasteurization and grow at refrigeration temperatures, causing spoilage problems as well and (iii) the pathogenic microorganisms, which are responsible for milk-borne diseases. Diseases, such as tuberculosis, brucellosis, and typhoid fever, result from the consumption of raw milk or dairy products made with milk that was insufficiently pasteurized or was poorly handled causing post-processing contamination. Bacterial pathogens, such as *Bacillus cereus*, *Listeria monocytogenes*, *Yersinia enterocolitica*, *Salmonella* spp., *Escherichia coli* O157:H7, and *Campylobacter jejuni*, are still of concern in raw milk and products. Among the pathogenic microorganisms present in raw milk, mold species of the genera *Aspergillus*, *Fusarium*, and *Penicillium* may produce health-hazardous mycotoxins. Figure 1 illustrates a representative overview of the microbial community of raw cow's milk.

### 1.2 The evolution of microbial identification techniques over the years: the impact of molecular tools on the detection of milk microbiome

Due to the significant importance of cow's milk worldwide, raw cow's milk microbiota has been thoroughly reviewed (Quigley et al., 2011, 2013a; Addis et al., 2016; Rainard, 2017; Derakhshani et al., 2018a; Yuan et al., 2019; Awasti and Anand, 2020; Fusco et al., 2020; Oikonomou et al., 2020; Owusu-Kwarteng et al., 2020; Parente et al., 2020) and associated to SCC (Oikonomou et al., 2014; Alekish, 2015; Nagasawa et al., 2019), mastitis and culture-negative mastitis samples (Kuehn et al., 2013; Metzger et al., 2018; Hoque et al., 2019; Taponen et al., 2019; Vasquez et al., 2019), intra-mammary infection (Falentin et al., 2016; Ganda et al., 2016, 2017; Zhang et al., 2016), lactation stage (Zhang et al., 2016; Petróczki et al., 2019), farm environment (Doyle et al., 2017; Skeie et al., 2019; Wu et al., 2019; Quintana et al., 2020), cow's genotype (Derakhshani et al., 2018b), cow's herd (Cremonesi et al., 2018; Petróczki et al., 2019; Albonico et al., 2020), weather conditions (Li et al., 2018), as well as storage temperature and duration (O'Connell et al., 2016).

For decades, the microbial ecology of milk and dairy products has been based on the study of microbial isolates through conventional microbiological techniques. These techniques rely on the detection, enumeration, and isolation of various microbial groups by using selective growth media and culture conditions. Several selective media have been developed, while some of them have been established as international standards through collaborative work between the International Organization of Standardization (ISO) and the International Dairy Federation (IDF) (IS017792/IDF180 2006; IS020128/IDF192 2006; ISO29981/IDF220 2010).

For the analysis and selective enumeration of pathogenic/spoilage bacteria and yeasts, such as *Listeria*, *Salmonella*, *B. cereus*, *E. coli* and other enterobacteria, and Candida, *Aspergillus niger* and *Saccharomyces cerevisiae*, respectively, chromogenic media have been developed, presenting high specificity and selectivity (Perry, 2017; Quiros et al., 2005; Zhao et al., 2016).

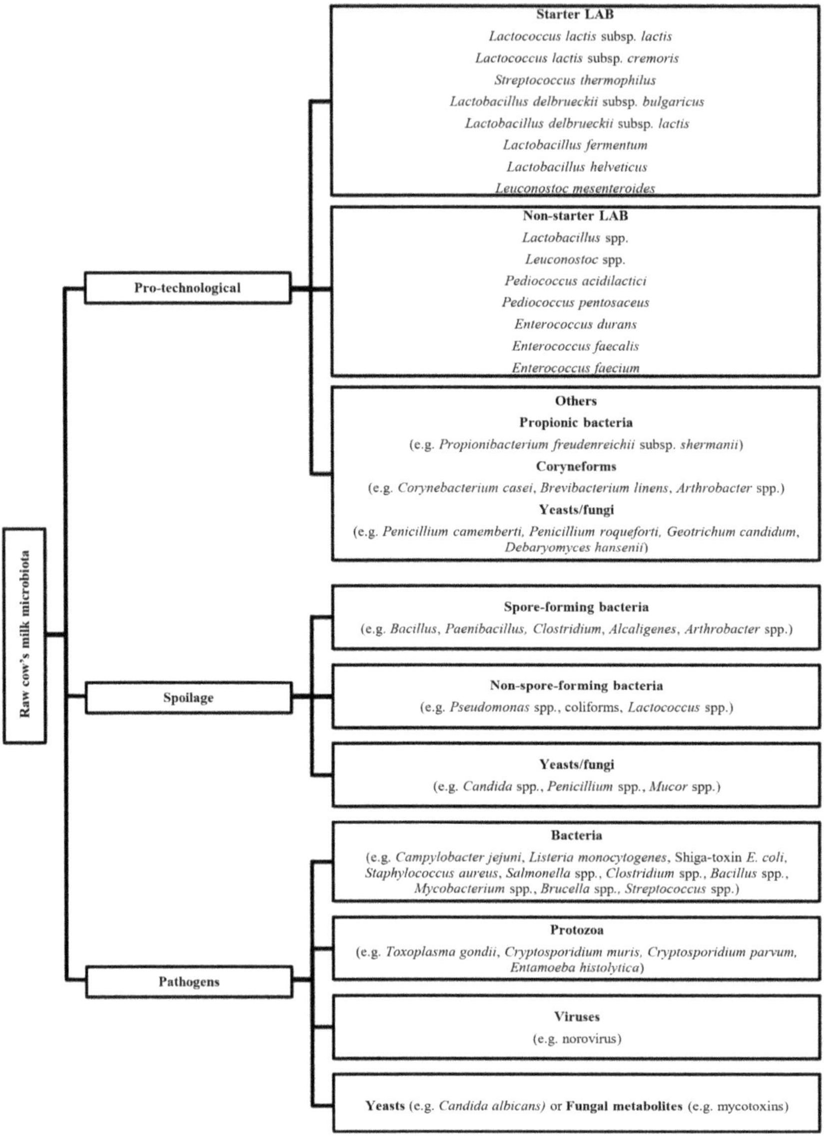


**Figure 1** The microbial community of raw cow's milk.

On the other hand, though, less attention has been paid to the development of such media for LAB and probiotics (Galat et al., 2016; Perry, 2017; Sohier et al., 2014). Despite the ease and low cost of conventional microbiological techniques, there are also several drawbacks, including the *de facto* limited selectivity of most growth media, the strain- and matrix-dependent efficiency, and the discrepancies between genotypic and phenotypic identifications

(Karimi et al., 2012; Van de Casteele et al., 2006; Zielińska et al., 2018). Furthermore, a major disadvantage of culture-dependent approaches is the failure to recover strains that cannot be cultured under the conditions used or strains that are viable but have entered a non-culturable state (viable but non-culturable cells – VBNC) (Fakruddin et al., 2013).

In the last two decades, however, research focused on the development and use of culture-independent approaches to characterize microbial food ecosystems without the need for enrichment or strain isolation thus leading to 'cultural' evolution (Cocolin and Ercolini, 2015). Among them, the recent application of high throughput sequencing (HTS) coupled with the omics-based methods, that is, amplicon sequencing, shotgun metagenomics, and metatranscriptomics, facilitates a comprehensive fingerprinting of the dairy microbiota. As attested, such studies reveal the more heterogeneous and complex, than previously thought, microbiota of raw milk (Fusco et al., 2020).

Amplicon sequencing is by far the most applied approach for metataxonomic studies. For the identification of the bacterial microbiota, selected hypervariable regions of the 16S rRNA gene are targeted with the V1-V3 and V3-V4 being the most commonly used. As to yeast/fungi, the internal transcribed spacers (ITS) DNA region ITS1-ITS2 has been most frequently used as a target, while the 18S and 28S rRNA genes are far less common. Due to the relatively short length of the regions targeted, the taxonomic resolution is often limited to the genus and, less frequently, to the species level. It should be stressed, however, that the identification at the species level or below is essential in the analysis of the microbiota of milk and dairy products. For example, several genera, like *Streptococcus*, *Staphylococcus*, and *Corynebacterium*, include both pro-technological and spoilage/pathogenic species. Therefore, the use of methods with the highest taxonomic resolution should be encouraged (Ferrocino and Cocolin, 2017; De Filippis et al., 2018; Yeluri Jonnala et al., 2018; Parente et al., 2020).

Despite the advantages of the omics approaches over the classical microbiological analysis, DNA-based techniques (both amplicon sequencing and shotgun metagenomics) fail to distinguish metabolically active/viable microorganisms from the inactive/dead ones. Therefore, RNA (metatranscriptomics), which is more easily degraded than DNA, or the use of dyes, such as propidium monoazide (PMA) and ethidium monoazide (EMA), prior to DNA extraction, is preferred, which prevent the PCR amplification of DNA from cells with a damaged membrane. Nevertheless, extensive benchmarking is required due to the differential ability of DNA dyes to penetrate cell membranes (Emerson et al., 2017; Parente et al., 2020).

As aforementioned, amplicon sequencing is the most widely used approach for metataxonomic analysis of microbial ecosystems. On the other hand, shotgun metagenomics studies of milk and dairy products are comparatively

**Table 1** Non-exhaustive list of the most recent studies based on culture-independent techniques used for the identification of the bacterial and yeast/fungal microbiota of raw cow's milk

| Identification method | Pro-technological taxa | Spoilage taxa | Pathogenic taxa | Reference |
|---|---|---|---|---|
| TTGE and DGGE | *Lactobacillus, E. faecium, E. durans, E. hirae, Lc. lactis, Candida, Kluyveromyces, Saccharomyces, Pichia* | *A. hydrophila, A. johnsonii, B. circulans, Clostridium lituseburense, Corynebacterium variabile, E. faecium, H. alvei, L. innocua, L. monocytogenes, Pseudomonas alcaligenes, Paenibacillus apiaries, Pseudomonas brenner, P. fluorescens, Pseudomonas stutzeri, P. synxantha, Staphylococcus epidermidis, Stenotrophomonas koreensis* | *Staphylococcus saprophyticus, Staphylococcus arlettae, Staphylococcus xylosus, L. monocytogenes, L. innocua, S. uberis, K. pneumoniae, S. marcescens, H. alvei* | Cocolin et al., 2002; Lafarge et al., 2004; Hagi et al., 2010; Raats et al., 2011 |
| RAPD | *Lactococcus, Enterococcus, Streptococcus, Lactobacillus, Leuconostoc, Pediococcus* | *A. hydrophila, Aeromonas media, Aeromonas piscicola, Acinetobacter bouvetii, A. guillouiae, Acinetobacter harbinensis, A. johnsonii, Acinetobacter lwoffii, Acinetobacter pakistanensis, B. thuringiensis, Brochothrix thermosphacta, Carnobacterium piscicola, C. freudii, Corynebacterium variabilis, Chryseobacterium balustinum, Chryseobacterium carnipullorum* | *H. alvei, S. marcescens, Staphylococcus pasteuri, Bacillus thuringiensis, Rodococcus erythropolis, Staphylococcus warneri* | Ercolini et al., 2009; Franciosi et al., 2009; Yuan et al., 2017 |
| MALDI-TOF MS | *E. faecium, E. durans, E. faecalis* | *P. fluorescens, Pseudomonas grimontii, Pseudomonas veronii* | *H. alvei, Pseudomonas syringae, Bacillus cereus, Bacillus pumilus, Bacillus licheniformis* | Vithanage et al., 2014; Dobranić et al., 2016 |

| | | | | |
|---|---|---|---|---|
| 16S metagenomics | *Streptococcus, Lactococcus, Leuconostoc, Lactobacillus, Enterococcus, Pediococcus, Bifidobacterium, Propionibacterium, Corynebacterium* | *Pseudomonas, Acinetobacter, Chryseobacterium, Flavobacterium, Hafnia, Serratia* | *Staphylococcus, Streptococcus, Bacillus, Clostridium, E. coli, Stenotrophomonas maltophilia* | Albonico et al., 2020; Skeie et al., 2019; Taponen et al., 2019; Vasquez et al., 2019; Wu et al., 2019; Zhang et al., 2019; Cremonesi et al., 2018; Li et al., 2018; Metzger et al., 2018; Porcellato et al., 2018; Bonsaglia et al., 2017; Doyle et al., 2017; Ganda et al., 2017; Kable et al., 2016; Oikonomou et al., 2014; Zhang et al., 2015; Quigley et al., 2013b |
| Shotgun metagenomics | *Streptococcus, Enterococcus, Corynebacterium, Lactococcus, Propionibacterium, Lactobacillus* | - | *Klebsiella, Escherichia, Enterobacter, Staphylococcus, Streptococcus Bacillus, Salmonella, Campylobacter* | Bhatt et al., 2012; Hoque et al., 2019 |

rare and metatranscriptomics studies are even less frequent. Due to the decreasing cost in sequencing and the improvement of bioinformatics tools and pipelines, both techniques are likely to become more regularly employed approaches in the near future (Yeluri Jonnala et al., 2018; Parente et al., 2020; Tilocca et al., 2020).

This chapter provides a comprehensive overview of the microbiota of raw cow's milk and its functionality as well as the methods employed for its fingerprinting. However, even though this overview cannot cover the numerous literature data accumulated so far, it does attempt to portray the key microorganisms found in the raw cow's milk ecosystem. Table 1 gives a representative overview from recent taxonomical studies performed on the bacterial and yeast/fungal microbiota of raw cow's milk using culture-independent approaches.

## 2 The microbial community of raw cow's milk

### *2.1 Raw cow's milk is a pool of pro-technological microbes*

The microbial composition of raw milk directly impacts the subsequent production of dairy products, as aforementioned, and thus, raw milk of high microbial quality is essential. The pro-technological microorganisms that are present in raw milk and also used in dairy fermented foods mostly comprise LAB, which are divided into two main groups.

In the first one, the so-called starter LAB (SLAB) rapidly convert milk lactose to lactic acid, which decreases the pH, and thus the safety and preservation of the final product are ensured, when good hygiene practices are applied as well. Apart from their principal role, SLAB also contribute to the organoleptic characteristics of the final product, that is, flavor, aroma, and texture. The LAB species that are mainly used as starter cultures in the production of dairy products (cheese, yogurt, and fermented milk) are *Lactococcus lactis* subsp. *lactis*, *Lactococcus lactis* subsp. *cremoris*, *Streptococcus thermophilus*, *Lactobacillus delbrueckii* subsp. *bulgaricus*, *Lactobacillus delbrueckii* subsp. *lactis*, *Lactobacillus fermentum*, *Lactobacillus helveticus*, and *Leuconostoc mesenteroides* (Leroy and De Vuyst, 2004; Bourdichon et al., 2012; Altieri et al., 2017).

The second group comprises LAB species that are implicated mainly in the ripening process during cheese-making and are referred to as non-starter LAB (NSLAB). NSLAB is a heterogeneous group that mostly includes *Lactobacillus* species, such as *Lactobacillus casei*, *Lactobacillus paracasei*, *Lactobacillus plantarum*, *Lactobacillus pentosus*, *Lactobacillus curvatus*, *Lactobacillus rhamnosus*, *Lb. fermentum*, *Lactobacillus buchneri*, *Lactobacillus parabuchneri*, and *Lactobacillus brevis*, as well as *Pediococcus acidilactici*, *Pediococcus pentosaceus*, *Enterococcus durans*, *Enterococcus faecalis*, *Enterococcus faecium*, and *Leuconostoc* spp. The addition of NSLAB as secondary cultures significantly contributes to the organoleptic characteristics of the final product through the production of key flavors and aromas, such as buttery, fruity, malty, nutty, chocolate-like, bitter almond, banana, sweet cherry, pungent and sulfur (Marilley and Casey, 2004; Smit et al., 2005).

Besides LAB, there are also some other groups of microorganisms of technological relevance, albeit not regarded as being as important as LAB. Among them, propionic acid bacteria (PAB), such as *Propionibacterium freudenreichii* subsp. *shermanii*, serve as starters in Swiss-type cheeses, for example, Emmental and Gruyere. Coryneform bacteria, such as *Corynebacterium casei*, *Brevibacterium linens*, and *Arthrobacter* spp. are generally regarded as being essential for the production of smear-ripened cheeses, for example, Camembert and Limburger. Finally, yeasts and molds, such as *P. camemberti*, *P. roqueforti*, *Geotrichum candidum*, and *Debaryomyces hansenii* are key members of the pro-technological microbiota of cheeses like Camembert, Brie, Roquefort, Gorgonzola, Stilton, and Danish Blue (Gobbetti et al., 2015, 2018).

### 2.1.1 The exploration of raw cow's milk microbiota using molecular techniques

Due to the extensive use of cow's milk in the dairy industry, raw cow's milk microbiota has been widely explored by both culture-dependent and -independent approaches. The typical composition of the raw cow's milk microbiota of a healthy udder obtained by conventional microbiological analysis mainly consists of LAB belonging to the genera *Lactococcus* (1–4 log CFU $mL^{-1}$), *Streptococcus* (1–4 log CFU $mL^{-1}$), *Lactobacillus* (2–4 log CFU $mL^{-1}$), *Leuconostoc* (1–3 log CFU $mL^{-1}$), and *Enterococcus* spp. (1–3 log CFU $mL^{-1}$) (Quigley et al., 2013a). However, culture-independent techniques revealed a much more diverse and complex microbial ecosystem than that described by conventional microbiological analysis, but, most importantly, these techniques paved the way to understand how a number of factors influence the microbial community of raw cow's milk.

In this perspective, denaturing gradient gel electrophoresis (DGGE) along with temporal temperature gel electrophoresis (TTGE) were used to monitor the dynamic changes of the bacterial communities in raw cow's milk samples before and after refrigeration. As was expected, refrigeration clearly impacts the microbial diversity of milk samples. However, it was interesting to observe that although several spoilage/pathogenic species were identified in both refrigerated and non-refrigerated milk samples, *Lc. lactis* was the main species identified mainly in the non-refrigerated samples, followed by *Leuconostoc carnosum, Leuconostoc lactis, E. durans, Enterococcus hirae, Propionibacterium acidipropionici*, and *Lactobacillus plantarum/Lactobacillus pentosus* (Lafarge et al., 2004). DGGE analysis has been also used to evaluate the influence of the feeding environment (from inside to outside grazing) on the raw cow's milk microbiota of the Holstein cows in Japan. Interestingly, DGGE profiles of the milk samples revealed that the genus *Lactobacillus* was predominant during the inside grazing, but after 1 week of outside grazing, the genus *Staphylococcus* was also identified as being dominant (Hagi et al., 2010).

When Delbès et al. (2007) analyzed the bacterial microbiota of raw cow's milk used for the production of the French Saint-Nectaire cheese using single-strand conformation polymorphism (SSCP), *Lc. lactis* was once again found to be the dominant LAB species, followed by *Lb. casei, E. faecalis, Lactobacillus kefiranofaciens*, and *S. thermophilus*. SSCP has been also used to study the potential link between milking practices and bacterial diversity of bulk cows' milk samples from various dairy farms in France (Verdier-Metz et al., 2009). The SSCP profiles classified the milk samples into three groups: group A was characterized mainly by Gram-positive non-LAB families, that is, *Corynebacterineae* and *Micrococcaceae*, and a high level of hygiene in milking practices, while groups B and C had similar profiles but different from that of group A. Both groups, B and C, were dominated by LAB, while less-intensive hygiene practices were applied. In particular, *Lc. lactis* and *Leuconostoc pseudomesenteroides* were mostly identified in groups B and C, respectively, while *Streptococcus uberis*, *Streptococcus parauberis*, and *S. thermophilus* were detected as well in both groups. Interestingly, Gram-negative bacteria belonging to the genus *Acinetobacter* were identified in both groups B and C.

More recently, Quigley et al. (2013b) used a combination of several culture-independent techniques, including flow cytometry, real-time quantitative polymerase chain reaction (real-time PCR), and 16S metagenomics analysis to assess the bacterial communities of raw and pasteurized cow's milk samples from industrial and research facilities around Ireland. It should be noted that the authors used the DNA dye EMA prior to DNA extraction to avoid the amplification of damaged cells. Based on the results of the 16S metagenomics analysis, Firmicutes was found to be the most abundant phylum in raw milk samples, followed by Proteobacteria, Bacteroidetes, and Actinobacteria. At the family level, *Streptococcaceae* was identified with the highest abundance, and, as expected, these reads were taxonomically assigned mainly to the genus *Lactococcus*. Moreover, a low fraction of the LAB genera *Leuconostoc* and *Streptococcus* was also apparent. The combination of the techniques employed also revealed the presence of a previously unrecognized bacterial microbiota in raw cow's milk. Most notably, the use of 16S metagenomics analysis resulted in several bacterial genera being identified in milk for the first time, among them the genus *Faecalibacterium*. The genus *Faecalibacterium* comprises to date only one species, namely *Faecalibacterium prausnitzii*, which was first discovered in human feces and subsequently in the feces of calves, pigs, and other agricultural species (Savin et al., 2019). Thus, it is not surprising that it was found in the raw cow's milk microbiota, as part of the external microbial contamination of the udder. Apart from the study of Quigley et al. (2013b), Oikonomou et al. (2014) and Doyle et al. (2017) also detected *F. prausnitzii* sequences in raw cow's milk using 16S metagenomics analysis. Interestingly, *F. prausnitzii* is one of the butyrate-producing bacteria in the colon, has anti-inflammatory properties, and its abundance in the gut is negatively correlated with obesity in humans (Savin et al., 2019). Therefore, raw cow's milk can be used as an alternative source to feces of new *F. prausnitzii* strains with potential probiotic applications.

Furthermore, Zhang et al. (2015) described the impact of different types of diet (high versus low concentrate diet) on the raw cow's milk microbiota using 16S metagenomics analysis. At the phylum level, the milk microbiota was dominated by Actinobacteria, Firmicutes, Proteobacteria, and Bacteroidetes regardless of the diet. However, at the genus level, diet-associated genera were observed despite the small number of animals enrolled in the study (four Holstein dairy cows). In detail, relative abundances of the genera *Enterococcus, Lactococcus*, and *Lactobacillus* were higher in the milk samples from the cows fed the high concentrate diet, while, on the other hand, relative abundances of the genus *Streptococcus* were higher in the milk from the cows fed the low concentrate diet. Nevertheless, a clear discrimination between the two groups of milk samples was not achieved using the principal coordinate analysis (PCoA).

Interesting results were also obtained by a large-scale study performed by Kable et al. (2016). The authors analyzed raw cow's milk samples from 899 tanker trucks in California during three seasons (spring, summer, and fall) searching for potential links between milk transport conditions and raw cow's milk microbiota using 16S metagenomics analysis. The results revealed, to an extent, different bacterial fingerprints among the three seasons considered, concerning the genera identified but most importantly the respective abundances. The averaged bacterial microbiota of raw milk samples collected in the fall was dominated by the genera *Streptococcus*, *Staphylococcus*, and *Enterococcus*, the one in the spring by *Streptococcus*, *Staphylococcus*, and *Corynebacterium*, and that of summer by *Streptococcus* and *Staphylococcus*.

Regarding the yeast/fungal population of raw cow's milk, the existing literature data are rather limited, despite the fact that several species are of significant importance in some cheese varieties. In one of the early studies, Cocolin et al. (2002) used a combination of classical microbiological analysis and DGGE fingerprinting to study the yeast diversity of raw cow's milk. Based on the culturing results, six *Candida* species, namely *Candida catenulata*, *Candida pararugosa*, *Candida parapsilosis*, *Candida zeylanoides*, *Candida pseudointermedia*, and *Candida rugosa* as well as *Kluyveromyces marxianus*, *Kluyveromyces lactis*, *S. cerevisiae*, *Pichia guilliermondii*, *Trichosporon mucoides*, and *Cryptococcus curvatus* were mainly identified. On the other hand, *Galactomyces* sp., *Candida kefyr*, *Candida humilis*, and *Saccharomyces bayanus/Saccharomyces pastorianus* were detected by the DGGE fingerprinting. In a more recent study, the yeast communities of raw cows' milk samples from various dairy farms located in different geographical regions of Quebec were analyzed using conventional microbiological techniques (Lavoie et al., 2012). Of the 339 yeast isolates, *Candida* was the most abundant genus and the most diverse as it comprised 11 species, followed by *Cryptoccocus* and *Pichia*. At the species level, *D. hansenii* was identified in the majority of the farms studied. Similarly, Buehler et al. (2017) analyzed bulk tank samples by classical microbiological analysis to identify the fungal species present. Although the majority of the isolates were identified as *Candida* spp., technologically important species were detected as well, that is, *D. hansenii* and *K. marxianus*. Fungal communities of raw cow's milk have also been associated with different altitudes across Valle d'Aosta region (North-Western Italian Alps) (Panelli et al., 2013). Bulk raw milk samples were collected from 1400 m, 1800 m, and 2200 m along with raw cows' milk samples from different herds at an altitude $< 700$ m (bottom of the valleys) and analyzed using multilocus sequencing with ITS1/ITS2, D1/D2, and β-tubulin being the targets employed. According to the results, 31 species were detected, among them several species with well-established positive contribution to cheesemaking, for example, *K. marxianus*, *D. hansenii*, *Galactomyces geotrichum*, and *Pichia membranifaciens*. Interestingly, *K. marxianus* was the only species that formed the core microbiota, and *Candida* was the most represented genus with a tendency to be detected at the highest altitudes.

## 2.2 Milk spoilers

### 2.2.1 The usual suspects of milk spoilage

As aforementioned, milk is also an ideal substrate for the growth of spoilage microorganisms. The microorganisms associated with spoilage of milk and dairy products can be either bacteria, such as aerobic spore-forming, for example, *Bacillus*, *Paenibacillus* spp.; anaerobic spore-forming, for example, *Clostridium*, *Alcaligenes*, *Arthrobacter* spp.; or non-spore forming, for example, *Pseudomonas*, *Lactococcus* spp., coliforms; or yeasts/fungi, for example, *Candida*, *Penicillium*, *Mucor* spp. (Lorenzo et al., 2018). The combination of insufficient hygiene practices during milking along with improper handling post-milking boosts spoilage microorganisms to thrive in raw milk, thus leading to low-quality dairy products (Postollec et al., 2012).

Many of the milk spoilers aforementioned are psychrotrophs (Ledenbach and Marshall, 2009; Perin et al., 2018). Nevertheless, under good hygiene practices, psychrotrophs represent approximately 10% of the raw cow's milk microbiota, compared to 70% in the case of low hygiene and poor milk handling (Perin et al., 2018). The main psychrotrophs in raw milk include Gram-negative genera, such as *Pseudomonas*, *Aeromonas*, *Serratia*, *Acinetobacter*, *Alcaligenes*, *Achromobacter*, *Enterobacter*, and *Flavobacterium* (Munsch-Alatossava and Alatossava, 2006; Samaržija et al., 2012; de Oliveira et al., 2015), while Gram-positive genera, namely *Bacillus*, *Clostridium*, *Corynebacterium*, *Microbacterium*, *Micrococcus*, *Streptococcus*, *Staphylococcus*, *Lactococcus*, and *Lactobacillus*, can be found to a lesser extent (Hantsis-Zacharov and Halpern, 2007; Vithanage et al., 2016). Psychrotrophs are mesophilic bacteria with optimal growth temperature between 15°C and 20°C; however, they can adapt to lower temperatures (3–7°C) very well (de Oliveira et al., 2015).

### 2.2.2 The impact of spoilers on milk and dairy products

The vast majority of milk spoilers belong to the genus *Pseudomonas* making approximately 70% of the total psychrotrophs found in milk (Ledenbach and Marshall, 2009; McPhee and Griffiths, 2011; Perin et al., 2018). The major metabolic activities of spoilage microorganisms include the production of heat-resistant proteolytic and lipolytic enzymes, which cause post-pasteurization milk deterioration (Scatamburlo et al., 2015; Ribeiro Júnior et al., 2018). In particular, psychrotrophs, such as *Pseudomonas*, *Acinetobacter*, and *Serratia* spp., produce lipases that hydrolyze the triglycerides of milk resulting in the release and accumulation of free fatty acids, for example, butyric, caproic, and caprylic acids, which generate rancid, bitter, and caustic flavors in milk and dairy products (Machado et al., 2017; Perin et al., 2018). Furthermore, the proteolytic activity of psychrotrophs leads to protein destabilization, causes gel formation mostly in ultra-high-temperature (UHT) milk (Machado et al., 2017) and induces a bitter, unclean taste, and reduces cheese yield (McPhee and Griffiths, 2011; Bautista, 2014). *Pseudomonas* species, such as *P. fluorescens*, *Pseudomonas gessardii*, *Pseudomonas lundensis*, *P. fragi*, and *Pseudomonas proteolytica*, show high proteolytic activity, while species of *Acinetobacter*, *Chryseobacterium*, and *Serratia* are considered less proteolytic. Moreover, although *Pseudomonas* spp. show high lipolytic activity as well, *Acinetobacter* species, such as *Acinetobacter guillouiae*, seem to be the most lipolytic ones (Von Neubeck et al., 2015; Yuan et al., 2018a). In addition to enzymatic activities, spoilage bacteria, such as *P. fluorescens*, *Bacillus licheniformis*, *Serratia liquefaciens*, and *Hafnia alvei*, have the ability to form biofilms on equipment surfaces, for example, milking machines and storage equipment, resulting in subsequent contamination and enzyme secretion during further processing (Machado et al., 2017; Yuan et al., 2018b).

Psychrotolerant coliforms belonging to the genera *Citrobacter*, *Enterobacter*, *Hafnia*, and *Serratia* have the capacity to catabolize triglycerides into free fatty acids, resulting in rancid flavors (Masiello et al., 2016), while *Klebsiella* is capable of producing exopolysaccharides provoking an unwanted ropy defect in pasteurized milk (Martin et al., 2018).

Although spoilage microorganisms, such as *Pseudomonas* spp., are the main actors in milk spoilage, spore-forming bacteria can also cause several adverse effects in milk and dairy products. The endospores produced by species belonging to the aerobic-spore former *Bacillus* can remain dormant for long periods and show resistance against extreme conditions, such as very low or high temperatures and starvation. The spores can be metabolically reactivated as soon as conditions are favorable; however, at this stage their inactivation by various means, for example, heat treatment and biocides, is feasible (Postollec et al., 2012). Due to their strong enzymatic activities, mesophilic strains of *Bacillus subtilis* and *B. cereus* induce rancid and bitter off-flavors in pasteurized milk (Bautista, 2014), while the latter is responsible for the 'sweet curdling' defects in milk cartons' surface, caused by its lecithinase and protease activities (Lücking et al., 2013; Bautista, 2014).

The anaerobic spore-forming *Clostridium* species, such as *Clostridium tyrobutyricum*, *Clostridium butyricum*, *Clostridium beijerinckii*, and *Clostridium sporogenes* can also survive pasteurization and consequently cause spoilage in dairy products, with *C. tyrobutyricum* being the most frequently detected. Clostridia are responsible for 'gas holes' formation, also known as 'late blowing,' mostly in hard and semi-hard cheeses, due to lactate fermentation into butyric acid, $CO_2$, and $H_2$, as well as rancid cheese flavors (Driehuis et al., 2016; André et al., 2017).

Despite the significant and positive contribution of LAB to the organoleptic characteristics of dairy products, they can also act as spoilage microorganisms in certain cases. For instance, some *Lactococcus* species that survive pasteurization, such as *Lc. lactis* subsp. *lactis*, can give a malty flavor to milk due to leucine catabolism into 3-methylbutanol (Lu and Wang, 2017). In addition, *Leuconostoc* spp. confer milk a sour taste and can produce $CO_2$ via protein catabolism, causing unfavorable hole formation during cheesemaking (Bautista, 2014).

Yeasts and fungi have also the ability to thrive in the nutrient-rich milk niche, thus affecting the organoleptic characteristics of the final products. Although several species belonging to the genera *Candida*, *Penicillium*, *Kluyveromyces*, *Galactomyces*, and *Mucor* are widely used in dairy production, other species can cause undesirable effects, such as off-odors, for example, production of alcohols, aldehydes, esters, as well as color and texture changes. For example, *G. geotrichum* and *Yarrowia lipolytica* are responsible for the toad skin and the surface browning of dairy products due to tyrosine decarboxylation, while *D.*

*hansenii* can cause off-odors by producing volatile compounds, for example, pentadiene (Garnier et al., 2017b; Awasti and Anand, 2020). On the other hand, fungi, such as *Mucor* and *Penicillium* spp., are responsible for the 'cat hair' defect and the development of dark hyphae on the product's surface, respectively (Martin et al., 2020). Figure 2 shows the defects of spoilers in milk and milk products.

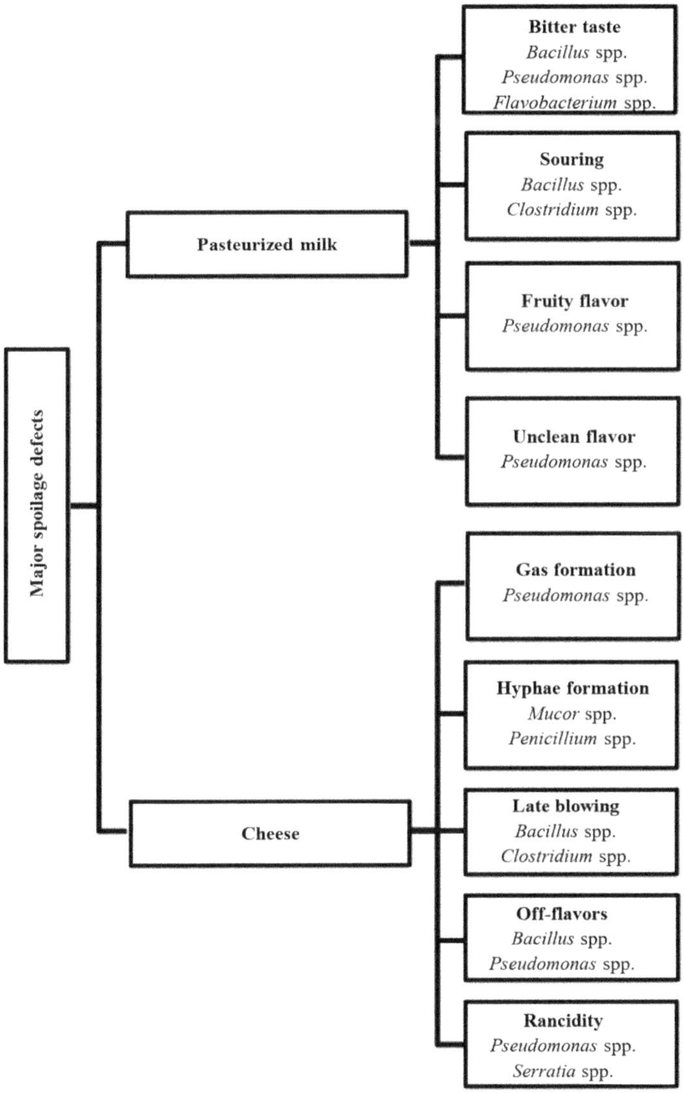


**Figure 2** Major spoilage defects reported in pasteurized milk and cheese.

### 2.2.3 Milk spoilers as a socioeconomic factor

Although spoilage microorganisms in raw cow's milk and their impact on milk and dairy products have been extensively studied, there are no comprehensive data on the resulting economic losses (Ghosh et al., 2016). Worldwide, the estimated amount of spoiled and wasted food losses, including fruits and vegetables, milk, and grain products, between production and consumption are approximately 1.3 billion tons per year. Moreover, it has been assessed that the global food waste due to microbial spoilage is approximately 25% (Petruzzi et al., 2017) and is caused mainly by bacteria (90–95% of the losses) and the rest by fungi (Awasti and Anand, 2020). Regarding milk and dairy products, there are some interesting reports from around the globe. In the United States, five billion liters of milk are discarded every year due to microbial spoilage (Ivy et al., 2012), while in Europe, losses in milk and dairy products during the processing stage can reach 48% (Ghosh et al., 2016). In many developing countries where food production is limited and preventive measures are not necessarily employed, food spoilage has a significant impact on food availability. It has been estimated that in these countries, food spoilage and waste, in general, are responsible for losses up to $US300 billion per year (Ghosh et al., 2016). The dairy sector losses in East Africa and Near East are close to $US90 million per year in total. Specifically, in Kenya, the dairy economic losses reach annually $US20 million, with 60 million liters of milk being wasted, while in Tanzania and Uganda, the annual economic losses can reach $US23 and $US14 million, respectively (Segrè et al., 2014).

### 2.2.4 Identification of milk spoilers using culture-dependent and -independent techniques

The biodiversity of psychrotrophs in raw cow's milk has been widely studied using both phenotypic and molecular methods. When raw cows' milk samples collected from a dairy plant in the United States were examined using culture-dependent (phenotypic and molecular) approaches, the largest percentage (90%) of the total isolated psychrotrophic bacteria were identified as *P. fluorescens* and *Pseudomonas putida* (Dogan and Boor, 2003). Similarly, the phenotypic characterization of psychrotrophs associated with raw cows' milk in Australia and Finland also demonstrated that *Pseudomonas* was the dominant genus and, in particular, the species *P. fluorescens*, *P. putida*, *Pseudomonas aeruginosa*, *Pseudomonas aureofaciens*, *Pseudomonas cepacia*, *Pseudomonas tolaasii*, *Pseudomonas synxantha*, *Pseudomonas marginalis*, *P. lundensis* with *P. fluorescens* being the main cause of spoilage in stored milk (Deeth et al., 2002; Munsch-Alatossava and Alatossava, 2006).

When TTGE and DDGE were employed to investigate the impact of 24 h refrigeration on the raw cow's milk microbiota in samples collected from farms and industrial dairy tanks in France, it was shown that counts of *Listeria innocua*, *L. monocytogenes*, *P. fluorescens*, *E. faecium*, *H. alvei*, *Serratia marcescens*, *Klebsiella pneumoniae*, *Kocuria rosea*, and *Aeromonas hydrophila* not only increased during storage but also limited the growth of *Lc. lactis*, which was the dominant species in all milk samples before refrigeration (Lafarge et al., 2004). In a similar study, DDGE revealed the prevalence of *Pseudomonas* and *Acinetobacter* spp. during 48 h refrigeration storage in raw cows' milk derived from farm bulk tanks and dairy plants' silo tanks in Israel (Raats et al., 2011).

The application of random amplification of polymorphic DNA (RAPD)-PCR and 16S rRNA gene sequencing in raw bulk milk collected from different farms in Italy revealed that *Pseudomonas* spp. followed by *H. alvei*, *S. marcescens*, and *Citrobacter freudii* were the most common contaminants (Ercolini et al., 2009). In a similar study, *P. fluorescens*, *P. fragi*, and *Pseudomonas psychrophila* were the dominant species identified in raw cows' milk samples produced in distinct Chinese farms, followed by *Acinetobacter johnsonii*, *Flavobacterium frigidimaris*, *Sphingobacterium kitahiroshimense*, and *S. liquefaciens* (Yuan et al., 2017).

Vithanage et al. (2014) confirmed the aforementioned findings by analyzing raw milk samples in Australia with high bacterial load, using 16S rRNA gene sequencing combined with classical and advanced phenotypic methods, such as the Biolog, API, and Microbact systems, along with matrix-assisted laser desorption/ionization-time of flight (MALDI-TOF) mass spectrometry (MS). Furthermore, the same group found that season considerably affected the diversity of psychrotrophs in milk collected from different Australian farms. In particular, the class Gammaproteobacteria, and specifically the genera *Pseudomonas*, *Acinetobacter*, and *Psychrobacter*, as well as several *Bacillus* species, such as *B. cereus*, *Bacillus weihenstephensis*, and *Bacillus circulans*, were widespread in milk collected during winter, while during spring and autumn, the genera *Hafnia*, *Serratia*, *Stenotrophomonas*, *Klebsiella*, and *Enterobacter* as well as species of the genera *Acinetobacter*, *Aeromonas*, *Arthrobacter*, *Brevundimonas*, *Microbacterium*, *Micrococcus*, *Pseudomonas*, *Pseudoalteromonas*, and *Rhodococcus* prevailed. In milk collected in summer, *Bacillus* species, including *Bacillus thurigiensis*, *B. licheniformis*, *Bacillus pumilus*, and *B. subtilis*, were predominant (Vithanage et al., 2016).

Interestingly, the combination of MALDI-TOF MS and 16S rRNA metagenomics mapped the psychrotrophs in raw cows' milk from distinct regions in New Zealand in a similar way. *Pseudomonas* was the dominant genus, with subdominant genera being *Acinetobacter*, *Bacillus*, *Buttiauxella*, *Carnobacterium*, *Lactococcus*, *Hafnia*, *Rahnella*, and *Serratia* (Zhang et al., 2019).

It is worth mentioning that although yeasts are found in raw milk at relatively high levels (2–5 log CFU $mL^{-1}$) (Quigley et al., 2013a; Awasti and Anand, 2020), their growth during milk storage is sporadic as psychrotrophs surpass rapidly the yeast population. Moreover, yeasts and fungi cannot survive pasteurization;

hence, any contamination of dairy products follows heat treatment and further processing (Garnier et al., 2017b). Even though yeasts and fungi have been studied to a much lesser extent compared to bacteria, it has been observed that the genera *Candida*, *Eurotium*, *Galactomyces*, *Debaryomyces*, *Yarrowia*, *Penicillium*, *Mucor*, and *Cladosporium* are responsible for the spoilage of dairy products (Keshri et al., 1998; Vacheyrou et al., 2011; Garnier et al., 2017a; Fröhlich-Wyder et al., 2019). Among the very few studies on yeast microbiota in raw milk, the genera *Penicillium*, *Pichia*, *Candida,* and *Debaryomyces* were the most frequently isolated from bulk tank milk samples from several farms in New York using classical microbiological analysis and molecular identification (ITS sequencing) (Buehler et al., 2017). Similarly, using the same approach in raw cow's milk samples deriving from the Italian Alps, *Candida* and *Klyveromyces* were the main genera identified (Panelli et al., 2013).

## 2.3 Pathogen intruders

### 2.3.1 Raw cow's milk as a source of foodborne pathogens: an incriminated food matrix

Although a number of diseases have been linked to the consumption of raw milk, the implementation of pasteurization has improved milk safety (Mungai et al., 2015) and, therefore, public health by preventing the spreading of pathogens (Martin et al., 2018). Before the adoption of pasteurization in 1938, the incidence of outbreaks associated with milk was estimated at 25% of the total food and waterborne outbreaks in the United States, whereas the respective percentage of milkborne outbreaks nowadays is estimated to be below 1% (Food and Drug Administration, 2017). Nevertheless, in recent years, there is a growing trend toward the consumption of raw milk (Lucey, 2015) due to its perceived health benefits (i.e. asthma and allergies), higher nutritional value, better taste (Oliver et al., 2009; Baars, 2013; Claeys et al., 2013), and the growing public demand toward natural products (Claeys et al., 2013). However, unpasteurized milk can be a vehicle for a number of pathogenic bacteria, hence, compromising public health. The consumption of contaminated milk may lead to diseases depending on the pathogenicity of the microorganism (or the toxicity of the toxin produced), the infection dose, the number of the ingested cells (or quantity of the toxins), and the health condition of the consumer. People belonging to susceptible groups (very young, elderly, pregnant, and immune-compromised) are at a higher risk of getting sick, although anyone can be affected (Claeys et al., 2013).

Milk can be a major source of foodborne pathogens of human health significance. Raw milk can harbor pathogens even when sourced from clinically healthy animals (Lucey, 2015) or has a low bacterial count (Jayarao et al., 2006). There are two main routes for the introduction of pathogens into the milk chain. The first route includes endogenous contamination by direct transfer of pathogens from the blood (systemic infection) of an infected animal or *via* infection in the udder (mastitis) (Verraes et al., 2015). Exogenous contamination, the second route, is due to microbial spread from feces, skin, milking equipment, farm personnel, and animal manure during or after milking (Verraes et al., 2015; Erkmen and Bozoglu, 2016).

Human pathogens that can potentially contaminate (raw) milk include spore-forming *Bacillus* spp. and *Clostridium* spp., pathogenic (Shiga toxin) *E. coli*, *Mycobacterium* spp., *Brucella* spp., *L. monocytogenes*, *Salmonella* spp., enterotoxin-producing *Staphylococcus aureus*, *C. jejuni*, *Coxiella burnetii*, *Streptococcus agalactiae*, and other streptococci as well as some mold species producing mycotoxins. Other pathogenic contaminants include species of the genera *Actinomyces*, for example, *Actinomyces radicidentis*; *Aeromonas*, for example, *Aeromonas trota*, *Aeromonas hydrophila*, *Aeromonas janda*, *Aeromonas caviae*, *Aeromonas sobria, and Aeromonas schubertii*; *Arcobacter*, for example, *Arcobacter butzleri*; *Corynebacterium*, for example, *Corynebacterium diphtheriae*; *Cronobacter*, for example, *Cronobacter sakazakii*; *Enterobacter*, for example, *Enterobacter cloacae* and *Enterobacter asburiae*; *Proteus*, for example, *Proteus mirabilis* and *Proteus vulgaris*; *Shigella,* for example, *Shigella dysenteriae*, *Shigella sonnei*, and *Shigella flexneri*; *Serratia*, for example, *Serratia marcescens*, as well as *H. alvei*, *K. pneumoniae*; *Stenotrophomonas maltophilia*; *Yersinia enterocolitica*; and *Yersinia pseudotuberculosis* (Fusco et al., 2020; Olajide and LaPointe, 2020). Protozoa, such as *Toxoplasma gondii*, *Cryptosporidium muris, Cryptosporidium parvum*, and *Entamoeba histolytica*, viruses (norovirus) and some yeast species, such as *Candida albicans*, can also be found (Olajide and LaPointe, 2020). Among the aforementioned pathogenic contaminants, the most common ones in order of decreasing frequency are *C. jejuni*, *Salmonella* spp., and Shiga-toxin *E. coli* (Claeys et al., 2013). In the United States, a survey during 2007 to 2012 revealed that among 81 outbreaks associated with the consumption of unpasteurized milk, the most commonly identified pathogen was *Campylobacter* spp. accounting for 81% of the outbreaks, followed by *E. coli* (13%), *Salmonella* spp. (3%), and *C. burnetii* (1%) (Mungai et al., 2015). Table 2 includes foodborne outbreaks of raw milk and dairy products recorded during the last 20 years.

### 2.3.2 Introduction of major milkborne pathogens into the food chain: food safety related issues

Although poultry is the leading cause of campylobacteriosis (An et al., 2018), raw milk is also considered one of the main possible sources of infection

**Table 2** Foodborne outbreaks of raw milk and dairy products recorded during the last 20 years

| Species | Country | Year | Implicated food | Reference |
|---|---|---|---|---|
| *B. abortus* vaccine strain RB51 | The USA (Texas) | 2017 | Raw milk | Cossaboom et al., 2018 |
| *C. jejuni* | The Netherlands | 2005/2007 | Raw milk | Heuvelink et al., 2009 |
| *C. jejuni* | The USA (California) | 2006/2007 | Pasteurized milk/ commercial raw milk and raw chocolate colostrum | Jay-Russell et al., 2013 |
| *C. jejuni* | The USA (Kansas) | 2007 | Raw milk/fresh cheese made from raw milk | Hunt et al., 2009 |
| *C. jejuni* | The USA (Multistate) | 2012 | Raw milk | Longenberger et al., 2013 |
| *C. jejuni* | Finland | 2012 | Raw milk | Jaakkonen et al., 2020 |
| *C. jejuni* | The USA (Pennsylvania) | 2013 | Raw milk | Weltman et al., 2013 |
| *C. jejuni* | The USA (Utah) | 2014 | Raw milk | Davis et al., 2016 |
| *C. jejuni* | Sweden | 2014 | Raw milk | Lahti et al., 2017 |
| *C. jejuni* | England | 2016 | Raw milk | Kenyon et al., 2020 |
| *C. jejuni* | The USA (Colorado) | 2016 | Raw milk | Burakoff et al., 2018 |
| *Coxiella burnetti* | The USA (Michigan) | 2011 | Raw milk | Signs et al., 2012 |
| *E. coli* O157 | Slovakia | - | Raw milk | Liptakova et al., 2004 |
| *E. coli* O157:H7 | The USA (Washington and Oregon) | 2005 | Raw milk | Bhat et al., 2007 |
| *E. coli* O157:H7 | The USA (Pacific Northwest) | 2005 | Raw milk | Denny et al., 2008 |
| *E. coli* O157:H7 | The USA (California) | 2006 | Raw milk | Schneider et al., 2008 |
| *E. coli* O157 | The USA (Connecticut) | 2008 | Raw milk | Guh et al., 2010 |
| *E. coli* O157 | Finland | 2012 | Raw milk | Jaakkonen et al., 2017 |
| *E. coli* (Shiga toxin) | England | 2014 | Raw milk | Butcher et al., 2016 |

| | | | | |
|---|---|---|---|---|
| *E. coli* O157:H7 | England | 2017 | Raw milk | Treacy et al., 2019 |
| *L. monocytogenes* | The USA (Multistate) | 2014 | Raw milk | Nichols et al., 2020 |
| *M. bovis* | Ireland | 2003–2005 | Raw milk from infected cow (tuberculous mastitis) | Doran et al., 2009 |
| *S.* Typhimurium | The USA (Multistate) | 2000 | Pasteurized milk (post-pasteurization) | Olsen et al., 2004 |
| *S.* Typhimurium | The USA (Multistate) | 2002 | Raw milk | Mazurek et al., 2004 |
| *S.* Typhimurium | The USA (Pennsylvania) | 2007 | Raw milk | Lind et al., 2007 |
| *S.* Agona | France, Spain, Greece | 2017 | Infant formula | EFSA and ECDC, 2018a |
| Staphylococcal enterotoxin | Japan (Osaka) | 2000 | Reconstituted milk | Ikeda et al., 2005 |
| Staphylococcal enterotoxin | Paraguay | 2007 | Ultrapasteurized milk | Weiler et al., 2011 |
| Staphylococcal enterotoxin | Austria | 2007 | Pasteurized milk, cocoa or vanilla milk | Schmid et al., 2009 |

(Fusco et al., 2020). *Campylobacter* harbors the intestinal tract of animals and, therefore, transmission occurs through cross-contamination, (Fusco et al., 2020) mostly via fecal contamination of the udders during the milking process (Kenyon et al., 2020). In addition, contamination due to insufficient pasteurization is also a recognized cause of campylobacteriosis outbreaks (Kenyon et al., 2020). Recently, An et al. (2018) found that among dairy cattle farms in Korea, the prevalence of *C. jejuni* was 60% (9/15), whereas it was found in 25.6% of the fecal samples in animal level (68/266). In addition, recent studies have demonstrated that the multi-stress-tolerant *C. jejuni* strains can survive high-temperature short-time (HTST) pasteurization (Oh et al., 2018).

*Salmonella* ssp. is an ubiquitous pathogen that can enter the milk chain from many sources (Fusco et al., 2020). Most outbreaks are related to raw milk consumption (Fusco et al., 2020), although contamination due to inadequate processing or post-pasteurization contamination can also occur (Adams and Moss, 2008). An outbreak due to multi-drug-resistant *Salmonella* post pasteurization has also been reported (Olsen et al., 2004).

Among the pathogenic *E. coli* that are associated with milk, Shiga-toxin producing *E. coli* O157:H7 (STEC) is of utmost importance due to its high virulence (Farrokh et al., 2013), although, other non-O157 serogroups have been isolated from raw cow's milk in Europe as well (EFSA and ECDC, 2018b). As with *Campylobacter* and *Salmonella*, STEC harbors the intestinal tract of dairy animals and can either contaminate milk through excretion from the udder due to sub-clinical mastitis or via the fecal route. STEC can persist in farms for years, and, consequently, cattle farms represent a reservoir for this pathogen (Hancock et al., 2001). Therefore, it is probably impossible to eradicate them, and, thus, preventive or curable measures are necessary to keep STEC at an acceptable level (Farrokh et al., 2013).

*L. monocytogenes* can contaminate raw milk mainly through the gastrointestinal tract (GIT) of animals and the environment (Ulusoy and Chirkena, 2019). Once the contamination occurs in the environment of a dairy farm, it can be easily transmitted from the environment to the animals or among animals (Hunt et al., 2012). Cross-contamination of the milking equipment through the feces is also possible (Hunt et al., 2012). Although pasteurization can kill *L. monocytogenes*, the presence of this pathogen in pasteurized milk can also occur due to post-process contamination or due to defects in technology during pasteurization, for example, inadequate temperature and technical errors (Ulusoy and Chirkena, 2019). Outbreaks of listeriosis have been principally linked to the consumption of dairy products, mainly fresh or soft cheeses made both from pasteurized or unpasteurized milk (Kousta et al., 2010; Fusco et al., 2020).

Food poisoning intoxication by *S. aureus* is attributed to the ingestion of food containing preformed enterotoxins (Kousta et al., 2010). Contamination of milk can occur through dairy cattle, in which *S. aureus* can cause mastitis. Contamination of raw milk or raw milk products by humans or the environment during handling/manufacture can also occur (Jørgensen et al., 2005a; Pexara et al., 2013). The production of enterotoxins is affected by a number of factors, although the density of *S. aureus* cells in milk (>$10^5$ CFU $mL^{-1}$) seems to be crucial for activating the corresponding gene regulators (Cousin et al., 2018). Even though pasteurization kills *S. aureus* cells, staphylococcal enterotoxins are thermostable, and once produced, they can withstand heat treatments and retain their biological activity (Jørgensen et al., 2005b). These toxins are also resistant against human gastrointestinal proteases (Balaban and Rasooly, 2000). Methicillin-resistant *S. aureus* (MRSA) isolated from cows' milk has also been found (Parisi et al., 2016; Keyvan et al., 2020).

*Streptococcus* species are major pathogens causing mastitis, alongside *S. aureus* and coliforms (Minst et al., 2012). Typical pathogenic species of the genus *Streptococcus* include *Streptococcus pneumoniae*, *Streptococcus pyogenes*, *S. agalactiae*, and *Streptococcus mutans* (Olajide and LaPointe, 2020). *S. agalactiae* is a highly contagious pathogen responsible for long-duration subclinical mastitis (Leelahapongsathon et al., 2016). It is cow-associated, well adapted to the mammalian glands (Minst et al., 2012), and causes invasive infections in humans (Bohnsack et al., 2004). Although control strategies implemented since 1960 have managed to reduce the incidence of mastitis due to *Streptococcus* spp. in several European countries and North America (Reyes et al., 2015), the last several years *S. agalactiae* has re-emerged in Scandinavian countries, such as Denmark and Norway (Reyes et al., 2015; Jørgensen et al., 2016). *Streptococcus dysgalactiae* subsp. *dysgalactiae* and *S. uberis* are considered environmental pathogens (Minst et al., 2012) also associated with mastitis (Olajide and LaPointe, 2020).

Brucellosis is one of the most widespread zoonotic diseases worldwide caused by *Brucella* spp. (Jansen et al., 2019). It is usually transmitted to humans through direct contact with infected animals or through the consumption of unpasteurized milk and dairy products (Kamal et al., 2013). Nevertheless, *Brucella* spp. are killed by pasteurization, therefore, pasteurized milk and dairy products thereof are safe for consumption (Adams and Moss, 2008; Jansen et al., 2019). In the EU, brucellosis has been eradicated in livestock by most of the member states, leading to an officially brucellosis-free status. As a result, foodborne brucellosis outbreaks are rare in large areas of the EU (EFSA and ECDC, 2018b). Although brucellosis has been eradicated from other countries as well, for example, Australia, the United States, and Canada, it still remains endemic in most areas of the world (De Santis et al., 2011) and a public health issue in the developing countries (Qasem et al., 2015).

Spore-forming bacteria associated with raw milk include predominantly *Bacillus* spp., such as *B. cereus* and *B. licheniformis*, and occasionally *Clostridium* species, such as *Clostridium perfringens*, and *Clostridium botulinum*. The count levels show seasonality, with higher levels during winter than in summer, due to the spore-contaminated bedding materials and silage (Erkmen and Bozoglu, 2016). Although pasteurization destroys vegetative cells of *B. cereus* and *C. botulinum*, it does not eliminate their heat-resistant spores (Claeys et al., 2013; Tirloni et al., 2017). On the contrary, pasteurization may promote germination of spores that are subsequently able to grow and produce toxins in pasteurized milk during storage (Claeys et al., 2013). Symptoms due to *B. cereus* food poisoning are usually mild, although fatal outbreaks have been also reported and can mainly be attributed to the secretion of enterotoxins causing diarrhea and emetic syndromes (Dierick et al., 2005). The emetic toxin of *B. cereus* is very heat-resistant; however, it has never been shown to be present in milk (Claeys et al., 2013). Although the prevalence of *C. botulinum* spores in raw milk is not known, contamination levels are probably low (Lindström et al., 2010).

*Mycobacterium* spp. and, in particular, *Mycobacterium bovis*, *Mycobacterium tuberculosis*, and *Mycobacterium avium* subsp. *paratuberculosis* are also potential pathogenic bacteria related to raw milk. *M. bovis* and *M. tuberculosis* are members of the *M. tuberculosis* complex and the etiological agents of bovine and human tuberculosis, respectively (Bolaños et al., 2017), whereas *M. paratuberculosis*, the etiological agent for paratuberculosis (Fusco et al., 2020). Foodborne tuberculosis can be transmitted through contaminated foods, primarily milk (Mariam, 2014). Reports on foodborne transmission are limited mainly to *M. bovis* contaminated foods (Sevilla et al., 2017), although detection of *M. paratuberculosis* in raw milk has been also reported (Fusco et al., 2020). In raw milk, *M. bovis* and atypical mycobacteria have been isolated in Tanzania, *M. bovis* and *Mycobacterium africanum* in Nigeria, *M. bovis* in Tunisia, whereas atypical *Mycobacterium* species in Turkey and Brazil (Bolaños et al., 2017). As with brucellosis, tuberculosis due to *M. bovis* is considered a rare disease in the EU due to the eradicating control programs in cow husbandry practices and milk pasteurization (EFSA and ECDC, 2018b). It should be stressed, however, that in 2017, 185 confirmed *M. bovis* tuberculosis cases were reported from the 27 EU member states. In general, infection due to *M. bovis* is more frequent in countries that lack eradicating programs and pasteurization treatments (Pérez-Lago et al., 2014). Nonetheless, a 15-month-old US boy died due to tuberculosis caused by *M. bovis* in March 2004 (Winters et al., 2005).

*C. burnetii* is also a zoonotic pathogen with ubiquitous prevalence in the environment and the etiological agent of Q fever (Pexara et al., 2018). Consumption of contaminated milk or dairy products is one of the main sources of foodborne transmission, although it can be also transmitted through aerosols and skin (Abdali et al., 2018). Infected animals shed the pathogen into the urine, feces, birth products, and milk (Capuano et al., 2012). It has been reported that infected animal can shed *C. burnetii* viable cells in milk for time longer than 1 year (Enright et al., 1957). Pasteurization is designed to reduce the levels of *C. burnetii*, which is the most heat-resistant vegetative pathogen, into levels that do not pose threat to public health (Martin et al., 2018).

Apart from bacteria, some yeasts and fungi species can be also considered as potential pathogens, mainly due to the production of health-hazardous mycotoxins. Mycotoxins are a group of secondary metabolites produced by fungi, mostly belonging to the genera *Aspergillus*, *Penicillium*, and *Fusarium*, under low water availability and unfavorable conditions for bacterial growth (Yiannikouris and Jouany, 2002; Becker-Algeri et al., 2016). Among them, aflatoxin $B_1$, produced mainly by *Aspergillus flavus* and *Aspergillus parasiticus* can be consumed by lactating mammalians, such as cows, through contaminated feed. The ingested aflatoxin $B_1$ is metabolized through hydroxylation in the liver by cytochrome P450 enzyme into aflatoxin $M_1$, which is then shed into milk (Prandini et al., 2009; Becker-Algeri et al., 2016). The presence of aflatoxin $M_1$ in milk, even in small quantities, is of concern, mainly because milk is widely consumed by children whose metabolic and immune systems are still developing, hence are more susceptible to the toxic effects of aflatoxins (Serraino et al., 2019). Throughout the world, Europe has the most extensive and detailed regulations for mycotoxins in foods. More specifically, the EU set the limit of the total aflatoxin levels to no more than 20 μg $kg^{-1}$ in lactating dairy feeds and 0.05 μg $kg^{-1}$ in milk (EU Regulation 466/2001).

### 2.3.3 Unravelling the pathogenic milk microbiota: recent trends in molecular techniques

With regard to health and safety issues, the detection of pathogens is of paramount importance. The most commonly applied techniques include the culture-dependent ones alongside molecular and immunological methods (Lazcka et al., 2007). Albeit classical microbiological analysis is the gold standard in food microbiology, molecular techniques are widely used for bacterial identification using several targets and even different protocols for the same target (Fusco and Quero, 2014). Conventional PCR-detection methods rely on targeting one or multiple genes, such as *invA* in *Salmonella* (Omar et al., 2018; Singh et al., 2018), *hlyA* (Paul et al., 2015) and *prfA* (Haghi et al., 2015) in *L. monocytogenes*, *nuc* in *S. aureus* (Faggiano et al., 2017; Keyvan et al., 2020) or *mecA* in MRSA *S. aureus* (Keyvan et al., 2020), *stx1*, *stx2*, and *eae* in *E. coli* (Faggiano et al., 2017; Jaakkonen et al., 2017, 2019), *hipO* in *C. jejuni* (Haghi et al., 2015), *bcsp31* (Haghi et al., 2015; Hoffman et al., 2016) and *omp25* (Hoffman et al., 2016) in *Brucella*, *com1*, *com2* (Haghi et al., 2015; Cornejo et al., 2020), and IS*1111* (Cornejo et al., 2020) in *C. burnetii* and *IS6110* for the *Mycobacterium* complex (Haghi et al., 2015).

In recent years, modern epidemiological studies also rely on advanced molecular tools for a more accurate fingerprinting of milk pathogens. In an outbreak linked to the consumption of raw cow milk contaminated with sorbitol-fermenting *E. coli* O157 in Finland (2012), the isolates were initially typed using pulsed-field gel electrophoresis (PFGE) and then subjected to whole-genome sequence (WGS) that diversified the pathogenic *E. coli* strains incriminated for the outbreak from the non-O157 ones on the basis of the presence of virulence genes (Jaakkonen et al., 2017). In a follow-up study of a milkborne campylobacteriosis outbreak due to the consumption of raw cows' milk in Finland (2012), the same authors analyzed the genomic diversity of *C. jejuni* isolates using a combination of PFGE, WGS, Multilocus Sequence Typing (MLST), and whole-genome MLST (wgMLST). According to the results, the outbreak caused by a single *C. jejuni* strain was persistent in bulk tank milk due to biofilm formation for a period of 7 months or longer (Jaakkonen et al., 2020). The significant importance of molecular techniques in outbreaks was also evident in the study of Butcher et al. (2016). The authors used the multilocus variable-number tandem repeat analysis (MLVA) alongside WGS as typing method to study an outbreak due to STEC O157 contamination in raw milk in England. Interestingly, based on the results of WGS, four additional cases were linked to this specific outbreak, and, ultimately, associated with the implicated farm. The same workflow, that is, molecular typing of bacterial isolates at first and then WGS, has been extensively applied to identify genomic variations or unravel epidemiological characteristics of bacterial isolates linked to milkborne outbreaks (Denny et al., 2008; Heuvelink et al., 2009; Longenberger et al., 2013; Lahti et al., 2017; Burakoff et al., 2018; Treacy et al., 2019; Kenyon et al., 2020).

Apart from their application in epidemiological outbreak studies, the aforementioned techniques have been also used to elucidate the persistence/prevalence of foodborne pathogens in milk and/or dairy farms. Recently, Jaakkonen et al. (2019) used a combination of typing techniques, that is, PFGE and MLST, to explore the genomic variations of STEC and *C. jejuni* in Finnish dairy farms and raw milk. The authors observed persistent contamination of the herds by the two pathogens and occasional STEC milk contamination during milking (Jaakkonen et al., 2019). PFGE along with WGS has been also used to assess the prevalence of *L. monocytogenes* in raw cows' milk from vending machines. Interestingly, it was confirmed that all *L. monocytogenes* belonged to a single clone (Ricchi et al., 2019).

The state-of-the-art approach of metagenomics has been widely used recently to shed light on the microbiota of raw milk, including pathogens, and the association of the microbiota with several factors, such as animal health, season, and animal diet. In detail, Hoque et al. (2019) employed shotgun metagenomics to analyze the bacterial microbiota of clinical mastitis and healthy cows' milk samples. Based on the results, the biodiversity of mastitis milk samples was higher compared to the healthy ones. At the genus level, *Escherichia*, *Pseudomonas*, *Staphylococcus*, *Streptococcus*, *Bacillus*, *Campylobacter*, *Klebsiella*, and *Salmonella* were the most abundant genera found in mastitis milk samples. However, due to the high taxonomic resolution of the data generated by shotgun metagenomics, the authors were able to identify bacteria species accurately. Among them, a number of species including *Klebsiella oxytoca*, *Klebsiella aerogenes*, *Pseudomonas mendocina*, *Pseudomonas thermotolerans*, and *Enterobacter cloacae* were identified for the first time in mastitis milk samples, and, according to the authors, this could possibly play a role in the mammary gland pathogenesis. Moreover, the functional profiles of the microbial communities in clinical mastitis and healthy cows' milk samples were also different. Most importantly, genes coding for epithelial cell signaling, epithelial cell invasion, legionellosis, and *Vibrio cholerae* pathogenic cycle were most abundant in the clinical mastitis metagenome samples.

In a similar study, Bhatt et al. (2012) analyzed subclinical mastitis milk samples from different cattle breeds using shotgun metagenomics. At the species level, *E. coli*, *P. aeruginosa*, *S. aureus*, *Salmonella enterica*, *B. cereus*, *K. pneumoniae*, and *S. flexneri* were found to be predominant with varying abundances depending on the breed. Furthermore, metabolic profiles of the metagenomes revealed the presence of genes associated with the 'virulence, disease and defense' subsystem of the System for Exchange of Excise Data (SEED) database. In this subsystem, genes related to antibiotic, multidrug, and methicillin resistance, as well as toxic compounds were predominant among all the metagenomes analyzed.

The microbiota of mastitis milk samples has also been evaluated using 16S metagenomics analysis. The most frequent bacteria detected include the genera *Streptococcus*, *Staphylococcus*, *Escherichia*, and *Salmonella*. However, other genera, such as *Klebsiella*, *Clostridium*, *Coxiella*, *Oscillospira*, and *Aerococcus* have been also identified in lower proportions (Oikonomou et al., 2012, 2014; Kuehn et al., 2013; Ganda et al., 2016; Metzger et al., 2018; Taponen et al., 2019; Vasquez et al., 2019).

Apart from mastitis, seasonal variations may also affect the microbiota of raw milk. Using 16S metagenomics, Kable et al. (2016) analyzed raw cows' milk samples from tanker trucks during three seasons (spring, summer, and fall). The results obtained revealed highly diverse bacterial communities, which exhibited seasonal variations. Interestingly, spring raw milk samples contained the highest number of operational taxonomic units, while the lowest ones were detected in the fall milk samples. The averaged bacterial microbiota of raw milk samples collected in the fall was dominated by the genera *Streptococcus*, *Staphylococcus*, *Corynebacterium*, *Pseudomonas*, and *Clostridium*; the one in the spring by *Streptococcus*, *Staphylococcus*, *Corynebacterium*, *Clostridium*, and *Acinetobacter*; and that of summer by *Streptococcus*, *Staphylococcus*, *Clostridium*, and *Corynebacterium*. Nevertheless, even with this complexity, a core microbiota was identified encompassing members of the Firmicutes, Actinobacteria, Bacteroidetes, Proteobacteria, and Tenericutes phyla. At the genus level, the core microbiota consisted of 17 genera with *Staphylococcus*, *Streptococcus*, and *Corynebacterium* being the most abundant. Endospore-forming genera *Clostridium* and *Bacillus* were also identified as part of the core microbiota, although at a lower proportion.

Similarly, Porcellato et al. (2018) performed a large-scale study analyzing milk samples every month during a 13-month period using 16S metagenomics. The milk samples were collected from two dairies located in two different geographical regions in Norway. It should be noted, however, that the selected dairies had similar climatic conditions but substantial variations in annual production volumes, that is, production < 40 million liters versus production > 80 million liters of milk. Although significant differences in microbial diversity were not detected between the two dairies, seasonal variations were observed concerning not only the genera identified but most importantly, the respective abundances. Most notably, the genera *Pseudomonas*, *Streptococcus*, and *Acinetobacter* were predominant in the winter milk and *Bacillus* in the summer one.

Finally, Zhang et al. (2015) investigated the impact of a high versus low concentrate diet on the raw cows' milk microbiota using 16S metagenomics analysis. At the phylum level, Proteobacteria was found to be more abundant in the milk samples from the cows fed the high concentrate diet; Firmicutes, Cyanobacteria, Tenericutes, and Fibrobacteres in the milk from the cows fed the low concentrate diet, while Actinobacteria, Bacteroidetes, and Acidobacteria were not influenced by the diet. Differences were also observed at the genus level. Most importantly, the relative abundances of the genera *Streptococcus*, *Brevundimonas*, and *Leucobacter*, as well as *Chryseobacterium*, *Streptococcus*, and *Corynebacterium* were found at higher proportions in high and low concentrate feeding, respectively. It should be noted though that despite the differences observed between the microbial fingerprints, the samples were not separated in the PCoA analysis.

## 3 Conclusion

Billions of people around the globe consume milk and dairy products every day, as part of a healthy and balanced diet. Due to its high nutritional value, milk comprises an ideal growth substrate for both beneficial and harmful microorganisms. These microbes, either dominant or subdominant, can have a number of roles, such as facilitating dairy fermentations, promoting health, causing spoilage, or even disease. The enormous evolution of molecular techniques and omics technologies over the last decades has enabled a significant breakthrough in the study of microbial communities. Among food ecosystems, the microbiota of milk is probably one of the most complex food ecosystems due to the multiplicity of sources of contamination. According to recent studies, there are several factors related to farming and storage of raw milk that have been shown to influence the diversity of the raw milk microbiota. However, still much remains to be done not only on the fingerprinting of the raw milk microbiota but also on the functionality of the microbiota and the interactions among the microbes. In this perspective, the decreasing costs of sequencing technologies and the availability of powerful bioinformatics tools and pipelines open new avenues for the study of food microbial ecosystems. Therefore, approaches, such as shotgun metagenomics and metatranscriptomics are likely to become more regularly employed. It should be stressed, however, that culture-dependent techniques cannot be completely replaced by culture-independent ones since cultivation and isolation of microorganisms are still deemed as strictly necessary. Overall, the combination of both approaches could provide fundamental insights into the milk microbiota.

## 4 Where to look for further information

Updated information regarding the milk microbiota, including milk-borne pathogens, can always be found on the official websites of the European Food Safety Authority (EFSA), European Centre for Disease Prevention (ECDC), Food and Drug Administration (FDA), Centers for Disease Control and Prevention (CDC), World Health Organization (WHO), Food and Agriculture Organization (FAO), International Dairy Federation (IDF) and National Dairy Council of US (NDC):

- https://www.efsa.europa.eu/en.
- https://www.ecdc.europa.eu/en.
- https://www.fda.gov.
- https://www.cdc.gov.
- https://www.who.int.
- http://www.fao.org.
- https://fil-idf.org.
- https://www.usdairy.com.

United States Department of Agriculture (USDA) published a report on raw milk consumption and food safety risks:

- https://www.ers.usda.gov/publications/pub-details/?pubid=91109.

The FAO's recommendations for the proper processing and handling of milk and milk products:

- http://www.fao.org/fileadmin/user_upload/livestockgov/documents/CXP_057e.pdf.

The FAO recently published a review on dairy market, referring to global milk production, dairy prices and international trade of milk products:

- http://www.fao.org/3/cb4230en/cb4230en.pdf?fbclid=IwAR2mR2yiq_ERQXHz6Kzr5CYfrB7GABXCNFL-Jz0oOMhhjsmwP9f6hSskheE.

In addition, the EFSA presents annual reports regarding the results of zoonoses monitoring activities:

- https://www.ecdc.europa.eu/en/all-topics-z/food-and-waterborne-diseases-and-zoonoses/surveillance-and-disease-data/eu-one-health.

In 2015, the EFSA published a scientific opinion regarding the bio-hazards associated with the consumption of raw drinking milk:

- https://www.efsa.europa.eu/en/efsajournal/pub/3940.

Information regarding the consumption raw milk can also be found in the following website of the FDA:

- https://www.fda.gov/food/buy-store-serve-safe-food/dangers-raw-milk-unpasteurized-milk-can-pose-serious-health-risk.

Outbreaks studies can be found in the following website of the CDC:

- https://www.cdc.gov/foodsafety/rawmilk/rawmilk-outbreaks.html.

Valuable scientific information can always be found in the "Morbidity and Mortality Weekly Report" series published by the CDC (https://www.cdc.gov/mmwr/index.html) and "Eurosurveillance" published by the ECDC: https://www.eurosurveillance.org).

The global burden of foodborne diseases has been recently estimated by WHO:

- https://apps.who.int/iris/bitstream/handle/10665/199350/9789241565165_eng.pdf?sequence=1&isAllowed=y.

## 5 References

Abdali, F., Hosseinzadeh, S., Berizi, E. and Shams, S. (2018). Prevalence of *Coxiella burnetii* in unpasteurized dairy products using nested PCR assay, *Iran. J. Microbiol.* 10(4), 220-226.

Adams, M. R. and Moss, M. O. (2008). *Food Microbiology* (3rd edn.), Cambridge: Royal Society of Chemistry Publishing.

Addis, M. F., Tanca, A., Uzzau, S., Oikonomou, G., Bicalho, R. C. and Moroni, P. (2016). The bovine milk microbiota: insights and perspectives from -omics studies, *Mol. Biosyst* 12(8), 2359-2372.

Albonico, F., Barelli, C., Albanese, D., Manica, M., Partel, E., Rosso, F., Ripellino, S., Pindo, M., Donati, C., Zecconi, A., Mortarino, M. and Hauffe, H. C. (2020). Raw milk and fecal microbiota of commercial Alpine dairy cows varies with herd, fat content and diet, *PLoS ONE* 15(8), e0237262.

Alekish, M. O. (2015). The association between the somatic cell count and isolated microorganisms during subclinical mastitis in heifers in Jordan, *Vet. Med.* 60(2), 71-76.

Altieri, C., Ciuffreda, E., Di Maggio, B. and Sinigaglia, M. (2017). Lactic acid bacteria as starter cultures. In: Speranza, B., Bevilacqua, A., Corbo, M. R. and Sinigaglia, M. (Eds), *Starter Cultures in Food Production*, Weinheim, Germany: John Wiley & Sons, Wiley-VCH Verlag GmbH & Co. KGaA, pp. 1-15.

An, J. U., Ho, H., Kim, J., Kim, W. H., Kim, J., Lee, S., Mun, S. H., Guk, J. H., Hong, S. and Cho, S. (2018). Dairy cattle, a potential reservoir of human campylobacteriosis: epidemiological and molecular characterization of *Campylobacter jejuni* from cattle farms, *Front. Microbiol.* 9, 3136.

André, S., Vallaeys, T. and Planchon, S. (2017). Spore-forming bacteria responsible for food spoilage, *Res. Microbiol.* 168(4), 379-387.

Awasti, N. and Anand, S. (2020). The role of yeast and molds in dairy industry: an update. In: Minj, J., Sudhakaran, V. A. and Kumari, A. (Eds), *Dairy Processing: Advanced Research to Applications*, Singapore: Springer Nature Singapore Pte Ltd, pp. 243-262.

Baars, T. (2013). Milk consumption, raw and general, in the discussion on health or hazard. *Nut. Food Res.* 1(2), 91-107.

Balaban, N. and Rasooly, A. (2000). Staphylococcal enterotoxins, *Int. J. Food Microbiol.* 61(1), 1-10.

Bautista, D. A. (2014). Problems caused by bacteria. In: Batt, C. A. (Translator) amd Tortorello, M. L. (Eds.). *Encyclopedia of Food Microbiology* (2nd edn.), San Diego: Academic Press, pp. 465-470.

Becker-Algeri, T. A., Castagnaro, D., de Bortoli, K., de Souza, C., Drunkler, D. A. and Badiale-Furlong, E. (2016). Mycotoxins in bovine milk and dairy products: a review, *J. Food Sci.* 81(3), R544-R552.

Bhat, M., Denny, J., MacDonald, K., Hofmann, J. and Jain, S. (2007). Escherichia coli O157:H7 infection associated with drinking raw milk–Washington and Oregon, November-December 2005, *MMWR Morb. Mortal. Wkly Rep.* 56(8), 165-167.

Bhatt, V. D., Ahir, V. B., Koringa, P. G., Jakhesara, S. J., Rank, D. N., Nauriyal, D. S., Kunjadia, A. P. and Joshi, C. G. (2012). Milk microbiome signatures of subclinical mastitis-affected cattle analysed by shotgun sequencing, *J. Appl. Microbiol.* 112(4), 639-650.

Bohnsack, J. F., Whiting, A. A., Martinez, G., Jones, N., Adderson, E. E., Detrick, S., Blaschke-Bonkowsky, A. J., Bisharat, N. and Gottschalk, M. (2004). Serotype III Streptococcus agalactiae from bovine milk and human neonatal infections, *Emerg. Infect. Dis.* 10(8), 1412–1419.

Bolaños, C. A. D., de Paula, C. L., Guerra, S. T., Franco, M. M. J. and Ribeiro, M. G. (2017). Diagnosis of mycobacteria in bovine milk: an overview, *Rev. Inst. Med. Trop. S. Paulo* 59, e40.

Bonsaglia, E. C. R., Gomes, M. S., Canisso, I. F., Zhou, Z., Lima, S. F., Rall, V. L. M., Oikonomou, G., Bicalho, R. C. and Lima, F. S. (2017). Milk microbiome and bacterial load following dry cow therapy without antibiotics in dairy cows with healthy mammary gland, *Sci. Rep.* 7(1), Article 8067.

Bourdichon, F., Casaregola, S., Farrokh, C., Frisvad, J. C., Gerds, M. L., Hammes, W. P., Harnett, J., Huys, G., Laulund, S., Ouwehand, A., Powell, I. B., Prajapati, J. B., Seto, Y., Ter Schure, E., Van Boven, A., Vankerckhoven, V., Zgoda, A., Tuijtelaars, S. and Hansen, E. B. (2012). Food fermentations: microorganisms with technological beneficial use, *Int. J. Food Microbiol.* 154(3), 87–97.

Buehler, A. J., Evanowski, R. L., Martin, N. H., Boor, K. J. and Wiedmann, M. (2017). Internal transcribed spacer (ITS) sequencing reveals considerable fungal diversity in dairy products, *J. Dairy Sci.* 100(11), 8814–8825.

Burakoff, A., Brown, K., Knutsen, J., Hopewell, C., Rowe, S., Bennett, C. and Cronquist, A. (2018). Outbreak of fluoroquinolone-resistant *Campylobacter jejuni* infections associated with raw milk consumption from a herdshare dairy–Colorado, 2016, *MMWR Morb. Mortal. Wkly Rep.* 67(5), 146–148.

Butcher, H., Elson, R., Chattaway, M. A., Featherstone, C. A., Willis, C., Jorgensen, F., Dallman, T. J., Jenkins, C., Mclauchlin, J., Beck, C. R. and Harrison, S. (2016). Whole genome sequencing improved case ascertainment in an outbreak of Shiga toxin-producing *Escherichia coli* O157 associated with raw drinking milk, *Epidemiol. Infect.* 144(13), 2812–2823.

Capuano, F., Mancusi, A., Casalinuovo, F., Perugini, A., Proroga, Y., Guarino, A. and Berri, M. (2012). Real-time PCR-based detection of *Coxiella burnetii* in cheeses, *Eur. Food Res. Technol.* 235(6), 1181–1186.

Claeys, W. L., Cardoen, S., Daube, G., De Block, J., Dewettinck, K., Dierick, K., De Zutter, L., Huyghebaert, A., Imberechts, H., Thiange, P., Vandenplas, Y. and Herman, L. (2013). Raw or heated cow milk consumption: review of risks and benefits, *Food Control* 31(1), 251–262.

Cocolin, L., Aggio, D., Manzano, M., Cantoni, C. and Comi, G. (2002). An application of PCR-DGGE analysis to profile the yeast populations in raw milk, *Int. Dairy J.* 12(5), 407–411.

Cocolin, L. and Ercolini, D. (2015). Zooming into food-associated microbial consortia: a 'cultural' evolution, *Curr. Opin. Food Sci.* 2, 43–50.

Cornejo, J., Araya, P., Ibáñez, D., Hormazabal, J. C., Retamal, P., Fresno, M., Herve, L. P. and Lapierre, L. (2020). Identification of *Coxiella burnetii* in tank raw cow milk: first findings from Chile, *Vector Borne Zoonotic Dis.* 20(3), 228–230.

Cossaboom, C. M., Kharod, G. A., Salzer, J. S., Tiller, R. V., Campbell, L. P., Wu, K., Negrón, M. E., Ayala, N., Evert, N., Radowicz, J., Shuford, J. and Stonecipher, S. (2018). Notes from the field: *Brucella abortus* vaccine strain RB51 infection and exposures associated with raw milk consumption-wise County, Texas, 2017, *M.M.W.R Morb. Mortal. Wkly Rep.* 67(9), 286.

Cousin, M. E., Härdi-Landerer, M. C., Völk, V. and Bodmer, M. (2018). Control of *Staphylococcus aureus* in dairy herds in a region with raw milk cheese production:

farmers' attitudes, knowledge, behaviour and belief in self-efficacy, *BMC Vet. Res.* 14(1), 46.

Cremonesi, P., Ceccarani, C., Curone, G., Severgnini, M., Pollera, C., Bronzo, V., Riva, F., Addis, M. F., Filipe, J., Amadori, M., Trevisi, E., Vigo, D., Moroni, P. and Castiglioni, B. (2018). Milk microbiome diversity and bacterial group prevalence in a comparison between healthy Holstein Friesian and Rendena cows, *PLoS ONE* 13(10), e0205054.

Dahlberg, J., Sun, L., Persson Waller, K., Östensson, K., McGuire, M., Agenäs, S. and Dicksved, J. (2019). Microbiota data from low biomass milk samples is markedly affected by laboratory and reagent contamination, *PLoS ONE* 14(6), e0218257.

Davis, K. R., Dunn, A. C., Burnett, C., McCullough, L., Dimond, M., Wagner, J., Smith, L., Carter, A., Willardson, S. and Nakashima, A. K. (2016). *Campylobacter jejuni* infections associated with raw milk consumption - Utah, 2014, *MMWR Morb. Mortal. Wkly Rep.* 65(12), 301–305.

Deeth, H. C., Khusniati, T., Datta, N. and Wallace, R. B. (2002). Spoilage patterns of skim milk and whole milks, *J. Dairy Res.* 69(2), 227–241.

De Filippis, F., Parente, E. and Ercolini, D. (2018). Recent past, present, and future of the food microbiome, *Annu. Rev. Food Sci. Technol.* 9, 589–608.

Delbès, C., Ali-Mandjee, L. and Montel, M. C. (2007). Monitoring bacterial communities in raw milk and cheese by culture-dependent and -independent 16S rRNA gene-based analyses, *Appl. Environ. Microbiol.* 73(6), 1882–1891.

Denny, J., Bhat, M. and Eckmann, K. (2008). Outbreak of *Escherichia coli* O157:H7 associated with raw milk consumption in the Pacific Northwest, *Foodborne Pathog. Dis.* 5(3), 321–328.

de Oliveira, G. B., Favarin, L., Luchese, R. H. and McIntosh, D. (2015). Psychrotrophic bacteria in milk: how much do we really know?, *Braz. J. Microbiol.* 46(2), 313–321.

Derakhshani, H., Fehr, K. B., Sepehri, S., Francoz, D., De Buck, J., Barkema, H. W., Plaizier, J. C. and Khafipour, E. (2018a). Invited review: microbiota of the bovine udder: contributing factors and potential implications for udder health and mastitis susceptibility, *J. Dairy Sci.* 101(12), 10605–10625.

Derakhshani, H., Plaizier, J. C., De Buck, J., Barkema, H. W. and Khafipour, E. (2018b). Association of bovine major histocompatibility complex (BoLA) gene polymorphism with colostrum and milk microbiota of dairy cows during the first week of lactation, *Microbiome* 6(1), Article 203.

De Santis, R., Ciammaruconi, A., Faggioni, G., Fillo, S., Gentile, B., Di Giannatale, E., Ancora, M. and Lista, F. (2011). High throughput MLVA-16 typing for *Brucella* based on the microfluidics technology, *BMC Microbiol.* 11(60), 60.

Dierick, K., Van Coillie, E., Swiecicka, I., Meyfroidt, G., Devlieger, H., Meulemans, A., Hoedemaekers, G., Fourie, L., Heyndrickx, M. and Mahillon, J. (2005). Fatal family outbreak of *Bacillus cereus*-associated food poisoning, *J. Clin. Microbiol.* 43(8), 4277–4279.

Dobranić, V., Kazazić, S., Filipović, I., Mikulec, N. and Zdolec, N. (2016). Composition of raw cow's milk microbiota and identification of enterococci by MALDI-TOF MS–short communication, *Vet. Arh.* 86(4), 581–590.

Dogan, B. and Boor, K. J. (2003). Genetic diversity and spoilage potentials among Pseudomonas spp. isolated from fluid milk products and dairy processing plants, *Appl. Environ. Microbiol.* 69(1), 130–138.

Doran, P., Carson, J., Costello, E. and More, S. J. (2009). An outbreak of tuberculosis affecting cattle and people on an Irish dairy farm, following the consumption of raw milk, *Ir. Vet. J.* 62(6), 390–397.
Doyle, C. J., Gleeson, D., O'Toole, P. W. and Cotter, P. D. (2017). High-throughput metataxonomic characterization of the raw milk microbiota identifies changes reflecting lactation stage and storage conditions, *Int. J. Food Microbiol.* 255, 1–6.
Driehuis, F., Hoolwerf, J. L. and Rademaker, J. L. W. (2016). Concurrence of spores of *Clostridium tyrobutyricum*, *Clostridium beijerinckii* and *Paenibacillus polymyxa* in silage, dairy cow faeces and raw milk, *Int. Dairy J.* 63, 70–77.
EFSA and ECDC (2018a). Multi-country outbreak of *Salmonella* Agona infections linked to infant formula, *EFSA Support. Publ.* EN-1365, 1–19.
EFSA and ECDC (2018b). The European Union summary report on trends and sources of zoonoses, zoonotic agents and food-borne outbreaks in 2017, *EFSA J.* 12(5500), 1–262.
Emerson, J. B., Adams, R. I., Roman, C. M. B., Brooks, B., Coil, D. A., Dahlhausen, K., Ganz, H. H., Hartmann, E. M., Hsu, T., Justice, N. B., Paulino-Lima, I. G., Luongo, J. C., Lymperopoulou, D. S., Gomez-Silvan, C., Rothschild-Mancinelli, B., Balk, M., Huttenhower, C., Nocker, A., Vaishampayan, P. and Rothschild, L. J. (2017). Schrodinger's microbes: tools for distinguishing the living from the dead in microbial ecosystems, *Microbiome* 5(1), Article 86.
Enright, J. B., Sadler, W. W. and Thomas, R. C. (1957). Pasteurization of milk containing the organism of Q fever, *Am. J. Public Health Nations Health* 47(6), 695–700.
Ercolini, D., Russo, F., Ferrocino, I. and Villani, F. (2009). Molecular identification of mesophilic and psychrotrophic bacteria from raw cow's milk, *Food Microbiol.* 26(2), 228–231.
Erkmen, O. and Bozoglu, T. F. (2016). *Food Microbiology: Principles into Practice*, West Sussex: John Wiley and Sons, Ltd.
EU (2001). Setting maximum levels for certain contaminants in foodstuff, Official Journal of European Commission, L077, Regulation 466/2001, 11–12.
Faggiano, G., Chieffi, D., Logrieco, A. F. and Fusco, V. (2017). Effect of refrigeration and probiotic adjunct on pathogenic and spoilage microorganisms in raw milk for direct human consumption, *J. Food Process. Preserv.* 42(2), e13499.
Fakruddin, M., Mannan, K. S. and Andrews, S. (2013). Viable but nonculturable bacteria: food safety and public health perspective, *ISRN Microbiol.* 2013, Article 703813.
Falentin, H., Rault, L., Nicolas, A., Bouchard, D. S., Lassalas, J., Lamberton, P., Aubry, J. M., Marnet, P. G., Le Loir, Y. and Even, S. (2016). Bovine teat microbiome analysis revealed reduced alpha diversity and significant changes in taxonomic profiles in quarters with a history of mastitis, *Front. Microbiol.* 7, Article 480.
Farrokh, C., Jordan, K., Auvray, F., Glass, K., Oppegaard, H., Raynaud, S., Thevenot, D., Condron, R., De Reu, K., Govaris, A., Heggum, K., Heyndrickx, M., Hummerjohann, J., Lindsay, D., Miszczycha, S., Moussiegt, S., Verstraete, K. and Cerf, O. (2013). Review of Shiga-toxin-producing *Escherichia coli* (STEC) and their significance in dairy production, *Int. J. Food Microbiol.* 162(2), 190–212.
Ferrocino, I. and Cocolin, L. (2017). Current perspectives in food-based studies exploiting multi-omics approaches, *Curr. Opin. Food Sci.* 13, 10–15.

Food and Drug Administration (2017). Grade "A" Pasteurized Milk Ordinance (PMO)-2017 Revision, United States Department of Health and Human Services, Public Health Service, Food and Drug Administration.

Franciosi, E., Settanni, L., Cavazza, A. and Poznanski, E. (2009). 'Biodiversity and technological potential of wild lactic acid bacteria from raw cows' milk, *Int. Dairy J.* 19(1), 3–11.

Fröhlich-Wyder, M. T., Arias-Roth, E. and Jakob, E. (2019). Cheese yeasts, *Yeast* 36(3), 129–141.

Fusco, V. and Quero, G. M. (2014). Culture-dependent and culture-independent nucleic-acid-based methods used in the microbial safety assessment of milk and dairy products, *Compr. Rev. Food Sci. Food Saf.* 13(4), 493–537.

Fusco, V., Chieffi, D., Fanelli, F., Logrieco, A. F., Cho, G. S., Kabisch, J., Böhnlein, C. and Franz, C. M. A. P. (2020). Microbial quality and safety of milk and milk products in the 21st century, *Compr. Rev. Food Sci. Food Saf.* 19(4), 2013–2049.

Galat, A., Dufresne, J., Combrisson, J., Thépaut, J., Boumghar-Bourtchai, L., Boyer, M. and Fourmestraux, C. (2016). Novel method based on chromogenic media for discrimination and selective enumeration of lactic acid bacteria in fermented milk products, *Food Microbiol.* 55, 86–94.

Ganda, E. K., Bisinotto, R. S., Lima, S. F., Kronauer, K., Decter, D. H., Oikonomou, G., Schukken, Y. H. and Bicalho, R. C. (2016). Longitudinal metagenomic profiling of bovine milk to assess the impact of intramammary treatment using a third-generation cephalosporin, *Sci. Rep.* 6, Article 37565.

Ganda, E. K., Gaeta, N., Sipka, A., Pomeroy, B., Oikonomou, G., Schukken, Y. H. and Bicalho, R. C. (2017). Normal milk microbiome is reestablished following experimental infection with *Escherichia coli* independent of intramammary antibiotic treatment with a third-generation cephalosporin in bovines, *Microbiome* 5(1), Article 74.

Garnier, L., Valence, F., Pawtowski, A., Auhustsinava-Galerne, L., Frotte, N., Baroncelli, R., Deniel, F., Coton, E. and Mounier, J. (2017a). Diversity of spoilage fungi associated with various French dairy products, *Int. J. Food Microbiol.* 241, 191–197.

Garnier, L., Valence, F. and Mournier, J. (2017b). Diversity and control of spoilage fungi in dairy products: an update, *Microorganisms* 5(3), 42.

Ghosh, P. R., Fawcett, D., Sharma, S. B. and Poinern, G. E. (2016). Progress towards sustainable utilisation and management of food wastes in the global economy, *Int. J. Food Sci.* 2016, 3563478.

Gobbetti, M., De Angelis, M., Di Cagno, R., Mancini, L. and Fox, P. F. (2015). Pros and cons for using non-starter lactic acid bacteria (NSLAB) as secondary/adjunct starters for cheese ripening, *Trends Food Sci. Technol.* 45(2), 167–178.

Gobbetti, M., Di Cagnoa, R., Calassob, M., Neviani, E., Fox, P. F. and De Angelis, M. (2018). Drivers that establish and assembly the lactic acid bacteria biota in cheeses, *Trends Food Sci. Technol.* 78, 244–254.

Guh, A., Phan, Q., Nelson, R., Purviance, K., Milardo, E., Kinney, S., Mshar, P., Kasacek, W. and Cartter, M. (2010). Outbreak of *Escherichia coli* O157 associated with raw milk, Connecticut, 2008, *Clin. Infect. Dis.* 51(12), 1411–1417.

Haghi, F., Zeighami, H., Naderi, G., Samei, A., Roudashti, S., Bahari, S. and Shirmast, P. (2015). Detection of major food-borne pathogens in raw milk samples from dairy bovine and ovine herds in Iran, *Small Rumin. Res.* 131, 136–140.

Hagi, T., Kobayashi, M. and Nomura, M. (2010). Molecular-based analysis of changes in indigenous milk microflora during the grazing period, *Biosci. Biotechnol. Biochem.* 74(3), 484–487.

Hancock, D., Besser, T., Lejeune, J., Davis, M. and Rice, D. (2001). The control of VTEC in the animal reservoir, *Int. J. Food Microbiol.* 66(1-2), 71-78.

Hantsis-Zacharov, E. and Halpern, M. (2007). Culturable psychrotrophic bacterial communities in raw milk and their proteolytic and lipolytic traits, *Appl. Environ. Microbiol.* 73(22), 7162-7168.

Heuvelink, A. E., van Heerwaarden, C., Zwartkruis-Nahuis, A., Tilburg, J. J. H. C., Bos, M. H., Heilmann, F. G., Hofhuis, A., Hoekstra, T. and de Boer, E. (2009). 'Two outbreaks of campylobacteriosis associated with the consumption of raw cows' milk, *Int. J. Food Microbiol.* 134(1-2), 70-74.

Hoffman, T., Rock, K., Mugizi, D. R., Muradrasoli, S., Lindahl-Rajala, E., Erume, J., Magnusson, U., Lundkvist, Å and Boqvist, S. (2016). Molecular detection and characterization of *Brucella* species in raw informally marketed milk from Uganda, *Infect. Ecol. Epidemiol.* 6, 32442.

Hoque, M. N., Istiaq, A., Clement, R. A., Sultana, M., Crandall, K. A., Siddiki, A. Z. and Hossain, M. A. (2019). Metagenomic deep sequencing reveals association of microbiome signature with functional biases in bovine mastitis, *Sci. Rep.* 9(1), Article 13536.

Hunt, K., Drummond, N., Murphy, M., Butler, F., Buckley, J. and Jordan, K. (2012). A case of bovine raw milk contamination with Listeria monocytogenes, *Ir. Vet. J.* 65(1), 13.

Hunt, D. C., Ocfemia, M. C. B., Neises, D., Hansen, G. and Aghoghovbia, S. T. (2009). Campylobacter jejuni infection associated with unpasteurized milk and cheese-Kansas, 2007, *MMWR Morb. Mortal. Wkly Rep.* 57(51), 1377-1379.

Ikeda, T., Tamate, N., Yamaguchi, K. and Makino, S. (2005). Mass outbreak of food poisoning disease caused by small amounts of staphylococcal enterotoxins A and H, *Appl. Environ. Microbiol.* 71(5), 2793-2795.

International Organization of Standardization/International Dairy Federation IS017792/ IDF (2006). Milk, milk products and mesophilic starter cultures-enumeration of citrate lactic acid bacteria–colony count technique at 25°C (vol. 180).

International Organization of Standardization/International Dairy Federation IS020128/ IDF (2006). Milk products–enumeration of presumptive lactobacillus acidophilus on a selective medium–colony count technique at 37°C (vol. 192).

International Organization of Standardization/International Dairy Federation ISO29981/ IDF (2010). Milk products–enumeration of presumptive bifidobacteria–colony count technique at 37°C (vol. 220).

Ivy, R. A., Ranieri, M. L., Martin, N. H., den Bakker, H. C., Xavier, B. M., Wiedmann, M. and Boor, K. J. (2012). Identification and characterization of psychrotolerant sporeformers associated with fluid milk production and processing, *Appl. Environ. Microbiol.* 78(6), 1853-1864.

Jaakkonen, A., Castro, H., Hallanvuo, S., Ranta, J., Rossi, M., Isidro, J., Lindström, M. and Hakkinen, M. (2019). Longitudinal study of shiga toxin-producign *Escherichia coli* and *Campylobacter jejuni* on Finnish dairy farms and in raw milk, *Appl. Environ. Microbiol.* 85(7), e02910-18.

Jaakkonen, A., Kivistö, R., Aarnio, M., Kalekivi, J. and Hakkinen, M. (2020). Persistent contamination of raw milk by *Campylobacter jejuni* ST-883, *PLoS ONE* 15(4), e0231810.

Jaakkonen, A., Salmenlinna, S., Rimhanen-Finne, R., Lundström, H., Heinikainen, S., Hakkinen, M. and Hallanvuo, S. (2017). Severe outbreak of sorbitol-fermenting *Escherichia coli* O157 via unpasteurized milk and farm visits, Finland 2012, *Zoonoses Public Health* 64(6), 468–475.

Jansen, W., Linard, C., Noll, M., Nöckler, K. and Al Dahouk, S. (2019). *Brucella*-positive raw milk cheese sold on the inner European market: a public health threat due to illegal import?, *Food Control* 100, 130–137.

Jay-Russell, M. T., Mandrell, R. E., Yuan, J., Bates, A., Manalac, R., Mohle-Boetani, J., Kimura, A., Lidgard, J. and Miller, W. G. (2013). Using major outer membrane protein typing as an epidemiological tool to investigate outbreaks caused by milk-borne *Campylobacter jejuni* isolates in California, *J. Clin. Microbiol.* 51(1), 195–201.

Jayarao, B. M., Donaldson, S. C., Straley, B. A., Sawant, A. A., Hegde, N. V. and Brown, J. L. (2006). A survey of foodborne pathogens in bulk tank milk and raw milk consumption among farm families in Pennsylvania, *J. Dairy Sci.* 89(7), 2451–2458.

Jørgensen, H. J., Mørk, T. and Rørvik, L. M. (2005a). The occurrence of *Staphylococcus aureus* on a farm with small-scale production of raw milk cheese, *J. Dairy Sci.* 88(11), 3810–3817.

Jørgensen, H. J., Mørk, T., Høgåsen, H. R. and Rørvik, L. M. (2005b). Enterotoxigenic *Staphylococcus aureus* in bulk milk in Norway, *J. Appl. Microbiol.* 99(1), 158 –166.

Jørgensen, H. J., Nordstoga, A. B., Sviland, S., Zadoks, R. N., Sølverød, L., Kvitle, B. and Mørk, T. (2016). *Streptococcus agalactiae* in the environment of bovine dairy herds–rewriting the textbooks?, *Vet. Microbiol.* 184, 64–72.

Kable, M. E., Srisengfa, Y., Laird, M., Zaragoza, J., McLeod, J., Heidenreich, J. and Marco, M. L. (2016). The core and seasonal microbiota of raw bovine milk in tanker trucks and the impact of transfer to a milk processing facility, *mBio* 7(4), e00836-16.

Kamal, I. H., Al Gashgari, B., Moselhy, S. S., Kumosani, T. A. and Abulnaja, K. O. (2013). Two-stage PCR assay for detection of human brucellosis in endemic areas, *BMC Infect. Dis.* 13(145), 145.

Karimi, R., Mortazavian, A. M. and Amiri-Rigi, A. (2012). Selective enumeration of probiotic microorganisms in cheese, *Food Microbiol.* 29(1), 1–9.

Kenyon, J., Inns, T., Aird, H., Swift, C., Astbury, J., Forester, E. and Decraene, V. (2020). *Campylobacter* outbreak associated with raw drinking milk, North West England, 2016, *Epidemiol. Infect.* 148, e13.

Keshri, G., Magan, N. and Voysey, P. (1998). Use of an electronic nose for the early detection and differentiation between spoilage fungi, *Lett. Appl. Microbiol.* 27(5), 261–264.

Keyvan, E., Yurdakul, O., Demirtas, A., Yalcin, H. and Bilgen, N. (2020). Identification of methicillin-resistant *Staphylococcus aureus* in bulk tank milk, *Food Sci. Technol.* 40(1), 150–156.

Kousta, M., Mataragas, M., Skandamis, P. and Drosinos, E. H. (2010). Prevalence and sources of cheese contamination with pathogens at farm and processing levels, *Food Control* 21(6), 805–815.

Kuehn, J. S., Gorden, P. J., Munro, D., Rong, R., Dong, Q., Plummer, P. J., Wang, C. and Phillips, G. J. (2013). Bacterial community profiling of milk samples as a means to understand culture-negative bovine clinical mastitis, *PLoS ONE* 8(4), e61959.

Lafarge, V., Ogier, J. C., Girard, V., Maladen, V., Leveau, J. Y., Gruss, A. and Delacroix-Buchet, A. (2004). Raw cow milk bacterial population shifts attributable to refrigeration, *Appl. Environ. Microbiol.* 70(9), 5644–5650.

Lahti, E., Rehn, M., Ockborn, G., Hansson, I., Ågren, J., Engvall, E. O. and Jernberg, C. (2017). Outbreak of campylobacteriosis following a dairy farm visit: confirmation by genotyping, *Foodborne Pathog. Dis.* 14(6), 326–332.

Lavoie, K., Touchette, M., St-Gelais, D. and Labrie, S. (2012). Characterization of the fungal microflora in raw milk and specialty cheeses of the province of Quebec, *Dairy Sci. Technol.* 92(5), 455–468.

Lazcka, O., Del Campo, F. J. and Muñoz, F. X. (2007). Pathogen detection: a perspective of traditional methods and biosensors, *Biosens. Bioelectron.* 22(7), 1205–1217.

Ledenbach, L. and Marshall, P. E. (2009). Microbiological spoilage of foods and beverages. In: Sperber, W. H. and Doyle, M. P. (Eds), *Compendium of the Microbiological Spoilage of Foods and Beverages, Food Microbiology and Food Safety*, New York: Springer, pp. 41–67.

Leelahapongsathon, K., Schukken, Y. H., Pinyopummintr, T. and Suriyasathaporn, W. (2016). Comparison of transmission dynamics between *Streptococcus uberis* and *Streptococcus agalactiae* intramammary infections, *J. Dairy Sci.* 99(2), 1418–1426.

Leroy, F. and De Vuyst, L. (2004). Lactic acid bacteria as functional starter cultures for the food fermentation industry, *Trends Food Sci. Technol.* 15(2), 67–78.

Li, N., Wang, Y., You, C., Rep, J., Chen, W., Zheng, H. and Liu, Z. (2018). Variation in raw milk microbiota Throughout 12 months and the impact of weather conditions, *Sci. Rep.* 8(1), Article 2371.

Lind, L., Reeser, J., Stayman, K., Deasy, M., Moll, M., Weltman, A., Urdaneta, V., Ostroff, S., Chirdon, W., Campagnolo, E. and Chen, T. (2007). Salmonella Typhimurium infection associated with raw milk and cheese consumption-Pennsylvania, 2007. *Morb. Mortal. Wkly Rep.* 56(44), 1161–1164.

Lindström, M., Myllykoski, J., Sivelä, S. and Korkeala, H. (2010). *Clostridium botulinum* in cattle and dairy products, *Crit. Rev. Food Sci. Nutr.* 50(4), 281–304.

Liptakova, A., Siegfried, L., Rosocha, J., Podracka, L., Bogyiova, E. and Kotulova, D. (2004). A family outbreak of haemolytic uraemic syndrome and haemorrhagic colitis caused by verocytotoxigenic *Escherichia coli* O157 from unpasteurised cow's milk in Slovakia, *Clin. Microbiol. Infect.* 10(6), 576–578.

Longenberger, A. H., Palumbo, A. J., Chu, A. K., Moll, M. E., Weltman, A. and Ostroff, S. M. (2013). Campylobacter jejuni infections associated with unpasteurized milk-multiple States, 2012, *Clin. Infect. Dis.* 57(2), 263–266.

Lorenzo, J. M., Munekata, P. E., Dominguez, R., Pateiro, M., Saraiva, J. A. and Franco, D. (2018). Chapter 3- main groups of microorganisms of relevance for food safety and stability: general aspects and overall description. In: Barba, F. J., Santana, A. S., Orlien, V. and Koubaa, M. (Eds), *Innovative Technologies for Food Preservation*, San Diego: Academic Press, pp. 53–107.

Lu, M. and Wang, N. (2017). Spoilage of milk and dairy products. In: Bevilacqua, A., Corbo, M. R. and Sinigaglia, M. (Eds), *The Microbiological Quality of Food*, Cambridge: Woodhead Publishing, pp. 151–178.

Lucey, J. A. (2015). Raw milk consumption: risks and benefits, *Nutr. Today* 50(4), 189–193.

Lücking, G., Stoecker, M., Atamer, Z., Hinrick, J. and Ehling-Schulz, M. (2013). Characterization of aerobic spore-forming bacteria associated with industrial dairy processing environments and product spoilage, *Int. J. Food Microbiol.* 166(2), 270–279.

Machado, S. G., Baglinière, F., Marchand, S., Van Coillie, E., Vanetti, M. C. D., De Block, J. and Heyndrickx, M. (2017). The biodiversity of the microbiota producing heat-resistant enzymes responsible for spoilage in processed bovine milk and dairy products, *Front. Microbiol.* 8, 302.

Mariam, S. H. (2014). Identification and survival studies of *Mycobacterium tuberculosis* within laboratory-fermented bovine milk, *BMC Res. Notes* 7(175), 175.

Marilley, L. and Casey, M. G. (2004). Flavours of cheese products: metabolic pathways, analytical tools and identification of producing strains, *Int. J. Food Microbiol.* 90(2), 139–159.

Martin, N. H., Boor, K. J. and Wiedmann, M. (2018). Symposium review: effect of post-pasteurization contamination on fluid milk quality, *J. Dairy Sci.* 101(1), 861–870.

Martin, N. H., Snyder, A. and Wiedmann, M. (2020). Yeasts and molds: spoilage mold in dairy products. This is an update of Sørhaug, T., Yeasts and Molds Spoilage Molds in Dairy Products. In: Fuquay, J. W. (Ed), *Encyclopedia of Dairy Sciences* (2nd edn.), San Diego: Academic Press, pp. 780–784.

Masiello, S. N., Martin, N. H., Trmčić, A., Wiedmann, M. and Boor, K. J. (2016). Identification and characterization of psychrotolerant coliform bacteria isolated from pasteurized fluid milk, *J. Dairy Sci.* 99(1), 130–140.

Mazurek, J., Salehi, E., Propes, D., Holt, J., Bannerman, T., Nicholson, L. M., Bundesen, M., Duffy, R. and Moolenaar, R. L. (2004). A multistate outbreak of *Salmonella enterica* serotype Typhimurium infection linked to raw milk consumption–Ohio, 2003, *J. Food Prot.* 67(10), 2165–2170.

McPhee, J. D. and Griffiths, M. W. (2011). Psychrotrophic bacteria *Pseudomonas* spp. In: Fuquay, J. W. (Ed), *Encyclopedia of Dairy Sciences* (2nd edn.), San Diego: Academic Press, pp. 379–383.

Metzger, S. A., Hernandez, L. L., Skarlupka, J. H., Walker, T. M., Suen, G. and Ruegg, P. L. (2018). A cohort study of the milk microbiota of healthy and inflamed bovine mammary glands From Dryoff Through 150 days in milk, *Front. Vet. Sci.* 5, 247.

Minst, K., Märtlbauer, E., Miller, T. and Meyer, C. (2012). Short communication: *Streptococcus* species isolated from mastitis milk samples in Germany and their resistance to antimicrobial agents, *J. Dairy Sci.* 95(12), 6957–6962.

Mungai, E. A., Behravesh, C. B. and Gould, L. H. (2015). Increased outbreaks associated with nonpasteurized milk, United States, 2007–2012, *Emerg. Infect. Dis.* 21(1), 119–122.

Munsch-Alatossava, P. and Alatossava, T. (2006). Phenotypic characterization of raw milk-associated psychrotrophic bacteria, *Microbiol. Res.* 161(4), 334–346.

Nagasawa, Y., Kiku, Y., Sugawara, K., Yabusaki, T., Oono, K., Fujii, K., Suzuki, T., Maehana, K. and Hayashi, T. (2019). The bacterial load in milk is associated with clinical severity in cases of bovine coliform mastitis, *J. Vet. Med. Sci.* 81(1), 107–112.

Nichols, M., Conrad, A., Whitlock, L., Stroika, S., Strain, E., Weltman, A., Johnson, L., DeMent, J., Reporter, R. and Williams, I. (2020). Short communication: multistate outbreak of *Listeria monocytogenes* infections retrospectively linked to unpasteurized milk using whole-genome sequencing, *J. Dairy Sci.* 103(1), 176–178.

O'Connell, A., Ruegg, P. L., Jordan, K., O'Brien, B. and Gleeson, D. (2016). The effect of storage temperature and duration on the microbial quality of bulk tank milk, *J. Dairy Sci.* 99(5), 3367–3374.

Oh, E., Chui, L., Bae, J., Li, V., Ma, A., Mutschall, S. K., Taboada, E. N., McMullen, L. M. and Jeon, B. (2018). Frequent implication of multistress-tolerant *Campylobacter jejuni* in human infections, *Emerg. Infect. Dis.* 24(6), 1037–1044.

Oikonomou, G., Addis, M. F., Chassard, C., Nader-Macias, M. E. F., Grant, I., Delbès, C., Bogni, C. I., Le Loir, Y. and Even, S. (2020). Milk microbiota: what are we exactly talking about?, *Front. Microbiol.* 11, Article 60.

Oikonomou, G., Bicalho, M. L., Meira, E., Rossi, R. E., Foditsch, C., Machado, V. S., Teixeira, A. G., Santisteban, C., Schukken, Y. H. and Bicalho, R. C. (2014). Microbiota of Cow's milk; distinguishing healthy, sub-clinically and clinically diseased quarters, *PLoS ONE* 9(1), e85904.

Oikonomou, G., Machado, V. S., Santisteban, C., Schukken, Y. H. and Bicalho, R. C. (2012) Microbial diversity of bovine mastitic milk as described by pyrosequencing of metagenomic 16s rDNA, *PLoS One* 7(10), e47671.

Olajide, A. M. and LaPointe, G. (2020). Raw milk associated microorganisms. In: McNamara, J. and McSweeney, P. L. H. (Eds), *Encyclopedia of Dairy Sciences* (3rd edn.), Netherlands: Elsevier.

Oliver, S. P., Boor, K. J., Murphy, S. C. and Murinda, S. E. (2009). Food safety hazards associated with consumption of raw milk, *Foodborne Pathog. Dis.* 6(7), 793–806.

Olsen, S. J., Ying, M., Davis, M. F., Deasy, M., Holland, B., Iampietro, L., Baysinger, C. M., Sassano, F., Polk, L. D., Gormley, B., Hung, M. J., Pilot, K., Orsini, M., Van Duyne, S., Rankin, S., Genese, C., Bresnitz, E. A., Smucker, J., Moll, M. and Sobel, J. (2004). Multidrug-resistant *Salmonella* Typhimurium infection from milk contaminated after pasteurization, *Emerg. Infect. Dis.* 10(5), 932–935.

Omar, D., Al-Ashmawy, M., Ramadan, H. and El-Sherbiny, M. (2018). Occurrence and PCR identification of *Salmonella* spp. from milk and dairy products in Mansoura, Egypt, *Int. Food Res. J.* 25(1), 446–452.

Owusu-Kwarteng, J., Akabanda, F., Agyei, D. and Jespersen, L. (2020). Microbial safety of milk production and fermented dairy products in Africa, *Microorganisms* 8(5), Article 752.

Panelli, S., Brambati, E., Bonacina, C. and Feligini, M. (2013). Diversity of fungal flora in raw milk from the Italian Alps in relation to pasture altitude, *Springerplus* 2, Article 405.

Parente, E., Ricciardi, A. and Zotta, T. (2020). The microbiota of dairy milk: a review, *Int. Dairy J.* 107, Article 104714.

Parisi, A., Caruso, M., Normanno, G., Latorre, L., Sottili, R., Miccolupo, A., Fraccalvieri, R. and Santagada, G. (2016). Prevalence, antimicrobial susceptibility and molecular typing of methicillin-Resistant *Staphylococcus aureus* (MRSA) in bulk tank milk from southern Italy, *Food Microbiol.* 58, 36–42.

Paul, M., Baranzoni, G. M., Albonetti, S. and Brewster, J. D. (2015). Direct, quantitative detection of *Listeria monocytogenes* in fresh raw whole milk by qPCR, *Int. Dairy J.* 41, 46–49.

Perez, P. F., Doré, J., Leclerc, M., Levenez, F., Benyacoub, J., Serrant, P., Segura-Roggero, I., Schiffrin, E. J. and Donnet-Hughes, A. (2007). Bacterial imprinting of the neonatal immune system: lessons from maternal cells?, *Pediatrics* 119(3), e724–e732.

Pérez-Lago, L., Navarro, Y. and García-de-Viedma, D. (2014). Current knowledge and pending challenges in zoonosis caused by *Mycobacterium bovis*: a review, *Res. Vet. Sci.* 97 (Suppl.), S94–S100.

Perin, L. M., Pereira, J. G., Bersot, L. S. and Nero, L. A. (2018). The microbiology of raw milk. In: Nero, L. A. and De Carvalho, A. F. (Eds), *Raw Milk: Balance Between Hazard and Benefits*, San Diego: Academic Press, pp. 45–64.

Perry, J. D. (2017). A decade of development of chromogenic culture media for clinical microbiology in an era of molecular diagnostics, *Clin. Microbiol. Rev.* 30(2), 449–479.

Petróczki, F. M., Tonamo, T. A., Béri, B. and Peles, F. (2019). The effect of breed and stage of lactation on the microbiological status of raw milk, *Acta agrar Debr.* 1(1), 37–45.
Petruzzi, L., Corbo, M. R. and Sinigaglia, M. (2017). Microbial spoilage of foods: fundamentals. In: Bevilacqua, A., Corbo, M. R. and Sinigaglia, M. (Eds), *The Microbiological Quality of Food*, Cambridge: Woodhead Publishing, pp. 1–21.
Pexara, A., Solomakos, N. and Govaris, A. (2013). Prevalence of methicillin-resistant *Staphylococcus aureus* in milk and dairy products, *Hellenic Vet. Med. Soc.* 64(1), 17–34.
Pexara, A., Solomakos, N. and Govaris, A. (2018). Q fever and prevalence of *Coxiella burnetii* in milk, *Trends Food Sci. Technol.* 71, 65–72.
Porcellato, D., Aspholm, M., Skeie, S. B., Monshaugen, M., Brendehaug, J. and Mellegård, H. (2018). Microbial diversity of consumption milk during processing and storage, *Int. J. Food Microbiol.* 266, 21–30.
Postollec, F., Mathot, A. G., Bernard, M., Divanach, M. L., Pavan, S. and Sohier, D. (2012). Tracking spore-forming bacteria in food: from natural biodiversity to selection by processes, *Int. J. Food Microbiol.* 158(1), 1–8.
Prandini, A., Tansini, G., Sigolo, S., Filippi, L., Laporta, M. and Piva, G. (2009). On the occurrence of aflatoxin $M_1$ in milk and dairy products, *Food Chem. Toxicol.* 47(5), 984–991.
Qasem, J. A., AlMomin, S., Al-Mouqati, S. A. and Kumar, V. (2015). Characterization and evaluation of an arbitrary primed polymerase chain reaction (PCR) product for the specific detection of *Brucella* species, *Saudi J. Biol. Sci.* 22(2), 220 –226.
Quigley, L., O'Sullivan, O., Stanton, C., Beresford, T. P., Ross, R. P., Fitzgerald, G. F. and Cotter, P. D. (2013a). The complex microbiota of raw milk, *FEMS Microbiol. Rev.* 37(5), 664–698.
Quigley, L., McCarthy, R., O'Sullivan, O., Beresford, T. P., Fitzgerald, G. F., Ross, R. P., Stanton, C. and Cotter, P. D. (2013b). The microbial content of raw and pasteurized cow milk as determined by molecular approaches, *J. Dairy Sci.* 96(8), 4928 –4937.
Quigley, L., O'Sullivan, O., Beresford, T. P., Ross, R. P., Fitzgerald, G. F. and Cotter, P. D. (2011). Molecular approaches to analysing the microbial composition of raw milk and raw milk cheese, *Int. J. Food Microbiol.* 150(2–3), 81–94.
Quintana, Á. R., Seseña, S., Garzón, A. and Arias, R. (2020). Factors affecting levels of airborne bacteria in dairy farms: a review, *Animals (Basel)* 10(3), Article 526.
Quiros, M., Wrent, P., Valderrama, M. J., de Siloniz, M. I. and Peinado, J. M. (2005). A beta-glucuronidase-based agar medium for the differential detection of the yeast Debaryomyces hansenii from foods, *J. Food Prot.* 68(4), 808–814.
Raats, D., Offek, M., Minz, D. and Halpern, M. (2011). Molecular analysis of bacterial communities in raw cow milk and the impact of refrigeration on its structure and dynamics, *Food Microbiol.* 28(3), 465–471.
Rainard, P. (2017). Mammary microbiota of dairy ruminants: fact or fiction?, *Rainard, Vet. Res.* 48, Article 25.
Regulation 853/2004/EC (2004). Regulation (EC) No 853/2004 of the European Parliament and of the Council of 29 April 2004 Laying Down Specific Hygiene Rules for Food of Animal Origin, L226, Brussels, pp. 22–82.
Reyes, J., Chaffer, M., Sanchez, J., Torres, G., Macias, D., Jaramillo, M., Duque, P. C., Ceballos, A. and Keefe, G. P. (2015). Evaluation of the efficacy of intramuscular versus

intramammary treatment of subclinical *Streptococcus agalactiae* mastitis in dairy cows in Colombia, *J. Dairy Sci.* 98(8), 5294–5303.

Ribeiro Júnior, J. C., de Oliveira, A. M., Silva, F. G., Tamanini, R., de Oliveira, A. L. M. and Beloti, V. (2018). The main spoilage-related psychrotrophic bacteria in refrigerated raw milk, *J. Dairy Sci.* 101(1), 75–83.

Ricchi, M., Scaltriti, E., Cammi, G., Garbarino, C., Arrigoni, N., Morganti, M. and Pongolini, S. (2019). Short communication: persistent contamination by *Listeria monocytogenes* of bovine raw milk investigated by whole-genome sequencing, *J. Dairy Sci.* 102(7), 6032–6036.

Samaržija, D., Zamberlin, Š. and Pogačić, T. (2012). Psychrotrophic bacteria and milk and dairy products quality, *Mljekarstvo* 62(2), 77–95.

Savin, K. W., Zawadzki, J., Auldist, M. J., Wang, J., Ram, D., Rochfort, S. and Cocks, B. G. (2019). *Faecalibacterium* diversity in dairy cow milk, *PLoS ONE* 14(8), e0221055.

Scatamburlo, T. M., Yamazi, A. K., Cavicchioli, V. Q., Pieri, F. A. and Nero, L. A. (2015). Spoilage potential of Pseudomonas species isolated from goat milk, *J. Dairy Sci.* 98(2), 759–764.

Schmid, D., Fretz, R., Winter, P., Mann, M., Höger, G., Stöger, A., Ruppitsch, W., Ladstätter, J., Mayer, N., de Martin, A. and Allerberger, F. (2009). Outbreak of staphylococcal food intoxication after consumption of pasteurized milk products, June 2007, Austria, *Wien. Klin. Wochenschr.* 121(3–4), 125–131.

Schneider, J., Mohle-Boetani, J., Vugia, D. and Menon, M. (2008). *Escherichia coli* 0157:H7 infections in children associated with raw milk and raw colostrum from cows-California, 2006, *MMWR Morb. Mortal. Wkly Rep.* 57(23), 625–628.

Segrè, A., Falasconi, L., Politano, A. and Vittuari, M. (2014). *Background Paper on the Economics of Food Loss and Waste (Unedited Working Paper)*, Rome: Food and Agriculture Organization.

Serraino, A., Bonilauri, P., Kerekes, K., Farkas, Z., Giacometti, F., Canever, A., Zambrini, A. V. and Ambrus, Á (2019). Occurrence of aflatoxin M1 in raw milk marketed in Italy: exposure assessment and risk characterization, *Front. Microbiol.* 10, Article 2516.

Sevilla, I. A., Molina, E., Tello, M., Elguezabal, N., Juste, R. A. and Garrido, J. M. (2017). Detection of mycobacteria by culture and DNA-based methods in animal-derived food products purchased at Spanish supermarkets, *Front. Microbiol.* 8, 1030.

Signs, K. A., Stobierski, M. G. and Gandhi, T. N. (2012). Q fever cluster among raw milk drinkers in Michigan, 2011, *Clin. Infect. Dis.* 55(10), 1387–1389.

Singh, P., Singh, R. V., Gupta, B., Tripathi, S. S., Tomar, K. S., Jain, S. and Sahni, Y. P. (2018). Prevalence study of *Salmonella* spp. in milk and milk products, *Asian J. Dairy Food Res.* 37(1), 7–12.

Skeie, S. B., Håland, M., Thorsen, I. M., Narvhus, J. and Porcellato, D. (2019). Bulk tank raw milk microbiota differs within and between farms: a moving goalpost challenging quality control, *J. Dairy Sci.* 102(3), 1959–1971.

Smit, G., Smit, B. A. and Engels, W. J. M. (2005). Flavour formation by lactic acid bacteria and biochemical flavour profiling of cheese products, *FEMS Microbiol. Rev.* 29(3), 591–610.

Sohier, D., Pavan, S., Riou, A., Combrisson, J. and Postollec, F. (2014). Evolution of microbiological analytical methods for dairy industry needs, *Front. Microbiol.* 5, 16–16.

Taponen, S., McGuinness, D., Hiitiö, H., Simojoki, H., Zadoks, R. and Pyörälä, S. (2019). Bovine milk microbiome: a more complex issue than expected, *Vet. Res.* 50(1), Article 44.

Tilocca, B., Costanzo, N., Morittu, V. M., Spina, A. A., Soggiu, A., Britti, D., Roncada, P. and Piras, C. (2020). Milk microbiota: characterization methods and role in cheese production, *J. Proteomics* 210, Article 103534.

Tirloni, E., Ghelardi, E., Celandroni, F., Bernardi, C. and Stella, S. (2017). Effect of dairy product environment on the growth of *Bacillus cereus*, *J. Dairy Sci.* 100(9), 7026-7034.

Treacy, J., Jenkins, C., Paranthaman, K., Jorgensen, F., Mueller-Doblies, D., Anjum, M., Kaindama, L., Hartman, H., Kirchner, M., Carson, T. and Kar-Purkayastha, I. (2019). Outbreak of shiga toxin-producing *Escherichia coli* O157:H7 linked to raw drinking milk resolved by rapid application of advanced pathogen characterisation methods, England, August to October 2017, *Euro Surveill.* 24(16), 1-7.

Ulusoy, B. H. and Chirkena, K. (2019). Two perspectives of *Listeria monocytogenes* hazards in dairy products: the prevalence and the antibiotic resistance, *Food Qual. Saf.* 3, 233-241.

Vacheyrou, M., Normand, A. C., Guyot, P., Cassagne, C., Piarroux, R. and Bouton, Y. (2011). Cultivable microbial communities in raw cow milk and potential transfers from stables of sixteen French farms, *Int. J. Food Microbiol.* 146(3), 253-262.

Van de Casteele, S., Vanheuverzwijn, T., Ruyssen, T., Van Assche, P., Swings, J. and Huys, G. (2006). Evaluation of culture media for selective enumeration of probiotic strains of lactobacilli and bifidobacteria in combination with yoghurt or cheese starters, *Int. Dairy J.* 16(12), 1470-1476.

Vasquez, A. K., Ganda, E. K., Capel, M. B., Eicker, S., Virkler, P. D., Bicalho, R. C. and Nydam, D. V. (2019). The microbiome of Escherichia coli and culture-negative nonsevere clinical mastitis: characterization and associations with linear score and milk production, *J. Dairy Sci.* 102(1), 578-594.

Verdier-Metz, I., Michel, V., Delbès, C. and Montel, M. C. (2009). Do milking practices influence the bacterial diversity of raw milk?, *Food Microbiol.* 26(3), 305-310.

Verraes, C., Vlaemynck, G., Van Weyenberg, S., De Zutter, L., Daube, G., Sindic, M., Uyttendaele, M. and Herman, L. (2015). A review of the microbiological hazards of dairy products made from raw milk, *Int. Dairy J.* 50, 32-44.

Vithanage, N. R., Dissanayake, M., Bolge, G., Palombo, E. A., Yeager, T. R. and Datta, N. (2016). Biodiversity of culturable psychrotrophic microbiota in raw milk attributable to refrigeration conditions, seasonality and their spoilage potential, *Int. Dairy J.* 57, 80-90.

Vithanage, N. R., Yeager, T. R., Jadhav, S. R., Palomboc, E. A. and Datta, N. (2014). Comparison of identification systems for psychrotrophic bacteria isolated from raw bovine milk, *Int. J. Food Microbiol.* 189, 26-38.

von Neubeck, M., Baur, C., Krewinkel, M., Stoeckel, M., Kranz, B., Stressler, T., Fischer, L., Hinrichs, J., Scherer, S. and Wenning, M. (2015). Biodiversity of refrigerated raw milk microbiota and their enzymatic spoilage potential, *Int. J. Food Microbiol.* 211, 57-65.

Weiler, N., Leotta, G. A., Zarate, M. N., Manfredi, E., Alvarez, M. E. and Rivas, M. (2011). Foodborne outbreak associated with consumption of ultrapasteurized milk in the Republic of Paraguay, *Rev. Argent. Microbiol.* 43(1), 33-36.

Weltman, A., Longenberger, A. H., Moll, M., Johnson, L., Martin, J. and and Beaudoin, A. (2013). Notes from the Field: recurrent outbreak of *Campylobacter jejuni* infections associated with a raw milk dairy-Pennsylvania, April–May 2013, *MMWR Morb. Mortal. Wkly Rep.* 62(34), 702.

Winters, A., Driver, C., Macaraig, M., Clark, C. and Munsiff, S. S. (2005). Human tuberculosis caused by *Mycobacterium bovis*-New York City, 2001–2004, *MMWR Morb. Mortal. Wkly Rep.* 54(24), 605–608.

Wu, H., Nguyen, Q. D., Tran, T. T. M., Tang, M. T., Tsuruta, T. and Nishino, N. (2019). Rumen fluid, feces, milk, water, feed, airborne dust, and bedding microbiota in dairy farms managed by automatic milking systems, *Anim. Sci. J.* 90(3), 445–452.

Yeluri Jonnala, B. R., McSweeney, P. L. H., Sheehan, J. J. and Cotter, P. D. (2018). Sequencing of the cheese microbiome and its relevance to industry, *Front. Microbiol.* 9, Article 1020.

Yiannikouris, A. and Jouany, J. P. (2002). Mycotoxins in feeds and their fate in animals: a review, *Anim. Res.* 51(2), 81–99.

Young, W., Hine, B. C., Wallace, O. A. M., Callaghan, M. and Bibiloni, R. (2015). Transfer of intestinal bacterial components to mammary secretions in the cow, *PeerJ.* 3, e888.

Yuan, L., Sadiq, F. A., Burmølle, M., Liu, T. and He, G. (2018a). Insights into bacterial milk spoilage with particular emphasis on the roles of heat-stable enzymes, Biofilms, and Quorum Sensing, *J. Food Prot.* 81(10), 1651–1660.

Yuan, L., Sadiq, F. A., Liu, T. J., Li, Y., Gu, J. S., Yang, H. Y. and He, G. Q. (2018b). Spoilage potential of psychrotrophic bacteria isolated from raw milk and the thermo-stability of their enzymes, *J. Zhejiang Univ. Sci. B* 19(8), 630–642.

Yuan, L., Sadiq, F. A., Burmølle, M., Wang, N. I. and He, G. (2019). Insights into psychrotrophic bacteria in raw milk: a review, *J. Food Prot.* 82(7), 1148–1159.

Yuan, L., Sadiq, F. A., Liu, T., Flint, S., Chen, J., Yang, H., Gu, J., Zhang, G. and He, G. (2017). Psychrotrophic bacterial populations in Chinese raw dairy milk, *LWT* 84, 409–418.

Zhao, L., de Hoog, G. S., Cornelissen, A., Lyu, Q., Mou, L., Liu, T., Cao, Y., Vatanshenassan, M. and Kang, Y. (2016). Prospective evaluation of the chromogenic medium CandiSelect 4 for differentiation and presumptive identification of non-*Candida albicans Candida* species, *Fungal Biol.* 120(2), 173–178.

Zhang, R., Huo, W., Zhu, W. and Mao, S. (2015). Characterization of bacterial community of raw milk from dairy cows during subacute ruminal acidosis challenge by high-throughput sequencing, *J. Sci. Food Agric.* 95(5), 1072–1079.

Zhang, Z., Li, X. P., Yang, F., Luo, J. Y., Wang, X. R., Liu, L. H. and Li, H. S. (2016). Influences of season, parity, lactation, udder area, milk yield, and clinical symptoms on intramammary infection in dairy cows, *J. Dairy Sci.* 99(8), 6484–6493.

Zhang, D., Palmer, J., Teh, K. H., Biggs, P. and Flint, S. (2019). '16S rDNA high-throughput sequencing and MALDI-TOF MS are complementary when studying psychrotrophic bacterial diversity of raw cows' milk, *Int. Dairy J.* 97, 86–91.

Zielińska, D., Ołdak, A., Rzepkowska, A. and Zieliński, K. (2018). Enumeration and identification of probiotic bacteria in food matrices. In: Holban, A. M. and Grumezescu, A. M. (Eds), *Advances in Biotechnology for Food Industry*, San Diego: Academic Press, pp. 167–196.

# Chapter 19

## Understanding the contribution of milk constituents to the texture of dairy products: liquid milk products

*Hilton C. Deeth, University of Queensland, Australia*



### 1 Introduction

The texture of liquid milk products is important for two major reasons: its effect on human sensory perception and the effects on processing. Major texture characteristics affecting the former are thickness or viscosity, consistency (e.g. presence of lumps) and mouthfeel; while the most important texture characteristic with regard to processing is viscosity, which affects processes such as heat transfer and pumping.

Several variables affect the texture of liquid milk products, of which the two major ones are the composition of the product and heating - which is particularly important in processing. Changes in texture also occur during storage, due to chemical, physicochemical and biological factors; some of these have a major effect on the physical and sensory properties of the product. Many of these changes are deemed to be normal while others result in defects.

Several reviews related to this topic have been published, including Vélez-Ruiz et al. (1997), Carr et al. (2002) and McCarthy and Singh (2009).

http://dx.doi.org/10.19103/AS.2022.0099.19

## 2 Viscosity of milk

Viscosity, or more generally, flow behavior of liquid dairy products is a major determinant of texture (Horne 1993). It is a measure of how easily a liquid flows when subjected to a shear force. Viscosity, $\eta$, is defined as the ratio of shear stress (t) (Pa) to shear rate ($\gamma$) ($s^{-1}$) and is measured in units of Pa s. The unit of centipoise (cp) is still used, the conversion factor being: 1 cp = $1 \times 10^{-3}$ Pa s or 1 mPa s.

An important compositional variable in viscosity is the volume fraction, $\phi$, both total and of individual components. It relates to the fraction of the product volume occupied by the hydrodynamic volume of the components. These volumes are determined by the voluminosities of the components, which were reported by Walstra et al. (1999) to be:

- Fat globules ~$1.11 \times 10^{-3}$ $m^3$ $kg^{-1}$
- Casein ~$3.9 \times 10^{-3}$ $m^3$ $kg^{-1}$
- Whey proteins ~$1.5 \times 10^{-3}$ $m^3$ $kg^{-1}$
- Lactose ~$1.0 \times 10^{-3}$ $m^3$ $kg^{-1}$

The viscosity of a product is largely determined by the total volume fraction of its components. The total volume fraction is a sum of the volume fractions of the individual components, that is, $\phi_{total} = \Sigma\phi_{i,}$ where $\phi_i$ is the volume fraction of component i. The volume fraction of an individual component is calculated by the formula $\phi_i = V_iC_{i,}$ where $V_i$ is the voluminosity ($m^3$ $kg^{-1}$) and $C_i$ is the concentration (kg $m^{-3}$) of component i. Thus, a 10% dispersion of micellar casein would have a volume fraction of 0.39, while a 10% solution of whey protein concentrate (WPC) would have a volume fraction of 0.15.

From the above data, it is apparent that low-molecular-weight materials such as lactose and minerals have a minor effect on viscosity, soluble whey proteins have much less effect than casein and fat globules also have less effect than casein. Therefore, the viscosity of a particular milk or cream depends largely on the concentration and physical state of the fat and proteins. If the whey proteins aggregate into small particles, they will exert more influence on viscosity than when they are in solution. A similar effect occurs when fat globules form aggregates; this may even result in converting a Newtonian to a non-Newtonian product (Mineshita et al. 1980).

In an investigation of the flow behavior of milk and cream, Morison et al. (2013) used mass fraction rather than volume fraction. They considered mass fraction to be more convenient for mixtures, as the volume fraction depends on the hydration of the molecules, which is influenced by concentration and competition for water from other molecules.

## 2.1 Flow behavior models

Liquid dairy products can be categorized as Newtonian or non-Newtonian. Newtonian liquids are defined by Equation 1:

$$\tau = \eta\gamma \tag{1}$$

where η, viscosity, is the slope of the line relating shear stress (τ) to shear rate (γ). The important property of a Newtonian liquid is that the viscosity is independent of the shear rate. Bovine milk and cream products with a total volume fraction up to ~0.6 are usually considered Newtonian, provided the shear rate is not too low and cold agglutination does not occur (see Section 5.2 for a description of cold agglutination). Rohm et al. (1996) reported raw milk to be non-Newtonian below a shear rate of 250 $s^{-1}$ but Newtonian at > 800 $s^{-1}$. Furthermore, some authors have reported on the non-Newtonian behavior of milk at temperatures below 40°C (e.g. Huppertz et al. 2020). According to Saif et al. (2004), caprine milk showed non-Newtonian behavior at temperatures below 60°C. At these temperatures, viscosity of all the types of caprine milk and cream varied exponentially with shear rate.

The viscosity of milk and cream products under Newtonian conditions at a particular temperature can be predicted by Eilers semi-empirical equation (Equation 2) (Walstra and Jenness 1984):

$$\eta = \eta_0 \left(1 + \frac{1.25\phi_{Total}}{1 - \phi_{Total} / \phi_{max}}\right)^2 \tag{2}$$

where $\eta_0$ is the viscosity of water plus milk salts; $\phi_{max}$ is the hypothetical volume fraction if all particles were packed closely together and is, according to Walstra and Jenness (1984), ~0.9 for fluid milk products but may be higher for evaporated milk. However, other values of $\phi_{max}$ have been reported. For example, Nöbel et al. (2012) used the lower figure of 0:68, Jeurnink and de Kruif (1993) used 0.74 and Van Vliet and Walstra (1980) cited 0.78 and 0.88.

de Kort et al. (2011) used Eilers equation to calculate the volume fraction of caseins in a micellar casein suspension, knowing the viscosity of the suspension. In that case, caseins were virtually the sole contributor to the total volume fraction.

The viscosity of water at 20°C is 1.005 mPa s, that of whole milk is 2.127 mPa s and skim milk is 1.790 mPa s (McCarthy and Singh 2009). As a rule of thumb, the viscosity of whole milk is about twice that of water.

When $\phi_{total}$ exceeds 0.6, viscosity increases steeply with $\phi_{total}$ and the rheological behavior becomes non-Newtonian (Walstra and Jenness 1984). Thus, concentrated liquid milk products with total solids (TS) of > ~ 30% exhibit non-Newtonian behavior which cannot be completely defined by Equation 1. The most common non-Newtonian flow behavior is shear thinning. While different models are used to describe non-Newtonian behavior, the Power Law model (Equation 3) is the simplest and most widely used:

$$\tau = K\gamma^n \quad (3)$$

where K is the consistency coefficient and n is the flow behavior index. The flow behavior of a product can thereby be described by its K and n values. Products with n = 1 are Newtonian and the consistency coefficient is the viscosity. Products with n < 1 are shear thinning or pseudoplastic and their apparent viscosity (viscosity at a particular shear rate) decreases with shear rate.

Two other models used to describe non-Newtonian behavior are the Bingham (Equation 4) and the Herschel-Bulkley models (Equation 5):

$$\tau = \tau_0 + \eta_p\gamma \quad (4)$$

$$\tau = \tau_0 + K\gamma^n \quad (5)$$

where $\tau_0$ represents the yield stress, the intercept on the shear stress axis of the shear stress–shear rate flow curve and $\eta_p$.is the plastic Bingham viscosity which is the slope of the linear flow curve beyond the yield stress. Both models incorporate a yield stress. If yield stress is zero, the Herschel-Bulkley model equates to the Power Law model.

Non-Newtonian behavior can be classified as time-independent or time-dependent. The two types of time-independent behavior are shear thinning or pseudoplastic and shear thickening or dilatant. Shear-thinning behavior is much more commonly encountered in liquid dairy products than shear thickening. The various types of flow curves representative of the four models are shown in Fig. 1.

An example of the application of the different models was presented by Xiang et al. (2011) for the flow behavior of reconstituted skim milk treated with pulsed electric field (PEF) technology. The rheological properties were measured at shear rates between 0 and 200 $s^{-1}$. The shear rate–shear stress flow curves exhibited a yield stress which eliminated Newtonian and Power Law as appropriate models. The Herschel-Bulkley model was found to give the best fit with the experimental data (Table 1).

A time-dependent flow behavior is thixotropy, where a product decreases in viscosity over time at a constant shear rate, possibly to an equilibrium level. The product may regain its viscosity when rested (Lewis and Heppell 2000). Cream exhibits thixotropic behavior (Everett 2007).

### 2.2 Effect of milk composition

The relationship between the texture of milk and its composition has been researched for over 100 years. The early work concerned unconcentrated

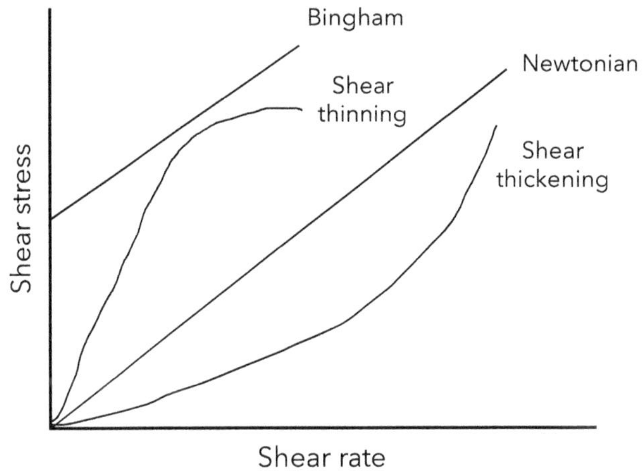


**Figure 1** Flow curves corresponding to different types of flow behavior.

**Table 1** Fitting of flow models for PEF-treated reconstituted skim milk (5% TS)

| Model | Yield stress (mPa) | Consistency index (K) (mPa s) | Flow behavior index (n) |
|---|---|---|---|
| Newtonian | Na | 1.80 | Nr |
| Power Law | Na | 5.08 | 0.777 |
| Bingham | 53.2 | 1.34 | Nr |
| Herschel-Bulkley | 39.6 | 1.46 | 0.988 |

na = not applicable; nr = not reported; data from Xiang et al. (2011).

milk, but more recent work has been on concentrated milk (Vélez-Ruiz et al. 1997) (see Section 3). Much of the research has been conducted on the effect of composition together with temperature; this is discussed in Section 2.4. In addition, some texture issues relating specifically to fat-containing products are addressed in Section 5.

Cox et al. (1959) reviewed six papers on the subject, published between 1913 and 1956. Each had developed empirical equations for the relationship between viscosity and milk components, but Cox et al. (1959) found 'appreciable differences' between them. They concluded that these papers had merely confirmed that within the ranges of composition found in natural milk, viscosity increases with fat percentage at constant TS percentage and increases with TS percentage at constant fat percentage.

Rohm et al. (1996) analyzed 95 samples of farm bulk milk and developed a linear relationship (Equation 6) which they claimed satisfactorily predicted the viscosity of milk based on fat and protein percentages:

$$\eta = 0.96 + 0.058F + 0.156P \quad (6)$$

where F is fat content (%) and P is protein content (%). As indicated above, protein has a much greater effect than fat on viscosity.

Minerals can also have an effect on the viscosity of liquid milk products. For example, the role of calcium in the viscosity of 9% micellar casein solutions was investigated by de Kort et al. (2011). Addition of the calcium chelators sodium hexametaphosphate (SHMP), disodium hydrogen phosphate ($Na_2HPO_4$), trisodium citrate (TSC) and sodium phytate (SP) caused increases in viscosity; SHMP caused the highest viscosity which was attributed to cross-linking of the proteins. The increase in viscosity appeared to be due to decreased ionic calcium activity causing swelling of the casein micelles, thus increasing their voluminosity. Minerals have also been shown to be significant in viscosity development in concentrated milk during storage (Bienvenue et al. 2003) (see Section 3.1.2).

Lactose generally has little significance in the texture of liquid dairy products because it is soluble and makes up a small fraction of the volume compared with proteins and fat. However, it can be significant when it reaches saturation point in a product and begins to crystallize. This occurs in sweetened condensed milk in which crystallization is initiated by seeding with lactose crystals and the size of the crystals is controlled (Walstra et al. 1999).

## 2.3 Effect of temperature

A major factor influencing flow behavior of liquid dairy products is temperature. All products show a decrease in viscosity/consistency coefficient with increasing temperature. Table 2 shows the effect with bovine whole and skim milk in the range 10–40°C and Table 3 shows the effect of heating (90°C for 10 min, commonly used in yogurt manufacture and as a preheat in powder manufacture) on bovine and buffalo milks with different fat contents (Khalifa and Ghanimah 2013). The effect of fat is apparent in the difference in viscosities between whole and skim milk in Table 2 and more explicitly in Table 3. Interestingly, buffalo milk

**Table 2** Apparent viscosities (mPa s) of bovine whole and skim milks measured at different temperatures

| Temperature (°C) | Whole milk | Skim milk |
|---|---|---|
| 10 | 2.23 | 1.90 |
| 20 | 2.01 | 1.66 |
| 30 | 1.81 | 1.48 |
| 40 | 1.60 | 1.31 |

Based on data by Khalifa and Ghanimah (2013); viscosity was measured on a Brookfield viscometer at shear rate of 60 $s^{-1}$.

**Table 3** Apparent viscosities (cp) of unheated and heated (90°C for 10 min) bovine and buffalo whole milks with different fat contents

| | Bovine milk | | Buffalo milk | |
|---|---|---|---|---|
| Fat content (%) | Unheated | Heated | Unheated | Heated |
| 3 | 2.04 | 2.46 | 2.20 | 2.77 |
| 4 | 2.12 | 2.51 | | |
| 5.5 | | | 2.37 | 2.85 |
| 7 | | | 2.46 | 2.93 |

Based on data by Khalifa and Ghanimah (2013); viscosity was measured on a Brookfield viscometer at shear rate of 60 $s^{-1}$ at 20°C.

at 3% fat had higher viscosity than the corresponding bovine milk, presumably due to the higher solids-not-fat content in the buffalo milk.

Viscosity decreased exponentially with increase in temperature (Bakshi and Smith 1984). It is generally considered that the Arrhenius model accurately describes the viscosity change with temperature (e.g. Veléz-Ruiz and Barbosa-Cánovas 1998; Herceg and Lelas 2005; Ho et al. 2019).

Ho et al. (2019) modeled the changes in the viscosity of liquid MPC (TS 19.8%) during heating at 100 or 120°C for 30 s. Heating caused a significant increase in viscosity in MPC systems, which followed second-order kinetics. The Arrhenius plot showed a distinct deviation in slope at 77.9°C which corresponded to the change in whey proteins, chiefly β-lactoglobulin (β-Lg), from unfolding to aggregation.

Jeurnink and de Kruif (1993) found that after heating skim milk at 85°C for 15 min, the viscosity predicted by Eilers equation (Equation 2) was too low. They attributed the difference to the denaturation of whey proteins. It was concluded that, since aggregated β-Lg did not lead to an increase in viscosity after heating, the unfolded (sticky) state of β-Lg was mainly responsible for the increased viscosity. This was proposed as the reason why the viscosity of heated skim milk did not conform to Eilers equation, which relies entirely on volume fractions. They found that the Batchelor equation (Equation 7), which takes account of the attraction between particles (in this case, casein micelles), was more applicable:

$$\eta = 1 + 2.50\phi + k_H\phi^2 \quad (7)$$

where $k_H$ is the so-called Huggins coefficient which accounts for the hydrodynamic interactions between the hard spheres and has a value of ~6.2.

## 2.4 Combined effect of temperature and milk composition

Several authors have developed semi-empirical models to describe the change in viscosity with temperature and other variables (Vélez-Ruiz et al. 1997). For example, Bakshi and Smith (1984) developed an empirical equation (Equation 8), relating viscosity to fat content and temperature. They used this equation to calculate the pumping requirements of milk with various fat contents and at different temperatures:

$$\ln \eta = -8.9 + 0.1F + 2721.5 / T \tag{8}$$

where F, fat % (w/w); T, temperature (°C).

Bertsch and Cerf (1983) produced similarly shaped curves for the viscosities of a range of milk products, with fat contents up to 15.5%, at temperatures from 70°C to 135°C, the highest of which were in the ultra-high-temperature (UHT) range. They also developed an equation (Equation 9) to relate viscosity to fat content and temperature, although more complex than Equation 8. Viscosities calculated from this equation for products at temperatures down to 50°C agreed well with published values:

$$\begin{aligned}\ln\eta = {} & 3.92\times10^{-5}T^2 - 1.951\times10^{2}T + 0.666 \\ & + F\left(-9.53\times10^{-6}T^2 + 1.674\times10^{-3}T - 4.37\times10^{-2}\right) \\ & + F^2\left(9.75\times10^{-7}T^2 - 1.7394\times10^{-4}T + 9.83\times10^{-3}\right)\end{aligned} \tag{9}$$

Similarly, Fernández-Martin (1972) formulated a complex equation relating viscosity to TS content and temperature. The author maintained that it could be used to predict the viscosity of milk at any temperature (0–80°C) with TS concentration of 0–30% ,with the ratios of fat to solids-not-fat of 0.01–0.4. A nomogram was constructed from which viscosity could be calculated for milk with any TS level and at any temperature in the above ranges.

The change in viscosity with temperature is particularly relevant to UHT processing. Temperatures change rapidly and different shear conditions prevail, making it difficult to predict what the viscosity of the product might be at any specific location in the plant. The viscosity strongly influences the Reynolds number and determines whether the flow regime is streamlined or turbulent. This in turn influences the rate of heat transfer and the distribution of residence times. For non-Newtonian fluids, the Reynolds Number, Re, at which flow changes from laminar to turbulent, depends on the value of the flow behavior index, n. For example, the critical Re at which the change to turbulent flow occurs, $Re_{crit}$, moves from 2100 for a Newtonian liquid (n = 1), to 2900 for a non-Newtonian liquid of n = 0.97 and to 4000 for a non-Newtonian liquid of n = 0.46, according to Equation 10 (Lewis and Heppell 2000):

$$Re_{crit} = 3258(1.61 - n) \tag{10}$$

## 3 Viscosity of concentrated milks

Concentrated milks are widely produced, mostly for production of milk powders. They are made by either evaporation or membrane filtration. The milk is usually heated before concentration (~90°C for 10 min) which denatures a large proportion of the whey proteins. The milk is concentrated to high TS (≥45%) as the concentration is more energy efficient than drying (Sutariya et al. 2017). However, the viscosity of the concentrate is a limiting factor in the level of TS achieved.

The flow behavior of concentrated milks changes from Newtonian at low TS to non-Newtonian at higher TS. The viscosity increases exponentially with increasing TS. Viscosity is an important factor in milk concentrates, because high viscosity concentrates are difficult to pump and atomize in the spray drier. In practice, concentrates are stored for only short times before they are dried, because they increase in viscosity during storage.

### *3.1 Composition and temperature*

Bloore and Boag (1981) reported that the viscosity of skim milk concentrate was influenced by five principal factors: Temperature, TS, protein content, age and preheat treatment. They developed an empirical equation (Equation 11) to describe the effects of the first three factors. Interestingly, the developed curves of viscosity versus temperature pass through a minimum, where the authors concluded that the increase in viscosity matches the reduction in viscosity with rising temperature:

$$\mathrm{Ln}\eta = 3.765 + 0.207\mathrm{TS} - 0.207\mathrm{T_c} + 0.044\mathrm{T_c}^2 + 0.075\mathrm{P} \tag{11}$$

where $T_c$ = (concentrate temperature, °C)/10; TS = (concentrate TS, %)/2; P, protein %, $\eta$ = viscosity measured 14 s after the concentrate is heated.

Mun et al. (1999) studied the flow curves for 1×–5× and 1×–5.3× UF-concentrated whole and skim milk, respectively. Skim milk concentrates up to 4× concentration (TS 19.4%) and whole milk concentrates up to 3× (TS 25.8%) showed Newtonian behavior, while the flow behavior of 5× and 5.3× skim milk concentrates (TS 22.9 and 27.2%) and 5× whole milk concentrates (TS 39.8%) fitted the Power Law model. The viscosity of the 5× concentrates decreased by about 50% over the shear rate used (from ~50 to ~1000 $s^{-1}$). The viscosity of the 5× whole milk concentrate was six times higher than the viscosity of 5× skim milk concentrate at low shear rates and 3.5 times higher at high shear rates. When the apparent viscosity at 146 $s^{-1}$ of the 5× skim and whole milk concentrates were plotted against casein content, the curves were close to each other up to about 8% protein, but thereafter the curves diverged, with viscosity of the whole milk concentrate rising more steeply than that of the skim milk concentrate.

Vélez-Ruiz and Barbosa-Cánovas (1998) reported on the flow properties of concentrated milk in relation to TS level, temperature and through four weeks of storage. They applied Newtonian, Power Law and Herschel-Bulkley models. The flow curves of concentrates with TS of 12.6% up to 22.3% were Newtonian, those of concentrates with TS between 22.3% and 30.5% fitted the Power Law model and those of concentrates with 42.4% and 48.6% best fitted the Herschel-Bulkley model. The milk concentrates with 42.4% and 48.6% TS exhibited yield stresses. The flow behavior index varied from 1.00 to 0.326 for concentrates with 17.2%–42.4% TS. The concentrate temperature (5°C, 15°C and 25°C) had a minimal effect on the flow behavior index but had a noticeable effect on the consistency coefficient. In contrast, storage had a significant effect on the flow behavior index but not on the consistency coefficient.

Morison et al. (2013) examined the effect of composition (TS 9%–50%) and temperature (20°C–60°C) on the flow behavior of concentrated milk and cream. Samples with TS > 30% exhibited non-Newtonian (shear thinning) behavior for which the Power Law model fitted well, 'within the uncertainty of the data'. The viscosity/consistency coefficient increased with TS content. For the non-Newtonian samples, the flow behavior index, n, decreased with increasing TS content. These authors used an empirical equation (Equation 12) which they found described the behavior of the liquids with changes in composition and temperature:

$$\ln \eta_{rel} = f(w) = \sum_i \frac{a_i w_i}{w_w} \qquad (12)$$

where $\eta_{rel}$ is the ratio of apparent viscosity to the viscosity of water at the same temperature, $a_i$ is a temperature-dependent constant for component i, $w_i$ is the mass fraction of component i and $w_w$ is the mass fraction of water (i.e. $w_w = 1 - \sum_i w_i$ ). The authors determined values for $a_i$ for each component (fat, protein and lactose) at 20°C–60°C.

### 3.1.1 Role of whey proteins

The viscosity of concentrates is influenced by the denaturation of whey proteins, as this causes aggregation of whey proteins and the interaction of whey proteins with casein micelles, increasing their voluminosity. The viscosity is increased more when the whey proteins associate with the casein micelles than when they form whey protein aggregates. According to Anema et al. (2014), a major factor influencing the distribution of the whey proteins is the pH at heating. They reported that heating at pH 6.5 produced concentrates (up to 40% TS) with higher viscosity than those heated at pH 7.1. During heating at pH 6.5, more whey proteins associate with casein micelles than when heating occurs at pH 7.1. However, the viscosities of milk concentrates with > 40% TS were similar, regardless of the pH of heating. Interestingly, flow behaviors of the concentrates prepared from preheated and non-preheated milk at pH levels from 6.5 to 7.1 showed different trends. There was a marked decrease in Power Law-derived consistency coefficients and a corresponding decrease in flow behavior index over the pH range in the unheated samples, but virtually no change in the heated samples up to pH 6.9 and only a small change at pH 7.1.

Sutariya et al. (2017) concentrated reconstituted skim milk to 50%–55% TS at pH 6.5 and 6.7, using preheating conditions of 80°C for 5 min and 90°C for 5 min. More whey protein denaturation occurred at the higher pH and higher preheat temperature and this corresponded to the viscosities of the concentrates as given by the Power Law-derived consistency index (K). The flow behavior index, n, decreased as K increased. Notably, the viscosities of the concentrates were only marginally affected by the distribution of denatured whey proteins between the serum and casein phases.

The effect of whey proteins and pH on 20% TS concentrates was investigated by de Souza et al. (2015). They used mixtures of WPC and MPC and measured viscosities on a Rapid Visco Analyzer using a heat treatment of 90°C for 5 min. The highest viscosities were recorded for the solutions with 100% whey protein isolate (WPI) at pH 7.0 and the lowest for solutions of 100% MPC at pH 6.0. The authors explained the results by the greater denaturation of the whey proteins at the higher pH and the difference in size of the whey protein aggregates, which would dominate at pH 7.0 and the κ-casein–whey protein complex, which would be dominant at pH 6.0.

Ho et al. (2019) modeled the changes in viscosity of liquid milk protein concentrate (MPC) during preheating at 100°C or 120°C for 30 s. Heating caused a significant increase in viscosity in MPC systems, which they found followed second-order kinetics. The Arrhenius plot showed a distinct deviation in slope at 77.9°C which was attributed to the change in whey proteins, chiefly β-lactoglobulin, from unfolding to aggregation.

### 3.1.2 Role of minerals

The soluble minerals in concentrated milk have also been shown to have a role in viscosity development during storage (Bienvenue et al. 2003). Concentrated skim milks, with 45% TS but different mineral content, were stored at 50°C. The apparent viscosity and Bingham yield stress increased during storage, most markedly after 4 h. The changes up to 4 h were largely reversible under high shear, but irreversible in samples stored for longer times. Particle size analysis showed irreversible fusion and aggregation of casein micelles during storage. Significantly, the apparent viscosity and casein micelle aggregation during storage were decreased by reduction of the soluble mineral content of concentrated milks, but increased by addition of minerals. This suggests that the soluble mineral content could be manipulated to control storage-induced changes in the flow behavior of concentrated milks. Bienvenue et al. (2003) confirmed the results of Aoki et al. (2002) that an increase in particle size was due to aggregation and fusion of casein micelles. They showed that the decrease in soluble mineral content caused a decrease in particle size, thus explaining the effect on viscosity.

Milk concentrated by reverse osmosis (RO), nanofiltration (NF) and ultrafiltration (UF) and then UHT processed show different rates of increase in viscosity during storage (Hinrichs 2000). The order of the rate of increase from fastest to slowest was RO > NF > UF. The order of heat stability was the same. Hinrichs (2000) attributed the different stabilities to differences in their mineral contents. To demonstrate the role of minerals, Hinrichs (2000) mixed RO concentrate (2.3×) with electrodialyzed RO concentrate to produce samples with a range of mineral (ash) contents. After 40 weeks' storage at room temperature, after UHT processing, samples with 1.15% and 1.5% ash had viscosities of 25 mPa s and 150 mPa s, respectively. On this basis, Hinrichs (2000) concluded that UHT UF concentrates should be more stable during storage than NF or RO concentrates due to their lower mineral contents.

Bienvenue et al. (2003) proposed the following mechanism for the increase in viscosity during storage of concentrated milk: during concentration, the pH decreases and the ionic strength increases, which leads to a decrease in the electrostatic repulsion between the already closely packed casein micelles. Additionally, some calcium phosphate is deposited on casein micelles, shielding their charge and interfering with the stabilizing κ-casein hairy layer. The increase in ionic calcium would further reduce the negative charge on the micelle and lead to flocculation. This hypothesis predicts that a decrease in ionic strength and ionic calcium would reduce the casein aggregation and subsequent viscosity increase, consistent with experimental findings.

## 3.2 Preheating

In their study of the viscosities of a range of concentrates at different temperatures, Morison et al. (2013) also compared the viscosity of concentrates made from unheated and heated (90°C for 20 min) milk and concluded that the most significant contribution to changes in the viscosity of the milk concentrates was the preheat treatment of the proteins.

Bloore and Boag (1981) investigated the aging effect by holding concentrates produced under the preheat conditions in Table 4 at 45°C for 300 s. They found that the viscosity of the concentrate increased around 200% from 150 s to 300 s after it left the evaporator. Increasing the preheat temperature increased the viscosity of the concentrate and this effect increased with the preheat holding time. The effect correlates well with the denaturation of whey proteins. Table 4

**Table 4** WPNI (mg whey protein nitrogen/mL) in skim milk concentrates resulting from four preheat treatments

| | Temperature | |
|---|---|---|
| Holding time | 80°C | 113°C |
| 10 s | 7.5 | 4.0 |
| 120 s | 5.7 | 0.9 |

From Bloore and Boag (1981).

shows the denaturation of whey proteins, as given by the Whey Protein Nitrogen Index (WPNI) for four different preheat conditions. The authors concluded that, for a given WPNI, the viscosity increase in concentrates during storage after spray drying was least for the high-temperature, short-time preheat treatment.

Aoki et al. (2002) studied the effect of storage at 5°C on the viscosity of concentrated (4.5×) skim milk prepared from milk preheated at ≥ 90°C for 10 min. After preheating at 90°C, the viscosity increased markedly and gelation occurred after 3 days' storage. Preheating at temperatures above 90°C and concentrating to higher TS levels accelerated the viscosity increase. Electron microscopy of casein micelles in concentrated skim milk during storage for 3 days showed the disappearance of the boundary between micelles and aggregation of casein micelles. The increase in the viscosity and gelation of the concentrate during storage was attributed to this aggregation.

In the manufacture of recombined sweetened condensed milk, the viscosity is greatly influenced by the preheat treatment applied to the milk prior to concentration and drying. Cheng et al. (2000) showed that the viscosity increased markedly when the preheat treatment caused > 50% denaturation of the whey proteins (as measured by WPNI). This corresponded to heating at 85°C for 30 min or 90°C for 2 min.

## 4 Texture of UHT milk

UHT milk is produced by a very-high-temperature, short-time process, commonly 138°C–142°C for 2 s–5 s, in order to make it 'commercially sterile', that is, containing no microorganisms which will grow under the normal conditions of storage. This enables UHT milk to be stored at room temperature for up to 12 months. This puts considerable demands on the stability of the product and it is therefore not surprising that some defects develop during storage. Significantly, two major changes in texture occur which can limit the shelf-life of the product. These are the development of sediment and age-thickening (increase in viscosity) which lead to age gelation, the latter being the most serious as it renders the milk unacceptable to consumers (Deeth and Lewis 2017).

Both changes are related to aggregation of casein micelles which increases the particle size so that the aggregates precipitate (sedimentation) or increase the viscosity (age-thickening), or form into a gel network (age-gelation). Gelation tends to occur when the apparent viscosity (measured at 20°C at 60 $s^{-1}$) reaches ≥ 10 mPa s (Kocak and Zadow 1985a).

## 4.1 Sediment formation

The exact cause of sediment formation is not known, but it is believed to be related to destabilization of casein micelles at low pH and high ionic calcium level (Gaur et al. 2018). Sedimentation is reduced when pH is increased or ionic calcium levels are reduced (Lewis et al. 2011). Addition of TSC or disodium hydrogen phosphate is most effective in reducing sediment formation, by decreasing ionic calcium levels and increasing pH. Caprine milk is particularly susceptible to sediment formation because it has high ionic calcium; it cannot be UHT processed without additives to reduce the ionic calcium level (Boumpa et al. 2008).

The mechanism of sedimentation is still unclear, but it involves destabilization of the casein micelle surface whereby the repulsion between caseins is reduced, or the micelles become cross-linked by some means. It has been hypothesized that one way in which repulsion between micelles is reduced is by the release of the negatively charged κ-casein from the surface. Furthermore, ionic calcium may form crosslinks. As pH is reduced during storage of UHT milk, ionic calcium content increases and hence more is available for crosslinking (Anema 2019).

## 4.2 Age-thickening and gelation

Almost all UHT milks will undergo age-thickening and then gelation, if stored for long enough at room temperature. The increase in viscosity during storage is usually gradual at first, but at some point becomes more rapid until gelation occurs (Kocak and Zadow 1985a). The gelation cannot be reversed. The increase in viscosity and gel formation occurs most readily during storage at 25–30°C. It does not occur at ≥40°C (Datta and Deeth 2001). The nature of the gel may be like a set yogurt, a protein coagulum such as occurs in rennet coagulation or of a custard-like consistency.

A major factor involved in age gelation is proteolysis of caseins, although it can occur via non-proteolytic physicochemical means. The proteases responsible for the proteolysis are of two major types: the native milk alkaline protease, plasmin; and the heat-resistant bacterial proteases resulting from growth of contaminating psychrotrophic bacteria such as *Pseudomonas* spp in raw milk. Both types of protease destabilize casein micelles: plasmin by preferentially attacking β-casein and to a lesser extent $\alpha_s$-caseins; and bacterial proteases by preferentially attacking κ-casein. Both types of protease are stable to some UHT heating, especially direct heat processing, but plasmin can be inactivated by some preheating conditions such as 90°C for 30–60 s (Newstead et al. 2006). High UHT sterilization conditions such as 150°C for 4 s considerably reduce the activity of bacterial proteases but may not inactivate them completely (Topçu et al. 2006). Control of the quality of the raw milk is the best defense against the presence of bacterial proteases in UHT milk.

One strategy used by some companies to delay age-thickening/gelation is to add polyphosphates such as SHMP (at ~0.1%). This does not inhibit proteolysis but prevents the aggregation of casein micelles which follows the initial proteolysis. In contrast to polyphosphate, monophosphates accelerate age-gelation (Kocak and Zadow 1985b). The difference between the effects of SHMP and orthophosphate in stored UHT milk have been shown by electron microscopy. The casein micelles in milk containing orthophosphate become slightly distorted, have thread-like tails and aggregate into interconnected chains at gelation. On the other hand, the micelles in UHT milk containing polyphosphate remain unchanged and separated during storage.

In some cases of age-thickening/gelation, proteolysis does not appear to be involved and a physicochemical mechanism is proposed (Harwalkar 1992). It occurs after long periods of storage, whereas proteolysis-related gelation occurs after shorter storage periods. Anema (2017) reported that the casein micelles in these gels were depleted in κ-casein and proposed that gelation occurs through the coalescence of such micelles and the formation of a gel when the concentration of these aggregates exceeds a threshold level. Age-thickening/gelation of UHT-concentrated milk appears to occur by a physicochemical pathway, as proteolysis is seldom involved (de Koning et al. 1985).

### 4.3 Age-thinning

During storage, UHT milk may initially decrease in viscosity, a thinning effect, before its viscosity increases (de Koning et al. 1985). However, this thinning is of little concern. In other UHT products such as custard and desserts, which are deliberately thickened with starch, thinning during storage is of concern. The thinning is caused by amylases which break down the starch.

The origin of the amylase may be thermophilic bacteria such as *Geobacillus stearothermophilus* and *Bacillus licheniformis*, which produce heat-stable α-amylases (Barefoot and Adams 1980). These spore-formers are common contaminants of milk and milk powders used in the manufacture of custard and desserts. The growth of psychrotrophs in raw milk before processing may also result in production of amylase. Another source of amylase is xanthan gum which is sometimes used as a stabilizer in custard. *Xanthomonas campestris*, the organism used to produce xanthan gum, produces an amylase which has a similar heat stability to the *Bacillus* α-amylases.

## 5 Texture of fat-containing products

The density of fat is lower than that of milk serum and therefore, fat will rise to the top when milk is left undisturbed during cold storage. The rate at which the fat rises (the terminal velocity) depends on the relative densities of the fat and the milk serum, the size of the fat particles and the viscosity of the milk serum, as given by Stokes Law (Equation 13):

$$v = \frac{d^2 g(\rho_s - \rho_f)}{18\mu} \tag{13}$$

where d = diameter of the fat particle; $\rho_s$ = density of the milk serum; $\rho_f$ = density of milk fat; $\mu$ = viscosity of milk serum; v = terminal velocity; and g = acceleration of gravity.

In most cases, it is desirable to reduce the rate of fat rising in milk. According to Stokes Law, this can be achieved by reducing the diameter of the fat particle or by increasing the viscosity. The process of homogenization, which is used for most milk produced for direct consumption, reduces the fat globule size from an average of about 3 μm to ≤ 1 μm. It also coats the new fat particles with a layer of protein, largely casein, which increases their density and reduces their ability to rise during storage. In some situations, the fat separates from the milk serum despite the milk having been homogenized. This seldom occurs in pasteurized milk, as its shelf life is only of the order of two weeks. However, it occurs more often in UHT milk, whose shelf life at room temperature can be up to 12 months.

The rate of fat separation from the serum in milk and cream is accelerated in some situations through the agglutination of fat globules. This can result in a cream layer which may or may not be easily redispersed into the product and presents as a defect. Defects related to the instability of the fat emulsion in milk or cream are discussed in this section.

### *5.1 Fat separation in UHT milk*

During ambient storage of UHT milk, a fat layer may develop on the surface. In most cases, this is barely noticeable by consumers, but in other cases, the fat may form as a solid layer which is not easily redispersed and appears as lumps of fat when the milk is poured from the container.

In most cases, the development of the fat layer is caused by inefficiency of homogenization (Hansen 1990). This can be due to the condition of the homogenizer valve or the conditions of homogenization. If homogenizer valves are not well maintained, fat globules may escape proper homogenization and cause fat separation. It has been shown that even if the average diameter of the fat globules is small (e.g. 0.5 μm), a small percentage of large globules can represent a large proportion of the fat in the product and lead to fat separation (Hillbrick et al. 1998). For this reason, some companies use the rule of thumb that the average diameter of the fat globules should be < 1 μm, but the percentage greater than 5 μm should be < 2% and the percentage greater than 2 μm should be < 5%. The tendency to fat separation is decreased by increasing the homogenization pressure up to about 35 MPa of combined pressure of the two stages (Hillbrick et al. 1998) and ensuring that the temperature of homogenization is ≥ 50°C – when all the fat is in the liquid state.

Coalescence of small fat particles during storage can also result in fat separation. The resulting larger particles then have a greater tendency to rise and separate from the serum. This can occur when direct UHT heating (either steam injection of steam infusion) is used after homogenization. To avoid this, downstream homogenization is practiced in direct UHT plants, as the direct heating process can cause aggregation of fat globules. This may be why directly processed UHT milks show less fat separation than indirectly processed milks which may be homogenized upstream of the sterilization section (Ramsey and Swartzel 1984). The fat particles may also coalesce during storage and cause fat separation, although the exact mechanism of this is not well understood. This type of fat separation is exacerbated by high storage temperatures (Ramsey and Swartzel 1984; AlKanhal et al. 1994).

An interesting cause of fat separation in UHT milk is age-gelation, which results from changes in the structure of the caseins, mostly initiated by proteolysis. Since casein micelles may constitute up to 93% of the new membrane material around the fat particles and cover some 75% of their surface (Walstra and Jenness 1984), the new particles act as mini-casein micelles and are susceptible to proteolysis by either the native plasmin or bacterial proteases and thus to gelation (see Section 4.2). The resulting gel consists of agglomerated fat-containing particles which rise to the surface and cause a fat layer during storage (Kohlmann et al. 1991; Zhang et al. 2007).

## 5.2 Cold agglutination

The fat globules in milk naturally rise in all non-homogenized milks to produce a cream layer. However, the rise is accelerated in cold-stored milk, which has not been severely heated, by a process known as cold agglutination. The agglutination is caused by the interaction of immunoglobulins, mostly immunoglobulin M (IgM) but also IgG, together with lipoproteins of skim milk membranes, also known as extracellular vesicles. The antibody–antigen-type binding is through carbohydrate moieties (Euber and Brunner 1984). The effect of temperature on this interaction was demonstrated by Hansen et al. (2019). Cold agglutination was clearly evident when the (raw) milk was stored at 5°C, was markedly reduced at 15°C and did not occur during storage at 20°C or

37°C. Re-cooling the milk stored at these last temperatures restored the cold agglutination ability if the milk was stored at low temperature, say 5°C.

Heat treatment reduces cold agglutination by inactivating IgM. Hansen et al. (2019) reported that heating at 69°C to 73°C for 20 s (with heat-up times of 2 min 30 s to 3 min 20 s) progressively inactivated IgM. However, Mainer et al. (1997) found that IgM was not completely inactivated by pasteurization at 72°C for 15 s, having a $D_{72}$ value of 205 s. Thus, it appears that pasteurization conditions can be designed to retain at least some cold agglutination in pasteurized milk. This is advantageous in markets such as those of organic milk, where non-homogenized milk is preferred and a cream-line is expected by consumers.

Homogenization eliminates cold agglutination. This has been shown to be due to the physical destruction of the skim milk membrane material, an integral part of the agglutination system (Euber and Brunner 1984).

Cold agglutination is significant at the temperature at which milk is separated into cream and skim. Cold separation enhances cold agglutination and formation of fat clusters, thereby increasing viscosity. On the other hand, separation at around 40°C reduces the potential for cold agglutination, as much of the agglutinin is left in the skim (McCarthy and Singh 2009).

### *5.3 Cream plug*

An undesirable texture defect which can occur in non-homogenized milk is a thick cream layer known as a cream plug (Hoyle and Burton 1984). The cream plug may contain up to 60% fat and can prevent milk being poured from a bottle. It is formed by coalescence of fat globules through partial churning. Free (non-globular) fat binds the globules together to form the compacted fat plug. A similar phenomenon occurs with pasteurized and UHT cream, where it is sometimes referred to as gelling (Te Whaiti and Fryer 1975; Streuper and Van Hooydonk 1986). The presence of free fat is common in all cases of cream plug; this becomes evident as an oily layer when the product is poured into hot coffee or water.

Factors which have been reported to induce cream plug formation include excessive pasteurization temperature, multiple pasteurization treatments, mechanical stresses such as pumping, agitation during handling and fluctuations in storage temperature (Hoyle and Burton 1984). Streuper and Van Hooydonk (1986) reported that the rate of cooling of cream was the most critical step in relation to cream plug formation, with cream subjected to rapid cooling being more susceptible to plug formation than slowly cooled cream; the authors suggested that this was related to the crystallization behavior of the fat inside the fat globule. All of these factors disrupt the milk fat globule membrane, which allows the coalescence of fat globules. Some authors have also reported a seasonal effect on cream plug formation, which is probably linked to changes in the milk fat composition (Streuper and Van Hooydonk 1986; Precht et al. 1987).

## 5.4 Stability of the texture of pasteurized and UHT whipping cream

Thickening (an increase in viscosity) of cream after pasteurization or UHT treatment is not an uncommon occurrence. It bears a close resemblance to cream plug formation, as discussed in Section 5.3, in that it is characterized by damage to the fat-globule membrane leading to production of free fat, causing partial fat coalescence and agglomeration. Solidification of pasteurized cream stored under refrigeration before making butter can occur if the fat content is high (≥42%) and the temperature is ≤ 30°C. Free fat content can be 50%–60% of the total fat. The temperature of separation has a major effect; Te Whaiti and Fryer (1975) reported much more free fat formation during separation at 35°C than at 70°C. This is consistent with a recommendation that physical stresses on the fat globule, such as separation and pumping, should be avoided at temperatures at which the fat is a mixture of both liquid and solid portions, that is, between ~14°C and 40°C (IDF 1996).

The thickening of UHT cream during storage occurs more rapidly at room temperature than it does under refrigeration and because of this, some producers of UHT whipping cream recommend refrigerated storage of their product. Unfortunately, this negates the major advantage of UHT-sterilized products, which is that they can be stored without the need for refrigeration. However, thickening can still occur at reduced temperatures. Leedham and Deacon (1970) reported clustering of milk fat globules in UHT-treated 18% cream during storage for 8 weeks at 15.5°C. This caused a viscosity increase from 10 mPa s to 17 000 mPa s.

Several strategies have been proposed to minimize thickening of UHT cream. Some of these strategies increase the stability of the cream during storage but have a detrimental effect on whipping time and overrun. Since shear-induced damage to the fat globule membrane is an important cause of cream thickening, physical processes such as homogenization and pumping are key issues. Some homogenization is necessary to prevent creaming during storage, but if the homogenization conditions are too severe, there may be insufficient protein to form an effective new membrane around the smaller fat globules with greatly increased surface area. This can then lead to partial coalescence of fat globules and lead to thickening. In general, cream is homogenized at much lower pressures than are used for milk. Pressure recommendations for UHT whipping cream are 3 MPa/1 MPa (2-stage; Hoffmann 2011), 2.5 MPa/1.5 MPa (Lorenzen et al. 1993) and 2.1 MPa/0.9 MPa (Kieseker and Zadow 1973). Lower homogenization pressures are recommended for high-fat cream rather than for lower-fat cream.

Fink and Kessler (1985) demonstrated that cream is destabilized by high-temperature heating. The proportion of large fat globules and production of extractable free fat increased with increased holding times, 0.8–90 s and temperatures from 105°C to 145°C. The large fat globules were formed mainly by aggregation of smaller fat globules. The high-temperature heating caused a loss of emulsion stability and formation of a 'gel-like cream layer'. Heat treatment at ≤105°C did not cause destabilization of the cream after processing or during storage.

Fat separation or creaming with release of a serum layer occurs in UHT cream if the homogenization is not completely effective. Partial coalescence exacerbates the fat separation, as the larger coalesced fat globules rise faster than the smaller globules. The supply of protein to coat the new globule surface is limited in cream and therefore, the stability of the emulsion to partial coalescence of the fat globules can be enhanced by the addition of protein. Addition of sodium caseinate or skim milk powder (SMP) reduces the partial coalescence but impairs the whipping properties of the cream (IDF 1996). Addition of calcium-chelating agents such as sodium citrate or carbonate reduces coalescence through a reduction in calcium-mediated linking of caseins attached to the (homogenized) fat globules and reduces fat separation. However, reduction of calcium impairs the whipping properties; in fact, addition of calcium lactate improves whipping properties (Zadow 1970; IDF 1996).

Stabilizing texture-modifying agents such as carrageenan, alginate and microcrystalline cellulose are commonly added to whipping cream to increase viscosity and thereby reduce creaming and age-thickening. One reason for adding them is to overcome seasonal effects on the quality of creams, which has been reported by several authors (e.g. Keogh 1978). In addition, emulsifying agents such as mono- and diglycerides and even fat globule membrane material are used to stabilize the emulsion. The emulsifying agents tend to displace the milk protein at the surface of the fat globules and so negate the negative effect of adding caseinate or SMP on whipping properties. Stabilizers and emulsifiers have an interactive effect in reducing viscosity during storage (Towler 1988). Several food ingredient supply companies produce proprietary mixes of stabilizers and emulsifiers specifically designed to increase the stability of UHT whipping cream.

A study by Kováčová et al. (2010) illustrates the effects of high-temperature heating, homogenization and carrageenan addition on the stability and flow behavior of whipping cream. Cream with 30% fat was pasteurized and UHT sterilized (136°C/1.5 s), with and without added κ-carrageenan (0.04%). The flow behavior was found to best fit the Hershel–Bulkley model; flow properties of the creams derived from the use of the model are shown in Table 5. Both the yield

**Table 5** Effect of κ-carrageenan and processing on the flow behavior of whipping cream (data from Kováčová et al. 2010)

| κ-Carrageenan added (%) | Processing | Yield stress (mPa) | Consistency index (mPa s) | Flow behavior index |
|---|---|---|---|---|
| 0 | Pasteurization | 4.3 | 15.7 | 0.98 |
| | UHT | 2.7 | 9.7 | 1.02 |
| | UHT+homogenization | 51.0 | 40.8 | 1.31 |
| 0.04% | Pasteurization | 166 | 249 | 0.68 |
| | UHT | 398 | 353 | 0.64 |
| | UHT+homogenization | NR | NR | NR |

NR = not recorded, did not meet Herschel-Bulkley model due to sigmoid shape of the flow curve. However, the flow curves for 'UHT + homogenization' were considerably higher (higher shear stress) than for the 'pasteurization' and 'UHT' samples.

stress and the consistency index/viscosity were lower for the UHT-processed cream than for the pasteurized cream, but when UHT processing was combined with (downstream) homogenization, both yield stress and consistency index of the cream were much higher than for the pasteurized cream. The latter values for the cream with added carrageenan are not shown in Table 5, as flow curves for this cream had a sigmoid shape and hence did not meet the Herschel-Bulkley model. However, the flow curves for 'UHT + homogenization' were considerably higher [higher shear stress] than for the 'pasteurization' and 'UHT' samples at all shear rates. Carrageenan addition considerably increased the viscosity of the creams. The viscosity of homogenized creams and those with added carrageenan decreased during 5 min shearing at 100 $s^{-1}$, that is, they displayed thixotropy. Both homogenization and carrageenan addition reduced the level of creaming. UHT processing combined with homogenization considerably increased the whipping time of the cream, compared with pasteurization or UHT processing without homogenization. The effects of homogenization were attributed to the formation of clusters of small fat globules and of carrageenan addition by formation of carrageenan-casein micelle aggregates. The thixotropic behavior was attributed to breaking of the fat globule clusters.

## 5.5 Rebodying of cream

Rebodying is closely related to cream plug formation and thickening, discussed in Sections 5.3 and 5.4. It relates to the thickening that occurs when cream with > 25% fat, which has been refrigerated, is heated to 30–35°C and subsequently cooled slowly. The term was coined by Hening and Dahlberg (1932), who first reported the phenomenon as a commercial process for increasing the viscosity of cream. It was seen at that time to be an important phenomenon as it improved the consumer appeal of cream. It can also be used to improve the whipping properties of UHT cream (Everett 2007). Following the initial report of Hening and Dahlberg (1932), five patents were taken out on various features of the process (Weise et al. 1939).

The temperature changes which cause rebodying can occur during transport or storage under non-ideal conditions. The thickening is caused by the partial coalescence of the fat globules brought about by the rupture of some of the fat globule membranes due to temperature fluctuation. The extent of rebodying is influenced by the rate of cooling, which determines the size of the fat crystals (Huppertz et al. 2020).

Rebodying does not occur with recombined cream unless the cream is stabilized with phospholipid-containing material. Weise et al. (1939) concluded that milk fat globule membrane material was the only milk component which could induce rebodying in recombined cream, while Oortwijn and Walstra (1982) reported that lecithin added at 1% had the same effect.

### 5.6 Feathering

This phenomenon is encountered in both milk and cream, both UHT and pasteurized. It is characterized by the formation of a curd (sometimes in the shape of a feather) floating of the surface when the milk or cream is poured into a hot beverage such as coffee. The effect is due to the precipitation of destabilized protein and fat in the acidic coffee solution. Electron microscopy of the feathered material produced by UHT cream (18%) showed that the fat globules appeared to be linked together by a matrix of casein, in which sub-micellar casein particles were prominent (Anderson et al. 1977).

The product which is most known for feathering is coffee cream or 'half-and-half', which has 12–18% fat. In a study of the effect of storage at 16–25°C on feathering by homogenized UHT cream with 18% fat, Anderson et al. (1977) found a decrease in feathering score (greater instability; determined using a sodium acetate solution at pH 5.0 to simulate a coffee solution) during storage for 10 weeks. Concomitant with this decrease in stability was a decrease in the cream's sensitivity to heat. Samples with low feathering scores were unstable to heat at temperatures around 30°C, but creams with high feathering scores were stable at temperatures up to 85°C. From chemical analyses of the creams, the authors concluded that the probable cause of the decreased stability of the creams was a gradual increase in the amount of casein and Ca attached to the fat phase of the cream. In the initial stages of storage, addition of sodium carbonate and sodium citrate (which reduce ionic calcium levels) improved the feathering score of the creams. This was also shown for evaporated milk by Whitaker (1931) who concluded that the susceptibility to feathering was related to the salt balance, that is, the ratio of calcium and magnesium to citrates and phosphates. Tracy and Ruehe (1930) also identified soluble calcium as important in feathering, both in the water used in making coffee and in the cream.

Homogenization has been known for some time to greatly affect the feathering of cream. Burgwald (1923) reported that the higher the homogenizing pressure used, the more susceptible a cream is to feathering. Geyer and Kessler (1989) investigated the effect of the position of the homogenizer and the heating conditions on the feathering of UHT coffee cream with 12% fat. They concluded that homogenization was a major cause of feathering and that homogenization both before and after heat treatment up to 135°C was preferred for minimizing feathering. Their homogenizing conditions were 20/6.5 MPa at 73°C and 12.5/2.5 MPa at 73°C, for the first and second homogenizations respectively. Furthermore, they advocated a heat treatment of 95°C for 7 min after the first homogenization, to denature the whey proteins. It was claimed that this ensured that the fat globules were coated with a layer of denatured whey proteins, which were much less likely to form aggregates and create feathering when the cream was added to hot coffee. They found that UHT heating above 135°C exacerbated feathering; this was attributed to chemical changes in not only the whey proteins but also caseins, leading to the formation of very large fat globule-protein aggregates which even a second homogenization cannot break up completely. In addition, they recommended an alternative heat treatment of 125–129°C for 3 min to produce a coffee cream which is stable during storage.

## 6 Perceived texture of milk

There has been considerable interest in the effect of milk composition on the perceived viscosity of low-fat milks by consumers. Quiñones et al. (1997) altered the true protein content of skim and 1% fat milk and found texture and appearance to be the most important sensory characteristics for milks with different total protein contents. However, the sensory scores for texture were found to be more strongly related to appearance, particularly whiteness, than to objective measurements of viscosity. There was a good correlation between the visual appearance of the milks and the perceived textural characteristics including mouthcoating and thickness. This is consistent with the results of Phillips et al. (1995b), that addition of 2% non-fat dried milk (NFDM) to skim milk had the same effect on the objective relative viscosity as 2% fat addition, but the milk with added NFDM did not have comparable visual characteristics. They concluded that NFDM would not be a good candidate for a fat replacer. In a parallel study, Phillips et al. (1995a) varied the fat content of milk from 0 to 2% and again found that the appearance of the milk had a greater effect on perceived mouthfeel compared with the viscosity of the milk. This led the authors to suggest that a successful fat substitute should alter the perceived color of the milk.

For 2% milk, the level of true protein level could be adjusted to 2.8% or 1.8% with no change in sensory scores, but a decrease to 0.9% true protein caused a decrease in sensory scores. An increase in true protein from 2.8% to 3.7% increased the scores for both appearance and texture descriptors. In this study, Quiñones et al. (1998) concluded that trained panelists were unable to detect a change in texture caused by the removal of up to 32% of the protein from either 3.3% or 2% fat milks that originally contained ~ 2.8% true protein.

Misawa et al. (2016) investigated the effect of casein as a percentage of total protein on the texture of milks containing 1% and 2% fat. They showed that casein percentage of true protein had a greater effect on the (objective) relative viscosity[1] than the true protein percentage. The relative viscosity increased with increasing casein as a percentage of true protein. Interestingly, sensory panelists reported increased 'throat cling and mouthcoating' with increased casein as a percentage of true protein in 2% milks.

Hydrocolloids are widely used for altering the texture of liquid dairy products. Wagoner et al. (2020) examined the effects of hydrocolloid choice on the texture perception of dairy beverages. They concluded that there was a significant effect of viscosity on the perceived sensory perception of 'consistency, creamy/oily, mouthcoating and residual mouthcoating' in milk beverages, but only a weak effect of hydrocolloid type.

## 7 Effects of non-thermal technologies

A brief discussion of the effects of some non-thermal technologies on the texture of liquid dairy products follows. It is apparent that some of these can be extremely useful for altering the viscosity of concentrates to facilitate further processing, such as pumping and atomization in spray driers.

### *7.1 Ultrasonication*

Ultrasonication of reconstituted MPC (rMPC), with TS ≥ 30% and reconstituted SMP (rSMP), with TS ≥ 46%, both in batch and flow-through systems, caused decreased viscosity which did not return to pre-sonication levels (Deshpande and Walsh 2018). Sonication with a batch system at 50°C for 30 s (using a 500 W sonicator at 70% amplitude; energy density 63 J $mL^{-1}$) caused a 54.6% decrease in viscosity of 44% TS rMPC and an 18.9% decrease in viscosity of a 46% TS rSMP. Overall, the decrease in viscosity ranged from 27.3% to 54.6% for rMPC and 18.7% to 44.3% for rSMP. Reductions in viscosity were also observed

1 Relative viscosity (RV) was measured using a capillary viscometer and calculated by the equation: $RV = ET_m/ET_w \times SG_m$, where $ET_m$ = efflux time for milk, $ET_w$ = efflux time for water, $SG_m$ = specific gravity of milk at the temperature of measurement.

with continuous ultrasonication at 192 W for 10–30.2 s (energy densities of 64–1.92 J $mL^{-1}$).

Reduction in viscosity of 10–17% was reported by Zisu et al. (2013) for medium-heat skim milk concentrates (TS 50–60%) using batch sonication (40–80 W for 1 min) and continuous sonication with an energy density of 4–7 J $mL^{-1}$. Sonication delayed age-thickening, but could not eliminate the thickening if it had reached an advanced stage. The greatest effect of sonication was observed when it was used during the entire evaporation process, as skim milk was concentrated to 55% solids and for 40 min thereafter; the sonicated concentrate showed little sign of thickening. An unsonicated control thickened significantly after 40 min. Continuous-flow ultrasonication at 20 kHz using a 1 kW sonicator delivering energy densities of 60–240 J $mL^{-1}$ reduced the viscosity of 33% TS WPC by up to 10%. The viscosity reduction was correlated with particle size reduction (Zisu et al. 2010).

Some studies have shown an increase in viscosity of reconstituted WPI and WPC following ultrasonication at 20 kHz in batch treatments for 15 min (Jambrak et al. 2008; Kresic et al. 2008). Ashokkumar et al. (2009) ultrasonicated 5% reconstituted WPC at 10 kHz for up to 60 min and found a decrease in viscosity up to 5 min (energy density 155 J $mL^{-1}$), but an increase thereafter until it was higher after 60 min than it was before sonication.

### *7.2 High-pressure processing*

The effects of high-pressure processing on some properties of milk were reviewed by Huppertz et al. (2002). Changes on the viscosity and particle size of raw whole caprine milk, due to treatment at 200–500 MPa for 25 min at 25°C, are shown in Table 6 (Nassar et al. 2019). Viscosity significantly increased with increased pressure. In this case, the change in viscosity with pressure was inversely related to the change in particle size.

**Table 6** Viscosity and particle size of caprine milk after high-pressure treatment at 200–500 MPa for 25 min at 25°C, measured after one day at 25°C and 4°C (data from Nassar et al. 2019)

| | Pressure (MPa) | | | |
|---|---|---|---|---|
| Temperature (°C) | 0 | 200 | 400 | 500 |
| *Viscosity (mPa s)* | | | | |
| 25 | 1.78 | 1.93 | 2.04 | 2.08 |
| 4 | 3.31 | 3.52 | 3.73 | 4.25 |
| *Particle size (nm)* | | | | |
| 25 | 329 | 313 | 289 | 211 |
| 4 | 302 | 278 | 230 | 201 |

The effect of high pressure in the range of 276–690 MPa on the flow and viscoelastic properties of concentrated milk was studied by Vélez-Ruiz et al. (1998). The rheological behavior of the milk concentrates was characterized as a function of TS (12.9–47.1%) and pressure. The viscosity increased with pressure and TS level. Concentrates with TS < 24% showed Newtonian behavior, while those with higher TS showed shear-thinning behavior. For those with higher TS, the relationship between apparent viscosity and concentration level was fitted by a single-term exponential model (Equation 14):

$$\ln \eta_{app} = A + bS \tag{14}$$

where $\eta_{app}$ is the apparent viscosity (MPa s), A and b are constants and S is the TS content (g $kg^{-1}$).

Pressure treatments above 414 MPa markedly increased viscosity. Treatments at pressures ≤ 276 MPa had little effect on viscosity, with only concentrates with TS ≥ 36% showing a small pressure-related increase.

The combined effects of pressure and TS on viscosity were modeled into a single logarithmic relationship (Equation 15):

$$\ln \eta_{app} = \beta_0 + \beta_1 S + \beta_2 P \tag{15}$$

where $\eta_{app}$ is the apparent viscosity (MPa s), $\beta_0$, $\beta_1$ and $\beta_2$ are constants, S is the TS content (g $kg^{-1}$) and P is pressure (MPa). The relationship is shown graphically in Fig. 2.

Pasteurized cream (35% fat) subjected to pressure at 450 MPa at 25°C for 30 min showed no significant change in flow behavior. However, the same pressure treatment of UHT cream with 35% fat markedly increased the viscosity; this was attributed to the formation of fat globule aggregates. Both pressure-treated creams were stable for 8 days at 4°C, showing no phase separation (Dumay et al. 1996).

## 7.3 PEF Technology

Xiang et al. (2011) showed that PEF treatment (electric field intensities of 15, 18 and 20 kV $cm^{-1}$ and number of pulses of 20, 40, or 60) changed the rheological properties of reconstituted skim milk (RSM) (TS 5%). The treated RSM showed shear-thinning, which increased with increasing electric field intensity and number of pulses. The flow curves were fitted to the Herschel–Bulkley model, mainly because they showed a yield stress. The increase in viscosity/consistency index was most noticeable at shear rates ≤ 50 $s^{-1}$; the apparent viscosities at 10 $s^{-1}$ increased about three-fold in the sample treated at the highest field intensity (20 kV $cm^{-1}$) with 60 pulses. The closeness to Newtonian behavior is indicated by the flow behavior index (n), which decreased from 0.99 to 0.96.

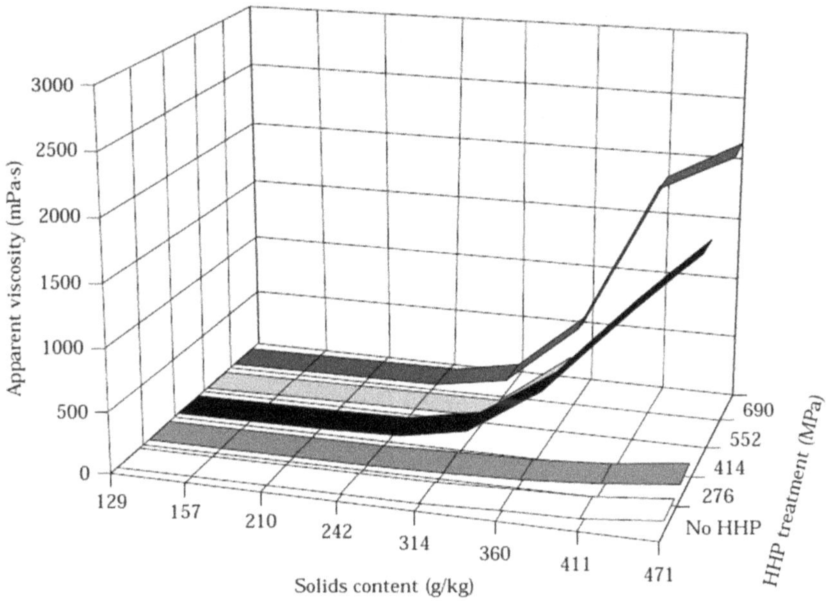


**Figure 2** Effect of concentration and pressure on the apparent viscosity (at 333 $s^{-1}$) of concentrated milk (From Vélez-Ruiz et al. 1998, with permission from Elsevier).

## 7.4 Shear-based technologies

### 7.4.1 High-pressure homogenization

Pereda et al. (2007) observed a significant decrease in viscosity of raw whole milk when treated with high-pressure homogenization (HPH) at 200 MPa at 30°C and 40°C inlet temperatures, but no significant change in viscosity with treatment at 300 MPa at 30°C and 40°C. Amador-Espejo et al. (2014) extended the range of HPH treatment and reported reduced viscosity of bovine whole milk at 200 MPa at 55°C, 65°C and 75°C inlet temperature, but treatment at 200 MPa at 85°C and at the 300 MPa at 55°C, 65°C, 75°C and 85°C increased the viscosity. The decrease in viscosity at the lower pressures and temperatures is probably due to the reduction in particle size, resulting from the partial dissociation of the casein micelles and migration of κ- and $a_{S1}$-caseins to the serum phase (Mohan et al. 2016). The increase in viscosity was attributed to the formation of aggregates of fat globules, resulting from the insufficiency of proteins to cover the surface of the newly created small fat globules.

### 7.4.2 High-pressure jet technology

A recently introduced high-pressure shear-based technology which has been applied to milk is high-pressure jet (HPJ) technology, which operates at pressures of up to 600 MPa. It has a diamond, sapphire, or ruby nozzle with a diameter of 75-400 μm, through which liquid is forced under pressure into a jet stream. The physical phenomena experienced by the liquid are similar to those in HPH, namely, shear caused by turbulence and cavitation (Mohan et al. 2016; Tran et al. 2018). Tran et al. (2018) treated homogenized pasteurized whole milk with HPJ at pressures of 125 MPa, 250 MPa, 375 MPa and 500 MPa. The viscosity of the treated milk, obtained from the Newtonian flow curves, was unchanged (at ~ 2 mPa s) after treatment at 125 MPa, increased at pressures of 250 MPa and 375 MPa, reaching a maximum of approximately 5.0 mPa s, followed by a decrease at 500 MPa to 3.7 mPa s. The changes in viscosity closely reflected the changes in particle size (D[4,3]). The increase in viscosity was not due to whey protein denaturation but probably to casein structural modifications and formation of casein-fat aggregates.

### 7.4.3 Tribomechanical micronization

In this technology, the product is sheared between two hard metal discs spun at high speed in opposite directions, in close proximity to each other. In a study by Herceg and Lelas (2005), WPC (WPC60 and WPC80) and SMP were treated at 22 000 rpm at an inlet temperature of 20.1°C. The outlet temperatures were 30.4°C for the WPCs and 33.6°C for the SMP. Aqueous dispersions (10.3%, 15.5% and 20.6%) prepared from untreated and treated WPC and SMP powders were prepared and analyzed for their rheological properties according to the Power law. Apparent viscosities at 1290 $s^{-1}$ were also recorded. All aqueous dispersions prepared from the treated powders showed higher viscosities than those from the untreated powders. The increases in viscosity were attributed to the significant reduction in the size of particles in the treated products. The authors state that during tribomechanical treatment, the casein micelles in SMP are destroyed; however, no data were provided to support this statement. Apparent viscosity data are shown in Table 7. The consistency coefficients followed the same trend.

## 7.5 Enzymatic modification

Several enzymes are now available for modification of the texture of dairy products (Ercili-Cura et al. 2015). A major enzyme which has been used to alter texture is transglutaminase (TG). A focus of the research has been in altering the texture of fermented milk gels. Treatment of milk before or after fermentation leads to increased firmness of set gels and increased viscosity of stirred milk gels. It also leads to reduced serum separation due to the increase in the proteins' water-holding capacity (Ercili-Cura et al. 2015). As fermented products are beyond the scope of this chapter, this application of TG is not discussed further.

**Table 7** Apparent viscosities (at 1290 $s^{-1}$) (mPa s) of aqueous dispersions of WPC60, WPC80 and SMP powders, untreated and treated by tribomechanical micronization

| Dispersions | Untreated | Tribomechanically treated |
|---|---|---|
| WPC60 10.3% | 11 | 12 |
| WPC60 15.5% | 16 | 19 |
| WPC60 20.6% | 25 | 29 |
| WPC80 10.3% | 21 | 23 |
| WPC80 15.5% | 87 | 96 |
| WPC80 20.6% | 725 | 897 |
| SMP 10.3% | 10 | 12 |
| SMP 15.5% | 14 | 16 |
| SMP 20.6% | 17 | 21 |

Data from Herceg and Lelas (2005).

Another reported application of TG is in reducing the viscosity of MPC dispersions to which calcium chelating agents have been added to increase their heat stability. A viscosity increase occurs on additions of calcium-chelating agents, which results from dissociation of casein micelles and an increase in serum calcium. Power et al. (2020) added SHMP to dispersions of MPC which had been reacted with TG and found the TG-MPC dispersions with SHMP had lower viscosity than the untreated MPC with added SHMP. The effects were attributed to the crosslinking by TG, limiting micelle dissociation on SHMP addition.

Another protein-modifying enzyme which has been shown to modify the functional properties of liquid dairy products is protein glutaminase (PG). This enzyme deamidates glutamine and asparagine to produce glutamic and aspartic acids, respectively and ammonia. Treatment of skim milk with PG increased its viscosity (Miwa et al. 2010). The effect was dependent on the degree of deamidation. With a high level (75%) of deamidation, the viscosity increased from ~1.7 mPa s to ~ 3.0 mPa s. Deamidation causes dissociation of casein micelles; at 75% deamidation, their size was reduced from 150–200 nm to ~40 nm.

## 8 Conclusions

The texture, particularly viscosity, of liquid dairy products is important with regard to both processability and perceived organoleptic characteristics. It is affected by the nature and concentration of the components and by processes which affect these components. Viscosity is mostly affected by the protein components, but fat plays a major role in some products. Lactose and minerals make minor direct contributions. Importantly, viscosity increases with the TS level and decreases with temperature. With many products, viscosity increases during storage. This is significant with concentrated milk, as an increase in viscosity before spray drying makes atomization difficult, while an increase in viscosity during storage of UHT milk often leads to gelation and the end of shelf life.

Most changes in texture of liquid dairy products can be traced back to alterations of two important structures, the casein micelle and the milk fat globule. Disruption of the casein micelle can lead to aggregation with concomitant increase in viscosity, while disruption of the milk fat globule, particularly its membrane, can lead to coalescence of fat globules, leading to fat separation and other textural defects in products such as cream. A knowledge of the causes of changes in texture can enable actions to be taken to minimize adverse texture changes and maximize favorable changes. Several non-thermal technologies are now available to assist with such processes.

## 9 Future research

Future research on this topic is expected to focus on the relationships between physical measures of texture and sensory perception of texture. Newer analytical procedures such as tribology together with established methods such as particle size analysis will be used by product development technologists to assist in optimizing the sensory aspects of the product.

From this chapter, it is apparent that there is considerable information on the relationship between the rheological characteristics of products and the components of conventional milk and milk products. However, with the advent of non-conventional products entering the dairy product space, this information will now be challenged. Such products include cell-based products where 'milk' or milk components are produced in cell culture; these may be chemically similar but have different effects on the texture of the product. A second new dairy product category is the 3D-printed product which is necessarily produced by extrusion of a liquid compounded product, while a third type of 'dairy product' is based on plant proteins. Each of these new product types offers opportunities for the application of established techniques and the development of new techniques for providing an understanding of the physical nature of texture and its relationship to the texture perceived when the food is consumed.

## 10 References

AlKanhal, H. A., Abu-Lehia, I. H. and Al-Saleh, A. A. (1994). Changes in the quality of fresh and recombined ultra-high temperature treated milk during storage. *Int. Dairy J.* 4(3), 221–236.

Amador-Espejo, G. G., Suàrez-Berencia, A., Juan, B., Bárcenas, M. E. and Trujillo, A. J. (2014). Effect of moderate inlet temperatures in ultra-high-pressure homogenization treatments on physicochemical and sensory characteristics of milk. *J. Dairy Sci.* 97(2), 659–671.

Anderson, M., Brooker, B. E., Cawston, T. E. and Cheeseman, G. C. (1977). Changes during storage in stability and composition of ultra-heat-treated aseptically-packed cream of 18% fat-content. *J. Dairy Res.* 44(1), 111–124.

Anema, S. G. (2017). Storage stability and age gelation of reconstituted ultra-high temperature skim milk. *Int. Dairy J.* 75, 56–67.

Anema, S. G. (2019). Age gelation, sedimentation, and creaming in UHT milk: a review. *Compr. Rev. Food Sci. Food Saf.* 18(1), 140–166.

Anema, S. G., Lowe, E. K., Lee, S. K. and Klostermeyer, H. (2014). Effect of the pH of skim milk at heating on milk concentrate viscosity. *Int. Dairy J.* 39(2), 336–343.

Aoki, T. Z., Iwashita, T., Hotanii, T., Shiokawa, M., Ohki, S. and Iwasaki, H. (2002). Increase in the viscosity of concentrated skim milk during storage at low temperature. *Milchwissenschaft* 57, 532–535.

Ashokkumar, M., Lee, J., Zisu, B., Bhaskharacharya, R., Palmer, M. and Kentish, S. (2009). Hot topic: sonication increases the heat stability of whey proteins. *J. Dairy Sci.* 92(11), 5353–5356.

Bakshi, A. S. and Smith, D. E. (1984). Effect of fat-content and temperature on viscosity in relation to pumping requirements of fluid milk-products. *J. Dairy Sci.* 67(6), 1157–1160.

Barefoot, S. F. and Adams, D. M. (1980). Amylase activity in sterile spoiled pudding. *J. Food Sci.* 45(6), 1658–1661.

Bertsch, A. J. and Cerf, O. (1983). Dynamic viscosities of milk and cream from 70°C to 135°C. *J. Dairy Res.* 50(2), 193–200.

Bienvenue, A., Jimenez-Flores, R. and Singh, H. (2003). Rheological properties of concentrated skim milk: importance of soluble minerals in the changes in viscosity during storage. *J. Dairy Sci.* 86(12), 3813–3821.

Bloore, C. G. and Boag, I. F. (1981). Some factors affecting the viscosity of concentrated skim milk. *N. Z. J. Dairy Sci. Technol.* 16, 143–154.

Boumpa, T., Tsioulpas, A., Grandison, A. S. and Lewis, M. J. (2008). Effects of phosphates and citrate on sediment formation in UHT goats' milk. *J. Dairy Res.* 75(2), 160–166.

Burgwald, L. H. (1923). Some factors which influence the feathering of cream in coffee. *J. Agric. Res.* 26, 541–546.

Carr, A. J., Southward, C. R. and Creamer, L. K. (2002). Protein hydration and viscosity of dairy fluids. In: Fox, P. F. and McSweeney, P. L. H. (Eds), *Advanced Dairy Chemistry Volume 1: Proteins*. New York: Kluwer Academic Publishers, pp. 1289–1323.

Cheng, L. J., Birkett, R. A., Augustin, M. A. and Clarke, P. T. (2000). Viscosity of sweetened condensed milk concentrates: effects of preheat treatment applied during powder manufacture. *Aust. J. Dairy Technol.* 55, 115–118.

Cox, C. P., Hosking, Z. D. and Posener, L. N. (1959). Relations between composition and viscosity of cows milk. *J. Dairy Res.* 26(2), 182–189.

Datta, N. and Deeth, H. C. (2001). Age gelation of UHT milk – a review. *Trans. Inst. Chem. Eng. C. Food Bioprod. Proc.* 79(4), 197–210.

de Koning, P. J., Kaper, J., Rollema, H. S. and Driessen, F. M. (1985). Age-thinning and gelation in unconcentrated and concentrated UHT-sterilized skim milk. Effect of native milk proteinase. *Neth. Milk Dairy J.* 39, 71–87.

de Kort, E., Minor, M., Snoeren, T., van Hooijdonk, T. and van der Linden, E. (2011). Effect of calcium chelators on physical changes in casein micelles in concentrated micellar casein solutions. *Int. Dairy J.* 21(12), 907–913.

de Souza, A. B., Costa, L. C. G., Stephani, R., de Oliveira, M. A. L., Perrone, I. T. and Costa, R. G. B. (2015). Evaluation of the viscosity profile obtained for dispersions containing different proportions of milk protein concentrate/whey protein concentrate during simulated conditions of thermal processing. *Lebensm. Wiss. Technol.* 64, 536–539.

Deeth, H. C. and Lewis, M. J. (2017). *High Temperature Processing of Milk and Milk Products*. Oxford: Wiley Blackwell.

Deshpande, V. K. and Walsh, M. K. (2018). Effect of sonication on the viscosity of reconstituted WPI and milk protein concentrate as influenced by solids concentration, temperature and sonication. *Int. Dairy J.* 78, 122–129.

Dumay, E., Lambert, C., Funtenberger, S. and Cheftel, J. C. (1996). Effects of high pressure on the physicochemical characteristics of dairy creams and model oil/water emulsions. *Lebensm. Wiss. Technol.* 29(7), 606–625.

Ercili-Cura, D., Huppertz, T. and Kelly, A. L. (2015). Enzymatic modification of dairy product texture. In: Chen, J. and Rosenthal, A. (Eds), *Modifying Food Texture Volume 1: Novel Ingredients and Processing Techniques*. Cambridge: Woodhead Publishing, pp. 71–97.

Euber, J. R. and Brunner, J. R. (1984). Re-examination of fat globule clustering and creaming in cow milk. *J. Dairy Sci.* 67(12), 2821–2832.

Everett, D. W. (2007). Cream products. In: Hui, Y. H. (Ed), *Handbook of Food Products Manufacturing*. Hoboken, NJ: John Wiley and Sons Inc, pp. 725–749.

Fernández-Martín, F. (1972). Influence of temperature and composition on some physical properties of milk and milk concentrates. 2. Viscosity. *J. Dairy Res.* 39(1), 75–82.

Fink, A. and Kessler, H. G. (1985). The effect of heating on the state of the emulsion in homogenized cream. *Milchwissenschaft* 40, 394–397.

Gaur, V., Schalk, J. and Anema, S. G. (2018). Sedimentation in UHT milk. *Int. Dairy J.* 78, 92–102.

Geyer, S. and Kessler, H. G. (1989). Effect of manufacturing methods on the stability to feathering of homogenized UHT coffee cream. *Milchwissenschaft* 44, 423–427.

Hansen, A. P. (1990). Age gelation and fat separation: two major problems in aseptic dairy products. *J. Dairy Sci.* 73(Suppl. 1), 91.

Hansen, S. F., Larsen, L. B. and Wiking, L. (2019). Thermal effects on IgM-milk fat globule-mediated agglutination. *J. Dairy Res.* 86(1), 108–113.

Harwalkar, V. R. (1992). Age gelation of sterilised milks. In: Fox, P. F. (Ed) *Advanced Dairy Chemistry, Volume 1 Proteins* (2nd edn.). London: Elsevier Applied Science Publishers, pp. 691–734.

Hening, J. C. and Dahlberg, A. C. (1932). A temperature treatment for increasing the viscosity of sweet cream. *N.Y. State Agric. Exp. Sta. Tech. Bull.* 197, 1932.

Hinrichs, J. (2000). UHT processed milk concentrates. *Lait* 80(1), 15–23.

Herceg, Z. and Lelas, V. (2005). The influence of temperature and solid matter content on the viscosity of whey protein concentrates and skim milk powder before and after tribomechanical treatment. *J. Food Eng.* 66(4), 433–438.

Hillbrick, G. C., McMahon, D. J. and Deeth, H. C. (1998). Electrical impedance particle size method (Coulter counter) detects the large fat globules in poorly homogenized UHT processed milk. *Aust. J. Dairy Technol.* 53, 17–21.

Ho, Q. T., Murphy, K. M., Drapala, K. P., Fenelon, M. A., O'Mahony, J. A., Tobin, J. T. and McCarthy, N. A. (2019). Modelling the changes in viscosity during thermal treatment of milk protein concentrate using kinetic data. *J. Food Eng.* 246, 179–191.

Hoffmann, W. (2011). Cream: Manufacture. In: Fuquay, J. W., Fox, P. F. and McSweeney, P. L. H. (Eds) *Encyclopedia of Dairy Sciences* (2nd edn.) (vol. 1). San Diego: Academic Press, pp. 912-919.

Horne, D. S. (1993). Viscosity of milk and its concentrates. In: Dickinson, E. and Walstra, P. (Eds), *Food Colloids and Polymers: Stability and Mechanical Properties*. Royal Society of Chemistry, pp. 260-265.

Hoyle, J. B. and Burton, H. (1984). The problem of cream plug in bottled unhomogenized pasteurized milk. *Int. J. Dairy Tech.* 37(2), 42-45.

Huppertz, T., Kelly, A. L. and Fox, P. F. (2002). Effects of high pressure on constituents and properties of milk. *Int. Dairy J.* 12(7), 561-572.

Huppertz, T., Uniacke-Lowe, T. and Kelly, A. L. (2020). Physical chemistry of milk fat globules. In: McSweeney, P. L. H., Fox, P. F. and O'Mahony, J. A. (Eds), *Advanced Dairy Chemistry Volume 2: Lipids*. Switzerland: Springer Nature, pp. 133-167.

IDF (1996). UHT cream. *Bull. Int. Dairy Fed.* 315, 3-34.

Jambrak, A. R., Mason, T. J., Lelas, V., Herceg, Z. and Herceg, I. L. (2008). Effect of ultrasound treatment on solubility and foaming properties of whey protein suspensions. *J. Food Eng.* 86(2), 281-287.

Jeurnink, T. J. M. and de Kruif, K. G. (1993). Changes in milk on heating–viscosity measurements. *J. Dairy Res.* 60(2), 139-150.

Keogh, M. K. (1978). Seasonal effects of colloid additives on some properties of UHT 35% fat cream. *Ir. J. Food Sci. Technol.* 2, 77-83.

Khalifa, M. Y. and Ghanimah, M. A. (2013). Studies on fluid milk viscosity as affected by some factors. *J. Biol. Chem. Environ. Sci.* 8(2), 57-66.

Kieseker, F. G. and Zadow, J. G. (1973). Factors influencing preparation of UHT whipping cream. *Aust. J. Dairy Technol.* 28, 165-169.

Kocak, H. R. and Zadow, G. (1985a). Age gelation of UHT whole milk as influenced by storage temperature. *Aust. J. Dairy Technol.* 40, 14-18.

Kocak, H. R. and Zadow, J. G. (1985b). Controlling age gelation of UHT milk with additives. *Aust. J. Dairy Technol.* 40, 58-64.

Kohlmann, K. L., Nielsen, S. S. and Ladisch, M. R. (1991). Effects of a low concentration of added plasmin on ultra-high temperature processed milk. *J. Dairy Sci.* 74(4), 1151-1156.

Kováčová, R., Štětina, J. and Čurda, L. (2010). Influence of processing and kappa-carrageenan on properties of whipping cream. *J. Food Eng.* 99(4), 471-478.

Kresic, G., Lelas, V., Jambrak, A. R., Herceg, Z. and Brncic, S. R. (2008). Influence of novel food processing technologies on the rheological and thermophysical properties of whey proteins. *J. Food Eng.* 87(1), 64-73.

Leedham, L. F. and Deacon, R. J. (1970). Microscopic appearance of fat globules in relation to viscosity of UHT cream. Proceedings of the 18th International Dairy Congress (vol. 195).

Lewis, M. J. and Heppell, N. (2000). *Continuous Thermal Processing of Foods: Pasteurization and UHT Sterilization*. Gaithersburg, MD: Aspen Publishers.

Lewis, M., Grandison, A., Lin, M.-J. and Tsioulpas, A. (2011). Ionic calcium and pH as predictors of stability of milk to UHT processing. *Milchwissenschaft* 66, 197-200.

Lorenzen, P. C., Precht, D. and Malmgren, B. (1993). Comparison of different UHT-systems for production of whipping cream with and without additives. Protein and fat globule modifications by heat treatment, homogenization and other technological means for high quality dairy products. *Int. Dairy Fed.* Special Issue 9303, 440-443.

Mainer, G., Sanchez, L., Ena, J. M. and Calvo, M. (1997). Kinetic and thermodynamic parameters for heat denaturation of bovine milk IgG, IgA and IgM. *J. Food Sci.* 62(5), 1034–1038.

McCarthy, O. J. and Singh, H. (2009). Physico-chemical properties of milk. In: McSweeney, P. and Fox, P. F. (Eds), *Advanced Dairy Chemistry, Volume 3: Lactose, Water, Salts and Minor Constituents* (3rd edn.). New York: Springer Science and Business Media, pp. 691–758.

Mineshita, T., Yamamoto, A., Nagano, K., Fujii, S. and Takada, H. (1980). Flow properties of milk in a capillary at low shear-stress. *J. Food Proc. Pres.* 4(1–2), 1–12.

Misawa, N., Barbano, D. M. and Drake, M. (2016). Influence of casein as a percentage of true protein and protein level on color and texture of milks containing 1 and 2% fat. *J. Dairy Sci.* 99(7), 5284–5304.

Miwa, N., Yokoyama, K., Wakabayashi, H. and Nio, N. (2010). Effect of deamidation by protein-glutaminase on physicochemical and functional properties of skim milk. *Int. Dairy J.* 20(6), 393–399.

Mohan, M. S., Ye, R. and Harte, F. (2016). Initial study on high pressure jet processing using a modified waterjet on physicochemical and rennet coagulation properties of pasteurized skim milk. *Int. Dairy J.* 55, 52–58.

Morison, K. R., Phelan, J. P. and Bloore, C. G. (2013). Viscosity and non-Newtonian behaviour of concentrated milk and cream. *Int. J. Food Prop.* 16(4), 882-894.

Mun, D. M., Hsieh, T. -T., Tiu, C. and Sutherland, B. J. (1999). Steady shear viscosity of ultrafiltered whole and skim milk retentates before and during fermentation. *J. Texture Studies* 30(5), 549-561.

Nassar, K. S., Zhang, S. W., Lu, J., Pang, X. Y., Ragab, E. S., Yue, Y. C. and Lv, J. P. (2019). Combined effects of high-pressure treatment and storage temperature on the physicochemical properties of caprine milk. *Int. Dairy J.* 96, 66–72.

Newstead, D. F., Paterson, G., Anema, S. G., Coker, C. J. and Wewala, A. R. (2006). Plasmin activity in direct-steam-injection UHT-processed reconstituted milk: effects of preheat treatment. *Int. Dairy J.* 16(6), 573–579.

Nöbel, S., Weidendorfer, K. and Hinrichs, J. (2012). Apparent voluminosity of casein micelles determined by rheometry. *J. Colloid Interface Sci.* 386(1), 174–180.

Oortwijn, H. and Walstra, P. (1982). The membranes of recombined fat globules. 4. Effects on properties of the recombined milks. *Neth. Milk Dairy J.* 36, 279–290.

Pereda, J., Ferragut, V., Quevedo, J. M., Guamis, B. and Trujillo, A. J. (2007). Effects of ultra-high-pressure homogenization on microbial and physicochemical shelf life of milk. *J. Dairy Sci.* 90(3), 1081–1093.

Phillips, L. G., McGiff, M. L., Barbano, D. M. and Lawless, H. T. (1995a). The influence of fat on the sensory properties, viscosity, and color of low-fat milk. *J. Dairy Sci.* 78(6), 1258–1266.

Phillips, L. G., McGiff, M. L., Barbano, D. M. and Lawless, H. T. (1995b). The influence of nonfat dry milk on the sensory properties, viscosity, and color of low-fat milks. *J. Dairy Sci.* 78(10), 2113–2118.

Power, O. M., Fenelon, M. A., O'Mahony, J. A. and McCarthy, N. A. (2020). Influence of sodium hexametaphosphate addition on the functional properties of milk protein concentrate solutions containing transglutaminase cross-linked proteins. *Int. Dairy J.* 104, Article Number 104641.

Precht, D., Peters, K. H. and Petersen, J. (1987). Improvement of quality of whipping cream by adding carrageenan and milk constituents. II. Influence of additives on the plug formation in pasteurized, unhomogenized cream. *Milchwissenschaft* 42, 635–639.

Quiñones, H. J., Barbano, D. M. and Phillips, L. G. (1997). Influence of protein standardization by ultrafiltration on the viscosity, color, and sensory properties of skim and 1% milk. *J. Dairy Sci.* 80(12), 3142–3151.

Quiñones, H. J., Barbano, D. M. and Phillips, L. G. (1998). Influence of protein standardization by ultrafiltration on the viscosity, color, and sensory properties of 2 and 3.3% milks. *J. Dairy Sci.* 81(4), 884–894.

Ramsey, J. A. and Swartzel, K. R. (1984). Effect of ultra-high temperature processing and storage conditions on rates of sedimentation and fat separation of aseptically packaged milk. *J. Food Sci.* 49(1), 257–262.

Rohm, H., Müller, A. and Hend-Milnera, I. (1996). Effects of composition on raw milk viscosity. *Milchwissenschaft* 51, 259–261.

Saif, S. M. H., Wang, S., Lan, Y. and Yin, G. (2004). Viscosity of caprine milk with respect to fat content, shear rate, and temperature. *Trans. ASAE* 47, 2023–2031.

Streuper, A. and Van Hooydonk, A. C. M. (1986). Heat-treatment of whipping cream. 2. Effect on cream plug formation. *Milchwissenschaft* 41, 547–552.

Sutariya, S. G., Huppertz, T. and Patel, H. A. (2017). Influence of milk pre-heating conditions on casein whey protein interactions and skim milk concentrate viscosity. *Int. Dairy J.* 69, 19–22.

Te Whaiti, I. E. and Fryer, T. F. (1975). Factors that determine the gelling of cream. *N. Z. J. Dairy Sci. Technol.* 10, 2–7.

Topçu, A., Numanoğlu, E. and Saldamli, İ (2006). Proteolysis and storage stability of UHT milk produced in Turkey. *Int. Dairy J.* 16(6), 633–638.

Towler, C. (1988). Effect of additives on the characteristics of ultra-high temperature (UHT) sterilized whipping cream. *N. Z. J. Dairy Sci. Technol.* 23, 109–125.

Tracy, P. H. and Ruehe, H. A. (1930). Factors affecting the heat coagulation of homogenized coffee cream. *Stat. Bull. Univ. Illinois, Agric. Exp. Sta.* 352, 569–578.

Tran, M., Roberts, R., Felix, T. L. and Harte, F. M. (2018). Effect of high-pressure-jet processing on the viscosity and foaming properties of pasteurized whole milk. *J. Dairy Sci.* 101(5), 3887–3899.

Van Vliet, T. and Walstra, P. (1980). Relationship between viscosity and fat-content of milk and cream. *J. Text. Stud.* 11(1), 65–68.

Vélez-Ruiz, J. F. and Barbosa-Cánovas, G. V. (1998). Rheological properties of concentrated milk as a function of concentration, temperature and storage time. *J. Food Eng.* 35(2), 177–190.

Vélez-Ruiz, J. F., Cánovas, G. V. B. and Peleg, M. (1997). Rheological properties of selected dairy products. *Crit. Rev. Food Sci. Nutr.* 37(4), 311–359.

Vélez-Ruiz, J. F., Swanson, B. G. and Barbosa-Cánovas, G. V. (1998). Flow and viscoelastic properties of concentrated milk treated by high hydrostatic pressure. *Lebensm. Wiss. Technol.* 31(2), 182–195.

Wagoner, T. B., Cakir-Fuller, E., Shingleton, R., Drake, M. and Foegeding, E. A. (2020). Viscosity drives texture perception of protein beverages more than hydrocolloid type. *J. Texture Stud.* 51(1), 78–91.

Walstra, P., Geurts, T. J., Noomen, A., Jellema, A. and van Boekel, M. A. J. S. (1999). *Dairy Technology: Principles of Milk Properties and Processes*. New York: Marcel Dekker.

Walstra, P. and Jenness, R. (1984). *Dairy Chemistry and Physics*. New York: John Wiley.

Weise, H. F., Nair, J. H. and Flemming, R. S. (1939). Improvement in the viscosity of pasteurized cream through subsequent heat treatment. *J. Dairy Sci.* 22(11), 875–881.

Whitaker, R. (1931). The feathering of evaporated milk in hot coffee. *J. Dairy Sci.* 14(2), 177–188.

Xiang, B. Y., Simpson, M. V., Ngadi, M. O. and Simpson, B. K. (2011). Flow behaviour and viscosity of reconstituted skimmed milk treated with pulsed electric field. *Biosyst. Eng.* 109(3), 228–234.

Zadow, J. G. (1970). A study on some variables in the manufacture of UHT whipping cream. Proceedings of the 18th International Dairy Congress, 194.

Zhang, S., Jin, Y. and Yun, Z. (2007). Effect of low temperature heat treatment on the activity of lipase VIII from *Pseudomonas* species in UHT milk. *Trans. Chin. Soc. Agric. Eng.* 23(8), 231–235.

Zisu, B., Bhaskaracharya, R., Kentish, S. and Ashokkumar, M. (2010). Ultrasonic processing of dairy systems in large scale reactors. *Ultrason. Sonochem.* 17(6), 1075–1081.

Zisu, B., Schleyer, M. and Chandrapala, J. (2013). Application of ultrasound to reduce viscosity and control the rate of age thickening of concentrated skim milk. *Int. Dairy J.* 31(1), 41–43.

# Chapter 20

## Understanding the contribution of milk constituents to the texture of dairy products: fermented products

*Georg Surber, Harald Rohm and Doris Jaros, Technische Universität Dresden, Germany*



### 1 Introduction

Fermented milk products originate most likely from the Middle East and the Balkans and were already being produced at around 10000 BC by spontaneous fermentation, followed later on by backslopping, where a part of the fermented product is used for the inoculation of the next batch of base milk (Robinson and Tamime, 2006). Today, the vast majority of fermented milk is manufactured industrially under controlled conditions and with specified starter cultures. Microorganisms that are considered as safe for use in foods are listed in the inventory of microbial food cultures of the International Dairy Federation (IDF, 2018). They usually carry the 'Qualified Presumption of Safety' (QPS) status issued by the European Food Safety Authority (EFSA), or the 'Generally Recognized as Safe' (GRAS) status issued by the United States Food and Drug Administration (FDA). The general Codex standard from the Food and Agriculture Organization of the United Nations and the World Health Organization (FAO/WHO, 2003) specifies the aspects of composition and the microorganisms used for specific types of fermented milk (Table 1). It is mainly

http://dx.doi.org/10.19103/AS.2022.0099.20

**Table 1** Production microorganisms and composition of specific fermented milks according to FAO/WHO (2003)

| Fermented product | Yoghurt | Alternate yoghurt | Acidophilus milk | Kefir | Koumiss |
|---|---|---|---|---|---|
| Microorganisms | *Streptococcus thermophilus* and *Lactobacillus delbrueckii* ssp. *bulgaricus* | *S. thermophilus* and any *Lactobacillus* species | *Lactobacillus acidophilus* | Starter culture prepared from kefir grains[a] | *Lactobacillus delbrueckii* ssp. *bulgaricus* and *Kluyveromyces marxianus* |
| Minimum protein [%] | 2.7 | 2.7 | 2.7 | 2.7 | –[b] |
| Maximum fat [%] | 15 | 15 | 15 | 10 | 10 |
| Minimum titrable acidity, expressed as lactic acid [%] | 0.6 | 0.6 | 0.6 | 0.6 | 0.7 |

[a] Kefir grains comprise lactic acid bacteria including *Lactobacillus kefiri*, species of the genera *Leuconostoc*, *Lactococcus*, *Acetobacter* and lactose (*Kluyveromyces marxianus*) and non-lactose fermenting yeasts (e.g. *Saccharomyces unisporus*, *S. cerevisiae* and *S. exiguus*).
[b] not specified.

lactic acid bacteria that are used, and a few specific yeasts that are capable of converting lactose to lactic acid. The respective products are either fermented solely with lactic acid bacteria (e.g. yoghurt) or in combination with yeasts (e.g. kefir or koumiss) and/or moulds (e.g. viili) (Bengoa et al., 2019; Narvhus, 2014). Apart from lactic acid, the microorganisms generate various other metabolites (e.g. acetic acid, diacetyl, acetaldehyde or sometimes ethyl alcohol) that are responsible for the typical product flavour.

Although the legislation differs worldwide, fermented milk should have a minimum of 2.7% protein and a maximum of 10% fat (FAO/WHO, 2003), while yoghurt or acidophilus milk can contain up to 15% fat (see Table 1). Concentrated variants such as strained or Greek-style yoghurt are expected to contain ≥ 5.6% protein. To label a product as 'yoghurt', legislation widely dictates that it is made using the 'classical' thermophilic yoghurt starter, namely *S. thermophilus* and *L. delbrueckii* ssp. *bulgaricus*. In case the *L. delbrueckii* ssp. *bulgaricus* is replaced by other *Lactobacillus* species, mainly for the sake of reducing post-acidification, the product is referred to as alternate yoghurt, sometimes labelled 'yoghurt mild'. The Codex Standard also defines drinkable fermented milk, which is usually made with mesophilic lactococci (e.g. sour milk or cultured buttermilk) and composite products, which consist of at least 40% of fermented milk mixed with water or whey. A list of less well-known fermented milks and concentrated variants around the world can be found elsewhere (Surono and Hosono, 2011; Tamime et al., 2014). When permitted by national legislation, hydrocolloids can be applied in plain fermented milk (FAO/WHO, 2003). For instance, in the United States, gelatin, starch and pectin are sometimes added to the base milk for yoghurt production, whereas in most European countries (France, Germany, etc.), the use of hydrocolloids in plain yoghurt is not allowed (Das et al., 2019).

The manufacture of fermented milk usually starts with the preparation of base milk. Base milk composition may sometimes be substantially altered by pre-concentration for concentrated fermented milk (Jørgensen et al., 2019), or by changing the fat content for low-fat or high-fat products (Narvhus et al., 2019). The common manufacturing steps can be summarised as follows:

- Base milk treatment including fat adjustment and/or dry matter fortification, homogenisation and heat treatment (hydrocolloids for texture enhancement may be added to base milk);
- Inoculation with a specific starter culture and subsequent fermentation at the specified temperature;
- Post-fermentation treatment such as shearing, heating and/or serum separation for concentration and cooling;
- Addition of fruit pre-mixes with or without hydrocolloids, sugar and flavour; and
- Packaging and chilled storage.

Figure 1 gives a rough overview of the main manufacturing steps to distinguish between product varieties. For details, see Mokoonlall et al. (2016), Sfakianakis and Tzia (2014) and Walstra et al. (2006); see de Oliveira (2014) for variants with extended shelf life.

The composition of base milk and its processing considerably affect product texture. According to the ISO (2020), texture refers to mechanical, geometrical and surface properties of a product. Texture descriptors are usually identified in sensory assessments, but instrumental methods are helpful to draw more or less standardised conclusions. Table 2 summarises the fundamental measures frequently assessed for the analysis of physical properties of fermented milk, such as the firmness of set gel, apparent viscosity of stirred product, water-holding capacity (WHC) and gel particle size. Some of the sensory attributes have been denoted as being multisensory. Creaminess, for instance, has been described to be associated with sensory perceptions of thickness, smoothness and mouth-coating (Dickinson, 2018).

This chapter discusses the impact of the main milk components - protein, fat and lactose - in context with manufacturing conditions, on the texture of fermented milk. Further important texture-related aspects such as the use of hydrocolloids and starter cultures with texturising properties through the synthesis of extracellular polysaccharides (EPS) produced *in situ* are also addressed. A case study points to the current challenges of finding unambiguous relationships between product texture and the specific properties of EPS.

## 2 Protein

Approximately 80% of the total protein in milk is casein, which is the main contributor to the texture of fermented milk. Casein, with its four sub-fractions, is present in conjunction with calcium phosphate and forms colloidal supramolecules of approximately 200 nm in diameter, known as casein micelles (see elsewhere in this book). When milk is slowly acidified through the formation of lactic acid by the action of microorganisms, colloidal calcium

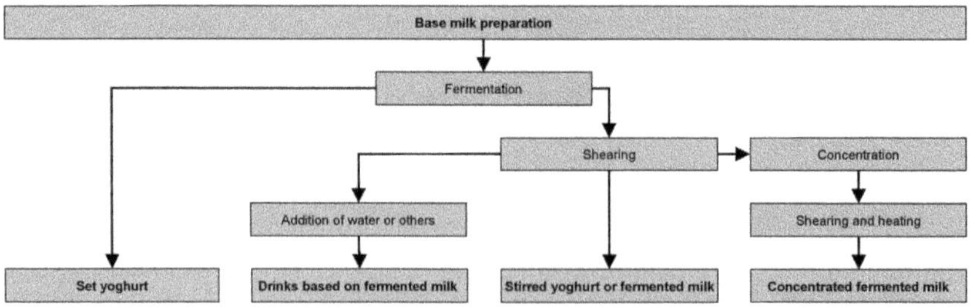


**Figure 1** Schematic overview of manufacturing steps for fermented milk.

**Table 2** Texture parameters of fermented milk frequently described in literature

| Parameter | Type of fermented milk | Reference |
|---|---|---|
| Shear rheometry | | |
| Thixotropic loop area | Stirred, Concentrated | (Krzeminski et al., 2011; Schmidt et al., 2016) |
| Apparent viscosity, Shear stress | Stirred | (Chua et al., 2017; Krzeminski et al., 2011) |
| Tribology | | |
| Friction coefficient | Set, Stirred, Concentrated | (Nguyen et al., 2017) |
| Oscillation rheometry | | |
| Stiffness as storage modulus from linear viscoelastic region, Phase angle, Critical stress and strain | Stirred, Concentrated | (Chua et al., 2017; Krzeminski et al., 2011; Zhang et al., 2016) |
| Stiffness at desired pH | Set | (Mende et al., 2012) |
| Penetration | | |
| Firmness (Force maximum) | Set, Stirred | (Buldo et al., 2016) |
| Laser diffraction | | |
| Sauter mean diameter, Diameter at 90th percentile | Stirred, Concentrated | (Krzeminski et al., 2011) |
| Micro- and Macrostructure | | |
| Confocal laser scanning microcopy, Scanning electron microscopy | Stirred, Concentrated | (Krzeminski et al., 2011; Remeuf et al., 2003) |
| Number, form and diameter of grains | Stirred | (Körzendörfer et al., 2019; Remeuf et al., 2003) |
| Serum release | | |
| Spontaneous or forced separation (water-holding capacity) | Set, Stirred, Concentrated | (Amatayakul et al., 2006b; Schmidt et al., 2016) |
| Sensory analysis | | |
| Thickness, smoothness, creaminess, ropiness, stickiness, graininess, mouth coating and others | Set, Stirred, Concentrated | (Jørgensen et al., 2015; Nguyen et al., 2017) |

phosphate dissociates from the micelles. This causes a destabilisation of the colloidal system, and beginning at pH ~ 5.4, reassociation occurs, allowing the subsequent formation of a three-dimensional network (Fig. 2) near the isoelectric point of casein (pH ~ 4.6) (McMahon et al., 2009). In the native state, the whey proteins remain dissolved in the milk serum and do not significantly contribute to the gel.

The heat treatment of base milk used for fermented products is usually much more intense than is necessary, for hygiene purposes (e.g. 95°C for 5 min). During heating at above 60°C, whey protein starts to denature, with

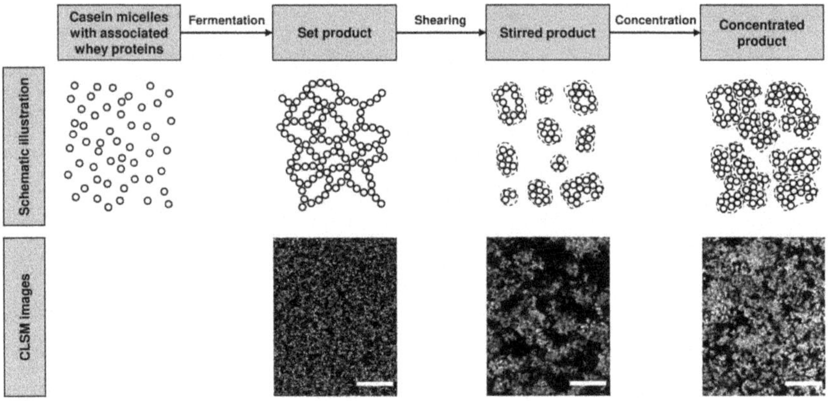


**Figure 2** Gelation of heated milk and formation of gel particle suspensions supported by confocal laser scanning microscopy (CLSM) images. Scale bar = 20 μm.

the degree of denaturation dependent on temperature and time. Reactive side groups, especially thiol groups, are exposed due to unfolding of the denatured whey protein, resulting in thiol-disulphide exchange reactions and some hydrophobic interactions mainly between β-lactoglobulin and κ- or $\alpha_{S2}$-casein (Anema, 2020). Consequently, a thin layer of denatured whey protein is formed around the surface of the casein micelle, which is responsible for the limited access of κ-casein for chymosin in cheese production. As a consequence, whey protein is incorporated into the gel network as additional bridging material and the pH at gelation is shifted towards the isoelectric point of whey proteins (pH 5.3 for β-lactoglobulin, the major whey protein; Anema, 2020). Furthermore, the capacity of the positively charged denatured whey protein to immobilise water is increased to almost that of casein (Remeuf et al., 2003). The consequent high degree of denaturation generally increases the WHC, firmness and stiffness of set gels (Anema, 2020; Lee and Lucey, 2010).

## 2.1 Fermentation temperature and post-processing

Fermentation at the optimum growth temperature of the respective starter culture, for instance at ~ 43°C for thermophilic lactic acid bacteria, results in a short production time and high hydrophobic interactions within the casein micelles (McMahon et al., 2009). As a consequence, casein voluminosity decreases with higher temperatures for casein micelle suspensions (3.5 mL/g at 45°C *vs.* 4.0 mL/g at 25°C), as demonstrated by Nöbel et al. (2016). A higher fermentation temperature results in a decreased WHC and lower firmness of set gels (Robinson et al., 2006; Wu et al., 2009). At a lower fermentation temperature, the generation time and hence the production time are prolonged, but the slower release of lactic acid causes an increase of interparticle junctions, resulting in a less coarse and more homogeneous network (Robinson et al., 2006).

For the production of stirred fermented milk, the gel is stirred or sheared and a gel particle suspension is obtained (see Fig. 2). The mechanical force applied during shearing, pumping, cooling and filling induces a significant viscosity reduction, as reviewed by Mokoonlall et al. (2016). It was shown that rebodying of the gel structure during cold storage depends on the intensity of mechanical stress, temperature and the sequence and combination of stirring operations (Guénard-Lampron et al., 2020).

## 2.2 Protein enrichment

Apart from using hydrocolloids, the texture of fermented milk is improved and texture losses during post-processing or transportation are reduced by increasing the protein content of base milk. This is achieved by either increasing the total dry matter without substantially changing the relative content of milk constituents, or by selectively raising the protein concentration. In the first approach, mainly (skimmed) milk powder or milk concentrate from evaporation or reverse osmosis is used for the fortification of base milk. The higher protein content enhances protein interactions and serum immobilisation in the gel network (Sodini et al., 2004). This causes a significant increase in the gel stiffness and firmness or apparent viscosity of set or stirred yoghurt, respectively (Mokoonlall et al., 2016; Sodini et al., 2004). For example, a dry matter increase from 12.3% to 15.7% (approx. 4.5–6.0% protein) increased the apparent viscosity of stirred yoghurt from 2.5 Pa·s to 4.5 Pa·s (Wu et al., 2009).

Milk protein concentrate ( ≤ 85% protein) or isolate ( > 85% protein) with a casein : whey protein ratio of ~ 4:1 is frequently used to raise the protein content of base milk without significantly increasing lactose or minerals, compared to (skimmed) milk powder. Enrichment of base milk with specific protein fractions allows the adjustment of the casein : whey protein ratio (Table 3). Whey protein concentrate and isolate generally improve texture and shorten fermentation time (Karam et al., 2013; Lesme et al., 2020); for example, gel stiffness is elevated by up to four times compared to the addition of skimmed milk powder. The recommended level of addition is to increase the protein by 1–2 percentage points. A higher amount of whey protein has been reported to evoke a metallic flavour and grainy texture (Karam et al., 2013). A high calcium content in the concentrate strengthens the effect because of a calcium-induced higher whey protein denaturation and consequently more interactions between casein and whey proteins (Anema, 2020; Lee and Lucey, 2010). The calcium content in whey protein products depends on the type of whey (sweet/acid), membrane pore size for protein separation and eventually salt reduction by means of ion exchange or electrodialysis.

**Table 3** Effect of specific protein fractions on texture of fermented milk as compared to the addition of skimmed milk powder

| Milk fraction | Type of fermented milk | Total protein [%] | Casein : whey protein ratio | Effect on texture | Reference |
|---|---|---|---|---|---|
| Whey protein concentrate | Set | 4.0 | 3:1, 2:1 and 1:1 | G′ at pH 4.6 increased by ~ 2 and ~ 4 times at 2:1 and 1:1, respectively. | (Zhao et al., 2016) |
| | Stirred | 4.5 | -[a] | Higher stiffness (nearly doubled), graininess and WHC. | (Buldo et al., 2016) |
| | Stirred | 4.5 | 1.8:1 | Two times higher apparent viscosity, but higher graininess and similar WHC. | (Remeuf et al., 2003) |
| Whey protein isolate | Stirred | 3.5, 4.5 and 6.0 | 1.5:1 and 1:1.5 | Larger particles (up to 4 times) and higher G′ (up to 3 times) with increasing whey protein content, apparent viscosity unaltered. Low-fat yogurt with 4.5% protein and 1.5:1 comparable to full-fat yoghurt. | (Krzeminski et al., 2011) |
| | Stirred | 4.0, 4.5 and 5.0 | 2.1:1, 1.5:1 and 1.2:1 | Higher firmness (up to 1.5 times) and yield stress (up to 2 times). Non-fat yoghurt with 0.5% whey protein isolate (4.0% protein) similar to full-fat yoghurt. | (Chua et al., 2017) |
| | Stirred | 4.3 - 4.8 | 3:1, 2:1 and 1.5:1 | Surface perimeter of grains (up to 2 times), stiffness (up to 3 times) and, syneresis (by one fifth) and yield stress (up to 5 times) decreased with increasing casein : whey protein ratio. | (Kücükcetin, 2008) |
| Milk protein isolate | Set | 3.5 | 4:1 | G′ at pH 4.6 increased by ~ 1.2 times. | (Peng et al., 2009) |
| Sodium caseinate | Set | 3.5 | 5.4:1 | G′ at pH 4.6 increased by ~ 1.5 times. | (Peng et al., 2009) |
| | Stirred | 4.0, 4.5 and 5.0 | 4.2:1, 3.9:1 and 3.5:1 | Similar G′ and yield stress. | (Chua et al., 2017) |
| Sodium caseinate or calcium caseinate | Stirred | 4.5 | 12:1 | Sodium caseinate: similar apparent viscosity, slightly higher graininess and lower WHC,<br>Calcium caseinate: lower viscosity. | (Remeuf et al., 2003) |
| Micellar casein concentrate | Set | 3.5 | 4.8:1 | Similar G′ at pH 4.6. | (Peng et al., 2009) |

[a] not determined.
G′, gel stiffness; WHC, water-holding capacity.

Whey protein also contributes to the texture when added as particulated whey protein aggregates. They can be distinguished into nano- ( < 1 μm) and microparticulated aggregates (1–10 μm), according to particle size (Lesme et al., 2020). Both are produced by acid precipitation and thermal denaturation of whey protein concentrate, and additional shearing is applied for microparticulated aggregates (Liu et al., 2016a). With experiments using a thiol-blocking agent, it was demonstrated that nanoparticulated aggregates are incorporated into the gel network via thiol–disulphide exchange reactions and hydrophobic interactions (Liu et al., 2016b). Hence, the stiffness as well as the apparent viscosity and WHC were higher compared to microparticulated aggregates or whey protein isolates in fermented milk (5% total protein, casein : whey protein ratio of one) (Liu et al., 2016a). The microparticulated aggregates, which are only partly incorporated into the network, provided a creaminess similar to milk fat (Ipsen, 2017). A native:denatured whey protein ratio of ~1 and preferably small particles is recommended, for better integration of the microparticulated aggregates in fermented milk (Ipsen, 2017). Recently, Hossain et al. (2020) reported on an alternative microparticulation technique, namely high-moisture hot extrusion. In their study, smaller microparticulated whey protein aggregates ( < 3 μm) were more effective in increasing the viscosity, firmness and creaminess of stirred yoghurt than the larger aggregates ( > 5 μm).

Casein was slightly less efficient than whey protein after increasing the protein concentration by 1–2 percentage points (see Table 3 and Karam et al., 2013; Lesme et al., 2020; Sodini et al., 2004). Sodium caseinate increased gel stiffness more than micellar casein, which was explained by the reduced buffering capacity of base milk with sodium caseinate (by 50%, between pH 5.8 and pH 4.1). This probably led to more solubilised colloidal calcium phosphate, and therefore, to an increase of the contact area of casein micelles (Peng et al., 2009). After size fractionation of casein micelles via microfiltration, concentrates with casein micelles of ~ 129 nm in diameter were reported to increase stiffness more than the casein micelles with ~ 183 nm in diameter. This was explained by a larger relative amount of κ-casein at the surface of smaller micelles, providing more interactions during acidification (Jørgensen et al., 2017). However, further research is needed to reveal how casein micelle size or specific casein fractions such as soluble κ-casein improve product texture.

Blends of protein fractions appear as a useful and interesting perspective for base milk fortification (Karam et al., 2013). For example, a blend of calcium caseinate and whey protein concentrate in a ratio of 2:1, for increasing the protein from 3.1% to 4.5%, showed the advantages of increased apparent viscosity and low syneresis without causing graininess, as compared to using only whey protein concentrate (Remeuf et al., 2003).

## 2.3 Concentrated fermented milk

Concentrated products usually have ≥ 8% protein (Tamime et al., 2014) and are produced from fermented milk by means of concentration with a separator or membrane filtration (see Fig. 2). This largely alters the texture, especially of low-fat products, since particle–particle interactions become more pronounced at higher protein concentrations (Mokoonlall et al., 2016). For example, the dependency of firmness on the protein content (4–12%) was described by a power law function (Körzendörfer et al., 2019). Increased particle interactions resulted in larger particles, which contributed to the higher stiffness and viscosity of concentrated products (8% protein) (Hahn et al., 2015), whereas smaller particles were responsible for a smoother (Jørgensen et al., 2015; Krzeminski et al., 2011) and more creamy perception (Cayot et al., 2008). Hence, post-fermentative steps such as shearing are helpful to eliminate graininess. To minimise structural losses upon shearing, mechanical treatments can be combined with heating to adjust the texture, as reviewed by Mokoonlall et al. (2016).

Concentrated products can also be made from base milk enriched to the desired protein content prior to fermentation. This was demonstrated with milk fortified with milk protein concentrate or micellar casein concentrate (~ 10% protein). Gel stiffness after shearing was similar compared to yoghurt obtained by post-fermentative concentration (Uduwerella et al., 2017). Using enriched base milk makes the separation of acid whey obsolete, but usually prolongs fermentation time because of the higher buffering capacity and mineral content of the concentrates (Peng et al., 2009). As high calcium content can cause a bitter taste, milk protein concentrate obtained by microfiltration in combination with diafiltration is more suitable to ensure low bitterness in products with ~ 8% protein (Schäfer et al., 2019). Low calcium content contributed to shorter fermentation time of high-protein yoghurt (9% protein). In combination with subsequent shearing by means of hydrodynamic cavitation, the texture was similar to yoghurt obtained by post-fermentative concentration (Meletharayil et al., 2016a). Gel stiffness of a product from skimmed milk enriched with milk protein concentrate (10% protein) could be reduced by applying acoustic cavitation during fermentation by means of ultrasound at 20 kHz, which helps in case the firmness is too high for further processing (Körzendörfer et al., 2019).

Whey protein has also been reported as suitable for concentrated products, as long as the casein : whey protein ratio remains higher than 2:1 and ~ 40–50% of the whey protein is in its native state. Heating to 95°C, on the other hand, causes a doubling of firmness, excessive graininess and a coarse microstructure in products with 8% protein (Jørgensen et al., 2015). Heating of base milk with > 4% of added whey protein may induce coagulation (Lee and Lucey, 2010). Microparticulated whey protein, preferably with particle size < 5 μm and a ratio of native:denatured whey proteins < 1, can be used as an inert filler in concentrated products (Ipsen, 2017). However, it seems that microparticulated whey protein prevents casein aggregation and creates serum channels, resulting in gels (8% protein) with low firmness, yield stress and WHC compared to gels without microparticulated whey protein (Silva and O'Mahony, 2018).

## 3 Fat

The size of fat globules dispersed in the aqueous phase of raw milk (~ 4% fat) is between ~ 0.1 and 15 μm (average: 3.3 μm) (Narvhus et al., 2019; Walstra et al., 2006). They are stabilised by the milk fat globule membrane, which consists of proteins and some lipids (triacyglycerols, diacylglycerols, monacylglycerols, sterols, sterol esters and phospholipids). Milk in its natural state tends to separate with the fat creaming up to the surface because of the obvious density difference, but also because of specific membrane properties. The coalescence of fat globules, which significantly increases creaming velocity, has also been reported (Walstra et al., 2006). As regards acid milk gels, the fat globules can be considered as being some sort of inert fillers which interrupt the network, and therefore reduce gel firmness (Schkoda et al., 2001).

To prevent creaming and to ensure physical stability, milk is usually homogenised to reduce the fat globule size by applying a pressure of 8–30 MPa at 50–60°C. The shear-induced reduction of the size of the fat globules to 0.1-1 μm significantly increases their specific surface area. The additional uncovered interface is coated by a thin layer of micellar casein, and in addition, by some whey protein (Walstra et al., 2006). This allows the fat globules to interact as pseudocasein particles with the bulk protein phase, and to link with the gel network upon acidification (Walstra et al., 2006). Fermented milk with smaller fat globules exhibits higher firmness, a smoother texture and a higher viscosity after stirring. Whiteness and water retention are also improved (Robinson et al., 2006; Truong et al., 2015; Walstra et al., 2006).

Applying either higher homogenisation temperature ( > 60°C) or heating milk prior to homogenisation may be helpful to improve texture, as the denatured whey protein at the surface of the fat globule interacts with casein (Walstra et al., 2006). Using atomic force microscopy, Obeid et al. (2020) showed that heating of whole milk at 140°C for 4 s followed by homogenisation (first stage: 20 MPa; second: 2 MPa) results in highly reactive denatured whey protein, which is responsible for a more branched protein network and stiffer gels upon chemical acidification compared to heating at 50°C for 10 min. On the other hand, Ji et al. (2011) found only slight differences in gel stiffness when the heat treatment was carried out before (370–430 Pa) or after homogenisation (400–450 Pa). Their most obvious finding was a higher resistance of gels against stress when milk was heated after homogenisation (yield stress of ~ 400 Pa *vs.* ~ 300 Pa at a fat globule size of ~ 0.2 μm). Narvhus et al. (2019) pointed out that complex interactions between the effects of both the increased degree of whey protein denaturation and the decreased fat globule size need to be studied in more detail.

A nearly linear increase of apparent viscosity and stiffness occurred in stirred yoghurt from fat-enriched milk with fat content between 3.5% and 12% (Krzeminski et al., 2011). A fat content of 3.5–6.0% was positively associated with sensory properties such as creaminess and smooth mouthfeel, whereas 12% fat induced a fatty mouthfeel perception (Tomaschunas et al., 2012). As regards gel formation, it has been reported that the protein phase still dominates gel texture as long as the fat content remains below 20% (Narvhus et al., 2019). However, in milk with > 9% fat, clusters of the globules may be formed upon one-stage homogenisation because of hydrophobic interactions and too little protein available to fully cover the newly formed fat surface (Narvhus et al., 2019; Walstra et al., 2006). These clusters can be disrupted again by applying a pressure of 3–5 MPa at the second homogenisation stage, as here the turbulent intensity is too low to considerably reduce globule size further (Walstra et al., 2006). Recently, Brighenti et al. (2018) showed that two-stage homogenisation at 25/5 MPa of milk with 12% fat decreased the Sauter mean diameter from 3.5 μm to 0.30 μm. The adsorption of protein on the newly created fat interfaces reduced the protein content of the bulk phase from 2.8 % in non-homogenised milk to 0.4 %.

There are also reports on the effects of homogenisation performed at elevated pressure, that is, at 50–350 MPa (Truong et al., 2015). Subjecting whole milk to 200 MPa or 300 MPa at 40°C reduced fat globule size to a Sauter mean diameter of ~ 0.2 μm (Serra et al., 2007). The respective set yoghurt showed higher WHC and increased firmness compared to yoghurt made from conventionally heated and homogenised milk (0.4 μm at 15 MPa). However, in a later study, they reported an increased graininess in stirred yoghurt made from milk homogenised at 200 MPa or 300 MPa, presumably because of predominantly casein–casein interactions (Serra et al., 2009).

When applying microfluidisation, the milk is divided into two streams, which are forced through narrow channels and then collide with each other and the wall in the interaction chamber (Truong et al., 2015). By microfluidising milk at 50–150 MPa, the stiffness of set gels increased to 156–170 Pa compared to microfluidising at 25 MPa (103 Pa) or conventional homogenisation at 20/5 Pa (57 Pa). The smaller fat globules were better incorporated into the protein network and the respective yoghurt showed an increased matrix density with lower porosity (Fig. 3 and Ciron et al., 2012). Microfluidisation at high pressure was suitable to increase apparent viscosity, smoothness and creaminess of non-fat or low-fat yoghurt to a level similar to that of full-fat yoghurt (Ciron et al., 2011, 2012).

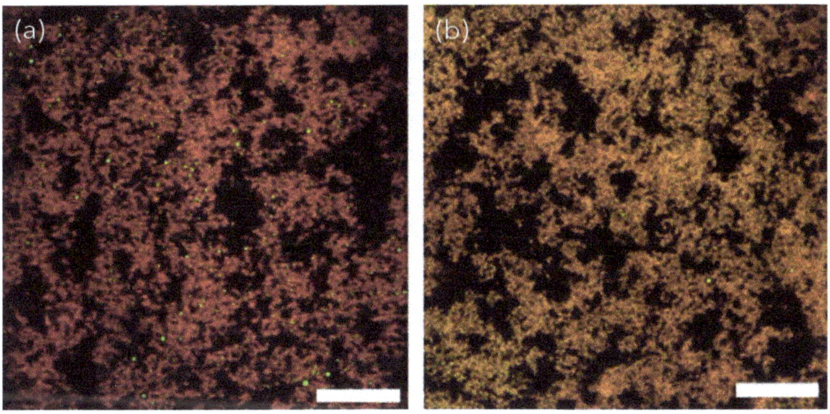


**Figure 3** Confocal laser scanning micrographs showing gross morphology of stirred yoghurts prepared with: low-fat milk (1.5%) homogenized at 20/5 MPa (a) or microfluidised at 150 MPa (b). Scale bar = 50 μm. Reproduced from Ciron et al. (2012) with permission from Elsevier.

Power ultrasound (20–100 kHz) was proposed as another tool to reduce fat globule size in milk (Truong et al., 2015). The combination of ultrasound and heat treatment (24 kHz at 72°C for 10 min) of base milk with 1.5% or 3.5% fat resulted in fat globules of ~ 0.4 μm median diameter. The resulting yoghurt therefore exhibited higher WHC and viscosity and showed smaller particles than yoghurt produced from milk heated at 90°C for 10 min (Riener et al., 2009).

## 4 Lactose

The role of lactose on the texture of fermented milk was recently summarised by Huppertz (2017). Apart from the primary function as a carbohydrate source for the lactic acid bacteria, residual lactose in yoghurt also enhances structure and texture through strengthening hydrophobic interactions between casein micelles (Huppertz, 2017). It was reported that a lactose content elevated to 6% or 11% in reconstituted milk protein concentrate (4% protein) resulted in higher gel stiffness and WHC (Meletharayil et al., 2016b). A lactose content of up to 15% in reconstituted skimmed milk (~ 4% protein) retarded denaturation of α-lactalbumin and β-lactoglobulin depending on heating conditions (80°C or 95°C) (Anema et al., 2006). It should be noted that increasing lactose content in yoghurt milk increases osmotic stress, which in turn may reduce the growth performance of lactic acid bacteria. A high lactose content may be likely in base milk for concentrated yoghurt, for instance when milk is fortified with milk powders of < 60% protein (Jørgensen et al., 2019).

Lactose-free fermented milk can be produced through the hydrolysis of lactose with β-galactosidase. Hydrolysis is either carried out prior to heating milk, or after heating simultaneously with fermentation. The latter approach is more frequently applied in the manufacture of fermented milk as preliminary lactose hydrolysis may reduce the acidification activity, either because of strain-dependent differences as regards the preferred carbon source and/or because of the increased osmotic pressure in hydrolysed milk (Dekker et al., 2019). However, Schmidt et al. (2016) stated that the acidification rate in either pro- or co-hydrolysed milk strongly depends on culture and substrate (non-fat regular or Greek-style). The fermentation time was lower in hydrolysed milk for three out of five starter cultures, indicating the importance of the selection of suitable strains.

## 5 Additional factors

### 5.1 Hydrocolloids

Except gelatin, hydrocolloids of practical relevance for fermented milk belong to a heterogeneous group of long-chain hydrophilic polysaccharides, which differ largely in their type and number of monosaccharides, branching and charge (Gyawali and Ibrahim, 2016). Hydrocolloids for fermented milk are specified in FAO/WHO (2003). They increase serum viscosity because of their high water-binding capacity and/or interact with casein or whey protein and hence improve texture. The protein-polysaccharide interactions can be referred to as co-solubility, incompatibility, segregation and complexation (Goh et al., 2020). These interactions are accompanied by electrostatic reactions, steric repulsion and depletion flocculation (Yousefi and Jafari, 2019).

Gelatin was frequently added in fermented milk. However, because of ethical reasons, it is being replaced more and more by polysaccharides (Table 4 and Das et al., 2019). A typical polysaccharide concentration is 0.04–0.1% for gellan, locust bean gum and guar gum, while higher amounts are used for pectin, xanthan (up to 0.5%) and modified starch, here up to 2% (Gyawali and Ibrahim, 2016). However, when locust bean gum, guar gum or carrageenan were added to the base milk in a concentration higher than ≥ 0.1%, they hindered gel formation, whereas xanthan and chemically modified tapioca starch increased stiffness of set yoghurt (Pang et al., 2015). Pang et al. (2019) demonstrated increased resistance against agitation for chemically modified waxy corn starch compared to the starch in its native state, which resulted in higher apparent viscosity and stiffness of stirred yoghurt (0.5% starch, ~ 3.3% protein). However, Nguyen et al. (2017) stated that modified starch, carrageenan or xanthan might also create undesirable mouthfeel such as chalkiness and lumpiness. Synergistic interactions between different hydrocolloids often provide advantages, as this was demonstrated by combining xanthan and locust bean gum (each 0.005%) together with a 0.5% whey protein isolate for stirred yoghurt (Pang et al., 2016).

Currently, an increased interest in the use of pectin for fermented milk and concentrated variants can be observed (Gyawali and Ibrahim, 2016). Depending on the amount of methyl esters, pectin is classified into low- (< 50%) and high-methoxy pectin (> 50%). High-methoxy pectin is usually preferred to ensure high viscosity in fermented milk, whereas low-methoxy pectin is often chosen to increase the firmness of set gels (Khubber et al., 2021; Krzeminski et al., 2014). The enhanced creaminess and low graininess induced by high-methoxy pectin were explained by a reduced self-association of whey protein, presumably induced by the predominant formation of whey protein–pectin complexes during heating (Krzeminski et al., 2014). These complexes possibly contributed to high WHC, for instance in strained yoghurt with 6–9% protein (Gyawali and Ibrahim, 2018).

### 5.2 Lactic acid bacteria

Lactic acid bacteria can be grouped based on their morphology into cocci and rods, and according to their optimum growth temperature, as mesophilic (~ 30°C) and thermophilic (~ 45°C). Homofermentative strains almost exclusively produce lactic acid, whereas heterofermentative strains produce lactic acid and other compounds such as $CO_2$, ethanol or acetic acid (ISO, 2010; Tamine et al., 2006).

In the industrial manufacture of fermented milk, acidification is usually carried out by multiple-strain or multiple-species starter cultures (ISO, 2010). For products such as kefir or koumiss, sometimes more traditional cultures are used; for example, complex communities of lactic acid and acetic acid bacteria with yeasts together in a matrix of proteins and polysaccharides, known as kefir grains (Bengoa et al., 2019). Apart from the primary purpose of quickly converting lactose to lactic acid, the different strains within a starter culture may have specific roles such as contributing to flavour or protecting against bacteriophages (Tamine et al., 2006). Starter cultures with the main role of acidifying milk are particularly used for liquid-like fermented milk, for example, cultured buttermilk and sour milk (Sodini et al., 2004). To limit post-acidification in yoghurt during cold storage, the ratio of 1:1 of *S. thermophilus* and *L. delbrueckii* ssp. *bulgaricus* is shifted to at least 3:1 (Zhang et al., 2016). In a recent review, Deshwal et al. (2021) reported on the use of bacteriocin-producing strains and on the possibility of strain improvement through random mutagenesis.

For the last three decades, strains capable of producing EPS have received increased attention as they naturally modify the texture of fermented milk by contributing to stiffness, viscosity, creaminess, shear resistance and WHC (Mende et al., 2016; Zeidan et al., 2017). However, the extent of texture enhancement is strain-dependent and related to intrinsic properties of the

**Table 4** Brief overview on the effect of hydrocolloids on the texture of low-fat fermented milk reported recently (in chronological order). Hydrocolloids were added to the base milk prior to acidification

| Reference | Type of fermented milk | Hydrocolloids (Amount [%]) | Effect on texture |
|---|---|---|---|
| Pang et al. (2016) | Stirred | Combinations of xanthan and locust bean gum (each 0.005), carrageenan (0.05) and starch (0.2) | A combination of 0.5% whey protein isolate with xanthan/ locust bean gum contributed to stirred gels with higher stiffness, but lower smoothness. The stiffness and WHC of these gels were similar to gels with starch and 0.5% whey protein isolate. A combination of 0.5% whey protein isolate with carrageenan considerably decreased stiffness and WHC, which was attributed to the strong interaction between carrageenan and milk proteins. |
| Nguyen et al. (2017) | Set, Stirred | Gelatin and modified starch (0.5–1.5) and xanthan and carrageenan ( ≤ 0.015) | Gelatin produced the highest firmness and the highest WHC of set gels, while xanthan and carrageenan were responsible for the highest viscosities of the respective stirred product. Modified starch, carrageenan and xanthan created mouthfeel sensations such as chalkiness and lumpiness. |
| Gyawali and Ibrahim (2018) | Concentrated | Two low-methoxyl pectins (0.05) | Pectins slightly increased the WHC of Greek-style yoghurt. The WHC was further increased by the addition of 1% whey protein concentrate (WPC80) together with pectin. |
| Bravo-Núñez et al. (2019) | Set, Stirred | Chemically modified and pregelatinised starch (2) | WHC, gel firmness and apparent viscosity in the following order: Hydroxypropylated starch > acetylated starch cross-linked with adipate > pregelatinized starch > acetylated starch cross-linked with phosphate. |
| Khubber et al. (2021) | Set | Low-methoxyl pectin (0.05–1.0) | WHC and gel stiffness increased with pectin concentration. |

WHC, water-holding capacity.

EPS, which can be distinguished into homo- (HoPS) and heteropolysaccharides (HePS). They are usually present in the serum phase, referred to as free EPS, and may be classified by the degree of ropiness they induce in the product. Some strains also produce capsular EPS, which are attached to the bacterial cell wall.

Most homopolysaccharides are produced from sucrose by extracellular enzymes, while the biosynthesis of HePS is more complex (Schmid, 2018). HoPS producers such as *Leuconostoc mesenteroides* are sometimes present in undefined traditional kefir cultures (Bengoa et al., 2019), or in starters for cultured buttermilk and some Nordic fermented milk (Surono and Hosono, 2011). There is a huge potential of HoPS in texturising milk, which has been demonstrated by adding dextran from *L. mesenteroides* to milk prior to acidification. The addition of 3% dextran increased gel stiffness and apparent viscosity by a factor of 2 and 3, respectively (Mende et al., 2013a). In a recent study, 6 out of 249 *L. mesenteroides* strains were found to increase the viscosity of milk supplemented with sucrose, and one strain was able to texturise milk without sucrose, suggesting that this strain synthesises HePS (Poulsen et al., 2020).

Heteropolysaccharides exhibit a wide diversity with respect to monosaccharide composition, branching, backbone stiffness and charge (Mende et al., 2016; Zeidan et al., 2017). They are produced in considerably lower amounts than HoPS (up to 240 mg/kg) mainly by the genera *Lactobacillus*, *Lactococcus*, or *Streptococcus* and play the major role in the texture enhancement of fermented milk (Amatayakul et al., 2006a; Folkenberg et al., 2006a; Schmidt et al., 2016; Surber et al., 2019). There is still uncertainty on whether concentration is important for the effects of HePS as their amount is strongly affected by several different intrinsic or extrinsic factors, for example, base milk composition, fermentation conditions and specific strains, making unambiguous statements difficult (Mende et al., 2016). In a systematic study where 0–300 mg/kg of an isolated ropy HePS from *S. thermophilus* ST-143 were added to milk prior to chemical acidification, a significant positive correlation was detected between HePS amount and gel stiffness or WHC of set and stirred gels (Mende et al., 2013b). These findings could be confirmed in an expanded study with isolates of ropy HePS from *S. thermophilus* DGCC7710; however, at a concentration > 300 mg/kg gel stiffness declined again (Nachtigall et al., 2021b).

The ability of HePS to evoke ropiness is easily accessible by pulling a spoon out of the product and judging the thread. Evidence of capsular HePS can be simply proven by negative staining with Indian ink and light microscopy (Surber et al., 2019). Ropy HePS are frequently reported to be responsible for high creaminess and high viscosity of fermented milk, while non-ropy HePS increase viscosity only moderately (Mende et al., 2016; Zeidan et al., 2017). In fermented milk with varying ropiness, the increase in extensional viscosity was more pronounced than the increase of apparent viscosity (Fig. 4 and Surber et al., 2020). Extensional viscosity was associated with a high thickness and stickiness of a model system (He et al., 2016), and might also be suitable to gain additional texture information of fermented milk when ropy HePS are present. Surber et al. (2020) also stated that higher relaxation times at higher ropiness indicate that stretchability is presumably induced by HePS–HePS interactions, most likely hydrophilic interactions (Zhou et al., 2019). Ropy HePS in stirred fermented milk lowered the ability of structure regeneration probably because of steric hindrance of the reassociation of the proteins. This hindrance, however, was less pronounced in case capsular HePS were also present (Folkenberg et al., 2006b; Surber et al., 2020). Probably for the same reason, ropy HePS contributed to smaller particles in stirred and concentrated fermented milk (Hahn et al., 2014; Nguyen et al., 2018; Surber et al., 2021a). In various studies, non-ropy HePS and capsular HePS seemed to be important for gel stiffness (Table 5), whereas other authors found a high stiffness also with ropy HePS (Mende et al., 2012, 2013b; Surber et al., 2019, 2021a). Capsular HePS primarily contribute to the WHC in set, stirred and concentrated fermented milk (Amatayakul et al., 2006a; Surber et al., 2021a).

The apparent viscosity of stirred yoghurt with ropy HePS decreased with the intensity of post-fermentative shearing (controlled by back-pressure) (Zhang et al., 2016). The firmness of cream cheese with ropy HePS was significantly lower when curd was homogenised at 30 MPa than at 0 or 15 MPa (Surber et al., 2021b). The authors hypothesised that the ropy HePS aggregated and separated from the protein upon shearing, indicating rather weak interactions

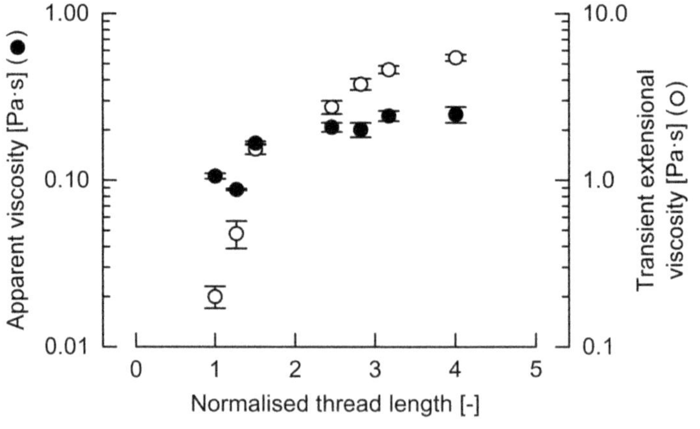


**Figure 4** Apparent shear viscosity (at a shear rate = 100/s) and transient extensional viscosity (at a Hencky strain = 4) of acid gel suspensions produced with *Streptococcus thermophilus* and *Lactococcus lactis* (~ 4% protein) *versus* thread length normalised to that of a HePS-negative strain (14 mm). Redrawn from Surber et al. (2020) with permission from John Wiley & Sons.

between HePS and protein, whereas the firmness of cream cheese with non-ropy HePS increased with curd homogenisation pressure. It is assumed that neutral HePS tend to separate from the protein, whereas charged HePS tend to associate or aggregate with protein (van de Velde et al., 2015), but a deeper understanding of the relationship between HePS–protein interactions and ropiness in fermented milk is still missing. In addition, a shear-induced alteration of molecular properties of the HePS should also be considered. A structural breakdown of ropy HePS is reported after microfluidisation of the aqueous solution, resulting in a decrease in viscosity (Nachtigall et al., 2019). Furthermore, the behaviour of capsular HePS upon excessive shearing of fermented milk is also unclear. For two of four *S. thermophilus*, the capsular HePS were removed from the cells with a laboratory-scale rotor–stator device in the fermentation media (Jaros et al., 2018).

## 6 Case study: exploring heteropolysaccharide structure-function relations

Lactic acid bacteria capable of producing EPS are widely used in the dairy industry. To take better advantage of the potential of specific HePS for improving the texture of fermented milk, the structure–function relations need to be reliably explored. In the last ~ 10 years, we have established cultivation methods and isolation procedures for obtaining HePS in an amount sufficient to study their intrinsic properties in aqueous solutions in detail. For this case study, we shortly summarise data on two strains of *S. thermophilus* (DGCC7710 and ST-143), which both produce neutral ropy HePS with a number of similar properties: molecular mass ($3.0 \cdot 10^6$ Da *vs.* $3.9 \cdot 10^6$ Da), intrinsic viscosity (5.4 dL/g *vs.* 5.2 dL/g) and hydrodynamic radius (62 nm *vs.* 69 nm), respectively (Nachtigall et al., 2019; and unpublished data). For the same HePS, lower molecular masses were reported by Pachekrepapol et al. (2017), which is probably related to differences in the isolation procedure. We recently summarised the impact that some of the isolation steps may have on HePS molecular mass (Nachtigall et al., 2021b).

In lab-scale fermentation of reconstituted skimmed milk (~ 4.6% protein), both DGCC7710 and ST-143 increased gel stiffness considerably (260 Pa and 305 Pa, respectively) compared to a HePS-negative strain (150 Pa), and lowered syneresis as determined by forced separation tests (Mende et al., 2012). Stirred gels from ST-143, however, showed a significantly lower apparent viscosity and higher syneresis than gels from DGCC7710.

A structural analysis of the HePS revealed similar monosaccharide composition and backbone length, but differences in glycosidic linkages and length of the side chain (Fig. 5); for example, the presence of $\beta(1\rightarrow4)$-linked monomers in the backbone and a dimeric side chain in ST-143 HePS. Both 1,4- and β-linkages were reported to result in lower flexibility of the

**Table 5** Overview of literature referring to the texture of fermented milk as affected by the type of extracellular heteropolysaccharides (HePS) of single strains (in chronological order)

| Reference | Strains and HePS types | Type of fermented milk | Specific effect on texture |
|---|---|---|---|
| (Hassan et al., 1996) | *S. thermophilus*<br>Non-ropy and capsular: S3855<br>Non-ropy: S4<br>*L. delbrueckii* ssp. *bulgaricus*<br>Non-ropy and capsular: L3<br>Ropy and capsular: RR | Stirred | Four times higher apparent viscosity and nine times higher hysteresis loop area for ropy HePS.<br>Higher apparent viscosity (1.3–2.4 times) and higher hysteresis loop area (1.4 - 2.0 times) for stirred gels with non-ropy and capsular HePS compared with non-ropy HePS. |
| (Purohit et al., 2009)[a] | *S. thermophilus*<br>Ropy: 5581, 4239, PH<br>HePS-negative: 5842<br>*L. delbrueckii* ssp. *bulgaricus*<br>Non-ropy: 3984<br>*Lactococcus lactis* ssp. *cremoris*<br>Ropy: JFR1 | Set, Stirred | Ropy HePS considerably contributed to lower G′ and slightly lower spontaneous syneresis of set gels.<br>Two times higher apparent viscosity of stirred gels with PH and JFR1 than with the other strains.<br>The hysteresis loop area was larger for stirred gels with ropy HePS than with non-ropy HePS, except for 5581 and JFR1. |
| (Gentès et al., 2011)[a] | *S. thermophilus*<br>Non-ropy: HC15<br>Ropy: 0131 and 2104<br>*L. delbrueckii* ssp. *bulgaricus*<br>Ropy: 291<br>Non-ropy: 210R, 11842, 702074 | Set, Stirred | Two to four times higher G′ for non-ropy HePS.<br>Higher apparent viscosity (up to 2 times) and WHC of stirred gels with ropy HePS. |
| (Kristo et al., 2011)[a] | *L. lactis* ssp. *cremoris*<br>Ropy: JFR1<br>HePS-negative | Set, Stirred | Ropy HePS contributed to 1.3 times higher G′ of set gels and 2 - 4 times higher apparent viscosity.<br>Structural breakdown after defined shear treatment was more pronounced for stirred gels with ropy HePS. |
| (Costa et al., 2012) | *L. lactis* ssp. *cremoris*<br>Ropy and capsular: DPC6532<br>HePS-negative | Set, Stirred | Ropy and capsular HePS contributed to 1.5 times higher G′ of set gels.<br>The apparent viscosity of stirred gels was 8 times higher. |

(*Continued*)

**Table 5** (*Continued*)

| Reference | Strains and HePS types | Type of fermented milk | Specific effect on texture |
|---|---|---|---|
| (Mende et al., 2012) | *S. thermophilus*<br>Non-ropy and capsular: DSM8713<br>Ropy and capsular: DGCC7710, ST-143<br>HePS-negative: DSM20259 | Set, Stirred | 1.7 - 2.0 times higher G′ for all HePS producers.<br>1.2 - 1.6 times higher apparent viscosity of stirred gels with ropy and capsular HePS.<br>1.3 - 1.6 times higher hysteresis loop area with ropy and capsular HePS.<br>The WHC of all gels was not clearly affected by a specific HePS type. |
| (Surber et al., 2021a) | *L. lactis*<br>Ropy and capsular: LL-1<br>Non-ropy: LL-2 | Set, Stirred, Concentrated | Higher G′ in set gels and higher WHC in stirred gels with ropy and capsular HePS.<br>Smaller particles (third less) and high sensory creaminess, yield stress (doubled) and serum retention of concentrated gels with ropy and capsular HePS. |

[a] Ability of strains to synthesise capsular HePS was not evaluated.
G′, gel stiffness; WHC, water-holding capacity.

backbone compared to 1,2-, 1,3- and α-linkages (Laws and Marshall, 2001; Zhou et al., 2019), and side chains were suspected to contribute to chain stiffness by hindering the orientation of the backbone (Tuinier et al., 2001). Information on the effect of backbone stiffness on texture is, however, scarce and controversially discussed in literature:

- A stiff backbone was linked to high apparent viscosity, firmness and whey retention of fermented milk (Gentès et al., 2011).
- The stiff backbone of HePS from ST-143 was linked to incompatibility between HePS and proteins, resulting in a less connected network, and hence, a lower gel stiffness compared to more flexible HePS from DGCC7785 (Khanal and Lucey, 2018).
- Other authors found no strong dependency of gel stiffness on backbone stiffness of neutral EPS (Girard and Schaffer-Lequart, 2007).

Considering this information, we hypothesise that both the linkage type and the side chain length contribute to a stiffer backbone of the ST-143 HePS, which in turn is responsible for a lower structure regeneration upon shearing, causing lower viscosity and higher syneresis of stirred gels. However, differences in more than one structural characteristic of HePS make unambiguous statements

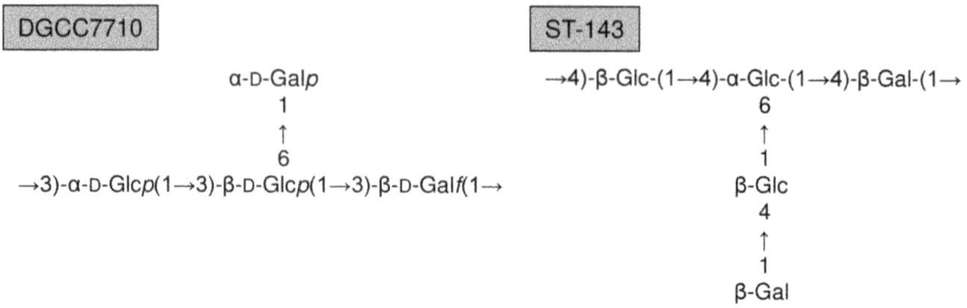


**Figure 5** Structures of HePS from *Streptococcus thermophilus* DGCC7710 (Nachtigall et al., 2019) and ST-143 (Pachekrepapol et al., 2018). Reproduced from Nachtigall et al. (2021a) under the permission of Creative Commons Attribution (CC BY) license.

difficult. Systematic studies with HePS that differ in only one structural feature are necessary to make predictions of texture effects possible.

## 7 Conclusion and future trends in research

Texture is a crucial quality issue of fermented milk and has therefore been studied extensively. There is, however, still a wide gap left open for research which is mainly owed to the complex interplay between milk constituents, starter cultures and manufacturing parameters. The following texture-related properties are most important for fermented milk: gel firmness or viscosity, creaminess and mouthfeel, WHC and smooth perception. Generally, an increase in dry matter helps to improve the texture, but at some point, it may also result in intermediates that are too firm for further processing, or in final products with texture defects.

Considering the individual milk constituents, the following fundamental statements can be given:

- Protein: the formation of the three-dimensional protein network upon acidification dominates the microstructure and hence the texture of fermented milk. The selection of appropriate ingredients for elevating the total protein content with or without adjusting the ratio of casein : whey protein is a crucial step.
- Fat: The main factors of influence are the absolute fat content and the size of the fat globules adjusted by homogenisation techniques. The impact of homogenisation on the composition of the fat globule membrane is also of importance.
- Lactose: Apart from being the main carbon source for the starter culture, the residual lactose also contributes to texture. In the production of lactose-free fermented milk, a higher osmotic pressure because of the elevated concentration of sugars and/or strain-dependent preferred carbon source needs to be considered.

- In addition, there are always various interactions between the different constituents that have to be considered, as well as the dependency of the texture of acidified milk on the processing steps.
- There is no fermented milk without starter cultures. Besides different acidification properties, there are mainly extracellular polysaccharide-producing lactic acid bacteria, which affect texture considerably.

Future trends in research are:

- Product composition: The demand for fermented milk with fewer non-dairy additives and products with a high protein content and excellent texture.
- Processing: Implementation of membrane filtration to obtain specific protein fractions and facilitating new opportunities for process and product development.
- Microbiology: The demand for an easy implementation of strains with well-predicted functionality. The use of new homopolysaccharide-producing genera such as *Leuconostoc* or *Weissella* in the production of fermented milk.

# 8 Where to look for further information

## *8.1 Further reading*

- Comprehensive introductions to fermented milk manufacture are Jørgensen et al. (2019), Mokoonlall et al. (2016) and Walstra et al. (2006).
- Reviews of milk fortification are Karam et al. (2013) and Lesme et al. (2020).
- Introduction to EPS from lactic acid bacteria is provided by Mende et al. (2016) and Zeidan et al. (2017).

## *8.2 Key journals and conferences*

- Several journals focus on fermented milk texture, for example, the *Journal of Dairy Science*, *International Dairy Journal* and also the *Journal of Texture Studies*.
- The American Dairy Science Association and the International Dairy Federation hold regular meetings.

## *8.3 Major international research projects*

- An overview of current or recently completed international projects is available at Community Research and Development Information Service of the European Union (https://cordis.europa.eu).
- Research centres, for example NIZO in the Netherlands, Teagasc in Ireland, Institut national de la recherche agronomique in France.

## 9 Acknowledgement

Part of the data presented in the case study was acquired under the support of IGF Projects (18318 BR, 20769 BR) of the FEI via AiF within the programme for promoting the Industrial Collective Research (IGF) of the German Ministry of Economic Affairs and Energy (BMWi), based on a resolution of the German Parliament and with support of the Deutsche Forschungsgemeinschaft (Bonn, Germany) under the grant number JA2033/1-2.

## 10 References

Amatayakul, T., Sherkat, F. and Shah, N. P. 2006a. Physical characteristics of set yoghurt made with altered casein to whey protein ratios and EPS-producing starter cultures at 9 and 14% total solids. *Food Hydrocolloids* 20(2-3), 314-324.

Amatayakul, T., Sherkat, F. and Shah, N. P. 2006b. Syneresis in set yogurt as affected by EPS starter cultures and levels of solids. *International Journal of Dairy Technology* 59(3), 216-221.

Anema, S. G. 2020. The whey proteins in milk: thermal denaturation, physical interactions, and effects on the functional properties of milk. In: Boland, M. and Singh, H. (Eds), *Milk Proteins*. Academic Press, San Diego, 325-384.

Anema, S. G., Lee, S. K. and Klostermeyer, H. 2006. Effect of protein, nonprotein-soluble components, and lactose concentrations on the irreversible thermal denaturation of β-lactoglobulin and α-lactalbumin in skim milk. *Journal of Agricultural and Food Chemistry* 54(19), 7339-7348.

Bengoa, A. A., Iraporda, C., Garrote, G. L. and Abraham, A. G. 2019. Kefir microorganisms: their role in grain assembly and health properties of fermented milk. *Journal of Applied Microbiology* 126(3), 686-700.

Bravo-Núñez, Á., Pando, V. and Gómez, M. 2019. Physically and chemically modified starches as texturisers of low-fat milk gels. *International Dairy Journal* 92, 21-27.

Brighenti, M., Govindasamy-Lucey, S., Jaeggi, J. J., Johnson, M. E. and Lucey, J. A. 2018. Effects of processing conditions on the texture and rheological properties of model acid gels and cream cheese. *Journal of Dairy Science* 101(8), 6762-6775.

Buldo, P., Benfeldt, C., Folkenberg, D. M., Jensen, H. B., Amigo, J. M., Sieuwerts, S., Thygesen, K., van den Berg, F. and Ipsen, R. 2016. The role of exopolysaccharide-producing cultures and whey protein ingredients in yoghurt. *LWT–Food Science and Technology* 72, 189-198.

Cayot, P., Schenker, F., Houzé, G., Sulmont-Rossé, C. and Colas, B. 2008. Creaminess in relation to consistency and particle size in stirred fat-free yogurt. *International Dairy Journal* 18(3), 303-311.

Chua, D., Deeth, H. C., Oh, H. E. and Bansal, N. 2017. Altering the casein to whey protein ratio to enhance structural characteristics and release of major yoghurt volatile aroma compounds of non-fat stirred yoghurts. *International Dairy Journal* 74, 63-73.

Ciron, C. I. E., Gee, V. L., Kelly, A. L. and Auty, M. A. E. 2011. Effect of microfluidization of heat-treated milk on rheology and sensory properties of reduced fat yoghurt. *Food Hydrocolloids* 25(6), 1470-1476.

Ciron, C. I. E., Gee, V. L., Kelly, A. L. and Auty, M. A. E. 2012. Modifying the microstructure of low-fat yoghurt by microfluidisation of milk at different pressures to enhance rheological and sensory properties. *Food Chemistry* 130(3), 510–519.

Costa, N. E., Wang, L., Auty, M. E., Hannon, J. A., McSweeney, P. L. H. H. and Beresford, T. P. 2012. Rheological, microscopic and primary chemical characterisation of the exopolysaccharide produced by *Lactococcus lactis* subsp. *cremoris* DPC6532. *Dairy Science and Technology* 92(3), 219–235.

Das, K., Choudhary, R. and Thompson-Witrick, K. A. 2019. Effects of new technology on the current manufacturing process of yogurt to increase the overall marketability of yogurt. *LWT* 108, 69–80.

de Oliveira, M. N. 2014. Fermented milks: fermented milks and yogurt. In: Batt, C. A. and Tortorello, M. L. (Eds), *Encyclopedia of Food Microbiology* (2nd edn.). Academic Press, Amsterdam, The Netherlands, 908–922.

Dekker, P. J. T., Koenders, D. and Bruins, M. J. 2019. Lactose-free dairy products: market developments, production, nutrition and health benefits. *Nutrients* 11(3), 1–14.

Deshwal, G. K., Tiwari, S., Kumar, A., Raman, R. K. and Kadyan, S. 2021. Review on factors affecting and control of post-acidification in yoghurt and related products. *Trends in Food Science and Technology* 109, 499–512.

Dickinson, E. 2018. On the road to understanding and control of creaminess perception in food colloids. *Food Hydrocolloids* 77, 372–385.

FAO/WHO 2003. *Codex alimentarius International Food Standards: Codex Standard for Fermented Milks CXS 243-2003*. Food and Agriculture Organisation of the United Nations/World Health Organisation, Rome, Italy.

Folkenberg, D. M., Dejmek, P., Skriver, A. and Ipsen, R. 2006a. Interactions between EPS-producing *Streptococcus thermophilus* strains in mixed yoghurt cultures. *The Journal of Dairy Research* 73(4), 385–393.

Folkenberg, D. M., Dejmek, P., Skriver, A., Skov Guldager, H. and Ipsen, R. 2006b. Sensory and rheological screening of exopolysaccharide producing strains of bacterial yoghurt cultures. *International Dairy Journal* 16(2), 111–118.

Gentès, M. C., St-Gelais, D. and Turgeon, S. L. 2011. Gel formation and rheological properties of fermented milk with in situ exopolysaccharide production by lactic acid bacteria. *Dairy Science and Technology* 91(5), 645–661.

Girard, M. and Schaffer-Lequart, C. 2007. Gelation and resistance to shearing of fermented milk: role of exopolysaccharides. *International Dairy Journal* 17(6), 666–673.

Goh, K. K. T., Teo, A., Sarkar, A. and Singh, H. 2020. Milk protein-polysaccharide interactions. In: Boland, M. and Singh, H. (Eds), *Milk Proteins*. Academic Press, San Diego, 499–535.

Guénard-Lampron, V., St-Gelais, D., Villeneuve, S. and Turgeon, S. L. 2020. Short communication: effect of stirring operations on changes in physical and rheological properties of nonfat yogurts during storage. *Journal of Dairy Science* 103(1), 210–214.

Gyawali, R. and Ibrahim, S. A. 2016. Effects of hydrocolloids and processing conditions on acid whey production with reference to Greek yogurt. *Trends in Food Science and Technology* 56, 61–76.

Gyawali, R. and Ibrahim, S. A. 2018. Addition of pectin and whey protein concentrate minimises the generation of acid whey in Greek-style yogurt. *Journal of Dairy Research* 85(2), 238–242.

Hahn, C., Müller, E., Wille, S., Weiss, J., Atamer, Z. and Hinrichs, J. 2014. Control of microgel particle growth in fresh cheese (concentrated fermented milk) with an exopolysaccharide-producing starter culture. *International Dairy Journal* 36(1), 46-54.

Hahn, C., Nöbel, S., Maisch, R., Rösingh, W., Weiss, J. and Hinrichs, J. 2015. Adjusting rheological properties of concentrated microgel suspensions by particle size distribution. *Food Hydrocolloids* 49, 183-191.

Hassan, A. N. N., Frank, J. F. F., Schmidt, K. A. A. and Shalabi, S. I. I. 1996. Rheological properties of yogurt made with encapsulated nonropy lactic cultures. *Journal of Dairy Science* 79(12), 2091-2097.

He, Q., Hort, J. and Wolf, B. 2016. Predicting sensory perceptions of thickened solutions based on rheological analysis. *Food Hydrocolloids* 61, 221-232.

Hossain, M. K., Keidel, J., Hensel, O. and Diakité, M. 2020. The impact of extruded microparticulated whey proteins in reduced-fat, plain-type stirred yogurt: characterization of physicochemical and sensory properties. *LWT* 134, 109976.

Huppertz, T. 2017. Lactose in yogurt. In: Shah, N. P. (Ed), *Yogurt in Health and Disease Prevention*. Academic Press, London, UK, 387-394.

IDF 2018. Inventory of microbial food cultures with safety demonstration in fermented food products. Bulletin of the IDF No. 495/2018. International Dairy Federation, Brussels, BE.

Ipsen, R. 2017. Microparticulated whey proteins for improving dairy product texture. *International Dairy Journal* 67, 73-79.

ISO 2010. Fermented milk products–bacterial starter cultures–standard of identity. ISO 27205:2010. International Organization for Standardization.

ISO 2020. Sensory analysis–methodology–texture profile. ISO 11036:2020. International Organization for Standardization.

Jaros, D., Mende, S., Häffele, F., Nachtigall, C., Nirschl, H. and Rohm, H. 2018. Shear treatment of starter culture medium improves separation behavior of *Streptococcus thermophilus* cells. *Engineering in Life Sciences* 18(1), 62-69.

Ji, Y. D., Lee, S. K. and Anema, S. G. 2011. Effect of heat treatments and homogenisation pressure on the acid gelation properties of recombined whole milk. *Food Chemistry* 129(2), 463-471.

Jørgensen, C. E., Abrahamsen, R. K., Rukke, E. O., Hoffmann, T. K., Johansen, A. G. and Skeie, S. B. 2019. Processing of high-protein yoghurt–a review. *International Dairy Journal* 88, 42-59.

Jørgensen, C. E., Abrahamsen, R. K., Rukke, E. O., Johansen, A. G., Schüller, R. B. and Skeie, S. B. 2015. Improving the structure and rheology of high protein, low fat yoghurt with undenatured whey proteins. *International Dairy Journal* 47, 6-18.

Jørgensen, C. E., Abrahamsen, R. K., Rukke, E. O., Johansen, A. G. and Skeie, S. B. 2017. Fractionation by microfiltration: effect of casein micelle size on composition and rheology of high protein, low fat set yoghurt. *International Dairy Journal* 74, 12-20.

Karam, M. C., Gaiani, C., Hosri, C., Burgain, J. and Scher, J. 2013. Effect of dairy powders fortification on yogurt textural and sensorial properties: a review. *Journal of Dairy Research* 80(4), 400-409.

Khanal, S. N. and Lucey, J. A. 2018. Effect of fermentation temperature on the properties of exopolysaccharides and the acid gelation behavior for milk fermented by *Streptococcus thermophilus* strains DGCC7785 and ST-143. *Journal of Dairy Science* 101(5), 3799-3811.

Khubber, S., Chaturvedi, K., Thakur, N., Sharma, N. and Yadav, S. K. 2021. Low-methoxyl pectin stabilizes low-fat set yoghurt and improves their physicochemical properties, rheology, microstructure and sensory liking. *Food Hydrocolloids* 111, 106240.

Körzendörfer, A., Schäfer, J., Hinrichs, J. and Nöbel, S. 2019. Power ultrasound as a tool to improve the processability of protein-enriched fermented milk gels for Greek yogurt manufacture. *Journal of Dairy Science* 102(9), 7826–7837.

Kristo, E., Miao, Z. and Corredig, M. 2011. The role of exopolysaccharide produced by *Lactococcus lactis* subsp. *cremoris* in structure formation and recovery of acid milk gels. *International Dairy Journal* 21(9), 656–662.

Krzeminski, A., Großhable, K. and Hinrichs, J. 2011. Structural properties of stirred yoghurt as influenced by whey proteins. *LWT–Food Science and Technology* 44(10), 2134–2140.

Krzeminski, A., Prell, K. A., Busch-Stockfisch, M., Weiss, J. and Hinrichs, J. 2014. Whey protein-pectin complexes as new texturising elements in fat-reduced yoghurt systems. *International Dairy Journal* 36(2), 118–127.

Kücükcetin, A. 2008. Effect of heat treatment of skim milk and final fermentation pH on graininess and roughness of stirred yogurt. *International Journal of Dairy Technology* 61(4), 385–390.

Laws, A. P. and Marshall, V. M. 2001. The relevance of exopolysaccharides to the rheological properties in milk fermented with ropy strains of lactic acid bacteria. *International Dairy Journal* 11(9), 709–721.

Lee, W. J. and Lucey, J. A. 2010. Formation and physical properties of yogurt. *Asian-Australasian Journal of Animal Sciences* 23(9), 1127–1136.

Lesme, H., Rannou, C., Famelart, M. H., Bouhallab, S. and Prost, C. 2020. Yogurts enriched with milk proteins: texture properties, aroma release and sensory perception. *Trends in Food Science and Technology* 98, 140–149.

Liu, G., Buldo, P., Greve, M. T., Nielsen, S. B., Nielsen, J. H. and Ipsen, R. 2016a. Effects of added whey protein aggregates on textural and microstructural properties of acidified milk model systems. *International Dairy Journal* 62, 43–52.

Liu, G., Jæger, T. C., Lund, M. N., Nielsen, S. B., Ray, C. A. and Ipsen, R. 2016b. Effects of disulphide bonds between added whey protein aggregates and other milk components on the rheological properties of acidified milk model systems. *International Dairy Journal* 59, 1–9.

McMahon, D. J., Du, H., McManus, W. R. and Larsen, K. M. 2009. Microstructural changes in casein supramolecules during acidification of skim milk. *Journal of Dairy Science* 92(12), 5854–5867.

Meletharayil, G. H., Metzger, L. E. and Patel, H. A. 2016a. Influence of hydrodynamic cavitation on the rheological properties and microstructure of formulated Greek-style yogurts. *Journal of Dairy Science* 99(11), 8537–8548.

Meletharayil, G. H., Patel, H. A., Metzger, L. E. and Huppertz, T. 2016b. Acid gelation of reconstituted milk protein concentrate suspensions: influence of lactose addition. *International Dairy Journal* 61, 107–113.

Mende, S., Mentner, C., Thomas, S., Rohm, H. and Jaros, D. 2012. Exopolysaccharide production by three different strains of *Streptococcus thermophilus* and its effect on physical properties of acidified milk. *Engineering in Life Sciences* 12(4), 466–474.

Mende, S., Peter, M., Bartels, K., Dong, T., Rohm, H. and Jaros, D. 2013a. Concentration dependent effects of dextran on the physical properties of acid milk gels. *Carbohydrate Polymers* 98(2), 1389–1396.

Mende, S., Peter, M., Bartels, K., Rohm, H. and Jaros, D. 2013b. Addition of purified exopolysaccharide isolates from *S. thermophilus* to milk and their impact on the rheology of acid gels. *Food Hydrocolloids* 32(1), 178-185.

Mende, S., Rohm, H. and Jaros, D. 2016. Influence of exopolysaccharides on the structure, texture, stability and sensory properties of yoghurt and related products. *International Dairy Journal* 52, 57-71.

Mokoonlall, A., Nöbel, S. and Hinrichs, J. 2016. Post-processing of fermented milk to stirred products: reviewing the effects on gel structure. *Trends in Food Science and Technology* 54, 26-36.

Nachtigall, C., Berger, C., Kovanović, T., Wefers, D., Jaros, D. and Rohm, H. 2019. Shear induced molecular changes of exopolysaccharides from lactic acid bacteria. *Food Hydrocolloids* 97, 105181.

Nachtigall, C., Rohm, H. and Jaros, D. 2021a. Degradation of exopolysaccharides from lactic acid bacteria by thermal, chemical, enzymatic and ultrasound stresses. *Foods* 10(2), 396.

Nachtigall, C., Surber, G., Bulla, J., Rohm, H. and Jaros, D. 2021b. Pilot scale isolation of exopolysaccharides from *Streptococcus thermophilus* DGCC7710: impact of methodical details on macromolecular properties and technofunctionality. *Engineering in Life Sciences* 21(3-4), 220-232.

Narvhus, J. A. 2014. Fermented milks: northern European fermented milks. In: Batt, C. A. and Tortorello, M. L. (Eds), *Encyclopedia of Food Microbiology* (2nd edn.). Academic Press, Amsterdam, The Netherlands, 895-899.

Narvhus, J. A., Østby, N. and Abrahamsen, R. K. 2019. Science and technology of cultured cream products: a review. *International Dairy Journal* 93, 57-71.

Nguyen, A. T. B., Nigen, M., Jimenez, L., Ait-Abderahim, H., Cunault, C., Marchesseau, S. and Picart-Palmade, L. 2018. A multi-scale approach to identify the role of heat treatment, milk protein composition and starter culture on the gel formation and the texture defects of acid milk gel. *Food Hydrocolloids* 85, 299-310.

Nguyen, P. T. M., Kravchuk, O., Bhandari, B. and Prakash, S. 2017. Effect of different hydrocolloids on texture, rheology, tribology and sensory perception of texture and mouthfeel of low-fat pot-set yoghurt. *Food Hydrocolloids* 72, 90-104.

Nöbel, S., Kern, C., Sonne, A., Bähler, B. and Hinrichs, J. 2016. Apparent voluminosity of casein micelles in the temperature range 35-70°C. *International Dairy Journal* 59, 80-84.

Obeid, S., Guyomarc'h, F., Tanguy, G., Leconte, N., Rousseau, F., Dolivet, A., Leduc, A., Wu, X., Cauty, C., Jan, G., Gaucheron, F. and Lopez, C. 2020. The adhesion of homogenized fat globules to proteins is increased by milk heat treatment and acidic pH: quantitative insights provided by AFM force spectroscopy. *Food Research International* 129, 108847.

Pachekrepapol, U., Lucey, J. A., Gong, Y., Naran, R. and Azadi, P. 2017. Characterization of the chemical structures and physical properties of exopolysaccharides produced by various *Streptococcus thermophilus* strains. *Journal of Dairy Science* 100(5), 3424-3435.

Pachekrepapol, U., Lucey, J. A. A., Gong, Y., Naran, R. and Azadi, P. 2018. Corrigendum to "Characterization of the chemical structures and physical properties of exopolysaccharides produced by various *Streptococcus thermophilus* strains" (*J. Dairy Sci.* 100:3424-3435). *Journal of Dairy Science* 101(6), 5668.

Pang, Z., Deeth, H. and Bansal, N. 2015. Effect of polysaccharides with different ionic charge on the rheological, microstructural and textural properties of acid milk gels. *Food Research International* 72, 62-73.

Pang, Z., Deeth, H., Prakash, S. and Bansal, N. 2016. Development of rheological and sensory properties of combinations of milk proteins and gelling polysaccharides as potential gelatin replacements in the manufacture of stirred acid milk gels and yogurt. *Journal of Food Engineering* 169, 27-37.

Pang, Z., Xu, R., Luo, T., Che, X., Bansal, N. and Liu, X. 2019. Physiochemical properties of modified starch under yogurt manufacturing conditions and its relation to the properties of yogurt. *Journal of Food Engineering* 245, 11-17.

Peng, Y., Serra, M., Horne, D. S. and Lucey, J. A. 2009. Effect of fortification with various types of milk proteins on the rheological properties and permeability of nonfat set yogurt. *Journal of Food Science* 74(9), C666-C673.

Poulsen, V. K., Koza, A., Al-Nakeeb, K. and Oeregaard, G. 2020. Screening for texturing *Leuconostoc* and genomics behind polysaccharide production. *FEMS Microbiology Letters* 367(20), 1-6.

Purohit, D. H., Hassan, A. N., Bhatia, E., Zhang, X. and Dwivedi, C. 2009. Rheological, sensorial, and chemopreventive properties of milk fermented with exopolysaccharide-producing lactic cultures. *Journal of Dairy Science* 92(3), 847-856.

Remeuf, F., Mohammed, S., Sodini, I. and Tissier, J. P. 2003. Preliminary observations on the effects of milk fortification and heating on microstructure and physical properties of stirred yogurt. *International Dairy Journal* 13(9), 773-782.

Riener, J., Noci, F., Cronin, D. A., Morgan, D. J. and Lyng, J. G. 2009. The effect of thermosonication of milk on selected physicochemical and microstructural properties of yoghurt gels during fermentation. *Food Chemistry* 114(3), 905-911.

Robinson, R. K., Lucey, J. A. and Tamime, A. Y. 2006. Manufacture of yoghurt. In: Tamime, A. Y. (Ed), *Fermented Milks*. Blackwell Publishing, Oxford, UK, 53-75.

Robinson, R. K. and Tamime, A. Y. 2006. Types of Fermented milks. In: Tamime, A. Y. (Ed), *Fermented Milks*. Blackwell Publishing, Oxford, UK, 1-10.

Schäfer, J., Sebald, K., Dunkel, A., Hofmann, T., Rosenthal, I., Schuster, R., Atamer, Z. and Hinrichs, J. 2019. A feasibility study on the pilot scale manufacture of fresh cheese from skim milk retentates without acid whey production: effect of calcium content on bitterness and texture. *International Dairy Journal* 93, 72-80.

Schkoda, P., Hechler, A. and Hinrichs, J. 2001. Improved texture of stirred fermented milk by integrating fat globules into the gel structure. *Milchwissenschaft* 56, 85-89.

Schmid, J. 2018. Recent insights in microbial exopolysaccharide biosynthesis and engineering strategies. *Current Opinion in Biotechnology* 53, 130-136.

Schmidt, C., Mende, S., Jaros, D. and Rohm, H. 2016. Fermented milk products: effects of lactose hydrolysis and fermentation conditions on the rheological properties. *Dairy Science and Technology* 96(2), 199-211.

Serra, M., Trujillo, A. J., Guamis, B. and Ferragut, V. 2009. Evaluation of physical properties during storage of set and stirred yogurts made from ultra-high pressure homogenization-treated milk. *Food Hydrocolloids* 23(1), 82-91.

Serra, M., Trujillo, A. J., Quevedo, J. M., Guamis, B. and Ferragut, V. 2007. Acid coagulation properties and suitability for yogurt production of cows' milk treated by high-pressure homogenisation. *International Dairy Journal* 17(7), 782-790.

Sfakianakis, P. and Tzia, C. 2014. Conventional and innovative processing of milk for yogurt manufacture; Development of texture and flavor: a review. *Foods* 3(1), 176-193.

Silva, J. V. C. and O'Mahony, J. A. 2018. Microparticulated whey protein addition modulates rheological and microstructural properties of high-protein acid milk gels. *International Dairy Journal* 78, 145–151.

Sodini, I., Remeuf, F., Haddad, S. and Corrieu, G. 2004. The relative effect of milk base, starter, and process on yogurt texture: a review. *Critical Reviews in Food Science and Nutrition* 44(2), 113–137.

Surber, G., Jaros, D. and Rohm, H. 2020. Shear and extensional rheology of acid milk gel suspensions with varying ropiness. *Journal of Texture Studies* 51(1), 111–119.

Surber, G., Mende, S., Jaros, D. and Rohm, H. 2019. Clustering of *Streptococcus thermophilus* strains to establish a relation between exopolysaccharide characteristics and gel properties of acidified milk. *Foods* 8(5), 146.

Surber, G., Schäper, C., Wefers, D., Rohm, H. and Jaros, D. 2021a. Exopolysaccharides from *Lactococcus lactis* affect manufacture, texture and sensory properties of concentrated acid milk gel suspensions (fresh cheese). *International Dairy Journal* 112, 104854.

Surber, G., Spiegel, T., Dang, B. P., Wolfschoon-Pombo, A., Rohm, H. and Jaros, D. 2021b. Cream cheese made with exopolysaccharide-producing *Lactococcus lactis*: impact of strain and curd homogenization pressure on texture and syneresis. *Journal of Food Engineering* 308, 110664.

Surono, I. S. and Hosono, A. 2011. Fermented milks | types and standards of identity. In: Fuquay, J. W. (Ed), *Encyclopedia of Dairy Sciences*. Academic Press, Amsterdam, The Netherlands, 470–476.

Tamime, A. Y., Hickey, M. and Muir, D. D. 2014. Strained fermented milks–a review of existing legislative provisions, survey of nutritional labelling of commercial products in selected markets and terminology of products in some selected countries. *International Journal of Dairy Technology* 67(3), 305–333.

Tamine, A. Y., Skriver, A. and Nilsson, L.-E. 2006. Starter cultures. In: Tamime, A. Y. (Ed), *Fermented Milks*. Blackwell Publishing, Oxford, UK, 11–52.

Tomaschunas, M., Hinrichs, J., Köhn, E. and Busch-Stockfisch, M. 2012. Effects of casein-to-whey protein ratio, fat and protein content on sensory properties of stirred yoghurt. *International Dairy Journal* 26(1), 31–35.

Truong, T., Palmer, M., Bansal, N. and Bhandari, B. 2015. *Effect of Milk Fat Globule Size on the Physical Functionality of Dairy Products*. Springer International Publishing, Cham, CH.

Tuinier, R., Van Casteren, W. H. M., Looijesteijn, P. J., Schols, H. A., Voragen, A. G. J. and Zoon, P. 2001. Effects of structural modifications on some physical characteristics of exopolysaccharides from *Lactococcus lactis*. *Biopolymers* 59(3), 160–166.

Uduwerella, G., Chandrapala, J. and Vasiljevic, T. 2017. Minimising generation of acid whey during Greek yoghurt manufacturing. *Journal of Dairy Research* 84(3), 346–354.

van de Velde, F., de Hoog, E. H. A., Oosterveld, A. and Tromp, R. H. 2015. Protein-polysaccharide interactions to alter texture. *Annual Review of Food Science and Technology* 6, 371–388.

Walstra, P., Wouters, J. T. M. and Geurts, T. J. 2006. *Dairy Science and Technology*. CRC Press Press, New York.

Wu, S., Li, D., Li, S., Bhandari, B., Yang, B., Chen, X. D. and Mao, Z. 2009. Effects of incubation temperature, starter culture level and total solids content on the rheological properties of yogurt. *International Journal of Food Engineering* 5(2), 1–17.

Yousefi, M. and Jafari, S. M. 2019. Recent advances in application of different hydrocolloids in dairy products to improve their techno-functional properties. *Trends in Food Science and Technology* 88, 468–483.

Zeidan, A. A., Poulsen, V. K., Janzen, T., Buldo, P., Derkx, P. M. F., Øregaard, G. and Neves, A. R. 2017. Polysaccharide production by lactic acid bacteria: from genes to industrial applications. *FEMS Microbiology Reviews* 41, S168–S200.

Zhang, L., Folkenberg, D. M., Amigo, J. M. and Ipsen, R. 2016. Effect of exopolysaccharide-producing starter cultures and post-fermentation mechanical treatment on textural properties and microstructure of low fat yoghurt. *International Dairy Journal* 53, 10–19.

Zhao, L. L., Wang, X. L., Tian, Q. and Mao, X. Y. 2016. Effect of casein to whey protein ratios on the protein interactions and coagulation properties of low-fat yogurt. *Journal of Dairy Science* 99(10), 7768–7775.

Zhou, Y., Cui, Y. and Qu, X. 2019. Exopolysaccharides of lactic acid bacteria: structure, bioactivity and associations: a review. *Carbohydrate Polymers* 207, 317–332.

# Chapter 21

## Understanding the contribution of milk constituents to the texture of dairy products: cheese

*Michel Britten, Agriculture and Agri-Food Canada, Canada*



### 1 Introduction

Milk is the primary source of nutrition for young mammals. It contains fat, proteins, minerals, lactose and vitamins to support the growth of the new-born. Fat is dispersed in milk as an emulsion with particle sizes ranging from 0.1 μm to 20 μm (average 3–4 μm) (Fox 2011). Fat globules are stabilised by a protective membrane, partially derived from the plasma membrane of the lactating cells (Lopez et al. 2015). Milk proteins are made of caseins and soluble or serum proteins. Caseins are a family of phosphoproteins representing ~80% of total milk protein. They have the ability to associate and form a colloidal dispersion in milk. The casein micelles contain thousands of casein molecules. They have a spherical shape, with diameters ranging from 50 nm to 500 nm (average ~150 nm) (Horne 2019). Milk is supersaturated with calcium and phosphate, and the formation of nanocluster complexes (~2.5 nm) within the casein micelle structure prevents precipitation (Gaucheron 2011). The rest of milk's minerals, milk sugar (lactose) and serum proteins are soluble in the aqueous phase of milk.

Milk has become an important part of the human diet. This is no surprise considering its high nutritional value. Furthermore, milk constituents have

http://dx.doi.org/10.19103/AS.2022.0099.21

excellent technological properties and can be easily processed into a wide variety of products. Milk's composition can be altered by simple physical processes such as centrifugation or filtration. Milk fat is reactive to shear stress, and mechanical treatments such as whipping and churning turn the fat globule dispersion into stable foams or spreads. Milk proteins are excellent emulsifiers and can also transform into gels. Gelation is obtained by acidification or enzymatic hydrolysis of casein, or by heat denaturation of serum proteins. The high nutritional and technological value of milk constituents has led to the development of a wide range of ingredients derived from milk (Britten and Remondetto 2018). These ingredients meet specific needs for food and non-food applications.

Cheese was one of the first manufactured food products. The primary objective of cheesemaking was to convert milk into a stable product via fermentation. Cheesemaking is a simple process, but the modification of a few parameters has a considerable impact on cheese characteristics (Fox et al. 2017). As a result, cheese exists in a great variety of textures and flavours. The structural properties of the cheese matrix are determined by the complex interactions between milk constituents (Lucey et al. 2003). Casein is the backbone of most cheeses and plays a critical role in cheese texture. Milk minerals are important for milk coagulation, and the portion bound to caseins has a direct influence on cheese texture. Inclusions of fat particles in the cheese matrix provide breaking points in the continuous casein network and contribute to the perception of creaminess and lubrication. Finally, serum proteins are sensitive to heat denaturation and, depending on the thermal history of the milk, can alter cheese composition and texture. This chapter focuses on the contributions of milk constituents to the development of the cheese matrix and its textural properties. After a quick review of the cheesemaking process, the specific contributions of the main milk constituents are presented.

## 2 Basic steps of cheesemaking

Cheesemaking is a dehydration process that is applied to milk to enhance its shelf life. The process variables must be optimised to control cheese composition, texture and flavour. Over the long history of cheesemaking, a great variety of cheeses have been produced worldwide. Despite this diversity, cheesemaking is based on a few fundamental steps: (1) milk standardisation; (2) acidification; (3) coagulation; (4) curd contraction and syneresis; (5) curd drainage and handling; and (6) cheese maturation (Fig. 1).

### *2.1 Milk standardisation*

Milk should be standardised to meet the target cheese composition. Since serum proteins are usually not recovered in cheese, the fat-to-protein ratio must

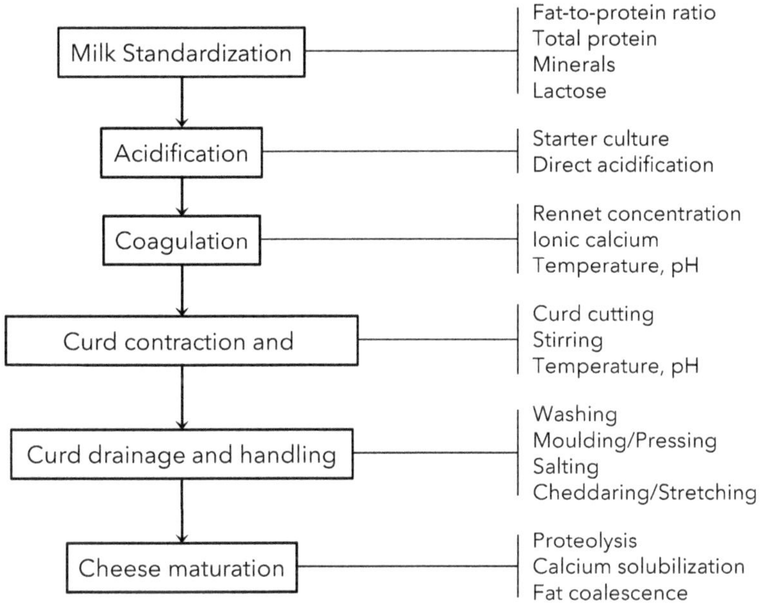


**Figure 1** Basic steps for cheese manufacture.

be lower than the target ratio in cheese. It is common practice to adjust the fat-to-casein ratio in milk to the expected fat-to-protein ratio in cheese, by adding skimmed milk or cream. To increase cheesemaking productivity, milk can be concentrated using ultrafiltration or microfiltration (Mistry and Maubois 2017). Alternatively, milk protein concentrate powders can be used to enrich cheese milk. However, concentrated milks are loaded with colloidal salts and special attention is required to ensure appropriate curd demineralisation during the cheesemaking process. Ultrafiltration, in diafiltration mode, has been proposed as a way to control the lactose content of cheese milk and reduce the risk of excessive acidification during maturation (Moynihan et al. 2016). After cheese milk standardisation, thermisation or pasteurisation is usually applied in order to reduce the viable bacterial load and control bacteria-associated enzymatic activities (Bintsis and Papademas 2017).

## *2.2 Acidification*

A starter culture is added to milk to produce lactic acid from lactose. Mesophilic, thermophilic or mixed cultures are chosen according to the type of cheese to be produced. The rate of acidification during cheesemaking is critical and depends on the amount and type of starter culture used, as well as on the environmental conditions. An appropriate acidification rate and decrease in pH are required to (1) prevent the growth of spoilage or pathogenic microorganisms; (2) control the activity of the coagulant during manufacture and ripening; (3) solubilise colloidal calcium phosphate and reduce the level of calcium and the ratio of colloidal to soluble calcium in the cheese curd; (4) promote syneresis and control cheese moisture content; and (5) influence the activity of enzymes during ripening (McSweeney 2007a). Direct acidification with lactic acid, citric acid or glucono-δ-lactone can be used as an alternative to acidification using starter cultures (Thybo et al. 2020). While direct acidification is more controllable than biological acidification, it does not contribute to the biochemical changes of cheese during maturation (Bintsis and Papademas 2017).

## 2.3 Coagulation

Milk coagulation is induced by acidification or by the proteolytic action of added rennet, which removes the caseinomacropeptide fragment from κ-casein. These changes reduce the net negative charge and eliminate the steric stabilisation layer at the surface of casein micelles. The destabilisation of casein micelles leads to the formation of a continuous and semi-solid network. The majority of cheeses are manufactured using rennet, and milk coagulation is triggered when a sufficient proportion of κ-casein is hydrolysed (McSweeney 2007b). The rate of aggregation of renneted micelles depends on the balance between attractive and repulsive interactions. Factors such as pH, temperature and the concentration of ionic calcium are known to influence rennet curd formation and firmness (Bintsis and Papademas 2017).

## 2.4 Curd contraction/syneresis

When the milk gel has attained sufficient firmness, it is cut in order to trigger the contraction and physical expulsion of whey. The extent of syneresis needs to be carefully controlled since it determines the cheese moisture content. Curd contraction results from the strengthening of protein-protein interactions and the rearrangement of the paracasein matrix. The principal processing variables affecting syneresis are cut size, temperature, rate of acid development and curd-stirring (McSweeney 2007c). Cutting the curd into smaller pieces promotes syneresis by increasing the surface area available for whey loss. A gradual increase of curd temperature (curd cooking) encourages gel rearrangement and increases permeability, which also promotes syneresis (Fagan et al. 2007). Acid production during cooking reduces the pH and the hydration of caseins, and enhances syneresis. It is important to choose a starter culture adapted to the cooking temperature to ensure sufficient acid production. Stirring during cooking is essential to facilitate heat transfer. It also prevents the curd pieces from matting together and it increases the collision rate between curd pieces. All these factors improve syneresis (McSweeney 2007c).

## 2.5 Curd drainage and handling

When the appropriate degree of syneresis is reached, the curd pieces are separated from the whey. The pH at drainage is important since it influences the cheese mineral content. Acidification during cheesemaking gradually solubilises colloidal calcium phosphate and the pH at drainage determines the proportion removed in the whey (Sheehan 2007).

After whey drainage, the curd is subjected to a variety of handling procedures such as washing, moulding, pressing, salting, cheddaring and stretching. Washing the curd, by adding water after partial or total whey drainage, is used to remove residual lactose and avoid over-acidification. The curd is usually transferred to moulds and pressure is applied to obtain the desired shape and weight of finished cheese. In most cases, cheese salting is achieved by immersion in saturated brine. Alternatively, dry salt can be rubbed on the surface of the cheese or added to milled curds. For some cheese varieties, cheddaring is a critical step for the development of the characteristic texture (Cheddar, Cheshire). After whey drainage, compacted curd blocks are piled up and turned periodically until the pH drops to ~5.3. The decrease in pH dissolves some colloidal calcium phosphate and makes the curd stretchable (Bintsis and Papademas 2017). Dry salting is performed after the blocks are milled to finger-sized pieces. The salted curd is then pressed to promote the fusion of curd particles and complete whey expulsion. In the manufacture of pasta-filata cheeses, the curd is stretched in hot water before brine salting. This treatment imparts structural changes and provides the desired textural and melting properties of the cheese.

## 2.6 Maturation

During maturation, various biochemical changes affect the flavour, aroma and texture of the cheese. Upon maturation of cheese, partial hydrolysis of paracasein takes place and fat globules coalesce, forming aggregates and pools of fat. The mineral equilibrium is altered during maturation and the concentration of soluble forms of calcium increases. The hydration of the cheese matrix also increases as a result of proteolysis and change in the mineral balance. Various other biochemical reactions occur during maturation, and these transformations are responsible for the finer aspects of cheese flavour (McSweeney 2004). The various factors, like the type and concentration of coagulant, indigenous milk enzymes, starter bacteria, secondary microflora, moisture and salt content, pH, residual lactose and ripening temperature affect the biochemical changes that take place during maturation and the final characteristics of cheese.

## 3 Contribution of caseins to cheese texture

The destabilisation of casein micelles and their aggregation into a continuous network is the fundamental process behind the development of cheese structure. The rearrangement of the casein matrix during cheesemaking and the changes that occur during maturation determine the textural properties of cheese. In this process, casein–casein interactions are altered and the conditions to be used for coagulation, curd processing and maturation are chosen to obtain the desired cheese texture.

### 3.1 Casein micelle structure

The casein micelle is an assembly of individual caseins and colloidal salts. The word 'casein' is used to describe a class of proteins obtained by recovering the precipitate formed after acidification of milk at a pH of 4.6 and a temperature of 20°C. Caseins are flexible proteins with an open conformation, which can be attributed to their high content of proline residues (O'Mahony and Fox 2013). The four main casein molecules ($\alpha_{s1}$-, $\alpha_{s2}$-, β- and κ-casein) differ in their amino acid composition and have a high level of structural heterogeneity due to post-translation phosphorylation and glycosylation (Holland and Boland 2014). Milk minerals are distributed in two fractions: the soluble fraction, which corresponds to the salts and ions present in the continuous phase of milk, and the colloidal fraction, which includes the minerals that are associated with the casein micelle and consists mainly of phosphorus and calcium (Gaucheron 2005).

The casein micelles are supramolecular aggregates, approximately 150 nm in diameter, resulting from the interactions between casein molecules, calcium phosphate and water. They contain ~4.0 g water per gram of protein and the minerals represent 6% to 8% of the dry matter (Silva et al. 2019). A first structural model was put forward more than 60 years ago (Waugh 1958) and since then, different models have been proposed in succession as the ability to analyse casein systems increased. These models fall into three general categories: core-coat, sub-micelles and internal structure (O'Mahony and Fox 2013). A model that is widely accepted (Dalgleish 2011, McMahon and Oommen 2008) describes the casein micelle as a flexible and relatively uniform protein matrix in which calcium phosphate nanoclusters are integrated and act as a cross-linking agent (De Kruif and Holt 2003). A slightly different model proposes that the micelle structure is governed by a balance between hydrophobic interactions and electrostatic repulsions between casein molecules (Horne 1998). In this 'dual-binding' model, hydrophobic interactions are the 'attractive' component between the caseins, and the micellar calcium phosphate also acts as a neutralising agent which reduces electrostatic repulsions. The concept of dual binding was used to explain the nature and type of casein–casein interactions involved in cheesemaking (Lucey 2002, Lucey et al. 2003). However, there is still debate regarding the exact nature of the attractive forces maintaining the integrity of casein micelles (Carver et al. 2017, Horne and Lucey 2017a).

### 3.2 Casein coagulation

Cheese structure develops from the coagulation of casein micelles under the action of rennet enzyme and acid production from starter culture. Rennet coagulation is a combination of three reactions: enzymatic hydrolysis, aggregation and rearrangement (Horne and Lucey 2017b). Chymosin hydrolyses κ-casein located on the surface of the casein micelles by cleaving the Phe (105)–Met(106) bond and releasing the caseinomacropeptide. Once sufficient κ-casein has been hydrolysed, electrostatic and steric repulsion between micelles is reduced and the attractive forces are dominant and lead to aggregation. The renneted micelles aggregate into clusters and chains that eventually form a continuous network that surrounds fat globules and serum pockets. After gel formation, fusion between particles continues and induces multiple rearrangements, such as the compaction of gel-forming elements, the increase in pore size and strand formation and breakage (Mellema et al. 2002). All these changes contribute to the gradual increase in gel firmness.

In addition to rennet and casein concentrations, the kinetics of milk coagulation is also affected by factors influencing attractive and repulsive forces between casein micelles, such as pH, calcium concentration and temperature. Milk gelation proceeds faster at lower pH due to increased rennet activity and reduction of charge repulsion between casein micelles. Compared to milk at natural pH (6.7), the gelation time was reduced by ~35% after acidification to pH 6.4 (Choi et al. 2007). Lower pH also induces the solubilisation of colloidal calcium phosphate (CCP), which increases the concentration of ionic calcium (Lucey 2011). The maximum gel firmness increases at a lower pH, with a peak observed at around 6.2. Any further decrease of milk pH will reduce gel firmness due to the loss of CCP cross-links and increased repulsions between the exposed phosphoserine residues (Lucey et al. 2003). The addition of calcium ($CaCl_2$) to milk reduces milk pH, which accelerates the κ-casein hydrolysis reaction. Adding calcium (<50 mmol $l^{-1}$), even at constant milk pH, increases the gel-firming rate and strength due to the neutralisation of negatively charged groups on the micelle surface and the formation of calcium bridges (Lucey 2011). The aggregation of renneted micelles is most affected by temperature. The optimum temperature for maximal gel firmness is around 30–35°C. At temperatures between 0°C and 10°C, casein hydrolysis still occurs but aggregation is inhibited (Bansal et al. 2007). At low temperature, hydrophobic interactions are reduced and β-casein chains are present on the surface of casein micelles, which increases steric stabilisation (Lucey 2002).

### 3.3 Casein network contraction and syneresis

After formation, the network of renneted casein is susceptible to further rearrangements. Renneted micelles form junctions with a limited number of neighbours, but reactive sites are still available on their surface (Fagan et al. 2017). Rearrangement into a more compact network would increase the number of bonds and decrease the total free energy of the system. Under cheesemaking conditions, syneresis is barely observed before cutting the curd; the gel is constrained and cannot shrink. However, at microscopic scale, local rearrangements of the network densify the casein strands and increase pore size. This phenomenon, called 'microsyneresis', is evidenced by the constant increase of permeability measured in constrained renneted gels (Dejmek and Walstra 2004). Cutting the curd releases the tensile strain and allows the gel to shrink and expel whey. Curd syneresis is a critical step in cheesemaking. It determines the retention of milk constituents, affecting both cheese composition and yields.

Whey release after cutting is proportional to the curd surface area, and reducing the size of particles promotes syneresis. Curd stirring also intensifies syneresis. It prevents sedimentation and increases the collision rate between particles. Furthermore, the deformation of curd particles induced by stirring provides the energy for the breaking of bonds within the casein strands, which contributes to network rearrangement and contraction. Acid production and an increase in temperature after gel cutting are the other factors known to encourage the rearrangement process and curd syneresis (Lucey 2002). As the pH decreases, the dissolution of CCP increases the liquid-like behaviour of the network, facilitating strand reorganisation and particle fusion (Mellema et al. 2002). A rise in temperature increases the thermal motion of casein particles and strands, thus enhancing network rearrangements (Lucey 2002). Higher temperature also strengthens the hydrophobic attractive forces, thereby contributing to contraction of the gel network (Giroux et al. 2014). During the cooking step of the cheesemaking process, the acidification rate and the temperature profile after curd cutting should be carefully controlled to meet the target cheese composition and properties.

### 3.4 Influence of curd handling on casein network structure

The additional processing steps after completion of curd syneresis alter the casein network and have direct impacts on cheese texture. The curd is first drained and washed (when appropriate) to eliminate free whey and reduce the residual lactose concentration. It is then pressed, knitted (cheddaring/stretching) and salted to achieve the target cheese composition and to impart the desired structure to the casein network.

After drainage, the cheese is pressed to force the fusion of curd particles into a cohesive mass and to eliminate most of the interstitial whey pockets. Pressing in moulds is used to shape the cheese loaves. A gradual increase in pressure is important, since too much pressure at the start can cause rapid closure of the rind and can prevent whey removal (Legg et al. 2017). A higher temperature during pressing promotes curd particle fusion and whey expulsion (Dejmek and Walstra 2004). As previously mentioned, a higher temperature promotes casein network rearrangements and increases the hydrophobic attractive forces. Pressing conditions determine the homogeneity and the final moisture content of the casein network. Poorly fused curd particles introduce weak points in the cheese matrix and produce a brittle texture. The hydration of the casein network, roughly corresponding to the moisture in the non-fat substances (MNFS), is an important factor influencing cheese texture (Marshall 1993). Cheese firmness is positively correlated with casein content and negatively correlated with moisture (Drake and Delahunty 2017).

Cheddaring is a texturising treatment applied to the drained curd. In the open-vat cheesemaking process, cheddaring is achieved by heaping the curd along the sides of the vat to allow fusion between particles (Legg et al. 2017). During this process, the pH gradually decreases and the drained mass of curd flows slightly under its own weight (Guinee 2016). Curd deformations orient the casein strands and allow the development of the fibrous or chicken breast-like texture. Stretching in hot water is another process for texturising cheese curd. Compared to cheddaring, the stretching process induces more drastic changes in the microstructure. Mechanical and thermal energy is transferred to the curd to transform the amorphous casein network into an oriented structure consisting of parallel-aligned casein fibres separated by long channels containing accumulated fat and serum (Everett and Auty 2017). The level of hydration and cross-linking of the casein network are critical for optimal stretching. Curd pH, temperature and calcium concentration determine the consistency of the curd and its ability to be plasticised during the process (De Angelis and Gobbetti 2011). Controlled acidification, coupled with a critical curd pH at the time of stretching, are the key technological parameters for pasta-filata cheeses. The fibrous structure of the casein network allows these cheeses to melt to a stringy, elastic and chewy consistency that is now considered the standard for pizza cheese.

Salt is added to cheese to act as a preservative and to add flavour (Guinee and Sutherland 2011). Salting also affects cheese moisture and casein hydration, two factors that influence textural properties. During salting in brine or in dry salt, cheese loses water due to the difference in osmotic pressure between the cheese aqueous phase and the brine, or the supersaturated salt solution on the cheese surface. At the same time, adding salt to cheese increases the hydration of the casein network due to a salting-in effect. The transfer of free water to the casein network reduces the expressible serum in pasta-filata cheeses and increases the viscous character of the matrix. As a general trend, increasing the salt concentration has been shown to increase cheese firmness, which is attributed to the loss of moisture and the increased voluminosity of the casein network (Guinee and Fox 2004). Adding salt at a relatively low concentration (salt-to-moisture ratio < 3%) assists in curd-knitting and enhances the textural attributes and melting properties of the cheese. However, an excessive concentration of salt (salt-to-moisture ratio > 6%) decreases casein network hydration through a salting-out effect and increases the elastic character of the cheese (Guinee 2004).

### *3.5 Influence of milk treatments on the casein network structure*

Before the cheesemaking process begins, milk composition is standardised to meet the target cheese composition. The fat-to-protein ratio is adjusted by adding skimmed milk or cream and the seasonal variation in protein content may be corrected by adding milk protein concentrate. Heat treatment is also applied to inactivate undesirable microorganisms and enzymes. Heating the milk denatures the whey proteins which interact with caseins. Cheese yield may increase as a result, but the denatured whey proteins also alter the casein network properties and cheese texture (see Section 6). Novel technologies, such as high-pressure treatment, have been proposed as alternatives to the use of heat treatments for microbial inactivation.

High-pressure treatment can be applied to milk in a batch or semi-continuous process at pressures of up to 800 MPa. Static pressurisation of milk promotes the dissolution of CCP and induces some disruption of the casein micelles (Huppertz 2016). The rennet coagulation time is reduced by high-pressure treatment and the gel is stronger. The casein network is also more compact, resulting in a cheese with increased initial firmness in comparison with cheese made from pasteurised milk (Buffa et al. 2001). These effects of high-pressure treatment are mainly attributable to the smaller size of the casein particles, which increases the casein cross-link density (Everett and Auty 2017). However, high-pressure treatment also promotes the denaturation of β-lactoglobulin (Huppertz et al. 2004). Binding of β-lactoglobulin on the surface of casein micelles can slow the aggregation of renneted micelles, limit casein network contraction and increase cheese moisture. As a consequence, when the denaturation rate of β-lactoglobulin is high, high-pressure treatment can reduce cheese firmness.

Other non-thermal processes have been applied to milk in order to reduce bacterial loads (Rashmi and Warrier 2020). High-pressure homogenisation, ultrasound and pulsed electric fields have been shown to have little effect on the casein micelles and the characteristics of the casein network in cheese (Huppertz 2016). The heat generated by these treatments, which induces some denaturation of whey proteins, is likely responsible for the slight differences observed in comparison with non-treated milk.

### *3.6 Influence of ageing on the casein network structure*

Immediately after production, cheese structure is heterogeneous, and depending on the curd-handling conditions, moisture and salt concentration gradients can be observed. In the first days of storage, the fusion of curd particles leads to a more coherent and homogeneous matrix. During maturation, biochemical and physicochemical changes occur, affecting the casein network and cheese texture. These changes are primarily mediated by the enzymatic proteolysis of casein, but also by a process of casein solvation and solubilisation.

The proteolytic hydrolysis of caseins into peptides is a driving force for changes in cheese texture during ageing. The breakdown of caseins leads to weakening of the casein network and a decrease in cheese firmness. Enzymes from several sources contribute to proteolysis. The rennet (chymosin) used for milk coagulation is partly retained in the cheese and is responsible for early cleavage of $\alpha_{s1}$-casein and a decrease of cheese firmness (Creamer and Olson 1982). Plasmin, the principal indigenous proteinase in milk, plays an important role as a ripening agent in high-cooked cheeses such as Emmental and Mozzarella. In these cheeses, residual chymosin is denatured by the heat treatments, but plasmin activity is increased due to thermal inactivation of the inhibitors of plasminogen (Farkye and Fox 1990). Other proteolytic enzymes involved in cheese softening are provided by starter and nonstarter microorganisms (Lamichhane et al. 2018). Proteolytic activity in cheese increases with increasing storage temperature and water activity. High-moisture cheeses are prone to proteolysis but increasing salt concentration retards proteolysis, because it depresses water activity (Guinee 2007). Optimal breakdown of the casein network is important, and insufficient proteolysis can cause a curdy texture in certain cheeses (Muthukumarappan and Swamy 2017).

During storage, rehydration of the casein network contributes to the softening of the cheese. The reabsorption of free water is affected by storage temperature. Caseins are relatively hydrophobic proteins and lower temperatures promote casein–water interactions at the expense of casein-casein interactions (Whetstine et al. 2007). Proteolysis also increases the water-binding capacity of the casein network. The hydrolysis of peptide bonds generates ionic groups with a strong affinity for water (Lucey et al. 2003). Ageing has been shown to induce the slow solubilisation of colloidal calcium phosphate (Hassan et al. 2004), which exposes charged groups on casein molecules and increases the mobility of the casein network. The softening of cheese texture is highly correlated with the solubilisation of the colloidal calcium content during storage (O'Mahony et al. 2005). Salt diffusion is another factor that enhances casein network hydration. A local increase in salt concentration during storage promotes casein solubilisation and hydration through a salting-in effect (Guinee 2004).

Changes in the casein network during ageing not only influence cheese texture at room temperature but also its melting properties. Increasing temperature alters the casein-casein interactions and can lead to cheese melting. Strengthening of hydrophobic interactions at high temperature induces contraction of the casein network, which provides fluidity to the cheese matrix by reducing the contact area between caseins (Lucey et al. 2003) and releasing free moisture (Guinee 2016). High-moisture cheeses tend to melt at lower temperature; however, pH and calcium concentration are also important factors in this process since they affect the charge repulsions between the caseins and the melting properties of the cheese (Lucey et al. 2003). During ripening, cheese meltability gradually increases as a result of casein hydrolysis and the exposure of charged groups resulting from the solubilisation of CCP (McMahon and Oberg 2017). The stretching characteristics of melted cheese improve in the first days of ripening, but are lost after a long period of storage, due to excessive proteolysis (Guinee 2016). The stretching characteristic requires that the casein molecules interact closely, and that the bonds holding them together break and reform quickly. These criteria are met through the relatively high concentration of intact caseins and the critical combination of pH and colloidal calcium concentration (Lucey et al. 2003).

## 4 Contribution of minerals to cheese texture

Although minerals account for a relatively small portion of milk solids, they play a critical role in cheesemaking. Their concentration and distribution in milk and the changes occurring during and after processing modulate the properties of the casein network and the textural properties of cheese.

### *4.1 Milk mineral equilibrium*

The concentrations of the different minerals in cow's milk is presented in Table 1. While sodium, potassium and chloride are almost totally dissolved in the aqueous phase of milk, portions of calcium (~69%), phosphorus (~46%), magnesium (~35%) and citrate (11%) are found in the colloidal phase of milk,

**Table 1** Mineral composition of cow milk[a]

| Mineral | Concentration ($mg \cdot kg^{-1}$) |
|---|---|
| Calcium | 1043–1283 |
| Magnesium | 97–146 |
| Phosphorus | 930–992 |
| Citrate | 1323–2079 |
| Sodium | 391–644 |
| Potassium | 1212–1681 |
| Chloride | 772–1207 |

[a] Adapted from Gaucheron (2005).

bound to the casein micelles (Gaucheron 2005). Insoluble minerals, present in the form of nanoclusters, act as cross-linking agents in the casein micelle structure (Holt 2004). Casein molecules contain phosphoserine residues which are the binding sites for calcium phosphate nanoclusters. Calcium and magnesium can also bind directly to phosphoserine, glutamic and aspartic residues (Bauland et al. 2020). Indeed, the concentration of CCP in milk is proportional to the casein concentration. It is common practice to express CCP as the concentration of insoluble calcium bound to casein. At natural milk pH and room temperature, the insoluble calcium concentration is ~30 $mg \cdot g^{-1}$ casein. In the soluble phase of milk, calcium forms complexes with phosphates and citrates, according to the association constants and the solubility of the salts (Gaucheron 2005). Almost 80% of soluble calcium is in the complexed form and the concentration of ionic calcium is ~2.0 mM.

Changes in the milk environment can have a major effect on the distribution of minerals. An alteration of pH or temperature and the addition of various salts are the main factors influencing the mineral balance in milk. As a result of decreasing pH, the acido-basic groups in milk (organic and inorganic phosphate, citrate and carboxylic residues) become more protonated, which induces the solubilisation of CCP to restore the equilibrium. In milk, the CCP is completely solubilised at a pH of about 5.2, and a lower pH (~3.5) is required to release calcium that is directly bound to the phosphoserine residues of casein (Le Graet and Brûlé 1993). The solubilisation of CCP is a major factor responsible for the high buffering capacity of milk. During fermentation, the lactic acid production that is required to reach a specific pH level will increase with increasing concentration of CCP in milk. The pH-induced changes of salt equilibria are irreversible, and the original structure of calcium phosphate nanoclusters cannot be restored after neutralisation of the acidified milk (Gaucheron 2011).

The milk salt balance is sensitive to temperature changes. Increasing temperature promotes the dissociation of the inorganic phosphate present in the aqueous phase of milk ($H_2PO_4^- \rightarrow HPO_4^{2-} + H^+$). The dissociated form of inorganic phosphate has a high affinity for calcium, which results in the formation of insoluble calcium phosphate salts (Gaucheron 2011). Conversely, the solubility of calcium phosphate increases when milk is cooled. Storage of milk at 4°C for 48 h has been shown to reduce the insoluble calcium concentration to 25.4 $mg \cdot g^{-1}$ casein, while heating milk at 90°C for 10 min increased the insoluble calcium concentration to 32.1 $mg \cdot g^{-1}$ casein (Rémillard and Britten 2011). Temperature-induced changes of the salt balance are reversible unless milk is heated at temperatures above 100°C for several minutes (Gaucheron 2011).

The mineral equilibrium in milk is altered by the addition of various salts. Calcium chloride is commonly added at a low concentration to cheese milk to enhance coagulation properties. The added calcium neutralises the negatively charged residues on caseins, increasing the aggregation of renneted micelles (Lucey and Fox 1993). Part of the added calcium reacts with soluble phosphates in the aqueous phase, resulting in the formation of insoluble calcium phosphates and the release of protons (Gaucheron 2011). The slight decrease in pH also promotes the aggregation of renneted micelles. When added to milk, sequestrants of calcium, such as citrate, EDTA or oxalate, reduce calcium phosphate saturation in the aqueous phase and shift the salt equilibria from the micellar phase to the aqueous phase (Horne 2009). The solubilisation of CCP reduces cross-links and increases electrostatic repulsion between caseins. A slight increase in pH, due to the binding of free protons by released phosphates, further increases electrostatic repulsions. Whereas at low concentration, calcium sequestrants loosen the casein micelle structure and increase hydration, at high concentration, calcium sequestrants can totally disrupt casein micelles (Gaucheron 2011). Adding NaCl to milk increases the concentration of soluble calcium and reduces the pH of milk. The added sodium participates in ion exchange reactions with the calcium and the protons bound to negatively charged residues of caseins. However, CCP nanoclusters are not thought to be affected by added sodium chloride (Cooke and McSweeney 2017).

### *4.2 Change in mineral equilibrium during cheesemaking*

The concentration and forms of calcium are important for the textural properties of cheese. According to Choi et al. (2008), it is the insoluble calcium that is bound to caseins, and not the total calcium, that contributes to the textural properties of cheese. However, it is difficult to isolate the effect of insoluble calcium, since it is often confounded with other characteristics known to influence cheese texture, such as the moisture-to-casein ratio, the extent of primary proteolysis and the cheese pH (Lucey et al. 2003). In general, reducing the insoluble calcium concentration in cheese reduces cheese firmness at room temperature and improves its melting properties (O'Mahony et al. 2005). This is due to the lower number of CCP cross-links, which increases hydration and bond mobility in the casein network (Cooke and McSweeney 2017). During cheesemaking, gradual acidification induces the dissociation of CCP and soluble calcium is removed with the whey. Important decisions that are made with respect to milk standardisation, renneting pH, cooking conditions and the whey drainage pH will determine the final concentration of insoluble calcium in cheese.

Using membrane separation technologies, milk can be concentrated to increase cheese plant productivity. Milk ultrafiltration also increases the recovery of whey protein, with a positive impact on cheese yield. However, during the process, the removal of milk permeate reduces the volume of water available to solubilise CCP during cheesemaking. In comparison with cheese whey, milk permeate contains much lower concentrations of calcium and phosphates. The concentration of calcium in permeate collected from milk ultrafiltered at natural pH varies between 250 $mg \cdot kg^{-1}$ and 350 $mg \cdot kg^{-1}$ depending on filtration temperature (Dumpler et al. 2017), while the calcium concentration in cheddar cheese whey (drainage pH = 6.35) is around 500 $mg \cdot kg^{-1}$ (Nelson et al. 2004). In order to meet the target concentration of insoluble calcium in cheese, the cheesemaking parameters, such as renneting and drainage pH or cooking conditions, need to be adjusted to increase the transfer of calcium to cheese whey. Alternatively, the calcium load of ultrafiltered milk (UF milk) can be reduced through the acidification of milk or the addition of calcium sequestrants before filtration (Soodam and Guinee 2018). Adding NaCl to highly concentrated milk has also been proposed as a way to modulate the calcium equilibrium in cheese made from highly concentrated milk (Mistry and Maubois 2017).

Acidification during cheesemaking is the main factor driving the solubilisation of micellar calcium. Acid-induced solubilisation is highly efficient when performed before the curd cutting and during the initial stages of syneresis, but its efficiency decreases significantly thereafter (Johnson and Lucey 2006). The curd matrix slows down the free diffusion of calcium to the surrounding whey, and the soluble calcium accumulates within the reduced volume of curd particles. As a result, CCP never completely solubilise in cheese, even when the pH falls below 5 (Cooke and McSweeney 2017). The pH at renneting is an important parameter which is positively correlated with the final concentration of insoluble calcium in cheese. Allowing more time for acidity development before renneting increases the solubilisation of CCP, but extends manufacturing time. Direct acidification with organic acids (Fernandez and Kosikowski 1986) or injection of carbon dioxide (Kelly et al. 2008) are appropriate alternatives, especially for cheeses made from UF milk. The use of organic acids with a strong affinity for calcium, such as citric acid, has been shown to further reduce the concentration of insoluble calcium in cheese (Thybo et al. 2020).

Cooking conditions also affect the solubilisation of CCP during cheesemaking through their effect on curd contraction. A rapid increase in temperature leads to the fast expulsion of serum, reducing the volume of water available in curd particles for further solubilisation of the CCP (Lucey and Fox 1993). The rate of syneresis and the rate of acidification are interrelated factors that determine the amount of CCP solubilised during the cooking phase of cheesemaking.

The pH at whey drainage influences the retention of calcium and is a distinct parameter responsible for texture differences between cheese varieties (Cooke and McSweeney 2017). A low pH at whey drainage results in a lower concentration of insoluble calcium in cheese. However, altering the drainage pH is less effective than altering renneting pH for controlling the insoluble calcium content in cheese (Johnson and Lucey 2006). The transfer of soluble calcium to whey depends on the diffusion of soluble species through the curd matrix, which is reduced in the late cooking phase. Furthermore, a decrease of the drainage pH is likely to reduce the final cheese pH - with potentially negative impacts on overall cheese quality. Curd-washing is primarily done to reduce residual lactose and lactic acid concentrations in cheese, but it has also been shown to reduce the concentration of insoluble calcium in cheese (Lee et al. 2011).

### *4.3 Change in mineral equilibrium during cheese ageing*

Changes in the textural attributes of cheese are observed during ageing. Both the proteolysis of caseins and the partial solubilisation of residual CCP bound to the casein network are involved in cheese softening during ripening. The texture changes in the early stages of ripening are due to solubilisation of the CCP (O'Mahony et al. 2005), while in the later stages, proteolysis becomes the dominant factor (Guinee et al. 2002). The rate and extent of solubilisation of residual CCP are affected by the change in pH, calcium crystal formation and proteolytic activity during ageing.

The solubilisation of CCP and the pH variations during ripening are interconnected. It has been reported that the pH of pasta-filata cheeses increases during ripening (Guinee et al. 2002). This effect is attributed to CCP solubilisation and the binding of protons by released phosphates. However, in cheeses containing active starter bacteria and significant quantities of residual lactose, the lactic acid production will reduce cheese pH and promote CCP solubilisation. In surface-ripened cheeses, moulds convert lactate to ammonia, which produces a pH gradient, with a higher pH at the surface (Abraham et al. 2007). The higher surface pH triggers calcium phosphate precipitation and the concomitant migration of soluble calcium phosphates from the centre to the outside, to restore the equilibrium (Le Graet et al. 1983). Solubilisation of CCP at the centre of the cheese reduces cross-linking and increases electrostatic repulsion, contributing to a soft and slightly runny interior.

The aqueous phase of cheese is supersaturated with various calcium salts. As a consequence, calcium phosphate or calcium citrate crystals can form in the aqueous phase of the cheese (Morris et al. 1988). The appearance of these calcium crystals coincides with a decrease in soluble calcium concentration, which would promote further solubilisation of CCP to balance calcium equilibrium (Lee et al. 2005). The growth of these crystals at the expense of CCP nanoclusters reduces cross-linking and increases the relaxation and mobility of the casein network (Lucey et al. 2003).

Proteolytic activity and the disappearance of CCP nanoclusters during ageing act synergistically to soften the casein network. Casein molecules interact with CCP via their phosphoseryl groups. It has been proposed that proteolysis could destabilise CCP nanoclusters and promote the formation of macrocrystals (Cooke and McSweeney 2017). The reduced number of CCP nanoclusters would decrease cross-linking and increase electrostatic repulsion, thus enhancing bond mobility in the casein network (Lucey et al. 2003). At the same time, the solubilisation of CCP in milk has been shown to increase the accessibility of caseins to proteolytic enzymes (Fox 1970), with a similar effect likely to occur in cheese (Lamichhane et al. 2018).

Changes in mineral equilibrium during cheesemaking and cheese ripening contribute to textural changes in the cheese. The solubilisation of CCP reduces cheese firmness and increases its meltability (O'Mahony et al. 2005). Although the residual concentration of insoluble calcium is a critical factor, it is involved in a complex interaction with other factors, such as the extent of proteolysis, the hydration of the casein network and the cheese pH. All these factors should be controlled by the appropriate cheesemaking and ripening conditions, in order to meet the target cheese textural properties.

## 5 Contribution of fat to cheese texture

In cheese, milk lipids are concentrated and integrated within the casein network. The texture of cheese is affected by both the concentration of fat and the structure of the fat inclusions. The interactions between the lipid phase and the casein network also contribute to the textural properties of the cheese. The original milk fat globule dispersion is altered during the cheesemaking process, which determines the impact of fat on cheese texture.

## 5.1 Fat dispersion in milk

Milk is an oil-in-water emulsion. The fat globules in bovine milk range in diameter from 0.1 to 20 μm, with a mean of 3–4 μm. Milk contains ~15 × $10^9$ globules $mL^{-1}$, with a total interfacial area of ~1.2–2.5 $m^2$ $g^{-1}$ fat (Fox 2011). These globules are surrounded by a lipoprotein membrane, the milk fat globule membrane, which protects them from coalescence and enzymatic degradation by lipases. Milk fat is composed mainly of triacylglycerols (about 98% of the total fat), with minor contributions from phospholipids, cholesterol, free fatty acids and mono- and diacylglycerols. Triacylglycerol consists of a glycerol backbone esterified with three fatty acid molecules. More than 400 different fatty acids have been identified in milk fat (Taylor and MacGibbon 2011a), but only 16 are present at significant concentrations (Table 2). Milk fat contains a large number of different types of triacylglycerols, which vary considerably in molecular weight and degree of unsaturation. As a result, milk fat melts over a wide temperature range (Taylor and MacGibbon 2011b). The proportion of solid fat is around 50% at 4°C and it gradually decreases with increasing temperature. Complete fusion of milk fat is observed at 40°C (Wright et al. 2011). Fatty acid composition is the main factor influencing the melting properties of milk fat. As a general rule, the melting point increases with the length of hydrocarbon chains and decreases

**Table 2** Proportion of major fatty acids of cow milk fat[a]

| Fatty Acid | Common Name | Proportion in Milk Fat (wt.%) |
|---|---|---|
| 4:0 | Butyric | 3.7 |
| 6:0 | Caproic | 2.4 |
| 8:0 | Caprylic | 1.5 |
| 10:0 | Capric | 3.2 |
| 12:0 | Lauric | 3.6 |
| 14:0 | Myristic | 11.1 |
| 14:1 | Myristoleic | 0.9 |
| 15:0 | Pentadecanoic | 1.2 |
| 16:0 | Palmitic | 28.3 |
| 16:1 | Palmitoleic | 1.6 |
| 18:0 | Stearic | 11.8 |
| 18:1 | Oleic | 18.8 |
| 18:1 | Vaccenic | 4.2 |
| 18:2 | Linoleic | 1.4 |
| 18:2c | Linoleic (conjugated) | 1.1 |
| 18:3 | Linolenic (alpha) | 0.9 |
| | Other fatty acids | 4.4 |

[a] Adapted from Taylor and MacGibbon (2011a).

with the number of double bonds. Palmitic acid (16:0) is the most abundant saturated fatty acid, while oleic acid (18:1) is the most abundant unsaturated fatty acid in milk fat. The fatty acid composition of milk fat is affected by factors such as the animal's diet and stage of lactation. In studies where the fatty acid profile was altered by diet, it was found that a higher oleic (18:1) to palmitic (16:0) acid ratio produced cheeses with lower firmness (Coppa et al. 2011, Frétin et al. 2019).

Milk handling before cheesemaking can affect the properties of fat and impact cheese texture. Mechanical stress can damage the milk fat globule membrane and alter the protection of milk fat (Michalski et al. 2002). The partially denuded surface is covered by milk protein, which changes the fat globule's interfacial properties and increases its susceptibility to lipolysis. Milk homogenisation can be used to intentionally destroy the original membrane during the manufacture of some cheeses where lipolysis is important for flavour development (Guinee and Kilcawley 2004). Homogenisation reduces the fat globule's size and increases the fat–serum interfacial area. Caseins are preferentially adsorbed onto the newly formed interface and the fat globules behave as pseudo-casein particles and become an integral part of the rennet gel during cheesemaking (Michalski et al. 2002). Milk homogenisation increases fat recovery in cheese, and the presence of homogenised fat in the casein network interferes with curd rearrangements and decreases syneresis. As a consequence, cheese moisture is increased, which tends to produce softer and creamier cheeses. The melting properties of cheese are also impaired by milk homogenisation, since casein-coated fat globules do not behave like 'real' casein particles when cheese is heated (Johnson 2011).

## 5.2 Change in fat dispersion during cheesemaking

During the coagulation of milk, native fat globules are considered non-interacting particles, which act as inert fillers (Vliet and Dentener-Kikkert 1982). They are mainly located in the serum pores of the casein network (Lopez et al. 2007). They do not interfere with the coagulation process and do not break the casein network. Depending on the membrane composition, damaged or homogenised fat can interact with renneted casein micelles during gel formation and act as an active filler. A slight increase in the curd-firming rate and in curd firmness was observed when homogenised cream was used to standardise cheese milk (Nair et al. 2000).

After coagulation, the curd is cut, stirred and brought to the cooking temperature. During the process, the curd grains contract and the whey is gradually expelled. The contraction of the casein network reduces the size of the serum pores, which induces some aggregation of fat globules (Lopez et al. 2007). At this point, the fat clusters prevent excessive contraction of the casein network and further elimination of serum from the pores. Fat clusters may also clog the curd channels and slow down the flow of serum. The presence of fat is known to reduce the rate of syneresis (Dejmek and Walstra 2004) and a positive correlation is observed between the fat content and the hydration of the casein network (non-fat substance) in cheese (Fenelon and Guinee 1999).

Pressing induces important changes in the cheese matrix. Under pressure, the milk fat globules are deformed, causing breakage of the protective membrane and coalescence of the aggregated globules. Fat coalescence is promoted by close contacts between fat globules. Hence, the extent of fat coalescence depends on the level of aggregation before pressing. Compared to large fat globules, small fat globules have a higher internal pressure, allowing them to better resist deformation and coalescence (Lopez et al. 2007). A lower degree of coalescence was observed in cheeses made from milk with small fat globules (Michalski et al. 2004). Extensive coalescence leads to the formation of fat pools in the cheese matrix. Free fat is not protected and is susceptible to oiling-off during cheese melting.

In the production of pasta-filata cheeses, the curd is subjected to a thermomechanical process in order to develop the characteristic fibrous structure. The curd is worked in hot water or brine at temperatures generally between 55°C and 65°C. In this temperature range, the fat is completely melted and hydrophobic interactions are strengthened, inducing the contraction of the casein network (Smith et al. 2018). Curd stretching in such conditions induces the formation of fat–serum channels within the cheese matrix. These fat–serum channels act as a low-viscosity lubricant which facilitates the displacement of casein fibres when the cheese is heated (Kindstedt et al. 2004). The structure of the fat is altered by the stretching process. Fat globule aggregation and coalescence are promoted, as is the proportion of free fat (Rowney et al. 2003). However, at high-shear work input, a reduction of free fat was observed (Sharma et al. 2017). High shear stress applied to plasticised curd is likely to promote the re-emulsification of free fat.

## 5.3 Change in fat dispersion during cheese ageing

During cheese ripening, the casein network relaxes and its water-binding capacity increases. As a result, the pockets of serum surrounding fat clusters gradually disappear (Lopez et al. 2007). With the reduction of free water, the fat inclusions come in direct contact with the casein network and may be considered active fillers. Active fillers reinforce a gel if the filler is stiffer than the gel network (Barden et al. 2015). The stiffness of the fat inclusions depends on the temperature, which controls the proportion of solid fat. At temperatures below 25°C, increasing the fat content was shown to increase cheese firmness (Yang et al. 2011), while the reverse trend was observed at higher temperatures (Ustunol et al. 1995). These observations seem to support the role of fat as an active filler in cheese. However, there is still debate regarding the nature of the interactions between fat inclusions and the casein network (Everett and Auty 2017). Microscopic observation of young cheese revealed the presence of fat globules surrounded by serum cavities, suggesting that there is no direct interaction with the casein matrix (Hassan and Awad 2005). It is likely that fat inclusions are present in both active and inert forms in the cheese matrix and that the proportion of the latter decreases during ageing.

The release of liquid fat provides lubrication between the layers of casein when the cheese matrix is deformed. Temperature has a considerable impact on cheese texture. With increasing temperature, the melting and coalescence of fat decrease friction and increase the perception of softness and creaminess (Liu et al. 2015). Since the fusion of milk fat is complete at ~40°C, its main effect on the melting properties of cheese is observed below this temperature. At temperatures above 40°C, the melting of cheese is driven by the contraction of the casein matrix and the release of moisture, which provide fluidity to the matrix (Lucey et al. 2003). Interestingly, the presence of fat improves the melting properties of the casein network. Fat inclusions reduce curd syneresis and promote the formation of serum pockets or serum channels (Lopez et al. 2007). As a result, more water is available for swelling of the casein network during ripening, and for moisture release when the cheese is heated. Cheese melting is often accompanied by the release of free oil. Excessive oiling-off gives the cheese a greasy appearance and is undesirable (Kindstedt et al. 2004). However, the formation of a thin film of free oil at the surface of cheese is required to slow down moisture evaporation during heating (Rudan and Barbano 1998). Surface dehydration during cheese melting can result in the formation of a tough skin, which limits flow and promotes scorching.

## 6 Contribution of serum proteins to cheese texture

Serum proteins correspond to the protein fraction remaining soluble when milk is acidified to a pH level of 4.6, and represent ~20% of true protein. β-lactoglobulin and α-lactalbumin, respectively, account for 50% and 30% of the total serum proteins. In their native state, the serum proteins have a globular structure, do not interact with casein micelles and are not sensitive to proteases. Unlike caseins, serum proteins are not concentrated during cheesemaking and are lost in the drained whey. The concentration of serum proteins in cheese is proportional to its moisture content. In cheese with 50% moisture, the serum protein concentration is less than ~0.3%. Various strategies have been proposed in order to increase the concentration of serum proteins in cheese. Milk ultrafiltration, milk heating and the addition of denatured whey proteins are the main practical approaches for increasing the recovery of serum protein during cheesemaking.

## 6.1 Milk ultrafiltration

Concentration of cheese milk by ultrafiltration (UF) is commonly done to increase the efficiency of the cheese plant and to contribute to a more consistent cheesemaking process (Soodam and Guinee 2018). Selective membranes with a molecular weight cut-off of around 10 kDa are used to concentrate caseins, fat, serum proteins and colloidal salts, while at the same time, lactose and soluble salts are removed with water in the permeate (Gavazzi-April et al. 2018). As a result, the concentration of serum protein in the aqueous phase of milk, and in the aqueous phase of cheese, increases proportionally with the concentration factor. To achieve total retention of serum protein in cheese, milk is concentrated until it reaches the composition of the drained curd before cheesemaking. This 'liquid pre-cheese' concept was first applied to Camembert, but many applications have been developed successfully for the manufacture of other cheese varieties (Mistry and Maubois 2017). Ultrafiltered milk (UF milk), concentrated two-fold or less, can usually be processed with standard cheesemaking equipment. However, the process needs to be adapted to manage the high buffering capacity of UF milk, the rapid increase in gel firmness after clotting, and the high concentration of colloidal calcium in cheese (Soodam and Guinee 2018). Cheese moisture usually decreases with increasing concentration of protein in cheese, but both protein and fat recovery have been shown to increase (Guinee et al. 2006).

Concentration of milk by UF can affect cheese quality, and some defects have been associated with high concentration factors, such as sandiness, crumbliness and excessive firmness (Mistry and Maubois 2017). These defects are mainly caused by the high calcium content of cheeses made from UF milk. Reducing the pH of milk prior to filtration, diafiltration during concentration, and addition of sodium chloride to milk concentrate have been shown to improve cheese quality (Guyomarc'h 2006). Since native serum proteins have little interaction with the casein network, increased retention in cheese is unlikely to have a direct effect on cheese texture. However, a high concentration of serum protein is associated with reduced proteolysis (Creamer et al. 1982) and increased firmness of mature cheeses. Serum proteins are resistant to proteolysis, and cheese protein breakdown decreases as serum protein is substituted for casein (De Koning et al. 1981). The reduced proteolysis during cheese ripening has also been attributed to the inhibition of rennet and plasmin activities by serum protein (Benfeldt 2006, Creamer et al. 1982).

Published by Burleigh Dodds Science Publishing Limited, 2022.

## 6.2 Heating milk

Milk can be thermally treated to eliminate pathogenic bacteria before cheesemaking. Heating milk in conditions more severe than pasteurisation can be used to denature serum proteins and increase their recovery in cheese. Temperatures exceeding ~65°C unfold serum proteins and expose the reactive groups that are buried in their native structure. The unfolded serum proteins form complexes with κ-casein on the surface of the casein micelles or in the aqueous phase of milk (Singh and Waungana 2001). The pH during the heat treatment is the main factor controlling the location of serum protein–κ-casein complexes (Donato and Guyomarc'h 2009). The proportion of complexes bound to the casein micelles increases with a decrease in the pH of milk before heating.

The heat denaturation of serum proteins in milk has a considerable impact on cheesemaking properties. Although it is exploited in the production of whey cheeses and smooth-textured lactic cheeses, the heat treatment of milk impairs rennet-induced coagulation. The rennet-clotting time increases and the curd-firming rate decreases with the intensity of the thermal treatment (Giroux et al. 2020). Increasing the milk pH before heating has been proposed as a way to reduce the amount of micelle-bound serum proteins and improve the coagulation properties of heated milk (Guyomarc'h 2006). However, similar deterioration of coagulation properties was observed for milk heated at different pH levels (Anema et al. 2007, Giroux et al. 2020). The poor coagulation properties of heated milk are attributable to the synergistic interactions between various factors (Kethireddipalli and Hill 2015): 1) the casein micelle surface is modified due to the attachment of serum proteins, which causes steric hindrance of the aggregation after renneting; 2) the soluble complexes bind to the casein micelles during renneting, thus interfering with the aggregation of casein micelles; and 3) mineral equilibrium is altered by heat treatment, with a reduction in the concentration of ionic calcium and an increase in precipitation of calcium phosphate onto casein micelles. Heat treatment of milk has also been shown to reduce curd syneresis and increase cheese moisture. The denaturation of serum proteins increases their water-binding capacity, and their attachment to casein micelles interferes with the rearrangement and contraction of the casein matrix (Giroux et al. 2020). Various approaches have been proposed to improve rennet coagulability of heated milk, such as increasing the coagulation temperature, acidifying milk before renneting and adding calcium chloride (Guinee 2016). The application of high-pressure treatment was shown to restore the coagulability of heated milk and to increase cheese yield (Huppertz et al. 2005).

Cheeses made from heated milk tend to have a porous matrix, a soft texture and a soggy appearance due to the increased moisture content and the binding of serum proteins to the caseins. When the cheesemaking parameters were held constant, the cheese moisture was shown to increase linearly with the denaturation rate of whey proteins (Giroux et al. 2020). The integration of denatured whey protein into the casein network reduces cheese cohesion and increases crumbliness. Denatured serum proteins also alter the melting properties of cheese. Despite the increase in cheese moisture, denatured serum proteins reduce the amount of free water released by the casein network and the flowability of cheese upon heating (Kethireddipalli and Hill 2015).

### *6.3 Adding denatured whey protein to milk*

To avoid the negative effect that heating has on the coagulation properties of milk, cheese whey can be heat-treated separately, before being added to cheese milk (Banks and Muir 1985). This approach prevents the formation of complexes between serum protein and κ-casein. Denatured whey protein aggregates are larger than the pores of the casein network and are recovered in the cheese matrix. The incorporation of denatured whey proteins has been shown to increase cheese moisture and reduce firmness (Perreault et al. 2017b). However, when compared to the 'heating milk' approach, adding denatured whey protein has a much lower impact on cheese moisture. For similar concentrations of denatured whey protein in cheese, the increase in moisture is four times lower with the addition of denatured whey proteins (Britten and Pouliot 2018, personal communication).

It is common practice to concentrate whey proteins by UF before heating and to apply shear stress during or after thermal treatment to control the characteristics of denatured whey protein aggregates (Guyomarc'h 2006). The conditions used to denature whey proteins influence the characteristics of whey protein aggregates (Nicolai et al. 2011). The composition (protein, lactose and mineral concentrations) and processing variables (pH adjustment, heating and shearing intensities) should be optimised to control the size distribution and the water-binding capacity of the aggregates. The addition of denatured whey protein to milk has been shown to reduce the rennet gel-firming rate and curd-contraction capacity (Perreault et al. 2016). Large whey protein aggregates, located in the pores of the casein matrix, are believed to sterically impede curd contraction (Guinee 2016). However, the smaller whey protein aggregates were found to have a more detrimental effect than the larger ones on rennet gel formation and contraction (Giroux et al. 2015). Smaller aggregates are believed to interact with renneted micelles and interfere with the formation and rearrangement of the casein network during cheesemaking (Perreault et al. 2017a).

Adding denatured whey proteins to cheese milk generally weakens the cheese body and decreases its firmness. Low-fat cheeses are excessively hard, and denatured whey proteins have been used to improve their texture (Stankey et al. 2017). Commercial whey protein ingredients are specifically designed for use in low-fat cheese applications (Hinrichs 2001). The whey proteins in these ingredients are microparticulated, allowing them to act as inert fillers in the cheese matrix and improve the melting properties through a ball-bearing effect (Schenkel et al. 2013). The increased moisture of cheeses with added denatured whey protein also reduces cheese firmness. A different study found that the increase in moisture content was the main factor responsible for the change in the elasticity of cheeses made with added denatured whey proteins (Perreault et al. 2017b). Large whey protein aggregates acted as active fillers and increased cheese elasticity when the moisture content was kept constant. The apparent contradiction between studies regarding the role of whey protein aggregates as active or inert fillers emphasises the importance of the denaturation conditions to control their characteristics.

## 7 Conclusion and future trends

During cheesemaking, milk constituents are modified and rearranged to produce cheeses with the desired characteristics. The texture of cheese is governed by its composition and the interactions between its constituents. Cheese diversity is impressive, and is likely to expand in the coming years. Cheese is eaten on its own and is also increasingly used as an ingredient in various food preparations. Specific textural and melting properties are required for optimal cheese performance for those applications. The adjustment of milk composition and the fine-tuning of cheesemaking parameters provide the flexibility to carefully control cheese composition and characteristics. Furthermore, new technologies, such as high-pressure processing, pulsed electric field and ultrasound treatment of milk, will offer new possibilities to modulate cheese properties.

## 8 Where to look for further information

A comprehensive coverage of the scientific aspects of cheese, emphasising fundamental principles, is found in *Fundamentals of Cheese Science* by Fox, Guinee, Cogan and McSweeney (2016). A detailed overview of the chemical, biochemical, microbiological and physicochemical aspects of cheese is found in *Cheese*, edited by McSweeney, Fox, Cotter and Everett (2017). Additional information on factors affecting cheese quality, and the specific cheesemaking processes and characteristics of a wide variety of cheeses are covered in *Global Cheesemaking Technology*, by Papademas and Bintsis (2017). Specific information on the rheological and textural properties of cheese and the relevant measurement techniques are presented in *Cheese Rheology and Texture* by Gunasekaran and Ak (2002).

- Fox, P. F., Guinee, T. P., Cogan, T. M. and McSweeney, P. L. H. 2016. *Fundamentals of Cheese Science*, Second Edition. Springer US.
- McSweeney, P. L. H., Fox, P. F., Cotter, P. D. and Everett, D. W. 2017. *Cheese: Chemistry, Physics and Microbiology*, Fourth Edition. Elsevier Inc.
- Papademas, P. and Bintsis, T. 2017. *Global Cheesemaking Technology: Cheese Quality and Characteristics*. Wiley.
- Gunasekaran, S. and Ak, M. M. 2002. *Cheese Rheology and Texture*. CRC Press.

## 9 References

Abraham, S., Cachon, R., Colas, B., Feron, G. and De Coninck, J. (2007). Eh and pH gradients in Camembert cheese during ripening: measurements using microelectrodes and correlations with texture. *International Dairy Journal* 17(8), 954–960.

Anema, S. G., Kim Lee, S. and Klostermeyer, H. (2007). Effect of pH at heat treatment on the hydrolysis of κ-casein and the gelation of skim milk by chymosin. *LWT - Food Science and Technology* 40(1), 99–106.

Banks, J. M. and Muir, D. D. (1985). Effect of incorporation of denatured whey protein on the yield and quality of Cheddar cheese. *International Journal of Dairy Technology* 38(1), 27–32.

Bansal, N., Fox, P. F. and McSweeney, P. L. H. (2007). Aggregation of rennet-altered casein micelles at low temperatures. *Journal of Agricultural and Food Chemistry* 55(8), 3120–3126.

Barden, L. M., Osborne, J. A., McMahon, D. J. and Foegeding, E. A. (2015). Investigating the filled gel model in Cheddar cheese through use of Sephadex beads. *Journal of Dairy Science* 98(3), 1502–1516.

Bauland, J., Famelart, M. H., Bouhallab, S., Jeantet, R., Roustel, S., Faiveley, M. and Croguennec, T. (2020). Addition of calcium and magnesium chlorides as simple means of varying bound and precipitated minerals in casein micelle: effect on enzymatic coagulation. *Journal of Dairy Science* 103(11), 9923–9935.

Benfeldt, C. (2006). Ultrafiltration of cheese milk: effect on plasmin activity and proteolysis during cheese ripening. *International Dairy Journal* 16(6), 600–608.

Bintsis, T. and Papademas, P. (2017). An overview of the cheesemaking process. In: *Global Cheesemaking Technology: Cheese Quality and Characteristics* Papademas, P. and Bintsis, T. (Eds). Chichester: Wiley-IEEE Press. pp. 120–156.

Britten, M. and Remondetto, G. (2018). Les ingredients laitiers. In: *Science et technologie du lait* Vuillemard, J.-C. (Ed). Québec, QC: Les Presses de l'Université Laval. pp. 389–446.

Buffa, M. N., Trujillo, A. J., Pavia, M. and Guamis, B. (2001). Changes in textural, microstructural, and colour characteristics during ripening of cheeses made from raw, pasteurized or high-pressure-treated goats' milk. *International Dairy Journal* 11(11–12), 927–934.

Carver, J. A., Thorn, D. C., Ecroyd, H. and Holt, C. (2017). Letter to the editor: a response to Horne and Lucey (2017). *Journal of Dairy Science* 100(7), 5121–5124.

Choi, J., Horne, D. S., Johnson, M. E. and Lucey, J. A. (2008). Effects of the concentration of insoluble calcium phosphate associated with casein micelles on the functionality of directly acidified cheese. *Journal of Dairy Science* 91(2), 513–522.

Choi, J., Horne, D. S. and Lucey, J. A. (2007). Effect of insoluble calcium concentration on rennet coagulation properties of milk. *Journal of Dairy Science* 90(6), 2612–2623.
Cooke, D. R. and McSweeney, P. L. H. (2017). From micelle to melt: the influence of calcium on physico-chemical properties of cheese. In: *Global Cheesemaking Technology: Cheese Quality and Characteristics* Papademas, P. and Bintsis, T. (Eds). Chichester: Wiley-IEEE Press. pp. 20–44.
Coppa, M., Ferlay, A., Monsallier, F., Verdier-Metz, I., Pradel, P., Didienne, R., Farruggia, A., Montel, M. C. and Martin, B. (2011). Milk fatty acid composition and cheese texture and appearance from cows fed hay or different grazing systems on upland pastures. *Journal of Dairy Science* 94(3), 1132–1145.
Creamer, L. K., Iyer, M. and Lelievre, J. (1982). Effect of various levels of rennet addition on characteristics of Cheddar cheese made from ultrafiltered milk. *New Zealand Journal of Dairy Science and Technology* 22(3), 205–214.
Creamer, L. K. and Olson, N. F. (1982). Rheological evaluation of maturing Cheddar cheese. *Journal of Food Science* 47(2), 631–636.
Dalgleish, D. G. (2011). On the structural models of bovine casein micelles - review and possible improvements. *Soft Matter* 7(6), 2265–2272.
De Angelis, M. and Gobbetti, M. (2011). Cheese | pasta-filata cheeses: traditional pasta-filata cheese. In: *Encyclopedia of Dairy Sciences* (2nd edn.) Fuquay, J. W. (Ed). San Diego: Academic Press. pp. 745–752.
De Koning, P. J., De Boer, R., Both, P. and Nooy, P. F. C. (1981). Comparison of proteolysis in a low-fat semi-hard type of cheese manufactured by standard and by ultrafiltration techniques. *Netherlands Milk and Dairy Journal* 35(1), 35–46.
De Kruif, C. G. and Holt, C. (2003). Casein micelle structure, functions and interactions. In: *Advanced Dairy Chemistry - 1 Proteins: Part A / Part, B* Fox, P. F. and McSweeney, P. L. H. (Eds). Boston, MA: Springer US. pp. 233–276.
Dejmek, P. and Walstra, P. (2004). The syneresis of rennet-coagulated curd. In: *Cheese: Chemistry, Physics and Microbiology* Fox, P. F., McSweeney, P. L. H., Cogan, T. M. and Guinee, T. P. (Eds). London: Elsevier Ltd., 71–103.
Donato, L. and Guyomarc'h, F. (2009). Formation and properties of the whey protein/κ-casein complexes in heated skim milk – a review. *Dairy Science and Technology* 89(1), 3–29.
Drake, M. A. and Delahunty, C. M. (2017) Chapter 20. Sensory character of cheese and its evaluation. In: *Cheese* (4th edn.) McSweeney, P. L. H., Fox, P. F., Cotter, P. D. and Everett, D. W. (Eds):San Diego: Academic Press. pp. 517–545.
Dumpler, J., Kieferle, I., Wohlschläger, H. and Kulozik, U. (2017). Milk ultrafiltrate analysis by ion chromatography and calcium activity for SMUF preparation for different scientific purposes and prediction of its supersaturation. *International Dairy Journal* 68, 60–69.
Everett, D. W. and Auty, M. A. E. (2017) Chapter 21.Cheese Microstructure1. In: *Cheese* (4th edn.) McSweeney, P. L. H., Fox, P. F., Cotter, P. D. and Everett, D. W. (Eds):San Diego: Academic Press. pp. 547–569.
Fagan, C. C., Castillo, M., Payne, F. A., O'Donnell, C. P. and O'Callaghan, D. J. (2007). Effect of cutting time, temperature, and calcium on curd moisture, whey fat losses, and curd yield by response surface methodology. *Journal of Dairy Science* 90(10), 4499–4512.
Fagan, C. C., O'Callaghan, D. J., Mateo, M. J. and Dejmek, P. (2017) Chapter 6. The syneresis of rennet-coagulated curd. In: *Cheese* (4th edn.) McSweeney, P. L. H., Fox, P. F., Cotter, P. D. and Everett, D. W. (Eds). San Diego: Academic Press. pp. 145–177.

Farkye, N. Y. and Fox, P. F. (1990). Observations on plasmin activity in cheese. *Journal of Dairy Research* 57(3), 413-418.

Fenelon, M. A. and Guinee, T. P. (1999). The effect of milk fat on Cheddar cheese yield and its prediction, using modifications of the Van Slyke cheese yield formula. *Journal of Dairy Science* 82(11), 2287-2299.

Fernandez, A. and Kosikowski, F. V. (1986). Physical properties of direct acidified mozzarella cheese from ultrafiltered whole milk retentates. *Journal of Dairy Science* 69(3), 643-648.

Fox, P. F. (1970). Influence of aggregation on the susceptibility of casein to proteolysis. *Journal of Dairy Research* 37(2), 173-180.

Fox, P. F. (2011). Milk lipids | fat globules in milk. In: *Encyclopedia of Dairy Sciences* (2nd edn.) Fuquay, J. W. (Ed). San Diego: Academic Press. pp. 675-679.

Fox, P. F., Cogan, T. M. and Guinee, T. P. (2017) Chapter 25.Factors that affect the quality of cheese. In: *Cheese* (4th edn.) McSweeney, P. L. H., Fox, P. F., Cotter, P. D. and Everett, D. W. (Eds). San Diego: Academic Press. pp. 617-641.

Frétin, M., Martin, B., Buchin, S., Desserre, B., Lavigne, R., Tixier, E., Cirié, C., Bord, C., Montel, M. C., Delbès, C. and Ferlay, A. (2019). Milk fat composition modifies the texture and appearance of Cantal-type cheeses but not their flavor. *Journal of Dairy Science* 102(2), 1131-1143.

Gaucheron, F. (2005). The minerals of milk. *Reproduction, Nutrition, Development* 45(4), 473-483.

Gaucheron, F. (2011). Milk salts: distribution and analysis. In: *Encyclopedia of Dairy Sciences* (2nd edn.) Fuquay, J. W. (Ed). San Diego: Elsevier Inc. pp. 908-916.

Gavazzi-April, C., Benoit, S., Doyen, A., Britten, M. and Pouliot, Y. (2018). Preparation of milk protein concentrates by ultrafiltration and continuous diafiltration: effect of process design on overall efficiency. *Journal of Dairy Science* 101(11), 9670-9679.

Giroux, H. J., Bouchard, C. and Britten, M. (2014). Combined effect of renneting pH, cooking temperature, and dry salting on the contraction kinetics of rennet-induced milk gels. *International Dairy Journal* 35(1), 70-74.

Giroux, H. J., Dupont, F., Villeneuve, G. and Britten, M. (2020). Effect of heating milk on whey protein denaturation and cheese-making properties. *International Dairy Journal* 111, 104831.

Giroux, H. J., Lanouette, G. and Britten, M. (2015). Effect of whey protein aggregates of various sizes on the formation and properties of rennet-induced milk gels. *Food Hydrocolloids* 45(0), 272-278.

Guinee, T. P. (2004). Salting and the role of salt in cheese. *International Journal of Dairy Technology* 57(2-3), 99-109.

Guinee, T. P. (2007). Salt in cheese: introduction: what are the functions of NaCl in cheese? In: *Cheese Problems Solved* McSweeney, P. L. H. (Ed). Cambridge, UK: Elsevier Ltd. pp. 80-81.

Guinee, T. P. (2016). Protein in cheese and cheese products: structure-function relationships. In: *Advanced Dairy Chemistry Proteins: Applied Aspects* (4th edn., vol. 1B) McSweeney, P. L. H. and O'Mahony, J. A. (Eds). New York: Springer. pp. 347-415.

Guinee, T. P., Feeney, E. P., Auty, M. A. E. and Fox, P. F. (2002). Effect of pH and calcium concentration on some textural and functional properties of Mozzarella cheese. *Journal of Dairy Science* 85(7), 1655-1669.

Guinee, T. P. and Fox, P. F. (2004). Salt in Cheese: Physical, Chemical and Biological Aspects. In: *Cheese: Chemistry, Physics and Microbiology* Fox, P. F., McSweeney, P. L. H., Cogan, T. M. and Guinee, T. P. (Eds). London: Academic Press. pp. 207–259.

Guinee, T. P. and Kilcawley, K. N. (2004). Cheese as an ingredient. In: *Cheese: Chemistry, Physics and Microbiology* Fox, P. F., McSweeney, P. L. H., Cogan, T. M. and Guinee, T. P. (Eds). London: Academic Press. pp. 395–428.

Guinee, T. P., O'Kennedy, B. T. and Kelly, P. M. (2006). Effect of milk protein standardization using different methods on the composition and yields of Cheddar cheese. *Journal of Dairy Science* 89(2), 468–482.

Guinee, T. P. and Sutherland, B. J. (2011). Cheese | salting of cheese. In: *Encyclopedia of Dairy Sciences* (2nd edn.) Fuquay, J. W. (Ed). San Diego: Academic Press. pp. 595–606.

Guyomarc'h, F. (2006). Formation of heat-induced protein aggregates in milk as a means to recover the whey protein fraction in cheese manufacture, and potential of heat-treating milk at alkaline pH values in order to keep its rennet coagulation properties: a review. *Lait* 86(1), 1–20.

Hassan, A., Johnson, M. E. and Lucey, J. A. (2004). Changes in the proportions of soluble and insoluble calcium during the ripening of cheddar cheese. *Journal of Dairy Science* 87(4), 854–862.

Hassan, A. N. and Awad, S. (2005). Application of exopolysaccharide-producing cultures in reduced-fat Cheddar cheese: cryo-scanning electron microscopy observations. *Journal of Dairy Science* 88(12), 4214–4220.

Hinrichs, J. (2001). Incorporation of whey proteins in cheese. *International Dairy Journal*, 11, 495–503.

Holland, J. W. and Boland, M. J. (2014). Chapter 5. Post-translational modifications of caseins. In: *Milk Proteins* (2nd edn.) Singh, H., Boland, M. and Thompson, A. (Eds). San Diego: Academic Press. pp. 141–168.

Holt, C. (2004). An equilibrium thermodynamic model of the sequestration of calcium phosphate by casein micelles and its application to the calculation of the partition of salts in milk. *European Biophysics Journal* 33(5), 421–434.

Horne, D. and Lucey, J. A. (2017a). Letter to the Editor: hydrophobic interactions in the caseins: challenging their dismissal by Holt, et al. (2013). *Journal of Dairy Science* 100(7), 5119–5120.

Horne, D. S. and Lucey, J. A. (2017b). Rennet-induced coagulation of milk. In: *Cheese: Chemistry, Physics and Microbiology* (4th edn.) Boland, M. and Singh, H. (Eds). London, UK: Elsevier Inc. pp. 115–143.

Horne, D. S. (1998). Casein interactions: casting light on the black boxes, the structure in dairy products. *International Dairy Journal* 8(3), 171–177.

Horne, D. S. (2009). Casein micelle structure and stability. In: *Milk Proteins* Thompson, A., Boland, M. and Singh, H. (Eds). London, UK: Elsevier Inc. pp. 133–162.

Horne, D. S. (2019). Casein micelle structure and stability. In: *Milk Proteins: From Expression to Food*. Elsevier. pp. 213–250.

Huppertz, T. (2016). Novel technologies for dairy processing: applications for non-cow milk. In: *Non-Bovine Milk and Milk Products*. Elsevier Inc. pp. 187–208.

Huppertz, T., Fox, P. F. and Kelly, A. L. (2004). High-pressure treatment of bovine milk: effects on casein micelles and whey proteins. *Journal of Dairy Research* 71(1), 97–106.

Huppertz, T., Hinz, K., Zobrist, M. R., Uniacke, T., Kelly, A. L. and Fox, P. F. (2005). Effects of high-pressure treatment on the rennet coagulation and cheese-making properties of heated milk. *Innovative Food Science and Emerging Technologies* 6(3), 279–285.

Johnson, M. E. (2011). Cheese | low-fat and reduced-fat cheese. In: *Encyclopedia of Dairy Sciences* (2nd edn.) Fuquay, J. W. (Ed). San Diego: Academic Press. pp. 833–842.

Johnson, M. E. and Lucey, J. A. (2006). Calcium: a key factor in controlling cheese functionality. *Australian Journal of Dairy Technology* 61(2), 147–153.

Kelly, A. L., Huppertz, T. and Sheehan, J. J. (2008). Pre-treatment of cheese milk: principles and developments. *Dairy Science and Technology* 88(4–5), 549–572.

Kethireddipalli, P. and Hill, A. R. (2015). Rennet coagulation and cheesemaking properties of thermally processed milk: overview and recent developments. *Journal of Agricultural and Food Chemistry* 63(43), 9389–9403.

Kindstedt, P., Carić, M. and Milanović, S. (2004). Pasta-filata cheeses. In: *Cheese: Chemistry, Physics and Microbiology* Fox, P. F., McSweeney, P. L. H., Cogan, T. M. and Guinee, T. P. (Eds). Academic Press. pp. 251–277.

Lamichhane, P., Kelly, A. L. and Sheehan, J. J. (2018). Symposium review: structure-function relationships in cheese. *Journal of Dairy Science Symposium Review: Structure-Function Relationships in Cheese* 101(3), 2692–2709.

Le Graet, Y. and Brûlé, G. (1993). Les équilibres minéraux du lait : influence du pH et de la force ionique. *Lait* 73(1), 51–60.

Le Graet, Y., Lepienne, A., Brule, G. and Ducruet, P. (1983). Migration du calcium et des phosphates inorganiques dans les fromages a pate molle de type Camembert au cours de l'affinage. *Lait* 63(629–630), 317–332.

Lee, M. R., Johnson, M. E., Govindasamy-Lucey, S., Jaeggi, J. J. and Lucey, J. A. (2011). Effect of different curd-washing methods on the insoluble Ca content and rheological properties of Colby cheese during ripening. *Journal of Dairy Science* 94(6), 2692–2700.

Lee, M. R., Johnson, M. E. and Lucey, J. A. (2005). Impact of modifications in acid development on the insoluble calcium content and rheological properties of cheddar cheese. *Journal of Dairy Science* 88(11), 3798–3809.

Legg, A. K., Carr, A. J., Bennett, R. J. and Johnston, K. A. (2017) Chapter 26.General aspects of cheese technology. In: *Cheese* (4th edn.) McSweeney, P. L. H., Fox, P. F., Cotter, P. D. and Everett, D. W. (Eds). San Diego: Academic Press. pp. 643–675.

Liu, K., Stieger, M., van der Linden, E. and van de Velde, F. (2015). Fat droplet characteristics affect rheological, tribological and sensory properties of food gels. *Food Hydrocolloids* 44, 244–259.

Lopez, C., Camier, B. and Gassi, J.-Y. (2007). Development of the milk fat microstructure during the manufacture and ripening of Emmental cheese observed by confocal laser scanning microscopy. *International Dairy Journal* 17(3), 235–247.

Lopez, C., Cauty, C. and Guyomarc'h, F. (2015). Organization of lipids in milks, infant milk formulas and various dairy products: role of technological processes and potential impacts. *Dairy Science and Technology* 95(6), 863–893.

Lucey, J. A. (2002). ADSA Foundation Scholar Award. Formation and physical properties of milk protein gels. *Journal of Dairy Science* 85(2), 281–294.

Lucey, J. A. (2011). Cheese: rennet-induced coagulation of milk. In: *Encyclopedia of Dairy Sciences* (2nd edn.) Fuquay, J. W. (Ed). San Diego: Elsevier Inc. pp. 579–584.

Lucey, J. A. and Fox, P. F. (1993). Importance of calcium and phosphate in cheese manufacture: a review. *Journal of Dairy Science* 76(6), 1714–1724.
Lucey, J. A., Johnson, M. E. and Horne, D. S. (2003). Invited review: perspectives on the basis of the rheology and texture properties of cheese. *Journal of Dairy Science* 86(9), 2725–2743.
Marshall, R. J. (1993). Physicochemical properties accounting for cheese texture. *Acta Psychologica* 84(1), 69–77.
McMahon, D. J. and Oberg, C. J. (2017) Chapter 40.Pasta-filata cheeses. In: *Cheese* (4th edn.) McSweeney, P. L. H., Fox, P. F., Cotter, P. D. and Everett, D. W. (Eds). San Diego: Academic Press. pp. 1041–1068.
McMahon, D. J. and Oommen, B. S. (2008). Supramolecular structure of the casein micelle. *Journal of Dairy Science* 91(5), 1709–1721.
McSweeney, P. L. H. (2004). Biochemistry of cheese ripening: introduction and overview. In: *Cheese: Chemistry, Physics and Microbiology* Fox, P. F., McSweeney, P. L. H., Cogan, T. M. and Guinee, T. P. (Eds). Academic Press. pp. 347–360.
McSweeney, P. L. H. (Ed) (2007a). Acidification: introduction. In: *Cheese Problems Solved*. Cambridge, UK: Elsevier Ltd. pp. 34–35.
McSweeney, P. L. H. (Ed) (2007b). Conversion of milk to curd: introduction. In: *Cheese Problems Solved*. Cambridge, UK: Elsevier Ltd. pp. 50–51.
McSweeney, P. L. H. (Ed) (2007c). What processing variables affect syneresis? In: *Cheese Problems Solved*. Cambridge, UK: Elsevier Ltd. pp. 75–76.
Mellema, M., Walstra, P., van Opheusden, J. H. J. and van Vliet, T. (2002). Effects of structural rearrangements on the rheology of rennet-induced casein particle gels. *Advances in Colloid and Interface Science* 98(1), 25–50.
Michalski, M. C., Camier, B., Briard, V., Leconte, N., Gassi, J. Y., Goudédranche, H., Michel, F. and Fauquant, J. (2004). The size of native milk fat globules affects physico-chemical and functional properties of Emmental cheese. *Lait* 84(4), 343–358.
Michalski, M. C., Cariou, R., Michel, F. and Garnier, C. (2002). Native vs. damaged milk fat globules: membrane properties affect the viscoelasticity of milk gels. *Journal of Dairy Science* 85(10), 2451–2461.
Mistry, V. V. and Maubois, J.-L. (2017) Chapter 27.Application of membrane separation technology to cheese production. In: *Cheese* (4th edn.). San Diego: Academic Press. pp. 677–697.
Morris, H. A., Holt, C., Brooker, B. E., Banks, J. M. and Manson, W. (1988). Inorganic constituents of cheese analysis of juice from a one-month-old Cheddar cheese and the use of light and electron microscopy to characterize the crystalline phases. *Journal of Dairy Research* 55(2), 255–268.
Moynihan, A. C., Govindasamy-Lucey, S., Molitor, M., Jaeggi, J. J., Johnson, M. E., McSweeney, P. L. H. and Lucey, J. A. (2016). Effect of standardizing the lactose content of cheese milk on the properties of low-moisture, part-skim Mozzarella cheese. *Journal of Dairy Science* 99(10), 7791–7802.
Muthukumarappan, K. and Swamy, G. J. (2017). Rheology, microstructure, and functionality of cheese. In: *Advances in Food Rheology and Its Applications*. Elsevier Inc. pp. 245–276.
Nair, M. G., Mistry, V. V. and Oommen, B. S. (2000). Yield and functionality of Cheddar cheese as influenced by homogenization of cream. *International Dairy Journal* 10(9), 647–657.

Nelson, B. K., Lynch, J. M. and Barbano, D. M. (2004). Impact of milk preacidification with CO2 on cheddar cheese composition and yield. *Journal of Dairy Science* 87(11), 3581-3589.

Nicolai, T., Britten, M. and Schmitt, C. (2011). β-Lactoglobulin and WPI aggregates: formation, structure and applications. *Food Hydrocolloids* 25(8), 1945-1962.

O'Mahony, J. A. and Fox, P. F. (2013). Milk proteins: introduction and historical aspects. In: *Advanced Dairy Chemistry: Proteins: Basic Aspects* (4th edn., vol. 1A) McSweeney, P. L. H. and Fox, P. F. (Eds):Boston, MA: Springer US. pp. 43-85.

O'Mahony, J. A., Lucey, J. A. and McSweeney, P. L. H. (2005). Chymosin-mediated proteolysis, calcium solubilization, and texture development during the ripening of Cheddar cheese. *Journal of Dairy Science* 88(9), 3101-3114.

Perreault, V., Morin, P., Pouliot, Y. and Britten, M. (2017a). Effect of denatured whey protein concentrate and its fractions on rennet-induced milk gels. *International Dairy Journal* 64, 48-55.

Perreault, V., Rémillard, N., Chabot, D., Morin, P., Pouliot, Y. and Britten, M. (2017b). Effect of denatured whey protein concentrate and its fractions on cheese composition and rheological properties. *Journal of Dairy Science* 100(7), 5139-5152.

Perreault, V., Turcotte, O., Morin, P., Pouliot, Y. and Britten, M. (2016). Combined effect of denatured whey protein concentrate level and fat level in milk on rennet gel properties. *International Dairy Journal* 55, 1-9.

Rashmi, K. G. and Warrier, A. S. (2020). Non-thermal processing of dairy foods. In: *Dairy Processing: Advanced Research to Applications*: Singapore: Springer. pp. 25-49.

Rémillard, N. and Britten, M. (2011). Quantitative determination of micellar calcium in milk and cheese using acid-base titration. *Milchwissenschaft* 66(2), 137-140.

Rowney, M. K., Roupas, P., Hickey, M. W. and Everett, D. W. (2003). The effect of compression, stretching, and cooking temperature on free oil formation in Mozzarella curd. *Journal of Dairy Science* 86(2), 449-456.

Rudan, M. A. and Barbano, D. M. (1998). A model of mozzarella cheese melting and browning during pizza baking. *Journal of Dairy Science* 81(8), 2312-2319.

Schenkel, P., Samudrala, R. and Hinrichs, J. (2013). The effect of adding whey protein particles as inert filler on thermophysical properties of fat-reduced semihard cheese type Gouda. *International Journal of Dairy Technology* 66(2), 220-230.

Sharma, P., Munro, P. A., Gillies, G., Wiles, P. G. and Dessev, T. T. (2017). Changes in creep behavior and microstructure of model Mozzarella cheese during working. *LWT - Food Science and Technology* 83, 184-192.

Sheehan, J. J. (2007). Why are certain cook temperatures used for certain cheeses? In: *Cheese Problems Solved* McSweeney, P. L. H. (Eds). Cambridge, UK: Elsevier Ltd. pp. 77-78.

Silva, N. N., Casanova, F., da Silva Pinto, M., de Carvalho, A. F. and Gaucheron, F. (2019). Casein micelles: from the monomers to the supramolecular structure. *Brazilian Journal of Food Technology* 22, 1-15.

Singh, H. and Waungana, A. (2001). Influence of heat treatment of milk on cheesemaking properties. *International Dairy Journal* 11(4-7), 543-551.

Smith, J. R., Carr, A. J., Golding, M. and Reid, D. (2018). Mozzarella cheese - a review of the structural development during processing. *Food Biophysics* 13(1), 1-10.

Soodam, K. and Guinee, T. P. (2018). The case for milk protein standardisation using membrane filtration for improving cheese consistency and quality. *International Journal of Dairy Technology* 71(2), 277-291.

Stankey, J. A., Lu, Y., Abdalla, A., Govindasamy-Lucey, S., Jaeggi, J. J., Ø Mikkelsen, B., Pedersen, K. T. and Andersen, C. B. (2017). Low-fat Cheddar cheese made using microparticulated whey proteins: effect on yield and cheese quality. *International Journal of Dairy Technology* 70(4), 481-491.

Taylor, M. W. and MacGibbon, A. K. H. (2011a). Milk lipids | fatty acids. In: *Encyclopedia of Dairy Sciences* (2nd edn.) Fuquay, J. W. (Ed). San Diego: Academic Press. pp. 655-659.

Taylor, M. W. and MacGibbon, A. K. H. (2011b). Milk Lipids | general characteristics. In: *Encyclopedia of Dairy Sciences* (2nd edn.) Fuquay, J. W. (Ed). San Diego: Academic Press. pp. 649-654.

Thybo, C. D., Lillevang, S. K., Skibsted, L. H. and Ahrné, L. (2020). Calcium balance during direct acidification of milk for Mozzarella cheese production. *LWT* 131, 1-8.

Ustunol, Z., Kawachi, K. and Steffe, J. (1995). Rheological properties of Cheddar cheese as influenced by fat reduction and ripening time. *Journal of Food Science* 60(6), 1208-1210.

Vliet, T. and Dentener-Kikkert, A. (1982). Influence of the composition of milk fat globule membrane on the rheological properties of acid milk gels. *Netherlands Milk and Dairy Journal* 36(3), 261-265.

Waugh, D. F. (1958). The interactions of αs-β- and κ-caseins in micelle formation. *Discuss. Faraday Soc* 25(0), 186-192.

Whetstine, M. E. C., Luck, P. J., Drake, M. A., Foegeding, E. A., Gerard, P. D. and Barbano, D. M. (2007). Characterization of flavor and texture development within large (291 kg) blocks of cheddar cheese. *Journal of Dairy Science* 90(7), 3091-3109.

Wright, A. J., Marangoni, A. G. and Hartel, R. W. (2011). Milk lipids | rheological properties and their modification. In: *Encyclopedia of Dairy Sciences* (2nd edn.) Fuquay, J. W. (Ed). San Diego: Academic Press. pp. 704-710.

Yang, X., Rogers, N. R., Berry, T. K. and Foegeding, E. A. (2011). Modeling the rheological properties of cheddar cheese with different fat contents at various temperatures. *Journal of Texture Studies* 42(5), 331-348.

# Index

www.ingramcontent.com/pod-product-compliance
Ingram Content Group UK Ltd.
Pitfield, Milton Keynes, MK11 3LW, UK
UKHW021823100926
8556IPUK00003B/132

9 781786 768193